The Art of Systems Engineering

A How-To guide for Systems Engineers

Anton F. Beck
Robert J. Monson, PhD

For those who inspired us,
For those who created with us,
For our most cherished of systems,
Mary and Kathy

TABLE OF CONTENTS

Introduction

This book is envisioned as a textbook that can be used to teach any student the Systems Engineering approach to problem solving, thereby enabling him or her to incorporate it into a personal problem solving toolkit. It is intended for use with a first semester "Introduction to Systems Engineering" course, typically comprising forty-five hours of instructional content. The final section of the text presents an interesting systems problem for the consideration of an instructor responsible for a second semester Systems Design course.

In general, we find it is very difficult to instruct students in the appropriate use of Systems Engineering tools and methodologies if those students do not possess sufficient experience in the workplace to understand the challenges of abstract problems and vague requirements. Due to this, when undertaking the Systems Engineering instruction at local universities, we attempted to utilize existing textbooks and approaches to instill in our students an appreciation for the Systems Engineering methodology and subsequently, to enhance their capabilities to a sufficient degree to allow them to transition into a Systems Engineering role within their organization.

This challenge of imparting appropriate Systems Engineering methodologies to students was further exacerbated by the difficulty we found in selecting a text that represented the actual methodology typically used by a Systems Engineer in a large-scale development program. While many high-quality texts exist on the subject of Systems Engineering, in general they are constructed in inclusive formats, which read more like encyclopedias of Systems Engineering knowledge, and less like texts focused on teaching the appropriate flow of Systems Engineering tasks.

Because of this, the authors elected to construct a unique and highly effective Systems Engineering instructional approach in their graduate school offerings at several universities. Those offerings are reflected in this text, and culminate in a large-scale proposal effort required of the students in order to make 'real' the Systems Engineering methodology and process.

In the first semester course, the students are introduced to a variety of Systems Engineering concepts and approaches, and they are required to apply them to relatively small-scale problems and sections of system development. While this can provide them with useful insights into the specific methodologies being described, often it results in a relatively disjointed educational process, as these students may not receive insight related to the value of the Systems Engineering tools being taught.

Our interest lies in producing Systems Engineering practitioners, and so we elected to undertake a unique development approach and the first and second semester instructional design. As previously stated, the first semester is focused on teaching the specific techniques and methodologies associated with Systems Engineering process, and the second semester is focused on providing a sophisticated problem

which acts as a backdrop for the actual application of Systems Engineering methods and processes by the students.
The last section of this text includes the proposal effort as provided by the authors in their graduate-level instructional efforts. This proposal package is designed to provide the students with a challenging problem, which must be solved within the time constraints of a single semester class. The statement of work (SOW) and the performance-based system specification (PBSS) are sufficiently vague, requiring that the students must spend considerable time in identifying and understanding the problems to be solved and the underlying knowledge that they are responsible to generate in order to adequately respond to the proposal. Additionally, the students, who are typically placed on a team of between seven and twelve students, find themselves in competition with several other teams in order to be selected for the award of the contract outlined in the request for proposal.

We find that this approach has been extremely effective in providing a real-world application of the Systems Engineering process for the students in class. Imagine, if you will, the students arriving on their first night of class for the second semester systems design course. They are immediately introduced to a team, with whom they will be working for the remainder of the semester. They are provided with the proposal documents as included in Volume 2 of this text, and they are told to proceed to create a proposal response that will be compliant with the request for proposal (RFP).

After several hours of discussion and wandering by the students, they begin to realize that they do not have sufficient knowledge to simply assign writing to each of the student members of the team. Instead, they must devise a method by which they can establish what information is known, establish what information must be known, to find a method in which they can derive that information, and ultimately to create a solution to the challenge outlined in the RFP.

In general, after several hours have passed in fruitless discussion and opining by the various members of the team, one or more members will suddenly realize that they already have tools at their disposal that will enable them to get their arms around such a vague and challenging problem. It is at this point that the students begin to realize why they were instructed in the various Systems Engineering methodologies the previous semester, and by direct application of those techniques they quickly discover that they can develop a very sophisticated response to the proposal in a relatively short period of time.

This insight and grounding of principles is the culmination of our educational approach in Systems Engineering, and we have found it to be a remarkably effective method by which to produce competent Systems Engineers for the workplace. In fact, our method showed such significant promise that we hired the students who demonstrated exceptional prowess in their proposal efforts, and they became highly successful Systems Engineers at the major defense contractor at which we were employed.

A Day in the Life of a Systems Engineer

I've been designated what? You mean I have to work with people? Where you stand is where you sit. It's obvious, why are you disagreeing with me? Who is running this place? Will it be like this until the end of time? Who changed the organization? How hard can this be? Can I survive this opportunity? Who is the customer? Complexity, ambiguity and uncertainty, they are everywhere; identify and manage them. Customers everywhere. This can't be real, nothing is this complicated! You can't be serious!

I've been designated what?

"Yes it is true, you have been designated as the Technical Director and/or Lead Systems Engineer on a proposal effort to integrate the command and control and communications infrastructure (C3) for the country of Euphoria. This is a two billion dollar business opportunity for the corporation, and a real career growth opportunity for you. Business Development has funded you for the past year to get a feeling for what is required to perform this job, and to create the system design and proposal for the customer. You have sixty days to do it. And oh, by the way, you will be working with several other companies as subcontractors (that Senior Management and business development have already selected), and you will have no direct personnel responsibility (no organization of your own), so that you can focus on the technical solution and get the technical proposal written. Give me a list of all the people that you need and we will see who we can free up from their current assignments. We are counting on you."

-Vice President of Systems Engineering

The above is a description of an interchange between a Lead Systems Engineer and the Vice President of Systems Engineering. It is based on a real program, led and pursued by real Systems Engineers, and was ultimately proposed and won.

What follows is a generalized description of the attributes of the environment that Lead Systems Engineers experience as they go about fulfilling their responsibilities in writing a winning proposal. The description will be seen through the Lead Systems Engineer's eyes. What we will see described is how it was, and not how it should be, and although the descriptions will be abstract, they will not be idealized descriptions of how things should have been. Let's see what we can learn by reading about a day in the life of a Systems Engineer.

You mean I have to work with people?

A good soldier is not violent.
A good fighter is not angry.

A good winner is not vengeful.
A good employer is humble.
This is known as the Virtue of not striving.
This is known as the ability to deal with people.
This since ancient times has been known as the ultimate unity with heaven.

- From the Tao Te Ching

The Ability to deal with People – their Diversity, Motivation, and Rewards

It is important for the Systems Engineer (particularly the Lead Systems Engineer - although it applies to all Systems Engineers) to understand the environment (the people and what motivates their behavior) within which he or she operates. Without this understanding the project and the Systems Engineer will most certainly fail.

A Systems Engineer, more so than in other engineering disciplines, must effectively interface with many organizations, both inside and outside of his or her own organization (corporation or governmental or other). Each of these organizational structures is populated with people in various roles and their attendant responsibilities, differing beliefs, differing organizational behavioral constraints and differing rewards.

The ability to deal with people reaches its most complex and most difficult level when a Systems Engineer is designated a Lead. The Lead (there will be different titles, for example Technical Director or Chief Engineer) will be responsible for the Systems Architecture of desired systems that may be valued in billions of dollars. Additionally, from a project perspective, he or she may be responsible for the work direction of large numbers of people (say from five to hundreds of people who may be spread geographically across the world), and the Lead Systems Engineer has no functional organization (no one reports to the Lead). That means that the Lead Systems Engineer cannot treat the members of the systems team as organizational subordinates and, consequently, has not been assigned the power to direct any of the team members to do anything! That implies that the Lead Systems Engineer will not be successful unless he or she is highly skilled at the art of dealing with the people that he or she encounters in his or her job functions. In large systems jobs, he or she will require some level of cultural understanding (particularly in jobs that span large geographic regions), as well as skills in Psychology and Sociology.

Diversity is the ability to deal with people that have differing interests, responsibilities, focus, goals, and rewards.

The technical people in the Systems Engineering team are generally highly skilled, self- actualized, independent-minded, in many cases extremely opinionated, and prone to respond negatively when given direction they feel is either arbitrary or dictatorial.

The non-technical people involved in the proposal, design, contract execution, and system lifecycle support are typically from Program Management, Proposal Management, Contracts, Subcontracts, Personnel, Legal, the Chief Executive Officer (CEO), and the Senior Management team's staff. Non-technical people involved have very little interest in technical matters. Their primary interests focus on the business, profitability, schedule, investment required, return on investment, and customer satisfaction.

The Lead Systems Engineer must satisfy the interests of the non-technical people in addition to performing his or her other duties in Systems Engineering. The ability to understand, convince, motivate, inspire, resolve conflicts and guide the development while keeping the project team focused and management team satisfied is an art. This challenging task must be accomplished for the system to be successfully proposed, to win the contract bid, and to successfully perform and sustain contract development over the lifecycle of the system.

It's obvious, why are you disagreeing with me?

The Organizational Environment - Collaborative with many differing Views and Opinions

The organizational environment that Systems Engineers typically face is a mature corporate industrial organization or a large government organization. Both organizational types constrain what a Systems Engineer can accomplish (policies, procedures, and internal processes) and provide facilities, domain experts, and support staff that enable the Systems Engineer to do the job.

The environment in which Systems Engineers operate is a collaborative environment; that is, it must effectively consider both technical and social complexities when creating new Systems Architecture and design or modifying an existing one. In a collaborative environment populated with people of differing viewpoints, there are differences of opinion on almost every topic. The Lead Systems Engineer must evaluate and reconcile many disagreements and differences of opinion without unduly ruffling the feathers of the individuals whose viewpoints are not adopted or used. At various times during the Systems Design process the complexities, ambiguities, and uncertainties may seem overwhelming.

Complexity, ambiguity and uncertainty are everywhere. Identify and manage them.

Both the types and the number of nodes in the Systems Architecture and the dynamics of their connections create the complexities found in systems. The nodes are typically organizational entities and the systems that support them, and each node spans multiple knowledge domains. The domains of knowledge (corpuses) and experience required to understand, describe, model, and simulate the behavior of the nodes and their interactions with other nodes in order to arrive at a Systems Architecture is typically more than are possessed by a single human being.

For example, nodes can be anything from organizational entities (defined by their policies procedures, processes, and command and control hierarchy and structure) to electrical, mechanical, data processing, communications, or other subsystems or systems. This implies that the Lead Systems Engineer must be able to identify the domain expertise required, and identify the domain experts required, to not only understand the systems requirements but to come up with an effective systems design. The Lead Systems Engineer must identify and bring together domain experts from many different domains to accomplish the creation of a complex systems design. Each of these domain experts will have a unique interpersonal style of communications, domain knowledge and emotional temperament. Some will have strong opinions and be somewhat intolerant of the other members and their opinions. Okay, all of them will.

The Lead Systems Engineer's challenge is to form a team that operates in a collaborative fashion to arrive at a systems design. However, that is just one aspect of collaboration that the Lead must create, orchestrate, and manage in order to achieve collaboration within the Systems Design team. Collaboration is also required outside of the Systems Engineering team and may span the rest of corporate structure and organizations external to the corporation, such as potential subcontractors, customers and government representatives of various agencies that touch their activities.

This implies that Systems Engineers and in particular, Leads, must possess the social skills necessary to effectively work with and manage other people in order to create systems that will work for their customers and make money for their companies.

The Systems Engineer must be able to draw on the other members of his or her team for specific domain knowledge, must stimulate discussion, must be patient while others expound on their topics, must tactfully terminate discussion that has drifted off the subject, must encourage brainstorming when appropriate, and must resolve conflicts by acting as a mediator or facilitator when there are strong differences of opinion between team members or between group of peoples comprised of subcontractors, suppliers and other people associated with the program.

The Systems Engineer must also be able consider multiple viewpoints, and must resist the cavalier dismissal of any viewpoint without a logical examination of its applicability to designing the system. Over time, the Systems Engineer develops the ability to recognize and understand multiple viewpoints from diverse technical and organizational associates, and to tolerate and encourage dialogue while avoiding rushed, premature conclusions by enduring ambiguity during the process. The skill and art that are developed with the help of thinking aids (frame works) that enable Systems Engineers to change their thought processes from everyday-life thinking to systems thinking is the essence of this text.

Where you stand is where you sit (Why viewpoints differ)

With that in mind, let us now consider the viewpoints of other organizations and in particular, specific jobs within our corporation such as Senior Management, Program Management, Contracts, Subcontracts, and Finance to try to understand why the individuals in these groups behave the way they do. We must understand what their motivations are, how they think, and how it affects the behavior of others, including the Systems Engineers. To accomplish this, we will examine the viewpoints of these positions, along with the viewpoints of the external organizations that are the customers for Systems Engineering developments.

Customers everywhere

Understanding the Systems Engineers' Internal Customers and Managing Expectations; the Key to Survival and Career Growth

In a typical industrial organization, there is a group of people that are designated as Senior Management. Senior Management consists of the Vice President or President (CEO) of the organization, Vice Presidents of Program Management, Proposal Management, Contracts, Subcontracts, Finance, Personnel, Legal, and Ethics that report to the CEO. The members of each of these groups have different responsibilities, viewpoints, and rewards as defined by their job functions. Although each of the organizations mentioned above may have the title of Vice President, the Systems Engineering Vice President has less influence in Senior Management meetings than the other Vice Presidents. This is because business issues such as profits, return on investment, customer satisfaction and long-term business goals of necessity have a higher weight in the Senior Management decision processes.

This often comes as a surprise to Systems Engineers and Engineers in general. It is often difficult for Systems Engineers and Engineers to realize and accept that business concerns and issues usually override technical issues. Sometimes Engineers get frustrated because they believe their technical opinions are ignored, and sometimes they go to great lengths to express their frustration. For example, there have been instances of ethics charges being leveled against Program Managers and Engineering Managers because the Engineers could not believe that the "best technical solution" was not proposed or used in the generation of a proposal or the execution of the contract. In many of the instances the charges were without substance and the result of an Engineer's frustration with getting either no explanation from their supervisor, or from getting an explanation that lacked credibility for why the "best technical solution" was not used.

Who is running this place? It's not Systems Engineering!!

Let us now look at the viewpoints, rewards, and responsibilities of the CEO and the Senior Management staff reporting to the CEO, and how their viewpoints, responsibilities and rewards affect the Lead Systems Engineer in the performance of his or her work. Understanding the viewpoints, responsibilities and rewards will enable the Systems Engineer to better understand what motives these people's behavior and preferences, and will enable the Lead Systems Engineer be more effective at winning new business and executing contracts. It will also increase the Systems Engineer's probability of long-term survival and career growth.

The CEO has overall responsibility for the fiscal performance of the entire organization and for meeting the goals of the organization, typically on an annual basis. These goals include profitability and other financial goals, organizational and governmental social goals, and marketing goals and strategies. The CEO must also develop and approve of business strategies, tactics, and salary plan allocations and will likely review the salary increases proposed (for Systems Engineers, too); particularly for those who have briefed the CEO or are responsible for the proposal or contract executions of very high value, such as one hundred million dollars and above.

The CEO is also responsible for interaction with the local community, the press (although a designated staff member may handle routine interactions with the press), and high-level members of other organizations, such as customers, the financial community and governmental organizations. The CEO, because of the breadth of responsibility and the frequent need for meetings and business travel, will have little time to delve deeply into the Systems Engineering issues or other technical issues. In addition, the CEO may lack the technical depth of knowledge required to ask incisive questions during a briefing that would aid in understanding its content.

If the CEO does not have the technical background required to ask incisive questions at technical briefings, he or she may ask few or no questions and may end technical briefings rapidly. The CEO's lack of time and lack of technical background will likely result in few technical briefings, and those may begin at a high technical level but turn into a Program Management briefing focused on cost and schedule!

This implies that the time allocated by the CEO to any issue, technical or otherwise, will be short. This also means that technical people who brief the CEO must tailor their briefings to be short and provide only the highest-level technical content, and to focus primarily on business issues such as cost and schedule. Lead Systems Engineers that don't understand the CEO's preferences should probably not brief the CEO, as CEOs typically have a couple of other behavioral attributes not mentioned yet. These include impatience and the ability to be exceedingly direct in expressing displeasure with the Lead Systems Engineer who is presenting the information. A Lead who experiences the CEO's expression of displeasure may experience it a second time, during his or her annual performance review.

> *"I'm late! I'm late! For a very important date! No time to say hello, goodbye! I'm late! I'm late! I'm late!"*
>
> *- White Rabbit from "Alice in Wonderland," By Lewis Carroll*

The viewpoints of a CEO can best be summarized by phrases frequently heard by members of the Senior Management staff or by the people who brief her. "Let's get to the bottom line, I don't have a lot of time for this, I have to go to another meeting, I have to catch an airplane to go to another meeting." Always remember that the CEO's viewpoint will be from a business perspective: profitability, customer satisfaction and the meeting of the goals by which her performance will be measured. If these goals are met, she will receive the maximum dollars plus base salary, thereby maximizing her financial gain. If the goals are not met, her personal financial outcome will not be maximized, her boss will chastise her, and she will be measured more often and with more granularity the next year.

It is rare for CEOs to survive in the organization if they repeatedly miss their goals. The behavior of the CEO and her staff will be emulated at all levels of management and within the organization, because it will become clear that similar behavior will be rewarded and behavior that is different will be held in disdain.

All the Senior Management staff that report to the CEO will (in her presence, at least,) emulate her behavior and few will be interested in the technical information. The only exception is likely to be the Vice President of Systems Engineering, because the CEO and the other Senior Management staff members rely on him to ensure the technical details necessary for the organization to be successful from a business perspective are handled.

That leaves only one chain of command that may provide a Systems Engineer the opportunity for an in-depth technical review of his work by technically experienced and competent people. That chain of command is the Engineering chain of command, comprised of the head of Systems Engineering, the Vice President of Systems Engineering and his staff, and the Systems Engineer's peers.

Can I survive this opportunity?

In addition to the responsibility of ensuring the program and system design requirements are met, the Vice President of Systems Engineering is responsible for meeting the schedule and financial constraints imposed by the contract and Senior Management. The Systems Engineering Vice President uses his management organization to ensure that reviews, such as the Systems Requirements Review (SSR), Preliminary Design Review (PDR), and Critical Design Review (CDR) are successfully performed and the artifacts that demonstrate their successful performances are created and made available for review by other organizations in a database. This database then can be used to create briefings for Senior Management, customers, and other outside organizations that may need to see the information.

The Vice President of Systems Engineering uses information from these databases to create presentations for the CEO and other members of the Senior Management team to provide them with summary indicators of the current status of the Systems projects ongoing, and to inform them of any current or impending changes to schedule, personnel, technical risk, and cost overruns or under runs.

The Vice President also serves as a translator, buffer, or referee between the Lead Systems Engineer and Senior Management. He is able to do this because he can operate comfortably and with confidence at multiple levels of detail and multiple technical domains while operating within the Senior Management team's domain. Because of these skills, he is able to advise lead Systems Engineers on the focus, breadth, depth, content, and the duration of presentations that the Lead may be required to give to CEO and/or members of her Senior Management team.

By virtue of the reviews, the Vice President of Systems Engineering and the Systems Engineering management organization ensure the success of either proposal or contract execution. They must make certain that the work is completed in such a manner that the outcome will be positive, (a proposal will be won or the contract will be executed) it is accomplished on schedule, it meets the requirements, and it is accomplished within the planned cost budgets.

Remember that the other organizations in the Corporation are primarily focused on financial and other business issues, including profitability, schedule compliance and customer satisfaction, and are therefore not interested in all the details on which the members of the Systems Engineering management team were briefed.

The Vice President of Systems Engineering is likely to have the most diverse set of viewpoints and experiences of anyone else on the Senior Management team, provided he worked as Systems Engineer previously, and had designed or participated in the design of systems or managed systems contracts at a major systems project level or better.

At the level of Vice President he is also likely to be on an executive compensation plan, salary and bonus dependent upon job performance. If he meets all of his goals, he will receive the maximum possible compensation and financial outcome. If he misses his goals, he will not maximize his personal financial outcome, will be chastised by his boss, the CEO, and will be closely scrutinized in the future. It is rare for a Vice President of Systems Engineering to survive in the organization if he repeatedly misses his goals, which typically encompass technical performance, schedule, and budgets.

Who is the customer?

As it turns out, everyone with a stake in the Systems program is a potential customer for the Lead Systems Engineer, in the sense that he needs to satisfy their needs for information by making convincing presentations, and by making sure that accurate readable reports are written and questions are answered intelligently and tactfully.

Some Systems Engineers must interact with customers, contractors, subcontractors, vendors and various government agencies. Each of them will have organizational structures similar to the Systems Engineer's organization structure. His ability to deal with people will be equally as important with these external organizations as with his own. In dealing with customers, he is likely to run into people with job descriptions that are not organically a part of the corporation for which he works. This is the role of a Systems Engineering and Technical Assistance contractor (SETA). This role is frequently occupied by experienced, mature Systems Engineers, and it is their role to advise the customer during the proposal and contract phase of a program regarding the proposing or performing of a contractor's technical approach. Customers frequently employ SETAs to get organizationally unbiased opinions on the direction and approaches being taken by the contractor to satisfy requirements. The Lead Systems Engineer frequently has the opportunity to brief the customer with SETAs present at the briefings.

Frequently, the SETAs will challenge the Lead System Engineer during the briefings. They will also be present when a customer is unhappy with the Development team's performance. This means that the SETAs are likely to meet the Corporate CEO and members of Senior Management. Having a CEO and/or members of the Senior Management staff receive poor performance briefings usually results in action items for the Development team. This is another issue on the Lead Systems Engineer's to-do list.

Till the end of time

The system's life cycle has several phases; pre-proposal, proposal, contract, and support. This requires the design, development, refined development, production and support of large systems (sustainment), with many replications of the designed system that are to be installed at many geographic locations.

It is interesting to consider the number of Systems Engineers participating in each phase, what they do, what their impact is on the design and the other people working on the system 'to be'.

The Pre-proposal Phase

This phase begins when Management evaluates a business opportunity and deems it to be worthy of the company's pursuit. We will not go into the details of their evaluation criteria (the details vary for each organization), other than to state that an attractive potential contract will present an opportunity of sufficient size, profitability and strategic market value. After the decision is made, a small business pursuit team is formed that includes at least one Systems Engineer and a Marketing person. During this phase, a Systems Engineer's job is to understand the customer's requirements by looking at corresponding open-source data and by having face-to-face discussions with people inside the customer's organization to better understand the users' needs.

The Marketing department typically initiates these face-to-face discussions. During the initial meeting the Systems Engineer presents technical capabilities, and how they might apply to a proposed solution to meet the customer's needs. After having gained a better understanding of the customer's requirements through this early contact, the Systems Engineer oftentimes defines and performs trade studies that are shared with the customer at subsequent meetings. This can take from weeks to years, depending on the complexity of the customer's needs, the estimated future contract value, and the System Engineer's company's budget for pursuit of the opportunity. The Systems Engineer's role, then, is to meet with the customer as frequently as possible to gain an understanding of the customer's requirements, to suggest possible ways to meet those requirements by presenting the customer with attractive alternative notional systems designs, and to instill in the customer the belief that the systems solution that his company's solution will be superior to those offered by all other competing companies.

The Systems Engineer is involved in an iterative (sequential contacts with the customer) and recursive (information from the previous meeting is used to gain better understanding of the requirements and notional system designs) process to discover the customer's requirements and to convince the customer that his company has the best solution.

When the customer's Request for Proposal (RFP) package (consisting of multiple documents that will be described later) finally arrives at the company's door, if the customer has a preferred solution in mind, the company may have a competitive advantage over its competition and System Engineer's life will change. The project's iterations will increase in time and scope, and work will increase to the point that the Systems Engineer must assemble a team to assist him.

Will this proposal ever end?

The Proposal Phase

The culmination of the Pre-proposal phase occurs when the RFP package arrives at the company. A Proposal team must be formed to respond to the RFP package. If the System Engineer assigned to the Pre-proposal phase assembled a team of two to six people during the earlier phase, he will have help in assembling the Proposal team and defining their roles and responsibilities during the Proposal Development effort.

During this phase the Systems Engineer's challenge is to:

Understand the customer's requirements
Think through every phase of the system's lifecycle
Describe the system to be designed without designing it in detail
Coordinate the technical team's work
Generate presentations to Management
Manage crises
Help define the statement of work
Help generate the proposal schedule and proposal outline in collaboration with the contract's Proposal Management team
Review and comment on the cost proposal
Mange the technical efforts of team members and subcontractors
Resolve technical ambiguities and uncertainties
Reconcile interpersonal conflicts
Integrate the technical proposal text

It would be marvelous if this were a challenge with an unlimited amount of time, but the time available is sixty to ninety days! In some ways the proposal effort is a project done in double time.

To illustrate this numerically, let us see what the ratio of the time to generate the proposal is to a notional systems life cycle of ten years. That ratio is 60/3650 = 0.0164, or 1.64% of the system's lifecycle! The understanding of the customer's requirements and the decisions to satisfy the requirements made during that time period will in large measure determine the success or failure of the proposal and the program. For example, if the requirements are not understood in enough detail, and from the customer's perspective, the design will not perform and the program will likely fail.

This is when the lead Systems Engineer's quality of life takes a downturn, because he needs to take what might be years of accumulated knowledge and understanding of the customer, the customer's requirements and solution preferences, and communicate them to the Proposal team in a very short period of time. How short is that time period? Let's do a little thought experiment to figure it out. First, assume that he generates the communication artifacts in seven days. Remember there are only sixty days to respond to the RFP. That means 7/60 = .116 or 11.6 percent of the time used to generate the proposal will be used to plan its generation! You will notice that the calculation for the percentage of time to communicate with the rest of the team used seven days instead of five. We also did not define how many hours a day the Systems Engineer worked. Why? To give the reader an idea of the intensity of the work environment and of the level of productivity needed to produce a proposal consisting of hundreds to thousands of pages in sixty days. Can 11.6% of the calendar time be spent planning proposal activities and tasks? Absolutely not, if past experience is any indicator.

The proposal cannot be accomplished on schedule if during 11.6% of the calendar time the Proposal team does nothing. Even less will be accomplished if they haphazardly begin by doing what they know how to do according to what they think needs to be done. This ensures that more money will be spent generating the proposal than is necessary. Moreover, it implies that when they find out what needs to be done, they will need to free their minds from what they have done, which is not an easy task.

So what does this all mean? This means that the lead Systems Engineer must anticipate the customer's issuance of the RFP and work as many hours as it takes to define the job for the proposal team without taking seven calendar days to do it. If he does this, his quality of life will decrease, as he is hardly ever home and doesn't sleep much. If he is really lucky, a draft RFP was presented weeks before the real RFP, so that the 11.6% of the time to write the proposal will not need to be spent communicating information to the Proposal team. Sounds like fun, right?

In addition, during this phase the Lead Systems Engineer must read and understand the customer's technical and business requirements with enough breadth and depth to assemble the technical team required to respond to the customer requirements, and to intelligently discuss the business requirements with Proposal Management, Program Management and Senior Management staff. What's more, the Lead Systems Engineer, together with Program Management and Proposal Management, develop a schedule and assign domain-qualified and experienced people to specific tasks, so the proposal may be written within the allotted time frame. It is during this period that his people skills will be stressed to the limit of his endurance, as the pressure to write the proposal within thirty to sixty days and the desire for regular real-time updates to all levels of management will relentlessly coexist.

All of the complexities and ambiguities inherent in the customer's requirements (there are typically several thousand requirements in a large systems job) and the social relationships among members of the Proposal team will be brought forward in a seemingly endless series of crises that are generated by the need to meet the proposal's delivery schedule. Almost every participant involved in the understanding of the customer's requirements and the generation of the proposal will need to have ambiguities resolved immediately, or the threat of missing the proposal due date will become a reality.

This can't be real, nothing is this complicated!

The RFP

So what is this thing called A Proposal, why is the delivery date so important, to whom is it delivered, and why is that important?

A proposal is a written story describing the system's design and the approach used to arrive at the design the customer is interested in purchasing. It is a document consisting of several documents that address the customer's technical, performance, business and proposal evaluation requirements as described and contained in a package called the Request For Proposal (RFP).

The customer creates the RFP that describes the customer's needs, and sends it to several companies to bid on what the customer believes may satisfy those needs. The customer is interested in getting the most delivered for the money that is spent. Generally, the RFP is provided to several different companies for competitive bidding. The companies that reply to the RFP will do so with a formal set of proposal documents addressing the customer's technical and business requirements, written to the specifications for the generation of the textual documents as described in the customer's RFP package. Timing is crucial due to the need for fairness in the process, and all responding companies must follow the same rules. Often, the company releasing the RFP does not know how to solve its problem, and by virtue of this, the noted requirements will be sparse, vague, and perhaps, misleading.

Since many large systems are procured by governmental agencies, we will examine how a governmental agency requests a bid for a system they are interested in buying. Large industrial organizations use similar processes. These types of organizations send RFPs to companies to acquire systems, expecting to receive multiple proposals for evaluation before they select the company that will receive the contract. Therefore, all aspects of what we learn by examining the procurement of a system by the federal government will be relevant to generating a response to a system RFP by any type of organization.

Specifically, we will look at the procurement of a system that does not yet exist, and must be designed, built, and tested. We will explore what is in an RFP package to get a better idea of the information types a Systems Engineer has to work with to generate a proposal. We will also describe how the material in the RFP package is generated by people in the customer's organization and others, such as SETA contractors that the customer may hire to help them.

What in the world does the customer want to buy? When does the customer want it? How many do they want?

The RFP Package

A large government RFP package contains a Request for Proposal (RFP), Statement of Objectives (SOO) or Statement of Work (SOW), Performance Based Systems Specification (PBSS) requirements or a Technical Requirements Document. The total page count in the RFP package can be in the hundreds of pages and the requirements can be in the thousands!

RFP

The RFP defines what the customer wants to buy in general terms, delivery schedules, the number of systems, place of delivery, pertinent paragraphs of various government contractual documents that are applied to the how the proposal is to be written

(Outline topical headings, text font, text size, etc.) and how the customer will evaluate each of the submitted proposals. The RFP is very specific (specific as natural language can be), because the customer expects to have more than one proposal to evaluate and needs a specific set of criteria to evaluate all of the proposals to ensure fairness and avoid lawsuits.

You can't be serious!

RFP Proposal Generation Writing Rules and Proposal Evaluation Criteria

The RFP contains two sections that require detailed literal compliance. The Lead Systems Engineer must be aware of each of them and ensure that every one is addressed in the technical proposal. Not complying with content of these sections will likely have the proposal declared non-responsive to the requirements. For example, if the allowed page count is exceeded, the reviewers will not read past the maximum number of pages specified in the RFP and as a result the bidder that submitted that proposal will receive a low technical evaluation score. Why? Because only part of the proposal was reviewed and so the reviewers were not able to evaluate all of the bidder's information, to ensure fairness. The reviewers have no choice in the matter, because the rules of the evaluation were spelled out in the RFP so that each bidder's proposal would be reviewed using the same evaluation criteria. In federal government procurements the sections of the RFP that contain the writing rules and the evaluation criteria are sections L and M.

Section L - Proposal Instructions

Section L - Proposal Instructions contains specific rules for writing the proposal, including font type, font size, paragraph headers, and numbers. This section also defines the type of documents to be submitted and the maximum number of pages of each document. A typical list is as follows:

Technical proposal 50 pages
Management proposal 50 pages
SEMP (Systems Engineering Management Plan) unlimited pages
SDP (Software Development plan) unlimited pages
System Test Plan (unlimited pages).

These instructions may seem trivial to the inexperienced and overconfident Systems Engineer; however, if they are violated, points will be deducted from the overall score in the evaluation of the proposal. Any point lost can mean the loss of the competition, and perhaps the loss of millions of dollars of work for the company.

Section M - Evaluation Criteria

Section M - Evaluation Criteria defines the rules that the proposal evaluators will follow. Different aspects of the proposal will have different factors associated with the weighting of the score of the Technical, Management and Past Experience documents in the submitted proposal package. For example, there may be more weight given to the Technical proposal than the Management proposal, and within each proposal volume various subsections may receive different weights. The prudent Systems Engineer will spend commensurate portions of his time on the different sections of the proposal to maximize his score potential.

Statement of Objectives (SOO)

The SOO describes the objectives of the system from Business Operation's viewpoint, providing a Concept of Operations. It does not typically give technical performance criteria, (hardware or software requirements) but it may contain specific values that relate to business operations, such as the number of transactions expected at a bank teller's window and the type of transactions and their frequency of occurrence (if you're really lucky.) The SOO is a high-level document and requires further definition to make sure the operations the customer wants the system to perform are included in the information given to the System Design team. The bidder has the responsibility of creating a Statement of Work (SOW) that defines the work to be done to build a system that will satisfy the operational objectives of the SOO.

The Statement of Work (SOW)

The Statement of Work (SOW) is sometimes provided instead of the SOO and it defines the work that needs to be done to build a system to satisfy the customer's requirements. It may be provided when the customer has a set of business rules that are invariant (e.g. air traffic control system) and wants a new or improved system architecture to help achieve more capable and efficient operations. However, when there is a SOO in the RFP package, the customer usually requires the delivery of a SOW as part of the proposal submission, so that the feasibility of the customer's approach to develop the system may be evaluated. Remember that in a proposal you are selling the customer on your solution as well as your ability to deliver it as planned.

Performance Based Systems Specification (PBSS)

A PBSS normally accompanies a SOO, as it defines the system's requirements in terms of the operational requirements or the business rules the system must meet. It is normally written at a higher level than a Technical Requirements Document (TRD) that is associated with an RFP for a system where the operational requirements are well understood and invariant.

An example of a PBSS requirement is:

The bridge automation system shall provide an automated toll collection capability for the bridges in the state of Euphoria that will provide twenty-four hour per day data on each vehicle that crosses any bridge in the state.

The customer usually provides a PBSS in the RFP that is at a very general and high level, and may ask the bidder, as part of the proposal submission, to deliver a much more detailed version. The reason for doing the submission of the updated PBSS to the customer during the proposal is so the customer can see how well the bidder understands the operational requirements during the Proposal Evaluation. Understanding the customer's operational requirements is key to designing a system that will meet the customer's needs.

Another reason for the customer to ask the bidders to update the PBSS is to gain insight into the operational requirements. Sophisticated and astute customers realize the value of independent views, as the PBSS may reveal aspects of their operations that have not been identified but may be critical to their success. The application of a technology that is unknown to a customer may enable a new and valuable operation that was not anticipated!

Technical Requirements Document (TRD)

The Technical Requirements Document (TRD) is usually associated with an RFP that has a SOW rather than a SOO. Where the operational rules (business rules) of the system are invariant and there is a set of similar alternative systems designs possible, the customer will include a TRD in the proposal package instead of a PBSS. It will contain requirements that are more focused on the details of system design specifics than on operations-oriented requirements.

For Example:

The system shall have 4 Input/ Output channels of 100 megahertz capacity and be compatible with XXX protocol.

Who changed the organization?

Assuming for the sake of discussion that the system's lifecycle is ten years long, and recognizing that average organizational regime changes in companies occur every two years, a Lead Systems Engineer who stays in his role for the duration of the lifecycle begins to realize that he is likely to face five different management organizations and inevitable changes in the system's team. This is yet another set of complexities, ambiguities, and uncertainties the Lead Systems Engineer must manage. It is quite likely that he will be required to the deliver similar briefings to subsequent Management teams. Notice the word 'similar', instead of the same briefings. Although the topics may be the same, the astute Lead Systems Engineer will accommodate the preferences of new Management teams by tailoring his briefings to align with them.

System Design and Production Phase

We have discovered that the Lead Systems Engineer assigned to lead a large systems design and production effort must possess:

Out standing people skills
The ability to deal with complexity, ambiguity, and uncertainty
An understanding of the iterative and recursive nature of the work involved in requirements understanding and system design
The system's lifecycle and phases and how they affect the Lead Systems Engineer
An understanding that the Proposal Development phase encompasses an extremely short time period with incredible amounts of work
The long-term effects of decisions on the system's lifecycle

We have not yet discussed in detail the System Design and Production Contract Phase or the Production Support Phase of the system's lifecycle. This is also a critical part of the Systems Engineering domain.

System Design and Production Phase / Production Support Phase

The lead Systems Engineer will be a candidate for the role of Chief Engineer during the Systems Design and Production Contract Phases of the program, although there is no guarantee she will get that role. A Chief Engineer is usually designated, and must make sure there are detailed and testable requirements, the design conforms to the requirements, and there are proper tests to verify and validate requirements. The Chief Engineer must also attend to customer reviews, conduct internal reviews, and help the Management team keep the customer satisfied.

During this phase a number the technical domains will be involved, such as communications, command-and-control and others. Some Systems Engineers that possess domain expertise in those areas will be assigned. The number assigned will be determined by the amount of work as originally defined in the cost proposal section and agreed upon during contract negotiations prior to the start of work.

Production Support Phase

The major design decisions were made during the proposal and contract phases. The systems that are deployed incorporate those choices made during those phases. Consequently, the number of people supporting the deployed systems is dependent upon the number of systems deployed, and more importantly, is defined by the details of the support contract.

During this phase of the system's lifecycle, there are likely to be few Systems Engineers involved, and the support staff members at each deployed system will probably be Field Engineers and Applications Engineers. However, the wise contractor knows that there is the potential for follow-on business during the Support phase, and so will assign a Systems Engineer, perhaps on a part time basis, to the Program Manager to help develop those opportunities.

During this phase it is likely that the lead Systems Engineer on the proposal and the contract will not be involved in the program. In general, he may be reassigned after the contract award to another new business pursuit, because his skills will be needed on new business pursuits and on other systems design development contracts.

So you still want to be a Systems Engineer?

It is an amazing accomplishment to complete a role as a Lead Systems Engineer. They must learn to thrive on complexity, ambiguity and uncertainty, all exacerbated by the relentless pressure of time. They can order and structure creative work and consolidate that work into a written, winning proposal. They secure and execute a successful contract by delivering the system on schedule, within budget, and performing according to requirements. Ultimately, they come back to work the next day and do it all over again. These are the true shapers of our future, and without their efforts the ability to produce complex, innovative and astonishing new systems would be virtually impossible.

Summary of the Text

The text is designed in a manner that allows the instructor to follow the design flow of the book to introduce the concepts of Systems Engineering to the students in the order in which they will typically be utilized. We begin by establishing the typical Engineering approach used. This includes the scientific method and other increasingly complex methods of problem-solving that have been shown to enhance the problem-solving process within the engineering domain. We also consider the concept of Bloom's Taxonomy, which establishes the method by which students gain and retain knowledge during the educational process. This taxonomy is the principal reason why we have constructed a practitioner-focused Systems Engineering curriculum, as it enables the students to learn and retain the important concepts of Systems Engineering in a manner that will enable them to apply the concepts within the professional workplace.

Next, we establish the need for Systems Engineers to develop a more holistic view of the problem than has been developed through their technical educational processes. This holistic view is essential in order to take advantage of the analytical skills developed by the typical technical professional, and to enhance their existing abstract and free-association reasoning capabilities. These concepts are presented in relation to developing a method of seeing more within the problem than is readily apparent. Many Systems Engineers have demonstrated that a superior ability to gather more information and insight from an existing problem will typically result in a vastly superior solution to that problem.

We then move into the development of multiple viewpoints of the problem, relying upon the systematic methods developed in large commercial corporations and the defense industry. Specifically, these viewpoints are related to the Department of Defense Architectural Framework (DoDAF) views, the development of which can provide the students with required viewpoints to enhance their understanding of the system.

In this text, students will learn to develop the AV-1, OV-1, Ov-2, OV-3, OV-4, and the OV-5. Use of these multiple viewpoints and the insights gained by their development will be crucial to Systems Engineers in their Requirements Development Phase. As these views are typically utilized as the basis for the majority of derived requirements in the system, the next step in this process is the extraction and recording of the requirements.

Once the known requirements have been developed, we explore the concept of establishing what we know about the problem and what we do not know about the problem. This results in an intellectual exercise in which we formulate how we gain the information that is not presently available in order to resolve the unknowns in our system. This includes modeling and simulation techniques, the modeling of discrete systems, and the modeling of dynamic systems.

We spend some time as well in discussing the unknown quantities in a given problem and the concept of uncertainty. This is of significant value to the typical Systems Engineering developer, as it is important to identify the gaps of knowledge that will necessarily exist in our approach to resolving the problem.

After completion of a revised requirements list, we move into establishing formal Systems Architecture for our design. Once this technical solution has been established, we can then establish the Physical Architecture necessary for the solution to be implemented in the real world.

In addition to this development process, we also include discussions of Risk Management and Specialty Engineering. In many organizations, Specialty Engineering includes concepts of reliability, maintainability and availability, usability, and other related concepts that will be useful to the Systems Engineering practitioner in understanding the limits of his or her solution.

We then discuss the concepts of verification and validation, ensuring that the solution as presented will be adequate for the satisfaction of the customer and organization. We also provide a brief summary of the Project Management process, in order to ensure that the reader gains a basic understanding of how a project plan is developed and documented once the system design has been completed.

The final section of the text describes the previously mentioned proposal effort and the approaches taken by the instructor in both the development and implementation of that proposal effort in a university setting. Topical appendices are also provided for the simplification of routine tasks in the Systems Engineering methodology, by providing templates for use by the students. Specifically, a verification and validation plan template is included for reference, and a Systems Engineering Management Plan (SEMP) also defines many important areas for consideration in a system's development.

Chapter 1: The Engineering Approach

To understand the position of Systems Engineering in the broader spectrum of problem solving and engineering, we will begin by examining the natural progression of understanding in the development of the Systems Engineering domain.

The Scientific Method

The goal of Science is to investigate and understand the natural world, to explain the events in the natural world, and to use those explanations to make useful predictions about the world in which we live. Scientists collect and organize information in a careful and orderly way, looking for patterns and connections between cause and effect. Therefore, science is an organized way of gathering evidence to learn about the workings of the natural world.

Science typically begins with an observation. This is the process of gathering information about events or processes in a way that allows us to draw conclusions about our environment. The information gathered is typically referred to as data. There are typically two types of data; quantitative data are specific numbers related to observational values, and qualitative data relates to descriptions and characteristics of a more abstract nature.

The scientific method follows a five-step process:

Observe
Form a hypothesis
Develop a controlled experiment
Conduct experiments, record and analyze results
Draw conclusions from results

A hypothesis is a potential explanation to a problem or unknown quantity. It is simply the conviction that has not yet been proven or disproven. A hypothesis must be stated in a way that is testable. The statement is considered testable if evidence can be collected that either does or does not support it.

The design of a controlled experiment is intended to allow the results from the experiments to be used to draw conclusions about the hypothesis. In many cases, a controlled experiment changes only a single variable at a time. All other variables are to remain unchanged or control. Additionally, an experiment is normally based upon the comparison between a control group and an experimental group. The difference between these groups is the single factor being measured. The control group serves as a baseline for the experiment while the experimental group shows the effect of the variable being tested.

There are at least two variables in any given experiment, including independent variables and dependent variables. The independent variables are those that are deliberately changed by the researcher. The dependent variables are those that are observed during the experiment.

Conducting, recording, and analyzing results is the process immediately following the experiment. The data that have been collected must be organized and analyzed to determine first whether the data gathered are reliable. Typically this relates to the repeatability of the experiment if it is conducted numerous times. Experiments that do not demonstrate reliability cannot be trusted to draw conclusions regarding the environment. Secondly, the data are analyzed to determine whether they support or negate the hypothesis. Ultimately, experiments will be repeated numerous times in order to establish the repeatability of the experiment prior to asserting any validity, which can be gathered from the results.

A theory is a broad and comprehensive statement of what is thought to be true about the environment. A theory must be supported by considerable evidence, and is formed after a hypothesis has been tested many times and supported in numerous cases.

As an example, assume we want to determine the effects of a certain fertilizer on the growth of plants grown in a greenhouse. We would like to ascertain if the plants will grow best with a week concentration of fertilizer, a medium concentration of fertilizer, or a high concentration of fertilizer. In order to do this, we will define a hypothesis: we predict that the plants will grow best with a medium concentration of fertilizer.

We then set up a controlled experiment. Assume we have two hundred plants that will be divided into four equally sized groups as follows:

Group 1 will contain fifty plants that will receive water.
Group 2 will contain fifty plants that will receive a week concentration of fertilizer and water.
Group 3 will contain fifty plants that will receive a medium concentration of fertilizer and water.
Group 4 will contain fifty plants that will receive a high concentration of fertilizer and water.

The plants will be watered daily, with an equal amount of water. They will be located in a randomized position within the greenhouse, and observed over a period of one month. The plants will ultimately be measured to see which ones grew the tallest.

In this experiment the control group contains fifty plants that are receiving only water. The experimental group is represented by the one hundred fifty plants that receive various concentrations of fertilizer.

After one month of watering the plants, the following data is obtained:

Group 1 grew to an average height of 15 cm.
Group 2 grew to an average height of 35 cm
Group 3 grew to an average height of 28 cm.
Group 4 grew to an average height of 10 cm.

When we analyze these results utilizing the scientific method, we find that the results do not support our original hypothesis that the plants would grow best with a medium concentration of fertilizer. We can draw a conclusion based upon these results that plants will grow best with a weak concentration of fertilizer.

The scientific method is very effective when utilized to create a basic understanding of an environment in which we do not already possess a great deal of understanding. This method is less useful when applied to the specific issue of problem solving. These challenges are typically better addressed by utilizing an engineering design method.

Engineering Design Methods

Engineering methods have been developed in order to directly address problems and provide solutions to those problems. Design problems typically have:

A goal to be achieved
Some constraints within which the goal must be achieved
Some criteria by which a solution can be measured

Constraints set specific quantitative targets or limits. Criteria are more flexible and might be used for judging between different design approaches, many of which might meet the specific design constraints.

Design problems are typically incompletely defined. This implies that there is no definitive formulation of the problem, and therefore any formulation may embody inconsistencies or inaccuracies that could dramatically impact the quality of the solution. In general, formulations of the problem will be solution dependent, thereby limiting our unbiased understanding of the problem and our ability to provide an optimal solution. Finally, there typically is no definitive solution to the problem.

Because of the nature of engineering problems, engineers tend to be solution-focused, whereas scientists tend to be a problem-focused. Engineers typically use solution conjectures as a way of exploring and better understanding the problem at hand. Engineers will actively manage and control the process of the development. By virtue of this, experienced designers will typically make well-planned, controlled exploration of the problem space. They will move between solution concepts and problem exploration seamlessly. They maintain a broad view across several sub-solution alternatives, and ultimately make engineering problem-solving look easy and intuitive.

A simple model of the engineering problem-solving process is shown in the following diagram, Figure 1.

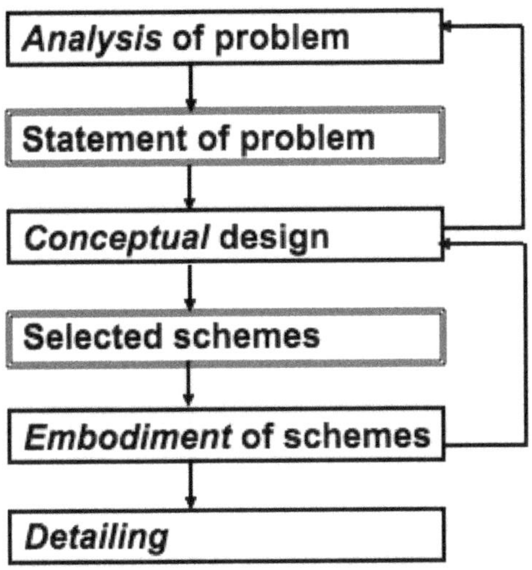

Figure 1: Common Problem-Solving Approach

In this problem-solving approach, two iterative loops are defined, which helps the engineer to clarify his understanding of the situation. The first of these loops includes a constant revisiting of the formulation of the problem, and the second focuses on developing numerous approaches for appropriate solutions. The engineer iterates amongst these two loops in order to provide an optimized solution.

In the VDI model, the problem is segregated into more manageable sub-problems and partial problems, which can then be addressed with aggregated solutions that will be assembled to create an overall solution. This approach is shown in the following diagram, Figure 2.

VDI model:

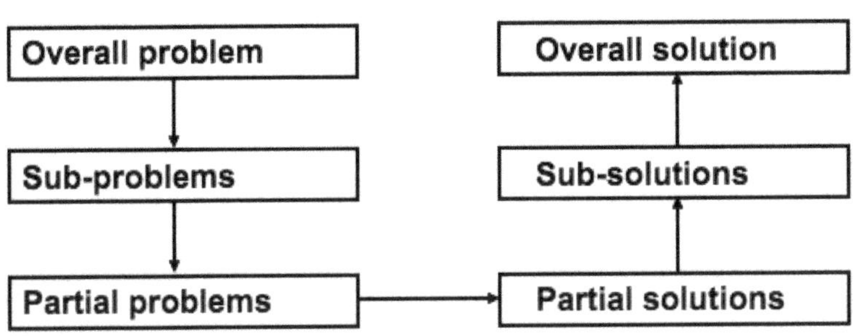

Figure 2: VDI Model

Cross proposes a similar problem-solving model to the VDI model, further adding a connection between the overall problem and overall solution generated. This connection is placed in order to assure that the solution meets the ultimate requirements of the proposed problem. This model is shown in the following diagram, Figure 3.

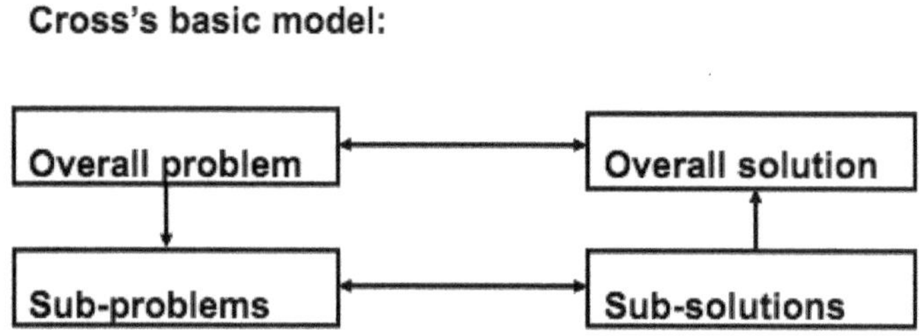

Figure 3: Cross' Model

Various methods have been proposed in a variety of publications that will generate numerous problem formulations and solution approaches. Among these techniques is brainstorming, which is a technique developed in order to avoid the phenomenon of groupthink and group-shift within the problem-solving team.

Many other methods have been proposed in order to generate a rich solution set for selection by the problem-solving team. These might include the generation of direct analogies, the use of personal analogies, the development of symbolic analogies and ultimately the use of fantasy or defining the optimal end state for the system.

Methods designed to enlarge the search are prevalent in the literature. These include methods developed to transform the problem such as magnification, modification, unification, subtraction, and addition. Random input will often generate unique and interesting solutions to a problem, such as when a random word is selected and the problem-solving team must define its application. Some techniques have been borrowed from the Black-Belt arena, including the Five Whys, and the development of a counter planning approach. One or more of these techniques might be beneficial to the development team.

Standardized methods for establishing an understanding of a challenge include the use of user scenarios, use cases, the development of an objectives tree, requirements development, quality function deployment, developing a morphological chart, utilizing weighted objectives, and applying value engineering to the challenge at hand. One or more of these techniques may be useful in gaining further insight into the problem set.

When we combine many of the previously mentioned examples and utilize several of the previously proposed models, we develop an understanding of how a problem might be approached to generate an efficient solution. We understand the transition from an overall problem to an overall solution requires a systematic process, and that process may include clarifying the objectives to be accomplished, identifying opportunities and approaches, and ultimately defining the details of those approaches to provide an acceptable solution. This is shown in the following diagram, Figure 4.

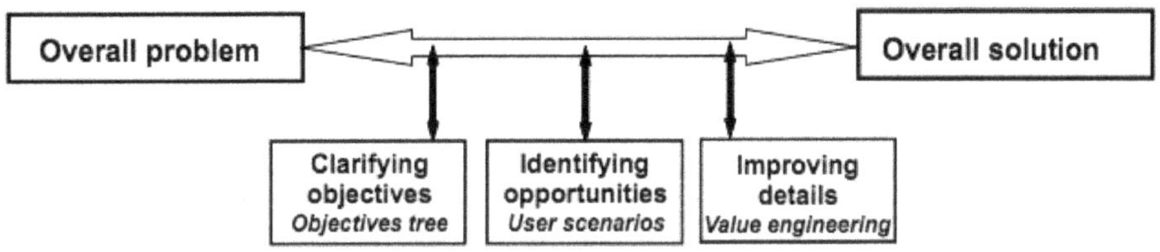

Figure 4: From Problem to Solution

An alternative complication has been proposed in the VDI model, where the problem is broken down into sub-problems. By virtue of this, we can utilize a simple approach in the generation of appropriate sub-problems. This concept is shown in the following diagram, Figure 5.

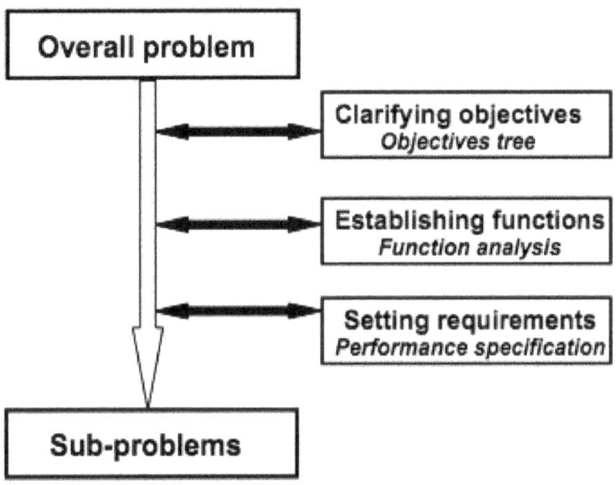

Figure 5: From Problem to Sub-problems

When these two approaches are combined, a complex diagram describing the problem-solving process is generated. This type of problem-solving model is often quite useful in establishing for the engineering team a practical and systematic method by which they may pursue appropriate solutions for their problem. This technique has been used in a wide variety of organizations, and is shown in the following diagram, Figure 6.

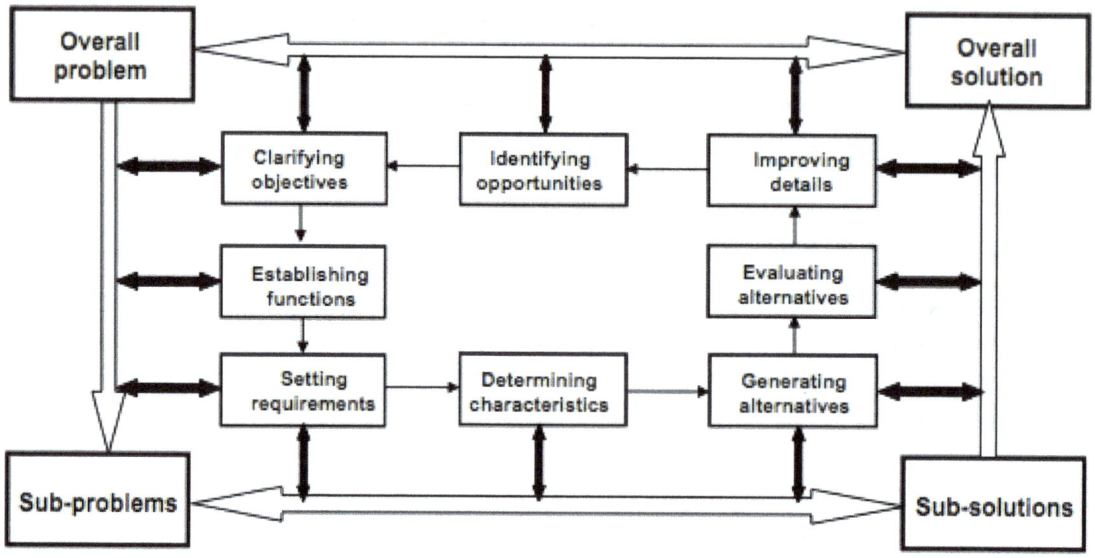

Figure 6: Aggregate Approach for Problem Solving

While this technique might be academically interesting, there are several factors associated with complex systems development that preclude the ability of the problem-solving team to utilize these concepts as independent paths to a solution. Complex systems typically demonstrate characteristics that require the problem be approached in a holistic and assembled manner. This will often limit the use of typical engineering design methods in the generation of appropriate systems solutions to complex problems. This challenge has led to the development of the Systems Engineering approach to problem solving as described in this text.

Bloom's Taxonomy

To understand Systems Engineering effectively, there is value in considering the methods by which cognition works in the human mind. That is to say that we are interested in how people learn new things and the process by which this learning takes place. By understanding these processes, we can better understand how to efficiently learn enough about our system to enable us to complete the system design.

As it turns out, this has been a question that many teachers have been asking for a long period of time; "Where do we begin in seeking to improve human thinking?"

Benjamin S. Bloom extensively contemplated the nature of thinking, eventually authoring or editing eighteen textbooks on the subject. Discussions that occurred during the 1948 convention of the American Psychological Association centered on asking Bloom to spearhead a group of educators who eventually undertook the ambitious task of classifying educational goals and objectives. Their intent was to develop a method of classification for thinking behaviors that were believed to be important in the process of learning.

Eventually this framework became a taxonomy of three specific domains:

The cognitive - knowledge-based domain, consisting of six levels
The affective – attitudinal-based domain, consisting of five levels
The psycho motor – skills-based domain, consisting of six levels

In 1956, eight years after the group first began, work on the cognitive domain was completed and the handbook commonly referred to as "Bloom's Taxonomy" was published. This volume was originally intended for university examiners but has since been transformed into a basic reference for all educators worldwide.

Bloom's taxonomy is a multi-tiered model of classifying thinking according to six cognitive levels of complexity. The lowest three levels are knowledge, comprehension, and application. The highest three levels are analysis, synthesis, and evaluation. The taxonomy is hierarchical; each level is subsumed by the higher levels.

Bloom identified six levels within the cognitive domain, from the simple recall or recognition of facts as the lowest level, through increasingly more complex and abstract mental levels. The highest order is classified as evaluation. A description of the six levels as well as verb examples that represent intellectual activity are listed in Figure 7.

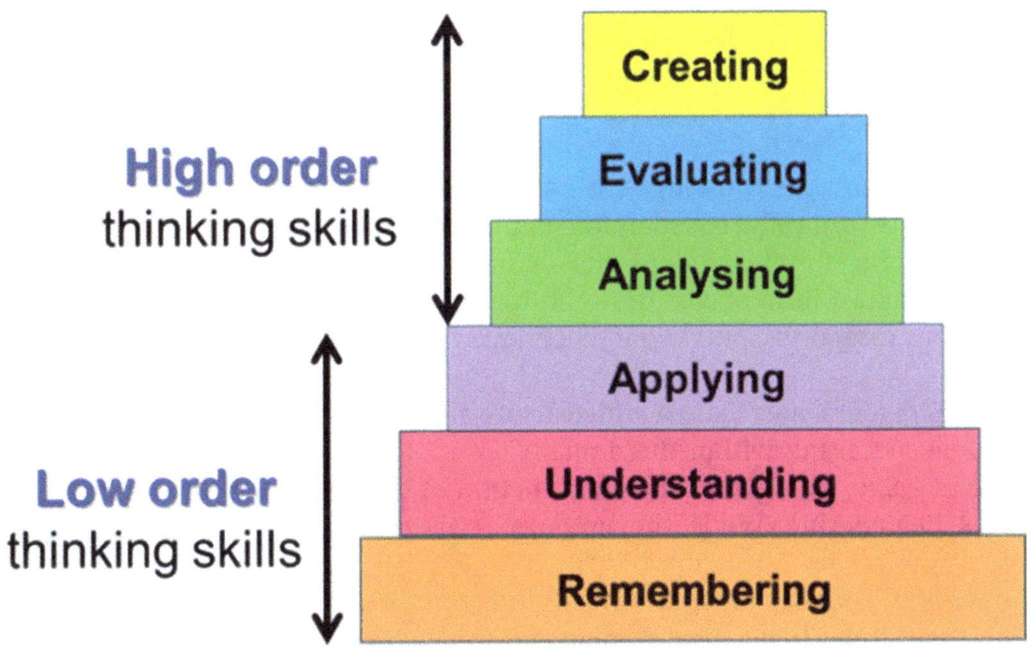

Figure 7: Bloom's Taxonomy Descriptors

Beginning from the lowest levels:

Knowledge is defined as the ability to remember previously learned material. This includes the recall of a wide range of material, from specific facts to abstract theories. All that is required of this level is the ability to remember information.

Comprehension is defined as the ability to understand the meaning intended by the material. This capability might be demonstrated by translating the material from one form to another, by interpreting the material, and by understanding the predictors generated by the material. Typically this requires more than the simple remembering of material, and results in the first level of understanding.

Application is the ability to use the learned material in a new manner. This might include the ability to apply rules, methods, concepts, principles, laws, and theories. To demonstrate application requires a higher level of understanding than those under comprehension.

Analysis is the ability to break down the material into its constituent parts so that the organizational structure maybe well understood. This might be demonstrated by the identification of the parts, analysis of the relationships between the parts, and recognition of the organizational principles involved. To demonstrate this level requires a higher intellectual comprehension because the individual must understand both the content and the structural form of the material.

Synthesis is the ability to combine various parts to form a new whole. This will typically involve the production of a unique result, a plan of operations, or a set of abstract relationships. This cognitive area includes the concept of creativity and innovation.

Evaluation is the ability to judge the value of the material for a given purpose. The judgments are typically based upon defined criteria. These may be internal or external criteria, and the student may actually determine the criteria to be utilized. This is the highest form of cognitive ability, because it requires all of the elements of the previous categories, plus value judgments from the individual.

Systems Engineers must learn a wide variety of new concepts and behaviors in order to become successful within the domain. An understanding of this taxonomy is of significant value, as it guides the construction of this text and approach to learning Systems Engineering. While the authors focus on understanding and imparting content to the practitioner, deconstruction of a proposal effort within this textbook will force the student to increase his or her understanding, moving to much higher levels of cognition. In general, students will reach the evaluation stage of cognition when completing a significant proposal utilizing the Systems Engineering concepts as described.

Chapter 2: The Systems Engineering Approach

Systems Engineering is a relatively recently developed domain of knowledge that is intended to improve the typical engineering organization's ability to deliver products that will meet customer needs more readily than those which have been developed in the past.
Systems Engineering does not require the development of skills in a new science or mathematical field. This makes it unique in engineering domains. By virtue of this, many different disciplines can develop capabilities that will result in excellent Systems Engineering approaches to problem-solving, although in general the majority of Systems Engineers are also Mechanical or Electrical or Software (or other) Engineers as well.

Simply stated, the system is an integrated composite of products, components, people and processes that will provide capabilities that will satisfy the needs of a customer.

Systems Engineering consists of two significant areas of interest: the technical knowledge domain in which the Systems Engineer operates, and Systems Engineering Management. Therefore, Systems Engineering is an interdisciplinary engineering management process that develops and verifies an integrated set of system solutions that will ultimately satisfy customer needs, whether implicit or explicit.

It is imperative for the engineering leader to utilize an effective subset of Systems Engineering tools and techniques in order to ensure efficiency and provide focus in the development of solutions for customer needs. This section of the text is not intended to be an exhaustive description of all elements of the Systems Engineering domain of knowledge, but rather a quick overview to provide some guidance for the interim leader to begin to incorporate Systems Engineering concepts into his or her product development process.

The Systems Engineering Approach

Many different models have been put forward with regard to the development of a system utilizing Systems Engineering principles. One of the most common has been adopted by INCOSE (the professional organization for Systems Engineers), which has been adapted from many authors in the domain including the Defense Acquisition University. In general, these process documents are quite complex and challenging to understand, but when boiled down to simple processes and a step-by-step methodology for product development, they are easily understandable and applicable for virtually any product development organization.

The V Diagram of Systems Engineering Development is illustrated in Figure 8.

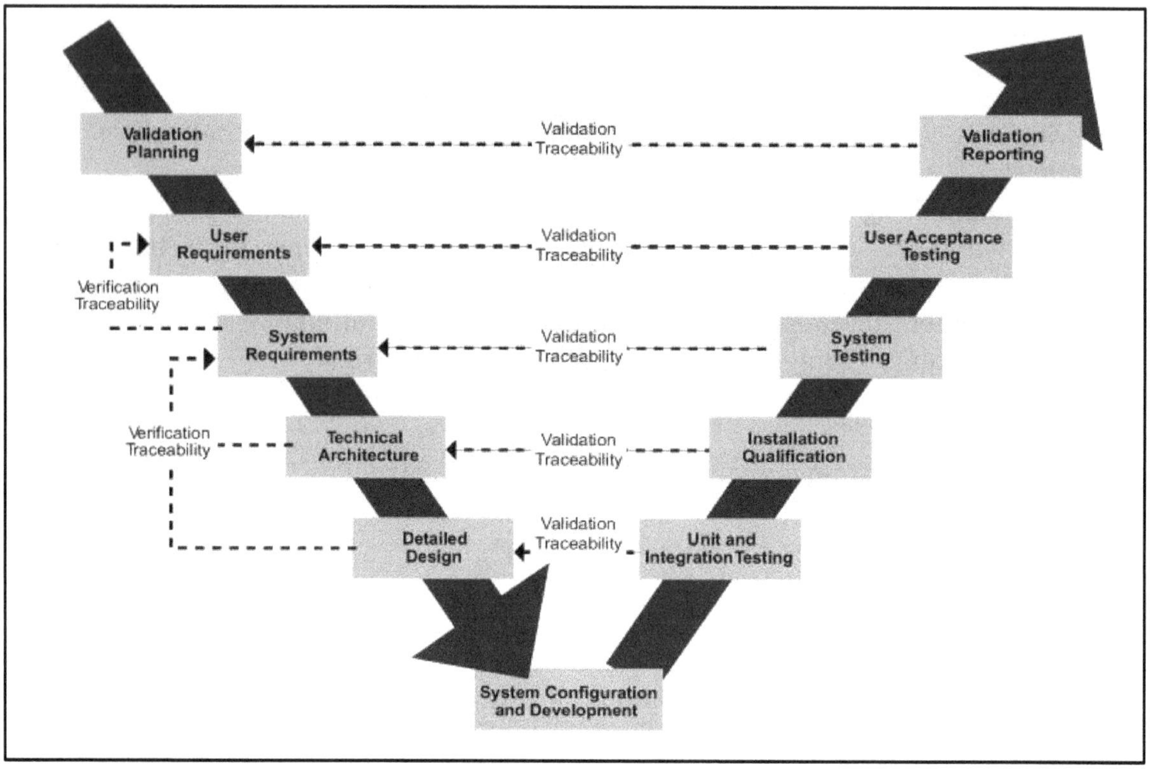

Figure 8: Systems V Diagram

This diagram demonstrates from a simplified perspective how the systems process will typically be applied in an engineering development project. Beginning from the upper left, the Systems Engineer will define users, user needs and ultimately (utilizing the tools in Systems Engineering methodologies), the requirements necessary for the project. Once the requirements have been developed, the Systems Engineer will continue to develop a systems architecture, which will ultimately be translated into a physical architecture and a detailed design, much as any standard engineering process is conducted today.

This simplified version was derived from a more complex version provided by the Defense Acquisition University. In this version, the entire program management cycle is included, and spans from the initiation of a program through the final retirement of the program products. This construction can be seen in Figure 9.

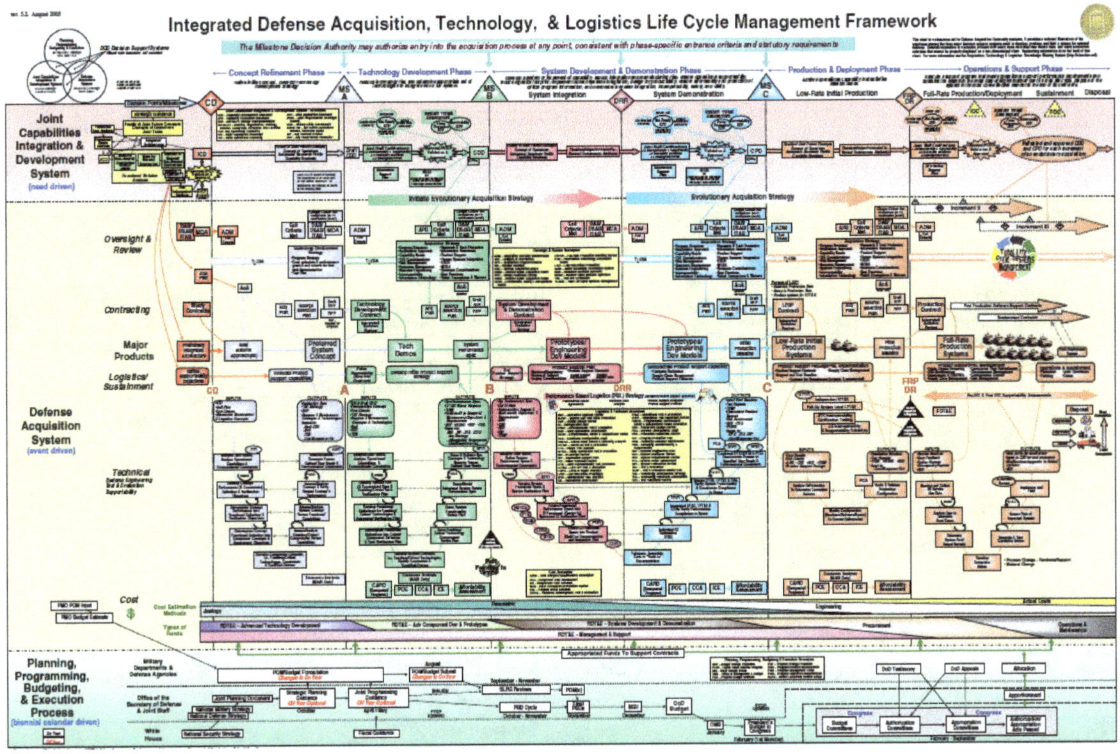

Figure 9: DAU Systems Process (www.dau.mil)

This is a highly complex diagram, and is often well beyond the needs of a typical engineering organization. However, upon reflection and observation one can readily understand that this diagram is nothing more than the combination of five independent System Development cycles in a longitudinal format. Therefore, we can investigate a much smaller section of this diagram and note the typical the diagram of systems development. This is shown in Figure 10.

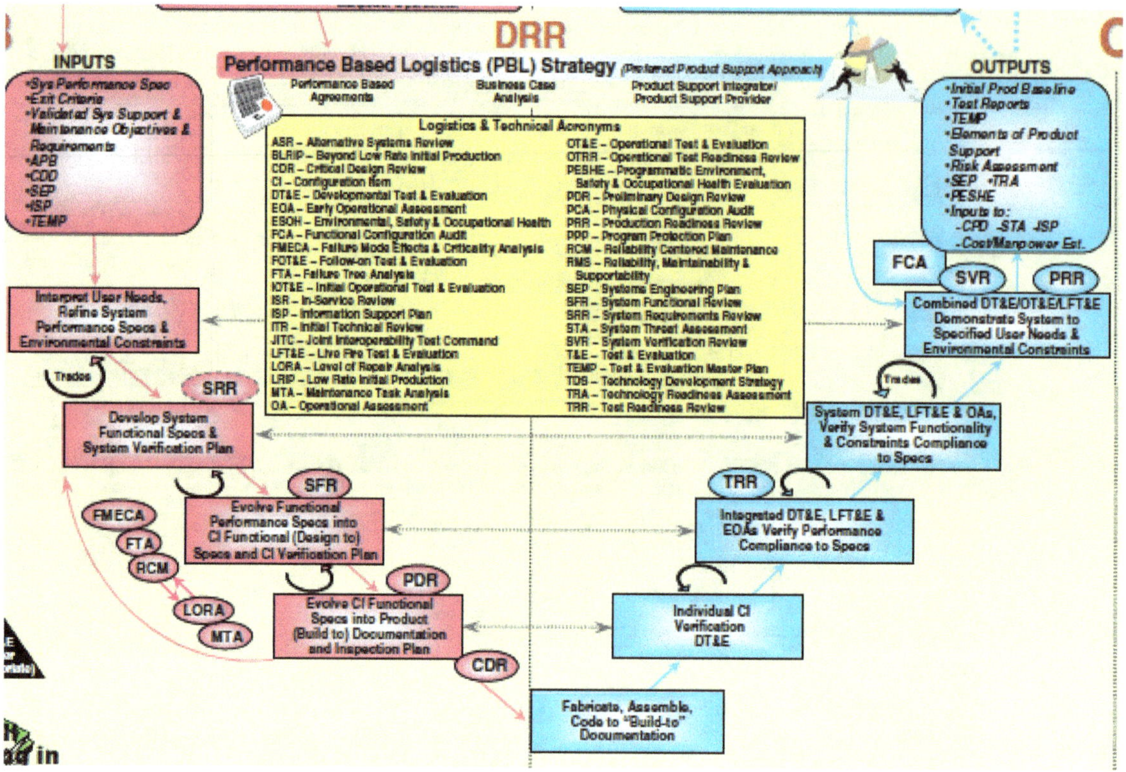

Figure 10: DAU Product Development Process

This process closely resembles the previously described model of development, so engineering leaders are urged to investigate this highly detailed model of the Systems Engineering process and choose those elements of greatest impact for their engineering organization. A subset of this complex system has been selected for discussion in this text, and has proven to be an effective methodology for small to midsize manufacturing and product development companies.

Systems Engineering in Context

To understand how Systems Engineering differs from Mechanical or Electrical Engineering, we will first examine a typical project schedule. This particular work breakdown structure has been developed as a generic WBS for a typical manufacturing entity. The reader will note that the system described contains mechanical, electrical, and software components.

A common approach to developing a new product with this content is shown in **Figure 11**.

Task Name
NPD Project Template
Gate#1 Project Approval
Develop Concept
Gate#2 Approval to Engineer Product
Product Design Phase
Mechanical Design
Develop Layout
Define Major Components
Develop Detail Design
Develop Assembly Drawings
Develop BOM
Check Drawings
Correct DWGs as needed
Release for Initial Pilot
Electrical Design
Develop Layout
Define Major Components
Develop Detail Design
Develop Assembly Drawings
Develop BOM
Check Drawings
Correct DWGs as needed
Release for Initial Pilot
Software development
Layout System
Develop ICDs
Define Major Components
Code Components
Initial testing
System Refinement
Preliminary release for Pilot
Gate#3 Approval to Build Prototype
Build Prototype
Execute In-House Testing (startup and verification)
Revise Design as Required
Gate#4 Approval to Release for Production
Production Release
Production support

Figure 11: New Product Development WBS

In order to incorporate Systems Engineering expertise into this WBS, we must add additional Systems Engineering tasks in two specific areas. These include the incorporation of systems development tasks prior to the detail engineering design of the system, as well as verification and validation tasks, which will be incorporated in place of testing. These tasks will be inserted in our project plan as shown in **Figure 12**.

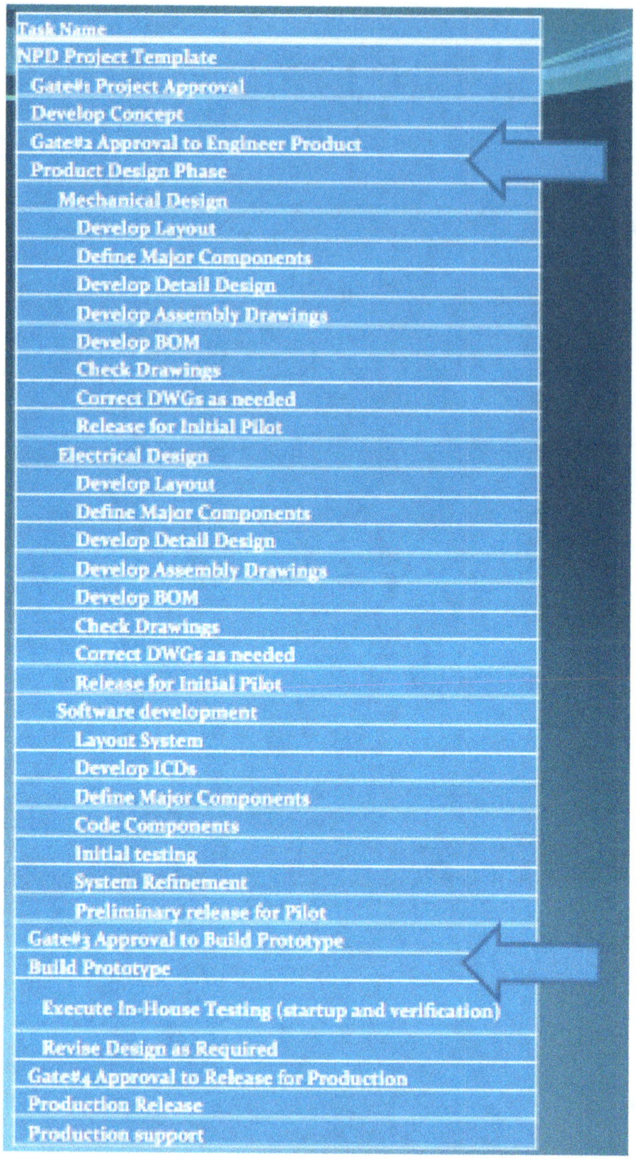

Figure 12: Where Systems Engineering Augments a Project

The tasks that will be incorporated representing the Systems Engineering development are as follows:

Product Design Phase
Detailed Project Planning
Develop Specification
Develop Schedule
Develop Budget/Risk and Charter
Systems Engineering
Identify Users
Identify User needs
Define development team
Develop AV-1
Establish general system requirements
Develop OV-2
Develop OV-5
Refine to black box level
Establish RVTM document
Populate requirements from views
Define System scope
Develop functional analysis
Compare OV-5 to functional analysis
Establish System Architecture
Define roles and responsibilities for physical architecture
Conduct Preliminary Design Review
Develop detail design
Mechanical Design

Figure 13: Systems Development Tasks

The additional tasks incorporated for the verification and validation process are as follows:

- Verification and Validation
  - Develop V&V plan
  - Execute In-House Testing (startup and verification)
    - Startup & Debug
    - Performance Testing
  - Revise per Test results
  - Field Beta Testing (Validation)
    - Tear Down Machine
    - Ship Machine
    - Install Machine
    - Field Testing of Machine
  - Revise Design as Required
    - Mechanical Revisions
    - Electrical Revisions
    - Conduct Critical Design review
- Gate#4 Approval to Release for Production

Figure 14: Verification and Validation Tasks

If a Gantt chart is utilized to demonstrate a generic project schedule with systems tasks, it would look similar to Figure 15. Note that the shaded areas represent the tasks added to a standard WBS in order to adopt a Systems Engineering perspective.

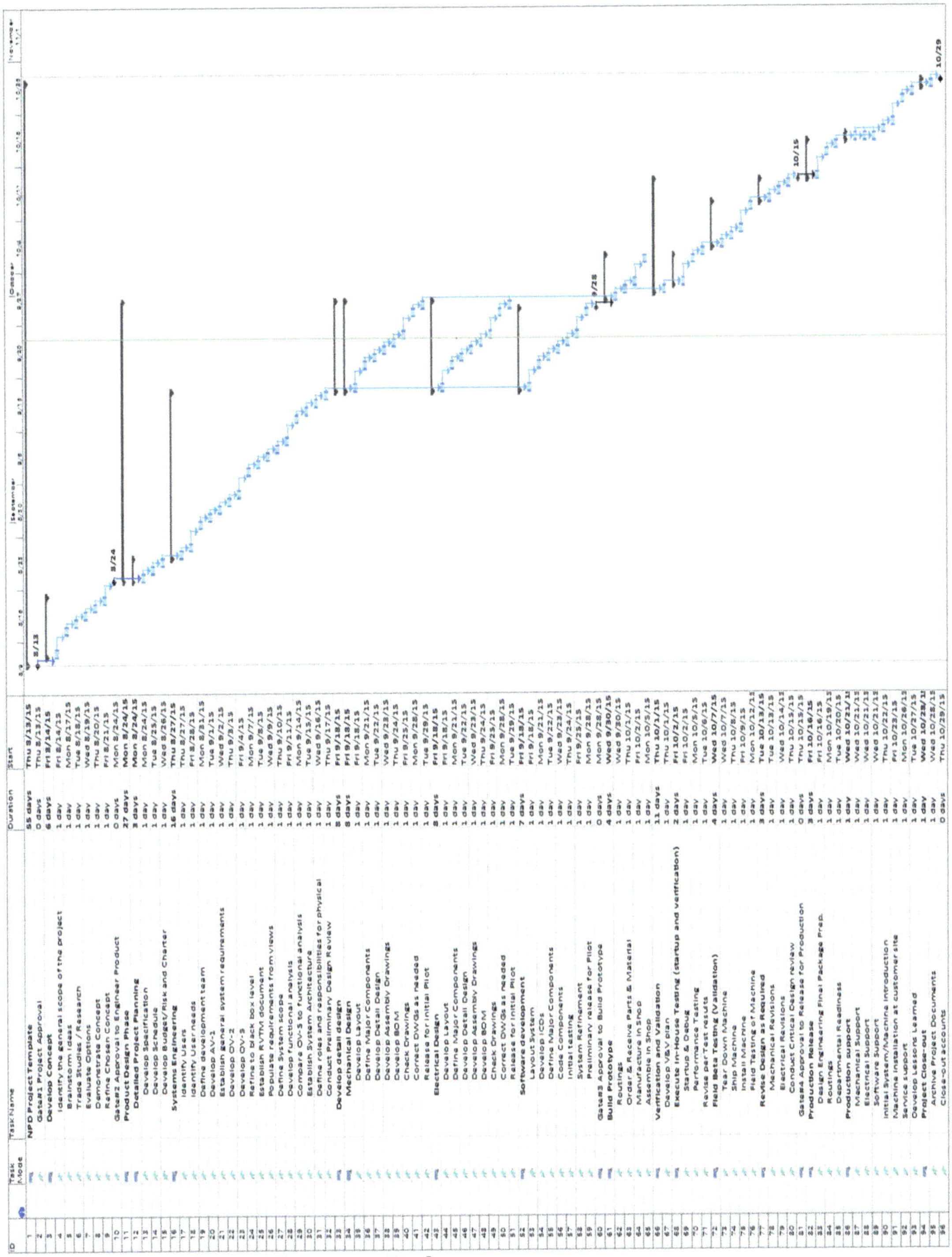

Figure 15: Gantt Chart with Systems Tasks

The Systems Engineering Process

There is a distinct difference between a classically trained engineer and a Systems Engineer. This distinction revolves around the ability of the Systems Engineer to develop a broader understanding of a problem and context within which the problem exists. Systems development in general requires a deeper understanding and a more reflective approach to formulating the problem than does classical engineering.

Therefore, Systems Engineers must consider more than the typical Engineering approaches to problem solving. In fact, many of the system tools are constructed in a manner that enables us to develop a more complete understanding of the issues of importance to the customer and developer. By more fully understanding the situation and problem, we can better address the problem and be more effective and efficient in the application of our engineering resources.

One considerable challenge that Engineers and Scientists face is the fact that we are typically intelligent and knowledgeable about the problem at hand. This knowledge and intelligence, when coupled with significant experience, often results in a quick jump to what we believe to be the optimal solution to the problem. In short, many engineers will look at a problem and almost immediately envision a solution to that problem. This is problematic for Systems Engineers, as it very quickly limits our ability to expand our understanding of the problem, and keeps us thinking 'inside the box', so to speak. Therefore, much of our development approach will be focused on keeping the Systems Engineer away from detail design as long as possible, until we have fully completed our systems architecture development. If we can accomplish this, we can remove initial bias and limitations from our final design, thereby creating a more focused and coherent solution to the problem not limited by our preconceived notions.

To understand the challenges we face, we will undertake a simple problem to illustrate these issues for our consideration. Imagine that you have been asked to develop a display for our local art museum, to put on display in our permanent collection a new and exciting acquisition. What information is necessary? When does the engineer fully understand the essence of the challenge?

Many Engineers would ask initial questions regarding the new acquisition. For example:

Is it a painting or a sculpture?
How much does it weigh?
What is its value?
How big is it?

If we establish that this is an approximately life-sized sculpture of a human being, carved from marble and weighing approximately six thousand pounds, most engineers have enough information to envision the final product.

If asked to document that final product, the most common solution would include a pedestal base reinforced to support and position the mass of the statue, which will be located on the floor of the museum and painted black. This is the extent of the system's development, and this is the solution generally provided.

Is this truly an adequate solution? Have we fully addressed the needs of the customer, and have we understood the true problem?

As a Systems Engineer, one tool that is utilized to enable us to better understand our problems is to look at our challenge from multiple viewpoints. These viewpoints might include different physical perspectives, different functional perspectives, knowledge domains, or a variety of other approaches. To explore these concepts, we will examine a photo of the statue in question.

Figure 16: Photo of Museum Acquisition

This new viewpoint immediately provides us with better understanding of the problem. Now we become more aware of the age of the acquisition, and perhaps the delicacy of the system that we must be aware of in our museum display. Does this fundamentally change the display we have previously designed? In most cases, we feel that the originally proposed design is still a superior solution, although we may incorporate slight enhancements and modifications to the design. But what if a new viewpoint is provided? What if we consider the historical provenance of this acquisition? This information is provided as follows:

Location: Louvre Museum, Paris, France

Artist/Maker: Unknown

Description, English: So-called "Venus de Milo" (Aphrodite from Melos). Parian marble, ca. 130-100 BC. Found in Melos in 1820.
Dimensions: H. 2.02 m (6 ft. 7 ½ in.)

Credit: Gift of the Marquis de Rivière to Louis XVIII of France, 1821
Accession number: Ma 399 (LL 299)
Current Location: Temporary - Department of Paintings, Sully, first floor, room 74

References: Hamiaux, M., Les Sculptures grecques, II, Paris, 1998, no. 52, pp. 41-44

This additional viewpoint often will prompt distinct changes in the design previously established. Now the designer will frequently focus on security issues, care for the statue, and potential insurance for liability issues. Therefore, by enhancing the understanding of our challenge with insight into the acquisition, we have dramatically enriched our approach to the problem. This is in essence what the Systems Engineering tools enable the Systems Engineer to accomplish.

Figure 17: Images of Venus de Milo

Before moving on from this challenge, let us consider many more viewpoints of the Venus de Milo statue, focusing on what insights these might provide us as Systems.

From these images, most Engineers would observe that lighting is clearly a significant factor, the background provided to the statue is of considerable importance, the ability of observers to see the statue from multiple angles is also of significant value. The statue is already exhibiting significant damage and decay, and therefore must be protected carefully when placed on display. Additionally, questions of security and safety for the display are often considered.

Again we see that the addition of information provided from different viewpoints can enrich our understanding of the problem significantly and therefore change the technical approach or solution.

If we take one final pass at increasing our understanding of this problem, we could observe the Venus de Milo on display at the Louvre in Paris, France. Upon observation, we would note several things:

The statute is located in a large salon.
No other artworks are located in the same salon.
The statue is raised a significant height above the floor.
The security barrier is approximately the same height as the pedestal.
A large amount of people comes to view the statue.
Many people would like their picture taken with the statue.
The statue is more of a destination than an appreciated artwork.

Upon reflection of this added data, it becomes clear that the challenge of placing this statue on display goes far beyond the design of a stable platform on the floor of the museum. We realize the real challenge is to create a setting that allows for effective movement of people throughout the museum and around the statute, free access to take photographs, and protection of the statue from the environment and the museum visitors. Due to the large number of sightseers, we must also provide additional cooling through the museum's heating and ventilation system.

If we had approached the problem based upon our initial assumptions and conclusions, it is unlikely we would have produced as an effective a solution as we ultimately arrived at after considering multiple viewpoints. Therefore, we will explore several methods that are typically used in Systems Engineering to develop multiple viewpoints of the problem at hand.

Chapter 3: The Art of Seeing

One of the biggest challenges facing the developing Systems Engineer is changing his or her patterns of thinking. Many of us who have worked our way through the technical educational system and academia have developed our analytical skills and capabilities at the expense of some of our abstract and free-association thought patterns.

It will become readily apparent to anyone studying Systems Engineering that new thought patterns and new approaches to problem solving must be developed. This ability to enhance the thinking processes is essential to fully understanding and embracing the holistic approach that is required for Systems Engineering.

A system, by definition, is a collection of related parts that interact in an organized way for a purpose. These three elements are of crucial interest to us:

Related parts
Interaction of parts
Organize for a purpose

Keeping these three factors in mind, we may enhance our understanding of the system simply by reflecting on the relationships between them. This creates insight related to systems and their development; however, if you want to understand a system you must study it in its entirety. Cutting it up into bits for study is likely to destroy its connectedness and the system itself.

Likewise, if you want to influence or control the behavior of a system, you must act on the system as a whole. Modifying it in one place in the hope that nothing will happen in another is impossible. This is what connectedness is all about. A Systems Engineer must become highly adept at understanding and visualizing this connectedness.

The system can be made up of many subsystems, such as the human body. The body is a system made up of many smaller subsystems such as the circulatory system, the digestive system, the skeletal system, and the nervous system.

This system can also be a part of a much larger system such as the human body, which may be part of a family, a community, a team or an organization.

A university is a system. The standard mental model of a university system is hierarchical. This is denoted by the concept of academic rank and the highly stringent organizational structure present in a university system. Most classrooms are an expression of this mental model.

Consider, for example, a typical university classroom. The instructor stands at the front of the classroom, possibly on a raised platform. The students are arrayed in orderly rows throughout the classroom seating area, all facing the instructor. There is a specific reasoning behind this construction. To the student, it is clear that within the classroom the authority resides with the instructor, exemplified by the raised position at the focal point of the classroom. Secondarily, the student is aware that the instructor is part of a much larger hierarchical system, meaning that the instructor is in turn outranked by individuals within the academic structure positioned higher than the instructor. This makes it very clear to the students that they are at the lowest rung of the ladder within the university system, and provides them with an understanding of where the credibility and authority lie within the system. This is of value to the instructor, as it immediately instills credibility and authority within the classroom environment.

However, to the Systems Engineer, the question is not one of authority, but rather; is this the optimal way in which to provide instruction at the university level? Many Systems Engineers have never even considered this question, including those who instruct at the university. This is largely due to the limitations introduced on our understanding by means of our mental models.

In general, we take in information from our environment and process it based upon our understanding of the source of the information. As an example, we see:

Events – things we see happening in the environment around us
Patterns of behavior – consistent and regular actions or events that occur within our environment over time
Systemic structure – permanent relationships and flows of information between different parts of the system.
Mental models - shared belief systems, ideas, assumptions, and goals of a given system that are usually unstated but are universally understood and accepted.

As we move from events to mental models, we also understand that this allows us to gain increasing leverage in our system's development. This is important to the System Engineer, as creating systems is highly unlikely to produce dramatic disruptive technologies for the organization. However, if one considers the change of mental models, which may be viewed based upon invalid assumptions or beliefs, we can often produce dramatic shifts to our understanding of the system to be developed.

Therefore, systems thinking looks deeper than at the surface events and even the patterns of behavior. It is critical that we begin to discover the mental models upon which we base our understanding of the system. This is seen in the Iceberg model of Figure 18.

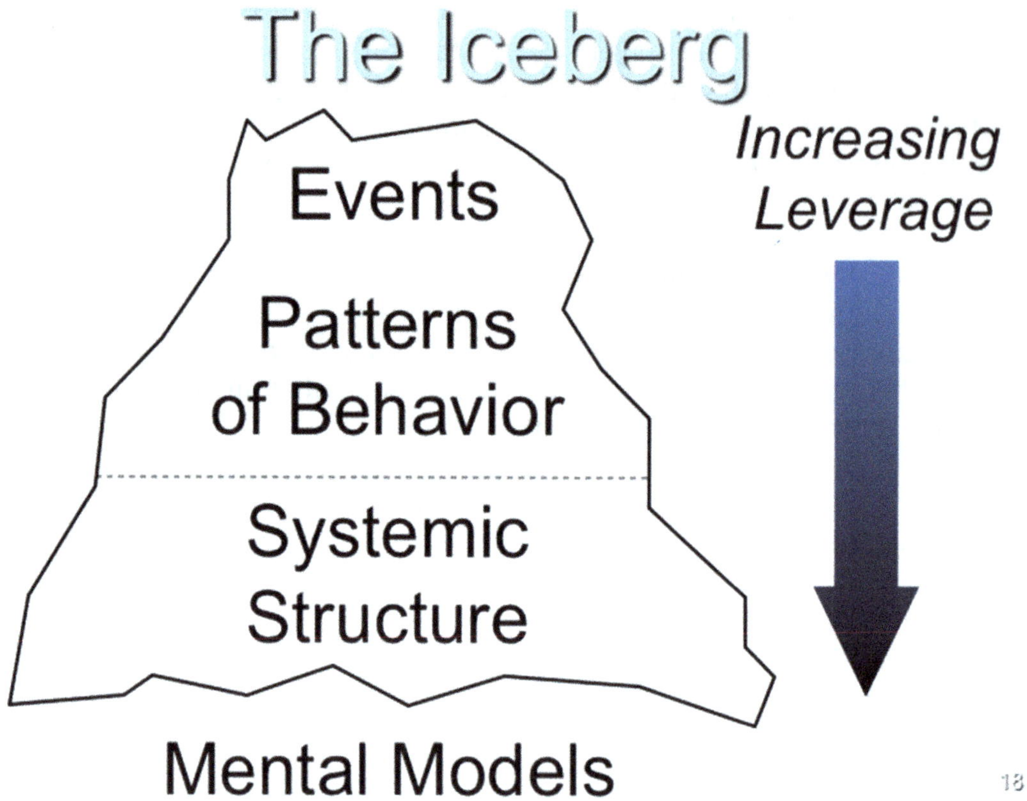

Figure 18: The Iceberg model

If we can successfully move our focus away from events and patterns of behavior towards systemic structure and the underlying mental models, we can avoid a variety of traditional problems faced by the system developer.

Sometimes we are stuck with mental models that create rules that don't really exist. While attending a meeting, have you ever asked yourself if the discussion is really true? Likewise, is the information based upon facts and data or is it an assumption held by many within the organization?

System tools have been developed to enable us to begin to see many of the mental models that can often be hidden very deeply within an organizational culture. In complex systems, it is very common that cause and effect will be separated by significant time and space. In addition, we often fixate on a specific part of the system, which can result in us missing the whole.

This has been best exemplified by the old story of five blind men who are asked to describe an elephant. Each feels the elephant and variously describes it as:

A long tube, like a snake
A heavy creature, like a tree
A thin and slender element, like a rope
A rough and hairy surface, like a brush

While each of these observations is correct, none of them embraces the whole of the system. It is critical that a Systems Engineer possesses this capability.

The essence of Systems Engineering is to find the leverage points for systemic change. A leverage point is where a small action can yield a large result. However, in our history our mental models have caused us to miss these leverage points, and we have erroneously developed:

Paved dirt roads in mountain areas that led to a decrease in safety
Low tar and low nicotine cigarettes that increased the intake of carcinogens
The US policy of fire suppression that increased the magnitude of forest fires
Seatbelts that resulted in more severe injuries to drivers
Airbags that resulted in more injuries to drivers

When one considers these developments, one realizes that we have created loops of misunderstanding. Clearly, the use of seatbelts and airbags has not contributed significantly to increasing driver safety. In fact, this result should have been expected. Consider, to the driver who is now located within a vehicle that contains ten to fifteen airbags, it seems much safer to drive at excessive speeds or in a distracted condition. A Systems Engineer should approach this challenge by looking at the systemic structure, and not the symptoms.

Systems tools have been developed to help us to see how our old mental models get in the way of new ideas. Consider a simple example, a wrangler who is responsible for the safe transport of mustangs across the western plains.

Figure 19: System Challenge

This is a particularly challenging job, typically resulting in very strenuous impacts to the human body, resulting in long-term degradation. There are imminent injuries, which may result in an inability to complete the task, as well as mentally and socially challenging tasks that must be accomplished in order to deliver the livestock as required. This is a very difficult job.

A Systems Engineer may approach this particular problem and suggest the following as capabilities to be considered:

Automate the tasks
Make the job less dependent on an animal
Make the job easier for the operator
Make the job more efficient
Save money
Save time

Engineers have the technical knowledge and desire to resolve this problem in a technologically intensive manner. This is shown in Figure 20.

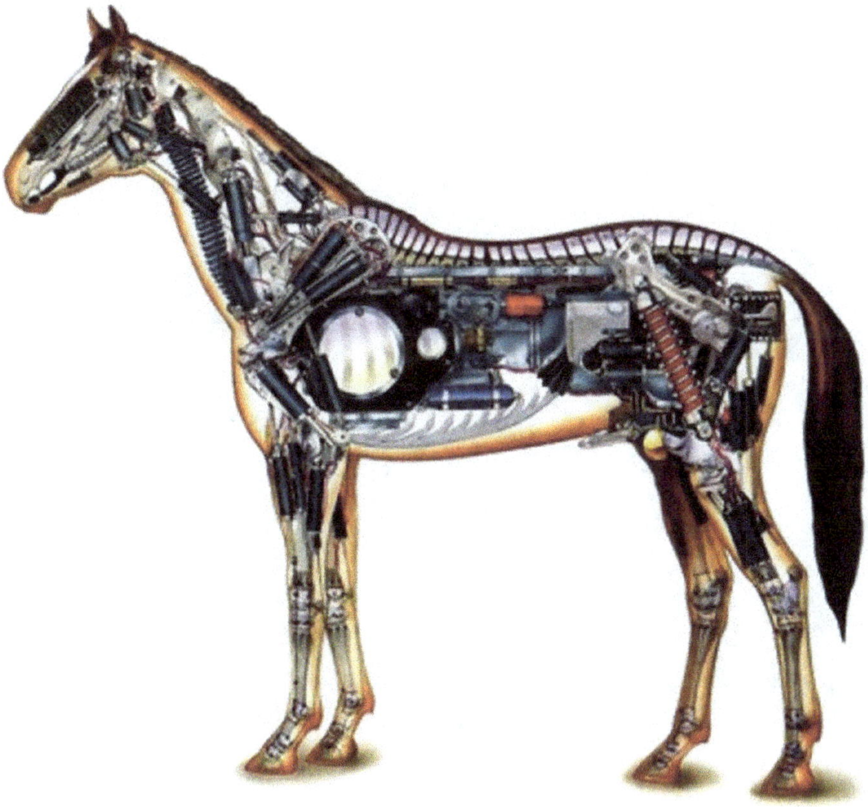

Figure 20: An Engineering Solution

Many of us are immediately considering a variety of solutions, several of which have been introduced to the market in the past. These include mechanical devices that walk like horses and are capable of carrying heavy loads. In fact, many of us may have considered this problem and refined our solution to the point of producing and aesthetically beautiful result. An example of this is shown in Figure 21.

Figure 21: An Aesthetically Pleasing Result

There fortunately is no shortage of engineers who are willing to take on such a challenge and to dedicate significant time and effort to complete the task. But upon reflection, perhaps we are missing the point. In this case, we have assumed that the solution will look similar to the current solution. This mental model is limiting our ability to think creatively and to produce highly disruptive solutions. In fact, if we reflect on the actual capabilities required of the system to solve this issue, there are much more obvious solutions available. However, they do require us to change our basic thinking in order to find simpler paths. An alternative solution is shown in Figure 22.

Figure 22: An Alternative Solution

Clearly, it would be much more efficient to consider the alternative solution before tackling the significant technological challenges of developing a horse-like robot. This is the principal reason why Systems Engineers must detach themselves from the solution and focus instead on what requirements or capabilities must be met.

To further illustrate the need for holistic thinking, we will examine the typical success and failure modes of projects within the technical domain. The results of several studies are included for consideration.

Key Elements of Project Outcomes (Standish Group CHAOS results, 1996)

Top Success Factors

- User involvement
- Executive Management support
- Clear requirements
- Proper planning
- Realistic expectations
- Smaller project milestones
- Competent staff
- Ownership
- Clear vision and objectives
- Hard working, focused staff
-

Top Failure Factors

- Lack of user input
- Incomplete requirements
- Changing requirements
- Lack of executive support
- Technology incompetence
- Lack of resources
- Unrealistic expectations
- Unclear objectives
- Unrealistic timeframes
- New technology

Critical Success / Failure Factors in Projects (Belassi & Tukel, 1996)

- Top management support throughout
- Proper goal definition and setting
- Accurate estimates
- End-user involvement
- Project Manager on site
- Appropriate team selected
- Testing and training

Patterns of Software Systems Failure and Success (Jones 1996)
Project Management Failure Factors

- No historical software measurement data
- Failure to use automated estimation tools
- Failure to use automated planning tools
- Failure to monitor progress of milestones
- Failure to use effective architecture
- Failure to use effective development methods
- Failure to include formal risk management
- Failure to use formal configuration management
- More than thirty percent creep in requirements

Principal reasons for DOD software project failures
Report of the Defense Science Board Task Force on Acquiring Defense Software Commercially (1994)

- Poor requirements definition
- Inadequate software process management
- Lack of integrated product teams
- Ineffective subcontract management
- Too little attention to software architecture
- Poorly defined, inadequately controlled interfaces
- Limited or no tailoring of military standards

Software Systems Failure Modes (Website: http://suraj.lums.edu.pk)
Hostile culture
    Punishing bad news messengers
Poor reporting structure
    Sponsors are not involved
The Success of Any Project is Linked to the People
    Take care of the people to succeed
Over-commitment
    Unclear expectations and understanding of the scope
Decision Making Escalation
    Movement becomes more important than progress
Political Pressures
    Hidden Agendas
Technology-focused developments
    Ignoring the Human
The lure of the leading edge
    It can be costly and heartbreaking
Complexity underestimated
    Simplified by individuals who are not informed
Inadequate consultations with major stakeholders
    User needs in the dark

Design by committee
  No major responsibility holder
Technical fix for a management problem
  Lack of whole picture thinking
Staff turnover
  Lack of engagement
Competency
  Decisions made by the wrong people
Receding deadlines
  Soft plans with malleable objectives and milestones
Inadequate testing and training
  Not having discipline in our development
Communication problems and work group problems
  As always

Most of these observations will come as no surprise to any experienced Project Manager. However, as might be expected of any Systems Engineer, it is not adequate to identify the issues as much as it is crucial to identify solutions to the existing problems. In order to do this, we will consider the construct of the human brain. We have previously mentioned that technical professionals have spent a great deal of time developing the analytical portion of their brains, however, if we consider a holistic viewpoint, this requires the abstract portion of the brain be utilized in the Systems Engineering domain. Therefore, we are interested in characteristics associated with both the left-brain and right-brain thinking patterns.

Left Brain Style

- Responds to verbal instructions
- Problem solves logically and sequentially
- Looks at differences
- Is planned and structured
- Prefers established, expected information
- Prefers talking and writing
- Prefers multiple choice tests
- Controls feelings
- Prefers ranked authority structures
- Sees cause and effect
- Is a splitter; distinction is important
- Draws on previously accumulated, organized information

Right Brain Style

- Responds to demonstrated instructions
- Problem solves with hunches, looks for patterns
- Looks at similarities
- Is fluid and spontaneous
- Prefers elusive, uncertain information

- Prefers drawing and manipulating objects
- Prefers open ended questions
- Is free with personal feelings
- Prefers collegial authority structures
- Is a lumper; connectedness is important
- Is analogic; sees correspondences, resemblances
- Draws on unbounded patterns, clustered around images

If we consider many of the common threads found in the previous Project Management reports, we begin to see some patterns emerging. It becomes clear that the majority of shortcomings brought about in the typical technical Project Management domain are caused by limited application of the holistic approach to Systems Engineering. This is apparent in the following list, derived from independent research studies:

Issue identified	**Cause of issue**
Hostile Culture	Lack of Right Brain
Poor reporting structure	Lack of Right Brain
The Success of Any Project is Linked to the People	Lack of Right Brain
Over-commitment	Lack of Right Brain
Decision Making Escalation	Left Brain Knee Jerk
Political Pressures	Lack of Right Brain
Technology-focused Dev – Ignoring the Human	Lack of Right Brain
The Lure of the Leading Edge	Lack of Right Brain
Complexity Underestimated	Lack of Left Brain
Inadequate Input from Stakeholders	Lack of Right Brain
Design by Committee	Lack of Right Brain
Technical Fix for Management Problem	Lack of Right Brain
Staff Turnover	Lack of Right Brain
Competency	Lack of Left Brain
Receding Deadlines	Lack of Right Brain
Inadequate Testing and Training	Lack of Right Brain
Communication Problems and Work Group Problems	Lack of Right Brain

Therefore, it seems apparent that there is significant value in enhancing both the left-brain and right-brain capabilities of the Systems Engineer. Several seemingly unrelated research studies suggest exactly the same conclusion.

The Art of Innovation (Kelley, 2001)

This text outlines the methodology utilized by IDEO Corporation to produce innovative projects. This organization is well known and recognized for its ability to innovate highly disruptive solutions to problems. Its methodology includes the following five steps. The appropriate brain constructs have also been associated with each of the steps.

1) Understand the market, client, technology, and constraints of the problem (left brain)
2) Observe real people in real-life (right brain)
3) Visualize new concepts and the customers who will use them (right brain)
4) Evaluate and refine the prototypes in rapid iterations (left brain)
5) Implement the new concept for commercialization (both sides)

Deep Survival (Gonzales, 2005)

In the text deep survival, Gonzalez investigates the principal causes for success and failure in critical survival situations. His conclusions are that those individuals practicing most or all of these skills are most likely to survive a critical situation. Those who are limited in their thinking often have catastrophic results. His taxonomy of thoughts and behaviors are:

1) Perceive
   Look, See, Believe. Be Observant and Cognizant
2) Stay Calm
   Use Humor, Use Fear to focus the Mind
3) Think/Analyze/Plan
   Get organized, set Small Achievable Objectives
4) Take Action
   Take bold and cautious Actions
5) Celebrate your Successes
   Take Joy in completing Tasks
6) Count your Blessings
   Be grateful, help others more than yourself
7) Play
   Singing, Mind Games, Poetry, Mathematics
8) See the Beauty
   Treat it like a Vision Quest
9) Believe you will succeed
   Develop a Deep Conviction of Success
10) Surrender
   Let go of your Fear, put away the Pain
11) Do whatever is necessary
   Be determined, have the Will and the Skill
12) Never give up
   Let nothing break your Spirit, there is always one more Thing you can do

Classifying these various behaviors by left and right brain activities:

Left Brain Activities

- Think/Analyze/Plan
- Take Action
- Celebrate your successes
- Believe you will succeed
- Do whatever is necessary
- Never give up

Right Brain Activities

- Perceive
- Stay Calm
- Count your blessings
- Play
- See the beauty
- Surrender

From these results, we can draw the conclusion that a holistic approach to survival will yield the most effective results. Overreliance on one or the other side of the brain can result in survival challenges.

Several related studies have also been identified that illustrate the value of a holistic approach to our educational process. In these studies, the use of music education has been connected to spatial reasoning or mathematical capabilities. In these studies, it is apparent that the students who demonstrated both concrete and abstract thinking modes were more successful within the educational system.

Results of the mixture of art and science education (various websites, 2007)

A study of thirty-seven student's IQ scores over an eight-month period

First group given no music lessons – 6% increase in spatial reasoning scores
Second group given music lessons – 46% increase in spatial reasoning scores

Medical School Acceptance

66% of music majors accepted (highest percentage)
44% of biochemistry majors accepted

After one year of music and visual arts education, first graders who underperformed in Kindergarten were twenty-two percent better at mathematical competency than their peers.

1993-95 study on three groups of preschoolers

1st group – no training – IQ ^ .5 points
2nd group – computer lessons – IQ ^ .35 points
3rd group – music lessons – IQ ^ 3.62 points

Students with musical performance experience

Scored 53 points higher on SAT verbal portion

Scored 39 points higher on SAT math portion than students without musical performance experience

Students with coursework in music appreciation

Scored 61 points higher on SAT verbal portion

Scored 42 points higher on SAT math portion than students without music appreciation courses

The Art and Science of Understanding a System

Success in any endeavor is first and foremost a matter of accurate seeing. Learning to really see what is in front of us allows us to better understand the world around us, and the essence of our problem more completely. This is a departure from linear thought, and results in a crossing of both artistic and scientific reasoning.

To solve a problem one has to:

Identify the problem exactly
Imagine the system and the problem
Imagine the solution to the problem
With your eyes you see what isn't working
You make it work in your mind, and you simply identify the difference

This is 'Whole-Brain' Reasoning

Whole Brain Reasoning does not imply that the capabilities of the typical engineer are of little use Systems Engineering. In fact, we require the nuts and bolts of observational analysis from the left side of the brain, followed by the free association image building from the right side of the brain. This is a departure from traditional linear-thinking or sequential problem-solving models. In essence, it is both an art and a science combining both sides of the brain.

Therefore, to the Systems Engineer, it becomes a matter of viewing the problem from appropriate viewpoints and asking questions that are appropriate given the problem at hand. As an example, imagine you went to your local hardware store in order to buy a new lawnmower. When you met with the associate at the store and requested advice on the purchase of a lawnmower you were met with a series of questions:

Do you want an 18-inch deck or 22-inch deck?
Is a mulching blade required?
Do you want variable rate propulsion?

Do you prefer side or rear exhaust?
Electric or gas powered preference?
Do you want a two or four-cycle engine?

The typical response is one of incredulity. The individual at the hardware store has asked for advice on the purchase of a lawnmower. This is likely due to a lack of expertise in the technical ins-and-outs of a lawnmower. For this reason, the questions should not be specifically technically oriented, but rather should be focused upon finding the capabilities necessary to satisfy the customer. Therefore, the Systems Engineer typically asks a different set of questions:

Why do you want to buy a lawnmower?
What is your yard like? What about trees? Hills?
Do you like to mow, or is it a chore? Is speed important?
How long do you need to own a lawnmower?
Who will do the mowing of your lawn?
Where is the nearest power outlet?

The second set of questions is far more likely to elicit practical responses from the customer, thereby enabling the system engineered to better define the capabilities required of the system design. As might be expected, this requires the Systems Engineer to depart from his typical analytical approach to a given problem and instead, expand his thinking into a holistic viewpoint that will better enable him to understand the problem.

Systems Engineering tools have been developed to enhance the thinking of the Systems Engineer in a manner that will lead to a more complete understanding of the system. These tools are specifically designed to force the Systems Engineer into alternative paths rather than those she normally would follow as a typical development engineer. Although many systems tools exist, this text will be limited to those tools the authors believe are most appropriate and beneficial to the average Systems Engineer.

An alternative way to consider the problem is this: a Systems Engineer must be able to take the information that is present for everyone, and come up with a completely different solution. This is essential in order to produce an innovative or a disruptive solution to the problem. Therefore, the effective Systems Engineer must become more adept at seeing what is present and drawing appropriate conclusions, rather than relying on past mental models and various assumptions in order to produce an average system solution. In fact, to identify a highly effective Systems Engineer who can exhibit these capabilities, one need to go no further than to consider the character of Sherlock Holmes. This astonishing detective demonstrated a keen ability to look at the same information available to every observer, and connect that information in a unique way that resulted in an optimal solution to the problem.

Chapter 4: DoDAF views

The Department of Defense has traditionally been involved in extremely large-scale and complex development projects. Due to this, the DOD is often at the forefront of new developments in Systems Engineering methodologies. One of the most powerful tools yet developed for use in Systems Engineering is the technique recommended for the production of an Architectural Framework. It is known as the Department of Defense Architectural Framework, or DoDAF.

This framework is intended to define a common approach for describing, presenting, and comparing enterprise architectures. It facilitates the use of common principles, assumptions and terminology in any description of the system to be designed. The principal objective of developing a framework of this sort is to ensure that architectural descriptions can be compared and related across organizational boundaries, national boundaries, and even within organizations.

One critical factor in applying this framework methodology is to focus upon the system to be developed as an element of an enterprise. This means it is no longer acceptable for us to design a machine or a device that is completely stand-alone and independent of all other systems with which it will come into contact. If we treat the system as an enterprise, we are no longer simply designing a new aircraft; we are in fact designing a new vehicle for use in the air system. This mindset is very effective for maintaining a coherence and pragmatic development strategy in most organizations.

Architecture is defined as the structure of components, their relationships, and the principles and guidelines governing their design and evolution over time. Architecture is the fundamental organization of a system embodied in its components, their relationships to each other and within the environment, and the principles guiding the system's design and evolution.

The intent of System Architecture is to allow appropriate make/buy decisions to be made prior to the development of systems or components, the ability to discriminate between options on a development project, the development of true and complete requirements, and the driving of one or more systems to a common use or purpose. System design, on the other hand, is used to develop the system components, build the system, and manage the changes and modifications to the system during its development cycle.

The enterprise's competitive edge and ultimate success are enabled by its ability to rapidly respond to changing business strategies and technologies. The competitive edge translates into higher levels of customer satisfaction, shorter work cycles, reduction in schedules, reductions in maintenance costs, and reduced development time, which all result in lower overall lifecycle costs.

Enterprise Architecture is the key facilitating ingredient providing a holistic view and a mechanism for enabling the design and development of innovative systems, as well as providing for a common communication platform and method to enhance the understanding of the enterprise. The overarching goal of Enterprise Context is to manage the complexity of the enterprise, online business strategies and implementations, and to facilitate innovations and rapid change in order to maintain business and technical advantages in the marketplace.

Any Architectural Framework is basically a tool. It should describe the methods for designing a system in terms of a set of building blocks, and for showing how the building blocks fit together. It should contain a set of tools and provide a common vocabulary.
The DoDAF work products are principally organized around three independent views, and also include a fourth summary view that is used to consolidate the other views. These views are described in **Figure 23**.

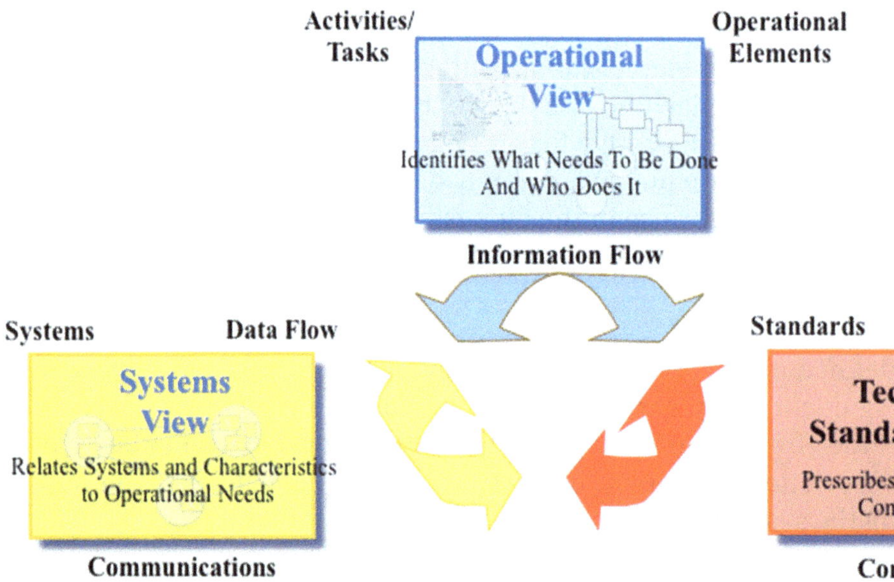

Figure 23: DoDAF Work Products

The DoDAF approach describes a set of twenty-six different work products that are used to ensure uniformity and standardization in the documentation and communication of the Architecture. The use of all of these work products (or views), is designed to enable the documentation of an entire Architecture, from requirements through implementation. In general, a smaller subset of work products are typically selected based upon the specific needs of the system being developed.

Therefore, the list of products is defined and categorized into four views:

All Views (AV): The summary information describing the Architecture's plans, scope, and definitions.

Operational View (OV): The behaviors and functions describing the way the system is used in operation or by the customer.
Systems View (SV): The system and applications supporting the mission functions
Technical Standards View (TV): The policies, standards, and constraints of the system

The entirety of the framework work products is listed in **Figure 24.**

Applicable View	Framework Product	Framework Product Name	General Description
All Views	AV-1	Overview and Summary Information	Scope, purpose, intended users, environment depicted, analytical findings
All Views	AV-2	Integrated Dictionary	Architecture data repository with definitions of all terms used in all products
Operational	OV-1	High-Level Operational Concept Graphic	High-level graphical/textual description of operational concept
Operational	OV-2	Operational Node Connectivity Description	Operational nodes, connectivity, and information exchange needlines between nodes
Operational	OV-3	Operational Information Exchange Matrix	Information exchanged between nodes and the relevant attributes of that exchange
Operational	OV-4	Organizational Relationships Chart	Organizational, role, or other relationships among organizations
Operational	OV-5	Operational Activity Model	Capabilities, operational activities, relationships among activities, inputs, and outputs; overlays can show cost, performing nodes, or other pertinent information
Operational	OV-6a	Operational Rules Model	One of three products used to describe operational activity—identifies business rules that constrain operation
Operational	OV-6b	Operational State Transition Description	One of three products used to describe operational activity—identifies business process responses to events
Operational	OV-6c	Operational Event-Trace Description	One of three products used to describe operational activity—traces actions in a scenario or sequence of events
Operational	OV-7	Logical Data Model	Documentation of the system data requirements and structural business process rules of the Operational View
Systems	SV-1	Systems Interface Description	Identification of systems nodes, systems, and system items and their interconnections, within and between nodes
Systems	SV-2	Systems Communications Description	Systems nodes, systems, and system items, and their related communications lay-downs
Systems	SV-3	Systems-Systems Matrix	Relationships among systems in a given architecture; can be designed to show relationships of interest, e.g., system-type interfaces, planned vs. existing interfaces, etc.
Systems	SV-4	Systems Functionality Description	Functions performed by systems and the system data flows among system functions
Systems	SV-5	Operational Activity to Systems Function Traceability Matrix	Mapping of systems back to capabilities or of system functions back to operational activities
Systems	SV-6	Systems Data Exchange Matrix	Provides details of system data elements being exchanged between systems and the attributes of that exchange
Systems	SV-7	Systems Performance Parameters Matrix	Performance characteristics of Systems View elements for the appropriate time frame(s)
Systems	SV-8	Systems Evolution Description	Planned incremental steps toward migrating a suite of systems to a more efficient suite, or toward evolving a current system to a future implementation
Systems	SV-9	Systems Technology Forecast	Emerging technologies and software/hardware products that are expected to be available in a given set of time frames and that will affect future development of the architecture
Systems	SV-10a	Systems Rules Model	One of three products used to describe system functionality—identifies constraints that are imposed on systems functionality due to some aspect of systems design or implementation
Systems	SV-10b	Systems State Transition Description	One of three products used to describe system functionality—identifies responses of a system to events
Systems	SV-10c	Systems Event-Trace Description	One of three products used to describe system functionality—identifies system-specific refinements of critical sequences of events described in the Operational View
Systems	SV-11	Physical Schema	Physical implementation of the Logical Data Model entities, e.g., message formats, file structures, physical schema
Technical	TV-1	Technical Standards Profile	Listing of standards that apply to Systems View elements in a given architecture
Technical	TV-2	Technical Standards Forecast	Description of emerging standards and potential impact on current Systems View elements, within a set of time frames

Figure 24: DoDAF Work Products

For most organizations, a reduced list of Architectural Framework Products is appropriate for product development. These views include:

AV– 1: Overview and summary information
AV– 2: Integrated dictionary
OV – 2: Operational node conductivity description
OV – 3: Operational information exchange matrix
OV - 5: Operational activity model

Please see http://aitc.aitcnet.org/dodaf for full descriptions of these products.

Developing the AV-1

The AV – 1 is a work product intended to describe the general overview and summary for the system to be developed. A typical AV-1 would include the following information:

Name – name of the system being developed
Architect – names of the developers of the system
Organizations involved– listing of major organizational stakeholders
When developed – date for outset of program
Purpose – describes a simply why the organization chooses to pursue this system development
Scope – describes in detail what is included and excluded from the system development
Mission – describes why the customer wants the system developed
Geographical considerations – describes where the system will be implemented
Rules, criteria, and conventions – establishes the context for the system development
Findings – used as a parking lot for issues of note throughout system development
Tools and file formats – establish baseline tool requirements for system development
Constraints – list any and all constraints necessary for successful completion of the system
Assumptions – describes any assumptions accepted by the development team

An example AV–1 for a simple power monitoring system is included in **Figure 25.**

The All View products are intended to capture the entirety of the operational, system, and technical views. They provide pertinent information to the entire Architecture, but do not represent a distinct view of the Architecture. The AV-1 sets the scope and context of the Architecture; the scope includes the subject area and timeframe of the Architecture, and the level to which the Architecture will be defined, and the context includes the setting and constraints on the system, which will indirectly define the Architecture. These views are a miscellaneous set of view products that supplement D3 Architectural views with either emergent or globally applicable Systems Engineering information.

The AV-1 is intended to describe the purpose, scope, and characteristics of the architecture. It contains much of the information that would also be found in a program's Systems Engineering Management Plan (SEMP). This view provides a high-level description of why and how the Architecture will be developed, which enables a focus to be placed on the Architecture effort. There is a well-defined process for creating the view:

Step 1: Determine the intended use of the Architecture; this will define the purpose of the effort.
Step 2: Determine the scope of the Architecture; this will define the Architecture views to be created and the product's identification.
Step 3: Determine the characteristics to be captured
Step 4: Determine the views and products to be built
Step 5: Use the Architecture's description for intended use

In general, the AV–1 is a very challenging document to create, as it is the System Development Team's first attempt to define and limit the scope of the effort to be undertaken. Therefore, it is common for the AV-1 to be revised and modified several times during the System Development cycle.

Special care must be taken to precisely define the purpose, scope, and mission for the development. This is crucial, as it defines the major constraints that set up the basic system Architecture for the development effort. Care must be taken to assure that terminology is precisely understood and defined, and that the documentation reflects the common understanding of the development effort.

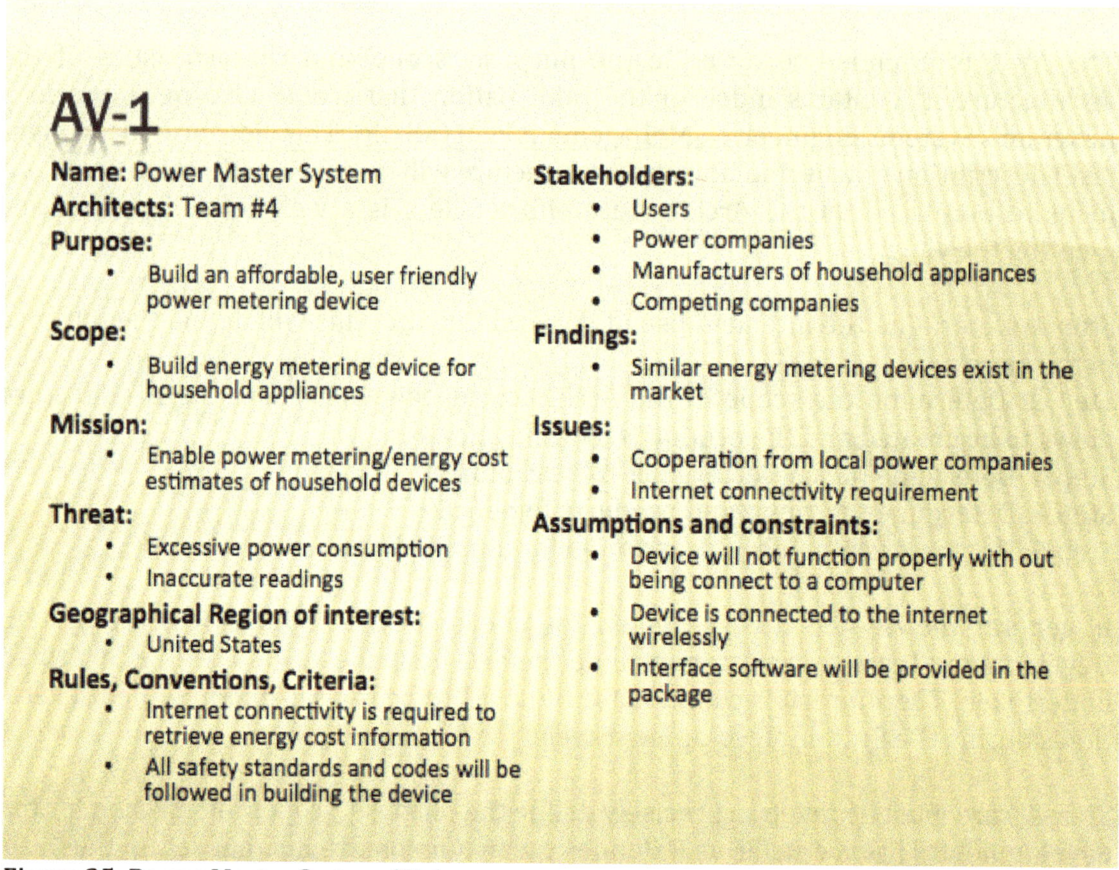

Figure 25: Power Master System AV-1

The AV-2 is an integrated dictionary. At a minimum, the integrated dictionary is a glossary with definitions of terms used in a given Architecture's description. Each labeled graphical item in the graphical representations should have a corresponding entry in the integrated dictionary. This provides a baseline of understanding for all members of the System Development team regarding the specific terminology and references used throughout the System Development.

The AV-2 is defined in line with the AV-1, and is used in two ways. First, the document defines the taxonomy of the Architecture that is being developed. Second, the AV-2 defines the Data dictionary that will be utilized throughout the development of the project. Careful attention to defining all potentially vague or nebulous terms will result in a higher probability of success for the System Development effort.

The Operational Views (OVs)

The Operational Views are intended to capture the operational concepts that described the execution of operational capabilities. Therefore, the OV is a description of tasks or activities, operational elements, and information flows required to accomplish or support the intended mission of the System Development effort. Additionally, operational views should address the inter-operability at the operational node and capability level of the system. This can be used to identify the required sharing of information and needs for specific operational information at specific operational locations. The Operational View products define:

Operational concepts for executing the operational capabilities
Operational nodes and connections to supporting organizations used to perform operational activities
Operational activities that process information
Information exchanges needed by the operational activities
Behavioral aspects of the operational Architecture

Developing the OV-2

The OV-2 is the operational node connectivity description. This view describes how the system interfaces with all other systems and components with which comes into contact. This may be limited to your enterprise, or may extend well beyond it. In general, it is constructed by placing the developing system at the center and connecting all other inputs and outputs to their corresponding systems or component sources. Each input and output will be described by a separate arrow, and will be numbered as a 'need line' for the diagram. These need lines are then described on the following page, the Data Elements Table, for the corresponding Architectural view.

Every operational node is an element that produces or consumes information, including operational or human roles, organization, logical or functional needs. Need-lines document the requirement to exchange information between the nodes. Note that this does not indicate how the information is transferred. A single need-line may have multiple operational information exchanges required. Only two need-lines can exist between any unique combination of operational nodes, and these need-lines are defined as one-way directional arrows.

Great care must be taken in establishing each of the need-lines, as after the Architectural View is completed, each of the defined need-lines will provide the System Development team with one or more requirements necessary for the completion of the system's development. The need-lines indicate the flow of information from one node to the other. The communications information flow is initially conceived in the mind as a functional flow of information and is transformed to requirements by inserting the word "shall" into the functional information flow statement.

The Power Master System OV-2 is included in **Figure 26**.

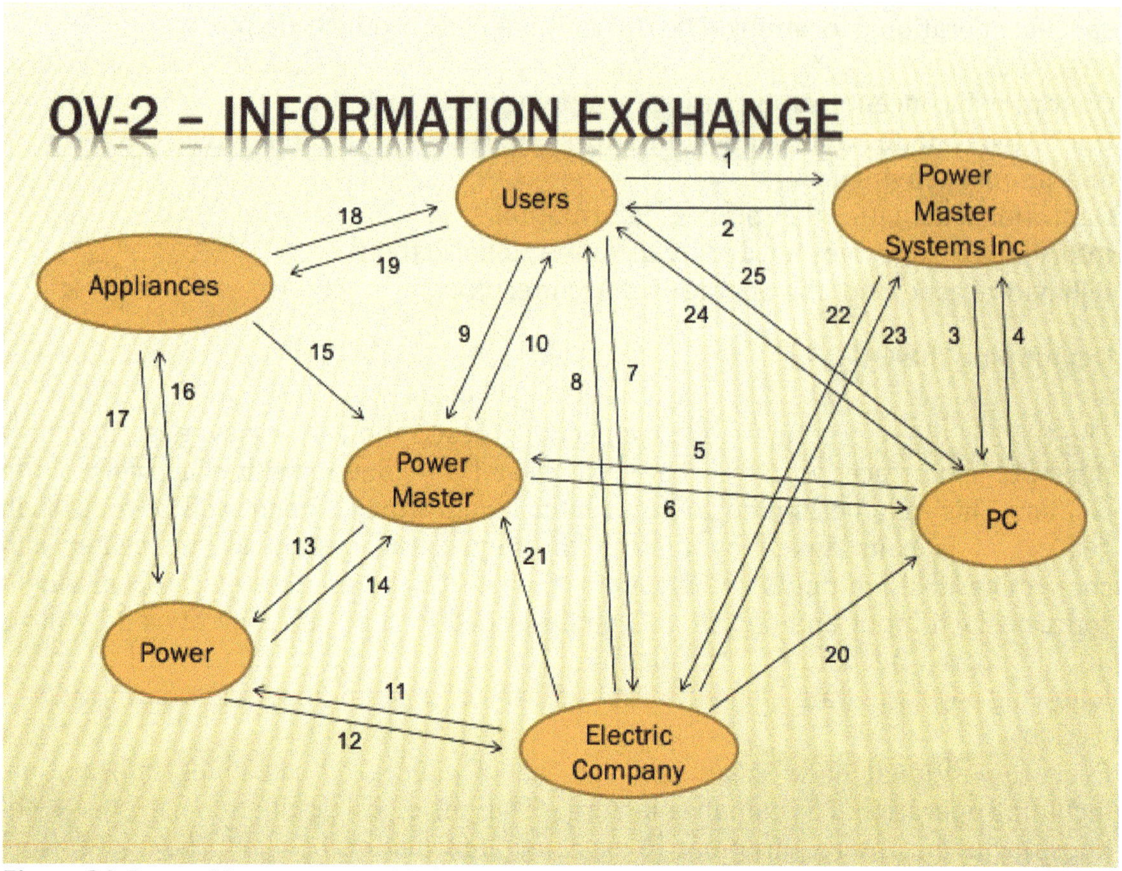

Figure 26: Power Master system OV-2

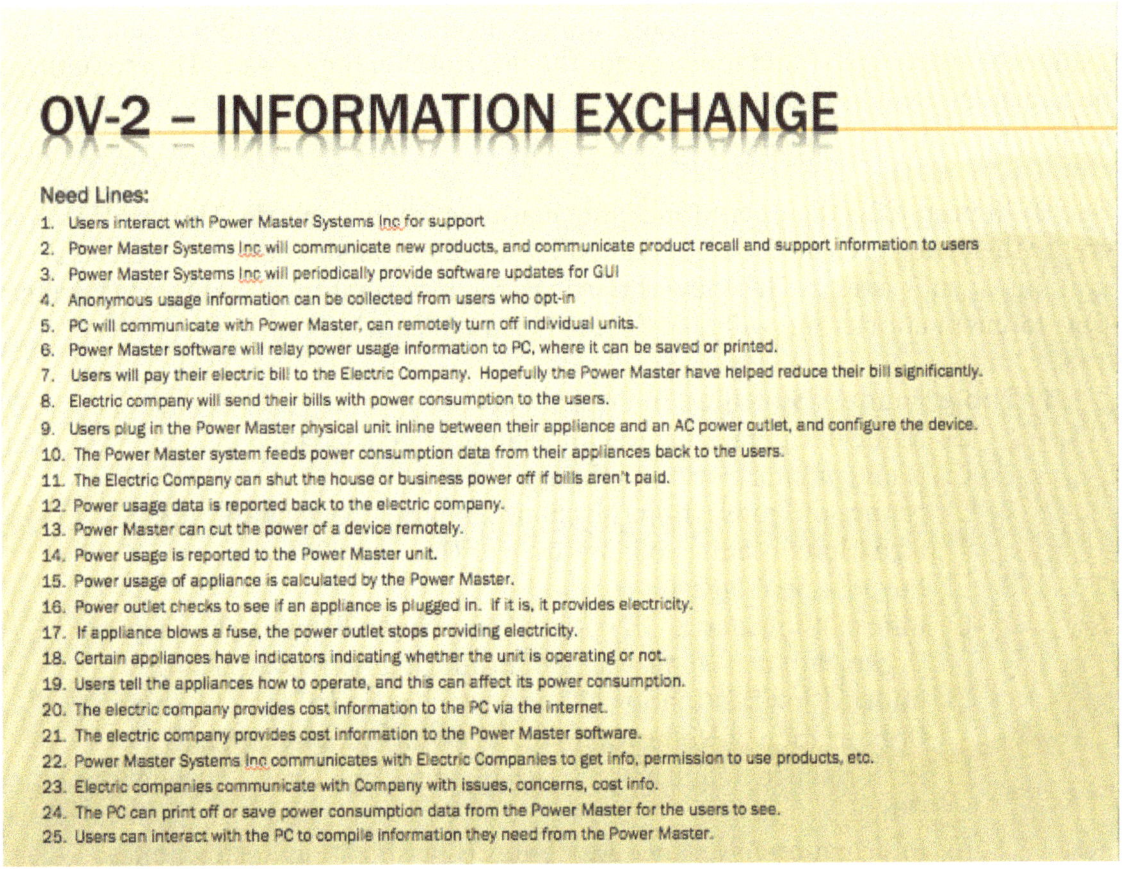

Figure 27: Power Master OV-2 Data Elements Table

Developing the OV-3

Levels of Abstraction and their Use

The OV-2 is the highest level of abstraction of narrative definitions of information flows, and is the starting point for the more detailed levels of abstraction of information flows that define detailed requirements, such as format and data rates. The progression is OV-2 (highest level of abstraction) to OV-3 (intermediate level of abstraction) to SV-6, with the SV-6 (design level abstraction) being the view that establishes the detailed requirements necessary for the design of the communications links between nodes. The approach is essentially a top-down approach that economizes the effort and matches it to the purpose for which it is developed.

For example, the presentation created for a meeting with the Marketing team and the customer might only require an OV-2 level of information, as it may provide enough information to convince the customer that further engagement with the vender is merited. It may also be enough to create a high-level model and simulation of node interconnectivity that will be used to acquire funding from management.

The layered abstraction approach of matching purpose to effort is key to utilizing Engineering staff effectively. Engineers like detail, have been trained to focus on detail, and will design to the most granular level of detail possible if left without guidance regarding the intended use for the information they create. This sometimes becomes evident in proposal activities, where engineers feel they must design a system for proposal in such detail that the design is ready for build.

The OV-3 product is the operational information exchange matrix. This is utilized to create the data words or packages that will be used to transfer data between adjacent systems or subsystems. This view is particularly important in data-intensive System Development efforts, and will take the form of a table including:

- **Table Headers Specified in Framework**:
  - ☐ Name of Operational Needline Supported (from OV-2)
  - ☐ Name of Information Exchange
  - ☐ Nature of Transaction (Mission/Scenario, Language, Content, Size/Units, Media, Collaborative or One-Way?)
  - ☐ Purpose or Triggering Event
  - ☐ Information Source (ID of Producing Node Element, Owning Organization of Node, Name of Producing Activity, UJTL ID)
  - ☐ Information Destination (ID of Receiving Node Element, Owning Organization of Node, Name of Receiving Activity, UJTL ID)
  - ☐ Performance Requirements (Frequency, Timeliness, Throughput, Other)
  - ☐ Information Assurance Attributes (Classification Restrictions, Criticality/Priority, Integrity Checks Required, Assured Authorization to Send/Receive)
  - ☐ Threats (Physical, Electronic, Political/Economic)
  - ☐ Operational Environment (Weather, Terrain, Policy/Doctrine Constraints)

Developing the OV-4

The OV-4 diagram is a view of the organization of the enterprise. This diagram will often be of limited usefulness as a stand-alone artifact, but can be quite useful when used in conjunction with other DoDAF artifacts.

Relationship of the OV-4 to the Integrated Architecture and Significance

OV-4s are a part of the Enterprise Integrated Architecture and are discovered after the OV-5 view lower level icons are appropriately defined and assigned to the affected OV-2 node(s). In many System Architectures a new activity in a specific node(s) requires an organizational change to the existing organizational Architecture in that node. For example, a new operator position, such as an Information Analyst, must be created to interpret the information received from the connection to the proposed system. Once the nature of the new position is discovered from an understanding of the specific OV-5 activity assigned to a node(s), its job description can be created as part of the OV-4. Therefore, the new system will impact the existing nodes to which it connects, and its Operational Organizational Architecture (OV-4) will be impacted by those connections.

Having spoken of the possibility of multiple Organizational Architectures due to the interconnection of the new system with the existing nodes, one needs to look at two examples of Organizational Architectures. The two of interest are the hierarchical top down traditional architecture found in annual reports and used in systems programs of less complexity, and the organizational architecture that is used in the execution of large 'systems of systems' contracts; the Integrated Product Team (IPT).

The OV-4 is the organizational construction that will be used for the operation of the system. It is useful for providing an understanding of work responsibilities and task accountabilities. An example is shown in **Figure 28**.

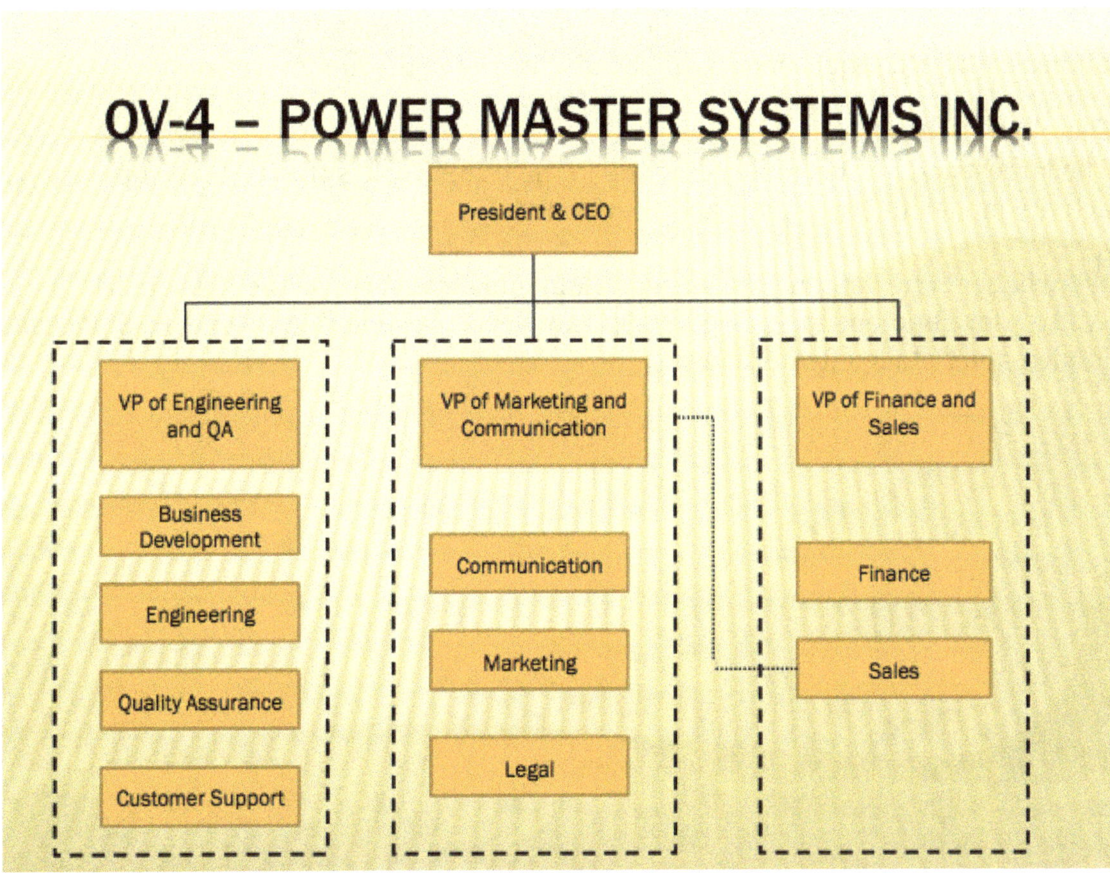

Figure 28: Power Master System OV-4

OV-4 Organizational Relationships Chart
This is an essential basic view, as it depicts the effects of external organizational connection to the Architecture of the system to be designed, and reflects the impact of the external organization on the new system.

Systems Requirements Discovery Methodologies

The two basic approaches to discovering system requirements are bottom-up and top-down. If we consider the development of a simple household toaster, we fulfill the system's requirements definition by defining its user needs and system requirements. This is an example of the top-down approach used in the generation of the OV -5 operational activity diagram.

A bottom-up approach may also be used, where all needed tasks are defined, agglomerated into subsystems and ultimately, into a system operational chart. This method is often used in the development of a Functional Analysis.

DoDAF enterprise architecture views enable the determination of the work or operations required to accomplish the objective task or capability. Either one or both of these techniques (bottom-up or top-down) may be used to determine requirements. When both are used, it is more likely that a more complete set of requirements will be discovered. As with any model creation, one either consciously or subconsciously makes assumptions that will fundamentally guarantee that not all of the real-world requirements will be discovered in the model world.

Developing the OV-5

The largest and most complex diagram in the framework is the OV-5, the operational activity model for the system to be developed.

We will begin by defining the various elements necessary for the household toaster system development, including Users, User Needs, System Requirements, States and Modes, and Detailed Requirements:

User Needs:

Toast Products	Quality	Timing
Safety	Efficiency	Maintainable
Size	Intuitive / User Friendly	
Adjustability	Aesthetics	

System Requirements:

120 VAC 50/60 hz	User Adjustment	Power Cord LG/Plug
UL listed	Indicators	Maintenance Access Panels
Portable	Safety Overrides GFCI	
Child Safe	Auto Shut-off	Art Deco / Chrome Finish / Pearl
Toasting Rate	Current Rate	Size / Wt

States:

- On
- Off
- Maintenance
- Storage
- Modes:
  - Toasting Bread
  - Toasting Bagel
  - Single Slot
  - Multiple Slot
  - Cooling
  - Adjustment Positions
  - Shutdown

Detailed System Requirements:

Number	Requirement	SuperReq	SubReq
1.0	**Usability**		
1.1	The system shall provide status indication		2.4
1.2	The system shall have an auto-shutoff	3.3	
1.3	The system shall be language independent	6.1	
2.0	**Operational**		
2.1	The system shall toast 1 product / minute		1.1
2.2	The system shall use 120 VAC		4.4
2.3	The system shall be portable		
2.4	The system shall allow user adjustment of toasting level	1.1	
3.0	**Security/Safety**		
3.1	The system shall use a GFCI	3.2	
3.2	The system shall be child-safe		3.1
3.3	The system shall fail to a power off condition		3.1
3.4	The system shall have a UL listing	3.2	
3.5	The system shall have a cord of 12 inches or less	3.2	
4.0	**Physical**		
4.1	The system shall have an art deco design		
4.2	The system shall have a chrome finish		
4.3	The system shall not exceed 12X12X12in		
4.4	The system shall use a standard LG Plug	2.2	
4.5	The system shall not provide hot surfaces to the user	3.2	
5.0	**Support**		
5.1	The system shall operate without maintenance for a period of three years		5.4
5.2	The system shall be replaceable, not repairable	5.2	
5.3	The system shall have maintenance panels	5.4	
5.4	The system shall require only cleaning maintenance beyond three years	5.1	5.3
6.0	**Training**		
6.1	The system shall require no training		6.1

Figure 29: Toaster System Development

The intention of establishing these factors early is to begin the development of the system utilizing the system V diagram. This begins with users and user needs as we progress down the left side of the V. The intention of this methodology is to begin with a given challenge or problem, and then to decompose that problem and continue decomposing until we ultimately end up with a large set of system requirements. This would be the equivalent of starting at the top of the following diagram, and decomposing a large number of requirements in the middle of the diagram. As we continue with detailed development we begin to address the individual requirements with Architecture artifacts, and then we build these up into subsystems and systems until ultimately, we create a singular solution for the problem at the bottom of the diagram.

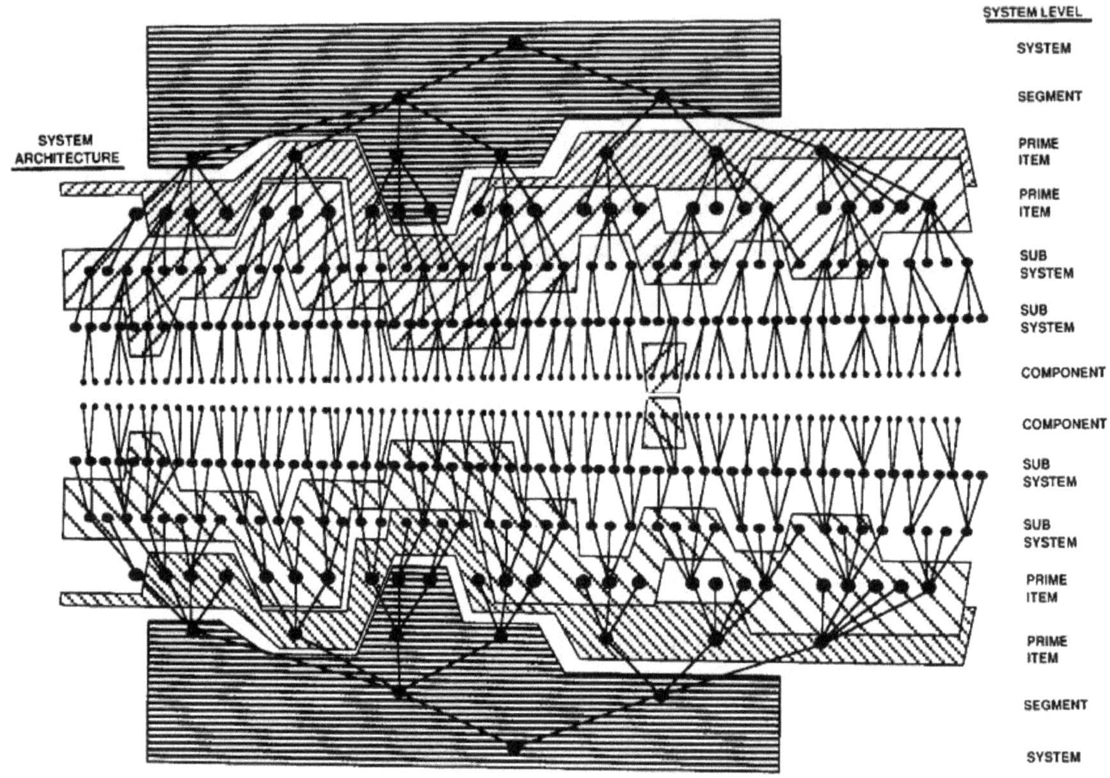

Figure 30: System Decomposition and Recomposition

The development of the OV-5, the operational activity model, is very similar to the Functional Analysis conducted in the Project Management section of this text. The principal difference between these two approaches in defining the operational activities is that the functional analysis is typically a bottom-up construction, whereas the OV-5 is a top-down construction of the same Architecture. We will be examining both of these in order to understand how they can be used in conjunction to produce more effective System Architecture.

Functional Analysis

A function is a process that receives inputs and transforms them into outputs. At the top level of detail, a Functional Architecture is an operational Architecture; a logical Architecture that defines what the system must do. Subsequent decompositions transform operational functions to systems functions that then can be allocated for implementation to hardware, software, and/or human implementation. The process of functional decomposition should be familiar, as it is used in creating the operational activity model, the OV-5. When we developed the OV-5 we used a process modeling tool, the IDEF0; this is the authors' preferred tool for functional analysis.

Using IDEF0 allows the Systems Engineer to discover the work to be done; that is, the operational functions required to reach the systems objective(s). These were defined in the earlier discussion of the OV-5 and its companion data elements table; data elements tables, or functionality statements, are transformed into requirements with the addition of the word "shall."

This tool is preferred because of its intuitive nature and its implementation convenience; that is, the use of PowerPoint or Visio to create the graphic process model representations, and the use of Excel spreadsheets to capture the operational functions.

Once the operational functions have been decomposed to a level of detail that allows them to be allocated to physical components, they must be allocated to physical configuration items to arrive at a physical Systems Architecture that supports the operational functions.

Functional Allocation

To create a physical system design, functions must be allocated to hardware, software, and/or people. A useful tool for assuring appropriate allocation of all required functions is a matrix that correlates functions with physical configuration items; illustrated in Figure 31. The exes in each box represent a specific part name and model number used during the process of developing a physically instantiated physical architecture.

Generic Physical resources (components)

Functions

	Analog-to-digital converter	Analytic software package	Computer memory	Storage algorithm	
Analog to digital conversion of video signal	X				
Generate average value of video signal		X			
save average value of video signal			X	X	
Function 4					
Function 5					

Figure 31: Mapping Functions to Physical Architecture

In addition to understanding the need for functional analysis of the system, it is also possible to decompose the overriding system requirements down into the logical subsystems being developed. One simple way to think about this is found in the development of an airplane. During the development cycle, each subsystem has a specific weight target to be hit, which is allocated in a manner as shown in Figure 32. If each group exceeds its allocation, the developer is in danger of producing an aircraft that cannot fly.

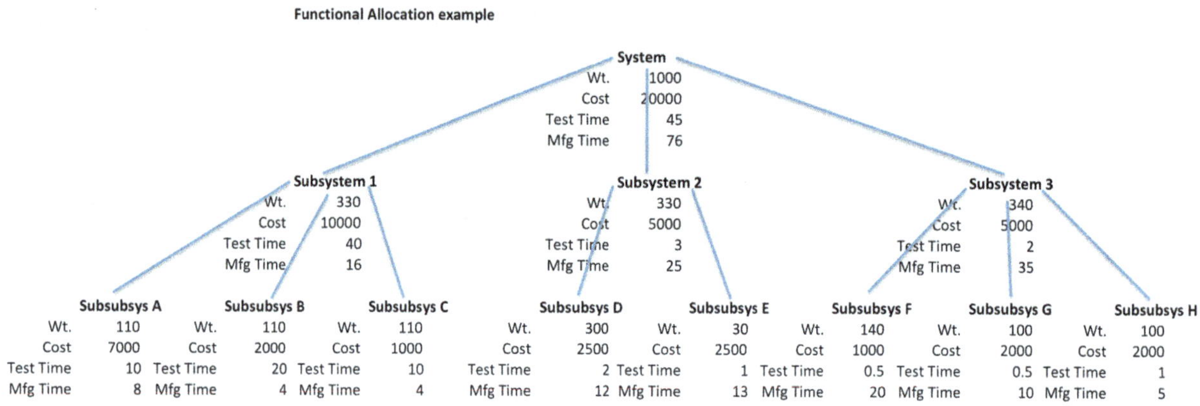

Figure 32: Functional Allocation

To conduct a Functional Analysis, the System Developer must consider the operation of the system under development and identify all needed capabilities by means of a verb/noun pairing. This provides a series of capabilities to be assembled into a holistic view of how the system shall operate. This can be accomplished by a free association exercise in which all of the capabilities are identified, allowing time later in the development to fill in any missing functions or capabilities.

From our simple household toaster example, imagine that we have identified a variety of capabilities that are communicated in this format:

Create setpoint
Clean cycle
Convert power
Update software
Override heat control
Shutdown power
Apply energy
Sense width
Toast product
Since complete
Shut down heat
Unload product
Loan product

Signal status
Input command
Request update
Reset manually
Clean device
Check load condition

These functions must then be assembled into a practical whole (an interconnected flow from operational start to finish), to provide an operational view of the toaster system.
This is shown in Figure 33.

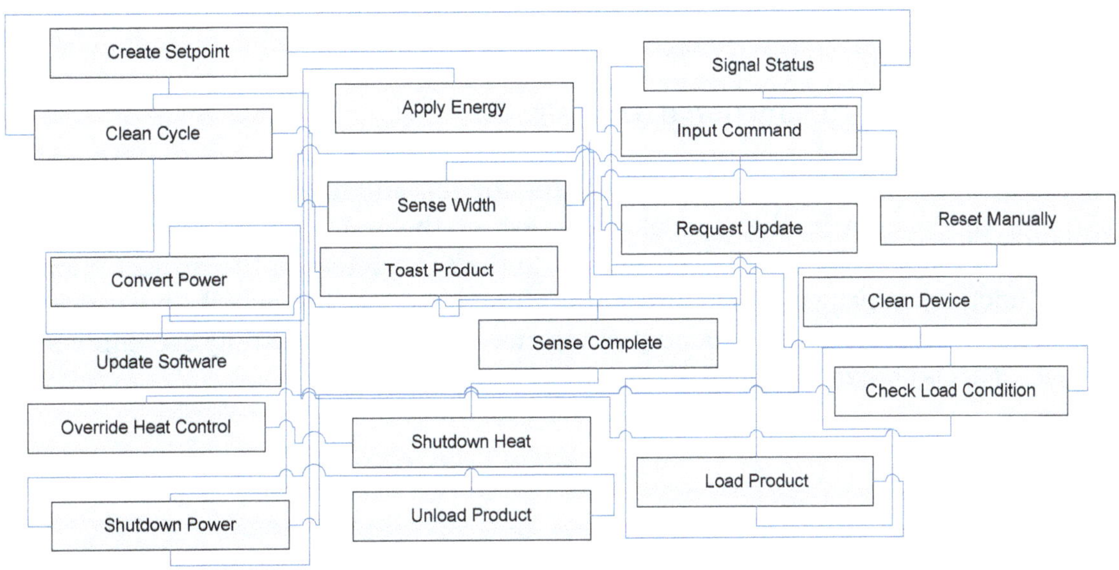

Figure 33: Functional Analysis of Toaster

Once all of the functions have been related one to another, the planner can untangle this diagram, producing what is generally constructed as a top left to bottom right flow of operations for the system. At this point it might become clear that there are several shortcomings that must be addressed, and additional functions may be added or superfluous functions may be deleted. The final outcome is to produce an understandable flow of operations for the proposed system. This is shown in Figure 34.

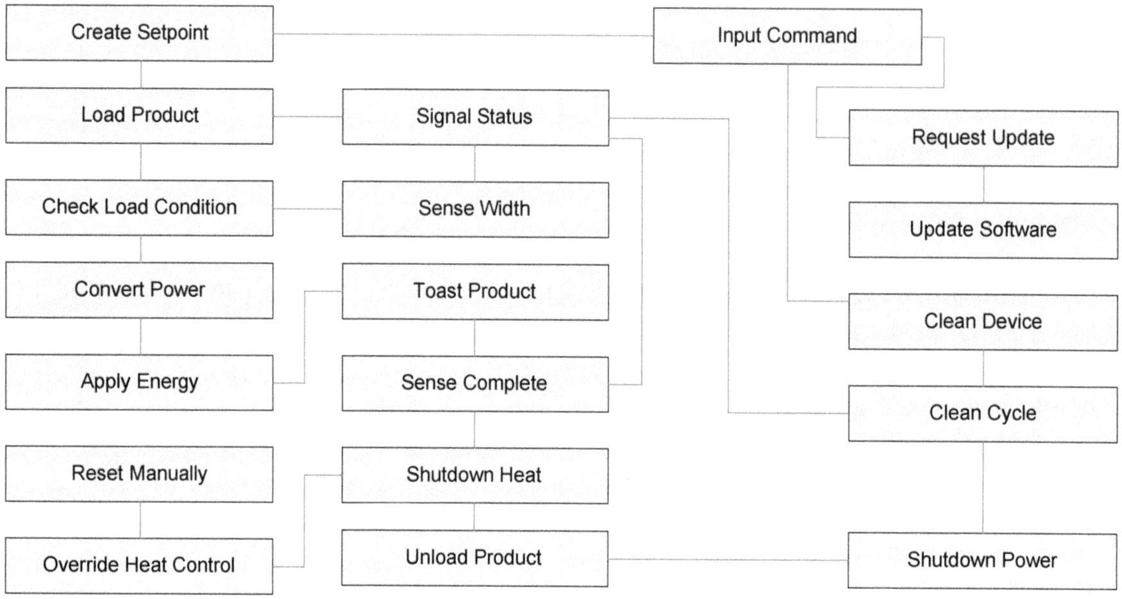

Figure 34: Simplified Functional Analysis

As can be seen in **Figure 34**, we begin at the top left and follow multiple flow paths through the system in the typical operation of the toaster. Once the simplified Functional Analysis has been diagrammed to the satisfaction of the project planner, these functions may be grouped into similar functional families, which will ultimately be used to define the major subcomponents necessary to successfully accomplish the project. This is shown in Figure 35.

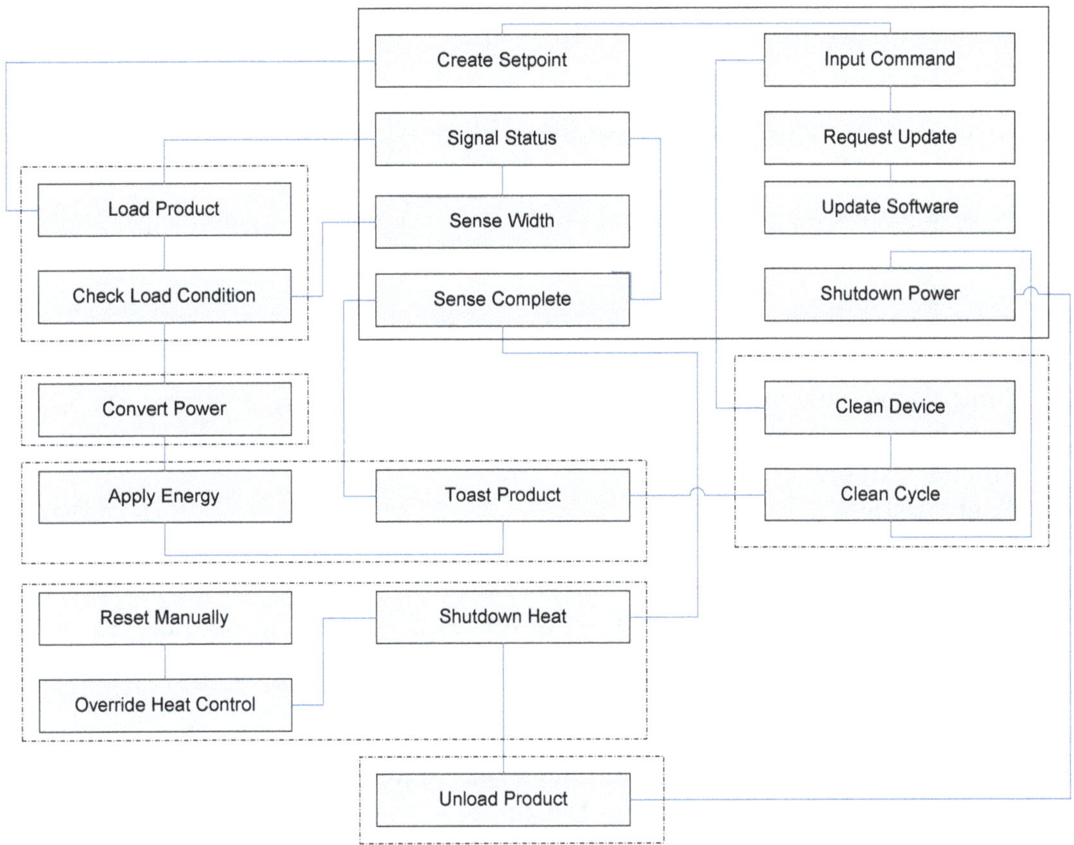

Figure 35: Grouped Functional Analysis

Once the functional analysis has been grouped, these groups can be collapsed into collectives or subsystems that must be provided by the system. This is of particular importance as it also defines the major subprojects that will be implemented during the execution of the project. As can be seen, we have identified the top level of our system, and the first and the second levels of decompositions that must be generated in order to complete the system. This is shown in Figure 36.

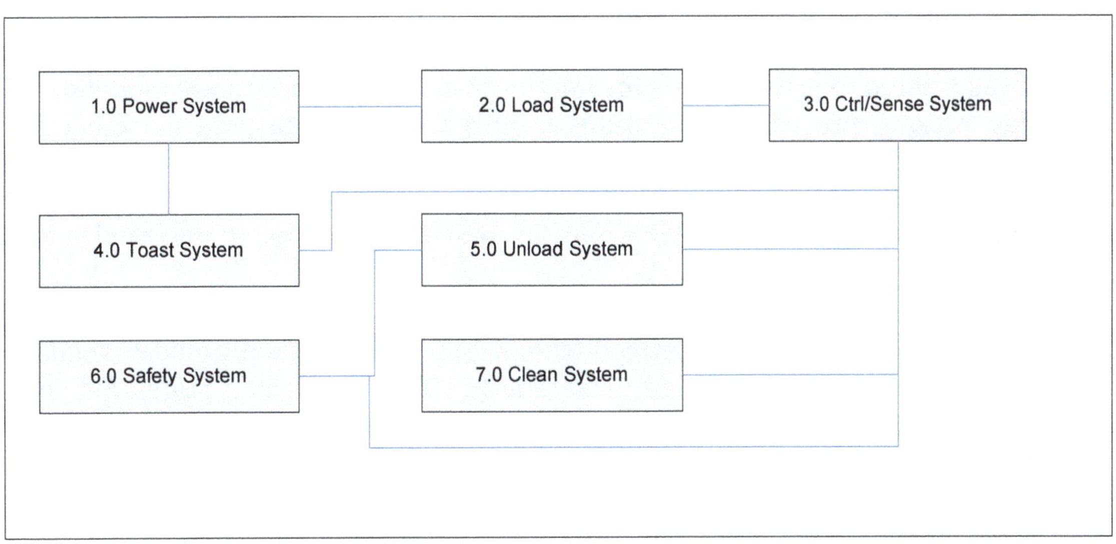

Figure 36: Subcomponents of Toaster

Once this effort has been completed, we are ready to begin the development of our System Architecture. This is generated from the major subsystems that must be present in the final system and the functions that are included within them. To clarify for our toaster project:

Toaster project
- Power system
- Load system
- Controls and sensing system
- Toasting system
- Unload system
- Safety system
- Cleaning system

In our functional analysis, we made a list of functional or operational needs for the system. We then related those needs one to another, which essentially established for us an operational flow diagram.

Conducting this exercise is of significant value, as we identified several functions the typical person would not think about at the outset of the project. We clarified further assumptions that may have limited or expanded our thinking on the subject, and we made some decisions to reduce our aperture on the program by limiting capabilities or market segment. However, we have several more steps to go before we have established effective System Architecture for our development.

To begin our OV-5, we will rely on the integrated definition for function modeling or IDEF0 modeling language. This is a federal information-processing standard, based upon a variety of software and hardware modeling languages currently in use. IDEF0 has been used by a wide variety of commercial and government entities in establishing Architectures for complex system design.

IDEF0 was created to provide a means for completely and consistently modeling the activities, actions, processes, operations required by a system and the functional relationships relating information or objects that support the integration of those functions. It also provides a generic but rigorous, precise and reasonably concise modeling technique that is useful in a wide variety of System Development projects.

IDEF0 is a method designed to model the decisions, actions, and activities of an organization or system. It was derived from a well-established graphical language; the structured analysis and design technique (SADT). The United States Air Force commissioned the developers of SADT to develop a function modeling method for analyzing and communicating the functional perspective of a system.

The principal characteristic of the IDEF0 modeling language is the cell modeling graphic representation. The 'box and arrow' graphics of the diagram show the function as a box, and interface to or from the function as arrows entering or leaving the box. To express functions, boxes operate simultaneously with other boxes, with the interface arrows constraining when and how operations are triggered and controlled. The basic syntax for the model is shown in Figure 37.

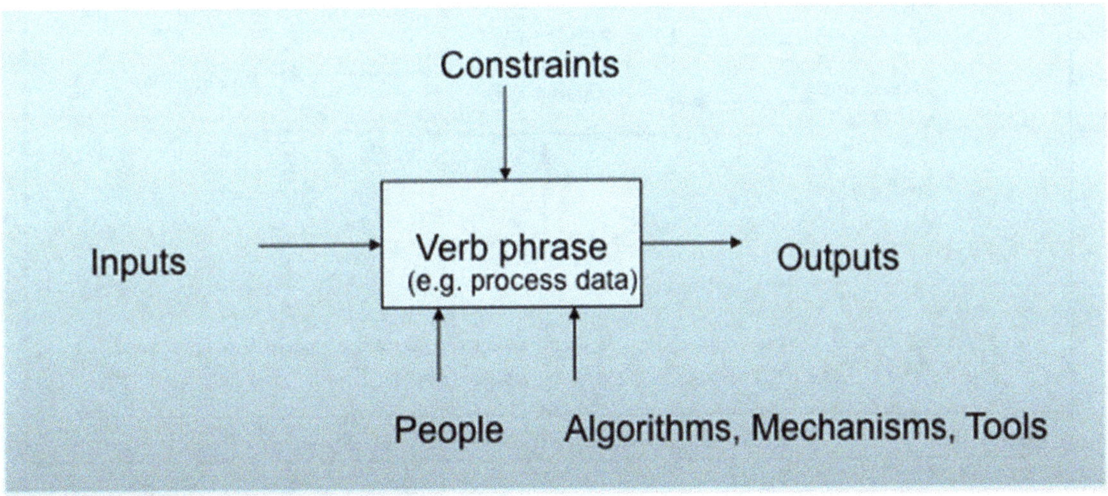

Figure 37: OV-5 Construction

At the highest level, this diagram is known as a context diagram. In this view, the system is entirely encompassed by the boundaries of the box in the diagram. As we continue to add hierarchical structure to the diagram we 'Drill Down' into the box, identifying the functions and capabilities that make up the box and add levels of detail to the system. Note that in functional analysis we started at the functions level and then aggregated the results to work towards a System Architecture from the bottom up; this method is a top down approach.

Thus the context diagram is decomposed into related dataflow diagrams that can be recomposed into the whole. This is shown in the following diagrams.

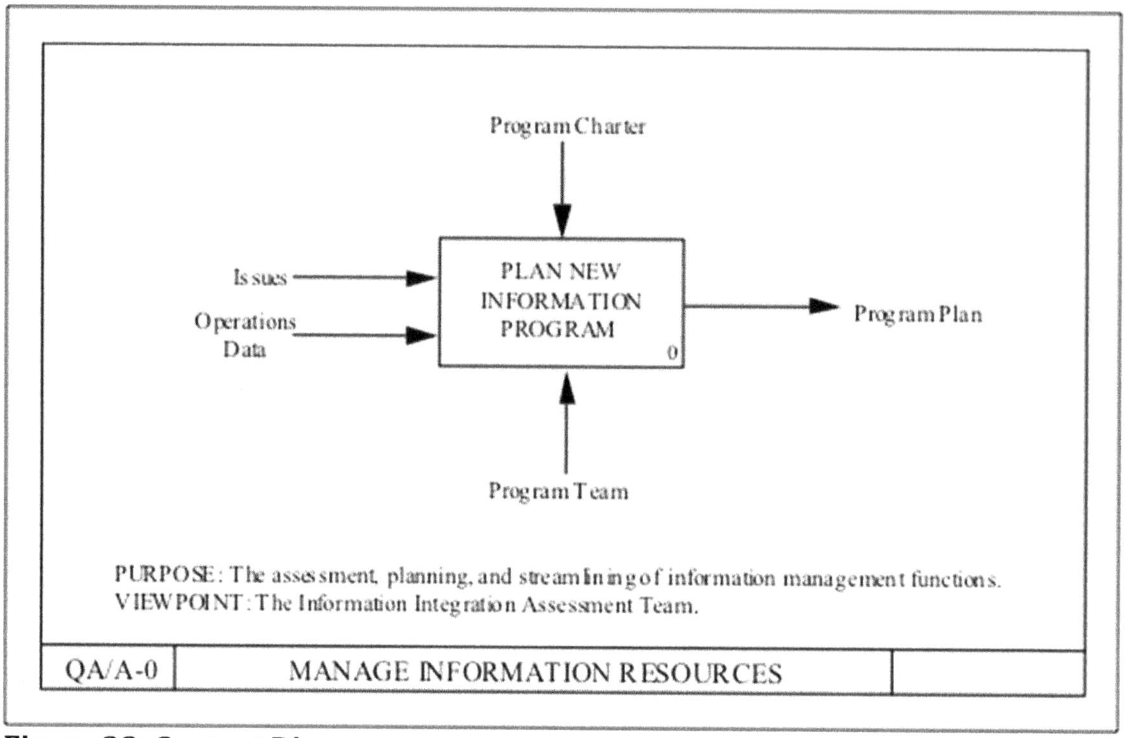

Figure 38: Context Diagram

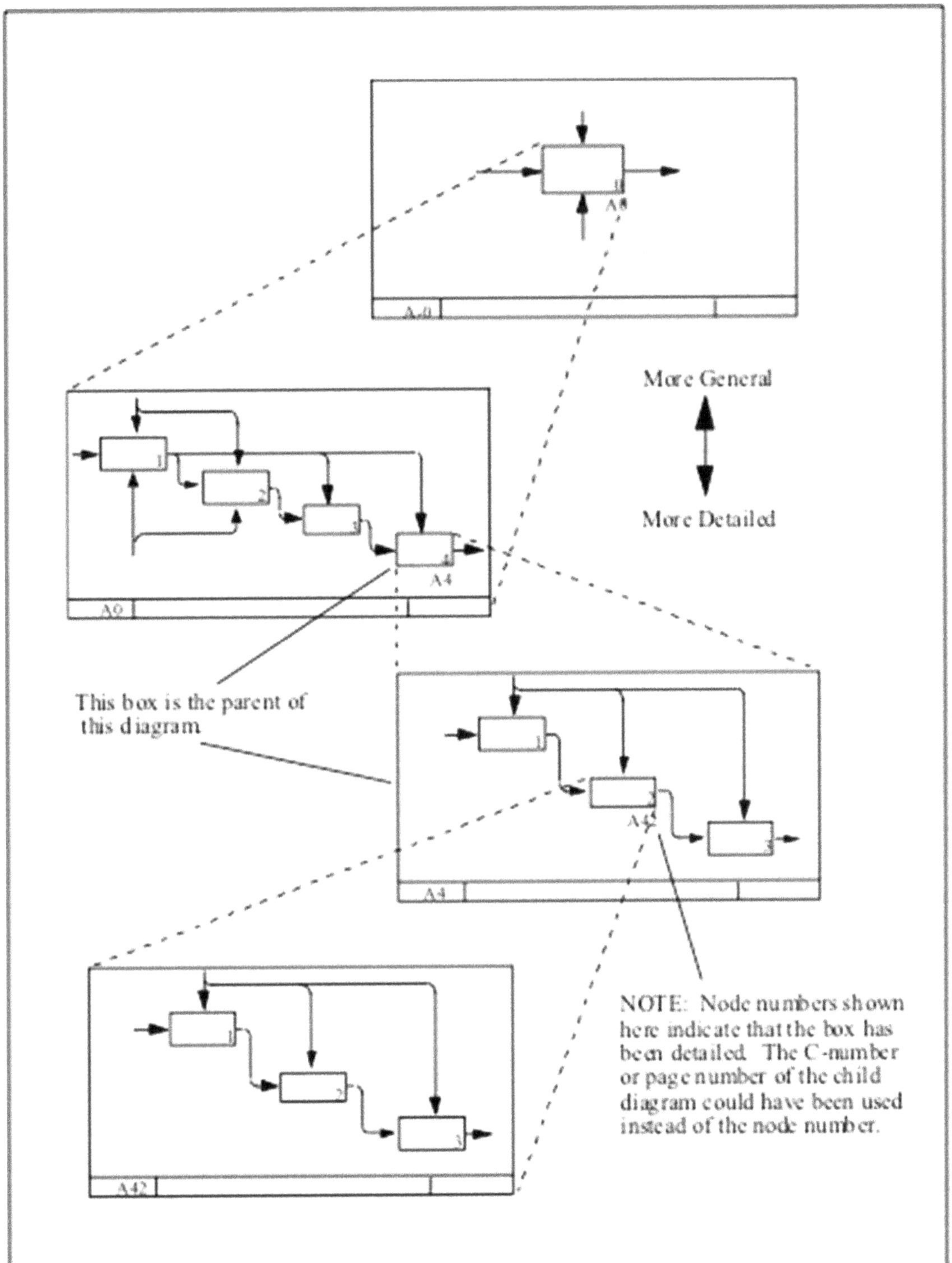

Figure 39: Decompositions of a System

These decompositions relate directly to our previous diagram and clarify how we are adding increasing levels of insight and detail to our system model, ultimately creating our System Architecture.

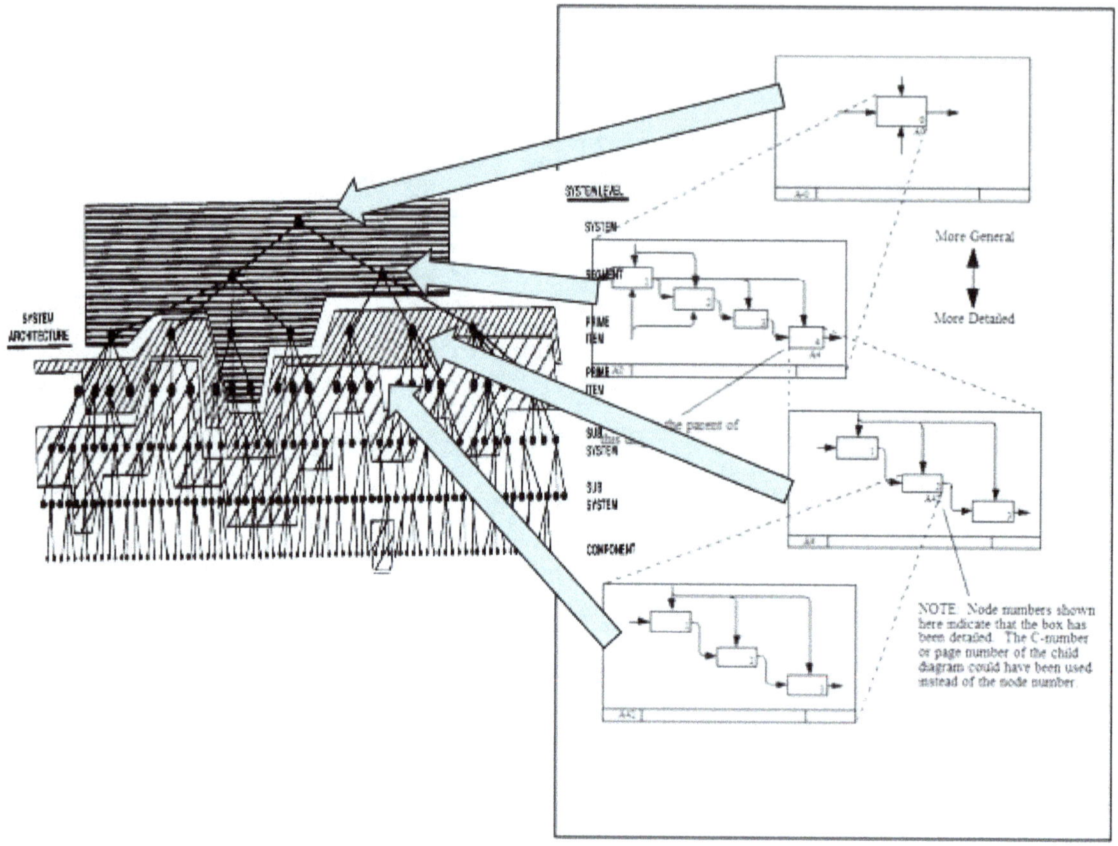

Figure 40: Decomposition leads to Requirements

This methodology allows us to take extremely complex and challenging systems and decompose them into manageable pieces until we create a System Architecture that may be addressed by Design Engineers. To illustrate this fact we will take a simple example: getting an individual to work in the morning.

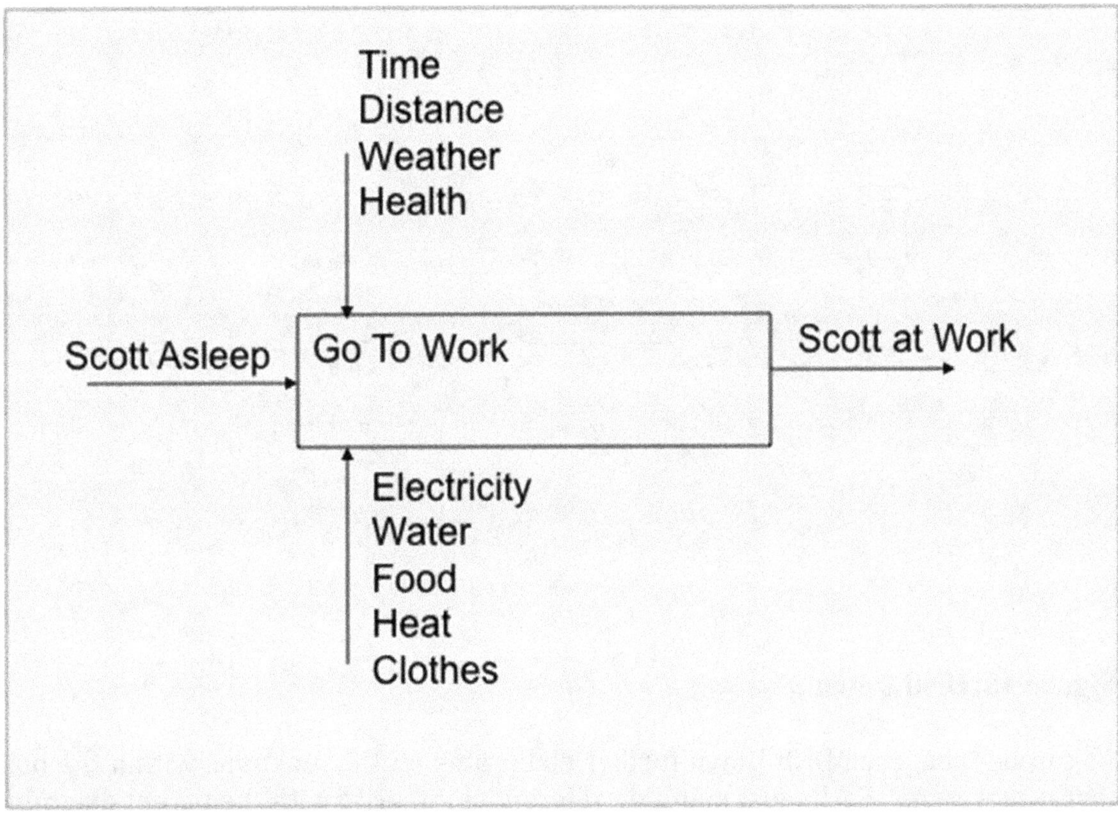

Figure 41: Context Diagram

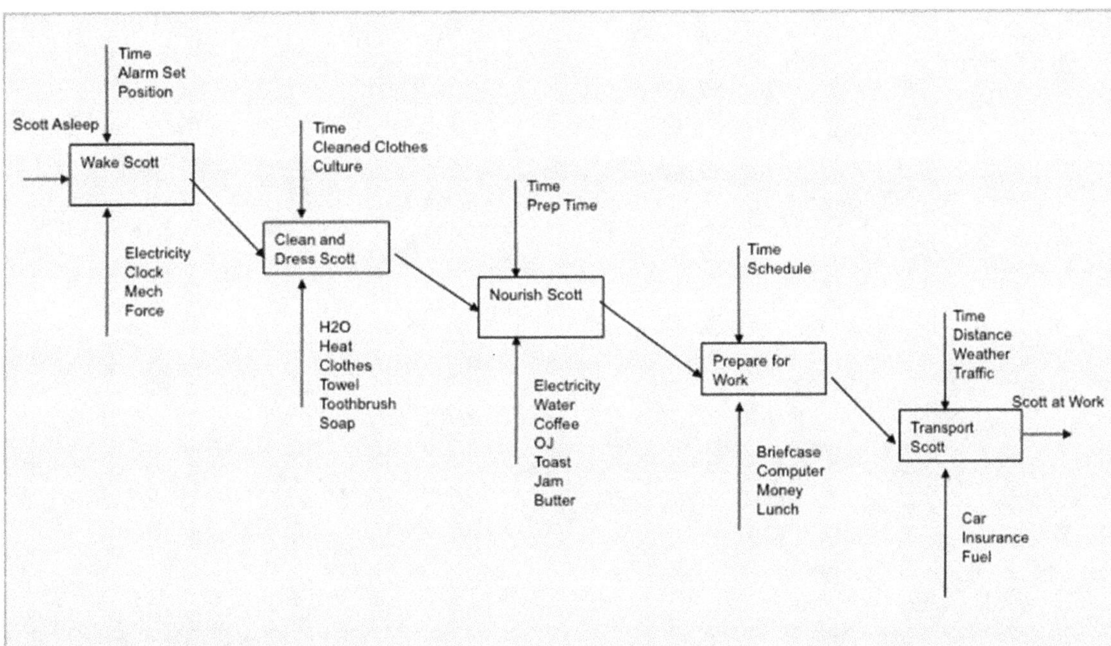

Figure 42: Drill Down 1

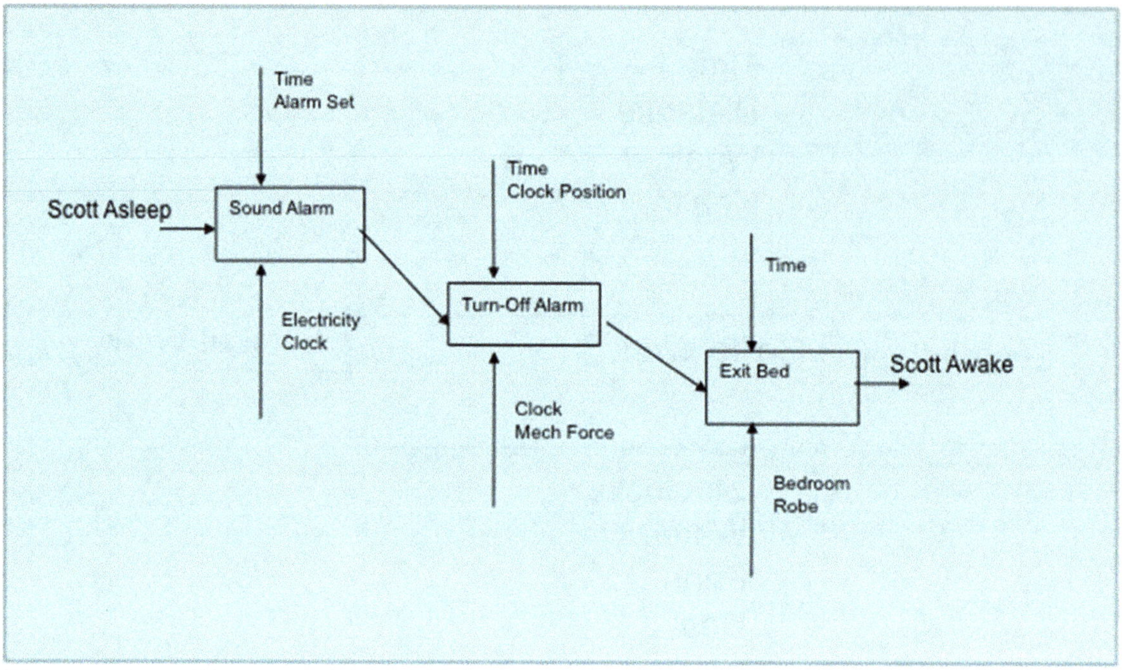

Figure 43: Drill Down 2

As can be seen, each Drill Down further elaborates on the functions within the box that it represents. Each must maintain referential integrity with the parent diagram, as shown in **Figure 44**.

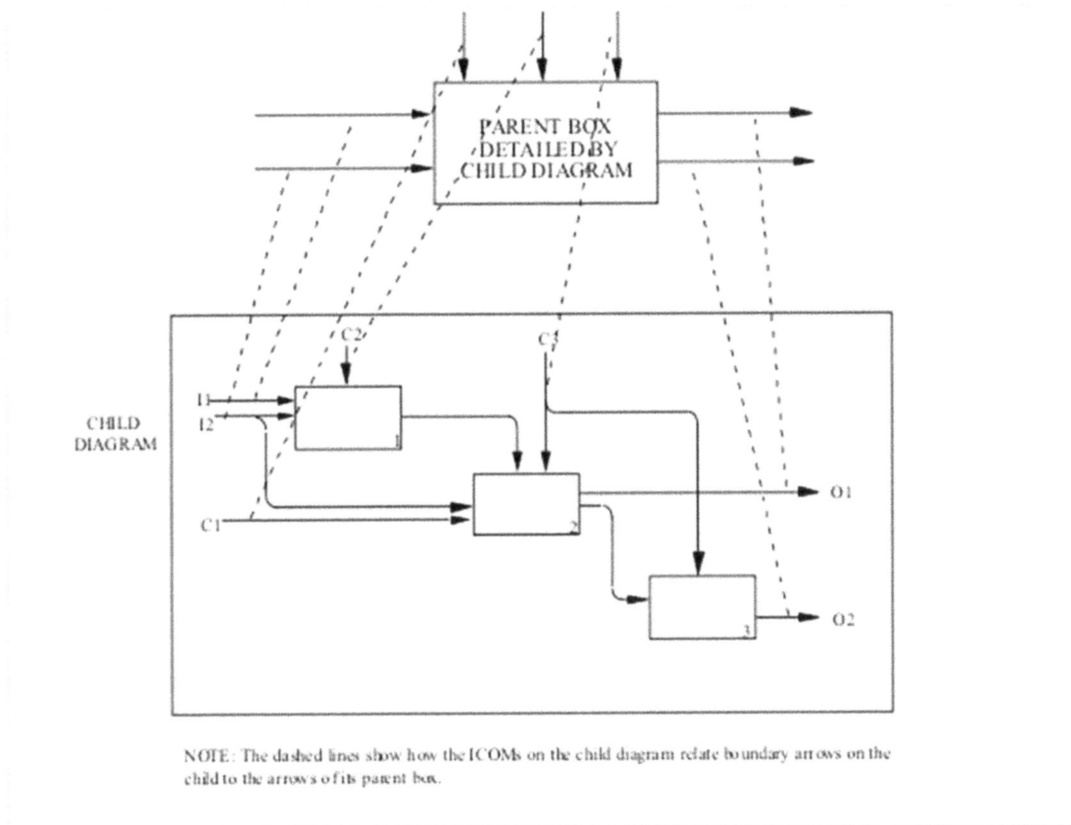

Figure 44: Referential Integrity of Diagrams

This effort can be continued until all of the boxes in the system are well understood or purchased components. We will now examine several differing problems, providing example OV-5 diagrams for review.

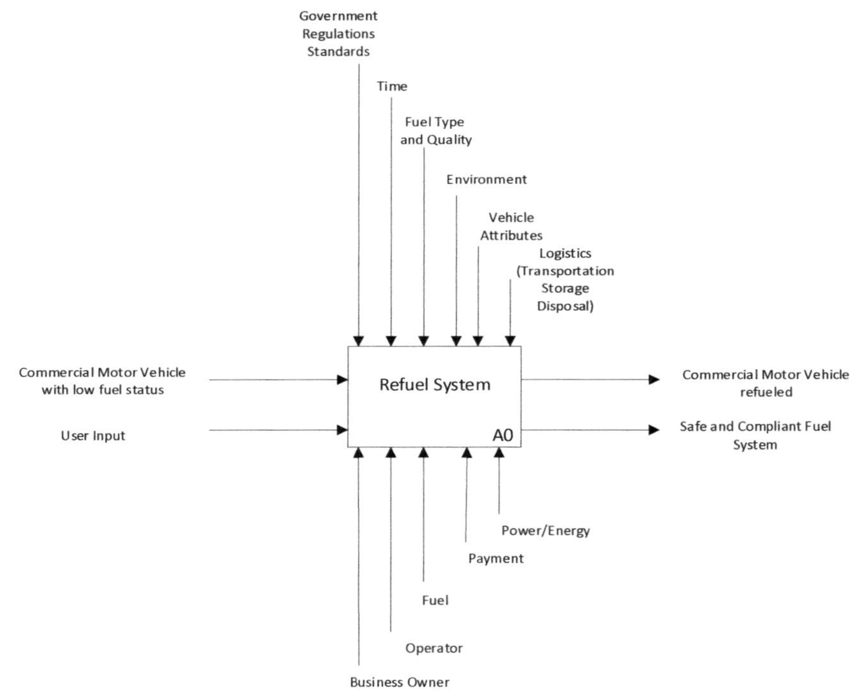

Mission: Optimize time spent performing the common task of refueling commercial vehicles while increasing user comfort.

Purpose: Build a system that streamlines the process of filling up a standard gas powered vehicle fuel tank.

Scope: Commercial Vehicles; US Geography

Benefits: 1) Time Savings 2) Service Quality 3) Safety

NODE: A-0	TITLE: Refuel System for Commercial Vehical	NO.: 27 Mar 17

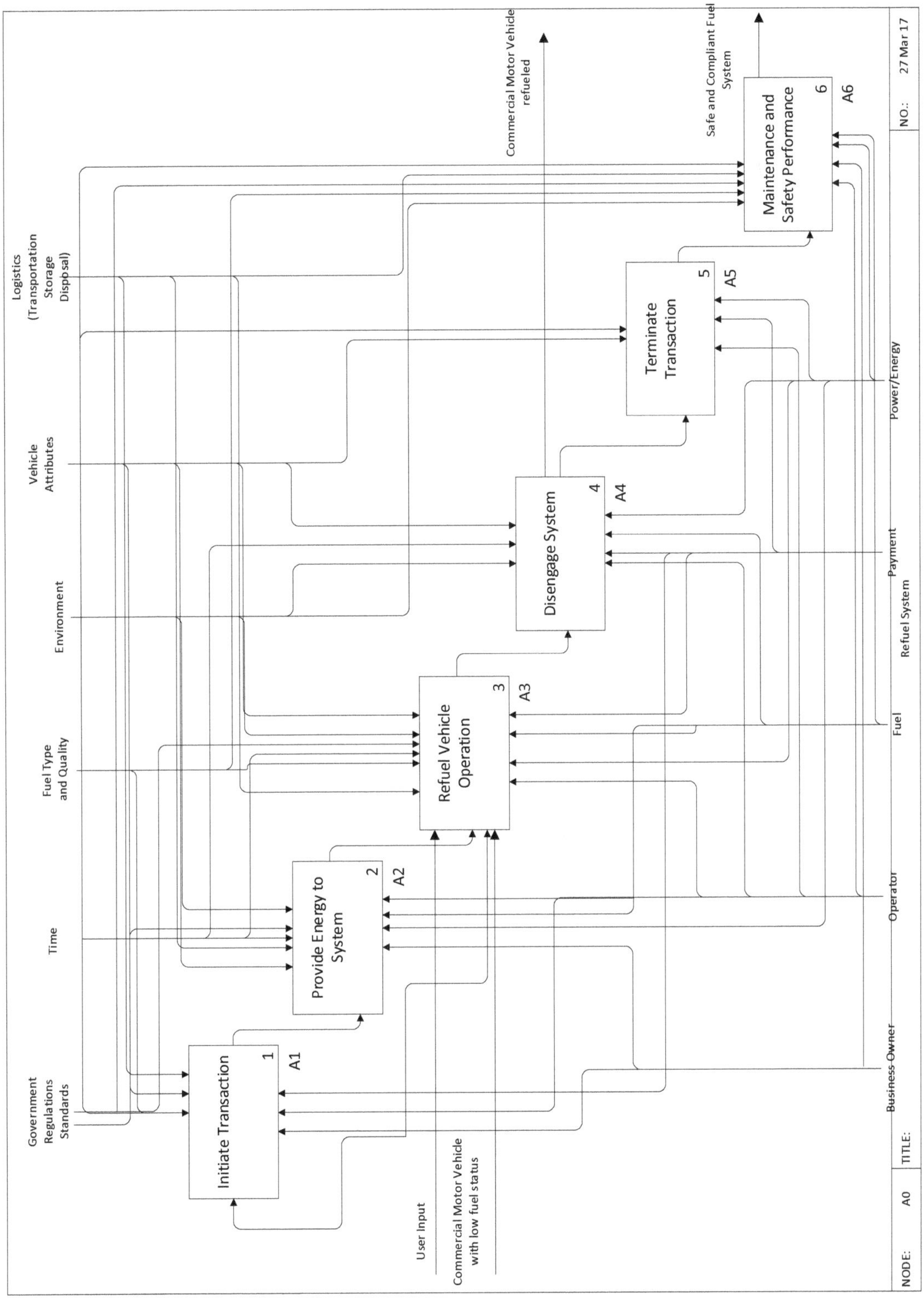
Government Regulations Standards
Time
Fuel Type and Quality
Environment
Vehicle Attributes
Logistics (Transportation Storage Disposal)
Commercial Motor Vehicle refueled
Safe and Compliant Fuel System
Initiate Transaction
1
A1
Provide Energy to System
2
A2
Refuel Vehicle Operation
3
A3
Disengage System
4
A4
Terminate Transaction
5
A5
Maintenance and Safety Performance
6
A6
User Input
Commercial Motor Vehicle with low fuel status
Business Owner
Operator
Fuel
Refuel System
Payment
Power/Energy
NODE:
A0
TITLE:
NO.:
27 Mar 17

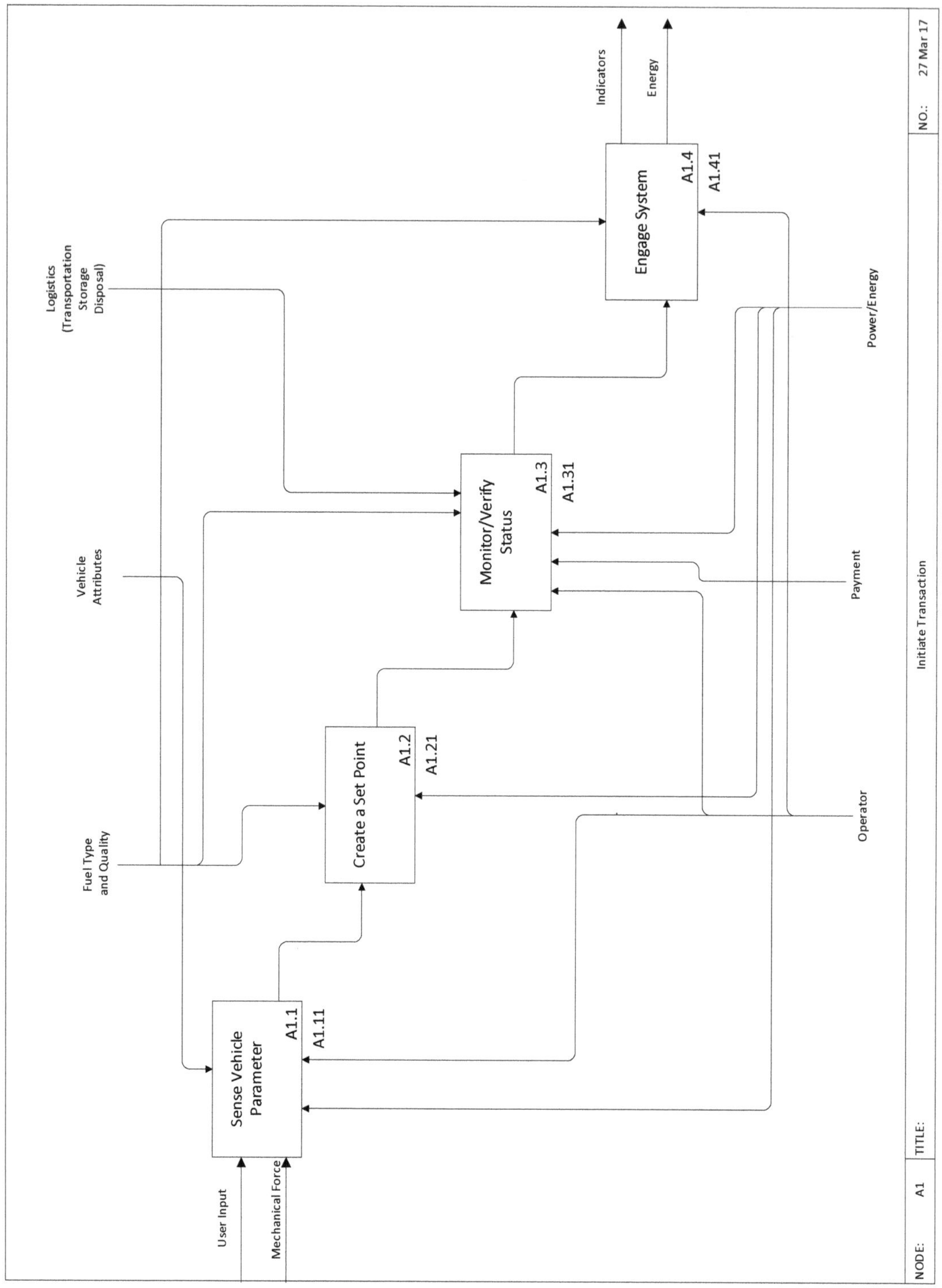
Logistics (Transportation Storage Disposal)
Vehicle Attributes
Fuel Type and Quality
Sense Vehicle Parameter
A1.1
A1.11
Create a Set Point
A1.2
A1.21
Monitor/Verify Status
A1.3
A1.31
Engage System
A1.4
A1.41
Indicators
Energy
User Input
Mechanical Force
Power/Energy
Payment
Operator
NODE: A1
TITLE: Initiate Transaction
NO.: 27 Mar 17

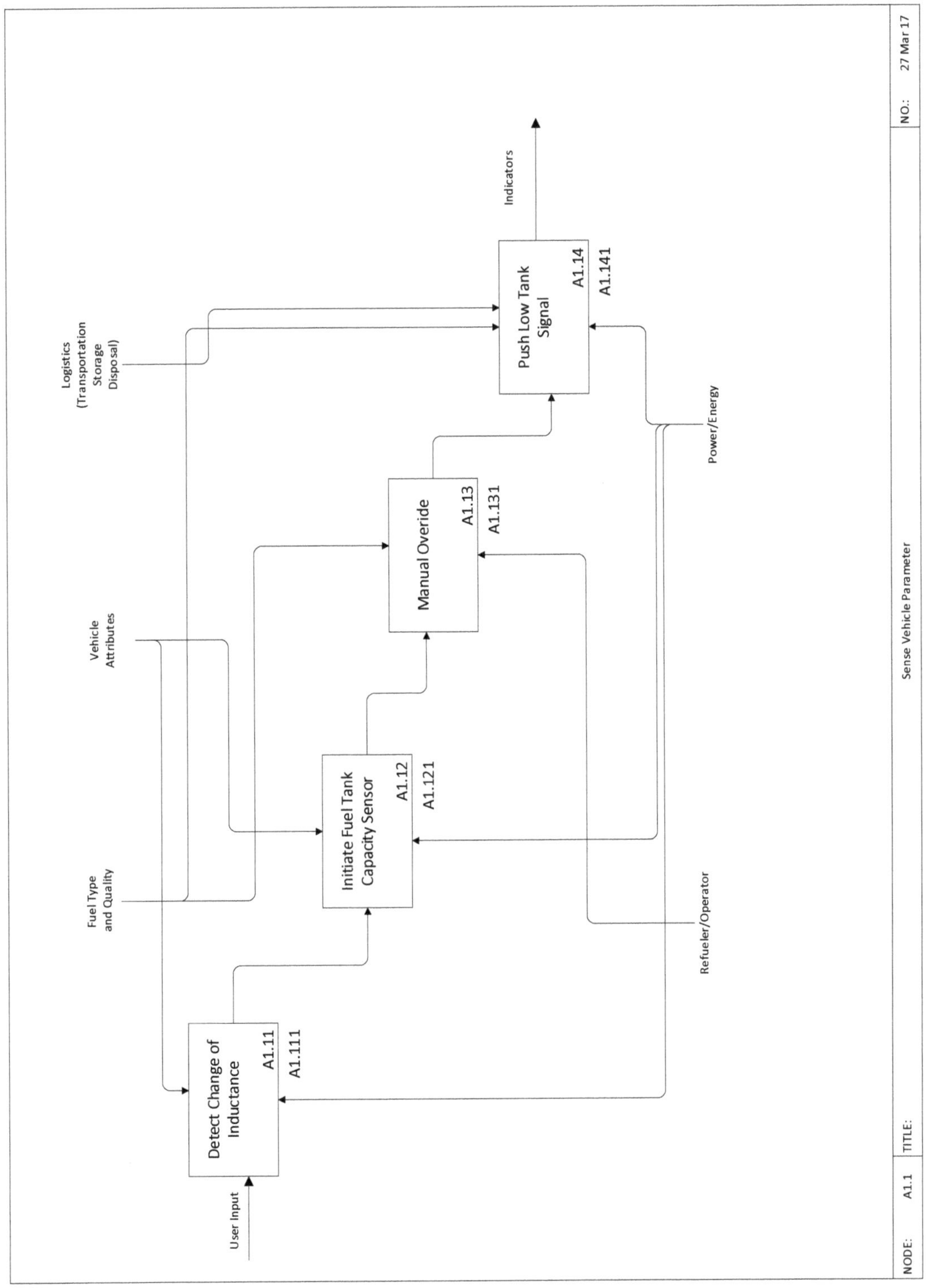
User Input
Detect Change of Inductance
A1.11
A1.111
Fuel Type and Quality
Initiate Fuel Tank Capacity Sensor
A1.12
A1.121
Vehicle Attributes
Manual Overide
A1.13
A1.131
Logistics (Transportation Storage Disposal)
Push Low Tank Signal
A1.14
A1.141
Indicators
Refueler/Operator
Power/Energy
NODE: A1.1
TITLE: Sense Vehicle Parameter
NO.: 27 Mar 17

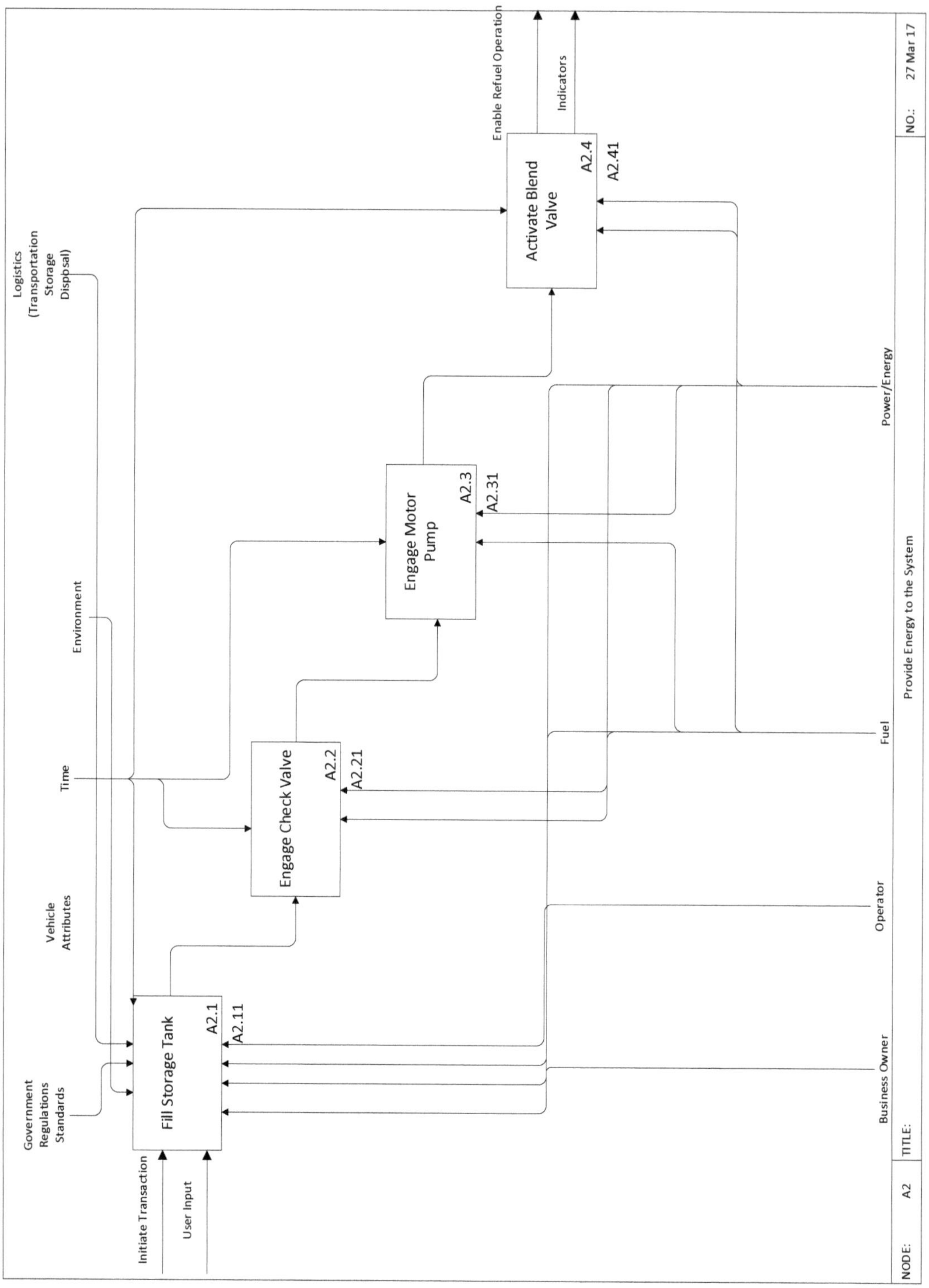

Fill Storage Tank
A2.1
A2.11
Engage Check Valve
A2.2
A2.21
Engage Motor Pump
A2.3
A2.31
Activate Blend Valve
A2.4
A2.41
Initiate Transaction
User Input
Government Regulations Standards
Vehicle Attributes
Time
Environment
Logistics (Transportation Storage Disposal)
Enable Refuel Operation
Indicators
Business Owner
Operator
Fuel
Power/Energy
NODE:
A2
TITLE:
Provide Energy to the System
NO.:
27 Mar 17

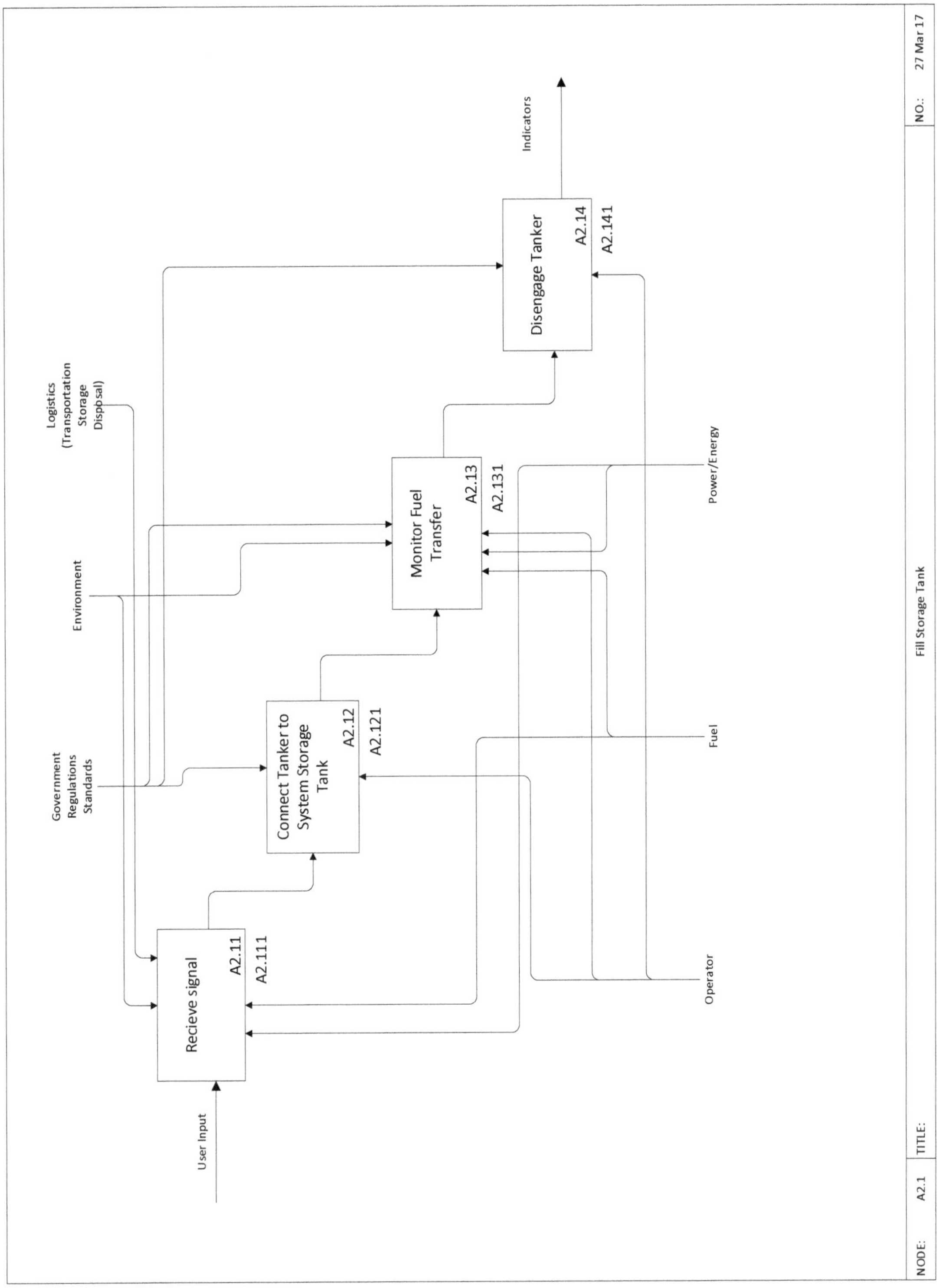

Recieve signal
A2.11
A2.111
Connect Tanker to System Storage Tank
A2.12
A2.121
Monitor Fuel Transfer
A2.13
A2.131
Disengage Tanker
A2.14
A2.141
User Input
Government Regulations Standards
Environment
Logistics (Transportation Storage Disposal)
Indicators
Power/Energy
Fuel
Operator
NODE:
A2.1
TITLE:
Fill Storage Tank
NO.:
27 Mar 17

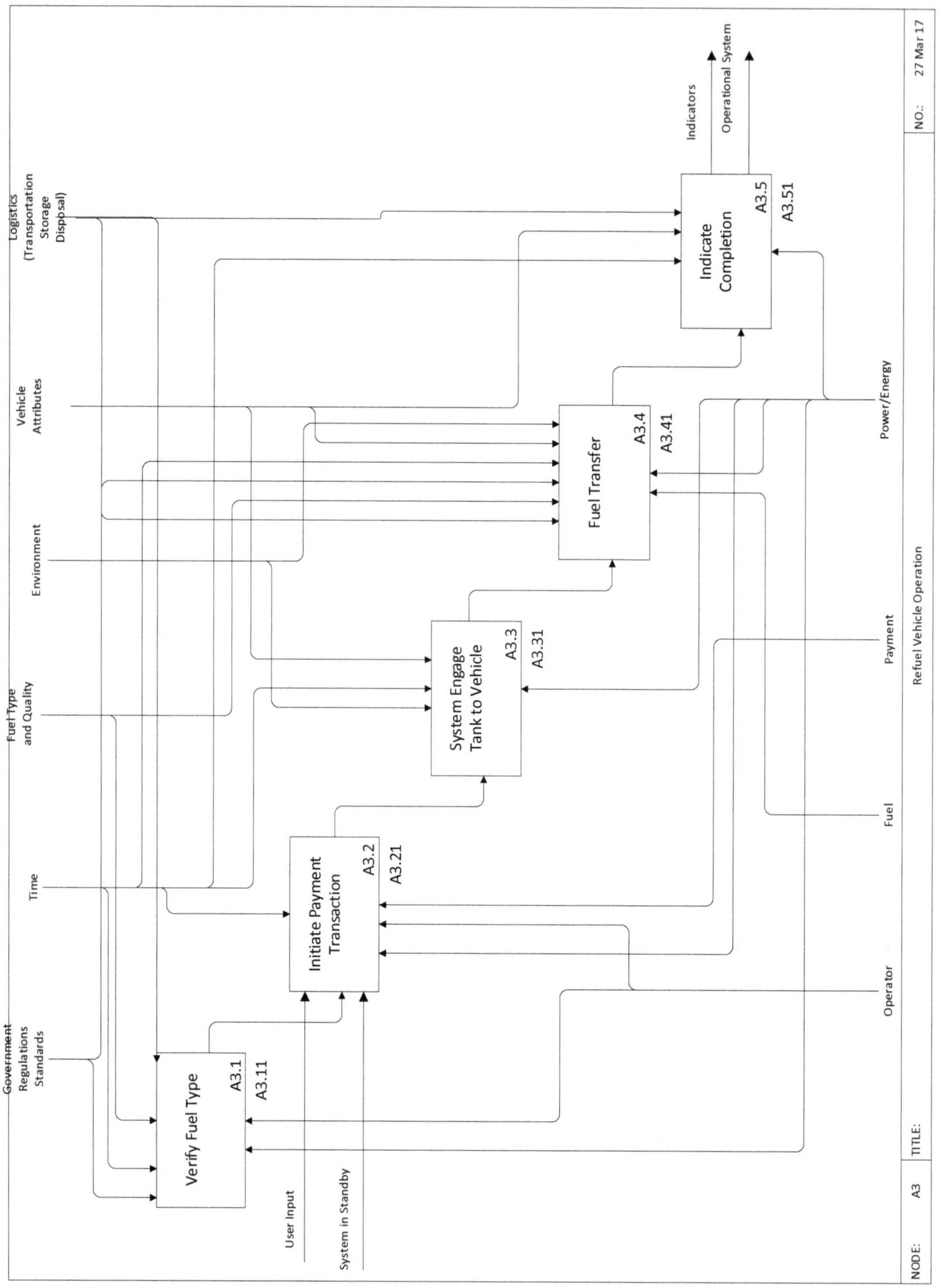
Government Regulations Standards
Time
Fuel Type and Quality
Environment
Vehicle Attributes
Logistics (Transportation Storage Disposal)
Verify Fuel Type
A3.1
A3.11
Initiate Payment Transaction
A3.2
A3.21
System Engage Tank to Vehicle
A3.3
A3.31
Fuel Transfer
A3.4
A3.41
Indicate Completion
A3.5
A3.51
Indicators
Operational System
User Input
System in Standby
Operator
Fuel
Payment
Power/Energy
NODE:
A3
TITLE:
Refuel Vehicle Operation
NO.:
27 Mar 17

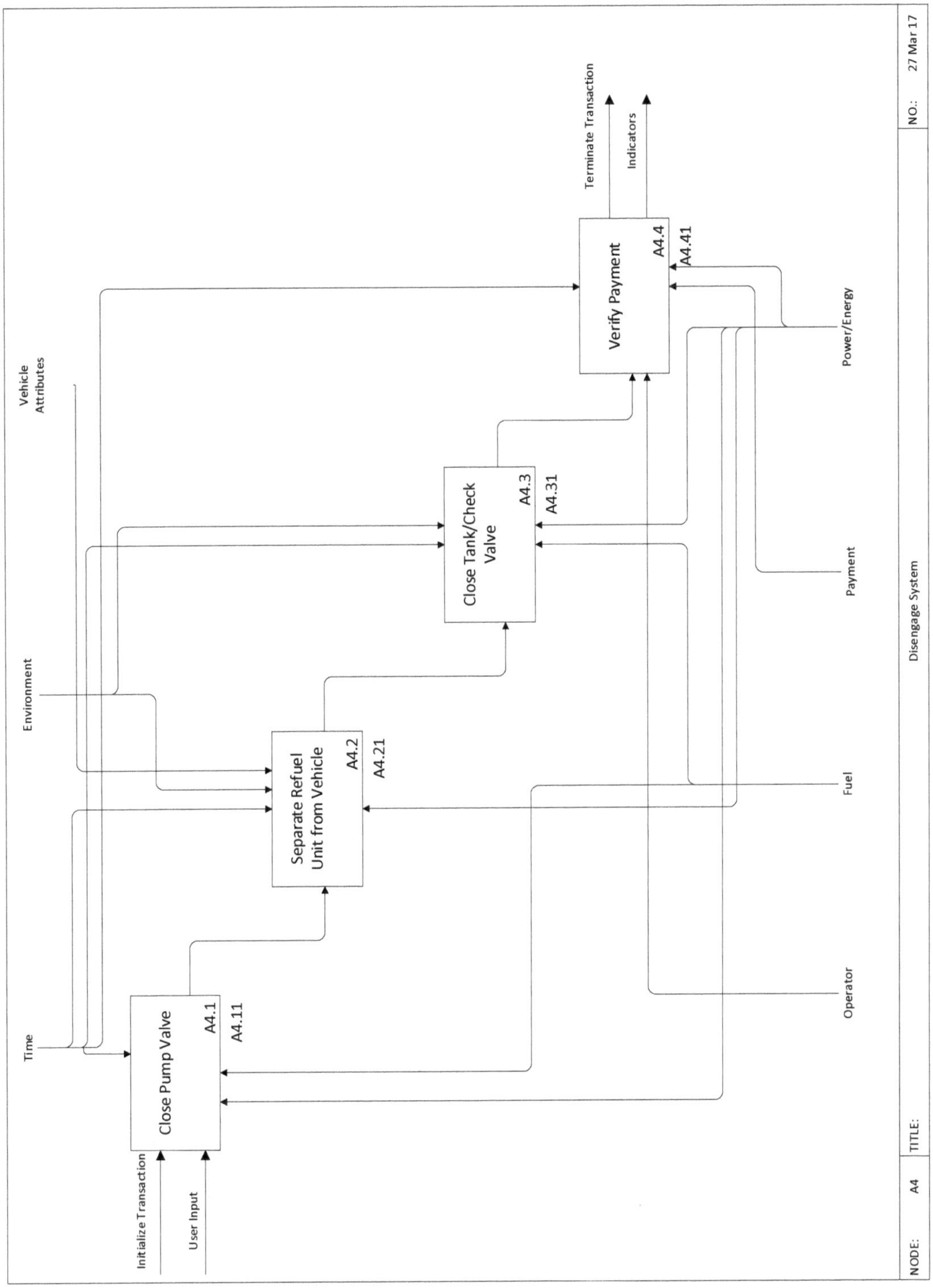

Close Pump Valve
A4.1
A4.11
Separate Refuel Unit from Vehicle
A4.2
A4.21
Close Tank/Check Valve
A4.3
A4.31
Verify Payment
A4.4
A4.41
Time
Environment
Vehicle Attributes
Initialize Transaction
User Input
Terminate Transaction
Indicators
Operator
Fuel
Payment
Power/Energy
NODE:
A4
TITLE:
Disengage System
NO.:
27 Mar 17

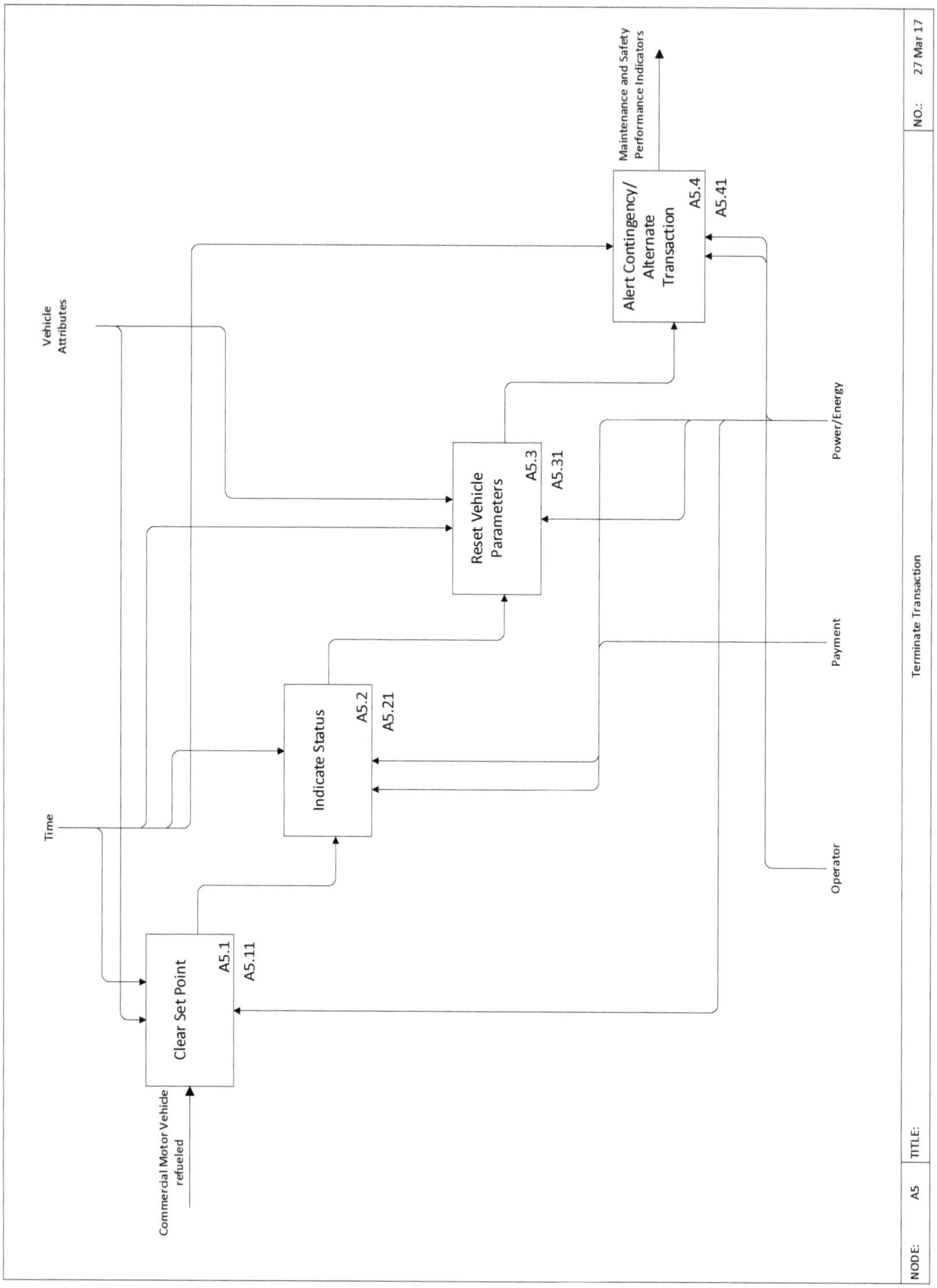

Commercial Motor Vehicle refueled
Time
Vehicle Attributes
Clear Set Point
A5.1
A5.11
Indicate Status
A5.2
A5.21
Reset Vehicle Parameters
A5.3
A5.31
Alert Contingency/ Alternate Transaction
A5.4
A5.41
Maintenance and Safety Performance Indicators
Operator
Payment
Power/Energy
NODE: A5
TITLE: Terminate Transaction
NO.: 27 Mar 17

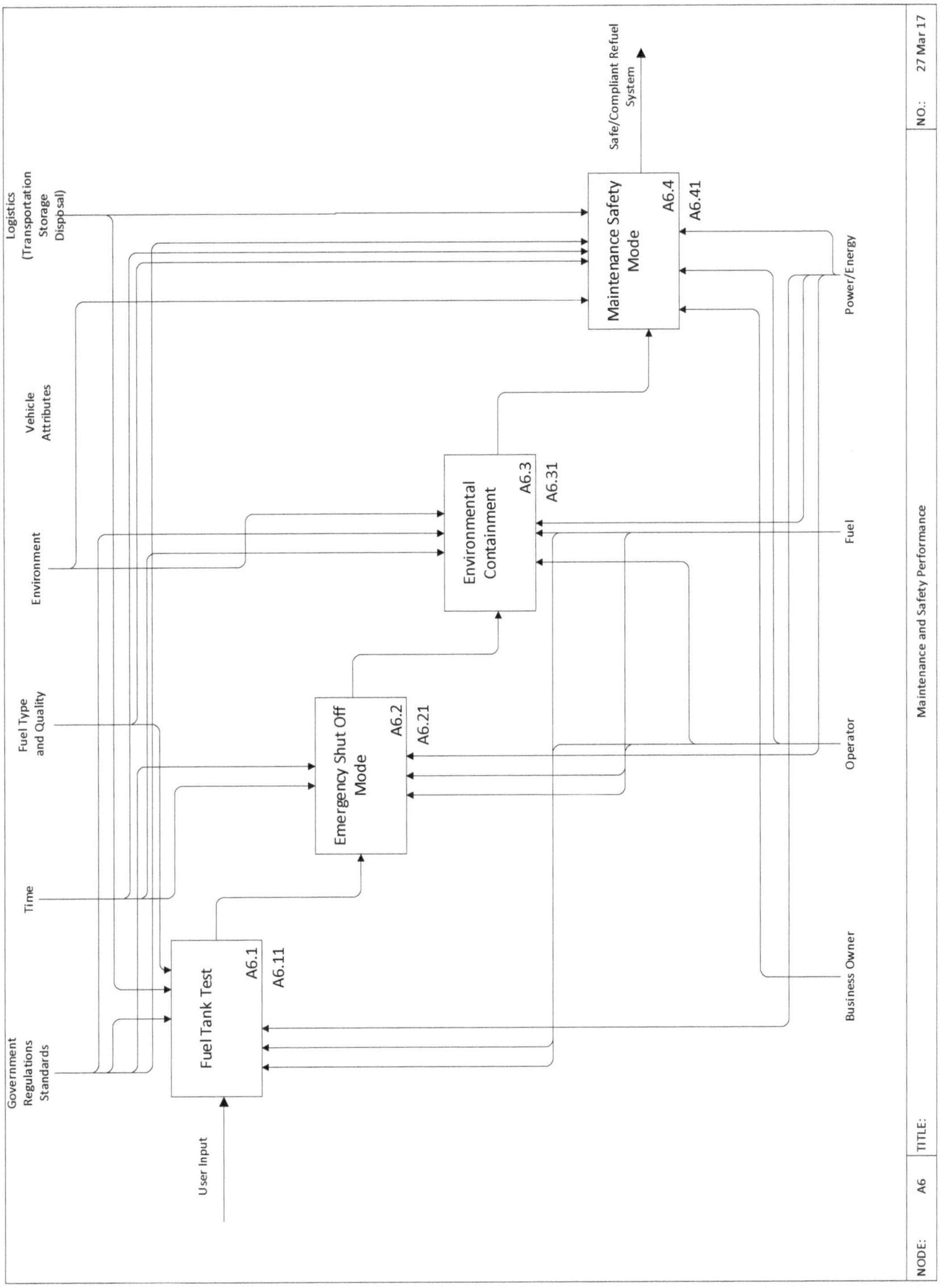

Figure 45: OV-5 Diagrams – Refueling System

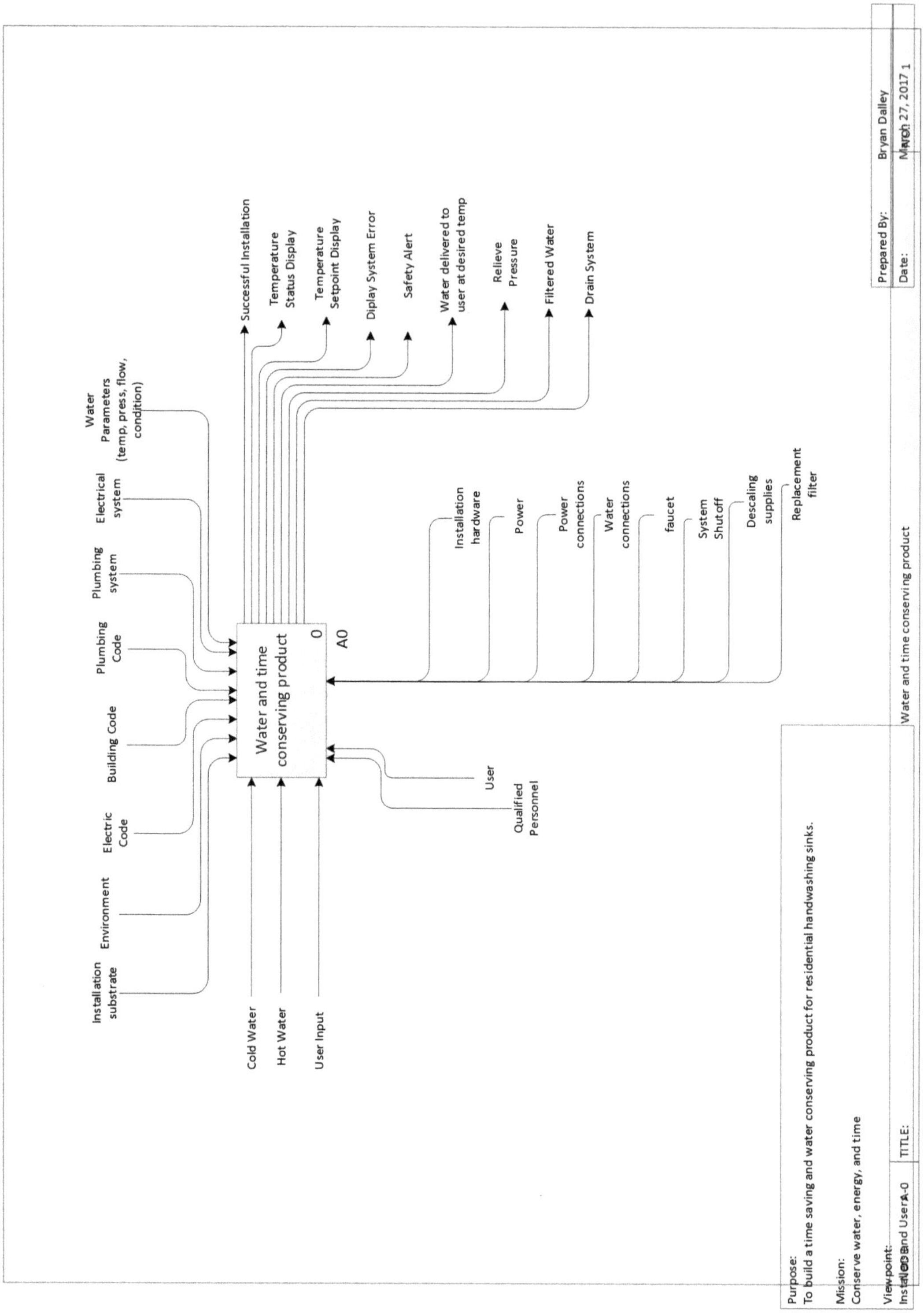
Water and time conserving product
0
A0
Installation substrate
Environment
Electric Code
Building Code
Plumbing Code
Plumbing system
Electrical system
Water Parameters (temp, press, flow, condition)
Cold Water
Hot Water
User Input
Successful Installation
Temperature Status Display
Temperature Setpoint Display
Diplay System Error
Safety Alert
Water delivered to user at desired temp
Relieve Pressure
Filtered Water
Drain System
Installation hardware
Power
Power connections
Water connections
faucet
System Shutoff
Descaling supplies
Replacement filter
User
Qualified Personnel
Purpose:
To build a time saving and water conserving product for residential handwashing sinks.
Mission:
Conserve water, energy, and time
Viewpoint:
Installer and User
NODE:
A-0
TITLE:
Water and time conserving product
Prepared By:
Bryan Dalley
Date:
March 27, 2017
NO.:
1

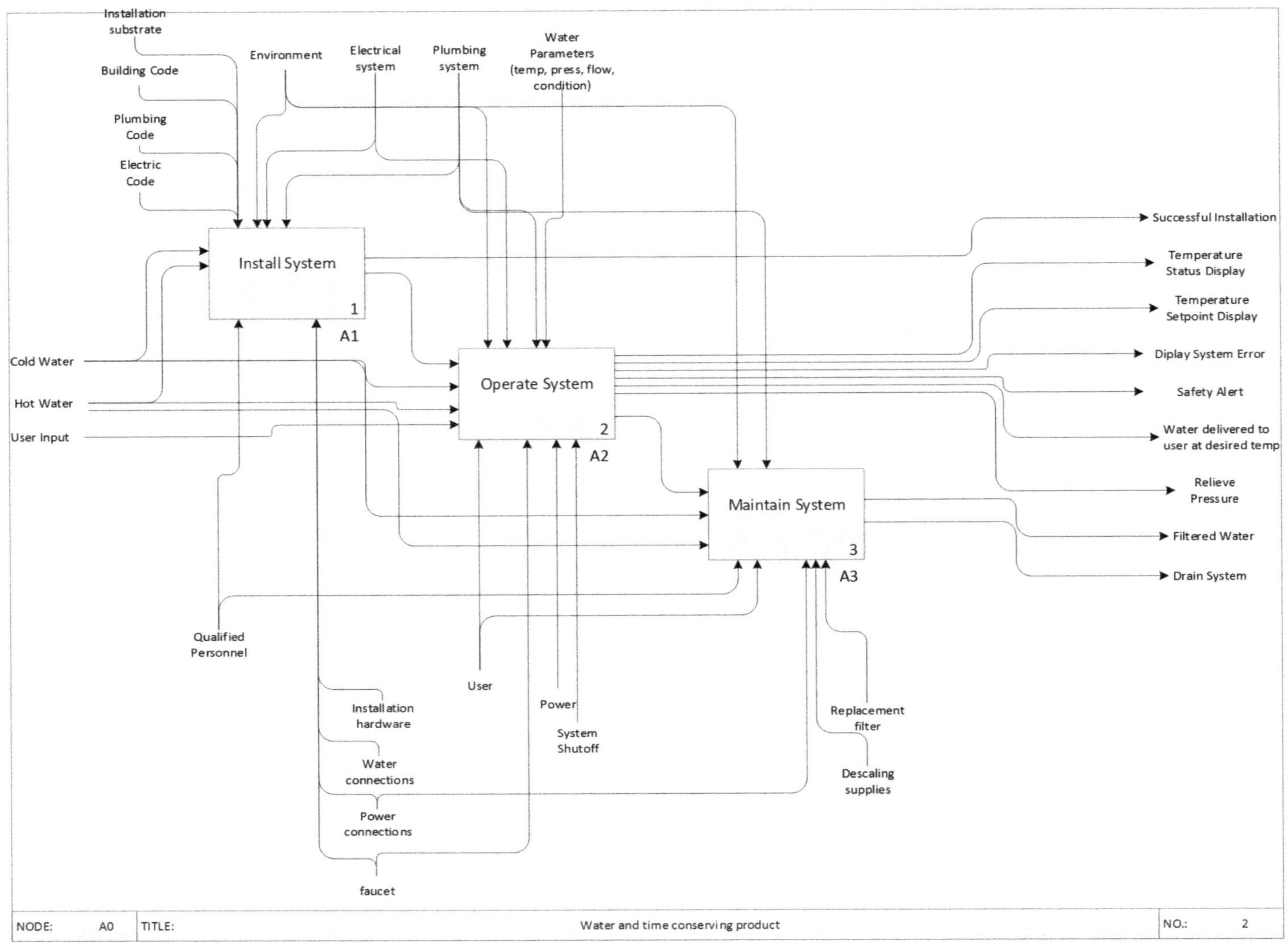
Installation substrate
Building Code
Plumbing Code
Electric Code
Environment
Electrical system
Plumbing system
Water Parameters (temp, press, flow, condition)
Install System
1
A1
Operate System
2
A2
Maintain System
3
A3
Cold Water
Hot Water
User Input
Successful Installation
Temperature Status Display
Temperature Setpoint Display
Diplay System Error
Safety Alert
Water delivered to user at desired temp
Relieve Pressure
Filtered Water
Drain System
Qualified Personnel
User
Power
System Shutoff
Installation hardware
Water connections
Power connections
faucet
Replacement filter
Descaling supplies
NODE: A0
TITLE: Water and time conserving product
NO.: 2

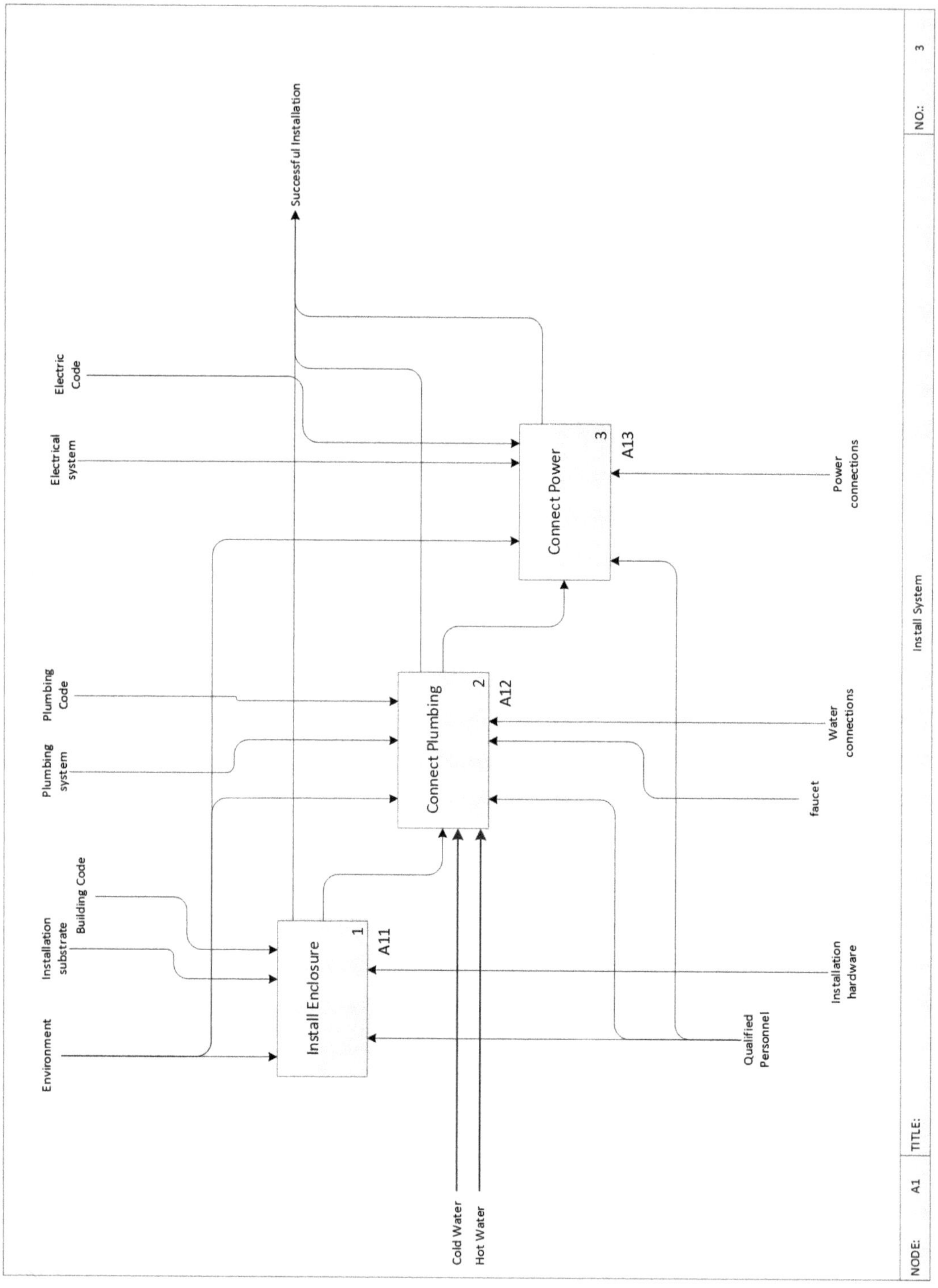

Environment
Installation substrate
Building Code
Plumbing system
Plumbing Code
Electrical system
Electric Code
Successful Installation
Install Enclosure
1
A11
Connect Plumbing
2
A12
Connect Power
3
A13
Cold Water
Hot Water
Qualified Personnel
Installation hardware
faucet
Water connections
Power connections
NODE:
A1
TITLE:
Install System
NO.:
3

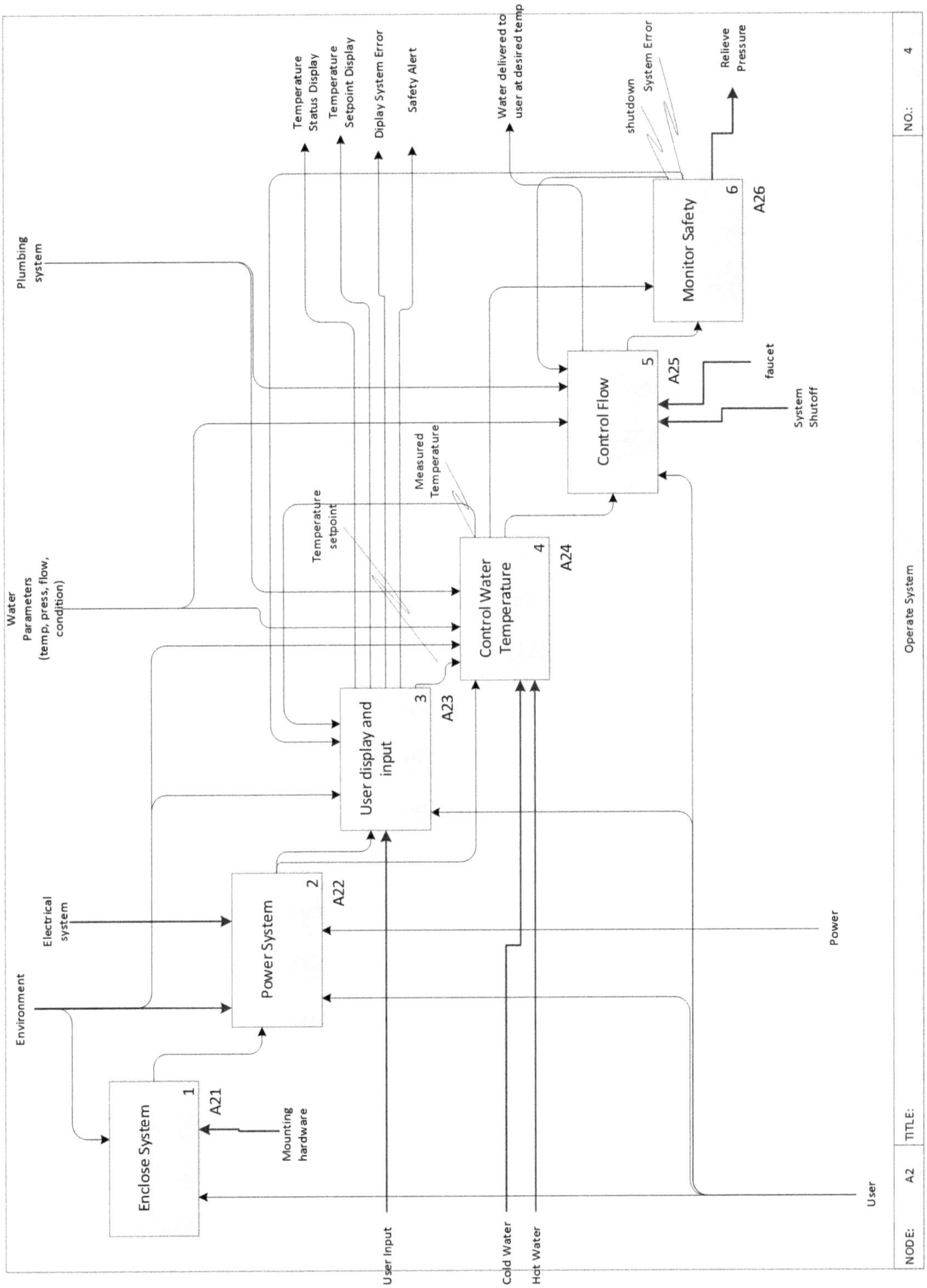
Enclose System
1
A21
Power System
2
A22
User display and input
3
A23
Control Water Temperature
4
A24
Control Flow
5
A25
Monitor Safety
6
A26
Environment
Electrical system
Water Parameters (temp, press, flow, condition)
Plumbing system
Mounting hardware
User Input
Cold Water
Hot Water
Power
User
faucet
System Shutoff
Temperature setpoint
Measured Temperature
Temperature Status Display
Temperature Setpoint Display
Diplay System Error
Safety Alert
Water delivered to user at desired temp
shutdown
System Error
Relieve Pressure
NODE:
A2
TITLE:
Operate System
NO.:
4

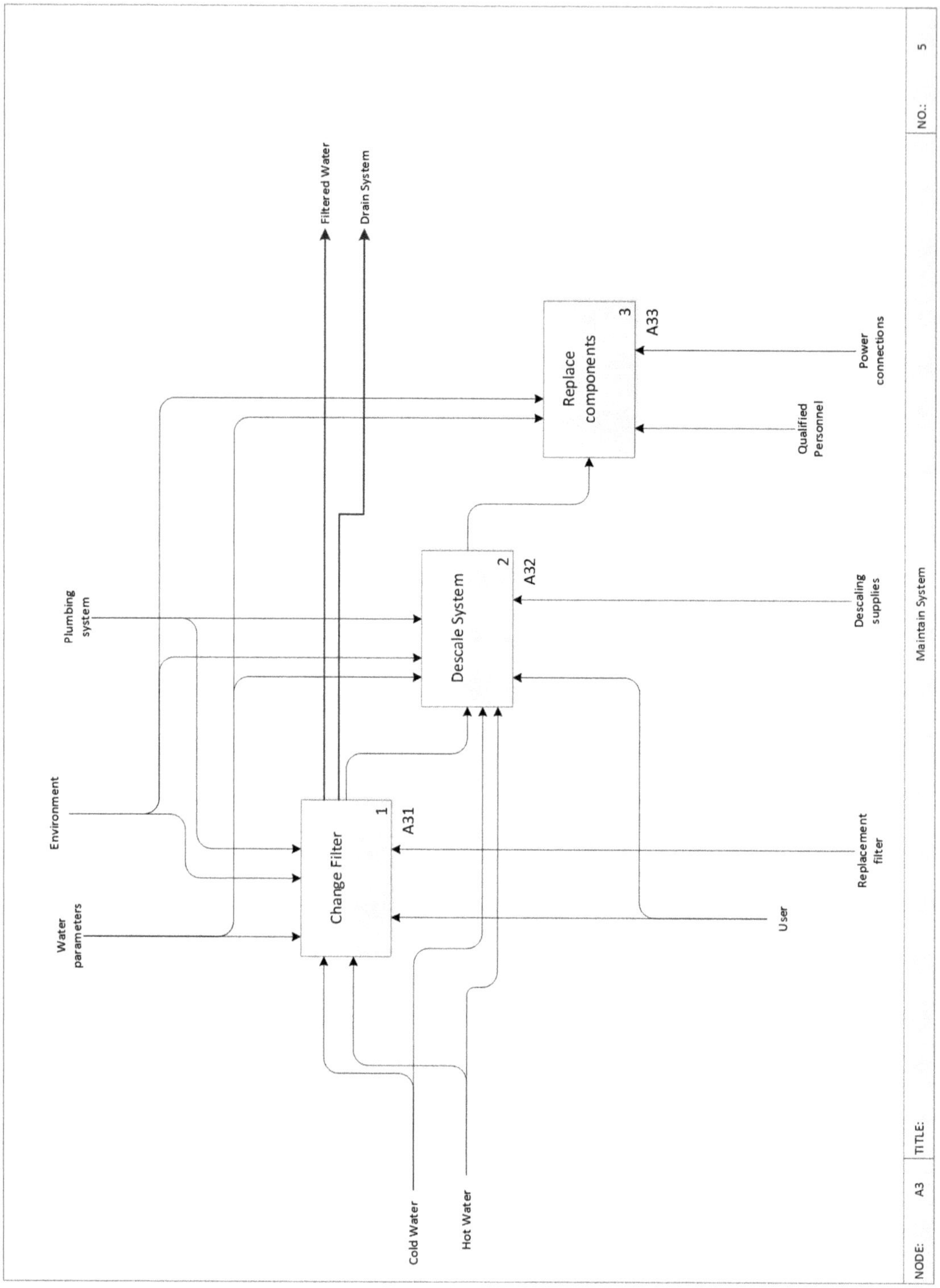
Water parameters
Environment
Plumbing system
Change Filter
1
A31
Descale System
2
A32
Replace components
3
A33
Filtered Water
Drain System
Cold Water
Hot Water
Replacement filter
User
Descaling supplies
Qualified Personnel
Power connections
NODE: A3
TITLE: Maintain System
NO.: 5

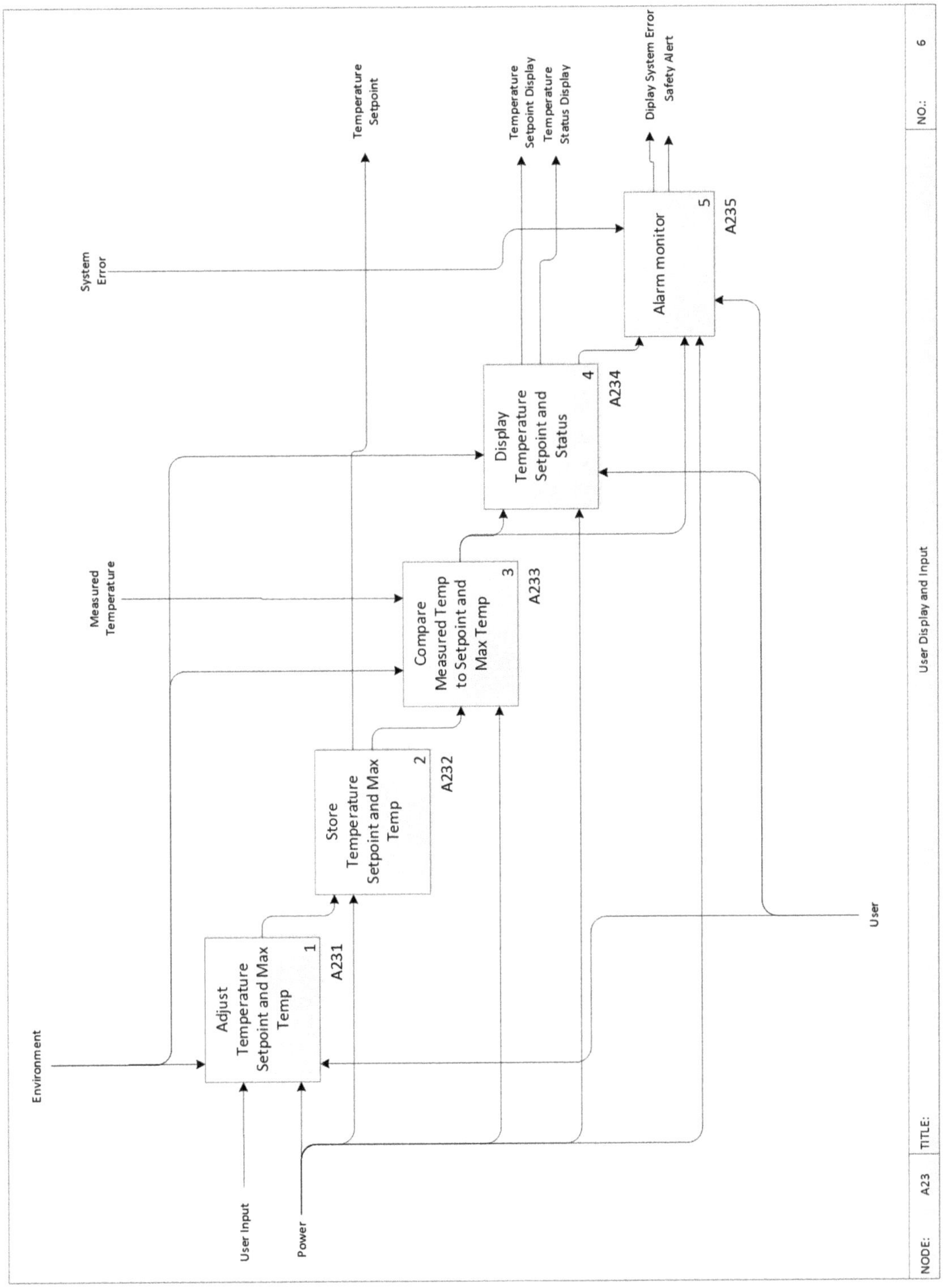

Environment
User Input
Power
Measured Temperature
System Error
Adjust Temperature Setpoint and Max Temp
1
A231
Store Temperature Setpoint and Max Temp
2
A232
Compare Measured Temp to Setpoint and Max Temp
3
A233
Display Temperature Setpoint and Status
4
A234
Alarm monitor
5
A235
Temperature Setpoint
Temperature Setpoint Display
Temperature Status Display
Diplay System Error
Safety Alert
User
NODE:
A23
TITLE:
User Display and Input
NO.:
6

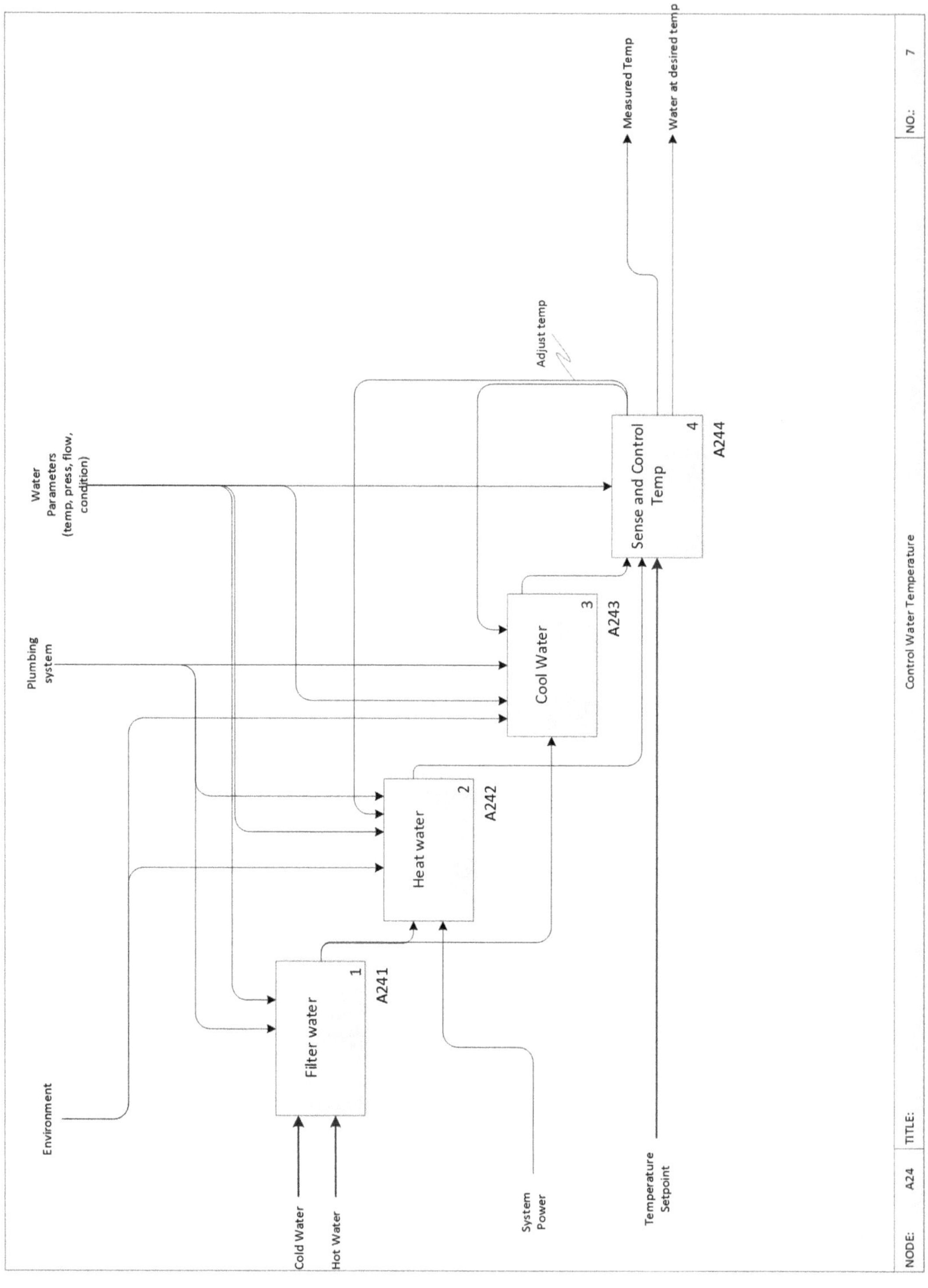

Figure 46: OV-5 Diagrams – Water System

These examples provide a great deal of details associated with the complexity of the proposed systems. Upon review, the reader should be able to grasp the intent of the developer by means of the information present on the drawings. If any operational element is unexplained or unclear, it is appropriate to add detail to the OV-5 diagrams. Please note that these diagrams should contain an appropriately numbered title block, and should demonstrate referential integrity. Careful adherence to the rules and methods of producing these diagrams can provide an extremely straightforward and easy communication tool for presenting your operational concepts to other members of the development team.

As a final example, we will look at a very simple device, demonstrating that these techniques of development of system architecture are applicable in a wide variety of domains, and are not limited only to very complex situations. In this case, we have chosen to diagram the operations of a small toy, a water pistol. This device is shown in **Figure 47**.

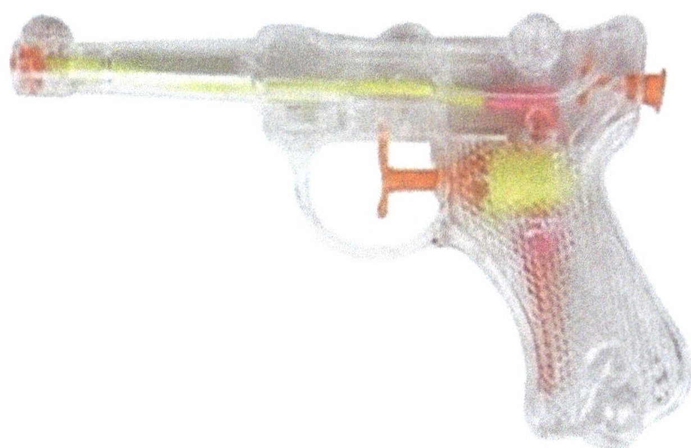

Figure 47: Water Pistol system

A simplified diagram is also included to clarify the operation of this system. The subsequent system architecture being developed is based upon this simplified operational diagram.

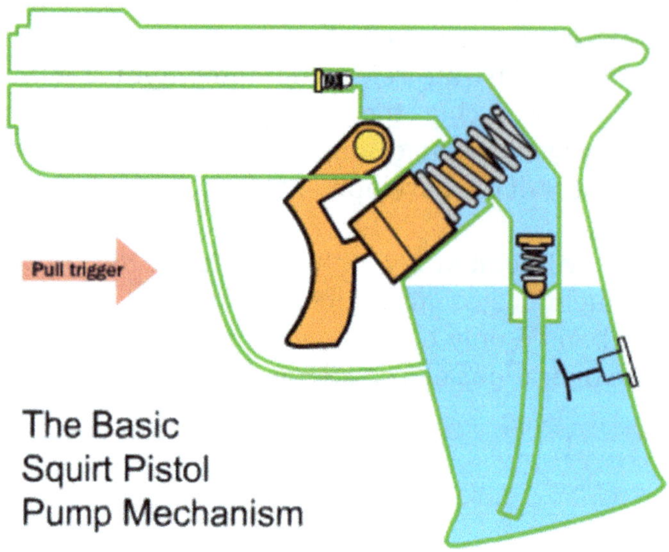

Figure 48: Operational Concept of Water Pistol

As we have previously done, we will first create a Functional Analysis of this system with an included consolidation of functions to provide a type of System Architecture for this device. This is shown in **Figure 49**.

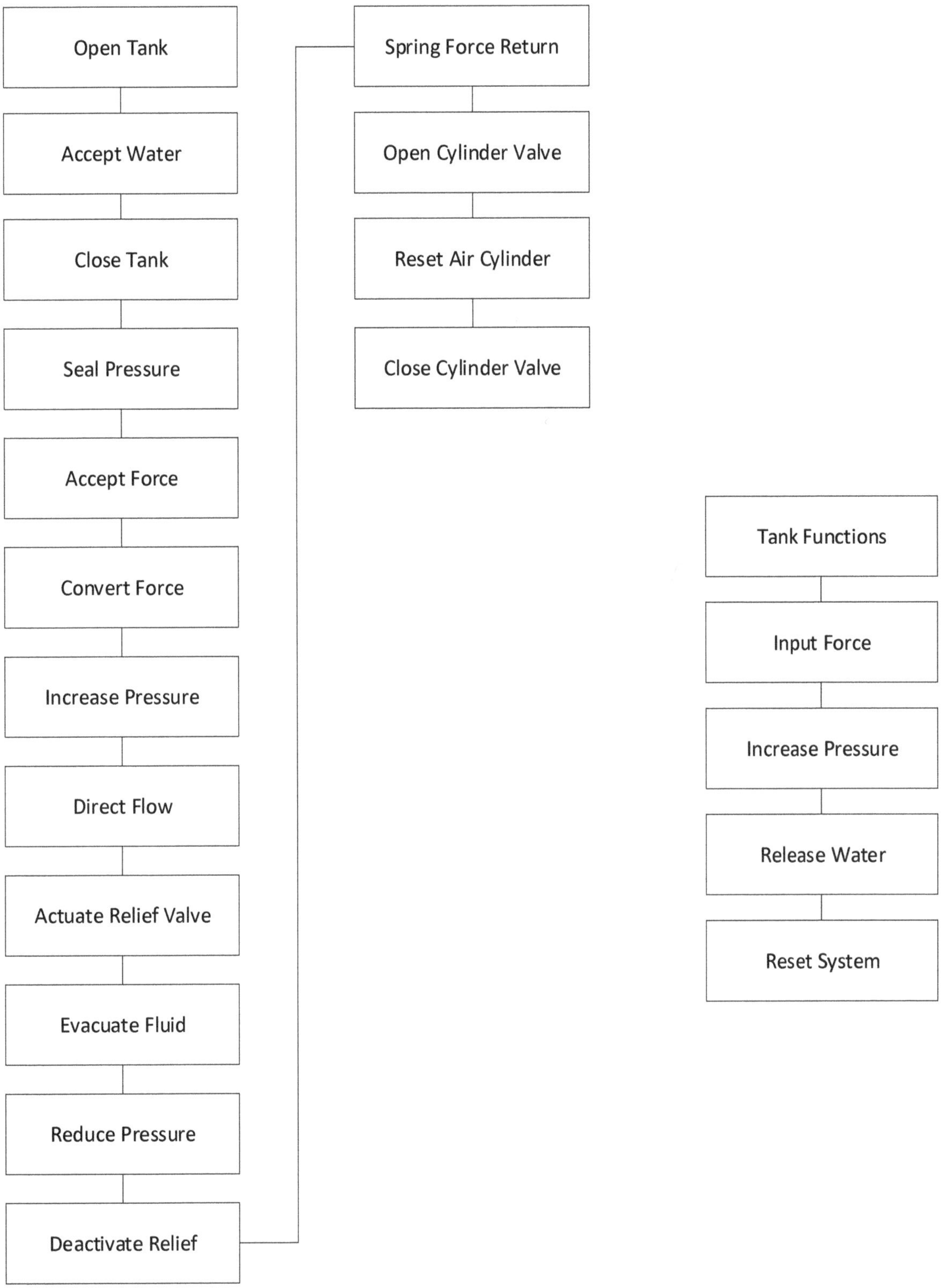

Figure 49: Functional Analysis of Water Pistol

We will also utilize our previously discussed techniques to develop the OV-5 in order to ascertain whether our Functional Analysis is complete and whether our System

Architecture is fully prepared. The context diagram and several Drill Downs are included in the following diagrams.

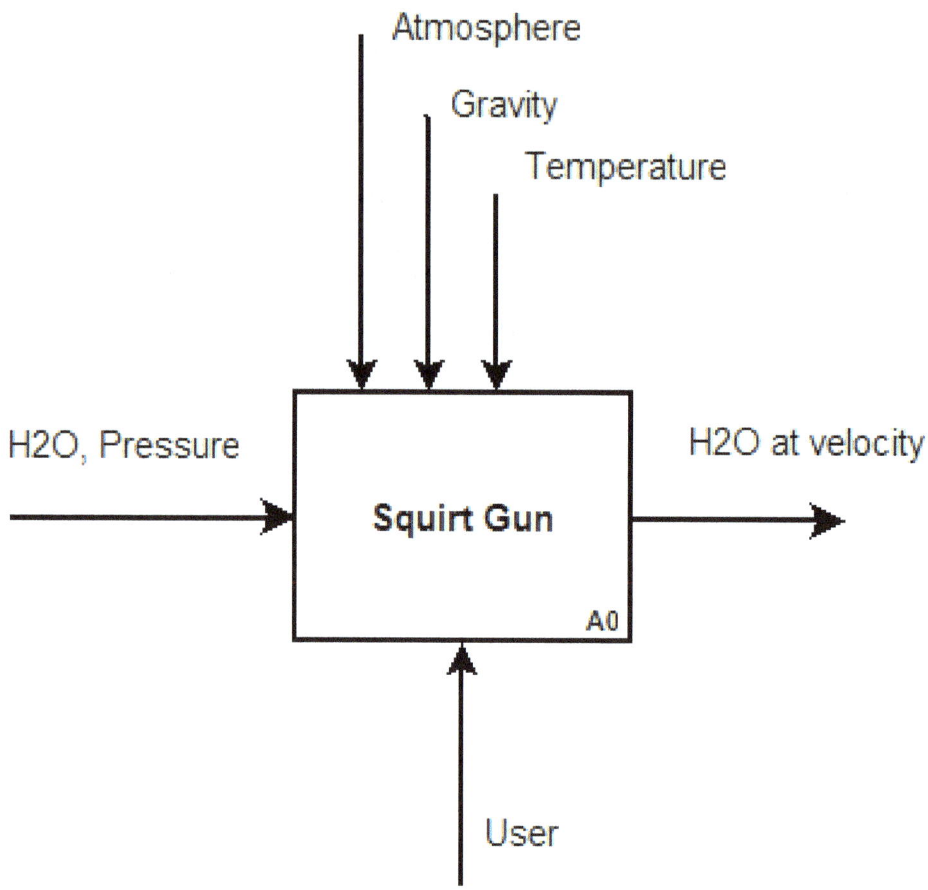

Figure 50: Context Diagram for Water Pistol

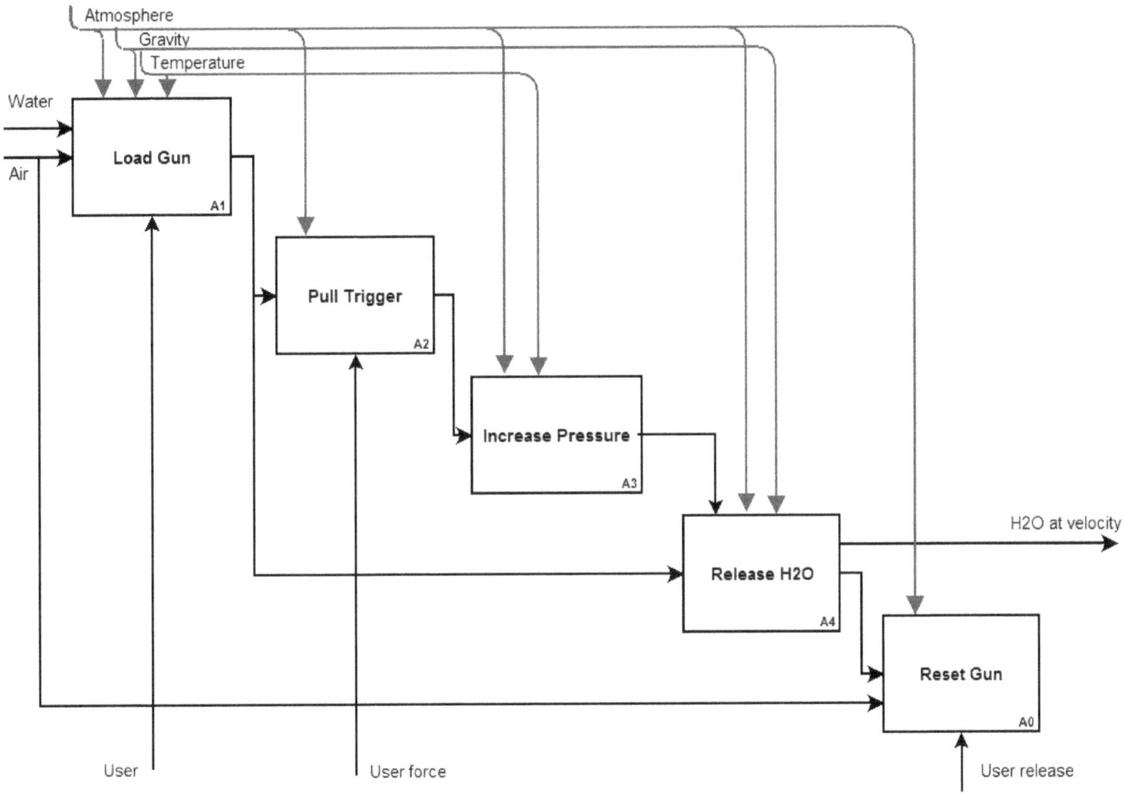

Figure 51: First Drill Down of Water Pistol

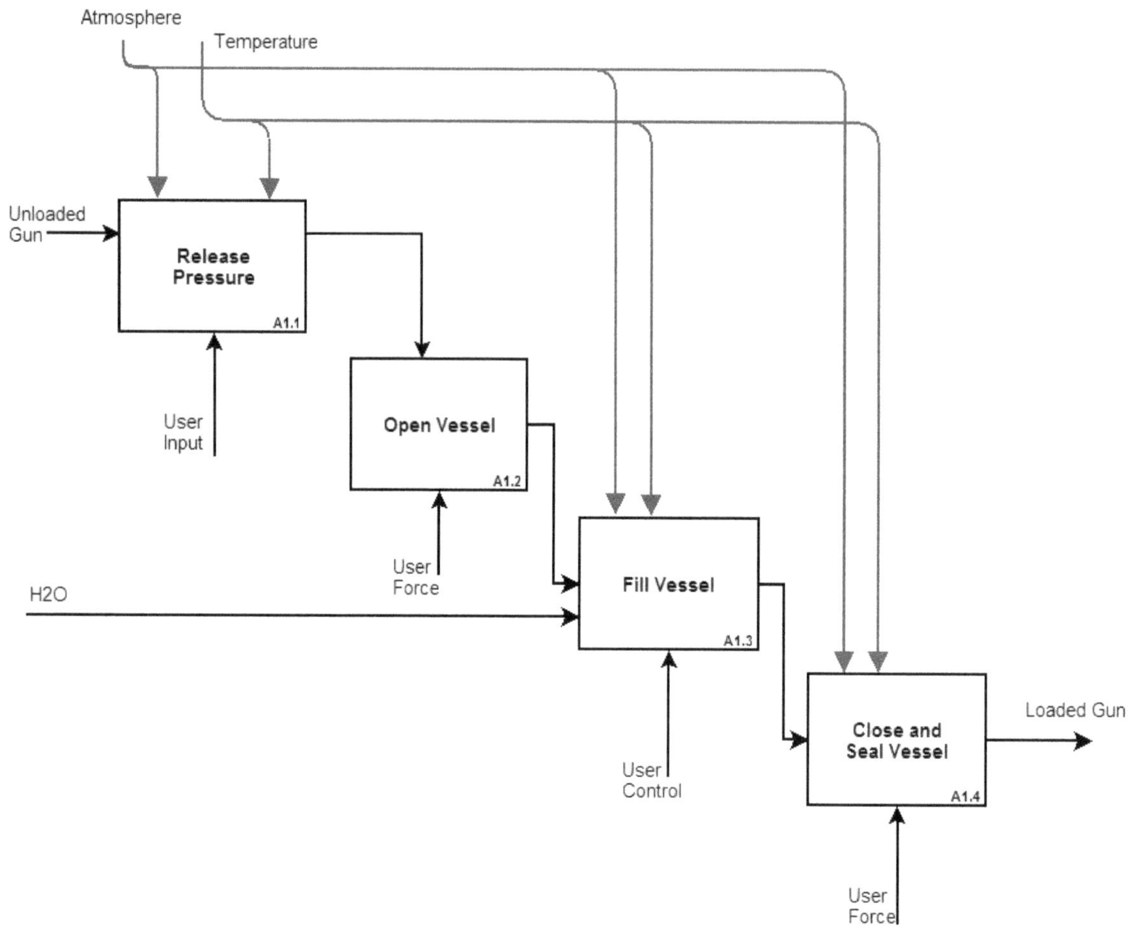

Figure 52: Second Drill Down of Water Pistol

The first Drill Down is remarkably similar to the System Architecture defined by the Functional Analysis. Had we completed all second Drill Down diagrams, we would have seen that each of the elements noted in the Functional Analysis appear in some form in the System Architecture on the OV-5. Utilizing these two techniques simultaneously enables us to better ensure that we have a complete description of the operational activities of the system.

Completion of the OV-5 establishes the System Architecture for the development effort. As might be expected, insights gained and conclusions developed through the generation of these work products often require us to go back and modify or rewrite previously defined work products for the system. In the end, we would like all our documents to be consistent so that they might be used as a record for future system developers who may be faced with a similar challenge, and as a justification of decisions made during the development effort.

Once the System Architecture is completed, it can be parsed out to the various Mechanical, Electrical, and Software Engineers who will be called upon to complete the Physical Architecture for the system. The System Architecture provides a generic operational diagram, and as long as the Component Engineers adhere to the logical constructs defined in the System Architecture, the manner in which the functions within the box are completed is of little concern to the Systems Engineer. Thus the Component Engineers are 'white-box' developers, in that they are concerned with what is inside each box and all of the details of the functionality. The Systems Engineers, on the other hand, are 'black-box' developers, as they are less concerned with the actual functionality that occurs within the box as long as it adheres to the logical construction defined in the System Architecture. This is one of the primary reasons why a Systems Engineer does not need to be an expert in software development or hydraulics in order to design a system containing those elements.

This is not to say that it is useless to have specific Engineering skills in a Systems Engineering endeavor. On the contrary, those skills are extremely useful, but when they are applied the Systems Engineer is acting in the role of a Component Engineer. As might be expected, the line drawn between these two domain areas is a gray area, and can often be quite difficult to define. Suffice it to say that the Systems Engineers and Component Engineers work hand-in-hand to assure the completion of an effective and coherent system.

Developing the OV-1

The OV-1 is an operational summary of the system. Usually provided in a graphical format, it is essentially a cartoon of the intended system to be developed. One can think of this work product as an abstract for a paper. It is developed late in the development cycle, but can be used as a common discussion factor for all members of the system's development team. Therefore, it is useful as a touchstone or guiding philosophy for many members of the development team. It will also be used as the introductory view for all individuals seeking a summary of the system's development. A sample for the power master system is included in **Figure 53**.

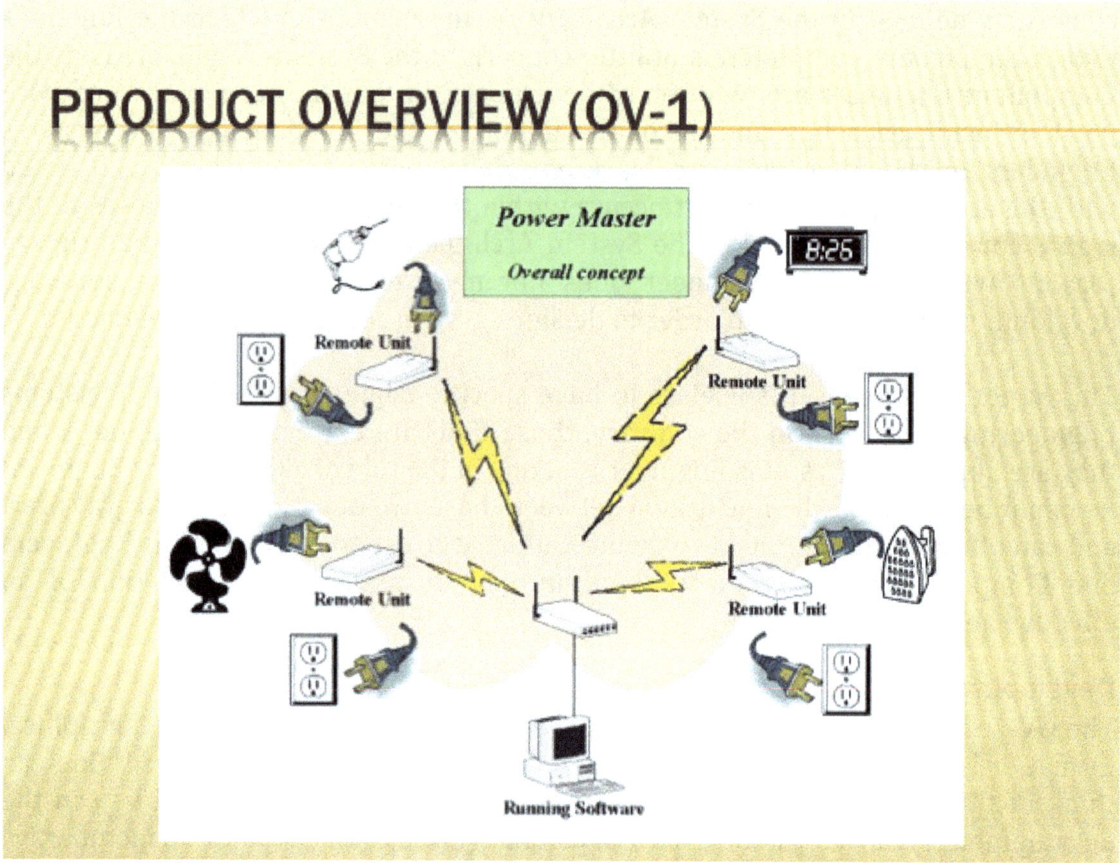

Figure 53: Power Master system OV-1

The Relationship between the Views

The DoDAF enterprise Architecture is an Integrated Architecture. This means that it is possible to identify ***where*** activities discovered in creating the OV-5 and the OV-4 are located in the OV-2's operational interconnectivity view. This allows the assessment of the impact that developing system's Architecture will have on external systems (external nodes), and what impact their requirements will have on the developing system's Architecture.

This is illustrated in Figure 54, which represents the system Architecture for a project called the Bureau of Inconsistency notification (BOPIN). The nodes contain the various activity alphanumeric identifiers that were discovered during the creation of the OV-5. Once one recognizes where these activities need to be performed, one can generate an idea that will later be verified by analysis, of what that activity will require in terms of configuration items in that node. This activity may be adding or modifying hardware or software, or implementing the organizational changes needed for the open system to interconnect with that particular node.

BOPIN OV-2

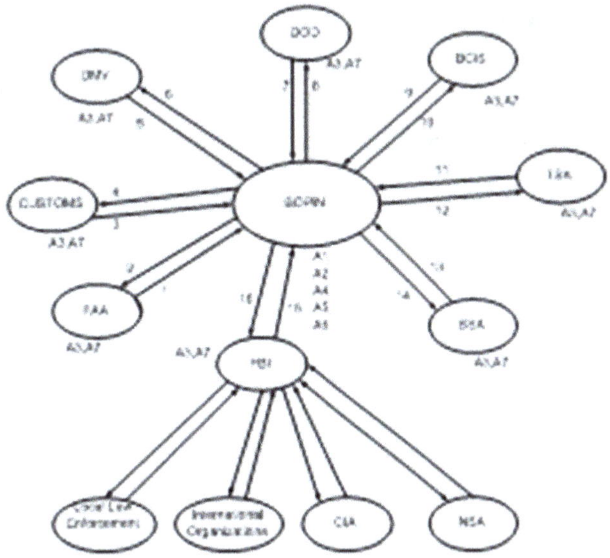

BOPIN

Figure 54: BOPIN OV-2

In reality, all of the system views are interconnected at various levels of understanding within the system architecture. Most system developers recognize the value of the operational interconnectivity diagram (OV-5) as the center of understanding of the system. However, when multiple individual diagrams are compared and contrasted, additional insights may be gained as previously described.

This can be of significant value to the developer, as it will enable a better understanding of implied constraints and challenges brought about by decisions made regarding the System Architecture. One view relating the various artifacts together is shown in Figure 55. Experienced system developers will discover and utilize additional relationships between the artifacts in order to gain insight into their system design.

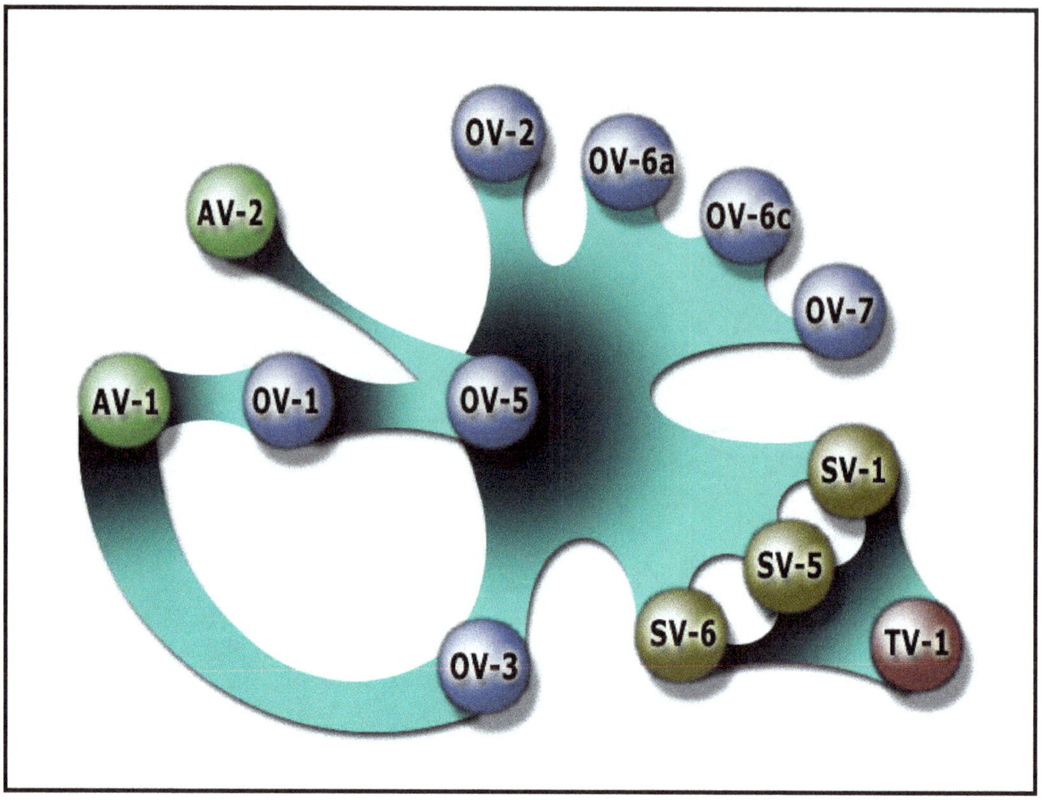

Figure 55: Relationship between DoDAF Views

Chapter 5: The Development of Requirements

One of the largest challenges in accomplishing System Development effectively is defining the problem to be solved and the criteria that will be used to establish if it has been solved adequately. Once we have defined our list of requirements, we have defined the criteria that will be used to assess the quality of the solution and its ability to meet the customer's needs.

Many engineering organizations maintain that a handful of requirements are adequate to complete any development project. While it is true that more requirements are typically more expensive to accomplish in the development process, the number of requirements is related to the quality of fit of solution. In other words, if adequate time is taken in the preparation and development of requirements, the validation of the system will be much more readily accomplished and the customer will be more satisfied.

Taking the example of our toaster, one might ask what an adequate number of requirements might be. If a rigorous requirements development is pursued with regard to this project, we might expect that there would be two hundred to three hundred requirements developed. If we were to consider a large-scale project such as the installation of a stadium or the development of a new tactical aircraft, we might expect that list to be much more extensive; perhaps twenty thousand requirements or more.

Of course it is unlikely that any requirements development group can simply sit down and begin writing a large number of requirements, and it is even less likely that those requirements will be evenly distributed across the needs of the program. Therefore, we have spent significant time developing a System Architecture, which will be used ultimately to develop the majority of the operational requirements for the system.

As previously stated, when we examine any work product for a system, we will observe many functional descriptions and need lines associated with those products. Every functional box and every need line will ultimately produce at least one requirement, and potentially more than one. This is likely to generate large numbers of requirements that would not otherwise be readily apparent to the development team.

Requirements Engineering is the process of establishing the services that the customer requires from a system and the constraints under which it will operate after development. Requirements may be functional or non-functional, where functional requirements describe system services or functions and non-functional requirements typically result in a constraint on the system or development process.

Most large-scale System Developments deal with problems that are so complex that they will never be fully understood until we have undergone a significant portion of the system's development. Insight is often achieved by means of the decomposition process, but this implies that requirements will normally be both incomplete and inconsistent. Of course, we seek to minimize these two characteristics in our Requirements Development Process.

This includes:

Requirements Elicitation
Requirements Analysis
Requirements Validation
Requirements Management

As we have previously stated, Systems Engineers are responsible for the development of requirements throughout the decomposition process. Responsibility for program development is noted in the following diagram.

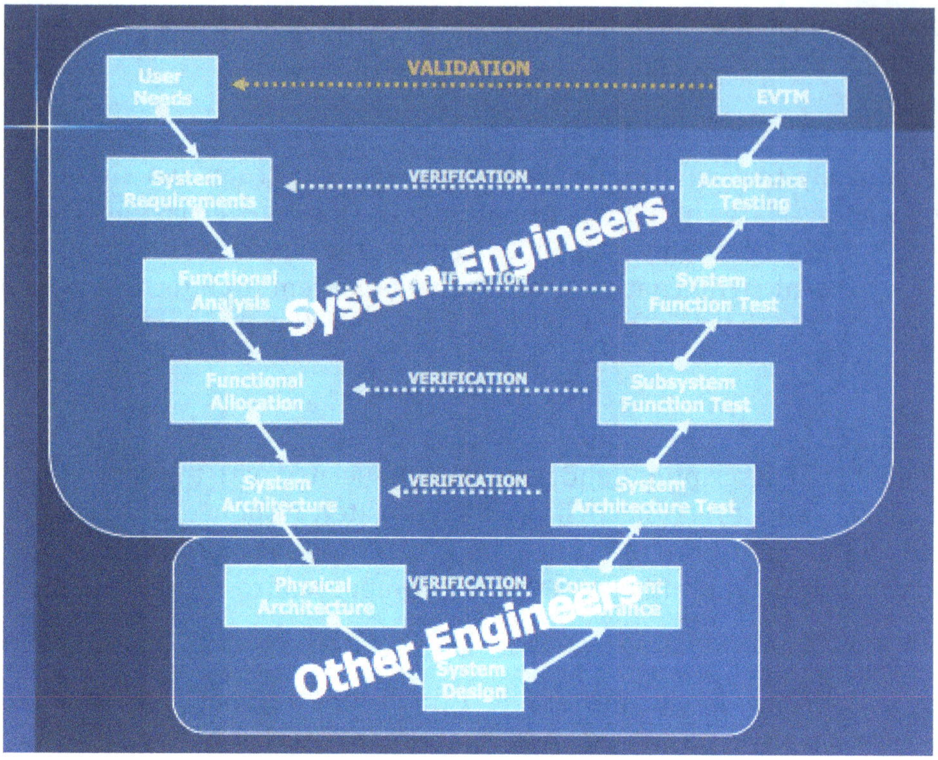

Figure 56: Development Responsibility

Although the responsibility for System Development will be passed from the System Engineer to the Component Engineers at the transition to Physical Architecture, requirements will typically be dynamic and potentially volatile until we accomplish and complete the detailed design of the system. This is largely due to the advances in knowledge that are created during the development process, and should not be considered a bad thing in the development. Minimization of this volatility will be of benefit to the organization with regard to the efficiency of resource use and waste.

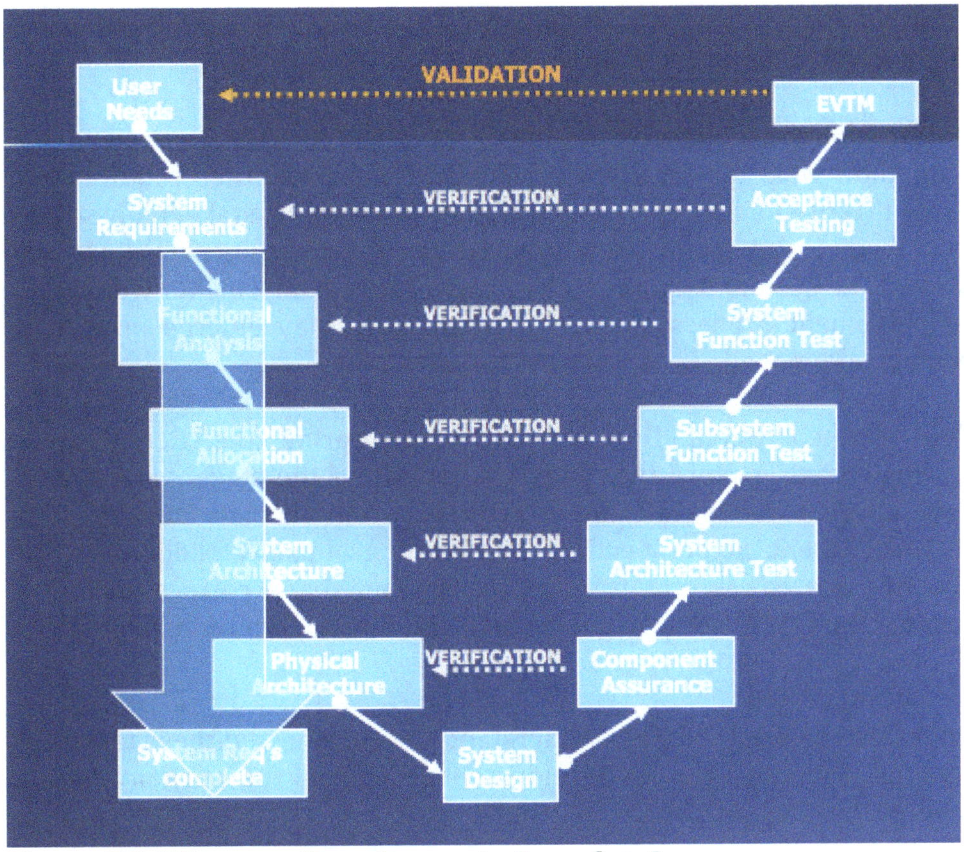

Figure 57: When are Requirements Complete?

In essence, during requirements definition, we are increasing the aggregation of our knowledge of the system in order to fully decompose the problem. By virtue of that, we hope to expose all of the details of the problem to our insight and understanding. We follow a simple process to advance the resolution of our understanding:

Requirements definition–
A statement in natural language plus diagrams of the services the system provides and its operational constraints. This is typically a document written for customers.

Requirements specification–
A structured document setting out detailed descriptions of the system services. Written as a contract between a client and contractor, and typically used to bridge the gap between the system architecture and physical architecture.

Design specification–
A detailed design description that can serve as the basis for a design or implementation; written for Component Engineers and Developers.

Requirements Elicitation typically involves technical staff working with customers to discern more about the application domain, the services expected and the customer's requirements of what the system shall provide, as well as systems operational constraints. This will involve all the stakeholders of the system including End-users, Managers, Engineers, Maintenance, Domain Experts, and Trade Unions.

As any System Developer will discover during this phase of the project, stakeholders typically do not know what they really want. They often express their requirements using their own terminology and language, and different stakeholders have potentially conflicting requirements. Organizational and political factors will influence many of the system requirements. Many of the requirements will change during the development process and new stakeholders may emerge or existing stakeholders may be removed.

The intention of the System Development team during the Requirements Development Phase is to explore the many viewpoints of the stakeholders of the system. These different viewpoints are examined within the structuring of the requirements to represent the perspectives of different groups of stakeholders. This multi—perspective analysis is important as there is no single correct manner in which to elicit and analyze system requirements, and any known requirement is of benefit.

Many techniques maybe used to better understand the system, and these tasks may be utilized both prior to and after the development of many of the work products in our DoDAF approach. They include:

Interviewing – Used to gain insight into the stakeholders' perspectives on the system; they are not particularly good for understanding the system's domain requirements.

Scenarios – Real-life examples of how a system may be used; they typically include a description of the starting situation, the normal flow of events, what may go wrong, and a final state when the scenario is completed. This analysis can be very useful in the development of operational activities and other products.

Use Cases – A scenario-based technique used to identify the actors in an interaction and the system operation with regard to the actors. Sequence diagrams are used to add detail to use cases showing the sequence of events in the system.

Ethnography – Social scientists spend a considerable time observing and analyzing how people actually work. In many cases people do not have the inclination or knowledge to be able to explain or articulate how they accomplish their tasks. Ethnographic studies have shown that work is usually richer and more complex than that suggested and simplified system models. These studies can help to add significant detail to vague or abstract areas in the System Architecture.

Modeling and Simulation – An abstract representations of a view of the system; a simulation is that model examined over time. Use of models and simulations often provide significant insight to the System Developers and define requirements and constraints that are not readily understood at the outset of the System Development. In fact, the development of our DoDAF work products is simply a form of modeling.

Collecting What We Know About the Problem

As we progress through the Systems Engineering development portion of the project, we are continuously refining and expanding our vision of the product. This vision is communicated to others by means of the defined System Architecture, the requirements list, and the Requirements Verification Traceability Matrix (RVTM). If we accomplish our tasks appropriately, this information is adequate to complete the System Development. The System Architecture can then be handed off to the Component Engineers for completion of the Physical Architecture and detailed design of the system.

However, in the midst of the system development cycle, often we will discover things about our system that draw into question our original assumptions, constraints or even our ability to successfully complete the project. Due to these shifts, we often will find scope changes to be necessary and appropriate in the development of the project. Additionally, we will also likely identify areas of knowledge that are undefined or unknown at the time. Part of the effort of the systems development team is to identify all of the information that we know about the system being developed, as well as to identify some of those areas where we are unsure or where data does not exist to support any conclusions.

Defining What We Do Not Know About the Problem

When documentation related to what we know about the system is complete, a Gap Analysis may be conducted to clarify and defined areas of uncertainty or unknowns on the project. This Gap Analysis is normally carried out by peripheral members of the development team, as those closest to the development are often clouded in their vision and may have a great deal of difficulty identifying gaps of knowledge or uncertainties.

A number of methodologies may be used to conduct the Gap Analysis. One such recommended method is a review of the System Architecture that examines each capability to be delivered with regard to the understanding of the System Architecture and logical construction of the capability. For example, in our water pistol example, an examination of the System Architecture may reveal uncertainty about the optimal placement for the relief valve that is used to release the water stream under pressure for the gun. A position might be defined based on personal preference or by identification of a model utilized to address this gap in knowledge. For example, two prototype water pistols with varying positions for the relief valve might be created, with each being tested for performance characteristics and the optimal position being defined by the test results.

In a different example, if we were asked to create a system to read the license plates of all cars passing over a given bridge in our locality, we may have very little idea how much data would be compiled in a twenty-four-hour period. We may choose to clarify this gap in our knowledge by utilizing Department of Transportation records, or perhaps setting up an experiment to count the number of vehicles passing over that bridge during a given period of time.

The Systems Engineer must identify which gaps need to be closed before moving forward with the system's design, weighing the amount of time and investment that is required to close the gaps. In some cases, modeling and simulation might be used, or prototyping or comparison to similar systems may also be adequate. In other cases, the time requirements and investment are too large to consider a closing of the gap, in which case the System Engineer may choose to accept the risk of a design decision in order to keep the program progressing.

Formulating How We Will Understand What We Do Not Know

Human beings think in terms of deterministic relationships and in terms of certainty. It's not natural for us to think in terms of multiple possibilities, and so we tend to go through life feeling very confident about what we think we know and what we believe, and what we think we can plan and execute. However, our constant companion in all our endeavors is chance, randomness, or what we perceive as randomness but is oftentimes one of many possible outcomes in considering a future course of action. In reality, all we have to work with are our perceptions and worldview models of reality, and our existing mental models. All these models have explicit and implicit assumptions. We are typically not aware of the implicit assumptions that we have used to create the models that we normally perceive as reality.

Occam's Razor

Occam's Razor is the modeling / simulating process we will use to convert the real world into a model world. Occam, a fourteenth century philosopher, coined a Latin phrase that translates to:

They should not be multiplied without good reason.

In other words, don't make things harder than they need to be.

This will become a guiding principle for us when defining what we do not know and how we will gain insight regarding our system development.

Reality comprehension

Human beings can only perceive reality; that is, they can perceive entities and associated entity attributes, and represent them symbolically. For example, Annie, a Systems Engineer, has omniscient powers of the perception of reality entities and their associated attributes. This means that she can perceive **all** of the entities and associated attributes in a real world situation, and demonstrate deductively that she has **<u>captured all of them</u>** for the particular reality situation of interest. The question is: How will she record them? What does the recording of entities and associated attributes of reality situation look like?

It turns out that Annie prefers to record entities and associated attributes in matrix format in an Excel spreadsheet like the one shown in Figure 58. In the spreadsheet, every entity has uniquely identified attributes that are associated with that entity. Annie has is named the spreadsheet the Reality spreadsheet R_{anem}. It represents the total and complete entity and associated attributes of a given reality situation.

	attributes index identifier										
entity idnetifier	a_1	a_2	a_3	a_4	a_5	a_6	.	.	.	.	a_n
e_1	a_{1e1}	a_{2e1}	a_{3e1}	a_{4e1}	a_{5e1}	a_{6e1}	.	.	.	.	a_{ne1}
e_2	a_{1e2}	a_{2e2}	a_{3e2}	a_{4e2}	a_{5e2}	a_{6e2}	.	.	.	.	a_{ne2}
e_3	a_{1e3}	a_{2e3}	a_{3e3}	a_{4e3}	a_{5e3}	a_{6e3}	.	.	.	.	a_{ne3}
e_4	a_{1e4}	a_{2e4}	a_{3e4}	a_{4e4}	a_{5e4}	a_{6e4}	.	.	.	.	a_{ne4}
e_5	a_{1e5}	a_{2e5}	a_{3e5}	a_{4e5}	a_{5e5}	a_{6e5}	.	.	.	.	a_{ne5}
e_6	a_{1e6}	a_{2e6}	a_{3e6}	a_{4e6}	a_{5e6}	a_{6e6}	.	.	.	.	a_{ne6}
e_7	a_{1e7}	a_{2e7}	a_{3e7}	a_{4e7}	a_{5e7}	a_{6e7}	.	.	.	.	a_{ne7}
.											
.											
.											
.											
.											
e_m	a_{1e7}	a_{2e7}	a_{3e7}	a_{4e7}	a_{5e7}	a_{6e7}	.	.	.	.	a_{nem}

Figure 58: Annie's Omniscient Total and Complete Reality model

Now let's step back to reality, and realize that Annie's is a mythical ideal Systems Engineer with omniscient perception capabilities. Real Systems Engineers are not omniscient and at best will perceive only a subset of the $\mathbf{R_{anem}}$ omniscient model created by Annie. A spreadsheet of a real Systems Engineer's perception of reality, or the Identified Reality Model, will be labeled $\mathbf{I_{anem}}$, and is shown in Figure 61.

	attributes index identifier							
entity idnetifier	a_1	a_2	a_3	a_4	a_5	a_6		
e_1	a_{1e1}	a_{2e1}	a_{3e1}	a_{4e1}				
e_2	a_{1e2}	a_{2e2}	a_{3e2}	a_{4e2}	a_{5e2}	a_{6e2}	.	.
e_3	a_{1e3}	a_{2e3}						
e_4	a_{1e4}	a_{2e4}	a_{3e4}	a_{4e4}				
e_5	a_{1e5}	a_{2e5}	a_{3e5}	a_{4e5}	a_{5e5}		.	.
e_6	a_{1e6}	a_{2e6}						
e_7	a_{1e7}	a_{2e7}	a_{3e7}	a_{4e7}	a_{5e7}	a_{6e7}	.	.

Figure 59: Identified Reality Model

In keeping with Occam's razor – nothing should be more complex than it needs to be – even before any assumptions are made to simplify the reality model, the number of entities and associated attributes shown in $\mathbf{I_{anem}}$ has been significantly reduced. The lack of omniscience in real Systems Engineers means that we have implicitly removed possibly significant entities from our Identified Reality Model.

We have epistemic ignorance of possible entities that may significantly impact information derived from an examination of the data produced by running simulation models based on our Identified Reality Model $\mathbf{I_{anem}}$. If the entities and associated attributes that we did not identify in our reality model are significant, we will discover this when we Validate the Data from a simulation model run by comparing it to real world measured data. Because multiple attributes may be missing from around the model, as seen in Figure 61, it may require some work to identify the entities of interest by discovering the causal relationships between entities and the differences between real data simulation and the run data.

There are several reasons to further simplify the Identified Reality Model and its entities and attributes, and create the Reality Model required to create desired simulation models. First, the developer may need a quick, high-level simulation model that takes very little time to develop. Second, simplification may be necessary to better understand the behavior of a complex system. Finally, a high-level simulation model and the data it contains may be needed for Marketing presentations, Advertising, Senior Management presentations, and presentations to obtain financing for a Project Development activity.

Figure 60 is a matrix that has a reduced set of entities that are reduced from Reality Model $\mathbf{I_{anem}}$. This platonic set, $\mathbf{I_p}$, is an idealized set with reduced or eliminated reality entities; the simplest and most topical Reality Model.

	attributes index identifier							
entity idnetifier	a_1	a_2	a_3	a_4	a_5	a_6		
e_1	a_{1e1}	a_{2e1}	a_{3e1}	a_{4e1}				
e_2	a_{1e2}	a_{2e2}	a_{3e2}				.	.
e_3	a_{1e3}							
e_4	a_{1e4}	a_{2e4}	a_{3e4}	a_{4e4}				
e_5	a_{1e5}						.	.

Figure 60: Platonic Idealized Model

The reductions in the entities and associated attributes were made by assuming that they were not needed for the purpose of the model. The specific entities and associated attributes were selected for the purpose of the simulation and to adhere to Occam's razor.

As an example, Engineers or those who have studied fundamental Physics are familiar with the formula F = ma. An Identified Reality Model for this formula is seen in Figure 63.

	attributes index identifier							
entity idnetifier	a_1	a_2	a_3					
e_1 envirioment	a_{1e1} air	a_{2e1}temperature 70	a_{3e1}pressure 14.7 psi					
e_2 mass	a_{1e2} length	a_{2e2} width	a_{3e2} depth					
e_3 veolcity	a_{1e3} 5 mph	a_{2e3} north						

Figure 61: Identified Reality Model for F=ma

F=ma is based on a Platonic Reality Model, $\mathbf{I_p}$. Its Identity and Associated Attributes Matrix are shown in Figure 62.

	attributes index identifier							
entity idnetifier	a_1	a_2	a_3					
e_1 envirioment	a_{1e1} vacuum							
e_2 mass	a_{1e2} point mass, 0 length, width and depth							
e_3 veolcity	a_{1e3} 0 to infinity							

Figure 62: Platonic Reality Model, all Identified Real-world Entities and Attributes Assumed Away

The formula is clearly a Platonic Idealization of reality. Its application to real-world problems will need to be tempered by taking into consideration the difference between real-world attributes and the Platonic Idealizations used in the creation of the model- formula. Even with such a simple example we can see that the assumption sample space we created has eight members that are entity and associated attribute sets, e1 environment, e2 mass, and e3 velocity. These are the simplifying or simplifying assumptions we have made, and caused the exclusion of real-world entities and associated attributes from our Reality Model. There may be others, implicit assumptions, that are unknown to us. This is important because if they are unknown we have no understanding of the impact of the 'assumed away' real-world attributes, and so we have no knowledge of the impact of their exclusion from our model of reality when we use it in designing systems.

For much more complicated real-world reality with more real-world attributes and interrelationships, Reality Models must identify a larger number of simplifying assumptions to create useable models and simulations. Assumptions will typically exclude more real-world attributes from our model, because more complex real-world situations will require more simplification to arrive at useful models that have reasonable simulation run times. In short, the more complicated the real-world situation, the greater the need for more simplifying assumptions and the exclusion of real-world attributes and relationships from our models and simulations. As with the simple model F=ma, and in any real world situation, there are entities and associated attributes of the real world that we may identify and assume away in a model's creation. These are explicit assumptions we use to exclude whatever subset of real-world attributes we decide to exclude in creating our model and our simulation runs.

We have most likely not identified all of the attributes that are occurring in the real-world situation we are trying to model. The explicit attributes required to create a Reality Model more representative of reality may be included or excluded by our assumptions. The number of all-inclusive attributes of reality to consider for inclusion in our models and simulations via the method of assumption is unknowable, except for in the simplest of real-world situations. This means that every Reality Model is provisional and subject to change with the discovery of an attribute, set of attributes or subset of attributes that must be included to make it a more valid model of reality.

Developing the Entity Associated Attributes Reality Matrix

The first step in coming up with a list of real world attributes and relationships for the reality situation of interest is the OODA loop. OODA stands for Observe, Orient, Decide, and Act.

Observe means to look at physical reality and identify its attributes and relationships as best you can.
Orient means to identify where you stand in relation to the information you observed. **Decide** means to identify what possible courses of action are available and what course of action should you take.
Act means to take a course of action and make some simplifying assumptions so that a reality model can be constructed and its simulation runs can be completed.

A useful framework for getting humans to inject reality into models is to begin with the AV-1 outline from the DoDAF Enterprise Architecture framework, paying particular attention to the assumptions and constraints outline's elements. These assumptions are often derived through the use of a stream of consciousness activity conducted by individuals or groups of individuals, where there is no initial attention paid to whether the assumptions made have any basis in physical reality. After the assumptions are made, rigorous, slow thinking should be applied to test whether they are based in physical reality. If not, they should be put aside in the creation of at least the first set of assumptions, upon which subsequent work in creating a system design is performed.

In general, the information created by a set of assumptions is a multidimensional array of a Physical Reality Model's entities and associated attributes that in practice can only be approximated. For example, consider the relationship in probability and statistical analysis of polling theory and practice. It is useful to consider a comparison between population sample space and a sample of the population sample space. Before the population sample is polled a questionnaire is created that is also based on a set of assumptions. Since these assumptions are usually not available in the public domain, it is impossible to tell whether they were based on empirical reality or on normative expectations.

A small sample of is taken of a huge population sample space by sending the polling questionnaire to a representative sample of the population. This requires another set of assumptions regarding the definition of a representative sample of the population sample space, and draws inferences and conclusions about likelihoods of future events or opinion. For example, the sample size of a population sample space is one thousand individuals, while the population sample space may be in the hundreds of millions of individuals. The poll will publish the error involved relative to the sample size chosen, which in the polling of one thousand individuals results in a +/- three percent error.

At this point the analogy between polling sampling theory and practice and the creation of Reality Models reaches a point of departure. The computation of the error between the model and reality is usually more difficult to compute than the polling error. The model error we are talking about is the error that is the difference between the model and reality.

There is more involved than making a simple error computation. Several simulation runs of a model are done, each run having a slightly different set of assumptions, with the data from each run compared to physical empirical data. When an acceptable degree of error is reached, the simulation runs are ended and the model is considered to be a faithful representation of reality. When it is impossible to compare model simulation run outputs to empirical real world data, one does not really know how close the model is to physical reality. Using an un-validated model when designing a system carries the risk of designing a system that may not work in an operational real-world environment. However, its use may still be justified if it provides more information regarding the real environment.

Whatever reality model we have constructed in our minds will never represent all of the variables in reality. It may be imperfect, but it may also be useful. Simplifying assumptions are always required to gain a useful understanding of complex phenomena and systems.

In addition, an assumption set has no correlation to reality. While it may be appropriate for mathematicians, it is a mind projection fallacy for Systems Engineers. This may result in the creation of a Reality Model that has no basis in reality. The firm adherence and advocacy of such a model is not a desirable attribute of a Systems Engineer. It is a behavioral demonstration of the belief in a normative reality model; how reality should be not how it is. Systems Engineers must accept what is, and design a solution that will operate in that real-world context.

This part of the book has described how to formulate 'how we understand what we do not know'. This is a methodology for creating and simplifying Reality Models. It has shown that ordinary human Systems Engineers can only comprehend the entities and associated attributes of the real world they can perceive.

In conclusion, perception is reality, and what you see is all there is; however, this is only true if it is not tested with simulation models runs that may reveal at some simulation iteration the assumed-away factor or unrecognized entity and associated attributes that are existent in reality. The model or simulation is only as good as the ability to make predictions based upon them.

With that in mind, let us now turn our attention to models and simulations.

How We Know What We Do Know (Empirical Data Measurements)

In the section "Formulating How We Will Understand What We Do Not Know," we described, by logical arguments, the impossibility of completely and exhaustively identifying all of the entities and associated attributes. So how do we identify appropriate entities and associated attributes? We do this by measurement; by measuring the attributes of the entities. These measurements are typically made by using time as an index; the unique identifier for a specific measurement. Let's look at a typical example of the measurements of several entities and their attributes. In Figure 63, we see a figure we have already encountered.

	attributes index identifier							
entity idnetifier	a_1	a_2	a_3	a_4	a_5	a_6		
e_1	a_{1e1}	a_{2e1}	a_{3e1}	a_{4e1}				
e_2	a_{1e2}	a_{2e2}	a_{3e2}				.	.
e_3	a_{1e3}							
e_4	a_{1e4}	a_{2e4}	a_{3e4}	a_{4e4}				
e_5	a_{1e5}						.	.

Figure 63: Platonic Idealized Set I_p

Suppose that we measured all of the attributes associated with entity e_1. What would a recording of these measurements look like? (Note that for convenience the measurement cells were populated using a random number generator.)

Figure 64 is an example of a recording of measurements taken regarding attributes of entity e_1.

	entity e_1 attributes measured Values			
	a_{1e1}	a_{2e1}	a_{3e1}	a_{4e1}
Time	0.50722	0.74535	0.16546	0.8142827
t_0	0.69429	0.38318	0.59192	0.0342998
t_1	0.14635	0.38016	0.79214	0.0386182
t_2	0.34397	0.36987	0.45287	0.4521102
t_3	0.1911	0.95754	0.51406	0.4989852
t_4	0.71769	0.48248	0.09279	0.7351563
t_5	0.76425	0.37646	0.71088	0.0472538
.				
.				
.				
t_n	0.98466	0.83271	0.8451	0.1116461

Figure 64: Measurement Data for Entity e_1

If one measured the attributes of all of the entities identified in Figure 60, a graphical representation of the data sets would look like the measurement matrix shown in Figure 64.

Figure 65 is a depiction of an ensemble of measurements of the entity attributes identified in Figure 60 Platonic idealized Set $\mathbf{I_p}$.

Measurement ensemble for e_1

	entity e_1 attributes measured Values			
	a_{1e1}	a_{2e1}	a_{3e1}	a_{4e1}
Time	0.50722	0.74535	0.16546	0.8142827
t_0	0.69429	0.38318	0.59192	0.0342998
t_1	0.14635	0.38016	0.79214	0.0386182
t_2	0.34397	0.36987	0.45287	0.4521102
t_3	0.1911	0.95754	0.51406	0.4989852
t_4	0.71769	0.48248	0.09279	0.7351563
t_5	0.76425	0.37646	0.71088	0.0472538
t_n	0.98466	0.83271	0.8451	0.1116461

Measurement ensemble for e_2

	entity e_1 attributes measured Values			
	a_{1e1}	a_{2e1}	a_{3e1}	a_{4e1}
Time	0.50722	0.74535	0.16546	0.8142827
t_0	0.69429	0.38318	0.59192	0.0342998
t_1	0.14635	0.38016	0.79214	0.0386182
t_2	0.34397	0.36987	0.45287	0.4521102
t_3	0.1911	0.95754	0.51406	0.4989852
t_4	0.71769	0.48248	0.09279	0.7351563
t_5	0.76425	0.37646	0.71088	0.0472538
t_n	0.98466	0.83271	0.8451	0.1116461

Measurement ensemble for e_3

	entity e_1 attributes measured Values			
	a_{1e1}	a_{2e1}	a_{3e1}	a_{4e1}
Time	0.50722	0.74535	0.16546	0.8142827
t_0	0.69429	0.38318	0.59192	0.0342998
t_1	0.14635	0.38016	0.79214	0.0386182
t_2	0.34397	0.36987	0.45287	0.4521102
t_3	0.1911	0.95754	0.51406	0.4989852
t_4	0.71769	0.48248	0.09279	0.7351563
t_5	0.76425	0.37646	0.71088	0.0472538
t_n	0.98466	0.83271	0.8451	0.1116461

Measurement ensemble for e_5

	entity e_1 attributes measured Values			
	a_{1e1}	a_{2e1}	a_{3e1}	a_{4e1}
Time	0.50722	0.74535	0.16546	0.8142827
t_0	0.69429	0.38318	0.59192	0.0342998
t_1	0.14635	0.38016	0.79214	0.0386182
t_2	0.34397	0.36987	0.45287	0.4521102
t_3	0.1911	0.95754	0.51406	0.4989852
t_4	0.71769	0.48248	0.09279	0.7351563
t_5	0.76425	0.37646	0.71088	0.0472538
t_n	0.98466	0.83271	0.8451	0.1116461

Measurement ensemble for e_4

	entity e_1 attributes measured Values			
	a_{1e1}	a_{2e1}	a_{3e1}	a_{4e1}
Time	0.50722	0.74535	0.16546	0.8142827
t_0	0.69429	0.38318	0.59192	0.0342998
t_1	0.14635	0.38016	0.79214	0.0386182
t_2	0.34397	0.36987	0.45287	0.4521102
t_3	0.1911	0.95754	0.51406	0.4989852
t_4	0.71769	0.48248	0.09279	0.7351563
t_5	0.76425	0.37646	0.71088	0.0472538
t_n	0.98466	0.83271	0.8451	0.1116461

Figure 65: Measurement Ensemble for Platonic Reality Entity Set represented by I_p

Figure 64 is representative of the data sets Systems Engineers encounter. Data sets are created by the measurement of real-world data. The challenge is to create models and determine relationships between the attributes (variables) from data sets that may or may not provide causal relationships between variables. The data is used to establish the way in which one variable relates to one another.

Before we get too far along let's take a look at Figure 60 Platonic idealized set $\mathbf{I_p}$, and count the possible ways in which the variable data could be arranged for analysis. Looking at Figure 60, we see that there are thirteen uniquely identified attributes or variables. Suppose that we have ten measurements for each of the thirteen attributes. Let's calculate the number of different ways the measurement data could be arranged to determine covariance (or example) within the data.

The number of ways the data may be arranged is of import. We remember that if we have M and N sets of data, the number of ways the data may be arranged is MxN. And for more sets of data the numbers may be estimated by calculating the product of those set members. For example, assuming three sets of data, MxNxL, and so on. Since we have thirteen sets of data with each data set having ten members, we can readily see that there are 10^{13} possible arrangements (10,000,000,000,000); a combinatorial explosion.

Avoidance of Combinatorial Explosions

Some thought must go into the entity and associated attribute selection for measurement. The intuitive inclination to measure as much data as possible is likely to cause combinatorial explosions that are likely to, if entered into, be costly, time-consuming, and unproductive.

Combinatorial explosions are also likely to occur during the testing phase of the program if the tests become more detailed than minimally required, either to verify a design or to validate design. This is also true for testing, which involves measurement of systems entity attributes that result from the satisfaction of compliance with requirements.

Models Derived from the Analysis of Empirical Data

The process used to derive models from measurement data implies that the engineer needs to fit functions to measurement data. In keeping with Occam's razor and to avoid combinatorial explosions, one initially tries to fit the data to linear functions that are the simplest and most commonly used. Covariance, correlation, linear regression and least-squares methods are then applied. If this is not fruitful the engineer fits nonlinear functions to the data set. Nonlinear functions fall into three categories: intrinsically linear, intrinsically nonlinear but linearizable, and intrinsically nonlinear but not linearizable.

Intrinsically Linear Functions

A linear function in the context of linear regression means the functions are a linear combination of basic functions. With the basic function $\{1,x\}$, the linear combinations have the form:

$$y=B_0 + B_1x$$

and define a straight line.

Intrinsically Nonlinear but Linearizable Functions

With the basic function $\{2,x\}$, the result is a curve rather than a straight line. Intrinsically linear functions are those that can be written in the form:

$$y=B_0f_0+ B_1f_1+.....+ B_nf_n$$

where the B_is are constants in the f_is are the basic functions. Many important nonlinear functions, such as the exponential function

$$y=Ae^{rx}$$

and the power function

$y=Ar^x$

where A and r are parameters to be fit, are not intrinsically linear. They can, however, be converted to intrinsically linear functions by suitable transformations, such as taking the logs of one or both variables. We call these functions intrinsically nonlinear but linearizable.

Nonlinear Non-transformable Functions

Some important functions are both nonlinear and non-transformable. An example of such a function is the Verhulst population function (aka logistic function) shown below in difference equation form:

X1=RX(1-X)

How would data sets generated by the above function look to a Systems Engineering Analyst trying to discover the function causing the data set, and how would they be used as Reality Models in the design of systems?

We use Excel to vary the parameters R and X and look at the data to discover what such a data set may look like.

Suppose the population of an animal species is given by the model (non-linear function):

X1=RX(1-X)

where X1 is the population one year and X is the population the preceding year and the birth rate, R, varies between 0 and 4.

This is an apparently simple equation and appears deterministic. It is hard to tell by looking at the analytical expression what the behavior of this "simple analytical model" will be over a range of values. This is the ideal application for an Excel spreadsheet, as it allows the exploration of the behavior of this expression over a wide range of values, iteratively and recursively.

To keep it simple, let us use a range of values for X and X1 between 0 and 1, with the actual population being 100,000 times the normalized values.

Let us now compute the eventual population for R=1, 2.6, 3.2, 3.5, and 3.57; and for each value of R calculate 20 years of population growth for X initial values of 0.1, 0.3, & 0.6.

We must create plots in Cartesian coordinates of eventual populations vice R-value for each initial x value (0.1, 0.3, 0.6), where the x-axis (horizontal axis) is the year axis and the y-axis (vertical axis) is the value of the eventual population. We will examine a subset of the plots. Figure 66 shows the population growth over time for R=1, and x=0.1. We can see that with a value of R=1 the population dies away over time.

X'=RX(1-X)	R	initial x value
0.100	1	0.1
0.090	1	
0.082	1	
0.075	1	
0.070	1	
0.065	1	
0.061	1	
0.057	1	
0.054	1	
0.051	1	
0.048	1	
0.046	1	
0.044	1	
0.042	1	
0.040	1	
0.038	1	
0.037	1	
0.036	1	
0.034	1	
0.033	1	
0.032	1	
0.031	1	
0.030	1	
0.029	1	
0.028	1	
0.028	1	
0.027	1	
0.026	1	
0.025	1	

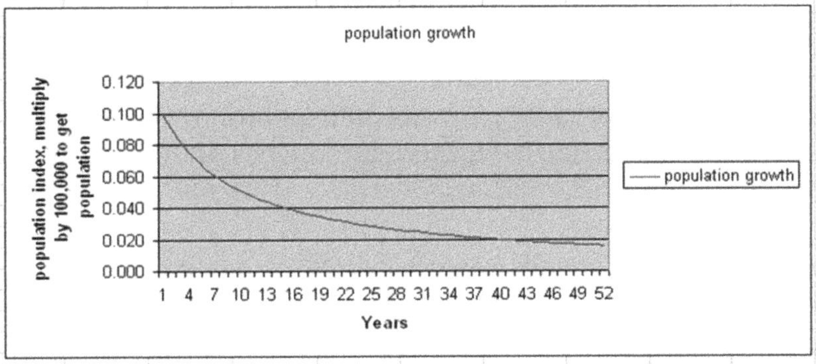

Figure 66: Population Growth over Time for R=1 and x=0.1

Figure 67 shows the population growth over time for R=2 and X=0.1. As we can see for the parameter values R=2 and x=0.1, the population grows to a stable value.

X'=RX(1-X	R	initial x value
0.100	2	0.1
0.180	2	
0.295	2	
0.416	2	
0.486	2	
0.500	2	
0.500	2	
0.500	2	
0.500	2	
0.500	2	
0.500	2	
0.500	2	
0.500	2	
0.500	2	
0.500	2	
0.500	2	
0.500	2	
0.500	2	
0.500	2	

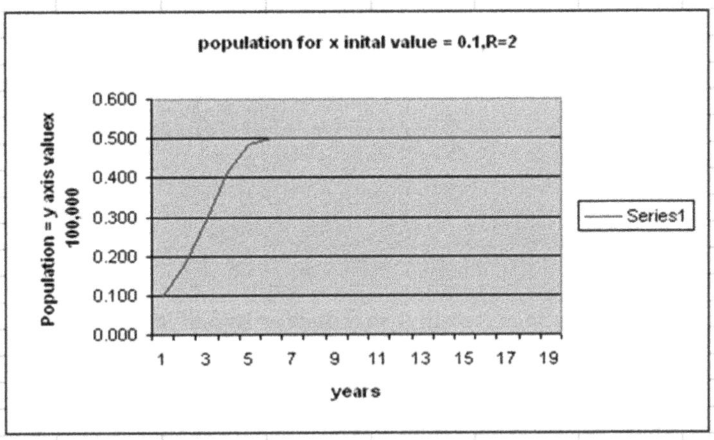

Figure 67: Population Growth over Time for R=2 and x=0.1

Now let's take a look and see what the data set looks like when R=3.57 and X=0.1. In Figure 68 we see chaotic behavior. If we look at the population data rather than the graph, we will see that the data looks like it is stochastic random data; however, it is data generated by the chaos created by the nonlinear function. Therefore, the apparently deterministic function is actually non-deterministic.

These examples demonstrate how difficult it is for a Systems Engineer Analyst to derive the function responsible for generating the data sets given that the underlying function is a nonlinear function.

Is there any way that the Systems Engineer Analysts can discover the underlying nonlinear function by analyzing empirical data sets derived from measurement? The least-squares methodology or the interpolation method may be tried, the modeler may try to fit quadratic cubic models to pieces of the data, or the modeler may use the heuristic approach.

If the Systems Engineer Analyst suspects that the data being analyzed is due to a logistics function, a logistics function can be implemented in Excel and parameters can be varied to see if the output of the Excel spreadsheet computations create data sets that are similar to the empirically collected reality data sets.

X^1=RX(1-	R	initial x value
0.100	3.57	0.1
0.321	3.57	
0.778	3.57	
0.616	3.57	
0.845	3.57	
0.468	3.57	
0.889	3.57	
0.353	3.57	
0.815	3.57	
0.538	3.57	
0.887	3.57	
0.357	3.57	
0.820	3.57	
0.528	3.57	
0.890	3.57	
0.350	3.57	
0.812	3.57	
0.544	3.57	
0.886	3.57	

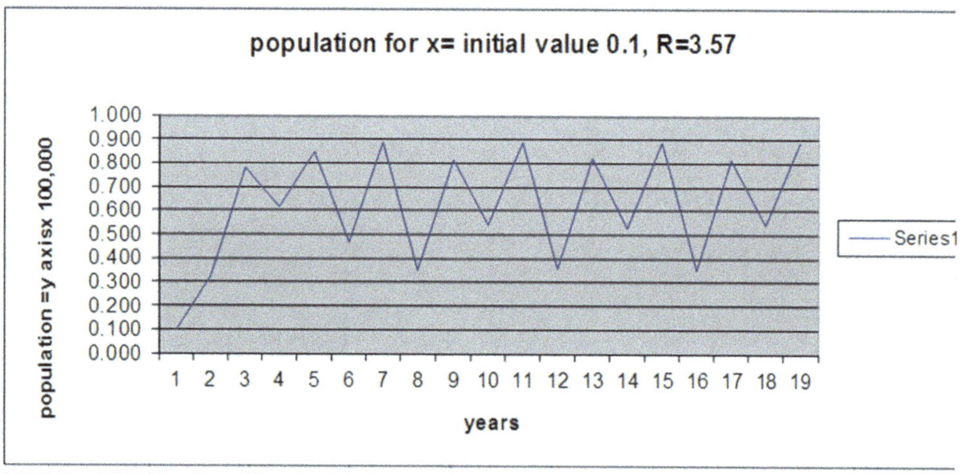

Figure 68: Population Growth over Time for R=3.57 and x=0.1

Chapter 6: Modeling and Simulation

According to Wikipedia, there are currently a total of seventy-seven simulation software packages available for engineers. Twenty-four are free simulation software packages and fifty-three are proprietary software simulation packages.

This course does not focus on learning how to use any one of them. It focuses on understanding the fundamentals of creating models and simulations using Microsoft PowerPoint and Microsoft Excel. PowerPoint is used for the creation of IDEF0 graphical representations of models and Excel for the creation of and the running of simulation models. Once the fundamentals of creating models and simulations are understood by using Power point and Excel, students will have a deeper understanding of the assumptions involved. Then, in the future, when creating models and simulations, they may be required to use any one of the seventy-seven currently available modeling and software simulation packages, or any new ones as they arise. This fundamental understanding will enable greater insight into the strengths and weaknesses inherent in any modeling endeavor.

One of the most important things to understand when creating models and simulations is that all models are abstract representations of reality, and as such, of necessity do not include, and cannot include a perfect representation of reality.

Modeling

If we ask several people for examples of models we would likely get a variety of responses including mathematical equations, toy trains, model airplanes, prototype cars, fashion models, or full-size mockups of a new automobile design. All these different interpretations of models have something in common; they are representations of a present, past, or future reality. They also have a purpose, and all required some resources for their creation.

The purpose of physical models can be representational, recreational, or may be to be a design aid. The physical models; trains or airplanes, for example, can be made from various materials requiring various skill sets and tooling, while the explicit purpose of a fashion models is to sell clothing and jewelry.

Systems Engineers will primarily be concerned with abstract models, the abstractions being symbolic representation of reality. These abstractions will be graphical representations, narrative descriptions, and mathematical symbolic relationships used to gain insight to and represent some aspects of the real world. **Figure 69** represents in a graphical format a description of the relationship between the real world, the model world, and the math world. These are the language quantities and relationships most often used in abstract models.

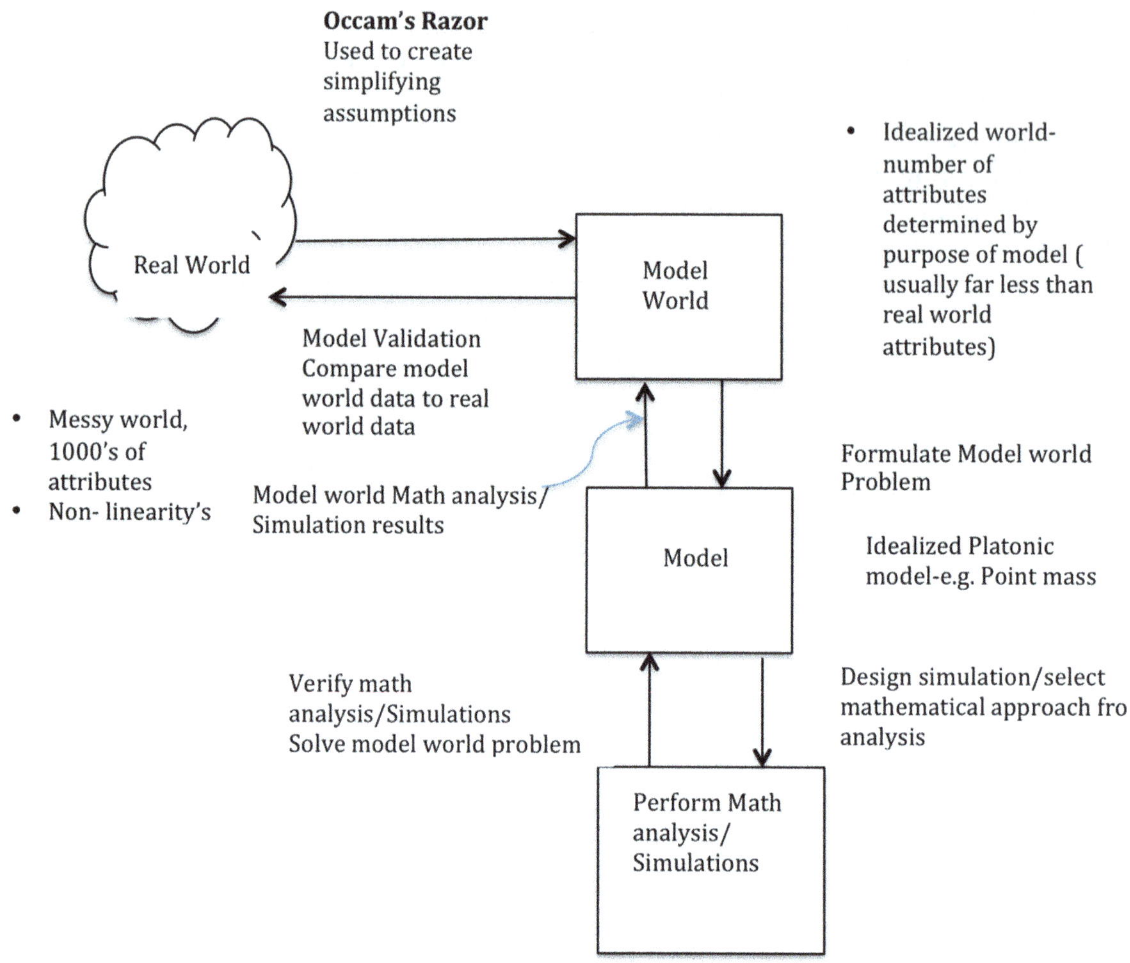

Figure 69: Occam's Razor

The problems that System Engineers must solve exist in the real world, which is a messy, nasty, complicated place. To illustrate this; the real world is portrayed as a cloud in **Figure 70**.

The first step in the modeling process is to simplify the real world to create a model world. As the figure shows, the model world leaves out much of the complexity of the real world. The original questions in the real world are translated to questions in the model world.

The second step is to construct a model of the problem in the model world. All the tools and techniques of mathematics and simulations are applied to the problem in the model world to provide answers in the model world posed problem.

The final step, Validation, determines how closely the mathematical analyses and/or model simulations we created to find solutions in the model answer the questions posed in the real world. To accomplish this, the information generated by mathematical simulation runs must be compared to real-world data.

Using the Validation information, the Systems Engineer must recall the purpose of the modeling and simulation activity and decide whether the model and the simulation meet the purpose, answer the question, or solve the real-world problem. If not, Occam's principles must be used to refine the model by making further assumptions and including more real-world attribute. The modeling process, including simulation runs, must be repeated, and as before, the outputs of the mathematical modeling and/or simulation runs must be compared to real-world data to see if the purpose of the modeling and simulation has been satisfied.

The process is repeated iteratively and recursively, in the sense that previous model run assumptions and attributes are in whole or in part reused in subsequent runs.

As one can see from the foregoing description, the creation of models and simulations and/or mathematical analyses is an iterative process that is performed until the modeler is satisfied that the models and simulations satisfy the intended purpose of the activity. This requires a comparison of the answers found in model-world problem space with real-world data.

Figure 70 is a graphical representation of the steps in the Mathematical Analysis/Modeling and Simulation effort described above.

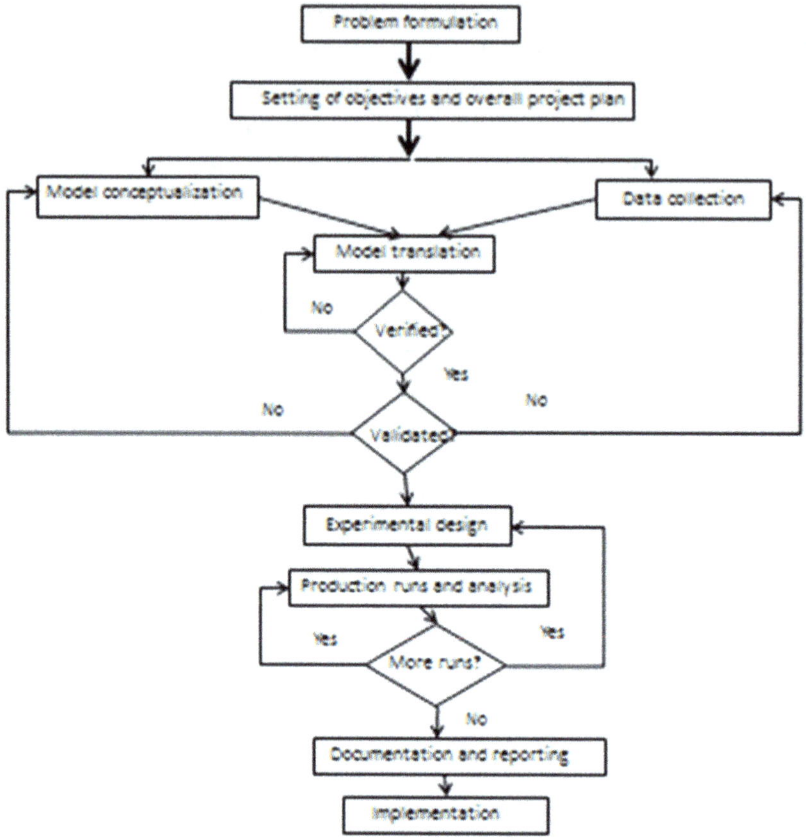

Figure 70: Graphical Representation of Modeling/Simulation Effort

Models and simulations

Models are abstract representations of reality. They are simplifications of the attributes of reality. In order to create them, simplifying assumptions must be made. What kind of assumptions are we talking about? Let's look at it from the perspective of three different domains: the Mathematician, Physicist, and Systems Engineering modeler.

The Mathematician makes assumptions and calls them axioms. Axioms are created by treating assumptions as "a priori"; or true without proof. From these axioms, the Mathematician uses deductive logic to create theories and theorems. They have no bearing on physical reality. They may not represent anything in physical reality and can be totally abstract creations of the human mind. What is fascinating is how often such Mathematical abstractions and formulations are useful in representing real-world quantitative relationships and values.

The Physicist takes in empirical data and uses it to draw inferences and generalizations. In essence, the Physicist uses inductive reasoning or logic to create what are known as physical laws, such as F=ma. What's interesting about the laws that are created using inductive logic is that they are provisional; new empirical data may cause modifications to previously defined laws or change them drastically.

Systems Engineers make assumptions and create models of reality to better understand customer needs and to identify requirements. A Systems Engineer is free to make any sort of assumptions she wishes to make, but they must have some bearing in physical reality. Systems Engineers create simulations to further define the problem space and to evaluate the dynamic response of existing or to-be-designed-and-built systems, using the physical laws determined by Physicists and mathematical techniques determined by Meta-mathematicians to be appropriate to the problem. Systems Engineers have one more complicating attribute that is usually not a factor in the technical activities of the Mathematician and Physicist: they deal with cognitive human beings who may be uncooperative, or who might be in conflict and competition with the purposes of the System Engineer and his company and/or organizational system.

Modeling and Simulation Tools

There are many different modeling and simulation tools available in the marketplace. The three most powerful Microsoft Office tools for Systems Engineers are Word, PowerPoint, and Excel. They are available to Systems Engineers as long as ta PC with Microsoft Office is nearby. This is important, as in many cases Systems Engineers are under severe time pressure to understand complex issues regarding a contract or bid that require a quick model and simulation run. PowerPoint is particularly useful in creating IDEF0 (Integration Definition for Function Modeling) activity models and Excel is an excellent tool to create rapid simulations that can be used in contract execution or new business pursuits. Excel has a set of mathematical formulations that allow the construction of deterministic mathematical analysis, stochastic mathematical analysis and the logic elements to create the related simulations.

IDEF0 modeling language

The IDEF0 Functional Modeling method is designed to model the decisions, actions, and activities of an organization or system. It was derived from the established graphic modeling language, Structured Analysis and Design Technique (SADT), developed by Douglas T. Ross and SofTech, Inc. In its original form, IDEF0 includes both a definition of a graphical modeling language (syntax and semantics) and a description of a comprehensive methodology for developing models. The US Air Force commissioned the SADT developers "to develop a function model method for analyzing and communicating the functional perspective of a system." IDEF0 should assist in organizing system analysis and promoting effective communication between the Analyst and customer through simplified graphical devices. Using PowerPoint as the graphical language and Excel as the mathematical tool for the creation of the data element tables and the simulation language, IDEF0 models may be created.

Modeling Discrete Systems

Particular Models of Interest for Systems Engineers
Systems Engineers will typically be concerned with: discrete time dynamical systems, discrete stochasticity, empirical modeling, continuous models, and continuous stochasticity.

Discrete Time Dynamical Systems
Discrete dynamical models are typically discrete time deterministic models. The mathematics that describe this behavior are called a discrete time dynamical system. A fundamental mathematical concept for working with discrete time models is the recurrence relation:

A function x on the nonnegative integers recursively defined if x(n) is a function of the values of x and some or all of the numbers less than n, that is,

$x(n)=f\,x(n-1), f\,x(n-2) \ldots.. x(2), x(1), x(0).$

Sometimes people prefer to approach discrete time dynamical problems with difference equations instead of recurrence relations. Basically, they are different ways of writing is the same equation and many textbooks use the terms interchangeably. A first-order difference equations is an equation of the form

$x(n)- x(n-1)= f(x(n-1)).$

Spreadsheets are an ideal tool for analyzing recurrence relations. It is easy to set up the equations, and the output is readily available in both numerical and graphical forms.

Spreadsheet Basics
A spreadsheet is a grid of rectangles called cells. In each cell, we can enter numbers, text or formulae, relating values in different cells to one another. To facilitate this, a cell has a unique address that allows one to refer to it and to its contents. The spreadsheet program used in in this book is from Microsoft Office Excel. It was selected for its familiarity, at least in some aspects of its use, and its extensive mathematical and logical functions.

Relative and Absolute Addressing
Oftentimes one sets a model then decides to change part of it. This is usually called creating "what if" scenarios. For example, what if the growth rates are decreased by five percent? If model parameters are incorporated with the model functions, a change in a parameter requires a spreadsheet to be extensively rebuilt. In order to avoid this, all the model parameters should be located in one section of the spreadsheet. Functions may then be written that that refer to the appropriate parameter cells instead of putting the parameters in the functions.

An Excel feature called "relative addressing" makes spreadsheet construction much easier and faster. It allows the specification of an absolute address for a cell, so that when it is copied the function address remains the same. In Excel, if a function refers to a cell G9 as G9 then whenever this function is copied, the new cell function also refers to G9. This is an absolute identity.

Discrete Stochasticity

Deterministic models do not take into account every uncertainty. In general, they are pretty good for large samples, but on a specific instance level they can provide inaccurate results. Models that take random events into account are called stochastic models. When building stochastic models, one must consider how to interpret noisy data in the output of the model and how to build models with stochastic compliments.

There are many different ways to incorporate random effects into models. One could assume that some intrinsic parameter of the model is random in nature or that the model output is random. An important distinction between stochastic models and deterministic models is that there is no single value for a stochastic model. A stochastic model takes on different values every time it is observed.

To understand a stochastic model we need to know what value it is most likely to be, and how far it is likely to be from its most likely value. One way to describe this information is to present the model as a distribution. Another is to define numbers measuring the center of the distribution; mean, median and the spread of the distribution, or the standard deviation, range, and interquartile range.

In considering the distribution of values, all possible values of a stochastic variable and the frequency of their occurrence are examined. The Statistician's term for this is the population distribution, with population referring to all possible values for stochastic model or parameter. It does not necessarily mean a collection of biological living organisms.

In general, planning the population distribution requires theoretical assumptions and mathematical analysis. Oftentimes, when creating stochastic reality models one cannot make justifiable theoretical assumptions or perform any analysis. Instead, one gathers experimental information via a process called sampling, or measuring a subset of the population distribution. This subset of the population distribution is called the sample distribution; it approximates and is representative of the population distribution.

A simple Stochastic Discrete Simulation Example

This example demonstrates the use of an Excel spreadsheet in the creation of a Monte Carlo simulation model.

The preferred example of discrete stochasticity is the Bernoulli experiment of flipping a coin.

Coin Flip Example

Problem Formulation

Background

You are a subject of the White Queen in Alice's Wonderland, and as a consequence, have no knowledge about the concept of flipping coins or the probability of occurrence of heads or tails in the process of the flipping of a coin. However, you are knowledgeable of the mysteries of discrete event simulation and you have a rudimentary knowledge of Excel. After your arrival in Wonderland, you were asked to make a bet on the flip of a coin. The coin had two sides; one side with a head on it and the other side with a tail. Since you have no experience with coin flips you decide to create a simulation using Excel as a simulation language and have formulated the problem statement as follows:

> **Problem Statement**
>
> Determine the probability of the occurrence of heads and tails in the flipping of a coin by performing fifty flips. At each flip calculate the probability of heads and tails and record the probability. Observe how the probability changes after each iteration of the simulation. Determine the probability of three heads in a row after three flips.

Objectives

Demonstrate the use of Excel as a discrete event simulation tool.

Demonstrate the calculation of the variation of the relative frequency of an event as function of the number of flips and the frequentist's view of probability.

Model Conceptualization

The physical realities of a coin flip

A physical coin has the following attributes: two sides, size, weight, mass, is made of a particular material, and is not the same on each side (the embossing process does not distribute the same amount of material on each side of the coin since the figures on each side of the coin are different)

Flipping Environment

The coin is flipped from a particular human being's hand, at a particular point in space and time, at a particular temperature, pressure, humidity, wind velocity, and is held in the hand in a particular orientation.

Abstracting the Essential Features of the Problem

Coin -The abstract coin will be two sided and the each side will be identical to the other side, except that we will call one side the head side and the other side the tail side to be able to uniquely identify each side.

Flipping Environment – The flipping process will not be affected by its environmental environment; it will not be affected by the real flipping environment described.

Model Translation

Problem: How do I go from the abstract model to something in an Excel spreadsheet that will generate the flip of a coin?

As I think about the process of flipping a coin I realize the following:

Each flip of the coin is a random event with a random outcome, heads or tails. This means that the outcome of either heads or tails is equally likely.
Therefore, I need to find a function in Excel that can generate either a head or tail outcome that is equally likely.
I can use the "Insert" command on the Excel tool bar and look at the functions I can insert into the spread sheet. Using "Rand()", will allow me insert a random value of any number between 0 and1. Hmmm, not exactly what I need.
I go to the "help" function, because I just realized that I can model heads with the number 1 and tails with the number 2! What I find is "RANDBETWEEN," which allows me to generate coin flips by randomly generating the numbers 1 and 2.

As I continue to think about the process of flipping a coin, I realize the following :

I can calculate the relative frequency for each coin flip by ($n_{heads}/n_{flip\ number}$) and I can calculate the relative frequency of tails in the same manner.
I can calculate the probability by using the first equation.
How can I tabulate the number of heads and tails so that I can compute ($n_{heads}/n_{flip\ number}$) for both heads and tails? Hmm, if I perform a logical test of the outcome of each output of the "RANDBETWEEN" function (the IF function in Excel) I can sort the occurrence of heads and tails by putting them in separate columns.
I need to keep track of the number of coin flips. How do I run the model? Press F9, and each depression will run the model with new random variables. I can press F10 to lock the model so that I can copy a specific run.

Model Verification

Is the computer program performing properly? If the input parameters and logical structure of the model are correctly represented in the computer verification, it has been completed. Common sense is used in completing this step.
Do the Excel spreadsheet results make sense? Does the probability of heads plus the probability of tails equal 1? If the answer is yes, the model may be considered verified.

Model Validation

Model validation is the determination that a model is an acceptably accurate representation of the real system.

If you physically flipped a coin for fifty times, what is the experienced probability of heads and tails? How does it compare to the simulation runs?

Validation is accomplished through the calibration of the model; an iterative process of comparing the model to the actual system behavior, and using the discrepancies between the two, and using the insights gained to improve the model. The process is repeated until the model accuracy is judged acceptable.

Event number n, (coin flip number)	coin flip performed =1, Toss not performed =0	Rand between 1,2 (1= HEADS,2 =TAILS	Number of tails	Number of heads	Probaility of tails after n flips	Probaility of heads after n flips	Probability of either heads or tails after n flips
1	1	0.221465	1	0	1.00	0.00	1.00
2	1	0.437569	1	0	1.00	0.00	1.00
3	1	0.616588	0	1	0.67	0.33	1.00
4	1	0.085556	1	0	0.75	0.25	1.00
5	1	0.151646	1	0	0.80	0.20	1.00
6	1	0.206053	1	0	0.83	0.17	1.00
7	1	0.379686	1	0	0.86	0.14	1.00
8	1	0.228125	1	0	0.88	0.13	1.00
9	1	0.599192	0	1	0.78	0.22	1.00
10	1	0.57333	0	1	0.70	0.30	1.00
11	1	0.038161	1	0	0.73	0.27	1.00
12	1	0.042866	1	0	0.75	0.25	1.00
13	1	0.867173	0	1	0.69	0.31	1.00
14	1	0.877877	0	1	0.64	0.36	1.00
15	1	0.065356	1	0	0.67	0.33	1.00
16	1	0.950527	0	1	0.63	0.38	1.00
17	1	0.390979	1	0	0.65	0.35	1.00
18	1	0.343513	1	0	0.67	0.33	1.00
19	1	0.91745	0	1	0.63	0.37	1.00
20	1	0.527862	0	1	0.60	0.40	1.00
21	1	0.815412	0	1	0.57	0.43	1.00
22	1	0.809526	0	1	0.55	0.45	1.00
23	1	0.292095	1	0	0.57	0.43	1.00
24	1	0.122681	1	0	0.58	0.42	1.00
25	1	0.404488	1	0	0.60	0.40	1.00
26	1	0.587878	0	1	0.58	0.42	1.00
27	1	0.69047	0	1	0.56	0.44	1.00
28	1	0.686282	0	1	0.54	0.46	1.00
29	1	0.469987	1	0	0.55	0.45	1.00
30	1	0.074172	1	0	0.57	0.43	1.00

Probability of tails after 5 flips	Probability of heads after 5 flips	Probability of either heads or tails after 5 flips	Probability of tails after 15 flips	Probability of heads after 15 flips	Probability of either heads or tails after 15 flips	Probability of tails after 50 flips	Probability of heads after 50 flips	Probability of either heads or tails after 50 flips
0.80	0.20	1.00	0.67	0.33	1.00	0.50	0.50	1.00

Figure 71: Coin Flip Simulation

In any modeling setting, exclude the details that are irrelevant to the given purpose or cannot be used given the constraints. We use Occam's razor so that we can reduce an unmanageable real-world problem to manifest manageable size in the model world. One must remember Occam's razor can be a sensitive and dangerous tool. If too much is cut out, the model solutions will have nothing to do with reality. However, if too little is cut out and the problem is difficult, we will not have met our initial purpose for creating a model that can be solved in a reasonable period of time. Occam's razor creates the assumptions that should be spelled out when reporting on a model that has been built.

Modeling Dynamic Systems

We have examined a variety of ways in which models are generated in order to further our understanding of the system. The most challenging models to generate are those describing dynamic systems. There are a wide variety of methodologies that have been developed regarding dynamic system simulation, and this text will not present an exhaustive list of techniques.

In general, Systems Engineers benefits from rapid and relevant models. While it is possible to create extremely high fidelity models, often these are generated at an expense that is unacceptable for the program. With this focus in mind, we present a methodology that enables the typical Systems Engineer to generate and simulate dynamic models in a very short period of time.

In their excellent text, Rosenberg and Karnopp present a methodology for modeling of dynamic systems known as Bond Graphs. This method has proven to be exceptionally flexible and will very quickly produce relevant results for Systems Engineers. We will first introduce the concept of Bond Graphs and the methods utilized to generate the state equations, and then we will provide several examples illustrating usefulness of this technique. Systems Engineers who require further details will benefit greatly by reading Rosenberg and Karnopp's text.

Models are intended to be simplified representations of a physical system. We use a model to predict what might happen when certain actions are taken within a particular system. We expect models to be sufficiently detailed to provide reasonably valid, or representative results. It is expected that these results will be of sufficient fidelity that will allow us to make decisions regarding our system. Therefore, any model we generate will mathematically represent reality. It should capture the essence of a decision-maker's problem and generate some balance between what detail is enough and what detail is unmanageable.

Modeling is the art of developing a mathematical description of a system that will represent that system. Simulation is the act of using the model generated over time in order to gain insight and understanding regarding the behavior of the system. Modeling and simulation are necessarily abstract, and will typically be incomplete.

One of the most insightful comments made regarding modeling and simulation is:

All models are wrong, but some are useful.

So in our consideration of modeling, we must recognize that truth is not what counts. What matters is whether a model is competent given the problem context.

For example, mathematical functions are commonly represented by models:

$$\sin X = X - \frac{X^3}{3!} + \frac{X^5}{5!} - \frac{X^7}{7!} + \ldots$$

We know in cases such as these that the mathematical function model may not be a very accurate representation of the object being modeled. However, within our region of interest or for the precision required to make our decision, the model captures the essence of the mathematical function. This might be perfectly adequate in order to make appropriate decisions regarding our system development. Why would this be important? The approximate mathematical function may allow much easier or faster calculations of equations. This can become of particular import when dealing with large-scale dynamic systems.

Simple models are not real. Real models are not simple. There is a fine line between modeling too much (more realistic but more complex,) and not enough (simpler but less accurate.) Therefore, we must consider several rules relating to modeling and simulation:

Never simulate unless you have to
Careful modeling is important to understanding the relevance of the results
Answers are only estimates within the specific context modeled
Know the question you want to answer before beginning the modeling effort

Dynamic model simulation is one of the most challenging numerical methods in use. Many different analytical methods have been developed over time, and many require the generation of appropriate state equations based upon both observation and experience. Some methods, such as Bond Graphs, create those equations in situ.

Bond Graphs

All physical systems will demonstrate the conservation of energy and the conservation of mass. The Bond Graph method is focused on the conservation of energy in a physical system. Since energy in a closed system is being conserved, it can only be affected by any one of three mechanisms:

Energy can be stored

Energy can be transported from one place to another
Energy can be transformed from one form to another

The total amount of energy present at any one place can only change if energy is flowing into or out of the system. In both cases, we understand that energy flows can be used to better understand the system. Those energy flows can be defined as the derivatives of energy with respect to time. In all physical systems, energy flows can be written as products of two different physical variables, one of which is extensive, and the other of which is intensive. In the case of coupled energy flows it may be necessary to describe a single energy flow as the sum of products of such variables.

The modeling of physical systems by means of Bond Graphs operates by utilizing a graphical description of energy flows. The energy flows are represented as directed arrows. Each directed arrow has two variables associated with it, which are responsible for the energy flow and are annotated above the arrow as an effort, and below the arrow as a flow. Flow points, and energy pushes. This is shown in the following diagram.

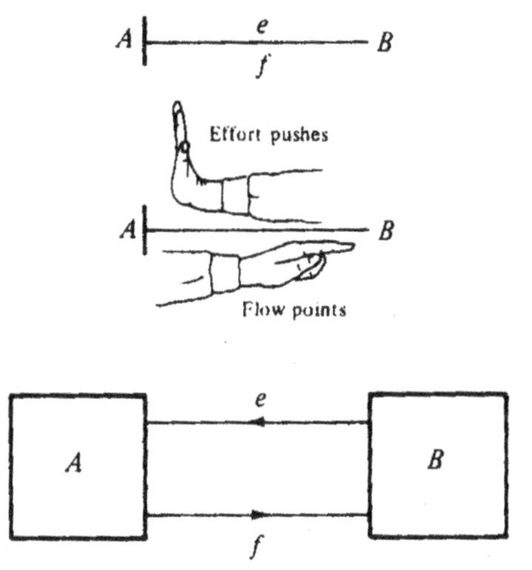

Figure 72: Energy Flows

These concepts are utilized as the basis for Bond Graph modeling. We will examine several of the basic construction components that are utilized in the development of a Bond Graph. The first elements of interest are the source of effort, or source of flow within the system. This concept is shown with its graphical representation in the following diagram.

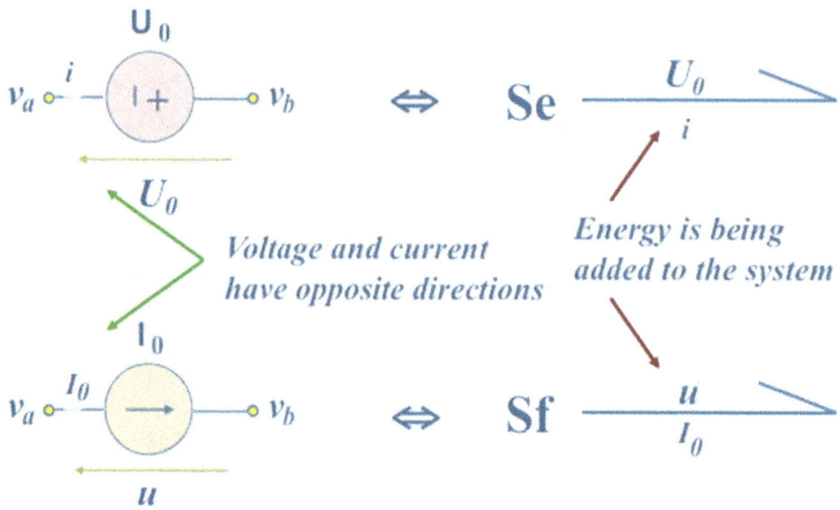

Figure 73: Energy Adders to a System

In general, the most commonly used components in the Bond Graph modeling method are elements that represent resistors, capacitors, and inductors. In the following diagram, these are represented as electrical elements. In the modeling of a mechanical system, these might represent resistance in the form of friction, capacitance or compliance in the form of a spring, and inductance or inertia in the form of a mass. Note the symbology used to describe these elements.

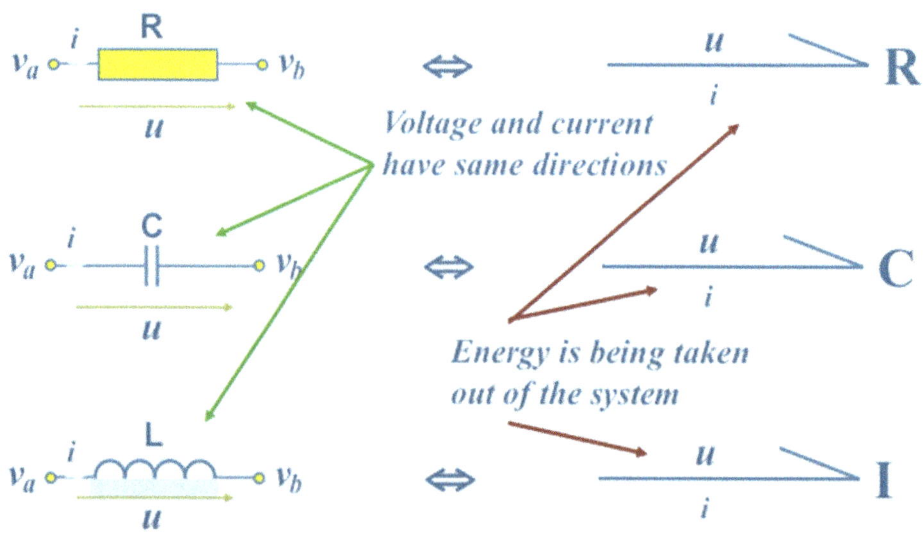

Figure 74: Electrical elements in modeling

In addition to these physical entities, there are also generated junctions. These junctions are one of two types; 0-junctions represent a common effort, 1-junctions represent a common flow. These are shown in Figure 75.

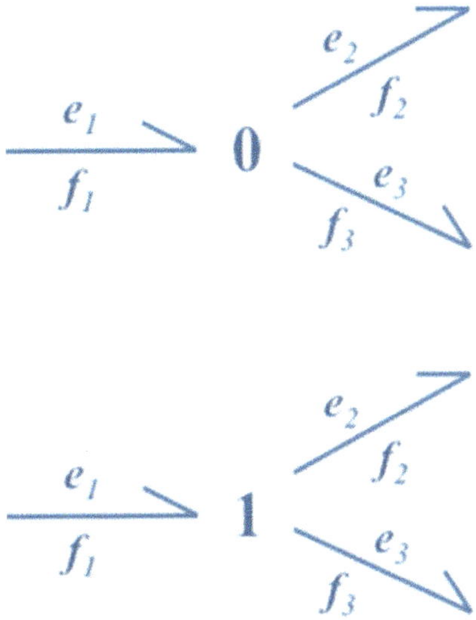

$$e_1 = e_2$$
$$e_2 = e_3$$
$$f_1 - f_2 - f_3 = 0$$

$$f_1 = f_2$$
$$f_2 = f_3$$
$$e_1 - e_2 - e_3 = 0$$

Figure 75: Junction definitions

The general rules that are utilized in order to generate a Bond Graph model are straightforward. Diagram the system to be modeled, and clarify each of the elements and the interconnections between the elements. Once this diagram has been completed, a Bond Graph can be generated.

In order to begin the model, a one junction is created for each physical element shown. The one bonds represents a common flow, and therefore they are interconnected by the use of 0 bonds. Each of the physical components is then attached to their respective 1 bond. The direction of the arrow is defined by the elements located at each bond. In the case of resistors, the arrow points toward the component. In the case of capacitors the arrow points toward the component. In the case of inductors, the arrow points toward the component. In the case of sources, the arrow points from the source. Each of these conventions relate to the flow of energy as expected.

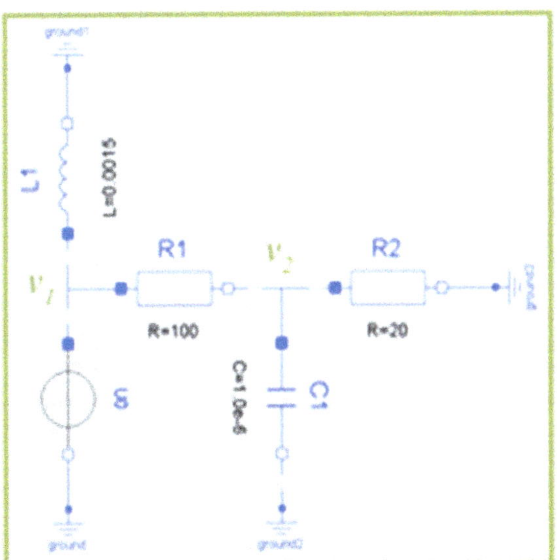

Figure 76: Example System

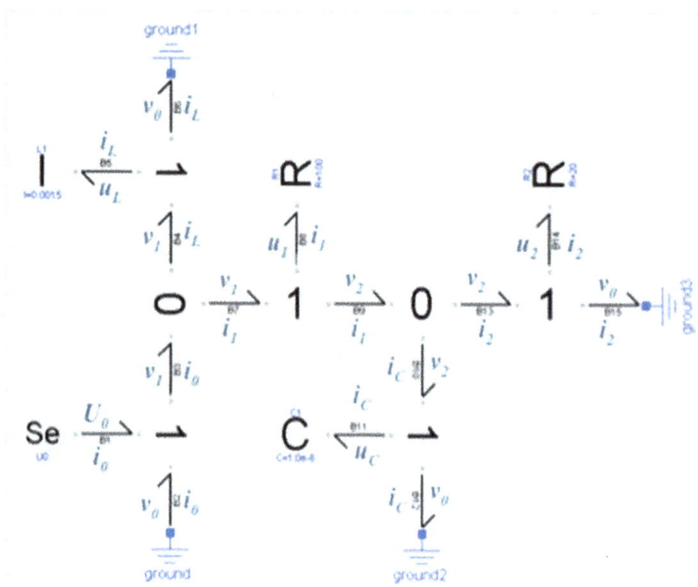

Figure 77: Bond Graph Model of System

Every bond defines the two separate variables, the effort and the flow. Therefore, we need two equations to calculate the values for these two independent variables. It is always possible to compute one of the two variables at each side of the bond. This is a very fortuitous occurrence, as it implies that we will always be able to solve the system of equations generated from a Bond Graph.

In order to understand how the equations will be generated, we assign causality to each of the links in our Bond Graph. The causality is represented by a straight bar (|), and denotes the side where the flow is being computed. Various rules are defined regarding the different components in a Bond Graph and are shown in the following diagrams.

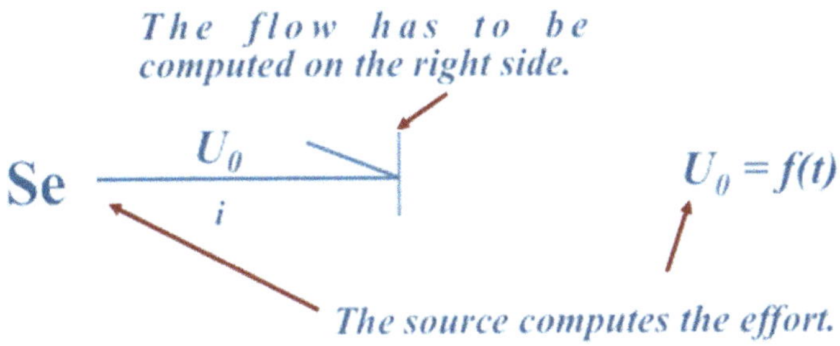

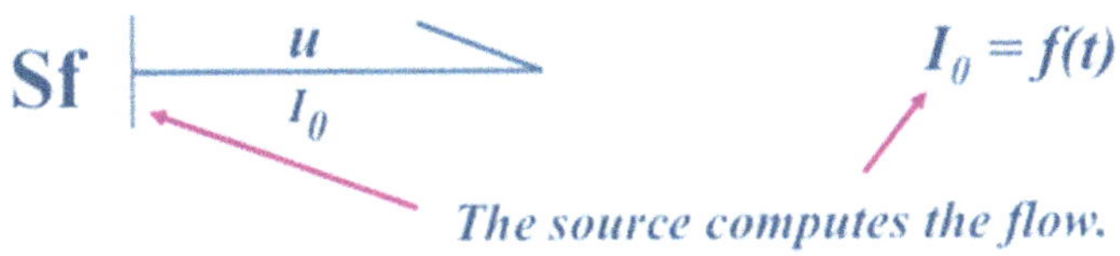

⇒ The causality of the sources is fixed.

$$u = R \cdot i \qquad\qquad i = u / R$$

⇒ The causality of resistors is free.

$$du/dt = i \;/ C \qquad\qquad di/dt = u \;/ I$$

⇒ The causality of the storage elements is determined by the desire to use ***integrators*** *instead of* ***differentiators****.*

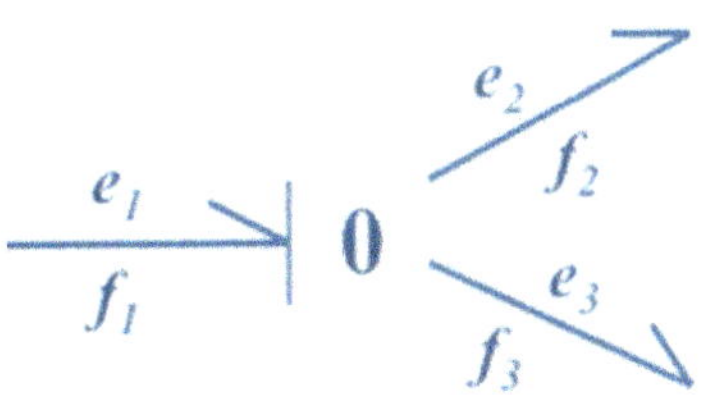

$$\Leftrightarrow \quad \begin{array}{l} e_2 = e_1 \\ e_3 = e_1 \\ f_1 = f_2 + f_3 \end{array}$$

Junctions of type 0 have only one flow equation, and therefore, they must have exactly one causality bar.

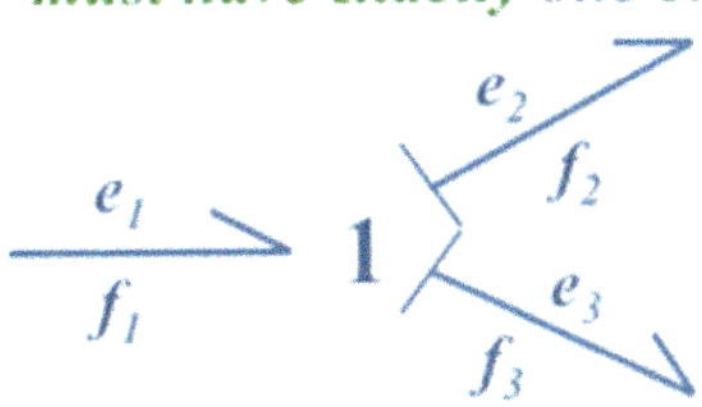

$$\Leftrightarrow \quad \begin{array}{l} f_2 = f_1 \\ f_3 = f_1 \\ e_1 = e_2 + e_3 \end{array}$$

Junctions of type 1 have only one effort equation, and therefore, they must have exactly (n-1) causality bars.

In the model shown in Figure 77, the diagram can be simplified in several ways. First, the grounds are removed from the system, as they are already represented by the common flow at the junction. Secondly, the junctions which do not have a connecting element associated with them can be collapsed and removed from the diagram. Therefore, the simplified model, with the addition of causality as defined by the previous rules, result in the diagram in Figure 78.

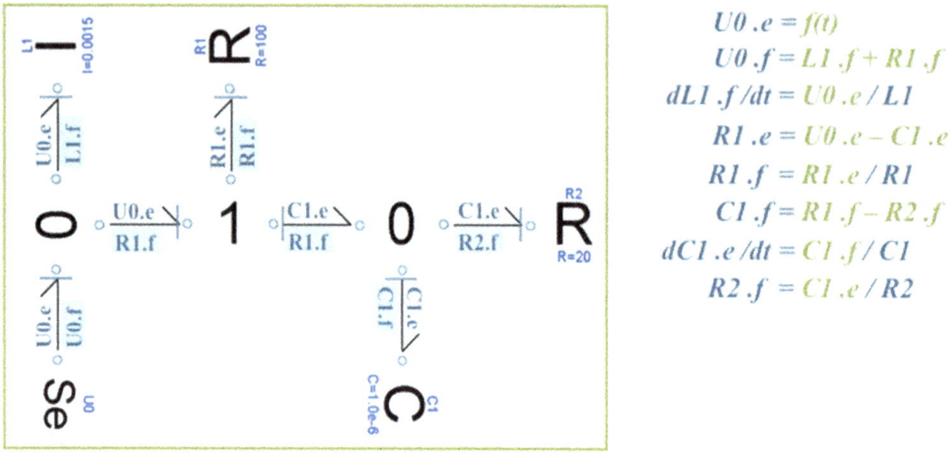

Figure 78: Bond Graph and State Equations

The efforts and flows are identified on this diagram, and the resulting state equations have been identified. By following this methodology, it is very simple and straightforward to generate a practical model of a dynamic system.

One of the most powerful modeling systems available to the Systems Engineer is the use of a spreadsheet upon completion of the state equations. As in the previous example, there are two first order differential relationships associated with this model. By virtue of this, we know that the model will likely generate nonlinear behavior, which is why we have undertaken the model in the first place. If the model were straightforward and resulted in a linear system, it would unlikely provide much of insight. However, in a case such as this, where it appears the system will generate complex behavior, these equations can be input into a simple spreadsheet and the resulting behavior of the system can be graphed. Several mechanical system models have been generated in order to illustrate these concepts.

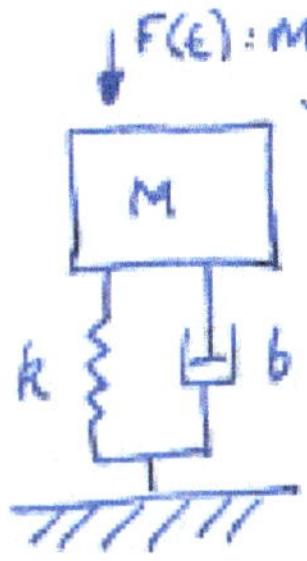

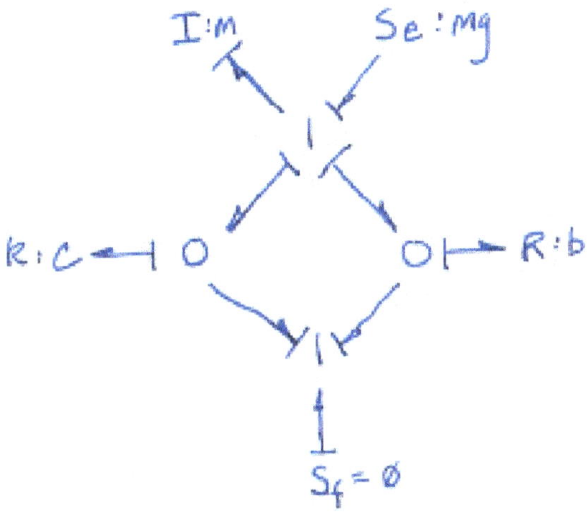

SIMPLIFY:

SINCE $S_f \Rightarrow \emptyset$, REMOVE 1 bond, 0 bonds

END WITH:

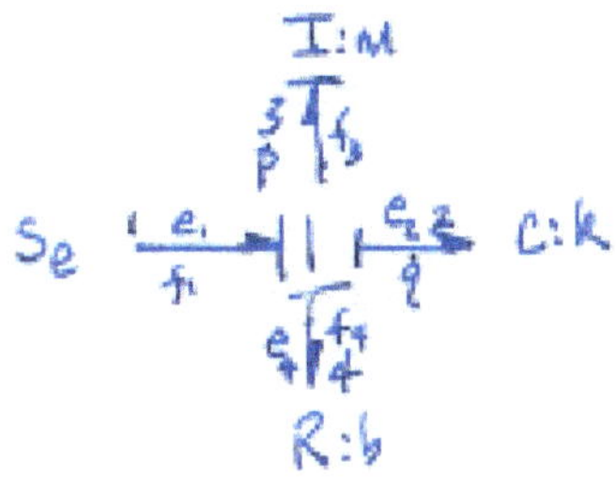

1) NUMBER BONDS

2) NOTE I & C ELEMENTS TO FIND $\dot{P}$ AND $\dot{Q}$

3) DERIVE STATE EQUATIONS

$$e_1 = SE_1$$
$$f_3 = P_3/M_3$$
$$e_2 = K_2 Q_2$$
$$e_4 = R_4 f_4 = R_4 \cdot f_3$$
$$e_3 = e_1 - e_2 - e_4$$

SUBSTITUTING:

$$e_1 = SE_1$$
$$f_3 = P_3/M_3$$
$$e_2 = K_2 Q_2$$
$$e_4 = R_4 P_3/M_3$$
$$e_3 = SE_1 - K_2 Q_2 - R_4 P_3/M_3$$

DISPLACEMENT $\frac{\partial Q}{\partial t} = f_2 = f_3 = f_4 \Rightarrow P_3/M_3$

MOMENTUM $\frac{\partial P}{\partial t} = e_3 \Rightarrow SE_1 - k_2 Q_2 - R_4 P_3/M_3$

Figure 79: Simple System Example 1

A simple mechanical system is defined in Figure 81. The Bond Graph is generated and simplified, causality is assigned, and the state equations are defined based upon the known associations within the Bond Graph. This enables us to define every effort and flow in the diagram, although there are only a few that we find to be of particular interest. In this case, they include the displacement (the integration of the velocity or flow variable) and the momentum (the integration of the inertia variable.) They also happen to be the only two first order differential relationships in the equations.

These equations can be quickly input into a spreadsheet, and the resulting dynamic behavior of the system can be calculated and displayed. This is shown in the following diagram. As will be obvious to anyone skilled in mechanical systems, this simple harmonic motion is the result expected from the dynamic system.

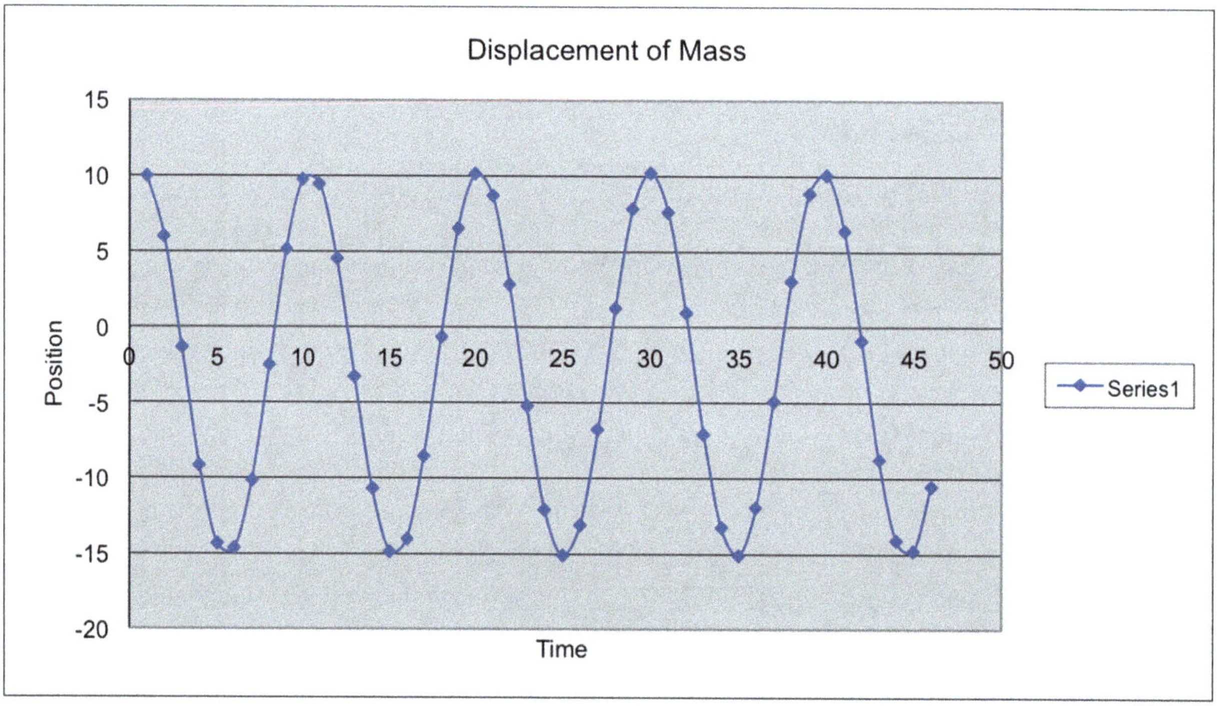

Figure 80: Displacement of Mass over Time

Several additional examples are included with sample results to illustrate the methods of Bond Graph modeling.

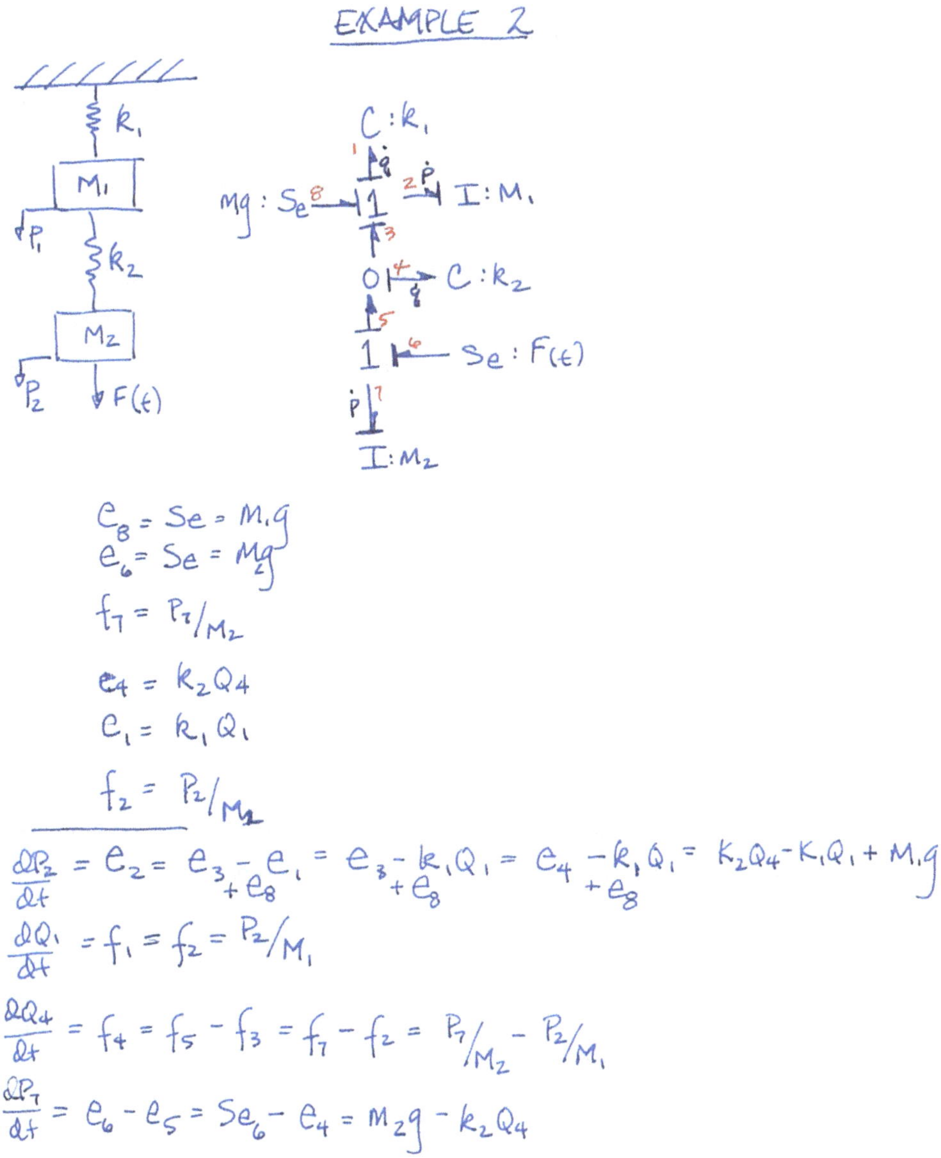

Figure 81: Simple System Model 2

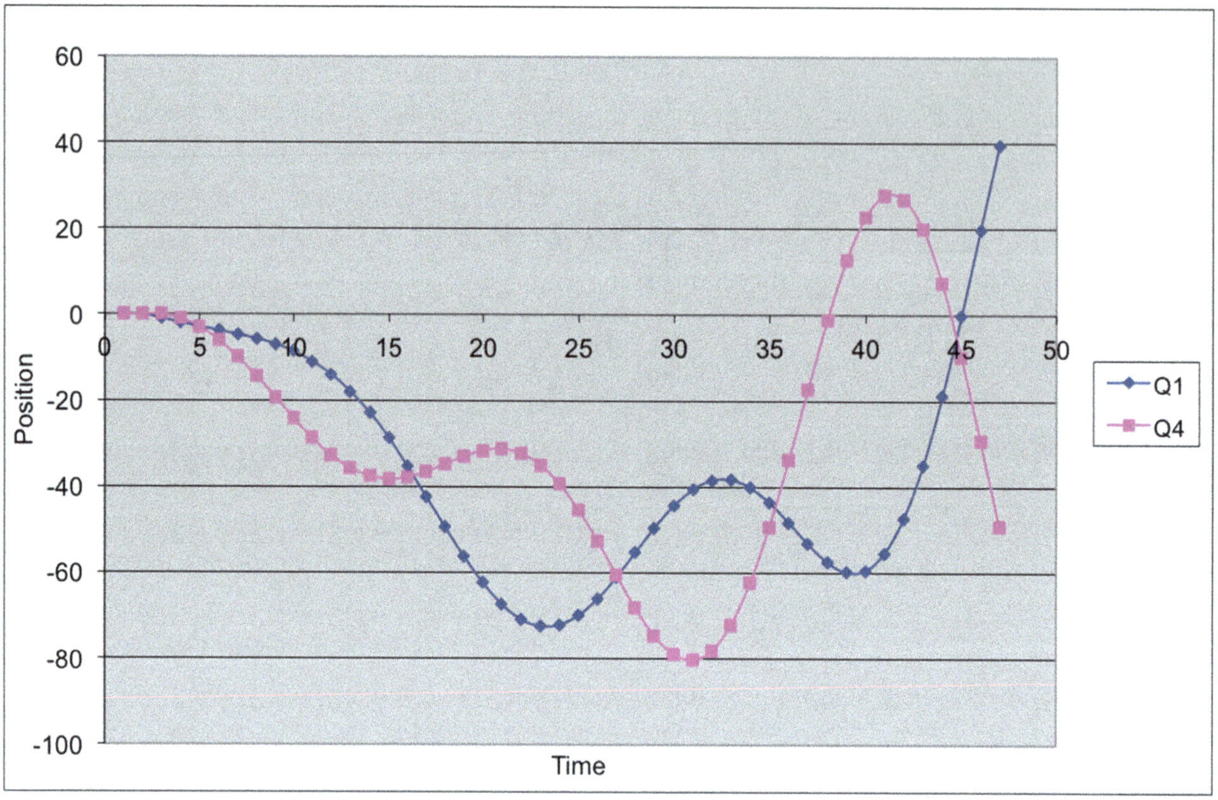

Figure 82: Displacement of Two Masses

This model demonstrates complex behavior, as the input for the lower weight is a dynamic input, and as the resulting behavior of the lower weight will introduce forces on the upper weight, no simple harmonic motion is generated. In fact, this system is rapidly going out of control, allowing the system developer to identify a potential problem with this arrangement prior to testing with a physical model. This model is representative of a car suspension with flexible tires in contact with the ground. The results would enable us to size our springs and shocks for the suspension, saving many hours of fabrication and testing of mechanical systems.

System Example 3

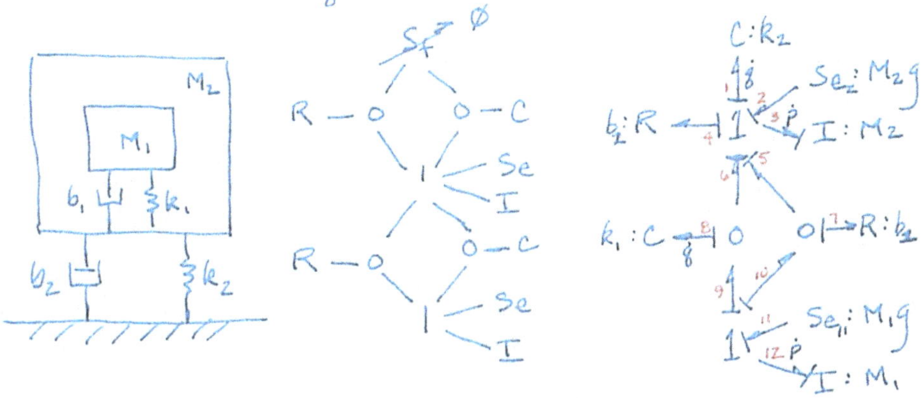

FIND: Q_1, Q_8, P_3, P_{12}

$$e_2 = Se_2 \qquad e_4 = b_2 f_3 = b_2 P_3/M_2$$

$$e_{11} = Se_{11} \qquad f_7 = f_{10} - f_5 = P_{12}/M_1 - P_3/M_2$$

$$f_3 = P_3/M_2$$

$$f_{12} = P_{12}/M_1$$

$$e_1 = k_2 Q_1$$

$$e_8 = k_1 Q_8$$

$$\frac{dQ_1}{dt} = f_1 = f_9 - f_8 = f_3 = P_3/M_2$$

$$\frac{dP_3}{dt} = e_3 = e_2 - e_1 - e_4 + e_5 + e_6 = e_2 - e_1 - e_4 + e_8 + e_7$$

$$= Se_2 - k_2 Q_1 - b_2 P_3/M_2 + k_1 Q_8 + b_1 (P_{12}/M_1 - P_3/M_2)$$

$$\frac{dQ_8}{dt} = f_8 = f_9 - f_6 = f_{12} - f_3 = P_{12}/M_1 - P_3/M_2$$

$$\frac{dP_{12}}{dt} = e_{12} = e_{11} - e_8 - e_{10} = Se_{11} - k_1 Q_8 - b_1 (P_{12}/M_1 - P_3/M_2)$$

Figure 83: Simple System Model 3

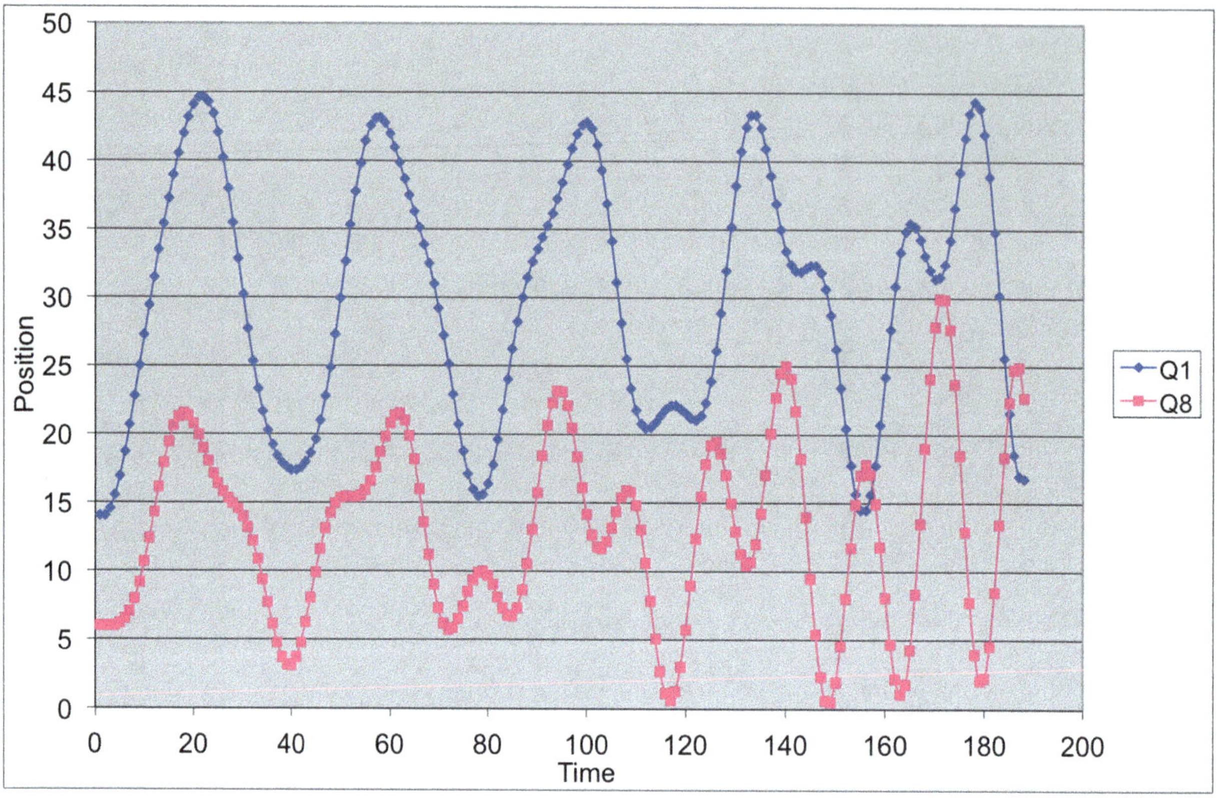

Figure 84: Simple System 3 Results

Earlier, we stated that it was important to understand what question needed to be answered in the modeling effort. In the case of simple system 3, we have isolated a cabinet inside a second isolated cabinet. This is a common method of shock and vibration management in the defense industry, and is utilized to protect sensitive components. In this particular case, the system was designed to withstand a shock event, and the question to be defined was whether in any situation the outer cabinet could be at top dead center while the inner cabinet was a bottom dead center (or any other configuration that would result in contact between the cabinets.) In this particular case, the Bond Graph model was sufficient to show that this design would result in a high probability of cabinet contact during a shock event, which resulted in significant modifications to the system design.

As can be seen by the previous examples, the use of Bond Graph modeling is flexible and straightforward. With the application of only a few minutes and a spreadsheet, the Systems Engineer can get a good feeling for the behavior of his or her dynamic system while there is still time to make needed modifications to the development. Many other methods are available to conduct dynamic modeling, and once again it is important to understand what question needs to be answered, and with what fidelity the model should be developed in order to begin to answer that question.

Establishing the Requirements List

There are several guidelines for writing good requirements:

1) Requirements should contain a single thought
2) Requirements use the word 'shall' in the definition
3) Requirements should form a complete sentence

As requirements are developed, they will be incorporated into a Requirements Verification Traceability Matrix (RVTM). This matrix is designed in order to allow requirements management to be conducted. There are several elements of note to incorporate into this particular matrix:

1) The requirement statement must be adherent to our guidelines.
2) The requirement must be verifiable.
3) The requirement list must be consistent and not contradictory.

Therefore, we will define our RVTM with several columns of interest. First, we will identify a number for each requirement. Secondly, we will incorporate the description of the requirement in its entirety. Third, we will define the traceability for the requirements (i.e. where it came from,) and the verification method. An example RVTM is included in **Figure 85**.

RVTM for Toaster Product

					Verification Method			
Requirement	**PBSS**	**OV-2**	**OV-5**	**Requirement**	**Inspection**	**Model**	**Test**	**Demonstrate**
1	1			The system shall operate on 120 VAC	X			
2	2			The system shall provide a manual shutoff	X			
3	3			The system shall not exceed 12X12X12 inches	X		X	
4	4			The system shall not mar countertops if slid on surface	X			
5	5			The system shall provide an indicator for operational status	X		X	
6	6			The system shall provide an audible alarm in the event of emergency shutdown	X			
7	7			The system shall autoshutdown in the event of overtemperature	X		X	
8	8			The system shall autoshutdown in the event of current draw over 10A	X	X		
9	9			The system shall weigh less than 4 lbs	X			
10	10			The system shall toast bread sices				X
11	11			The system shall toast bagel slices				X
12		1		The system shall allow user adjustment of setpoint				X
13		1		The system shall indicate setpoint for user				X
14		1		The system shall utilize an LG plug for power input				X
15		1		The system shall provide a push button start for the user				X
16		1		The system shall provide automatic unloading of product after completion of toast cycle				X
17		1		The system shall accept user loading of product				X
18		2		The system shall unload toast to an area at least 2 inches from heated surfaces				X
19		2		The system shall provide user audible notification of unloading event				X

Figure 85: Toaster RVTM

As noted previously, it is essential for the Systems Engineer to be able to generate many appropriate requirements for the system under development. We have discussed the generation of a variety of system artifacts that will enable us to generate an adequate list of system requirements.

As previously stated, every need line on the various system artifacts and every function or capability will represent one or more requirements. Therefore, in the development of our RVTM, we expect that every capability and every function or capability will be represented one or more times in the requirements list. When one examines a typical OV-2 diagram, it is apparent that we could expect to generate several dozen requirements from that single artifact. When examining a typical OV-5, it is apparent that we should be able to generate several hundred requirements with little difficulty. The systematic recording and traceability of an extensive requirements list is a crucial skill that Systems Engineers must develop. This requires a highly organized and detail oriented approach to the development of requirements, and also allows the system developer to assure himself that all operational capabilities are represented in the requirements list. Therefore, the RVTM becomes a repository and ultimately, the reporting artifact for not only the requirements, but also the traceability and verification methodology for each of those requirements.

Defining the Technical Solution

Upon completion of the system's Architecture, requirements list and RVTM, and when the Systems Engineer believes that the appropriate gaps in knowledge have been sufficiently closed or clarified, we are ready to finalize our documentation before proceeding to the Physical Architecture.

It is appropriate at this stage of the system's development process to conduct a meeting with as many stakeholders as are readily available to ensure the proposed concept of operations is in alignment with the stakeholders' identified needs. Often, during this meeting, the stakeholders will find it easier to clarify their needs and objectives with regard to the system. This may require some realignment of the concept of operations if new insights are presented. It is prudent to conduct this prior to beginning the development of the Physical Architecture to reduce potential rework of the detail design.

At this point, all information must be consolidated, including the Work Products and Requirements List, with a review of these documents for coherence and lack of conflicting requirements. This phase often requires revision of the System Architecture or adjustment to the Requirements List. When this has been completed, the System Architecture may be transferred to the Component Engineers for translation to Physical Architecture and detailed design. This is accomplished in a handoff meeting, where the Component Engineers are introduced to the System Architecture and the Systems Engineering team clarifies all uncertainties and questions. In many cases, the members of both teams are identical, but the role played by specific individuals on the team may shift as the focus of the work progresses through the development cycle.

Chapter 7: Defining the physical architecture

Once handoff has occurred, the Component Engineering team utilizes the System Architecture as a blueprint for the Physical Architecture definition. Paying close attention to the construction of each functional block on the Architecture diagrams, the Component Engineers establish physical designs to accomplish the capabilities outlined by the Systems Engineering group.

The Component Engineers continue to develop the project utilizing standard Engineering practices and procedures to complete the decomposition phase of the V diagram and begin the re-composition phase. Their output completes the detailed design documentation, as well as the initial verification of the components and subsystems produced.

This concludes the System Development portion of the Systems Engineering cycle. The remaining responsibility of the Systems Engineering group constitutes the verification and validation of the resulting system produced.

Physical Architecture Development

Physical Design Definition

The Physical Architecture of a system is a hierarchical description of the resources that comprise the system. It is composed of configuration items (CI), hardware, software, people, facilities, procedures, and documents.

Physical Design Abstractions

Generic Physical Architecture

As one goes about the process of requirements discovery the mind may spontaneously think about the generic nature of what the physical architecture may look like, at a very high level. For example, after reflecting on several requirements regarding the processing of input data from a sensor or several sensors, one may think, "Yes, this architecture will need a processor and memory and some sort of sensor(s)". This will be a high level conceptual view that does not permit the creation of numerical models and simulations. For example, on Project Bentley there may be general talk about the need for processors and software and people, no specific information regarding the numbers of people needed, detailed job descriptions, software attributes, processor performance levels, and processing throughput. This would be deemed a generic Physical Architecture.

Instantiated Physical Architecture

Instantiated Physical Architecture provides detailed data regarding hardware, software, people, documents, and interfaces. For example, detailed data regarding hardware might include memory size, clock rates, CPU or GPU, or both. It might also include interface capability, specific hardware brands, model numbers and detailed descriptions. Detailed data regarding software might include functional capabilities, memory usage, functionality, portability, API (Application Interface definitions), product names, and model numbers, for example. Detailed data regarding people might include organizational descriptions, detailed job descriptions, and identified subject matter experts. Detailed document descriptions might include plans, procedures, handbooks, and maintenance manuals. Detailed data regarding interfaces might include Internet Protocols, and OSI model approach.

Generic and Instantiated Architectures - Differences

Generic Physical Architecture

A Generic Physical Architecture defines the hierarchy in general terms. For example, it may be defined as processors with associated software, or a person in a building.

Instantiated Physical Architecture

An Instantiated Physical Architecture defines the specifics of processors, software, buildings, people, facilities, procedures, and documents in enough detail to permit performance modeling of the system related to the requirements being addressed.

Systems Engineering Roles in Physical Architecture Development

Systems Engineering's intent is to create and use representative instantiations for the generic components that are sufficient to model and run simulations to determine that the performance requirements of the system are met, and to ensure that the requirements decomposition makes sense. In addition, the intent is to ensure multiple (at least two, preferably three) Candidate Physical Architectures are created and trade-off studies are performed to select the best (based on specific evaluation criteria) Architecture that meets operational requirements.

The Systems Engineer uses generic and instantiated architectures to test the validity of the decomposition of the requirements. The Systems Engineer uses models and simulations as tools to verify the requirements decomposition process makes sense. Systems Engineers do not do the detailed design of the subsystems or the configuration items; but rather, ensures that subsystems and configuration item specifications are provided to configuration item designers.

Methods for Developing Alternatives for Generic and Instantiated Physical Architectures

The Systems Engineer uses two methods to ensure all of the systems functions are allocated to a configuration item in the Generic Physical Architecture, and subsequently, in the Instantiated Physical Architecture. Two steps that precede this are a mapping of requirements to the systems functions, and a mapping of functions to the generic physical architectures shown in Figures 88 and 89. All four of these steps use Excel for the matrix mappings.

#	Requirement description	Function a	Function b	*Function c*
	Requirement 1	X		
	Requirement 2			*X*
	Requirement 3	X	x	

Figure 86: Function Mapping

Generic Physical Resources (Components)

Functions

	Analog-to-digital converter	Analytic software package	Computer memory	Storage algorithm	
Analog to digital conversion of video signal	X				
Generate average value of video signal		X			
save average value of video signal			X	X	
Function 4					
Function 5					

Figure 87: Mapping Functions to Generic Physical Components

Morphological Box

The Systems Engineer makes sure that a morphological box is used to develop Instantiated Physical Architecture or Architectures. A morphological box is a matrix created in Excel in which the columns or rows represent the components in the Generic Physical Architecture, and the boxes in a given column or row represent all the choices for filling that generic component need. An example of a morphological box is shown in Figure 88. The information in the boxes below the column headers represents candidates for the design of the screwdriver. The selected options from the set of boxes that best fit the requirements for the screwdriver result in the Instantiated Architecture.

Handle size	Handle material	Screw driving element material	Driving point type	Blade length
1.5 " diameter	Wood	Steel	flat	3 "
,5" " diameter	Plastic	Plastic	Philips	4"

2 “diameter	Aluminum	aluminum	torx	8”
1” diameter	Plastic	Steel	square	6”

Figure 88: Morphological Box

The morphological box is used to ensure that multiple alternatives are considered for Instantiated Physical Architectures. It must be remembered that the Physical Architecture provides resources for every function identified in the functional Architecture.

A Process for Creating a Physical Architecture

We will use the IDEF0 modeling language to graphically demonstrate how to create a Physically Instantiated Architecture, given a set of requirements and functional Architecture. Figure 89 is high-level view of the Instantiated Physical Architecture development process.

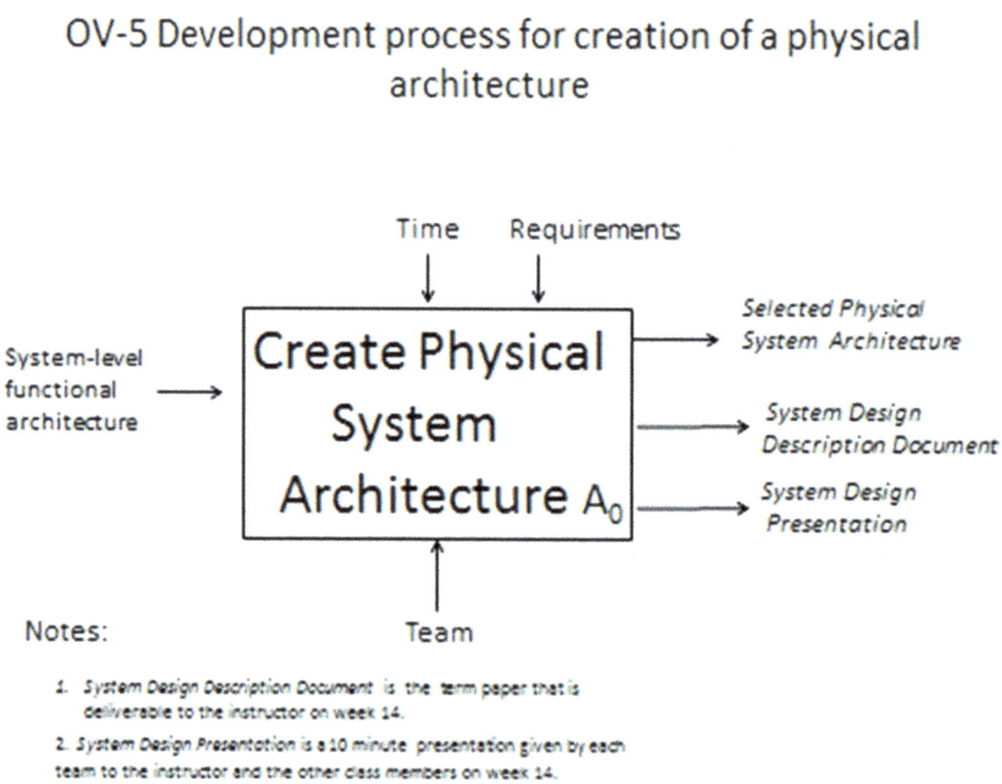

Figure 89: High Level View - Physical Architecture Development

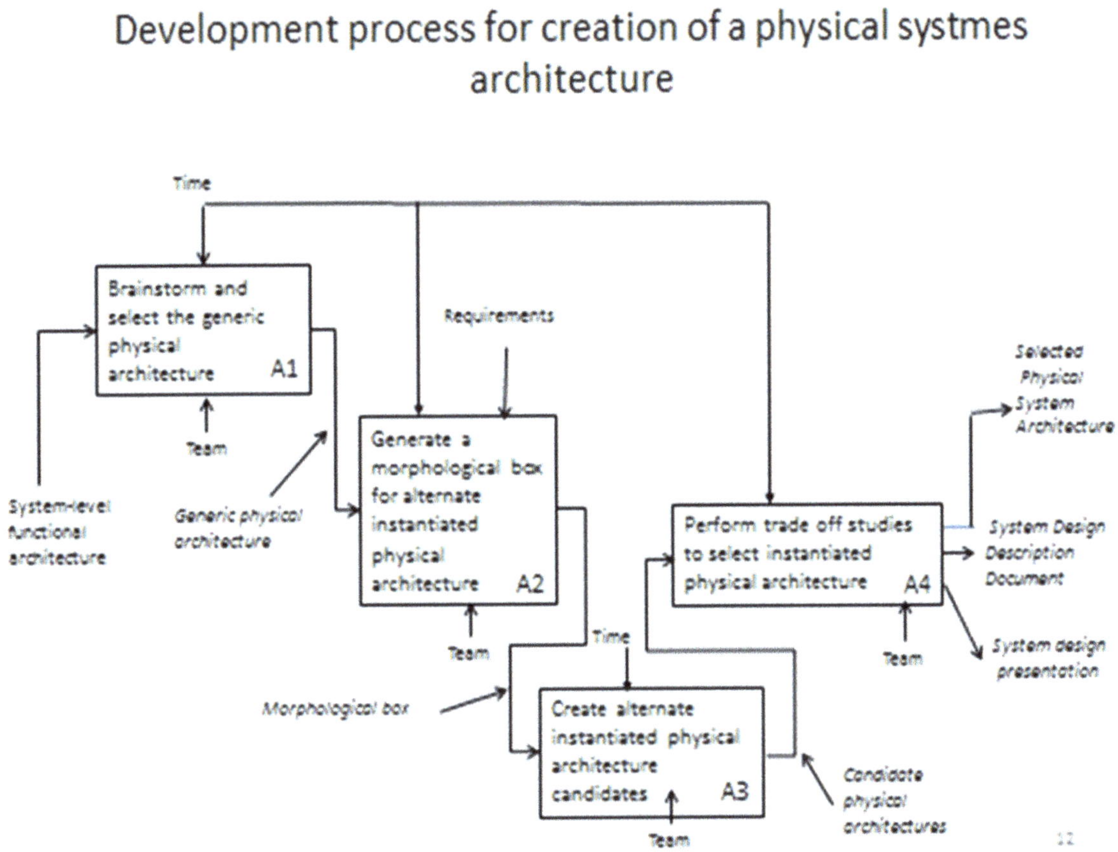

Figure 90: Steps to Create an Instantiated Physical Architecture

The First Step - Brainstorming/Brain Writing

The first step in creating Instantiated Physical Architecture is to brainstorm a generic physical architecture. Some ideas have probably occurred to the Systems Engineers during the requirements and functional Architecture phases of the system engineering activities. When these ideas occur during the requirements discovery phase and the functional Architecture phase, they are vague, generic ideas regarding the physical components needed to implement or build the Architecture. In the event that these ideas do not occur as part of the initial Systems Engineering phases, there are techniques for creating the Generic Physical Architecture. There are multiple techniques for brainstorming or brain writing activities. They are individual, group, individual/group, structured, and unstructured techniques. The following section provides more detail regarding those techniques.

Brainstorming/Brain Writing Details

This technique is utilized to generate a wide range of alternatives, and is commonly applied when many ideas are needed. The group operates in an evaluation-free period where criticism and discussion is in prohibited. The rationale for this approach is that even the most preposterous idea may stimulate the generation of a really superior idea. The more ideas generated, the better chance of finding a high quality solution. Analogies are often used in Systems Engineering because building upon our experiences with previous systems has a great deal of creative power.

A second approach, one of attribute listing, is based on the concept that Physical Architectures may all be traced to modifications of previous Architectures. The following tables provide more detail regarding the techniques and their applicability.

Creativity Techniques

Brainwriting /Brainstorming

Brainstorming/Brain writing categories	Examples
Brainstorming 1-A group of individuals generates ideas via verbal discussion, no defined procedures used	Unstructured group discussion
Brainstorming 2-A group of individuals generates ideas via verbal discussion within the bounds of predefined procedures	Classical brainstorming
Brainwriting/Brainstorming 1- a group of individuals generates ideas by a predefined written and verbal procedures	Brain writing game

Table 2- Brainwriting brainstorming categories- continued

Figure 91: Creativity Techniques

The Second Step - The Morphological Box

The second step is to create a morphological box to assist in generating a set of creative Candidate Instantiated Architectures to analyze. A morphological box is a matrix: columns represent the components of the Generic Physical Architecture, and the cells in a given column represent alternate choices for filling that generic component.

To identify all of the possible instantiations for a component in the Generic Architecture one looks down the column through all the rows to see the possibilities. Each option should have well-defined performance and cost characteristics.

The Generic Physical Architecture provides common designators for physical resources in a hierarchical decomposition that partitions the system in greater detail. Although the Generic Architecture has no substance in the sense of specific physical items, the structure is still very important because some Instantiated Physical Architectures can be eliminated from consideration merely on the basis of the division of the system components of the generic level.

Recommended number of instantiations

When performing a system design there should be at least two Candidate Physical Systems Architectures, so that a trade-off study may be done to pick the one that most capably meets the Evaluation/Selection Matrix.

Evaluation/Selection Matrix

An Evaluation/Selection Matrix is used to select the best Candidate Instantiated ystem design. The matrix uses weighted scores to compare each system component against evaluation criteria that include the user's operational requirements (user), cost, weight, size, power, and any other criteria that apply to the specific context.

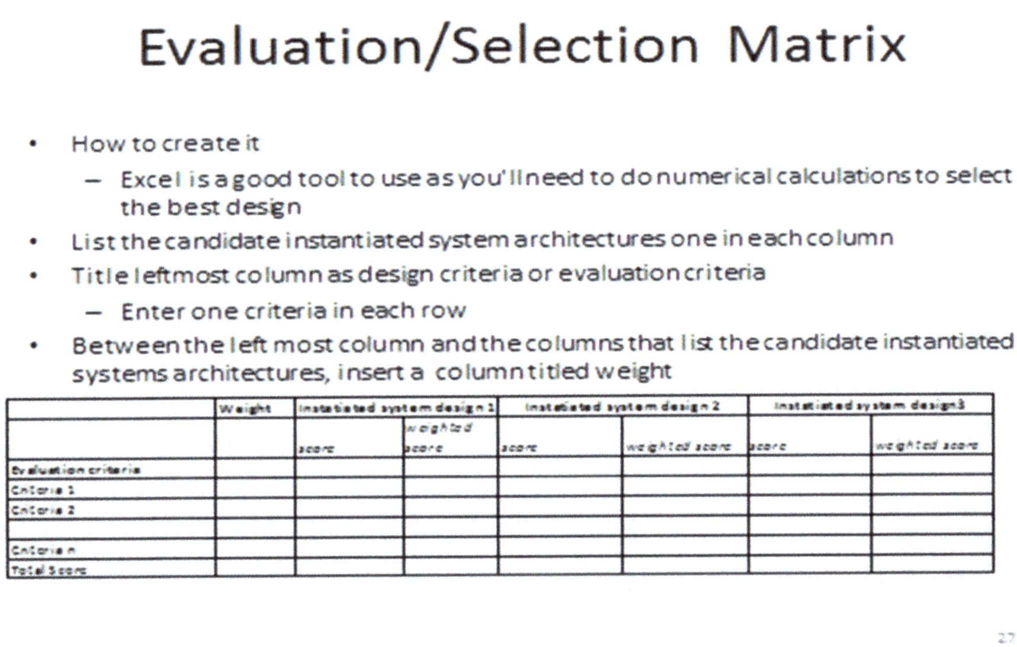

	Weight	Instatiated system design 1		Instatiated system design 2		Instatiated system design3	
		score	weighted score	score	weighted score	score	weighted score
Evaluation criteria							
Criteria 1							
Criteria 2							
Criteria n							
Total Score							

Figure 92: Evaluation/Selection Matrix

Instantiated Physical Design Evaluation/Selection Matrix

Example candidate selection criteria matrix
Instantiated physical architecture 1 - selected because it has the highest score

Notes:

1. A score of five 5 is the highest score and 1 is a lowest score
2. A score of 5 for the weight is the highest score signifying the most important criteria, and a score 1 signifies the least important waiting criteria.
3. This evaluation matrix shows that design 1 is the best fit, and designed 2 is the worst fit.
4. This is a very high level example. A real system can evaluation matrix contain many more rows

		Instantiated Physical Architecture 1		Instantiated Physical Architecture 2		Instantiated Physical Architecture 3	
Design criteria	Weight	score	Weighted score	score	Weighted score	score	Weighted score
Operational requirements	5	5	25	4	20	2	10
Cost	3	2	6	3	9	4	12
Size	2	4	8	2	4	5	10
Weight	5	2	10	2	10	3	15
Power	5	5	25	3	15	4	20
Total score			74		58		67

Table 2-Instantiated physical architecture evaluation matrix example

Figure 93: Instantiated Physical Architecture Evaluation Matrix

Utilizing this method, we can readily create a Generic Physical Architecture, an Instantiated Physical Architecture, and justification for the selections made from candidate physical components. This traceability might become useful in the future, as justifications or selection criteria might be questioned regarding the chosen design.

Chapter 8: Risk Management

I'm sure we've all heard the term Risk Management, either in school or at work. But what does Risk Management really mean? More fundamentally, what is Risk?

The definition of risk from Dictionary.com:

noun
1. Exposure to the chance of injury or loss; a hazard or dangerous chance: *It's not worth the risk.*
2. *Insurance.*
   a. The hazard or chance of loss.
   b. The degree of probability of such loss.
   c. The amount that the insurance company may lose.
   d. A person or thing with reference to the hazard involved in insuring him, her, or it.

e. The type of loss, as life, fire, marine disaster, or earthquake, against which an insurance policy is drawn.

verb (used with object)

3. To expose to the chance of injury or loss; hazard: *to risk one's life.*

4. To venture upon; take or run the chance of: *to risk a fall in climbing; to risk a war.*

Idioms

5. **At risk,**

a. In a dangerous situation or status; in jeopardy: *families at risk in the area of the weakened dam.*

b. Under financial or legal obligation; held responsible: *Are individual investors at risk for the debt part of the real estate venture?*

6. **Take /run a risk,** to expose oneself to the chance of injury or loss; put oneself in danger; hazard; venture.

Defining the Unknowns and Uncertainty

Origin of Risk

The key to the above definitions and examples of risk narratives is to understand that "risk" implies the chance of loss. In general, risk is composed of two components; the chance of occurrence of an event and the loss, or damage caused by the occurrence of the chance event. "Chance" is another expression of the uncertainty of the occurrence of future events. Notice that now we are using the word "uncertainty." The use of the word "uncertainty" really isn't clearing up the meaning of the word "chance"; that is, providing greater understanding or better comprehension of what we mean by "chance."

Merriam Webster's definition of **uncertainty**:

- The quality or state of being uncertain
- Something that is doubtful or unknown: something that is uncertain

So, the words "chance" and "uncertainty" in essence mean something that is doubtful or unknown. The question now arises: are there different types or classes of uncertainty? It is a question that is rarely asked. Typically, we react to the word "uncertainty" by immediately going to the measure of uncertainty, the Probability Calculus, to determine the likelihood of the occurrence of possible future uncertain events.

That clears everything up, right? Not really.

The word "uncertainty" needs some refinement in order for us to understand what we are talking about when we use the word. Most of us never think of the distinctions, or the different classes of uncertainties. We immediately jump to its measure, the Probability Calculus.

Are there different types of uncertainties? If so, what are the distinctions?

There are two different types of uncertainty; aleatory and epistemic. The measure of each uncertainty type is Probability Calculus. Let us now look at the distinctions between these two types and see what they may mean to the design of systems.

Aleatory and Epistemic Uncertainty

When classical determinism was in fashion, "uncertainty" only meant a lack of knowledge; that is to say, only epistemic uncertainty. Uncertainty, (The lack of knowledge) is a deficiency which might be remedied in principle by further learning experiments.

For example; flipping a coin is an example of epistemic uncertainty. An uncertainty is caused by the lack of precise knowledge of the three-space position of the coin flipper's hand, the relation of coin to the coin flipper's hand, and the center of density of the coin. Papers exist that provide a deterministic analysis of the outcome of coin flips provided perfect information in regard to the above. However, that information is generally not available. Exact measurements of position (noiseless, error free) measurement will never be perfect.

Error or noise in measuring position will likely exhibit a normal distribution. So the coin flip must be treated, if we wish to measure whether the outcome of the flip will be heads or tails, with the measure of a certainty called Probability Theory. Rather than a deterministic analysis of a coin flip, then, a probabilistic determination is made. The answer to the probability of either side of the coin occurring if n, the number of flips, goes to infinity, is obviously 0.5 or fifty percent.

From Quantum Mechanics and Chaos Theory we now know that there is much more uncertainty in characterizing the physical world than epistemic uncertainty; uncertainty arising from the lack of knowledge. There is an intrinsic uncertainty associated with the need for a statistical perspective in order to comprehend the physical world. This kind of uncertainty associated with randomness is called "aleatory" uncertainty, after the Latin word for dice. Aleatory uncertainty may be estimated, but cannot be reduced by advances in theoretical or observational science. This is very different than epistemic uncertainty, which arises from imperfect information and knowledge, which is potentially reducible by new data acquisition (new information), even though for some observations this may take generations.

The two types of uncertainty reflect the underlying duality of probability. On the one hand, probability has a statistical aspect, concerned with stochastic laws of random processes, and on the other hand, probability has this epistemic aspect concerned with assessing degrees of belief in the propositions with no statistical basis. In the absence of statistics, these beliefs may be based on logic and our personal judgment.

Recognizing that different types of uncertainty exist is important because of the practical consequences of decision-making in the presence of uncertainty.

In order to dramatize the importance of recognizing the difference between the epistemic and aleatory uncertainty, let us look at what the recognition of this difference means in relation to ground shaking that arises in earthquakes. Let us call the estimate of the severity of ground shaking G, from earthquake of given magnitude M, and arising at some distance R. This estimate is an essential component for establishing design criteria for seismic loads and is based on a parametric attenuation relation:

$G=f(M,R)$

This equation is derived from the regression analysis of data recordings of ground motions at various earthquake-monitoring stations. The application of such design criteria to a single particular site is based on the assumption that the spatial scatter is a good guide to the temporal variability of the ground motion at a particular site.

This implies that an ergodic assumption was made regarding the spatial and temporal nature of the recorded data on which the regression attenuation function was based. For example, for an earthquake of a given magnitude, the rupture geometry and dynamics are not fully known. Differences in ground shaking due to wave propagation path or site effects also contribute to the aleatory uncertainty due to the presence of random heterogeneities in the geological structure. However, all the uncertainty to the ground motion is not due to aleatory factors. A good part of the uncertainty is epistemic, which should be evermore reducible in principal, as more data are acquired on earthquake instances at a given earthquake-monitoring site.

Epistemic uncertainty is due to incomplete knowledge about the physics governing the earthquake rupture and wave propagation process, and insufficient information about local geological conditions. These deficiencies should be rectified through further seismological monitoring, and geological and geotechnical field investigations.

Why is all of this important the Systems Engineer designing a new system? One of the reasons it is important has to do with the focus of energy and application of resources. Without a distinction, the recognition between aleatory and epistemic uncertainties, energy and resources might be applied to further efforts to increase knowledge of aleatory uncertainties that are inherent in the physical world. Such efforts may only result in the expenditure of resources and efforts and not improve or increase knowledge.

More specifically risk reduction efforts on reducing aleatory uncertainties from a measurement perspective using probability will not be productive. Now that we have examined and discussed the differences between the two types of uncertainties and their effects on risk reduction, let us look at the human inclinations regarding risk. Humans love certainty and will created even when it's an illusion. The following paragraphs discuss these tendencies and describe some of their effects.

As humans, we seem to have a need for certainty. We like it so much that authorities (Parents, organizations, politicians, insurance companies, for example) create illusory certainty to satisfy our need. Parents use certainty to prevent their children from trying to learn firsthand about possible dangers (An effective risk reduction strategy). Young children are prepared to believe in values, rules and stories without question, which facilitates their integration into their social group and culture. Social conventions, whether learned from one's family or wider culture, are like elementary perception, a source of the illusion of certainty.

Illusory certainty is part of our perceptual, emotional and cultural inheritance. It can provide us with images of our environment that are useful, although not always correct, as well as feelings of comfort and safety.

The illusion of certainty can make us believe that our job performance is solely due to our personal skill, a particularly dangerous illusion. This can be a threat to job survival, because it hides from our consciousness, (Our constant invisible companion regarding job performance) the effect of chance!

Let's reflect a little about the split between skill and chance by looking at various job titles and try to estimate, from what we know about each of the job functions, the weighting between skill and chance, and job performance and job survival. To do this we will create a matrix, as shown in Figure 94.

Job title	Skill	Chance-negative event	Comments
Infantry soldier-in combat situation	high	high	Chance event can negate skill-e.g. lucky shot by enemy
Dentist	High	Low	Sill is likely to be dominant
Systems Engineer- in lead role on high dollar value proposal effort	High	High	Less competent competitor writes winning proposal at better price
Options trader- several successful trades	High	High	Winning streak ends

Figure 94: Job skill vs. Risk comparison

Each of the job descriptions in the matrix has a skill component and a chance component. What criteria can we create to measure the degree to which skill and chance influence each job type? This is a difficult question to answer. Do we count the number of successes of job executions to total number of job instances to determine the spilt between skill and chance? I f so where do we set the threshold (A ratio of greater than 0.5, or maybe .75 is probably a reasonable metric)?

Looking at the matrix, it seems reasonable that the job that has the highest likelihood of having skill be the predominant component is the job of Dentist. However, experience of knowing individuals in all of the jobs in the matrix believe that sole reason for their successful job performance is skill. That is until they experience the effects of chance events, or "unforeseen bad luck".

In one of the author's experiences with an options trader, it was all skill, (from the options trader's viewpoint) until the string of chance events that the options trader was riding on ended. Bewilderment then set in, but unfortunately, the realization that success was due to both skill and chance was not perceived by the trader. The illusion of certainty in his personal skill level was not shaken.

Throughout history, humans have created belief systems that promise certainty. Certainty has become a consumer product. Insurance companies, investment advisors, election campaigns, and the medical industry market certainty all over the world, for example.

As Systems Engineers we need to be able to objectively assess the uncertainties in our proposals and projects. We need to be able to determine whether those uncertainties are risks to the successful implementations of our proposals and projects. As you can see from our lifelong exposure to illusory certainties, our need for certainty, and the continual marketing of certainty, this is not an easy task.

Risk definition

The word risk can be used as a noun or a verb. When used as a **noun** it has the following meaning: a situation involving exposure to danger. When used as a **verb**: to expose (someone or something valued) to danger, harm, or loss. Notice no mention of probability is identified in the definition of risk.

One of the key responsibilities of system engineer is to be able to identify and aggregate risks (technical risk, schedule risk, operational risk, and cost risk) for all phases of a systems program and to mitigate the risks (reduce, eliminate, or avoid). That includes all phases of system program, for example: concept formulation, requirements determination, system design, systems build, and systems tests (verification and validation). Typically, the system engineer is also responsible for determining the probability of occurrence of each risk and must be able to compute the impact to the program should the risk be realized.

An example method of risk identification

The authors have used a method that was identified by Albert Einstein, "Ein kleine gedanken' (Translation: a small thought experiment) to ascertain risk. So what do we mean by a small thought experiment? We actually mean 'to think through'. The task steps required to go from concept formulation, through test and evaluation in the creation of a system, defining the maturity of the technology involved, the nature of the program, the experience of the people involved in the program, the nature of the customer organization that is responsible for the program, and the experience that the performing organization as with the technology and with the customer program organization. For each of the major areas just described a list of possible events is created; both potential positive events and potential negative events. For example, let us consider a simple case of sawing a wooden board in half. Prior to sawing the board, the place to saw the board is marked with a pencil.

Let's now do a thought experiment regarding the marking of the board. There are three possible events outcomes regarding marking the board accurately (i.e. exactly in half). It can be marked too short, exactly in half and marked too long. Consequently; cut too short, exactly in half, or too long. Each of the previous activities has three possible outcomes. In each case only one of the possible outcomes is the desired positive outcome. So what does this mean regarding risk identification? It means that we have identified four risks to the program (marked too short, marked too long and cut too short, cut too long). Now the question is, how do we assess the probability of these occurrences?

How to Identify Risk Probability

At this point, we have identified four risks; marked too short, marked too long, cut too short, and cut too long. However there were three possibilities for marking and cutting. So with no prior knowledge regarding the skill of the persons doing the marking and the sawing, one applies Laplace's principle of indifference. If there are n possible outcomes, and no further knowledge, it must be assumed that the probability of any of the n outcomes is $1/n$.

However, the probability of failure (marking the board and cutting the board exactly) is 2/3 for each activity. Now let's assume that John, an experienced board marker, has a track record of mismarking one out of 10,000 markings. Let us assume that Sam, an experienced sawyer also has one out of 10,000 inaccurate saw cuts. We now have additional information that allows us to compute probabilities of failure that are far different from those that we computed before we knew the skill levels and success rates of the marker and sawyer. Let us further assume that the cost of market board is $10 and the cost of a sawn board is $100. With that information and the probabilities that we have calculated we can now create a risk identification and mitigation matrix. This shown in Figure 95.

ID number	Risk Description before mitigation	Probability of occurrence Before mitigation	Cost	Expected value (cost) Before mitigation	Mitigation description	Probability of occurrence after mitigation	Cost	Expected value after mitigation
1	The risk is the board will be marked too shoirt	0.66	$ 10.00	$ 6.60	Use John to mark the board.	0.0001	$ 10.00	$ 0.001
2	The risk is the board will be marked too long	0.66	$ 10.00	$ 6.60	Use John to mark the board.	0.0001	$ 10.00	$ 0.001
3	The risk is the board will be cut too short	0.66	$ 100.00	$ 66.00	Use Sam to cut the board.	0.0001	$ 100.00	$ 0.010
4	The risk is the board will be cut too long	0.66	$ 100.00	$ 66.00	Use Sam to cut the board.	0.0001	$ 100.00	$ 0.010
			Total	$ 145.20			Total	$ 0.022
		Risk reduction	**$ 145.18**					

Figure 95: Risk Identification and Mitigation matrix

We have just described a methodology for risk identification and for risk probability of occurrence. We use the concept of a little thought experiment, thinking through the program, to identify risks and the simple example that we created. When we didn't have any knowledge other than the total number of ways that negative outcomes could be produced, we used Laplace's principle of indifference to identify the probability that the specific risk would be realized. After we discovered that John and Sam would be assigned to, respectively, marking and sawing, we used the probabilities that they had empirically demonstrated in their history of marking and sawing.

Until we go through the process just described, all we can say about the success of the program is that it's uncertain. We are not sure if it will perform the functions it is designed to perform, that it will be delivered on time, and that program will meet its cost goals. We used Laplace's principle of indifference, a frequency interpretation of probability. There are other interpretations that can be used to determine the probability of occurrence and they are described in the following section. The one that is chosen for use will depend on the level of knowledge we have.

Probability interpretations

There are three (3) major interpretations: propensity, degree of belief, and frequency.

Propensities

Propensities are the properties of an object, such as the physical symmetry of a die. If a die is constructed to be perfectly symmetrical, the probability of rolling a 6 is 1/6. The reference to a physical design mechanism or trait that determines the risk of an event is the essence of the propensity interpretation of probability. The propensity interpretation differs from the subjective interpretation because what matters is the die's design. It is not sufficient that someone's subjective probability satisfies the laws of probability.

Degree of belief

In situations that are new, where there is no prior data, such as when a procedure or process has never been done before, what is often times requested is the probability of success. In those cases one makes a conjecture, usually intuitive of some numerical value (say 80%) probability that the process or procedure or new design will work. This may be as good as it gets as no one else may have any other information that is any better. What's important to realize here is that if there is an 80% probability of success, there must also be a 20% probability of failure. That is in keeping with the law probability that says the sum of the probability of success and the probability of failure must equal one.

An example of such a situation is: Christiaan Barnard's account of his first encounter with Louis Warshkansky, who was then the soon to become first man to have a heart transplant. Washkansky was in bed reading a book. Barnard introduced himself and explained the operation. Dr. Barnard told Louis there was a chance to get back to normal life again. Louis wife asked, "What chance do you give him?" the doctor replied "80%". Eighteen days after the operation, Washkansky died. Barnard's 80% reflected a degree of belief or subjective probability. In the subjective view, uncertainties can be transformed into risks as long as they satisfy the laws of probability – the exhaustive an exclusive set of alternatives such as survival and death add up to one.

So Barnard's 80% of survival, although unstated, meant that there was a 20% chance the patient would not survive.

Frequencies

For a frequentist, a probability must be based on a large number of observations and is defined as the relative frequency of an event in a specified reference class. Frequentist's are not interested in what someone believes about an outcome about rolling a die, nor would they study the design of the die to determine the probability of rolling a 6.

They would determine the probability empirically by rolling the die many times and computing the relative frequency with which the outcome was a 6. Frequentist's are the most cautious about moving from uncertainties to risks. For them, risks refer only to situations for which a large body of data exists (Life insurance companies life expectancy tables are an example).

For the system engineer, risk management is a crucial element of any system development. To understand the methodology of accomplishing risk management, we must of course consider the aforementioned characteristics of risk and uncertainty, and utilize this understanding to create a rational risk register that will describe for us to vulnerabilities that exist on our system development.

There are several actors of interest for us to understand:

Issues: Anything that requires the project manager's attention
Problems: anything or an event that has gone wrong and requires action to correct
Hazard: A situation or condition that is likely to cause you harm or loss underserved and circumstances
Crisis: an unplanned event or possible situation which triggers a real or perceived threat to safety, health or to the environment or to an organizational repetition
Risk: A potential problem or event that may occur and if it occurs may cause harm or opportunity

Risk is not a certainty, but it does imply if a course of action or inaction is taken under conditions of uncertainty, the risk takers may be exposed to possible loss or gain regarding their desired outcome.

Risk tolerance is the amount of risk that is acceptable to the individual, organization or client. It will certainly be different for all of the individuals working in that organization. This must be considered when assessing risk on any system development, as when that information is communicated to different parties within the organization it must be tempered with regard to their particular risk tolerance.

Why do people take risks? At times it is related to the potential for profit or rewards, excitements, benefits, and potentially stupidity. We cannot completely aluminate risk from any system development, although there are methods by which we can mitigate those risks which are present.

To define risk: it is a future event or series of events with a probability of occurrence and the potential for impact to the objectives of the program. This implies that there are two components to assessing risk, the probability and the impact to the program.

The risk management methodology contains a five-step program. These steps include:

Risk identification
Qualitative risk analysis
Quantitative risk analysis
Risk response planning
Risk monitoring and control

Risk identification is one of the most challenging aspects of a risk program. This requires that the system engineer consider all of the complexities of the program and ask themselves several questions:

What are the things that could potentially go wrong on this program?
What are the possible causes for those things going wrong?
If things do go wrong, what could the potential consequences be?
What are the things that could potentially impact the program positively?

These risks are typically described in a narrative form, and must define the specific concern in a manner that will reduce the nebulous nature of the statement in order to provide specificity to the developer. These risks are then logged in a table, which will be utilized throughout the life of the system development.

Qualitative risk assessment is frequently used to ascertain those areas of greatest concern for the system development team. This type of risk analysis will require the assignment of high, medium, or low, to the probability of the risk occurring as well as a high, medium, or low, to the impact on the development project. This methodology is often used for executive summaries of the program, but lacks precision and predictive capabilities.

Quantitative risk assessments require a more stringent analysis on the part of the system developer. This method will typically require the system developer to ascertain the probability of occurrence by defining a percentage of likelihood with a precision defined by the knowledge of the risk. Secondly, the consequences of that risk occurring must be considered and an estimate is created regarding the financial impact on the project objectives. These two factors are then multiplied to define the potential risk exposure on the project.

Risk response planning includes the potential mitigations that will be implemented regarding the risks identified and quantified. There are several methods commonly used for risk response, which include:

Risk avoidance
Risk transference
Risk mitigation
Risk deferral
Risk acceptance

In general, the risks identified will be identified by number and mapped onto what is commonly referred to as a heat map in the risk management process. The qualitative assessment is shown in Figure 96.

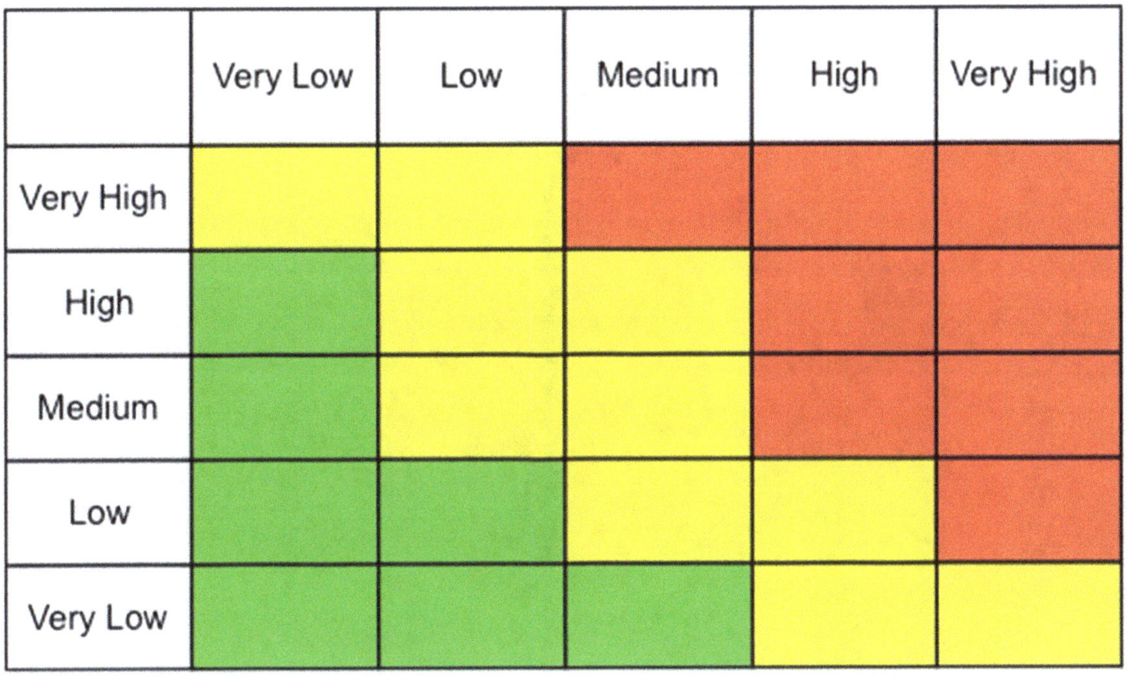

Figure 96: Qualitative Risk matrix

It is of course potentially possible that some of the risks will represent a negative impact on the program and others will present a positive impact. In some organizations, these are identified as threats and opportunities, and are shown on adjacent heat maps as shown in Figure 97.

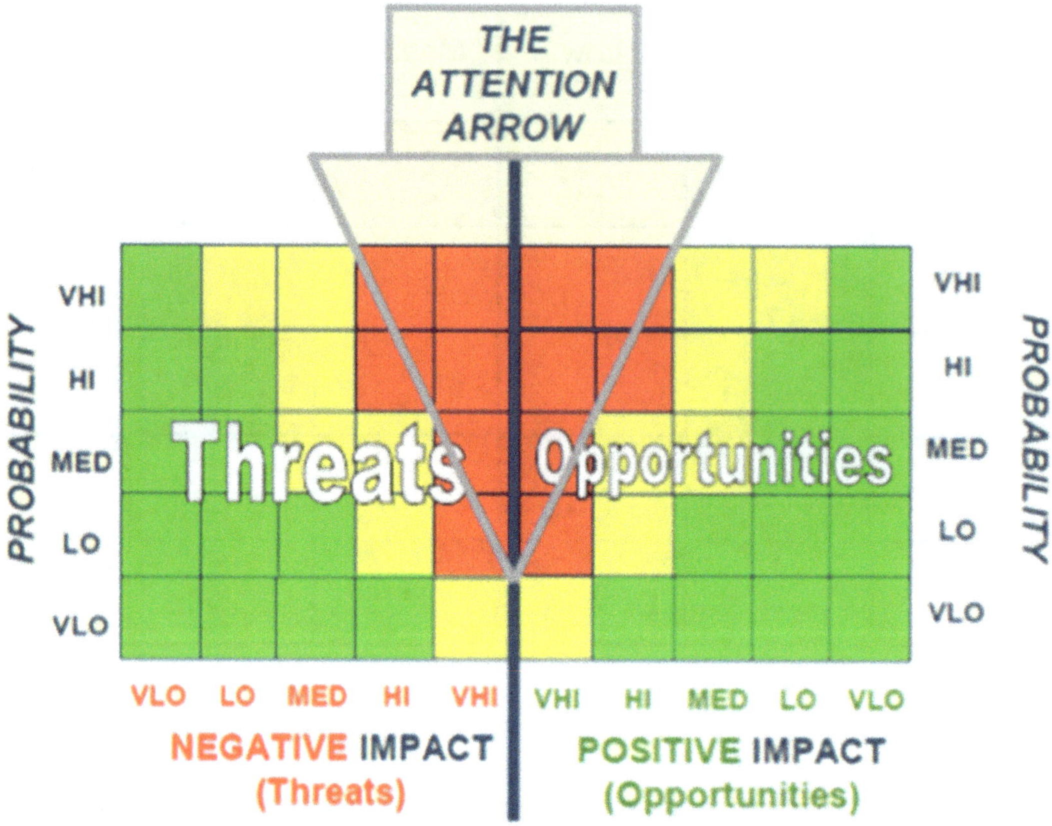

Figure 97: Threats and Opportunities matrix

Quantitative risk management is typically tracked via spreadsheet, where each risk is identified, assigned to a specific point of contact, the probability and impact are identified which enables the calculation of an estimated monetary value. This estimated monetary value is then summed for the program, which enables an assessment of the financial exposure existing on the program. An example matrix is shown in Figure 98.

	Description	POC	Probability	Impact	EstMonValue	Actions
1	Heating Method doesn't work	T. Jones	0.5	$1,500.00	$750.00	
2	Heat at outer case causes safety issues for people	R. Williams	0.75	$2,000.00	$1,500.00	Add insulation around canister
3	Risk of cord getting into toaster and burning off insulation	R. Williams	0.2	$3,500.00	$700.00	Reduce length of cord to 15 inches
4	May miss schedule due to delivery of heating core delay	M. Sharpe	0.85	$12,000.00	$10,200.00	
5	Customer may not like surface finish	E. Smith	0.05	$100,000.00	$5,000.00	
6	UL Listing test	R. Miller	0.35	$15,000.00	$5,250.00	Use UL Consultant in case of failure
7	Power supply noise below required threshold	F. Meeker	0.8	-$12,000.00	-$9,600.00	Remove filter from design if it works
	TOTAL RISK				$13,800.00	

Figure 98: Risk Matrix / Register

The quantitative risk management requires a greater degree of analysis and insight into the risks identified for the program. Due to this, once the risks have been identified they can be effectively normalized for use in the risk heat map. This is useful in order to establish the relative importance of those risks for management.

To accomplish this, one must examine the lowest and highest probabilities identified, and evenly spread those probabilities across the Matrix fields. Secondly, the impacts are assessed for maximum and minimum and these are used as guides to create the impact buckets that will be used for the risk assessment. An example of this is shown in the following quantitative heat map.

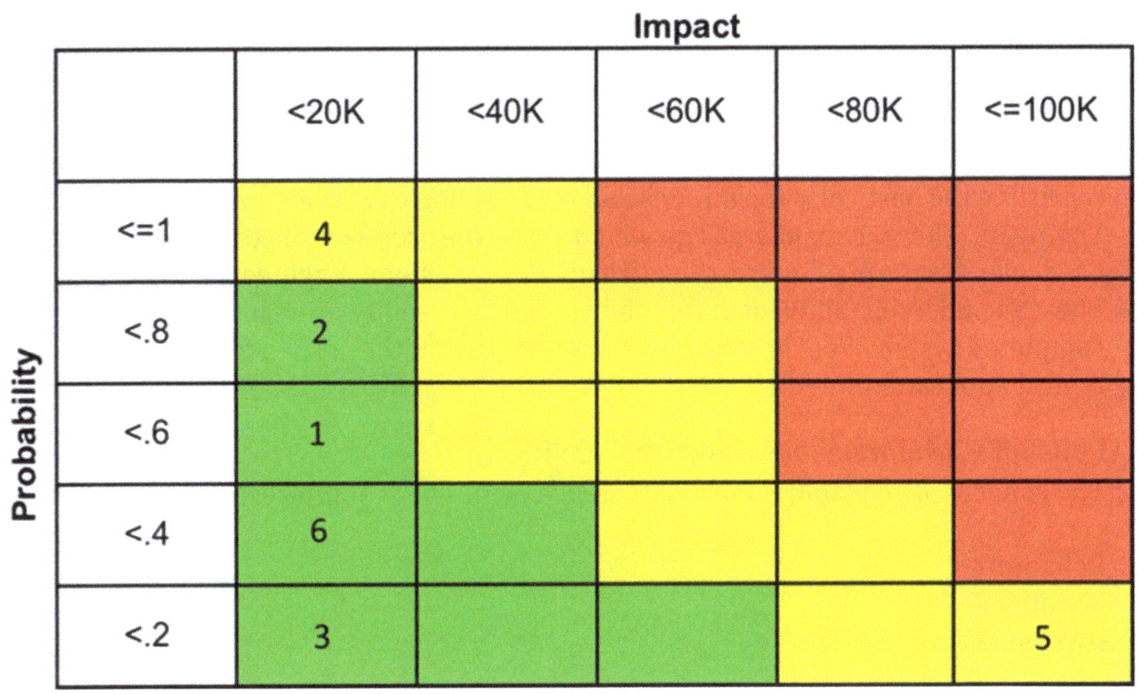

Figure 99: Risk Quantitative Heat Map

Figure 99 identifies the mapping of the risks identified in the previous risk matrix. In this particular case, attention would be paid to risk number five, as this provides the greatest threat to the success of the system development. In general, those risks identified in the yellow or red region are typically targeted for mitigation strategies in order to reduce the exposure of the system development.

Risk management is the process of monitoring and adjusting these risks on a systematic basis during the program development cycle. In most cases this translates to a review of the risk register during regular project meetings, where each risk is assessed to define if any mitigation strategies have been implemented or if occurrences on the project have changed either the probability or impact of any of the risks. The risk register and the corresponding heat map are then updated and communicated during the regular business rhythm within the organization.

Chapter 9: Specialty Engineering

Specialty engineering refers to those highly specialized domains of knowledge that are necessary for the successful completion of a complex project. These include at least the following domain areas:

Reliability, Maintainability, Availability (RMA)
Human factors
System usability
System safety
System security
Electromagnetic effects
Facilities
Logistics

Each of these domain areas requires the development of a great deal of expertise regarding the particular techniques and methodologies adopted by Systems Engineers. These domain areas go well beyond the scope of this text, however to particular areas are of interest to virtually every Systems Engineering development. These include; reliability, maintainability, and availability as well as usability of the completed system. For these reasons, we will discuss these specific areas and their typical application.

Reliability, Maintainability and Availability

The concept of availability includes three specific areas of interest. These areas are:

Reliability
Maintainability
Availability

Reliability is the probability that a device or a system will operate without failure for a given. And under given operating conditions. It is described by the equation:

$$R(t) = e^{-\lambda t}$$

where λ is failure rate

Maintainability is the probability that a device or a system that has failed will be restored to operational effectiveness within a specified time. It is defined by the following equation:

$$M(t) = 1 - e^{-\mu t}$$

where μ is repair (restoration) rate

Availability is the proportion of the system uptime to the total time, which includes uptime and downtime, over a long period of time. These relationships can be seen most readily in the following diagrams.

System Operational States

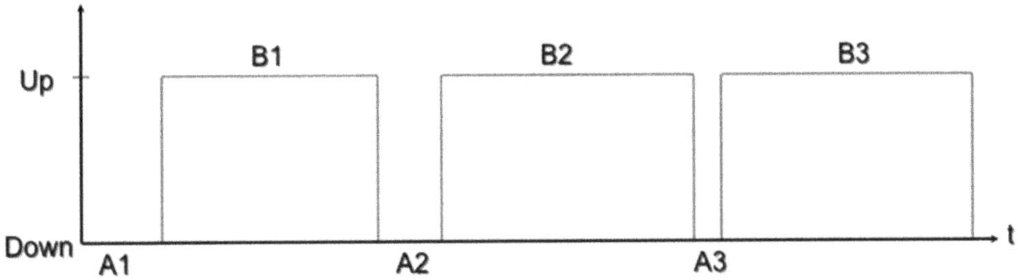

Up: System up and running
Down: System under repair

Figure 100: System Operational States

Mean Time To Fail (MTTF)

MTTF is defined as the mean time of the occurrence of the first failure after entering service.

$$MTTF = \frac{B1 + B2 + B3}{3}$$

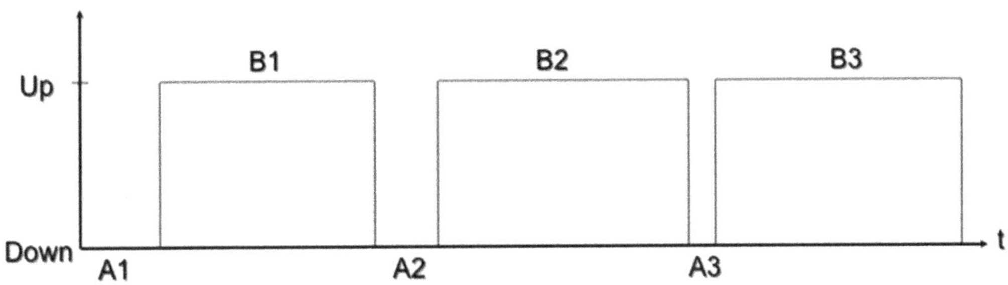

Figure 101: Mean Time To Fail

Mean Time Before Failure (MTBF)

MTBF is defined as the mean time between successive failures.

$$MTBF = \frac{(A1 + B1) + (A2 + B2) + (A3 + B3)}{3}$$

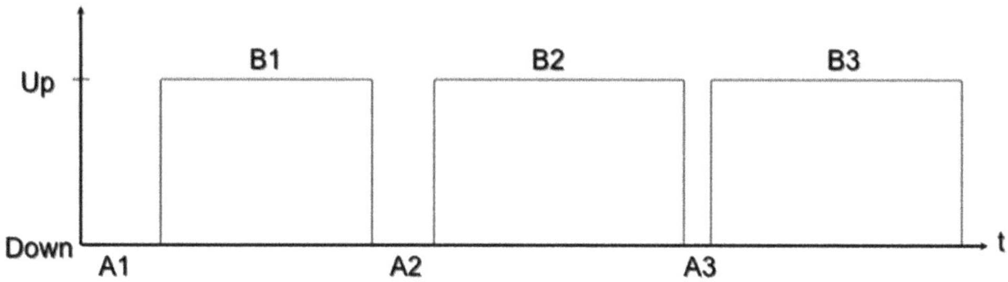

Figure 102: Mean Time Between Failures

Mean Time To Repair (MTTR)

MTTR is defined as the mean time of restoring a device or system to operation condition.

$$MTTR = \frac{A1 + A2 + A3}{3}$$

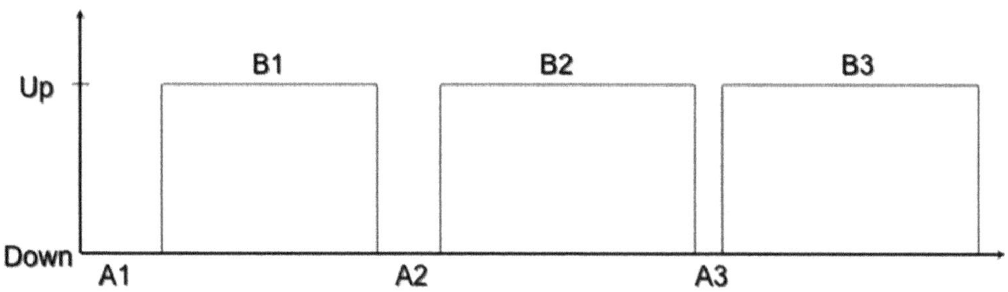

Figure 103: Mean Time To Repair

Reliability/Maintainability Measures

Reliability R(t)

$$(\text{Failure Rate})\ \lambda = 1 / \text{MTBF}$$
$$R(t) = e^{-\lambda t}$$

Maintainability M(t)

$$(\text{Maintenance Rate})\ \mu = 1 / \text{MTTR}$$
$$M(t) = 1 - e^{-\mu t}$$

Figure 104: Reliability and Maintainability measures

Failure rate is a function of time, and is considered a constant during the useful life of the device or system. If a piece of equipment has a failure rate lambda, it's probability of survival from time zero to t, or reliability R(t) is given by:

$$R(t) = \exp\left[-\int_0^t \lambda(u)du\right]$$

If the failure rate lambda is constant, the expression reduces to:

$$R(t) = e^{-\lambda t}$$

To illustrate, we will consider a simple example:

A system has 4000 components with a failure rate of 0.02% per 1000 hours. Calculate λ and MTBF.

λ = (0.02 / 100) * (1 / 1000) * 4000 = 8 * 10-4 failures/hour

MTBF = 1 / (8 * 10-4) = 1250 hours

A second example is provided for review:

A first generation computer contains 10000 components each with λ = 0.5%/(1000 hours). What is the period of 99% reliability?

MTBF = t / (1 – R(t)) = t / (1 – 0.99)
t = MTBF * 0.01 = 0.01 / λav

Where λav is the average failure rate
N = No. of components = 10000
λ = failure rate of a component
= 0.5% / (1000 hours) = 0.005/1000 = 5 * 10-6 per hour

Therefore, λav = N λ = 10000 * 5 * 10-6 = 5 * 10-2 per hour

Therefore, t = 0.01 / (5 * 10-2) = 12 minutes

Calculating Reliability

The reliability of a series system is expressed as the product of the reliabilities of the individual contributors to the series:

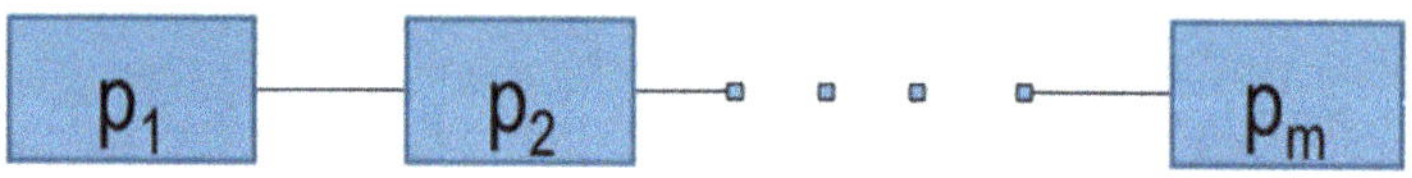

$$\text{Reliability} = \prod_{i=1}^{m} p_i$$

$$R_e = R_1 * R_2 * R_3 \ldots R_n$$

Figure 105: Series elements reliability

The reliability of a parallel system is equal to one minus the product of the unreliability's of the individual contributors to the system:

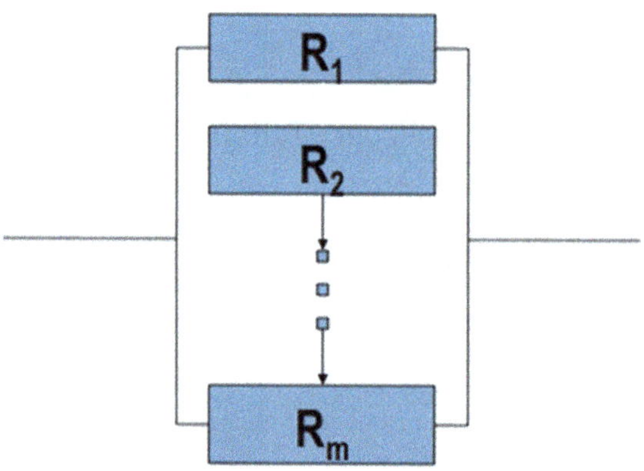

$$\text{Reliability} = 1 - \prod_{i=1}^{m}(1-p_i)$$

$$R_e = 1- (1 - R_1)(1 - R_2)(1 - R_3) \ldots (1 - R_n)$$

Figure 106: Parallel elements reliability

Therefore one can calculate the reliability of any system by simplifying the diagram from parallel and series relationship until all that remains are series components. This is shown in .

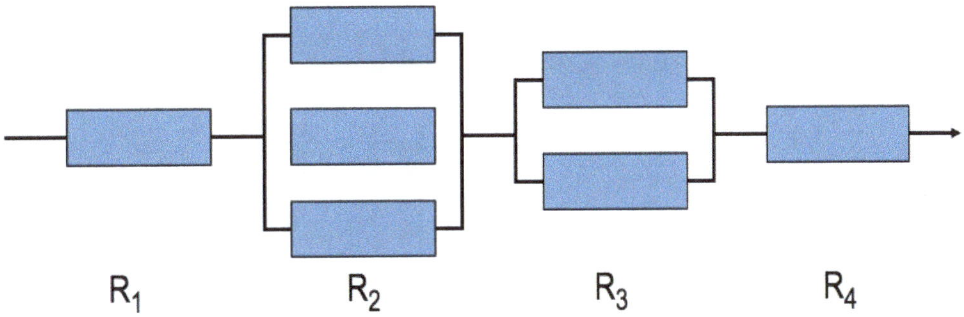

Figure 107: Complex system reliability

Any system can be expressed as a combination of series and parallel segments (subsystems). Consequently, by finding the reliability of the distinct Segments of the system, the overall reliability of the system can then be found as the product of the segment reliabilities, that is, for the system above with segment reliabilities of R1, R2, R3 and R4, the overall system reliability is:

System Reliability = (R1)(R2)(R3)(R4)

As an exercise, we will examine a complex system to calculate the reliability:

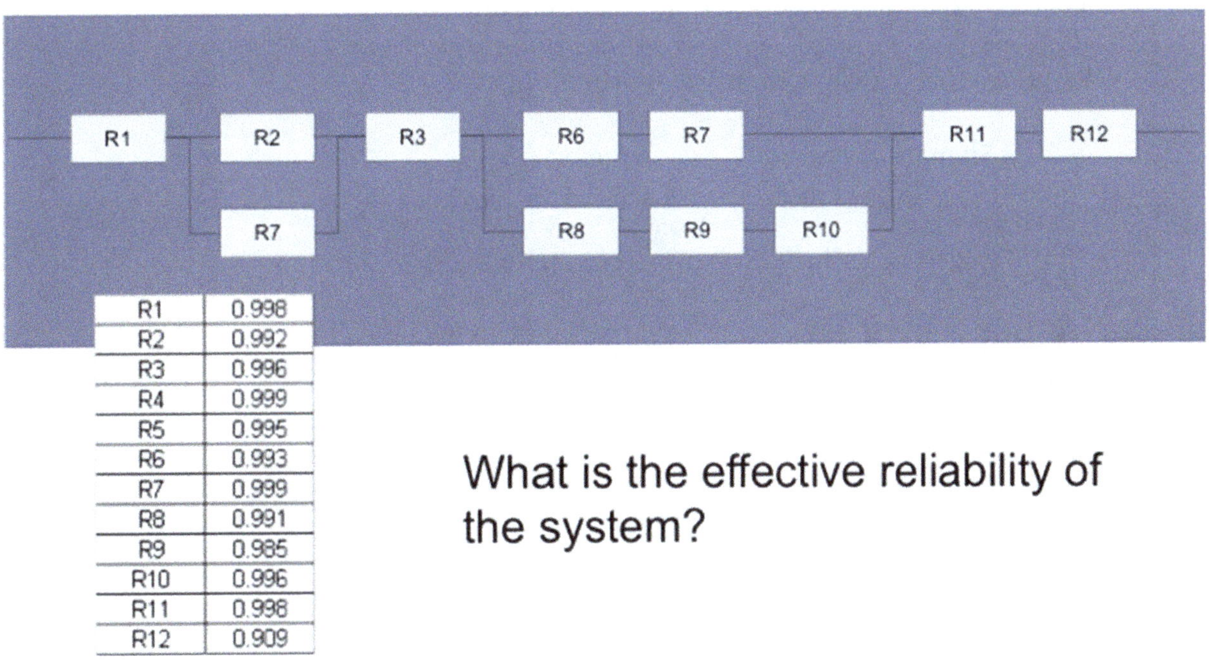

R1	0.998
R2	0.992
R3	0.996
R4	0.999
R5	0.995
R6	0.993
R7	0.999
R8	0.991
R9	0.985
R10	0.996
R11	0.998
R12	0.909

Figure 108: Reliability example

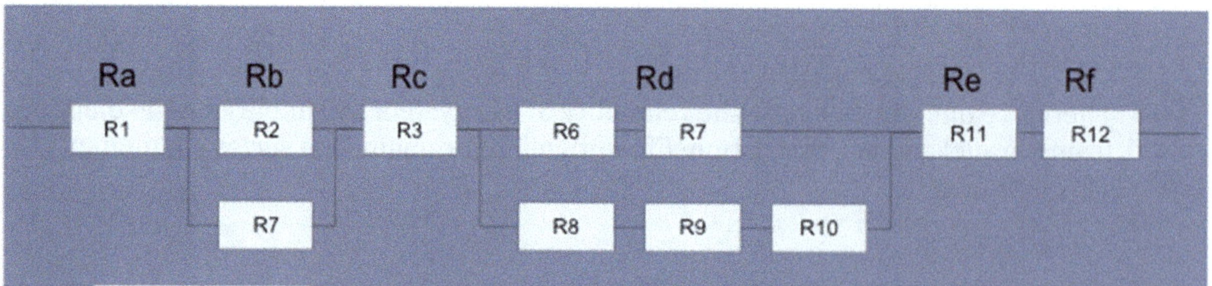

R1	0.998
R2	0.992
R3	0.996
R4	0.999
R5	0.995
R6	0.993
R7	0.999
R8	0.991
R9	0.985
R10	0.996
R11	0.998
R12	0.909

Ra	0.998
Rb	0.999992
Rc	0.996
Rd	0.999778
Re	0.998
Rf	0.909
Reff	0.901539

Figure 109: Reliability solution

To gain some insight into the level of reliability that might be desired in a given application, we can examine the effective sigma levels and defects from statistics.

2 Sigma = 308,537 DPMO
3 Sigma = 66,807 DPMO
4 Sigma = 6,210 DPMO
5 Sigma = 233 DPMO
6 Sigma = 3.4 DPMO (99.999966% good)
σ process capability defects per million opportunities

What does this translate to in real world numbers?

3 Sigma

- 20,000 lost pieces of mail per hour
- 5,000 surgery mistakes per week
- 2 long or short airplane landings at a major airport per day
- 200,000 wrong drug prescriptions per year
- 15 minutes of unsafe water per day

6 Sigma

- 7 lost pieces of mail per hour
- 1.7 surgery mistakes per week
- 1 long or short airplane landings at a major airport every 5 years
- 68 wrong drug prescriptions per year
- <1 second of unsafe water per day

- We currently observe 1 failure in 2,000,000 in commercial airliner flight – Much better than 6 sigma

Figure 110: Effective failures in various applications

Therefore, a seven-step methodology is presented to enable the effective calculation of system reliabilities:

1) Define the product and its functional operation. Use functional block diagrams to describe the systems. Define failure and success in unambiguous terms.

2) Use reliability block diagrams to describe the relationships of the various system elements (e.g. series, parallel, etc.)

3) Develop a reliability model of the system.

4) Collect part and subsystem reliability data. Some of the information may be available from existing data sources. Special tests may be required to acquire other information.

5) Adjust data to fit the special operating conditions of your system. Use care to assure that your adjustments have a scientific basis and are not merely reflections of personal opinions.

6) Predict reliability using a mathematical model.

7) Verify your prediction with field data. Modify your models and predictions accordingly.

In general, reliability calculations are carried out by Systems Engineers who specialize in their use. These engineers typically utilize publicly available tables of reliability data for common components, as well as formulating test methodologies and accelerated test schemes that will allow them to make reasonable predictions of reliability of components.

Percolation Modeling and System Reliability

New technologies are constantly emerging in the field of electronics. One of those breakthroughs was in the development of a transparent conductive film to be used in touchscreen applications. This particular transparent conductive film was comprised of a substrate and an irregular/randomized network of silver nanowires on the order of 35 nm in diameter and 60um in length.

These nanowires were applied to the film during a roll-to-roll processing step, and although their distribution was not completely random, they were effectively in an unbiased randomized position on the surface of the film. This implies that to maintain conductivity from point A to point B within the film substrate, that the random position of these wires must form a conductive network. Due to the ergodic nature of the distribution, the potential always exists for an interruption to the conductivity in the nanowire network.

For this reason, it was of interest to the material producing organization to understand the potential for failure within their circuit designs. This required the ability to model the conductive network effectively, and to utilize the output from many simulations of that model in order to make a prediction of network reliability.

In a typical square meter of transparent conductive film, there are located more than 10 billion wires. This means that every square meter of film has more than 1500 kilometers of wire located within it. This would imply that there are more than 15 billion the connections between wires per square meter of film. While that number is remarkably large, the fact that very small traces are utilized in electronics manufacture suggests that the potential for open circuits always exists. An example of this is shown in Figure 111.

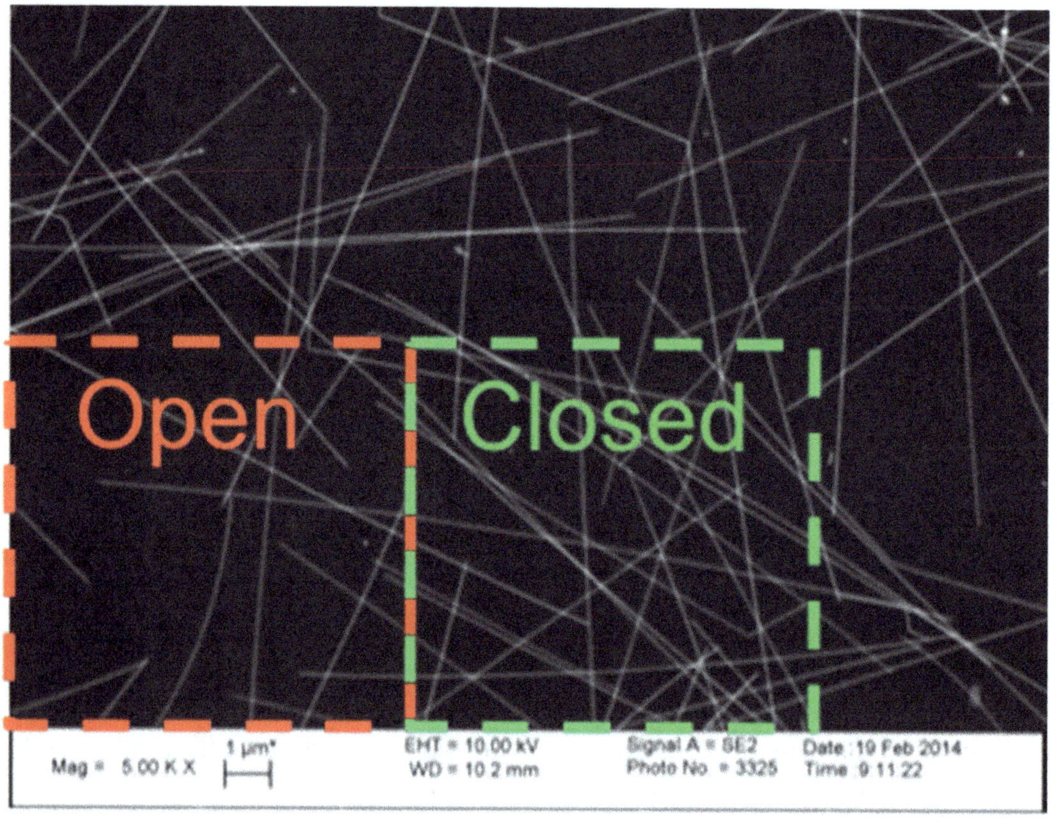

Figure 111: AgNW network example

In general, the transparent conductive film is created as a uniform conductive surface, and circuits are created within the surface of the film by the application of laser etching. Once the laser etching has been completed, very long narrow traces are typically located within the film, creating a worst-case situation for conductivity failures.

Empirical methods of assessing the circuit reliability were developed in order to provide a measure of predictability for the customers of this conductive material.

This empirical model was based upon the assumption that a well understood unit cell percolation reliability could be utilized with a standard series reliability calculation in order to predict the effective trace length that could be used with a particular film and circuit geometry. This is shown in Figure 112.

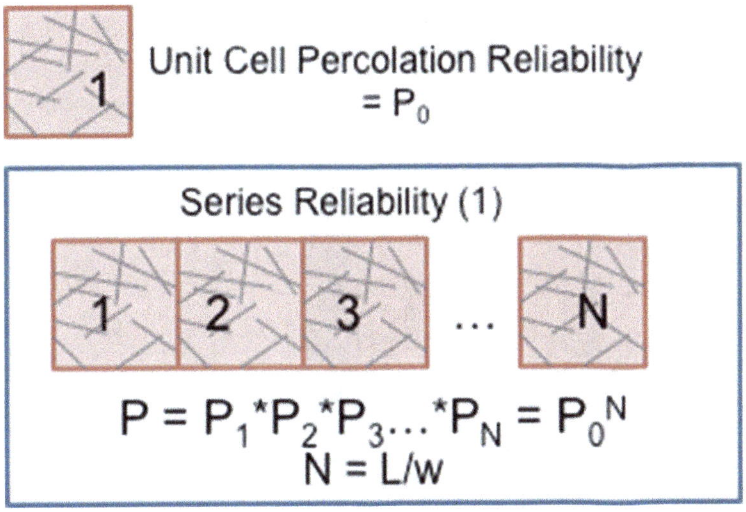

Figure 112: Circuit reliability approach

Traces in series have probability of success related to length:

P2 = P1 (L2/L1)
L2 = L1*Log(P2)/Log(P1)

Example: P1 = 50 at L1 = 0.5 m length, probability of success, P2, at L2 = 2.0 m, is:

P2 = 0.50^(2/0.5) = 0.5^4 = 0.0625 or 6.25%

Example: What length can this trace width yield 90% success (assume 2200 mm for L1)?

L2 = 2200 mm * Log(0.9)/Log(0.5) = 334 mm

This predictive reliability for the circuit was then tested by empirical method, with the creation of a serpentine test pattern that provided a long circuit of appropriate geometry for comparison to predicted values. This geometry is shown in Figure 113.

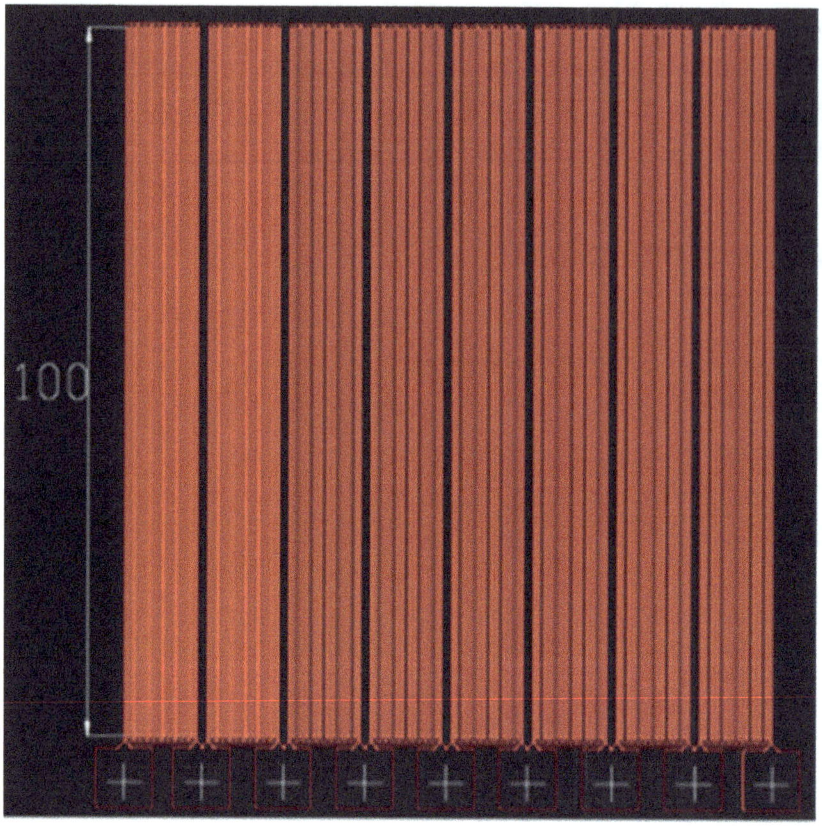

Figure 113: Serpentine test pattern

The testing of the system was conducted over a range of samples in order to produce recommended design rules for the customers based upon empirical results. The data was compiled into a tabular form in order to clarify for the customer how to properly utilize the film while compensating for natural limitations in reliability. These design rules are shown in .

Touch Panel Application	Trace Width (µ)	Trace Length (mm)
Discrete Button	500	100-200
Backgammon	600-2000	1000-3000
Bars & Stripes	400	4000-6000
Single Layer Multi Touch	50-75	3000-6000

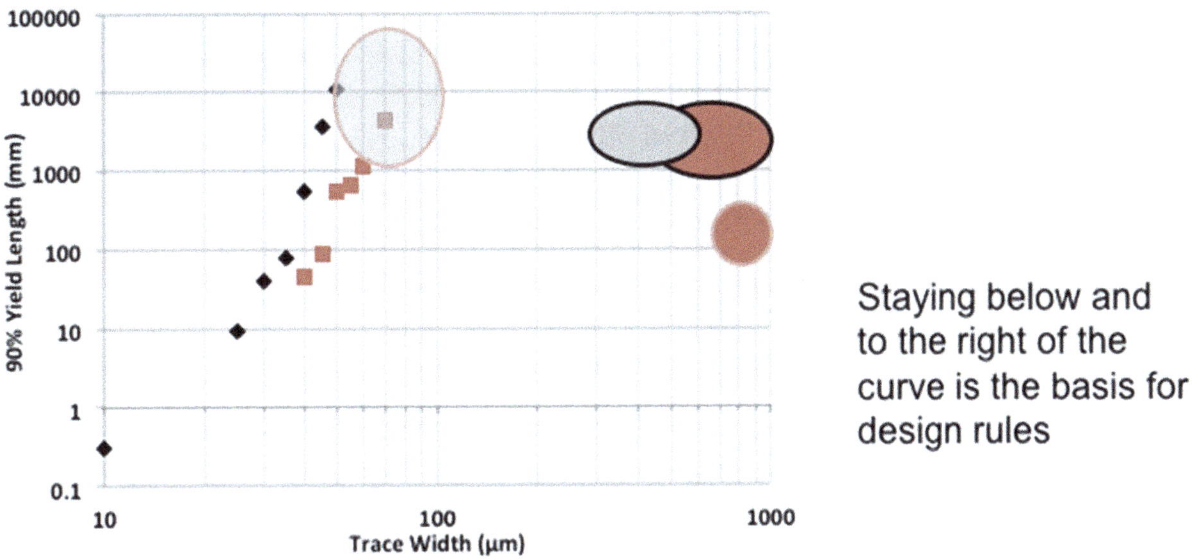

Figure 114: Design rules for AgNW 100 product

Unfortunately, these design rules lacked precision and they are based upon a very limited set of physical tests. The customer, however, relies upon this predictive data in order to generate many millions of dollars worth of product. It is not acceptable to the customer to have estimates of reliability differ from reality by more than a few percentage points. Bus, it became apparent that the system needed to be modeled in order to apply it across many millions of samples.

The process of percolation modeling requires the development of several methodologies to enable prediction for the system. These include:

Randomly dropping wires into a fixed region
Analyzing conductivity across the square
Establishing reliability of the square
Computing the system reliability

Wire distributions can be generated by a line-based approach, where a center point is defined for a wire and a slope is assigned and then the wire is extended symmetrically from this point. A grid pattern method may also be used, defining a first endpoint and then creating a second end point at a generated distance from the first. Last, a pixel-based approach may be utilized. In this case, and endpoint is defined, Then a wire is generated at a random angle and random length from that starting point. Each of these methods has processing implications. The first of these is shown in Figure 115.

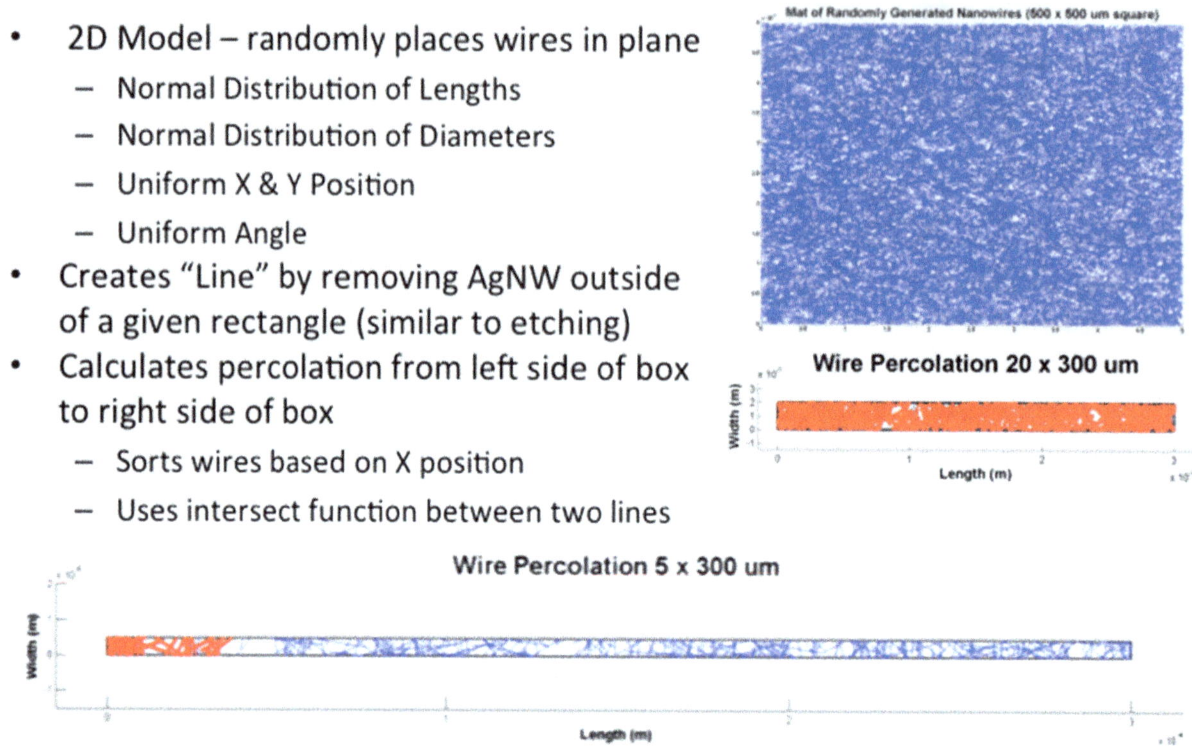

Figure 115: Line-based approach to modeling

The second methodology is a essentially a lattice method for creating a conductive network. This method can rapidly generate a manageable network with predictive interconnectivity, but unfortunately it does not appear to represent reality to a sufficient extent. This methodology is shown in Figure 116.

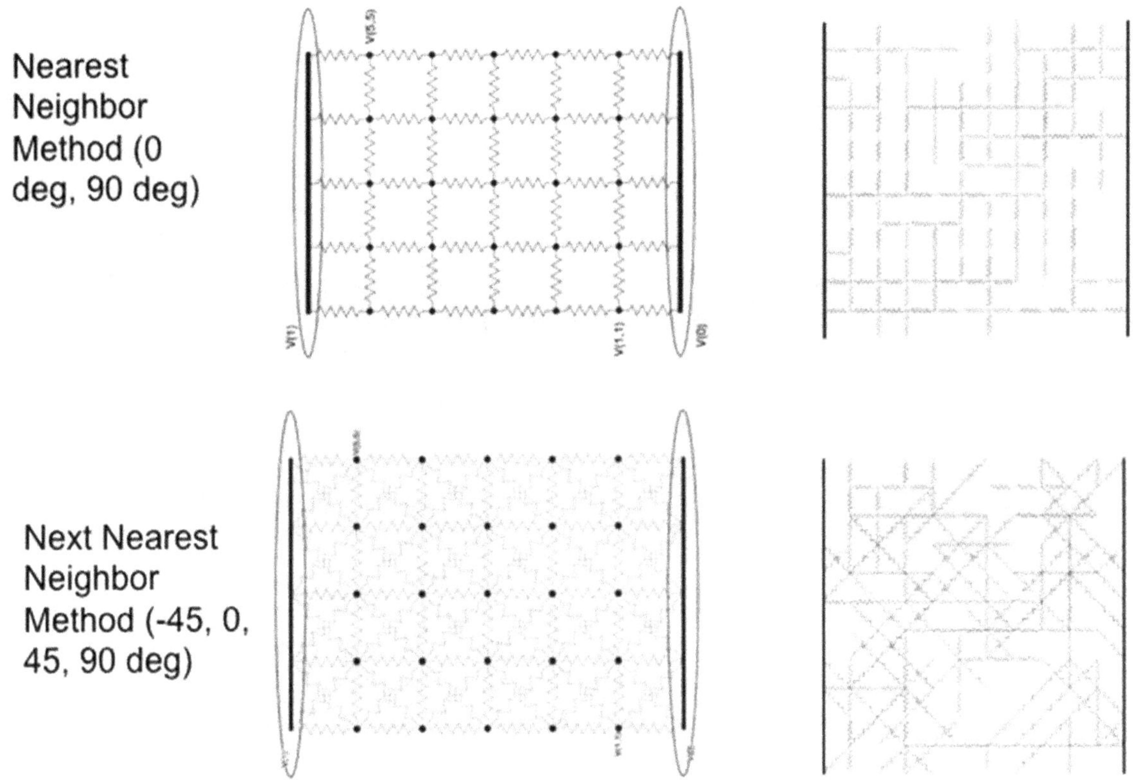

Figure 116: Lattice method of modeling

The pixel-based approach to modeling most closely represents the generation of a random network of wires as occurs during the physical process of creating transparent conductive film. Several examples of differing distributions utilizing this method are shown in Figure 117.

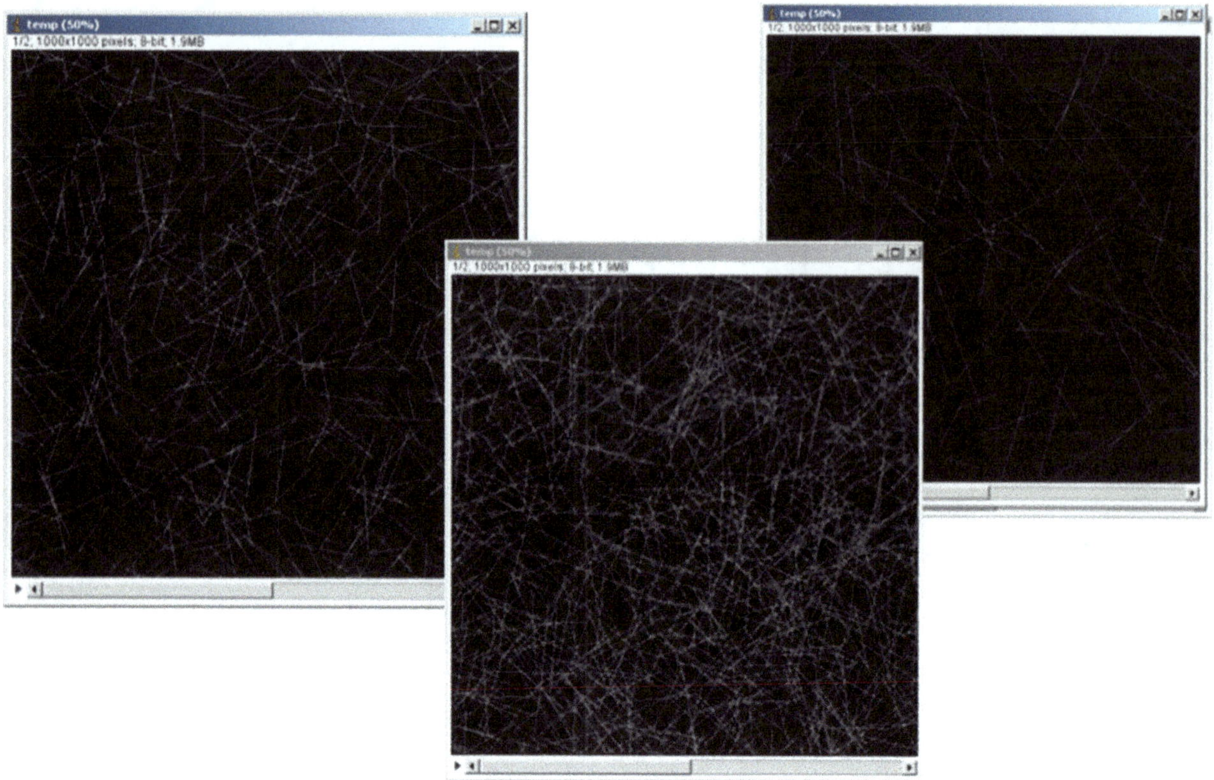

Figure 117: Pixel based method of modeling

In order to assure that this model was representative of reality, we made several defining assumptions. First, that wire diameters and lengths are typically represented by a normal distribution within the sample. We also ensured that there were no practical edge effects built into the system by conducting our experiment on a largely oversampled data set (Figure 118). Last, we validated the dispersion by comparing the generated model to actual microscope images to ensure similar appearance and wire count.

- We purposely create a border of a single wire length to assure the distribution is regular within the analysis box

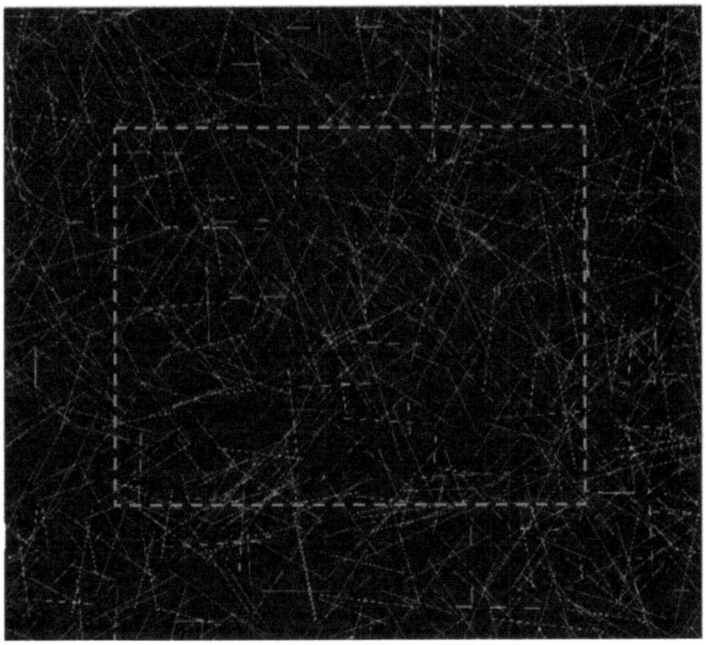

Figure 118: Oversampled image

In the generation of this model, we've incorporated a large number of variables that would enable the model to be highly flexible and allow the rapid testing of proposed configurations of randomized nanowire networks. Some of those variables included wire diameter and variation, wire length and variation, and material laydown percentage.

The face validity of the dispersion was compared to actual microscopic images of the material, and the percent solids lay down from actual film was utilized to calculate the model lay down, which ultimately resulted in a good correlation between the model and empirical samples.

Several different methods of establishing conductivity across the square were identified and developed. Ultimately, the most effective method from a processing standpoint included the following steps:

Skeletonizing the image
Common agglomeration analysis
  Defines all agglomerates that extend from one edge to another in the square
  Much less computationally demanding
  Loses multiple pathways within the same agglomerate
Record the number of independent pathways across the image in both directions

When a model of this type was originally created, individual squares of model data were created and conductivity was checked across those squares. After approximately one month of processing the model, an open circuit was discovered in a typical wire lay down geometry. This corresponded to approximately 1.5 failures in 1 million sample squares. To replicate these results and establish a statistical basis would require at least a year of processing time to ensure reliable conclusions. This was considered a brute force approach, and a much more elegant approach was desired.

After some consideration, we established a methodology that would enable us to predict the reliability of a finished circuit by utilizing the predicted reliability of individual squares in two dimensions. This methodology allowed us to analyze 1 million 50 μ squares in a period of 136 hours. Increasing the square size to 100 μ resulted in an increase of processing time to 330 hours, a factor of 2.4. This was far more attractive than the previous models proposed.

To accomplish this reliability prediction, we followed a simple process:

Define the number of paths through the fixed region
Define the expected reliability of the square
Repeat for the orthogonal directions of the square
Use the expected reliability of many samples, (approximately 1,000,000) to define the dispersion of the results, which allows in amalgamated prediction of reliability.

For example:

On X 100 μ squares, we average 14.45 horizontal connections across the square with a standard deviation of 3.09. This gives us a Z score of 4.67. Therefore, the horizontal reliability for the square is 0.9999990. (The Z-score implies the reliability value) An identical calculation is carried out in the opposing direction for the square. This is critical as most transparent conductive film shows differing conductivities in different directions within the film substrate.

We can then create a geometric construction of the desired electronic circuit made up of unitized squares equivalent to those modeled. An example of this circuit is shown in Figure 119.

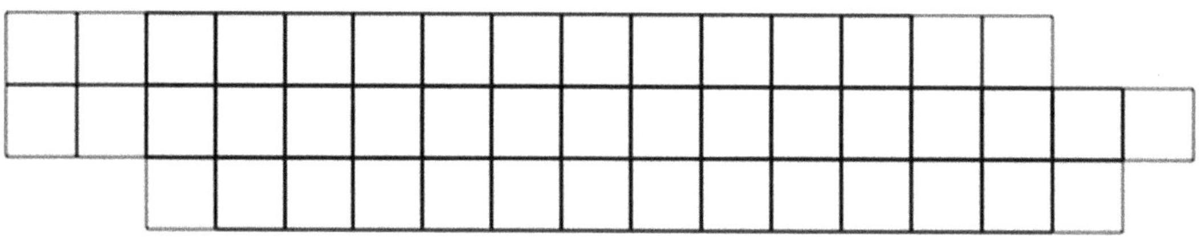

Figure 119: Example circuit built up of modeled squares

Utilizing parallel and series reliability calculations enables the calculation of a system reliability. This is shown in Figure 120.

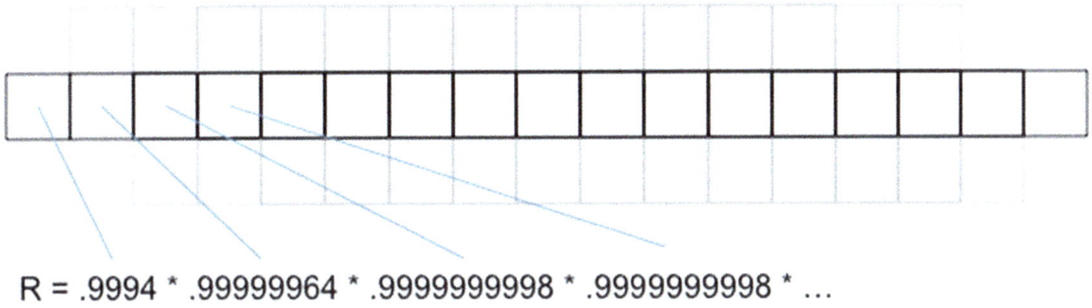

Figure 120: Calculating System Reliability

We can combine these geometries to establish a horizontal reliability for a given trace width and a vertical reliability for a given trace width. This results in a simple system reliability value defined by:

#squares = Length of segment/length of square

$$\boldsymbol{\Sigma\ R_{sys} = R_h \wedge \#\ squares_h * R_v \wedge \#\ squares_v}$$

This directly predicts production yield for a given material and design.

These predictions are based entirely on Z scores. Also known as standard score, it is the number of standard deviations from the mean, and it can be used to calculate the probability of the event occurring given the distribution

Z-score of 1 => 15.8% chance (1:6)
Z-score of 3 => 0.135% chance (1:741)
Z-score of 4.5 => 0.000003% (1:294331)
Z-score of 4.76 is 1 in a million

What is the acceptable threshold for a design and material? This is entirely dependent upon the customer and the desired yield of the electronic devices. However, it should be apparent to the reader that this information can be utilized to predict a wide variety of characteristics about the material and the electronic devices. An example of this is shown in Figure 121.

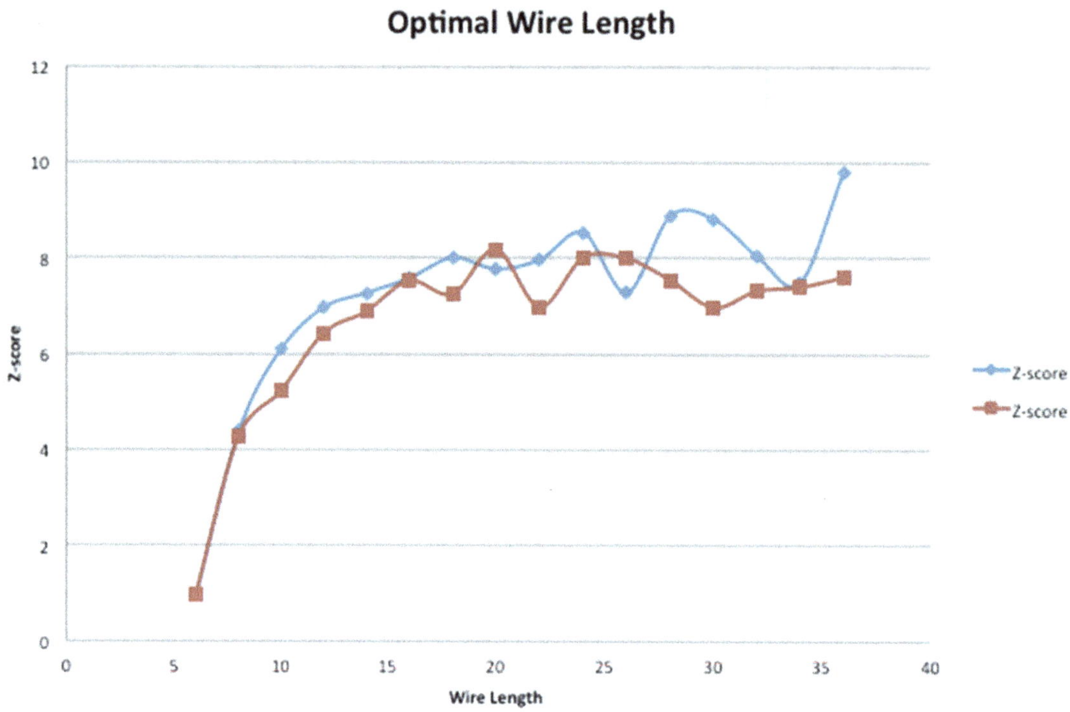

Figure 121: Defining optimum wire length based upon Z-score

Although this example differs significantly from a standard reliability calculation as previously discussed, the methods demonstrate the power of understanding the impact of reliability on the success of a system development. In addition, it is clear that the use of this modeling methodology increased the robustness of the predicted reliability significantly, while also decreasing the required processing time to a more manageable level.

Usability

"Usability is like oxygen– you don't notice it until it's missing." –Anonymous

Usability is the degree to which your product is both easy to use and an effective interface for the people who use it. It is a characteristic of a product, and defines whether that product is efficient, effective and satisfying for those who use it. It is also the name for the techniques and methodologies developed in Systems Engineering to assist in the creation of usable products.

There are two international standards that define usability and human – centric design:

ISO 9241 – 11: Usability refers to the extent to which a product can be used by specified users to achieve specified goals with defectiveness, efficiency and satisfaction in a specified context of use.

ISO 13407: human – centered design is characterized by:
The active involvement of users and a clear understanding of user and test requirements An appropriate allocation of functions between users and technology
The iteration of design solutions
Multidisciplinary design

The best way to ensure usability is to design it into your product from the beginning. User centric design involves translating user and business needs into solutions that will meet those needs effectively and efficiently. Properly designed systems fit their context so well that they are easy-to-use and beneficial to the user.

Effective and usable products never happen by chance. Typically, they are achieved by a design that is based on an understanding of the natural physical, psychological, social, and emotional characteristics of the human beings that will be operating those systems. It takes into account the type of tasks that will be accomplished in the work environment, the constraints of the technology being developed, and results in the creation of an interactive experience that allows the human user to fully utilize the system being developed.

Well-designed products are much easier to use, and are generally more useful to the user then poorly designed products. Good design will increase productivity, job satisfaction and user acceptance of the designed system. The design will also focus your limited resources towards building those products that will satisfy the goals of the user most readily, and will reduce the amount of waste. This waste can be thought of as the money spent on developing a product that will not be used effectively or accepted by the customer. There is a long track record of effective designs with poor usability that failed in the marketplace. One of these is shown in Figure 122.

Figure 122: Joint Strike Fighter competing designs

In the joint strike fighter design competition, two competing companies developed fighter aircraft for the program. These two aircraft had stringent requirements on flight characteristics, mission capability, cost-effectiveness, sustainment and many others. In fact, this program resulted in the largest awarded contract in the history of mankind. As one might expect, this would generally cause any organization to take great notice with regard to meeting the requirements of the design competition. In Figure 122 the two competing aircraft are shown. While both of these aircraft had roughly equivalent capabilities, it was clear from the outset that one aircraft had a significant advantage.

The aircraft on the right side of the image was designed in order to resemble previously designed fighter aircraft, taking advantage me effectiveness of previous designs. This required a significant engineering gamble, as the vertical takeoff requirement of this aircraft typically required the movement of the engine exhaust port in order to achieve vertical thrust. In this particular case, the aircraft on the left used the conventional thrust technique, while the aircraft on the right used a completely innovative solution to accomplish the same capability need, a ducted fan.

When one considers the assessment of capabilities of these two aircraft, the importance of usability becomes clearer. It will be necessary at some point to rely upon fighter pilots to communicate to the competing organizations regarding the effectiveness of their respective aircraft. As might be expected, fighter pilots walking out to these two aircraft are immediately attracted to the one on the right, while the aircraft on the left appears ungainly and unattractive. By virtue of this, the fighter pilots already have a higher degree of acceptance of the aircraft on the right, simply because they find the aircraft to be more attractive. This is an important lesson for any system developer; one must know the audience from which the assessment of capability will be delivered.

A well-designed user-interface is manageable and effective, enabling users to complete their work successfully, efficiently, and with a feeling of competence. The most effective user-interfaces have been designed based upon principles of human interface design. The following principles are a consolidated list from a wide range of published sources and are based in a long history of human computer interaction research, cognitive psychology, and design best practices.

Systems that exhibit a high degree of usability will demonstrate the following characteristics:

Utility
The system should provide the necessary capabilities and address the practical needs of the users. The information and functions provided to the user must be relevant to the users task and environment.

Consistency
The system developer should follow appropriate standards and conventions for the development of the system. Within any given application or family of products, make sure that the terminology, documentation, commands and actions are used consistently.

Use knowledge of the world and knowledge in the head
The designer should utilize commonly listed concepts, terms, metaphors, analogies, and follow typical real-world conventions to presents information in a natural and logical order. Methodologies and sequences should follow generally accepted practices within the industry. As an example, the cold water faucet is typically located to the right of the spout, whereas the hot water faucet is located to the left. There is very little justification for violating this simple knowledge of the world, and doing so will often result in a less acceptable interface.

Simplicity
Keep the most commonly used elements of the system visible, And keep those that are less frequently used easily accessible. Design a system that is usable by the typical operator with little or no instruction. This will greatly facilitate the acceptance of your system in general, and will increase the feeling of competency of the user. Display data in a clear and efficient manner to the user.

Communication
Provide effective feedback to the user so that she sees the result of her actions and can predict what is occurring within the system. Use organization to reinforce meaning: Put related things in one location, and keep unrelated things separated.

Ensure that any and all instructions are concise and focused supporting the needs of the operator.

Robustness
Allow for variations in users and input to the system. Prevent the operator from making serious errors whenever possible, and limit the access to potentially destructive activities. Provide clear and concise messages that describe any problem encountered, and suggest an appropriate solution for recovery from an error. Provide off–ramps to enable a user to abandon an unwanted path or group of actions. The ability to reverse actions without repercussions is also a typical desire of operators.

Efficiency
Allow frequent users to advance their knowledge and skill, providing shortcuts that may increase the efficiency. Make users the initiators of actions rather than reactors, this will increase the user's sense that they are in charge of the system.

Reduce cognitive requirements
Make the operators work easy, simple, fast and enjoyable. Automate any repetitive workload. Keep information displays brief and simple, consolidating and summarizing data when ever possible. Do not require the user to remember significant amounts of information, as the human ability to recall lists of items precisely is severely limited. All of these factors are focused on reducing uncertainty for the operator.

To accomplish effective usability, the system developer must conduct a significant amount of user research in order to understand the Interface needs of the operator the system is being designed for. There are many different approaches that have been developed for effective user research and they are typically categorized as:

Context of use methods
Ethnographic methods
Field study methods
Focus groups

Context of use methods include the domain of contextual inquiry. This is a structured interview method is used to gain information and descriptions about the context of use of the system, where users are asked a set of standard questions and then observed as they work in a contextual environment. The results of a contextual inquiry are used to define requirements and to learn what is important to the users in their approach to the system.

Context of use analysis is the act of collecting and analyzing information regarding the users, the user tasks, and constraints regarding the system. The data for this type of analysis is typically gathered by using operator interviews, workshops, surveys, site visits, artifact analysis, focus groups, observational studies and contextual inquiry. This type of a study is an essential input to the problem definition and requirements development for a highly usable system.

Often these types of studies are conducted over long periods of time, in order to understand the longitudinal effects of changes in the products for user community. A good example of this is the typical user-interface provided on mobile phones. While at first it was a new foreign-language to most users, the community rapidly developed a familiarity with typical gestures and motions used in these interfaces, which would allow changes in the development of future systems based upon this knowledge.

Many usability studies focus on the usability of the initial use of a system, which is related to the system learnability. Studies related to the use of systems after the user has become familiar with them are known as longitudinal studies. Both of these will generally result in insights that will be valuable to the interface developer in the system development process.

Ethnography is the process of gathering information about users and tasks directly from users in their normal environment. Traditional ethnography focuses on long-term studies that may span weeks or even months. Information is typically collected through observation, interviews, observer logs, artifact collection, and work product analysis.

Ethnography is a valuable approach, typically due to the fact that many users cannot effectively describe how they accomplish their jobs. Many of the behaviors and tasks become so ingrained that the user is not really conscious of accomplishing them during the normal workday. Due to this, if the system developer wants to capture all effective actions required in the use of a system, and ethnography may provide a means in which that may be accomplished.

A field study is a method of collecting data about users and their needs, and involves observation and interviewing. In this case, focus is on tasks and task flows required of the operators, inefficiencies in those flows and the organizational environment presented to the user. Typically researchers involved in a field study will observe operators as they work, recording observations about particular activities or tasks being accomplished, and utilizing questions to advance their understanding of the work accomplished by the operator. Field studies have been used to great advantage in advancing the efficiencies of manufacturing floors, and are commonly referred to as time studies.

A focus group is a discussion where a moderator will work with a group of participants utilizing a set of questions related to the system being developed. These are often used early in the product development stages as a means to enhance the requirements list and to provide meaningful feedback to the system developers about the users and products in a contextual environment. Often prototypes are utilized in this type of study to gain further insights into the requirements of the user group.

Once the user research has been completed, requirements must be defined that relate directly to the future usability of the system. There are a wide variety of methods utilized by the system engineer defining and refining those requirements related to usability. These include:

Affinity diagramming
This is a participatory method that is used where a wide variety of concepts are recorded for use by the team, and sorted into related groups and categories. This method has been used to help clarify complicated problems by organizing qualitative data to reveal themes associated with that data. These things can often be utilized to define a priority for the requirements to be utilized in the design for usability.

Competitor analysis
In this method, competing products or systems are examined and their characteristics are recorded. Those characteristics will typically be related to design advantages exhibited by the product, as well as deficiencies identified by users. This information can be utilized to identify the desired characteristics of the system being developed.

Future workshop
This is a participatory activity in which a group of people focuses on developing solutions to vague or challenging problems. In the initial phases, the problems are identified and categorized and then the participants focus on developing a future scenario in which those problems can be reduced or eliminated. This is similar to defining an optimized future state for a system.

Quality function deployment
This is a methodology that has been developed for use by a participatory group in order to effectively prioritize the needs specified by multiple users, the appropriate actions to be taken to address those needs, and the design characteristics that are desired in the final system development. This is a highly complex system that will generally yield very insightful results regarding the requirements most desired by the operators.

Scenarios and use cases
Use cases are typically focused on the finding specific tasks or group of tasks that must be accomplished by the user of the system. They will typically include the functional steps required to operate the system being developed. Scenarios differ from use cases in that they also include the actors in their assessments. As a simple example, the use case for a withdrawal of money at the automated teller machine would include the functional steps making a request, verifying identity, specifying the amount, checking the available funds, and dispensing of the withdrawal. The scenario may include other situations like utilizing the system in the rain or while

operating a motor vehicle. Each of these methods are valuable approaches used to gain insight into the requirements of a usable system.

The usability of a system requires significant attention by the system developer early in the development process. However, most of these development areas are vague and nebulous, and require specialized skills in order to effectively accomplish the development of a highly usable system. It is recommended that any system developer focused upon usability should familiarize themselves with the current state-of-the-art in the usability profession, as common strategies and approaches for dealing with specific challenges are maturing in the workplace, and the use of these existing systems may be of significant advantage to the system engineer.

Chapter 10: Verification

The verification and validation of a system is the second major development area for the Systems Engineer after system development. The verification and validation process constitutes the right-hand side of the V diagram of the systems process, and represents further re-composition of the system including the checking and verifying that we have completed the system as intended, and in a manner that will satisfy the customer.

Verification and validation are comprised of three related concepts:

Verification –
The process of evaluating a system or component to determine whether the products of a given development phase satisfy the conditions imposed at the start of that face. Did we build the product correctly?

Validation –
The process of evaluating a system or component during or at the end of the development process to determine whether it satisfies the specified requirements. Does the product do what it should?

Certification–
A written guarantee that a system or component compliance with its specified requirements and is acceptable for operational use. Is an outside expert happy enough with the product to stake his/her reputation on it?

Verification and validation are typically conducted by the Systems Engineer in order to produce an efficient and effective design result. A fast design process only helps if you get it right. If you get it wrong, you spend far more money fixing problems because you have progressed farther into the design phase before you address them.

We must put verification and validation into context to fully understand the impact on the success of a system design. This will also allow us to better understand the difficulty of accomplishing verification and validation on complex systems. First and foremost, we must understand:

No computer code / physical design can be proved to have no errors.
No documentation package can be proved to have no errors.

Verification and validation is typically carried out not to assure ourselves that our system is perfect and in no need of future attention, but rather that we are reasonably confident that the system will operate in the manner that we have planned for a specified period of time.

We frequently do this throughout the development cycle, perhaps without giving it a great deal of thought. For example, when we are seeking to create a system that will improve upon an existing system, we may utilize a model of that system in order to simplify our approach.

If you are designing a heatsink, you may rely on thermodynamic equations and fin theory to begin your system design. However, you may not give a great deal of thought to the validation of the equations with regard to real-world behaviors. In reality, these equations are not particularly predictive of the physical behavior of most heatsinks. Without understanding that we are relying on a numerical representation of reality in our modeling of the system, we may be missing an important element of the validation of the system. Therefore we must validate the translation from the physical world to the numerical description before we begin any development at all. If we assume our model is correct, and it is not, we will likely not produce a system of value to the customer.

Once the model of the system is complete, it can be used to design the physical entity for the heatsink product. If we were to compare the physical product with what we intended to design (based upon our model), we would be verifying that our design reflects the behavior predicted by model. Ultimately, we will examine the heatsink in operation, which would be the validation that the design is effective in the operational context.

So in very simple terms we can clarify the difference between verification and validation:

Verification – Did we do what we said we would do?
Validation – Did we do what the customer needed?

Typically verification and validation is conducted at a level commensurate with the criticality of the system designed. For example, if our system affects critical performance that may result in a loss of life, or loss of prestige, or financial or social loss, we would likely verify and validate at a higher integrity level then a system that did not demonstrate these characteristics. Many companies that do not produce mission-critical systems do a very low level of verification and validation. Companies that do have a more direct impact on safety or health may do a very rigorous verification and validation. An example of the former might be a software company that produces simple applications. An example of the latter might be an automotive company.

As discussed during the system development phase, we have defined a large number of system requirements that are constructed in a specifically verifiable form. These were eventually entered into our RVTM, which creates for us a convenient mechanism by which to verify our system.

Verification is usually conducted throughout the system development phase as each design phase is completed in the system development. Typically verification will take one of four potential forms, as shown in our traceability matrix. We will use one or more of these modes to check each of our requirements in order to complete our verification process. The four modes include:

Inspection – Do the characteristics of the component match the specification?
Analysis – Does a model predict that the system will perform as required?
Test – The component is placed in a simulated or actual operational use to check performance.
Demonstration – The entire system is demonstrated for the customer to ensure satisfaction.

Verification and validation (V&V) are, by definition, on–going activities. There is likely no endpoint to verification or validation challenges. This implies that it is possible to continue verification and validation long beyond any practical return. Therefore, in order to manage the expense incurred in the V&V process, we must examine the impact of cost and time with regard to our confidence level. This is shown in **Figure 123**. Based upon the testing integrity required, we will define appropriate limits for our verification and validation processes.

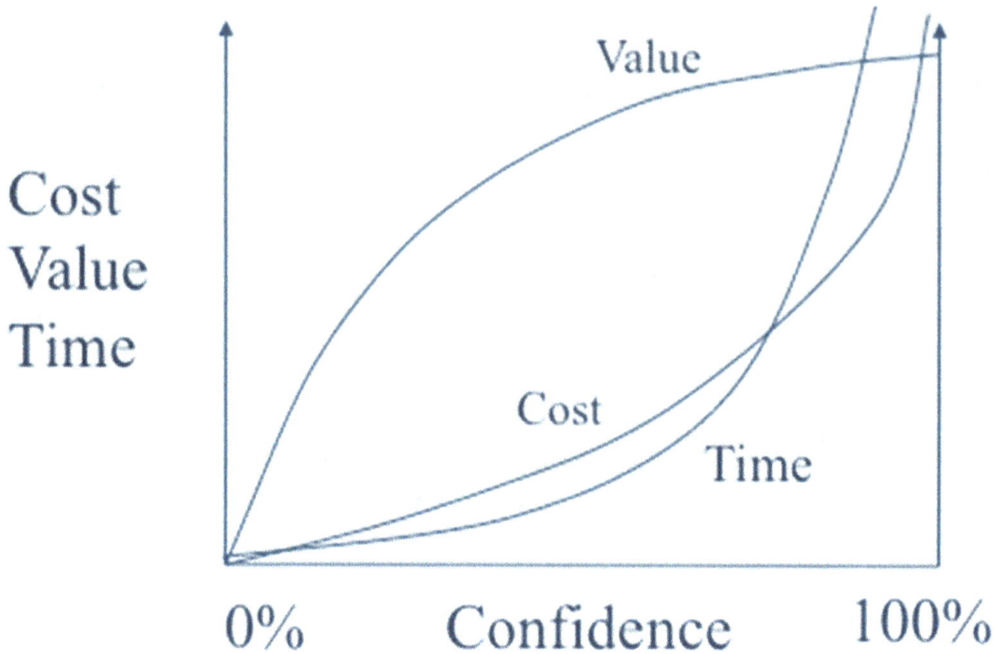

Figure 123: Confidence of V&V

Verification

In order to accomplish Verification, we need to ascertain whether the system has been created or built correctly. To begin, careful planning is required to get the most out of any testing and inspection process. Test planning is about defining standards for the testing process rather than describing the product tests.

The testing process will begin with unit testing, then module testing, then subsystem testing, then systems testing, and finally culminating in an acceptance test. This is representative of moving back up the right-hand side of the systems V diagram. The structure of a formal test plan should include:
The testing process
Requirements traceability
Tested items
Testing schedule
Test recording procedures
Hardware and software requirements
Constraints

Inspections and Testing are complementary and not contradictory Verification techniques. Inspections will check conformance of the system with the specification, but not conformance of the system with the customer's real requirements. Additionally, inspections cannot check non-functional characteristics such as performance, usability, and reliability.

In order to speed up the inspection or Verification of a system, check lists of common errors are typically used to help focus the inspection. Experienced professionals are also used, relying upon their experience and expertise to judge the systems have been completed correctly.

Testing can take many different forms. For example, many organizations conduct hardware testing; essentially shake and bake testing of the system both statically and in operation. Some may also conduct stress and life testing, running a highly stressed system to find points of weakness or failure. Margin testing is also utilized, which overdrives a system in order to find the margin of safety. System-level testing focuses on usability tests and checks that the system can, in fact, be operated in the manner intended within reasonable time limits.

White–box testing examines all elements of a component or subsystem to ensure that each of the internal capabilities performs as expected. Black–box testing tests each of the functional blocks within the System Architecture to ensure that the system operates correctly. Typically, the process begins with white-box testing of all components, and progresses through black box testing of consolidated components until the fully assembled system is both Verified and Validated. Ultimately, the entire system is tested as a black-box solution for the customer.

Validation focuses on the customer's acceptance of the product. In general, this is accomplished by demonstration of the system in operation. This type of testing is largely focused on demonstrating each of the established user needs from the outset of the System Development process.

As previously stated, V&V is a big can of worms. Since we cannot prove that the system, specification or requirements are complete or correct, we must instead come to some conclusion with regard to what confidence level we feel is appropriate prior to shipping the product. The general state of knowledge is that we will keep testing until it takes a long time to find the next defect. The definition of long time is dependent upon the organization.

Ultra dependable systems are becoming more commonplace. These are systems that we expect will never fail. For example, banking systems, healthcare systems, and transport systems fall into this classification. In general, a very rigorous V&V approach will be applied, and the testing will be conducted ten times longer than the acceptable mean time between failures (MTBF). This will likely ensure that our system will demonstrate ultra-dependable behavior.

For the assistance of the reader, a sample Verification and Validation template has been included in Appendix A.

Chapter 11: Validation

While Verification assesses whether we developed the system we intended to develop, Validation determines if the Systems Engineering process has developed the right system. This is assessed based upon the needs originally expressed by the stakeholders. Validation is carried out by the Systems Engineers, based upon what they believe the customer's needs were at the outset of the development project. Operational validity is the matching of the capabilities of a designed system to the originating concept of operations, and is typically only accomplished after the entire system has been verified. However, conceptual validity, requirements validity, and design validity are also important considerations that can be addressed earlier in the development cycle.

Conceptual validity is the correlation between the specified needs of the users and the operational concept. Systems Engineers may establish this early in the design process based upon the customer's stated needs, although it should be assumed that this is an incomplete process that is likely to change throughout the development cycle.

As systems develop, it is likely that the knowledge levels of the developers and stakeholders will expand. Thus, stakeholders' needs are likely to change and by virtue of that, operational concepts must also change in alignment with those needs.

Operational validity is the assessment of whether the operations accomplished by the use of the system will meet the needs of the stakeholders. Requirements validity is the correlation between the concept of operations and the originating requirements list. In requirements validity it is assumed that the concept of operations is an effective solution to the stakeholders' needs. The validation is accomplished by establishing that the requirements have been completed effectively in relation to operational needs. Keep in mind, however, that throughout the development of requirements we have been adamant that they should be stated in design independent terms, which results in a large challenge for the validation of the system. Therefore, the validation process must assess the requirements validity as well as the fulfillment of the originating user needs in the development process.

In a simplified sense, the validation process of the system development effort focuses on whether or not the system will meet the needs of the stakeholders in a manner that will result in their purchasing the system. Often this results in a demonstration of the system, with user acceptance of that system after demonstration. While many similarities exist in the validation efforts, no single checklist or method has been defined that will address all validation needs for the System Developer. In general, the System Developer should be able to prove that the following questions have been answered:

Does the system satisfy the users' needs?
Does the system operate in a manner consistent with the users' expectations?

Can the user is effectively use the system to accomplish the desired tasks?
Does the system demonstrate a reasonable cost, design, and sustainment approach for the user?
Does the system represent the original concept of operations?
Is the customer satisfied with the system?

The Validation Process

Establish a Validation Plan

Identify the scope of the validation for the system based upon the needs and requirements. This activity is carried out concurrently with system definition and system requirements development.
Identify needs according to their origin, especially those that will limit validation success.
Define appropriate validation techniques to be applied to the identified needs, Such as inspection, analysis, demonstration, and modeling.
Select the validation actions that will be taken based on the system design, objectives and constraints.
Optimize the validation plan to define the most appropriate approach to ensure all needs are validated effectively.

Perform the Validation Actions

Detail each validation action, note the predicted results, the validation technique utilized, and the validation means required for the action.
Acquire all necessary components to conduct the validation.
Carry out the validation procedures recording appropriate results.

Analyze Results

Analyze the obtained results and compare them to the predicted results.
Conduct gaps analysis for those results that do not correlate to expected values.
Define whether further actions need to be taken regarding the potential issue for the generation of a change request for the requirements list.

Maintain Process Control

Update the validation plan according to the progress made with regard to the system development and requirements modifications.
Coordinate validation activities with regard to the schedule and budget.

The final output of the validation process should include the development of validation artifacts. This includes the original validation plan as specified at the outset of the action. A completed validation report and verification matrix will be provided that identifies each validation action, applied techniques, methods of execution, and results obtained. Validation procedures and methodologies will be identified in the event later replication of the validation process must be conducted. Any non-compliance, non-conformance, or trouble reports will also be included in the final validation report.

Technical Performance Measurements

Performance metrics are defined early in the development cycle, and are applied throughout the entire system development process. Technical performance measurements are defined at the start of a program, and will likely include the plan to track the progress of the selected technical parameters. They are typically used to provide a red flag to Systems Engineers of potential technical problems and an assessment of the impacts of proposed changes and system performance. They are also utilized in decision-making and trade-off analysis throughout the system development process. Generally, TPM parameters that are tracked are the cost drivers of the efforts, elements that lie on the critical path, and those that represent technical risk items. System performance requirements are both testable and measurable, which implies that they are verifiable quantities.

Validation includes two specific metrics of interest to the Systems Engineer. These are the Measure of Effectiveness (MOEs) and the Key Performance Parameters (KPPs). An example project, the Card-Craft System, is included as an example of this approach.

The Card-Craft System is focused on providing the ability to produce secure identification cards for a user in an efficient and unobtrusive manner. The OV-1 of the Card-Craft System is included in Figure 124.

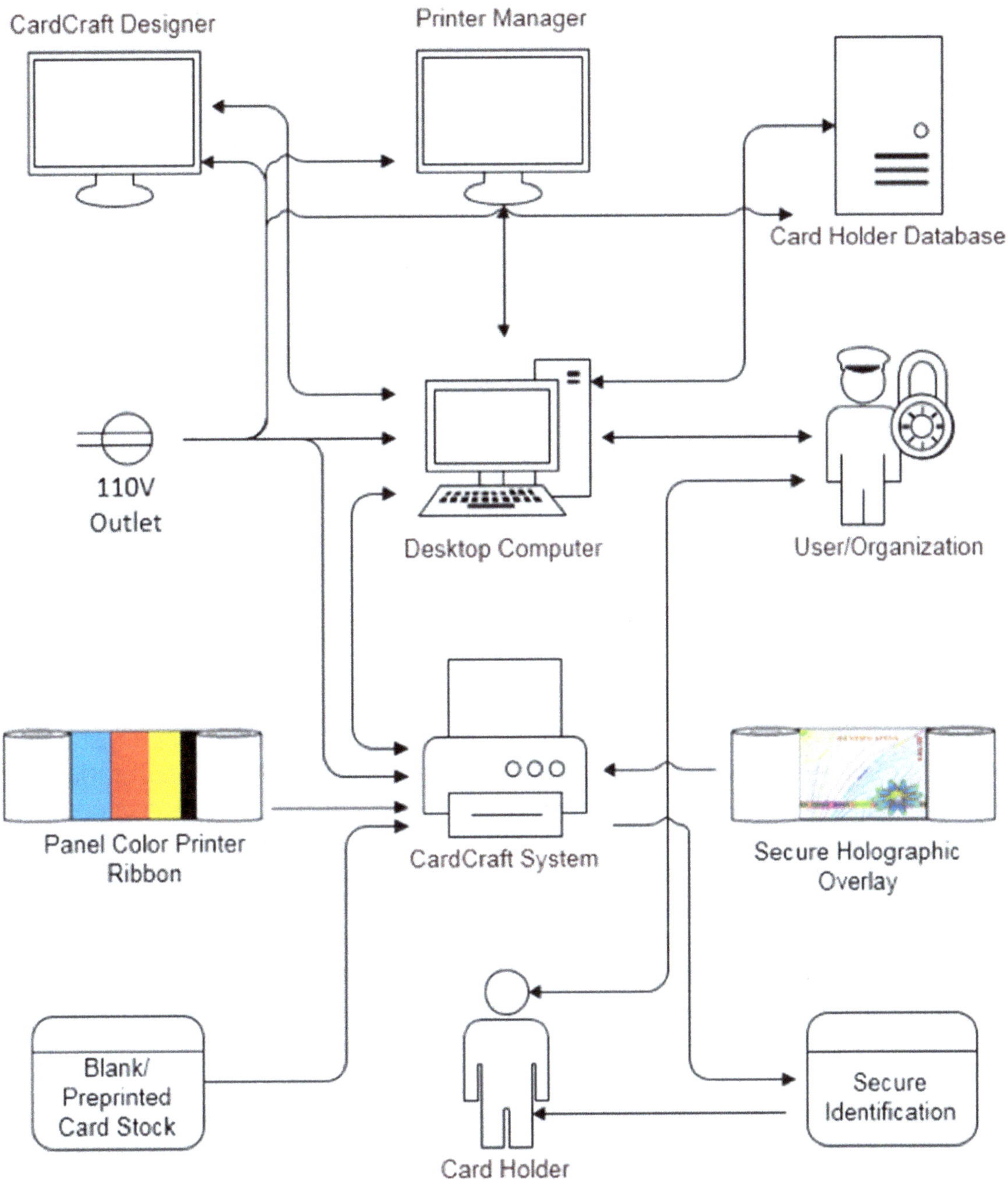

Figure 124: CardCraft System OV-1

The system has an extensive list of requirements defined; a sample of which is shown in the Card-Craft RVTM in Figure 125.

Requirement	Req ID	Subsystem/Area	Comply		Verification Phase			Verification Method		
			Y	N	D	Q	A	I	T	A
System shall comply with all UL USA standards	1	Agency Requirements					x	x	x	
System shall comply with FCC I.T.E. Class A USA	2	Agency Requirements					x	x	x	
FCC Radio ID Number USA	3	Agency Requirements					x	x	x	
Holographic and Clear versions of both high and standard durability overlay shall exisit	4	Consumables			x			x		
The printer shall support YMCK, YMCK-K, YMCKT-KT color ribbons	5	Consumables			x			x	x	
The printer and laminate shall have the ability to print and laminate .020" to .030" plus or minus 10% (0.030" nominal) PVC, PC/PVC Composite, PET(G)/PVC Composite, and Translucent PVC/Teslin cards	6	Consumables			x	x			x	
All Overlays shall resist delamination- High Durability Overlay applied to cards shall have: A grade ≥ 3.0 when tested using ANSI INCITS 322-2002 Tape method	7	Consumables-Durability				x	x		x	
Overlay materials shall be tested with all printers against the ISO 24789 card life protocol tests for NID. (expected testing will be conducted externally - i.e. Eclipse lab)	8	Consumables-Durability				x	x		x	
Final card testing shall be conducted externally - i.e. Eclipse lab)	9	Consumables-Durability					x		x	
Overlay flash shall not exceed 0.5mm	10	Consumables-Quality				x	x		x	
Unused Overlay and Laminate shall not exceed 5%	11	Consumables-Quality			x	x	x	x	x	
Resolve the card bowing exceeding .060" ISO 7810 Card Warpage Spec	12	Consumables-Quality			x	x	x		x	
Enclosure plastics Shall marked with recyclable symbol, per Resin Identification Code	13	Environmental Standards			x			x		
The Printer and Laminator shall be designed for use on a desk of table top	14	Hardware			x			x		
The Printer and Laminator shall weigh less than 20 pounds	15	Hardware			x	x		x		
Height of The Printer and Laminator shall not exceed 18 inches	16	Hardware			x			x		
Depth of the printer and lamnator shall not exceed 24 inches	17	Hardware			x			x		
Input and output hopper capacity shall be 25 cards.	18	Hardware			x			x		
Reject hopper capacity shall be 10 cards	19	Hardware			x			x	x	
Card feeder shall feed one card at a time	20	Hardware			x	x	x		x	
a mulit-LED shall display status	21	Hardware				x		x	x	
CardCraft System shall produce audiable notifiactions when errors are encountered	22	Hardware				x			x	
CardCraft shall have a display lit LCD screen with membrane buttons	23	Hardware			x			x		
CardCraft System throughput shall be no less than 75 cards per hour with overaly	24	Hardware				x	x		x	
Consumables shall be held in a removable Laminator Supply Cartridge	25	Hardware-Laminator			x			x		
Laminator shall be headted with a halogen element	26	Hardware-Laminator			x			x		
Laminator roller shall be constructed of hard coat anodized aluminum	27	Hardware-Laminator			x			x		
Heating element shall be placed in laminator roller	28	Hardware-Laminator			x					
Laminator heating monotired via IR sensor 1/4 the lenght of the roller	29	Hardware-Laminator				x		x	x	
Laminator shall consistnatly hold a lamination temperature range of 160 to 200 C	30	Hardware-Laminator				x	x		x	x
Cards shall be able to pass the laminator with out contact from lamination roller	31	Hardware-Laminator			x			x		

Figure 125: Card-Craft RVTM (Sample)

Measures Of Effectiveness

The MOEs (Measures of Effectiveness) are developed with the customer's requirements are be used to demonstrate that the designed system will satisfy the customer's needs. Measures of effectiveness are a measure of how well an operational task is accomplished through using the system. The data used to measure the effect comes from the use of the system in its expected environment. Employees often have green/yellow/red limits that are established early in the development program and are tracked to maintain a green status. These are also useful in establishing if the measure is achievable. The MOE can be stated either as a question or as a declarative statement, and is often stated in terms of accomplishment gained per cost incurred.

The Card-Craft MOEs are listed in Figure 126. The MOEs outline the requirement or requirements to be tested, and the testing method that is to be used.

MOE	Requirement	Quantitative Measures	Measurement Method	Scoring
The CardCraft System produces personalized cards with a secure overlay quickly	24 Hardware, 42 Hardware Printer	System is able to produce 300 CPH with out and overlay and 75CPH with and overlay	Run a 300 card job without and overlay and a 75 card job with an overlay. Measure run time.	CardCraft System must meet specifications outline in requirements 24 and 42
The system provides alerts when supplies are "empty"	21, 23 Hardware	Audible and visual alerts indicate errors	Allow CardCraft system to run out of supplies in card hopper, printer ribbon, overlay, as well as manually induce a double feed card jam ensuring audible and visual alerts are seen and herd.	All errors must generate audible and visual alerts
Personalized cards are durable	8 Consumables-Durability	ISO 24789 card life protocol tests for NID	Complete ISO 24789 National ID test sequence	All test cards shall pass the ISO 24789 test sequence with minimal wear and shall be functional with the card holder clearly identifiable
CardCraft System is able to run for long periods with minimal manual interventions	85 Usability-Error Handling	Card jams preventing automatic clearing of the machine	Personalize 2400 cards (300 CPH X 8H)	≤2 manual interventions - Pass >2 manual interventions - Fail
User operates system with only basic instruction manual	86, 87, 91 Warranty and Service	User can successfully accomplish basic production level tasks on a 'Production Operation Use Checklist'	Production operation use check list encompasses basic production operation of system. Provide new user with basic instruction manual and record items that are successfully and unsuccessfully completed	≥90% - Highly Successful ≥80%, <90% - Successful <80% - Fail
CardCraft System produces high quality cards with consistent image and overlay placement without visual defects	74 Software-Usability, 102 Hardware Printer, 103 Hardware-Laminator	Secure Overlay placement ± 0.015" Long Axis, ± 0.015" Short Axis and Image placement ± 0.015" Long Axis, ± 0.005" Short Axis	Personalize 500 cards, measure image and overlay variability on 10 cards at 25%, 50%, 75% completion of the 500 card run	Secure Overlay placement >0.015" any axis = Fail : Image placement > 0.015" Long Axis = Fail, > 0.005" Short Axis = fail
System connects to Windows 7 and XP	60 Software-Controller	Connection successful / unsuccessful	Test connection from devices with the two different operating systems. Operating systems to test are Windows7 and XP.	CardCraft Systems shall work with Windows 7 and XP operating systems
The CardCraft System produces flat personalized cards with a secure overlay	12, Consumables-Quality, 32Hardware-Laminator	Measured card warpage less than 0.06"	Personalize 500 cards, measure warpage on 10 cards at 25%, 50%, 75% of total run. Warpage measured by placing cards on a flat reference surface. Highest point will be recorded.	Card Warpage ≤0.06" - Pass >0.06" - Fail

Figure 126: Card-Craft MOEs

Key Performance Parameters

The KPPs (Key Performance Parameters) are key specifications that must be met to make the completed system operational. They should be defined with a threshold and an objective value, must be testable, and shall outline the measurement methods along with quantitative measures. KPPS are used during development to ensure the system meets or exceeds them at completion. However, KPPs are not to be used in trade off analysis, as they flow directly from the operational requirements. If a completed system cannot meet KPPs it will be deemed non-operational.

The Card-Craft System KPPs are listed in Figure 127.

ID	KPP	Quantitative Measures	Measurement Method	Score
1	Image Placement	± 0.015" Long Axis, ± 0.005" Short Axis	Personalize 500 cards, measure image and overlay variability on 10 cards at 25%, 50%, 75% completion of the 500 card run	>0.015"% - Fail:≤0.015" - Pass Long Axis; >0.005"% - Fail:≤0.005" - Pass Short Axis
2	Secure Overlay Placement	± 0.015" Long Axis, ± 0.015" Short Axis	Personalize 500 cards, measure image and overlay variability on 10 cards at 25%, 50%, 75% completion of the 500 card run	>0.015"% - Fail:≤0.015" - Pass
3	Printer Manager Communication	15 seconds delay maximum	Committing setting and changed in the printer managers shall take no longer than 15 seconds to change setting and refreeze screen with updated values	≤15seconds - Pass:>15 seconds – Fail
4	User Interface	99% Uptime	Power system and use continuously for 5 simulated work days of 8 hours	≥99% - Pass:<99% - Fail
5	System Availability	95% Uptime	Power system and use continuously for 5 simulated work days of 8 hours	≥95% - Pass:<95% - Fail
6	Secure Overlay Adhesion	A grade of greater than 3	Five Cards will be selected at random and subjected to ANSI INCITS 322-2002 Tape method	>15% - Fail:≤15% - Pass

Figure 127: Card-Craft KPPs

Chapter 12: Project Management

Project Management abilities are the least effectively taught and most critical skills that successful Engineers must possess. These skills transcend the usual Engineering or Business Management abilities taught to leaders, as they become life skills that can be used on a day-to-day basis and are useful outside the Engineering domain. They have become critical because research suggests virtually all white-collar work is project-based. In addition to this work focus, most technical professionals also seek to plan their personal lives around a project orientation. Therefore, it is imperative that any effective leader in an Engineering organization demonstrates appropriate and pragmatic Project Management skills.

Many taxonomies exist that describe the project lifecycle; the Project Management Institute (PMI) adopts one of the most common. This lifecycle incorporates a five-phase approach to project understanding, shown in Figure 130.

Figure 128: The Project Life Cycle

The first phase of a project is Project Initiation. This is focused on shaping the project in a manner that will be appropriate for the individual, the project team, and the organization for which the project is being completed. This implies that the project leader is well acquainted with the organizational goals and strategy and potentially many of the tactics being implemented within the organization. With this knowledge, the practitioner can plan an appropriate project, which will fit well with the expectations and needs of the organization. This alignment is critical to ensure the effective application of resources within the organization, and often a minimum investment of time is applied to this phase. This can be problematic in later phases, as the project being accomplished may be less suitable than other projects for the future success of the organization.

The second phase of the project lifecycle is Project Planning. This is the heart and soul of most Project Managers' day-to-day existence. It involves defining the specification, schedule and budget for the project deliverables, in order to lay out appropriate resource needs, timelines and funding that are required to complete the project as specified. If a Project Manager was only to become well versed in a single phase of Project Management, Project Planning is the most suitable. With that being said, Project Planning is typically not adequate to expect project success, and the Project Manager must continue to further develop skills until a level of mastery in all phases of the project lifecycle is achieved.

The third and fourth phases of the project lifecycle are typically concurrent in any given development environment. Phase three focuses on Cost Control, which is the establishment of appropriate metrics and status methods for the project. It is required to maintain updates and communications for other members of the organization, and to accurately project the likely outcome of the project at any given time. Effective Cost Control is highly useful and predictive, and it enables the Project Manager to take action quickly in the event of a downturn in project status. Additionally, utilizing a well-established methodology such as Earned Value allows easy communication across project boundaries due to a consistent and practical status system.

The fourth phase of the project lifecycle is Project Execution. This focuses on the ability of the Project Manager to engage the workforce and to enable that workforce to accomplish the objectives and deliverables set out in the project plan. This involves the application of the soft skills of management, which include teaching, learning, communication, motivation, relationship building and other related skills that are often grouped under the classification of emotional intelligence. While the project-planning phase is the most significant for the future success of the project, an ability to completely execute a project effectively is a skill not easily mastered by many technical professionals. This skill is typically not valued highly by those professionals because it requires a focus on the people as opposed to the technical aspects, something which many project professionals are ill-equipped to handle. Therefore, effective Project Management will require the effective execution of the project in addition to appropriate planning and monitoring.

The last phase of the project lifecycle is known as Project Closeout or Closure. This phase is intended to require the Project Manager to revisit the original project plans and compare them with the actual executed history of the project in order to measure the accuracy of the original projections and estimates. This assessment is expected to improve the future planning and estimating capabilities of the project leaders that will manage multiple projects throughout their career. Often this phase gets ignored when resources become tight, and project leaders move on to their next challenges. This is unfortunate, as project planners are then unable to assess their previous estimation capabilities, thereby limiting their readiness to more effectively plan and execute projects in the future. Effective Project Managers will ensure that an appropriate project postmortem and Closeout is accomplished in order to achieve these goals.

Phase 1: Project Initiation

Project Initiation is a critical aspect in the development of effective projects within the organization. Engineering leadership should take great care to assure that Project Initiation is accomplished on all projects, as this sets the stage for project alignment and the effective support of programs within the organization.

Defining Projects:

A Project is a temporary endeavor undertaken to create a new product or service (Project Management Book of Knowledge - PMBOK.)

Project Management:

Project Management is the application of knowledge, skills, tools and techniques to project activities in order to meet or exceed stakeholder needs and expectations from a project (PMBOK).

Project Management Largely Consists of Two Parts:

Project Planning

The Requirements of the work
The Quantity of the work
The Resources needed for the work

Project Control and Execution

Measuring and assessing status
Making adjustments to the plan
Managing the engagement of the team

Appropriate initiation of any project requires considerable thought and reflection on the organization for which the work is to be performed. The project leader must consider what strategic imperatives are currently underway within the organization, and which competitive priorities might be most appropriately applied in the execution of the project. Once the project leader has a clear understanding of this alignment, it is appropriate to move forward in the development of the initial project plan.

To do so, a project leader begins at the definition of the goals to be achieved by the project. These will differ from the organizational goals, but should very likely be in alignment with or in support of those business goals. Therefore, to begin our plan we must clearly define our goals in a narrative form.

Define the overall intent of your project – **The Goal**

As an example, if we are responsible for the development of a new product and a chocolate chip cookie company, we must discern the ultimate goal of our project. We might be tempted to suggest that it is to dramatically increase our market share, or increase our profitability. This type of goal can be problematic, for the same reason that the application of a business strategy is challenging for an engineering organization. It is very difficult to know which recipe might be best suited to increase market share, although one might instead change that focus a bit and produce an effective goal statement for the project team.

Goal statement: Provide a new chocolate chip cookie product that will taste better than the competitors' products.

This goal statement is effective because it clearly delineates the criteria used to define the appropriate project plan to accomplish the ultimate business goal of increasing market share. Our next step, then, is to begin to define how this will be accomplished. As the reader will recall, this is synonymous with defining the strategy of the project, and how this strategy aligns with the organizational strategy. This is typically referred to as the project mission and is delineated in the mission statement.

To begin to understand how we might approach this question, we might consider additional viewpoints, such as:
How do you intend to accomplish this work? What is your general approach? What key skills and abilities will make your approach the correct one that will benefit the organization the most?

Consideration and reflection upon these questions will help to define for the project leader the competencies or skill areas of the resources that can best be applied on the project. Utilizing our example of a chocolate chip cookie company, we know that our previous strategy was to consult the finest chefs for recommendations of recipes and techniques and to use the finest ingredients available in the preparation of our cookies. Since this is the case, and we have been able to effectively apply these skills in the past, it seems appropriate for us to continue to focus on areas where we have demonstrated a significant capability or differentiator over our competition. Please also note that if the intention is to right a sinking ship, it may be appropriate to make a significant departure from past strategies in order to become more competitive or to remain competitive in the marketplace.

In this case we might choose to answer those questions by focusing on a general approach to increase the quality of ingredients that are utilized in our cookies. For example, a portion of our project may include research into what makes a cookie better tasting to the consumer, as well as research into what consumers believe defines a better quality ingredient. Once we have established this understanding we can define our mission statement for the project, which is a sentence or two describing how we will accomplish the goals. Another way to think of this is that the mission of the project should describe simply why your customer wants you to accomplish the project.

Project Mission: Produce an innovative cookie that the average consumer believes taste better than the competitors' cookies.

It is important to understand that this mission statement should reflect one of the five competitive priorities previously defined, including cost, quality, dependability, or flexibility for innovation. In this case we have elected to take an innovative approach to produce the best tasting cookie, which previously did not exist. This might also imply that we could charge a premium for this product.

Next, we will define a statement of work for the effort to be undertaken. This statement of work should represent an objective view of the mission to be accomplished. By objective, we mean that the final statement must be both measurable and achievable in order to establish that we have completed the deliverables we have set out to complete. Some care must be taken not to write this in a goal orientation and to assure that the final result is provable as having been completed.

Goal: A vision or intent that may or may not be achievable

Objective: A measurable and achievable goal

Statement of work: Develop and document a new chocolate chip cookie product utilizing higher milk-fat chocolate and all natural dairy ingredients to produce the best tasting cookie for our customer.

Last, the purpose of the project should be defined. This should describe why the organization is undertaking the completion of this project.

Purpose: To provide a new chocolate chip cookie product that will result in the development of a new market of high quality cookies with superior flavor for our company.

Phase 2: Project Planning

Project Planning is a crucial component of the project lifecycle. There are many different published methods regarding planning, and several will be discussed here. Fundamentally, it comes down to a very simple understanding of what needs to be accomplished in the Project Planning process. There are three fundamental questions that must be answered:

What are you going to do?
How long is it going to take?
How much will it cost?

If these three questions are answered effectively, any individual can make an appropriate business decision based upon knowledge of the organization and this scoping of the project. Therefore, we will spend our time focusing on how to address these specific needs.

In order to answer the question, "What are you going to do?" we must scope the project appropriately, provide requirements that must be fulfilled for each of the appropriate stakeholders and ultimately derive a specification describing the fit and function of the system being developed.

The specification will describe every aspect of the system under development, but we also want to identify the higher-level deliverables that must be provided by the system in order to provide a complete description of what will be accomplished on the project.

There are several approaches that can be used in order to develop descriptions of what will be accomplished, but prior to that a significant effort that must be undertaken in order to define requirements necessary to achieve all of the stakeholder's needs under the project development effort.

Requirements Definition

A requirement is something that the product must do or a quality that the product must have. A requirement exists either because the type of product demands certain functions or qualities, or the client wants that requirement to be part of the delivered product.

Why focus on requirements?

To ensure project success, 8% - 14% of project costs should be invested in requirements development
Requirements define the project scope
Make / Buy decisions can be made based upon the requirements
Requirements can define how we implement the project
Requirements reveal where further research is necessary
Requirements help define what we do not know
This allows the requirements to drive the solution - not the reverse

Reasons for Good Requirements

Save time and money in project development
Increase efficiency of communications
Set appropriate expectations and parameters for the project
Prevent duplication of effort by project team members
Focus project resources on key milestones and maintain coherence
Spend less time on revising product or service requirements during or after project implementation
Fewer contract changes with external vendors and associated costs
Better product or service quality
Performance accountability – to be able to do what we said we would do

Requirements Definition

A requirement is something that the product must do or a quality that the product must have. A requirement exists either because the type of product demands certain functions or qualities, or the client wants that requirement to be part of the delivered product.

Why focus on requirements?

To ensure project success, 8% - 14% of project costs should be invested in requirements development

Requirements define the project scope

Make / Buy decisions can be made based upon the requirements

Requirements can define how we implement the project

Requirements reveal where further research is necessary

Requirements help define what we do not know

This allows the requirements to drive the solution - not the reverse

Reasons for Good Requirements

Save time and money in project development

Increase efficiency of communications

Set appropriate expectations and parameters for the project

Prevent duplication of effort by project team members

Focus project resources on key milestones and maintain coherence

Spend less time on revising product or service requirements during or after project implementation

Fewer contract changes with external vendors and associated costs

Better product or service quality

Performance accountability – to be able to do what we said we would do

Many different methods have been defined to enable us to begin to develop requirements. In general, there is no ideal method that will ensure that all the requirements or even the most critical requirements have been identified. Utilizing multiple techniques such as those described in the following paragraphs will be very useful in establishing a large percentage of those requirements that will be of interest to the project team and customer.

Requirement Definition Techniques

Interviews/Surveys/Focus Groups

This method is typically used when little is known about the expectations and needs of the customers and stakeholders. The basic premise of this methodology is to talk with people utilizing a structured interview technique in order to define:

What do they need from the developed project?
What do they want from the developed project?
What are they concerned about with regard to the effort?
What other elements of interest come to mind when they are considering the project?
What capabilities are implied that must be present in a successful system?

This methodology can be quite time-consuming, but allows its user to discover interesting perspectives and innovations that might not readily come to mind without the input of those most familiar with the application, the stakeholders.

Mind Mapping

Mind Mapping is a methodology that begins with our project focus located in the center of the mind map. From that center, we connect all related topics that might be of interest to us with regard to the project. This allows us to create a series of spokes around our central concept that will bring to mind many considerations we might not have perceived prior to this exercise. In addition to this, we can continue to connect spokes outward from the related ideas to more fully develop an understanding of what challenges need to be addressed in the project. This can be seen in Figure 131.

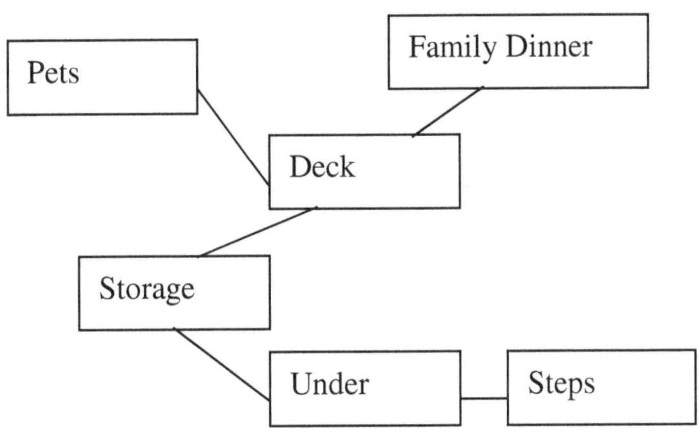

Figure 129: Mind Map of a Deck Project

Workflow Analysis

In workflow analysis, establishing the workflow throughout the lifecycle of the system being developed generates its requirements. This will essentially result in a Start – to – Finish Block diagram that is used to operate the system. From this diagram, a developer can carefully examine the relationships and establish whether any of the steps may be eliminated or if any challenges are present in the current workflow.

Additionally, this workflow can be connected to other systems within the organization, allowing the developer to explore the relationships amongst these interrelated systems. From the workflow analysis, each box and each connector will typically result in one or more requirements being identified.

A Cake Example produces a workflow:

Decide on Cake
Find Recipes
Check Inventories
Go to Store
Buy Groceries
Prepare Kitchen Components
Mix Cake
Grease Pan
Bake Cake
Cool Cake
Make Frosting
Frost Cake
Serve

From this example, it becomes clear that the project plan must include recipes and buying groceries in order to complete the overall project. This type of workflow analysis can help to create boundaries around the project enabling the stakeholders of the project to have a common understanding of what will be accomplished and what will not be accomplished within the project development.

Systems Analysis

Systems analysis is commonly used in complex systems development. The preferred method typically utilizes an object-oriented model of the system functionality in order to fully understand the interrelationships of the various elements of the system as well as the capabilities that must be delivered in the final product. This is similar to dataflow modeling in software development or event response modeling as is commonly used in electrical engineering.

As further detail is added to the system model, each of the boxes can be expanded into multiple functions within the box, which also generate additional understanding and requirements. In the Systems Engineering field, this type of systems analysis is what generates the lion's share of requirements for the system developer. Further details on this technique can be found in the Systems Engineering section of this text. An example of a top-level systems analysis is shown in Figure 132.

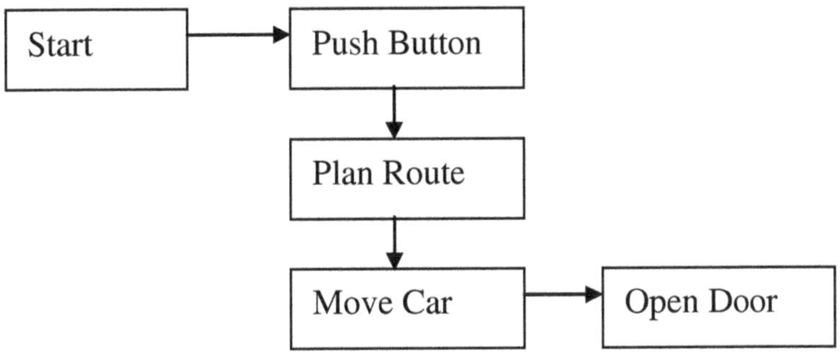

Figure 130: Systems Analysis of an Elevator

Prototyping

Prototyping is undertaken in order to address one of the fundamental shortcomings of customers in the project development process. The problem is in specifically identifying what they need or want from the project. Often this is due to an inability to see what is currently not existent, and it limits the customers' perceptions only to what they already have. The use of prototyping can generate new products and new approaches that will enable customers to better understand what you intend to accomplish and may provide additional requirements based upon their observations regarding the prototype.

The prototype can be a display-oriented or functionally oriented device, and can be rudimentary or elaborate depending on the technology availability and the criticality of the prototype results. In some cases extremely high fidelity prototypes can be created in CAD systems in order to provide the customer a realistic view of what the finished product will look like.

Additionally, prototypes may be used where modeling or simulation could become too demanding or expensive in the development of a usable end product.

Scenarios / Use Cases

Scenarios are activity-based descriptions or models of the system's intended use. Use cases are similar to scenarios except that they also include the actors associated with the activities. Both of these approaches might be very useful in understanding the interrelationships between the systems as well as between the actors in the system in a typical scenario. Additional insights and requirements may be generated for the system by walking through all of the activities and relationships.

Market Research

It is often of benefit to the project leader to conduct significant research into the market or the project's technology area of interest. This research might take the form of:

Internet searches

Publications review

Reviews and discussions with business contacts

Many project leaders are surprised to discover that in many cases other parties, sometimes in unrelated industries, have already accomplished the work they are contemplating. Some research is typically required to better understand the challenges faced by the project team prior to project launch.

Templates/Project Warehouses

In many cases it is well worthwhile to examine past efforts undertaken by the organization or records available in the public domain from other organizations that have accomplished similar projects. These documents can describe standard requirements that are utilized in that industry, and save considerable time in requirements development. These documents might include:

RFP – request for proposal*

Project introduction/kickoff documents

Business objectives

Background information supplied with the project

Scope descriptions from project sponsors

Business terms and conditions from contracting agents or vendors

Requirements documents included in business agreements

Proposal instructions included in an RFP

Evaluation criteria included in an RFP

*Note that this document is usually not available at the project level; this is more of a program level activity.

Performance Based Requirements

Performance-based requirements are typically the most useful for the project development team. These are requirements that are measurable and achievable where the performance measure is included in the description of the requirement. This requires that care be taken in the development of requirements to ensure the resulting requirement description is appropriately constructed and has enough detail to provide a performance-based profile.

Several rules are utilized to generate effective requirements. These include:

1) The requirement statement is a complete sentence.

2) The requirement statement uses the word 'shall' in its construction.
3) The requirement is limited to a single thought.

Example performance-based requirements will focus on issues such as cost, quality, speed, efficiency, weight, and size to name a few.

Example performance-based requirements might include:

General: Must print fast.

Performance-Based: Must print three inches per second.

General: The software system shall retain history.

Performance-Based: The system shall retain a minimum of sixty weeks of history and a minimum of four weeks of schedules.

General: Training shall be provided.

Performance-Based: Three training sessions shall be provided to thirty employees onsite at XYZ Corporation.

Types of Requirements

Functional - What does the product or service need to do?
Action requirements
Workflow steps
Examples:
Sum total hours worked
Calculate payroll
Print paychecks

Technical - Conformance to standards
Conformance to IEEE 802.11(b) Standard for radio frequency
System type
900 MHz vs. 2.4 GHz direct sequence spread spectrum (DSSS) system.

Usability - Describes how the functional requirement should be perceived
User-Friendly
Graphic user interface
Wizard-driven screens
Reliability: Mean-time between failures
Speed: Number of seconds/minutes/hours to complete
Accuracy: Percent accuracy or allowable errors
Number of steps to complete

Reliability: Mean-time between failures
Speed: Number of seconds/minutes/hours to complete
Accuracy: Percent accuracy or allowable errors
Number of steps to complete

Operational/Environmental - Describe the system environment

Components must tolerate external temperatures from –20 degrees to 110 degrees F. Equipment must withstand 95% humidity without deterioration in performance as measured by MTBF

Security
Confidentiality
    Encryption, authorization, authentication; privacy policies
Accessibility
Legal considerations

Physical Characteristics
Equipment size limitations
Portability
Weight limitations
Materials – durability
Company brand requirements

Support
Maintainability
Instruction manuals if to be maintained internally
Knowledge articles for internal support centers
Spare parts
Developer kits
Service Level Requirements
Repair turn-around (e.g. Must be repaired within 24 hours of service call)
Service representative availability (e.g. 24x7)

Training
Train the trainer
Train specified number of business or technical users
Describe level of participant skills
Number of hours/days
Specify content to be covered at a high level

It is generally considered good practice to ensure that some requirements appear in each of these specific areas after requirements definition is complete. Once all appropriate requirements have been defined, a completed statement of work and detailed specification can be completed. An example statement of work document has been included for reference.

Statement of Work Worksheet

Project Title:	
Project Leader:	Date Written:
Project Purpose:	
Background:	
Project Description:	

Deliverables:
Measures of Success:

Alternate Statement of Work Worksheet
Project Name:

____

Project Mission (strategy):

____ Objective

Statement:

Specification:

____
____
____
____
____
____
____
____
____
____
____
____
____
____
____
____
____
____
____
____
____
____
____
____
____
____
____
____
____

Project Specifications

The development of a detailed specification is intended to convey to all interested parties exactly what will be accomplished on the project, and what the characteristics of those deliverables will be. This is essentially how we will satisfy all the needs and requirements through the development project. This will also support the development of the top-down deliverables for the project.

The specification is intended to convey 'What we will do', and not describe how it will be done. In general there is a very fine line between these two points, but it is important to draw a distinction between functionality and implementation.

Specifications should be as concise as possible, but should also effectively communicate the look and feel of the project deliverables, the functions or services provided by the deliverables, and the key performance parameters (KPP) for the project. The specifications should also convey the success criteria for the project.

To develop a specification one can utilize several techniques.

1) Consider the high-level project deliverables:
    Define each tangible and measurable component
    Define the components' notable characteristics
    Establish several requirements in each of the domain areas

2) Establish the functional hierarchy
    What is the top-level deliverable?
    From what subcomponents may this be made?
    What elements comprise the subcomponents?

Imagine we are defining a specification for a vacation we are planning in the coming year.

What are the deliverables?
    An enjoyable vacation

Subcomponents of this include:
    Travel to the site
    Housing at the site
    Meals
    Entertainment/chores
    Travel back home

Our specification should reflect each of these areas.

Our next task is to add additional details to each of these subcomponents as follows:

Travel to the site
- Plan from home to the airport
- Plan for a flight to the destination locale
- Plan for transport from the airport to the destination
- Established luggage plan for all of these
- Consider meals as required during the transport

Housing
- Plan for seven nights
- Plan to remain at the same hotel
- Make sure the quality of lodging will meet our standards

Once this is completed, we can continue to add additional detail in order to complete our specification.

Project will include travel to the destination site, including:
- Reserved cab ride to airport for a party of four
- Reserved flight from MSP to ABC for a party of four
- Reserve midsize rental car at ABC for eight days
- Luggage to be limited to one carry-on per member of party
- Luggage will be checked from MSP to ABC
- Meals will be purchased at MSP and ABC for party of four

Project will include housing at the destination, including:
- Reserved two rooms of double occupancy at four-star hotel
- Seven-night stay
- Etcetera

Simply by increasing the level of detail included in the description, a thorough specification can be developed. In general, detail should be added until the project becomes specific enough that it might be understood by the majority of readers of the project plan. While it is true that some of the individuals associated with the project may purposely misconstrue elements of the specification, it is unrealistic to expect an all-inclusive specification that will provide an airtight description of the project to be completed. The final project specification should take the form of a written document describing the requirements and specifications for the system to be developed. This may look like the following:

Requirements for a household toaster (partial list)
The toaster shall operate on 120 VAC 60 Hz power
The toaster shall allow for user adjustment
The toaster shall have a power cord utilizing an LG plug
The toaster shall be well tested
The toaster shall provide indicators for the user

The toaster shall have maintenance access panels
The toaster shall be portable
The toaster shall contain a safety override ground fault circuit interrupter

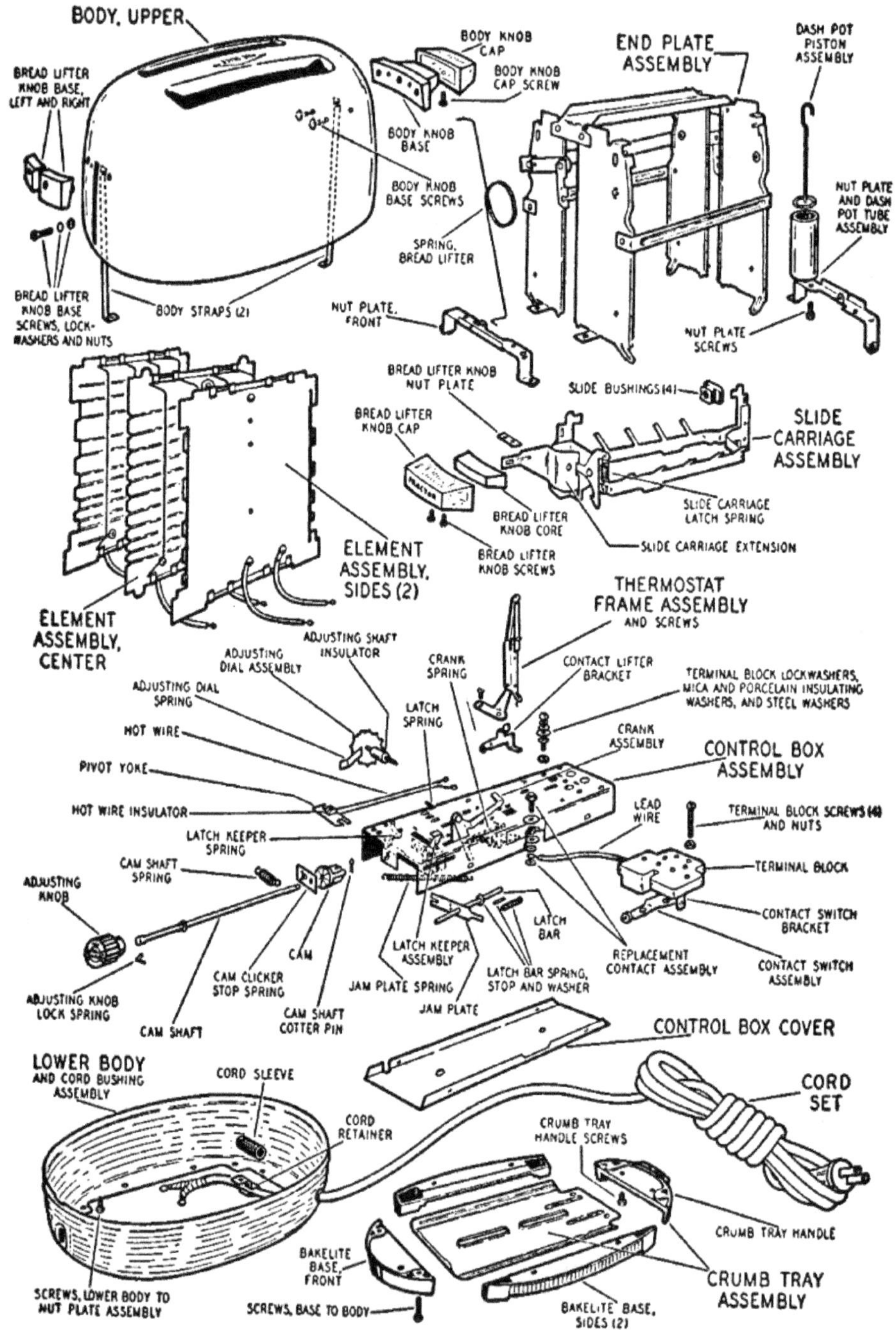

Figure 131: Graphical Toaster Specification

At this point we have defined one of the three factors of interest to the Project Manager; namely, 'What we will do'. Next we move on to the development of the project schedule.

The Project Schedule

In order to develop a project schedule, project leaders must shift their thinking from the tangible deliverables of the project to a hierarchy of tasks that must be completed in order to accomplish the needs of the project. Different methods are suggested for the development of this hierarchy of tasks. The author prefers the approach of Functional Analysis to define not only the capabilities of the system but also for the identification of the tasks that must be accomplished in order to produce those capabilities.

To conduct a Functional Analysis, the project planner must consider the operation of the system under development, and identify all needed capabilities by means of a verb/noun pairing. This is intended to provide a series of capabilities that must be assembled into a holistic view of how the system shall operate. This can be accomplished by a free association exercise in which all of the capabilities are identified, allowing time later in the development to fill in any missing functions or capabilities.

If we consider a simple example, the development of a countertop toaster for our home, one can imagine that we have identified a variety of capabilities that are communicated in this format:

Create setpoint
Clean cycle
Convert power
Update software
Override heat control
Shutdown power
Apply energy
Sense width
Toast product
Since complete
Shut down heat
Unload product
Loan product
Signal status
Input command
Request update
Reset manually

Clean device
Check load condition

These functions must then be assembled into a practical whole, which will provide an Operational View of the toaster system. This is shown in **Figure 132**.

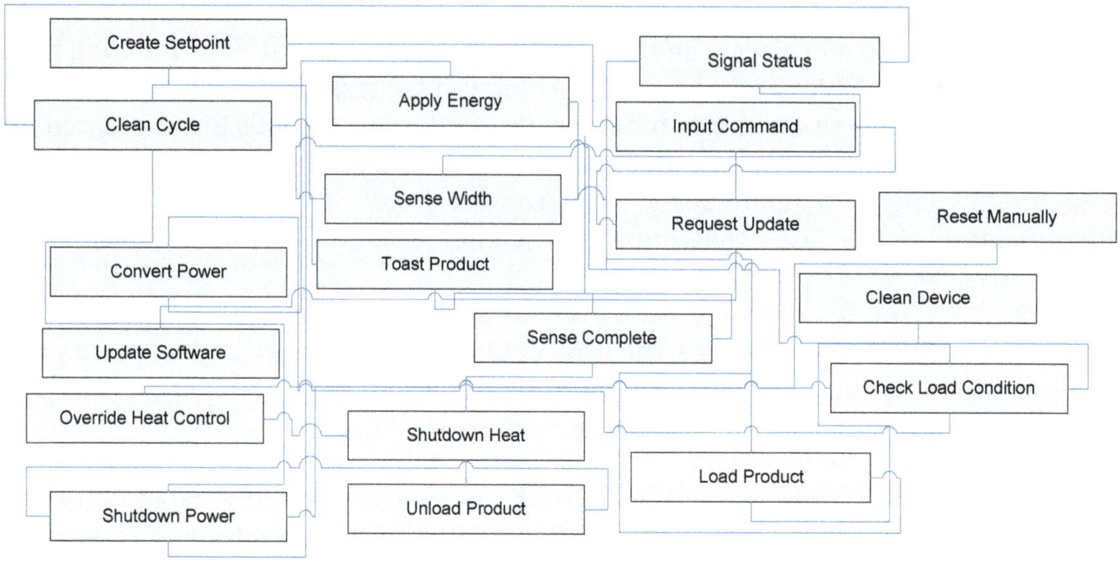

Figure 132: Functional Analysis of Toaster

Once all of the functions have been related one to another, the planner can untangle this diagram, producing what is generally constructed as a top left to bottom right flow of operations for the system. At this point it might become clear that there are several shortcomings that must be addressed, and additional functions may be added or superfluous functions may be deleted. The final outcome is to produce an understandable flow of operations for the proposed system. This is shown in **Figure 133**.

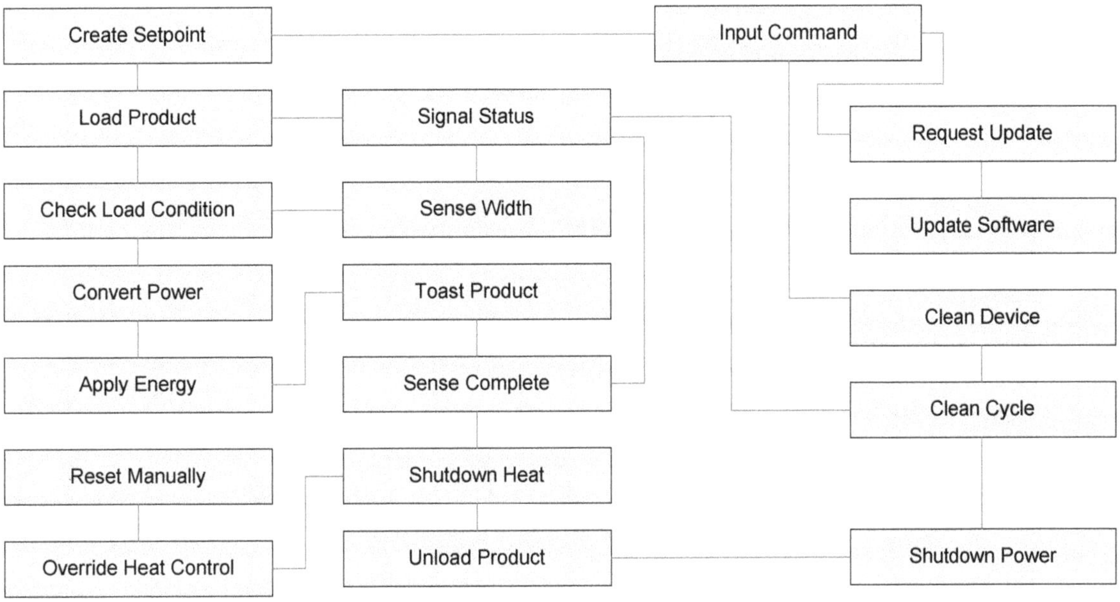

Figure 133: Simplified Functional Analysis

Once the Simplified Functional Analysis has been diagrammed to the satisfaction of the project planner, these functions may be grouped into similar functional families, which will ultimately be used to define the major subcomponents necessary to successfully accomplish the project. This is shown in **Figure 134**.

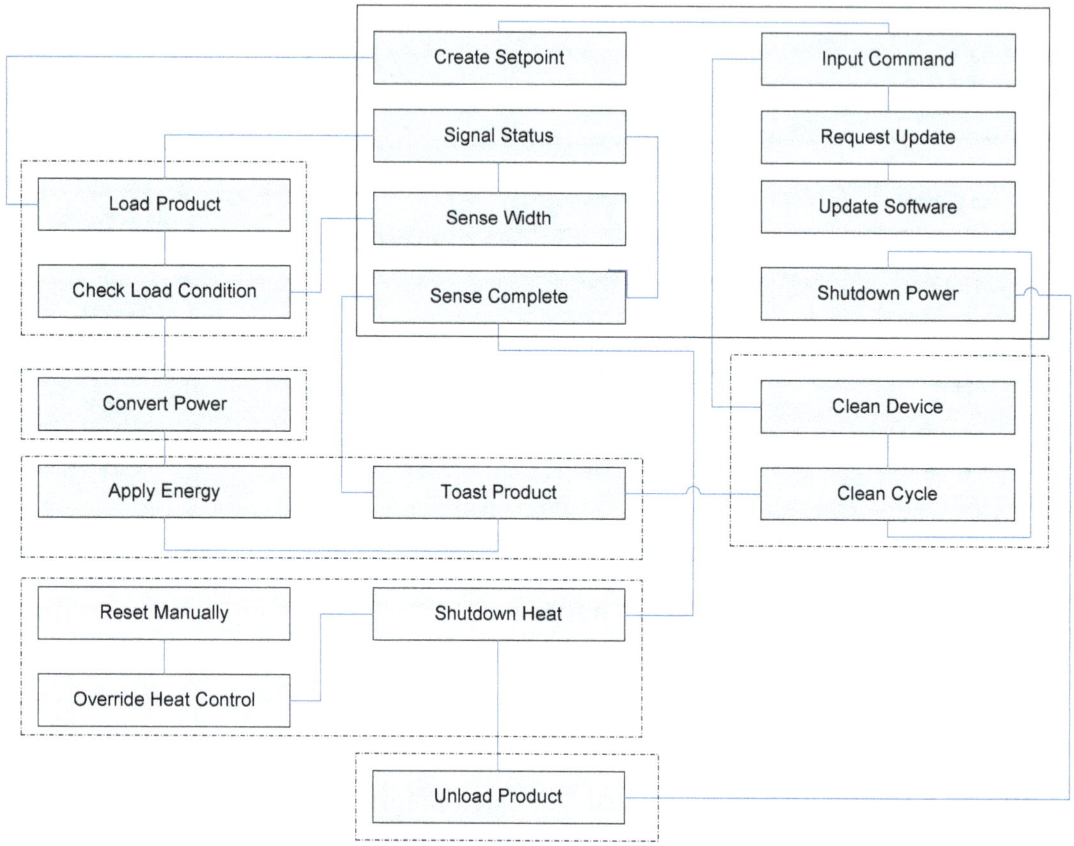

Figure 134: Grouped Functional Analysis

Once the Functional Analysis has been grouped, these groups can be collapsed into collectives or subcomponents that must be provided by the project. This is of particular importance as it defines the major subprojects that will be implemented during the execution of the project, and will coincide with the major headings in our hierarchical task listing. As can be seen, we have identified the top level of our system, the first level of indentation and the second level of functions that must be generated in order to complete the project. This construct will be extremely useful in the development of our work breakdown structure (WBS), which is the list of hierarchical tasks needed for the completion of the project. The WBS is also the most significant input to our schedule. This is shown in **Figure 135**.

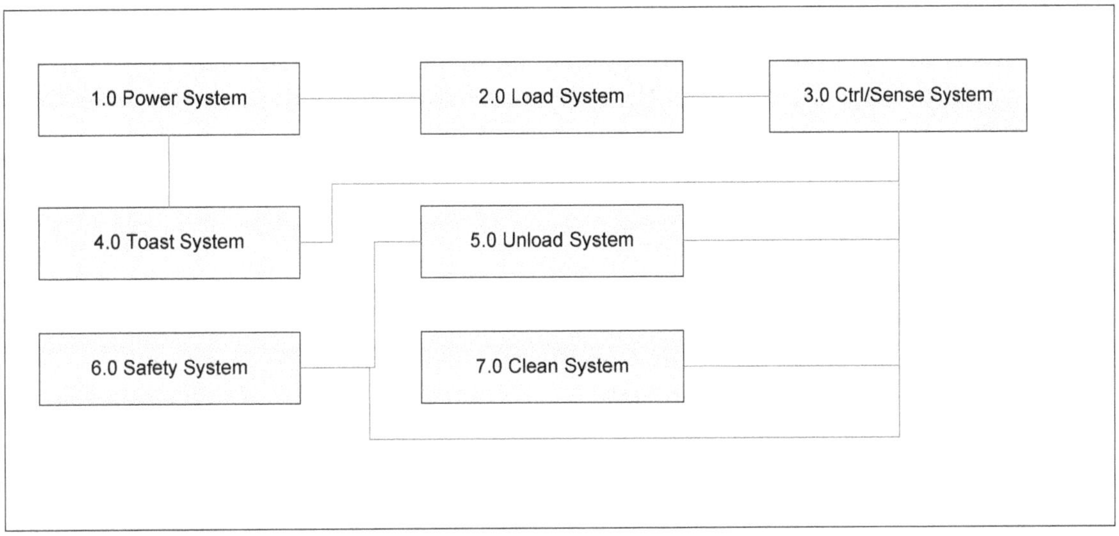

Figure 135: Subcomponents of Toaster

Once this effort has been completed, we are ready to begin the development of our hierarchical task list. This list will be utilized as an input to our schedule to define all necessary tasks that must be accomplished on the project. To clarify for our toaster project:

Toaster project

- Power system
- Load system
- Controls and sensing system
- Toasting system
- Unload system
- Safety system
- Cleaning system

To begin the generation of our work breakdown structure, we must translate these needs into tasks:

Toaster project

- Develop power system
- Develop load system
- Develop controls and sensing system
- Develop toasting system
- Develop unload system
- Develop safety system
- Develop cleaning system
- Integrate systems
- Prototype toaster
- Test toaster
- Release of System

Note that additional tasks have been added to the list to accomplish the integration and testing of the system being developed. The next step is to add additional details to each of the subprojects in order to describe all those tasks necessary to accomplish that milestone. For example:

Develop power system
- Layout power system
- Generate electrical drawings
- Generate wiring diagram
- Generate installation drawings
- Generate bill of materials
- Create documentation package
- Release power system

Additional detail can be added to each of these indented task groupings to create a completed hierarchical task list. Additional detail is typically added to the task list until the final task descriptions become self-explanatory to the reader. It may require many levels of indentation, but the further we delve into the project schedule, the more robust our project plan becomes.

We can see this in an example of making a chocolate cake:

Chocolate cake project
- Planning
    - Locate recipes
        - Select cake recipe
            - Select Grandma's chocolate cake recipe
        - Select frosting recipe
            - Select Four Seasons' chocolate glaze
- Logistics
    - Reserve kitchen for Saturday
    - Clear calendar for Saturday

As can be seen, the same technique is utilized in order to establish the entire hierarchical list of tasks. This is also known as the work breakdown structure, or WBS. Once this list of tasks is complete, we can incorporate it into a variety of project planning software packages. This WBS is inserted into the task description column, as shown in **Figure 136**. This format of project plan is known as the Gantt chart, and is characterized by a hierarchical list of tasks associated with a calendar.

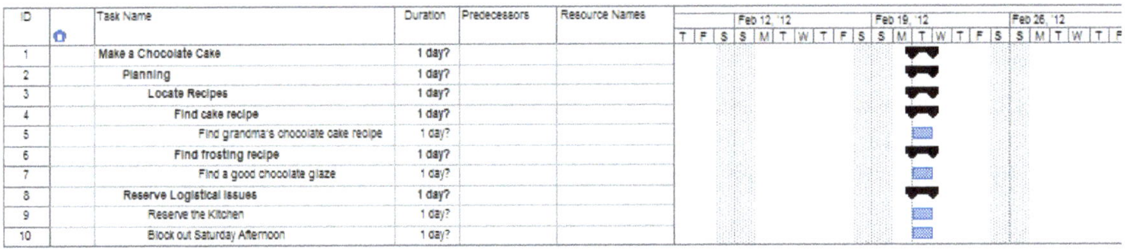

ID	Task Name	Duration	Predecessors	Resource Names
1	Make a Chocolate Cake	1 day?		
2	Planning	1 day?		
3	Locate Recipes	1 day?		
4	Find cake recipe	1 day?		
5	Find grandma's chocolate cake recipe	1 day?		
6	Find frosting recipe	1 day?		
7	Find a good chocolate glaze	1 day?		
8	Reserve Logistical Issues	1 day?		
9	Reserve the Kitchen	1 day?		
10	Block out Saturday Afternoon	1 day?		

Figure 136: WBS input to Gantt Chart

Note that on the Gantt chart, there are ten independent tasks. Of these ten tasks, only four are shown as bars in the diagram. These are also coincidently the tasks that are indented the furthest to the right throughout the project plan. As it turns out, these are the only tasks that require work to be completed. All other levels of tasks that have been out-dented from these tasks are simply summary tasks and are used as placeholders in the project plan (they define the hierarchy).

What this means is that we have only four tasks or work packages that must be completed to complete this scheduled project. Software packages such as Microsoft project and ProjectLibre take care of this identification automatically for us.

Our next challenge is to identify the relationship between the various tasks on the project plan. This is normally accomplished by linking the tasks one to another in order to clarify the order in which these tasks should be accomplished. The most common method links the finish of a task to the start of the following task. Other methods such as the start of a task may be linked to the start of another task as well. Ultimately, this linking enables us to establish the total duration of the project. This is shown in **Figure 137**.

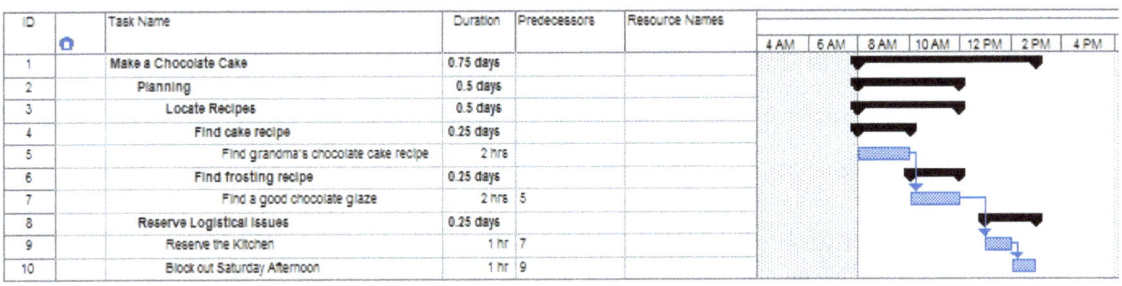

ID	Task Name	Duration	Predecessors	Resource Names
1	Make a Chocolate Cake	0.75 days		
2	Planning	0.5 days		
3	Locate Recipes	0.5 days		
4	Find cake recipe	0.25 days		
5	Find grandma's chocolate cake recipe	2 hrs		
6	Find frosting recipe	0.25 days		
7	Find a good chocolate glaze	2 hrs	5	
8	Reserve Logistical Issues	0.25 days		
9	Reserve the Kitchen	1 hr	7	
10	Block out Saturday Afternoon	1 hr	9	

Figure 137: Linked Project Schedule

Durations for each of the work package tasks must be defined at this point. These are input into the Gantt chart, and result in a chronological view of the project to be completed. Only the work package tasks are assigned durations and linked. All other tasks are summary tasks and simply follow the positioning of the work packages. Durations are estimated based upon the planner's knowledge of the work to be completed. In cases where the planner does not have sufficient knowledge to feel comfortable with these estimates, the aid of a subject matter expert is used. In some cases, if completed project plans are available with actuals that are similar to the tasks being planned, those actuals might be used as a basis of estimate.

At this point, resources are assigned to the work packages in order to establish what skill areas (and therefore what billing rates) are necessary to accomplish those tasks. This is shown in **Figure 138**.

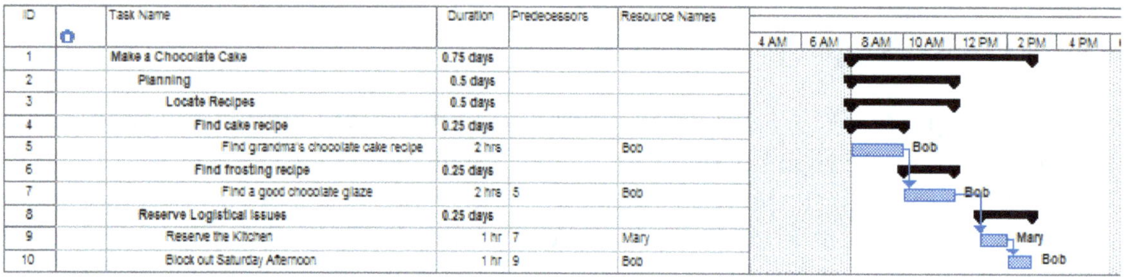

Figure 138: Resource Assignments

Resources are only assigned to the work packages. This ensures that resources are only paid for once in the project's final roll up of the budget. If resources are added to the summary tasks, the project leader will pay for those resources multiple times during the execution of the project.

Indentation levels of the task descriptions are related to the hierarchical task construction, whereas the indentations on the schedule portion of the Gantt chart are chronological in nature. These levels of indentation are entirely independent of each other.

Utilizing this methodology, we have successfully generated a detailed project plan to make a chocolate cake. There are multiple methods that can be used to accomplish this, and any means by which the project planner can generate an effective work breakdown structure and Gantt chart are considered reasonable in the generation of the project plan.

Networking Method

A second method of plan generation is the networking method. Imagine you have been provided a list of tasks to be completed, in no particular order. Begin by assigning a letter to each of the tasks:

(A) Bake ham
(B) Make sauce
(C) Warm buns
(D) Cut ham
(E) Preheat oven
(F) Buy food
(G) Get recipes
(H) Make salad

Build a network of the tasks based upon their interrelationships with each other. Then assign a duration value between the tasks representative of the effort required for completion. This is an event-based system, and allows easy modification of the network as tasks are added or deleted. This network is shown in **Figure 139**.

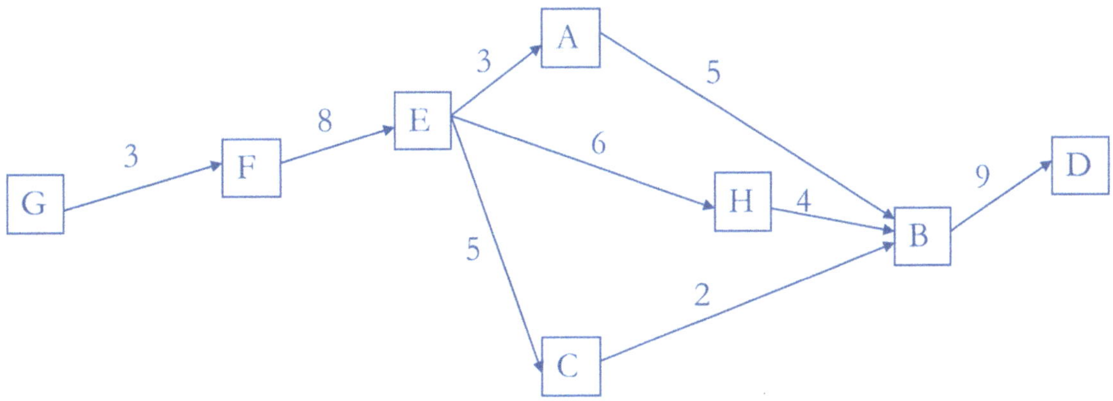

Figure 139: Network Diagram

The network diagram has a significant advantage in that it can help us to identify the critical path. This critical path is of significant importance to us as it represents the longest path from the start of the project to the end of the project. Those tasks that fall along the critical path are critical due to the fact that any delay in the critical path results in a commensurate delay in the delivery of the project. The critical path in this case is shown in **Figure 140**, and is comprised of tasks G-F-E-H-B-D.

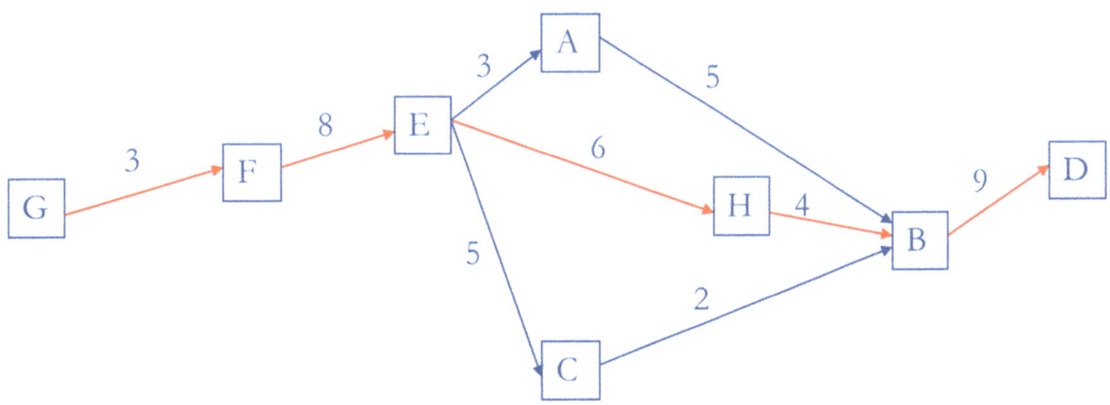

Figure 140: Network with Critical Path

In this instance the critical path contains ten units of time between tasks E and task B. The alternative paths require eight and seven units respectively. This means there are several units of float associated with each of these pads, which actually means that the eight units of work may float anywhere within the ten units allotted and still finish without impacting the completion of the project.

The network method of scheduling is often used by project planners due to the perception that it requires less work than the production of the Gantt chart. In reality these two approaches contain the same information for the project plan. In fact, if you choose to transition from a Gantt chart to network chart, the methodology is quite straightforward. First we will begin with the Gantt chart, as shown in **Figure 141**.

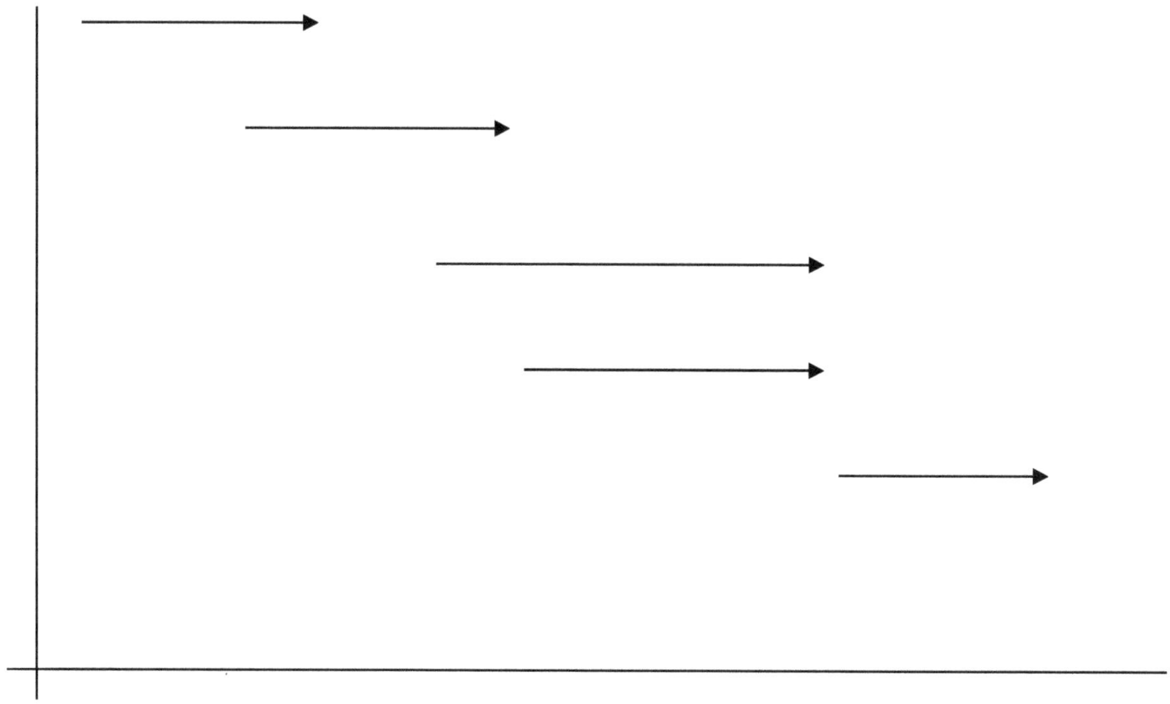

Figure 141: Gantt Chart

Next we will assign an event to the start and finish of every task.

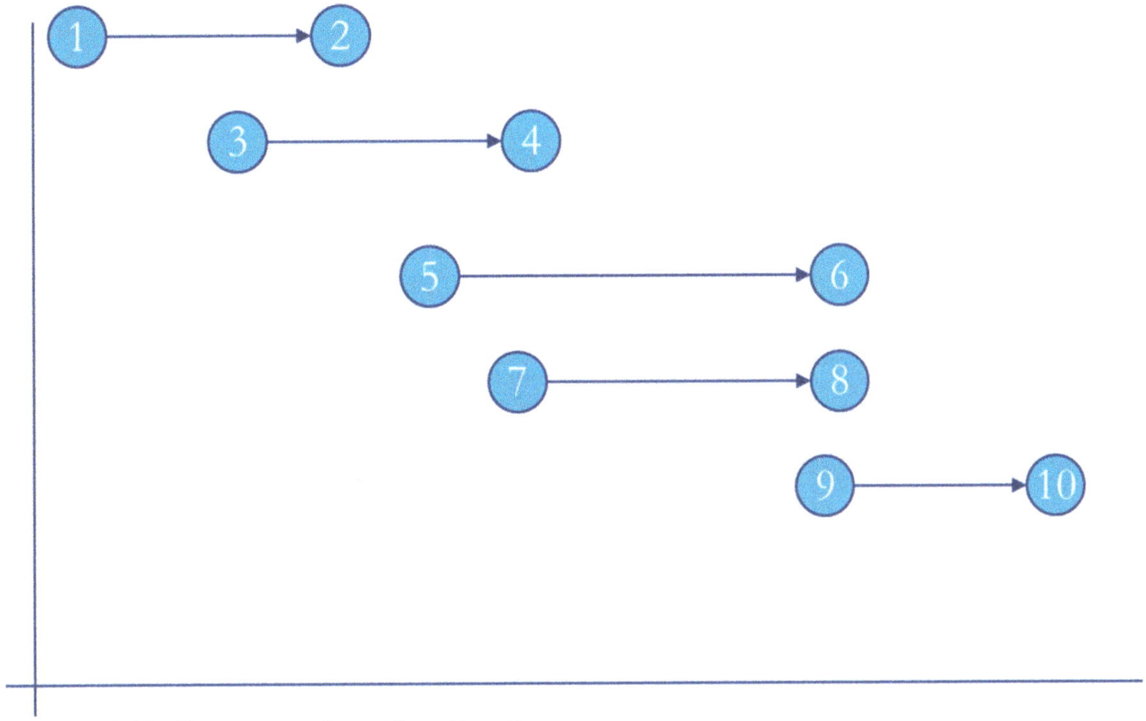

Figure 142: Events assigned to Gantt

We then assign relationships between the events representative of the relationships between the tasks on the Gantt chart.

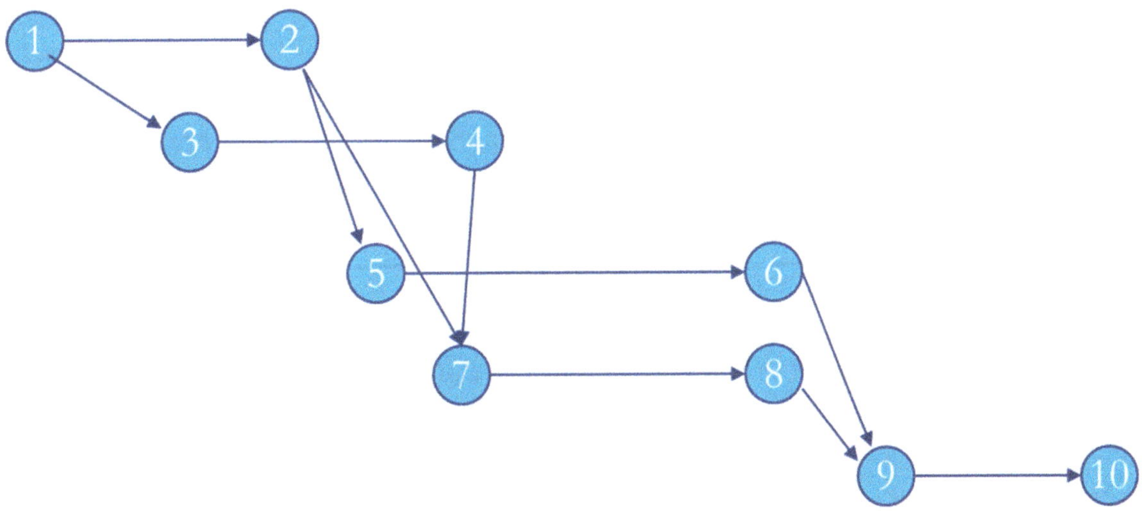

Figure 143: Network Diagram

Durations are added to the links and the critical path is identified.

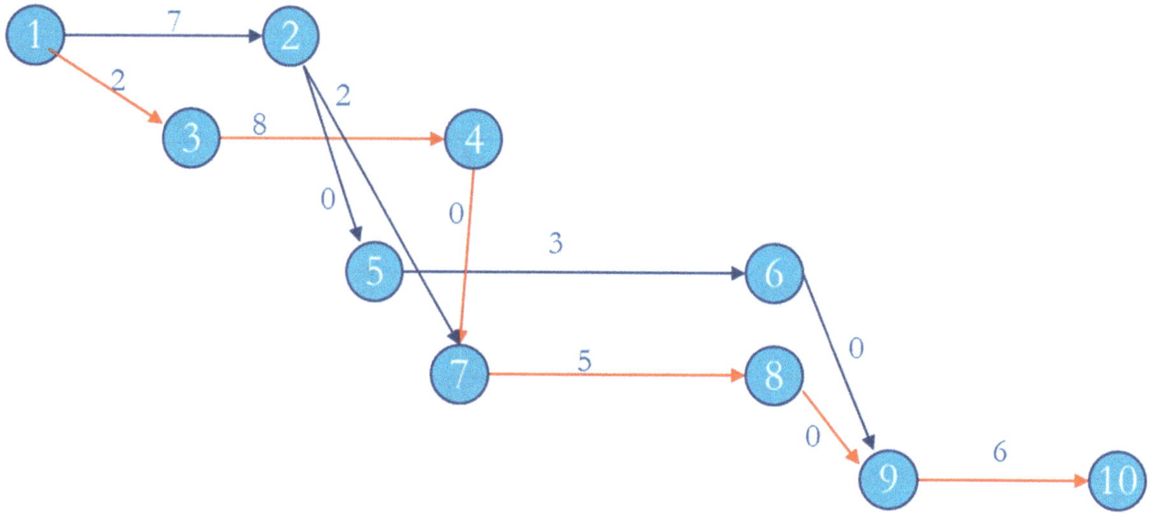

Figure 144: Network Diagram with Critical Path

The Preferred Method of Schedule Generation

The method that is utilized in many large companies for the generation of an appropriate schedule for a project plan includes the development of an Integrated Master Plan (IMP) and an Integrated Master Schedule (IMS). This method is relatively straightforward.

There are three steps to completing the integrated master plan.

1) Identify events
2) Identify accomplishments
3) Identify criteria

To complete a integrated master schedule, expand the task hierarchy to include all of the work packages necessary to accomplish the criteria, accomplishments and events. A simple software project is included as an example:

IMP Level	IMP ID	Task Name
Event	5.0	− Internal Integration & Pre-Qual Test Complete
Accomplishment	5.1	− GNC Software Builds Complete
Criteria	5.1.1	+ GNC Software Build 1 Complete
Criteria	5.1.2	+ GNC Software Build 2 Complete
Criteria	5.1.3	+ GNC Software Build 3 Complete

Figure 145: Integrated Master Plan

The addition of tasks will complete our integrated master schedule in much the same way that we developed our Gantt chart. The completed integrated master schedule is shown in **Figure 146**.

IMP Level	IMP ID	Task Name	CWBS	Duration	Early Start	Early Finish
Event	**5.0**	**− Internal Integration & Pre-Qual Test Completed**		**189 d**	**9/1/09**	**5/21/10**
Accomplishment	**5.1**	**− GNC Software Builds Completed**	**1.1.6.5**	**189 d**	**9/1/09**	**5/21/10**
Criteria	**5.1.2**	**− GNC Software Build 1 Completed**	**1.1.6.5a**	**63 d**	**9/1/09**	**11/26/09**
Task	5.1.2	Determine GNC SW Build 1 Requirements	1.1.6.5a	10 d	9/1/09	9/14/09
Task	5.1.2	Conduct GNC SW Build 1 Design	1.1.6.5a	18 d	9/15/09	10/8/09
Task	5.1.2	Code GNC SW Build 1	1.1.6.5a	25 d	10/9/09	11/12/09
Task	5.1.2	Conduct GNC SW Build 1 Unit Test	1.1.6.5a	10 d	11/13/09	11/26/09
Criteria	**5.1.3**	**− GNC Software Build 2 Completed**	**1.1.6.5b**	**63 d**	**11/27/09**	**2/23/10**
Task	5.1.3	Determine GNC SW Build 2 Requirements	1.1.6.5b	10 d	11/27/09	12/10/09
Task	5.1.3	Conduct GNC SW Build 2 Design	1.1.6.5b	18 d	12/11/09	1/5/10
Task	5.1.3	Code GNC SW Build 2	1.1.6.5b	25 d	1/6/10	2/9/10
Task	5.1.3	Conduct GNC SW Build 2 Unit Test	1.1.6.5b	10 d	2/10/10	2/23/10
Criteria	**5.1.4**	**− GNC Software Build 3 Completed**	**1.1.6.5c**	**63 d**	**2/24/10**	**5/21/10**
Task	5.1.4	Determine GNC SW Build 3 Requirements	1.1.6.5c	10 d	2/24/10	3/9/10
Task	5.1.4	Conduct GNC SW Build 3 Design	1.1.6.5c	18 d	3/10/10	4/2/10
Task	5.1.4	Code GNC SW Build 3	1.1.6.5c	25 d	4/5/10	5/7/10
Task	5.1.4	Conduct GNC SW Build 3 Unit Test	1.1.6.5c	10 d	5/10/10	5/21/10

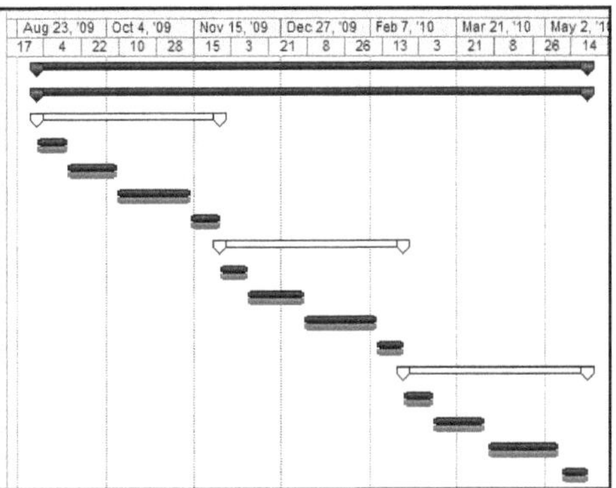

Figure 146: Integrated Master Schedule

This method is very straightforward and develops a highly detailed schedule that provides a coherent view of the project to be performed. As a final example in this section, we will provide the definition of a project schedule for a holiday dinner.

Project name: Holiday Dinner
Requirements:
Ten guests
Turkey with stuffing
Side dishes: potatoes, carrots, rolls, cranberries
Pre-meal wine
Dessert: apple and pumpkin pie

To define Events:
Holiday Meal

To define Accomplishments:
Holiday meal
- Invite guests
- Prepare for meal
- Prepare meal
- Serve meal

Serve dessert
Reset home

To define Criteria:

Holiday meal

- Invite guests
- Prepare for meal
  - Clean house
  - Prepare kitchen
  - Prepare dessert
- Prepare meal
  - Prepare turkey
  - Prepare side dishes
  - Set dining table
- Serve meal
  - Serve wine
  - Serve dinner
  - Clear dinner
- Serve dessert
  - Serve dessert
  - Clear dinner
- Reset home
  - Clean kitchen
  - Clean remainder of home that has been used

We then detail the task structure, creating the IMS (Integrated Master Schedule):

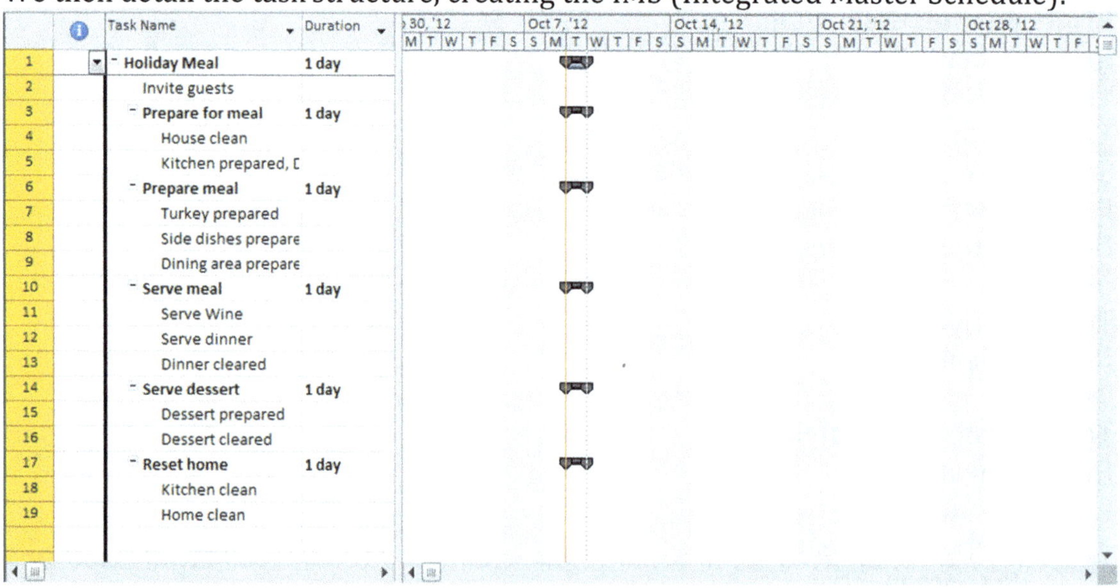

Figure 147: IMS

Add the task durations and link the tasks together:

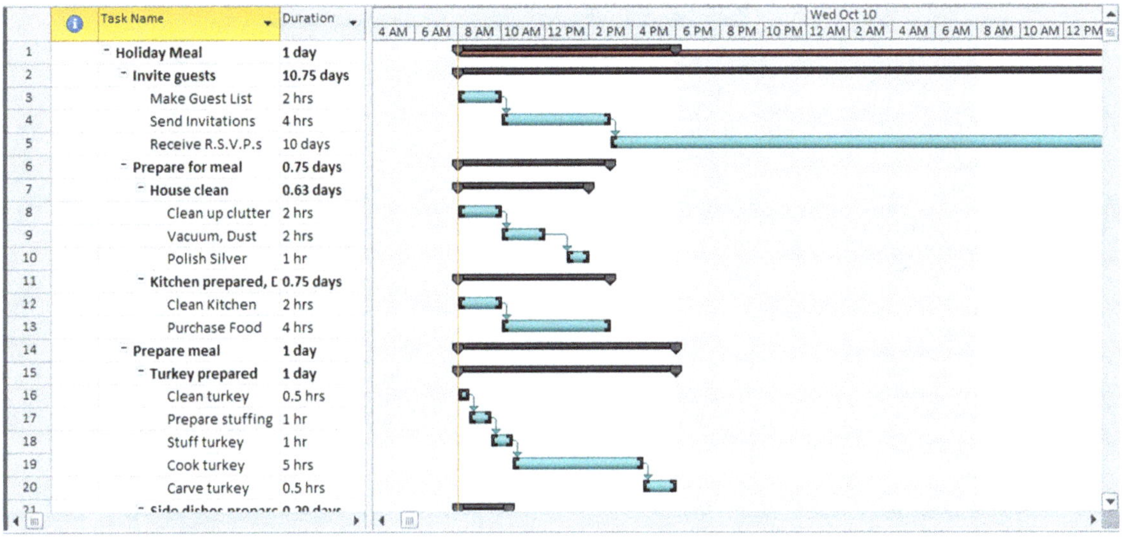

Figure 148: IMS with links

Link the subprojects together:

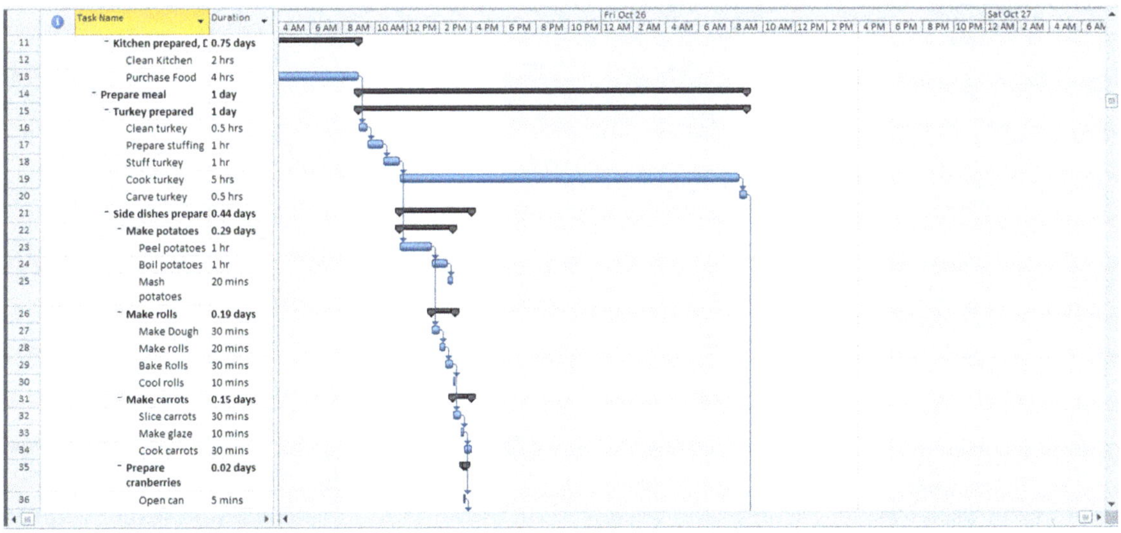

Figure 149: Linked IMS

Then assign resources and complete your project schedule.

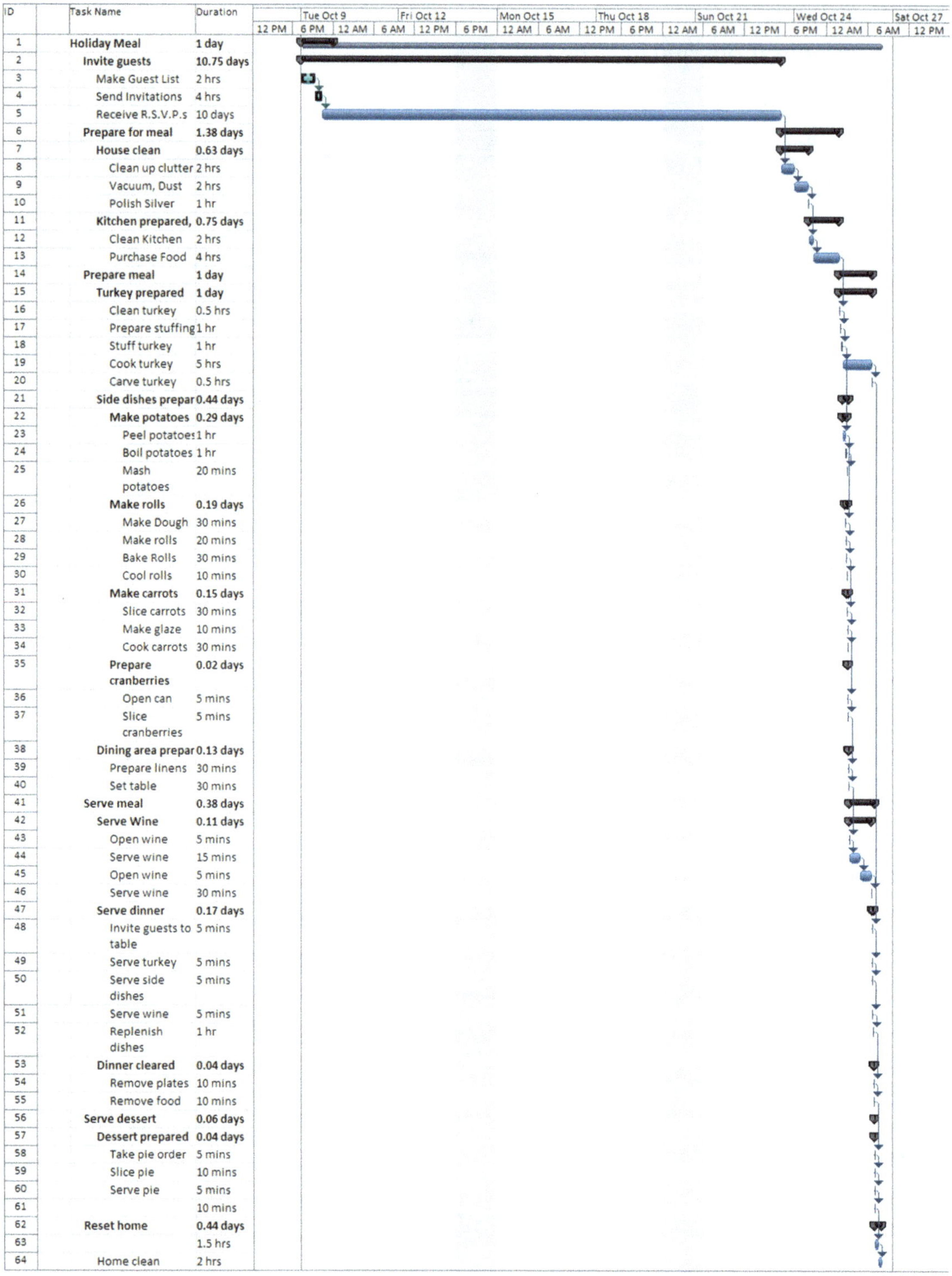

ID	Task Name	Duration
1	**Holiday Meal**	**1 day**
2	**Invite guests**	**10.75 days**
3	Make Guest List	2 hrs
4	Send Invitations	4 hrs
5	Receive R.S.V.P.s	10 days
6	**Prepare for meal**	**1.38 days**
7	**House clean**	**0.63 days**
8	Clean up clutter	2 hrs
9	Vacuum, Dust	2 hrs
10	Polish Silver	1 hr
11	**Kitchen prepared,**	**0.75 days**
12	Clean Kitchen	2 hrs
13	Purchase Food	4 hrs
14	**Prepare meal**	**1 day**
15	**Turkey prepared**	**1 day**
16	Clean turkey	0.5 hrs
17	Prepare stuffing	1 hr
18	Stuff turkey	1 hr
19	Cook turkey	5 hrs
20	Carve turkey	0.5 hrs
21	**Side dishes prepar**	**0.44 days**
22	**Make potatoes**	**0.29 days**
23	Peel potatoes	1 hr
24	Boil potatoes	1 hr
25	Mash potatoes	20 mins
26	**Make rolls**	**0.19 days**
27	Make Dough	30 mins
28	Make rolls	20 mins
29	Bake Rolls	30 mins
30	Cool rolls	10 mins
31	**Make carrots**	**0.15 days**
32	Slice carrots	30 mins
33	Make glaze	10 mins
34	Cook carrots	30 mins
35	**Prepare cranberries**	**0.02 days**
36	Open can	5 mins
37	Slice cranberries	5 mins
38	**Dining area prepar**	**0.13 days**
39	Prepare linens	30 mins
40	Set table	30 mins
41	**Serve meal**	**0.38 days**
42	**Serve Wine**	**0.11 days**
43	Open wine	5 mins
44	Serve wine	15 mins
45	Open wine	5 mins
46	Serve wine	30 mins
47	**Serve dinner**	**0.17 days**
48	Invite guests to table	5 mins
49	Serve turkey	5 mins
50	Serve side dishes	5 mins
51	Serve wine	5 mins
52	Replenish dishes	1 hr
53	**Dinner cleared**	**0.04 days**
54	Remove plates	10 mins
55	Remove food	10 mins
56	**Serve dessert**	**0.06 days**
57	**Dessert prepared**	**0.04 days**
58	Take pie order	5 mins
59	Slice pie	10 mins
60	Serve pie	5 mins
61		10 mins
62	**Reset home**	**0.44 days**
63		1.5 hrs
64	Home clean	2 hrs

Figure 150: Completed Project Schedule (IMS)

Evaluating Risk

Risk assessment is critical to the success of the Project Manager. It is common for the average Project Manager to begin the plan with a high likelihood of failure. This is largely due to the planning of a project irrespective of other work occurring within the organization. Risk assessment allows us to define the potential challenges that will arise during the project execution phase and account for them in our planning.

To define risk, consider each task as a separate project. We can postulate that a given task repeated by a wide variety of people will result in a variety of outcomes. In this case, we are interested in the completion time of the task. We see by the following diagram that the resulting outcome of the task can be modeled by a normal distribution.

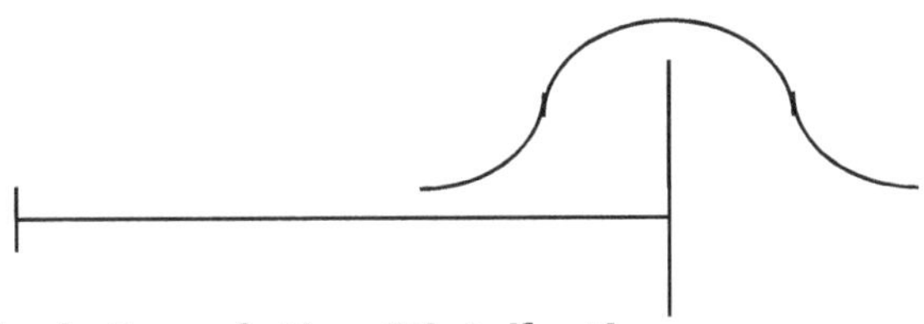

Figure 151: Task Completion Distribution

The issue of risk assessment is then related to defining the approximate shape of the distribution about the nominal task end. This shape will define for us the likelihood of completing the task as expected. Taken as a whole, all tasks within a project can be assessed in this manner, allowing a confidence interval to be calculated for the entirety of a project.

The information to define this distribution can be collected in a straightforward manner as shown in the following table.

Task #	Nominal Duration	Worst Case Duration
1	4d	5d
2	1d	1d
3	5d	40d
4	6w	7w
5	10d	11d

As expected, the nominal duration is the intended schedule time to complete the work. The worst-case duration is defined to assess the approximate width of the standard deviation. It is surprising to observe that several of the tasks show a distribution width that exceeds the task length. This is an insight into the psyche of the project planner, as he or she does not desire to communicate that the task schedule is thirty-five days plus or minus one year. This creates a significant credibility issue for the planner, and is therefore mentally suppressed.

We desire to communicate a schedule with realistic risks associated, and therefore a worst-case scenario is chosen as the appropriate method, bypassing this blind spot in the planner's viewpoint. As we observe from the above table, the risk of certain tasks is quite low and predictable. Other tasks show a great deal of risk, such as task three. These of course are the tasks requiring the attention of the Project Manager, as the greatest variability in response will be observed in this area.

For example, consider a project made up of the following tasks:

Task #	Nominal Duration	Worst-Case Duration
1	10d	15d
2	10d	20d
3	10d	10d
4	10d	13d

The project at the outset shows expected times of completion:

$T_{nominal} = 10d + 10d + 10d + 10d = 40\ d$

$T_{worst\ case} = 15d + 20d + 10d + 13d = 58\ d$

This provides us with an idea of the standard deviation, assuming the worst case shows a 3-sigma distribution.

$\sigma = 6\ d$

Therefore, if we are asked if we are confident that we will complete the project in forty-six days we statistically predict that this encompasses a nominal value of 40 d plus a one-sigma area of the distribution. This corresponds to a confidence interval as follows:

$C.I._{46days} = 67\%$

In the event 26 days have passed (about half the management allowed time) and the first two tasks are completed, we can recalculate the confidence interval.

$T_{nominal} = 10d + 10d = 20\ d$

$T_{worst\ case} = 10d + 13d = 23\ d$

$\sigma = 1\ d$

We must consider our confidence now of completing the remaining tasks within 22 days:

$C.I._{22days} = 20 \text{ days} + 2 \text{ sigma} = 93\%$

Therefore, although we are approximately half completed with the project, and half completed with the tasks, our confidence interval has increased due to the lower risk tasks still remaining to be completed. If we are able to shift the higher risk tasks further forward into the project, or if we are able to focus directly upon these tasks, we can have the greatest impact on the overall project delivery.

Imagine that in the above example, the tasks are identical but the first two tasks are completed within 25 days.

Task #	Nominal Duration	Worst Case Duration
1	10d	15d
2	10d	20d
3	10d	10d
4	10d	13d

$T_{nominal} = 10d + 10d = 20 \text{ d}$

$T_{worst\ case} = 10d + 13d = 23 \text{ d}$

$\sigma = 1 \text{ d}$

We must consider our confidence now of completing the task within 23 days:

$C.I._{23days} = 20 \text{ days} + 3 \text{ sigma} = 99.7\%$

Therefore it is very important for the project manager to mix the risks within the project appropriately to provide for lower risks as the project proceeds. Additionally, it should be noted that based on this model, all projects are not including risk calculations are currently planned with a confidence interval of only 50%.

Once we have completed the integrated master schedule, we have completed the second task in our project planning challenge, namely we have defined 'how long it will take'. Next we move on to developing the budget.

Budget Definition

To define an appropriate budget for our project, we must first make sure that our Integrated Master Schedule makes sense. To accomplish this, we will meet with the Functional Managers to ensure the schedule we have generated has realistic task identification and durations that will produce a reasonable expectation of being able to complete the project as planned. We meet with Functional Managers because they are the subject matter experts for the areas of interest, and it is very likely they will help us to produce a more accurate and representative schedule. What's more, since the Functional Managers will assign their personnel to this project, this is our first step at socialization of our upcoming project plan.

For each of the work package tasks within our Integrated Master Schedule, we will assign costs as appropriate for that task. The resources that are utilized to accomplish that task and their associated billing rates typically represent these costs. These billing rates are multiplied by the duration of the tasks to generate the actual cost of those resources accomplishing that task.

Indirect labor hours
Fixed costs
Materials
Purchased parts
Subcontracts
Freight
Travel costs
Other miscellaneous costs (10-15% add-on)

These can be incorporated into a simple project budget worksheet, such as included here for reference.

Cost Estimating Worksheet

Project Title:				Project Manager:		
Activity	Schedule	Budget				
No.: Name/Description	Hours	Labor hr x rate	Material	Other	Total	Resource Manager
Prepared by:				Date:		

As the budget is completed, we have accomplished what we originally set out to achieve.
We have defined:

What will we do?
How long will it take?
How much will it cost?

Our only remaining task is to complete the Project Charter. The charter document is a single document that incorporates all of the work accomplished to date on the project plan.

The Project Charter Document is used in a similar manner as a contract within the organization. The information is filled out at the completion of the project planning process, and defines exactly what will be accomplished on the project, what will not be accomplished on the project, the expected costs and the expected timing of the project. Space is provided to gain approval signatures from stakeholders on the project and the intent of the project charter is to use the document both to open and close the project.

An example Project Charter Document is included to provide some insight into the technical information required. We have already defined all the information required to complete the document, but the charter acts as a single repository for the information.

Project Charter

Prepared By:	
Version:	
Date:	

Sponsor:	
Project Manager:	

Approval Signatures

Name ____

Title ____Date______________________________ ______________

Name ____

Title ____Date______________________________ ______________

Final Customer Sign Off*

Name____

Title ____Date______________________________ ______________

Name____

Title ____Date______________________________ ______________

Key Stakeholders			
Project Sponsor:		Phone:	
Division:		Department:	

* Should be completed as a final signoff to the project.

Project Manager:		Phone:	
Division:		Department:	
Colleges / Departments Affected:			

Summary:	How does this project contribute to the overall Vision of the Organization? With which strategy does this project align?

Problem/ Opportunity	What is the problem or opportunity that this project poses to correct or take advantage? (Not the project benefits)

Benefits	Identify the Tangible and Intangible benefits that will be derived from this project implementation.
Tangible: Intangible:	

Critical Success Factors	Critical success factors are those business factors that should be accomplished in order to consider this project a success.
1.	
2.	
3.	
4.	

In Scope	Describe the items that are in scope for the successful completion of this project. Any items not listed are not understood as being in scope.

Out of Scope	Describe the items that are out of scope for the successful completion of this project. Any items not listed are not understood as being out of scope.

Risks / Opportunities	What risks or opportunities are associated with completing this project, or not completing this project? What risks or opportunities can be identified when considering resources needed to complete this project?

Costs	What are the costs associated with this project *(Cost are usually estimated with a confidence rating at the beginning and continue to firm up during the planning process.)*

Costs:				
Hardware	$ -		Consulting	$ -
Software	$ -		Internal Labor	$ -
Support Contract	$ -			
Other	$ -			
Expense Total	$ -		**Labor Total**	$ -
	Total Cost:	$ -		
Initial Cost	$ -			
Monthly Cost of Ownership	$ -			
Monthly Benefit	$ -			
Cummulative ROI	#DIV/0!			
**Based on a three year rate of return*				

ROI is calculated as 1-[{Total Cost + (Cost of Ownership * 36)} / (Monthly Benefit * 36)]

Interdependencies	In order to complete this project, does it require any other projects or tasks outside of the Project Team's control to complete in order to be successful?

Milestones	In this section the user can identify the top 4 milestones and their proposed completion dates.	
Milestone	Description	Completion Date
1.		
2.		
3.		
4.		

Technology	What technology will be used to deliver the solution?

Team Composition	How will the Project Team be composed?

Stakeholders	List all of the individuals, departments or users who will be impacted by the outcome of the project.

Resources	What resources are needed to complete this project in scope and on time? Resources include people, skills and equipment. Include an estimate to the time a resource is needed.

Assumptions	What assumptions are being made about the resources, budget, schedule, etc.?

Constraints	List major constraints that are on the project team and the completion of this project. Be sure to include milestone dates, resource availability and budget.

To fully understand the Project Management lifecycle, there are additional phases of the project that must be understood. These include Project Control, Project Execution and Project Close-out. Many textbooks exist regarding these phases of the project, and the finer points of several of these areas go beyond the scope of this text.

Phase 3: Project Control

Details of Project Control are included in the Earned Value Management section of this text. In that section, a method is presented that allows appropriate status definition and reporting on the project during the execution phase. That chapter outlines the details that are adequate for the majority of users to understand how to track and report on a project, and discusses the benefits of making reasonable predictions regarding the outcome of the project during the project execution.

Phase 4: Project Execution

Project Execution is probably the most convoluted and challenging phase of any project endeavor. This is the portion of the project in which the project leader must rely upon his or her ability to provide an engaging project plan that will result in the employees investing their efforts in the successful completion of the plan. This covers a wide variety of topics, including individual motivation, communications, social styles and personality styles, group behavior, sociology, psychology and anthropology. In general, technical professionals gain these skills through application of the project techniques previously mentioned, and often must learn by observation in the real world. It is of great benefit to the project leader, however, to investigate the wide variety of publications available to assist in this area of project execution.

The Role of Complexity and Chaos in Project Management

Discussions with individuals involved in the Project Management function will often produce the same list of Project Management problems. These problems will typically include items such as the following:

- Creeping Scope (Performance)
- Unrealistic Expectations
- Lack of Planning
- Lack of Control
- Poor Tracking
- Anomalies occurring during the Project
- Chaos throughout the project
- Others ...

Given the opportunity, the list could be extended to several pages. What is interesting, however, is the common thread in all the problems listed; that is, the appearance of Chaos within the structure of the project. This Chaos is frequently credited with the translation of a project with an exquisite project plan into a jumbled mess that is virtually unmanageable.

Unfortunately, it is not possible to tell you that your projects will be entirely manageable. In fact, what is closer to the truth is the assertion that all projects exist within a Chaotic environment, and will experience Chaotic periods. That's the bad news. The good news is that Chaos is understandable and manageable on a system level. You can manage a project in a Chaotic environment if you are willing to allow the project to act in a natural manner.

What is Chaos and Complexity?

Chaos is a relatively new branch of mathematics that has received a great deal of attention in recent years. One of the more straightforward explanations of the phenomena may be found in Bak (1996). In this profoundly named text, Bak describes from an empirical perspective how the realm of Chaos exists all around us. Bak notes that Chaotic models can be used to describe earthquake patterns, the use of English words in this textbook, and the pattern of behavior of the U.S. stock market.

Some have taken issue with Bak's work, stating that his model is nothing more than an empirical description of complex behavior, and is therefore of little use to the practitioner. One need only consider the fact that the majority of our natural sciences knowledge was developed by similar means to recognize the potential fallacy of this argument.

Bak proposes a model of Chaos that exists in the physical domain between order and randomness. This is shown in Figure 152.

Order Chaos
Random

Figure 152: Chaos in the Physical World

Obviously there is some hope, as chaos is not a random process. It is clear that chaos is more difficult to understand than an ordered model. In fact, chaos is a phenomenon that exhibits the behavior of a complex system. The system is so complex that we are unsure of all the constructs contributing to the behavior.

This type of model reflects the real world of Project Management. In short, Project Management is impacted by a wide variety of sources within an organization, many of which are beyond our reach or awareness. Thus, the study of Chaos will enable us to better understand the overall system of Project Management.

A Simple Chaotic Model: The Sandpile Model

It is beyond the scope of this text to describe in detail how Chaos and complexity function. It is sufficient for the understanding of the Project Manager to investigate the behavior of two simple Chaotic models. The first of these models is the Sandpile model. The second model is the Metered Traffic model.

The Sandpile is described in detail by Bak (1999, *How Nature Works*. Springer. New York) and can be implemented with a simple deck of cards. An exercise of this type will enable the Project Manager to very quickly grasp the behavior of a Chaotic system. First, we must define some of the behaviors of a complex system:

They operate under a set of simple rules
They self-organize into a state of criticality
Systems in a state of high criticality exhibit catastrophes

To model a simple sand pile, place 9 playing cards on a table in a three by three matrix as shown in Figure 153.

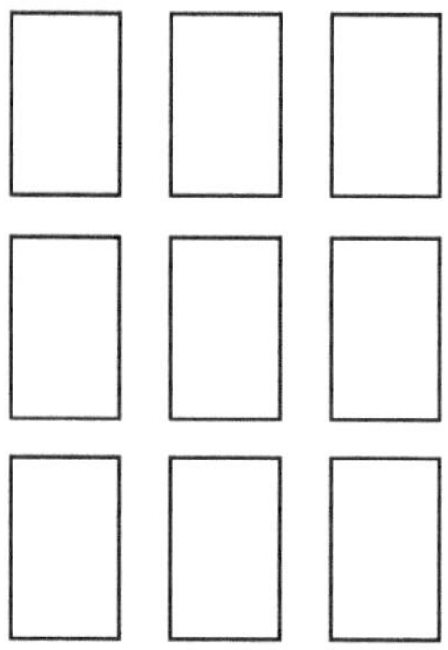

Figure 153: Sand pile Matrix

The 3X3 matrix will represent a Sandpile, and the addition of single cards to the system will represent a single grain of sand being dropped onto the Sandpile. As might be expected, when enough grains of sand are dropped into the matrix, some avalanching will occur. This effect is described with a simple set of operational rules for the simulation. These rules are as follows:

1. Introduce a card randomly to the matrix
2. If there are four cards in any pile, these cards avalanche into the N,S,E,and W piles adjacent to the original pile. This is shown for the center pile in Figure 154.
3. If a pile is on the edge of the matrix, cards moving out of the matrix are considered to have exited the system.
4. Record how many cards moved during the process.
5. Continue introducing cards into the matrix until you have exhausted the original deck.

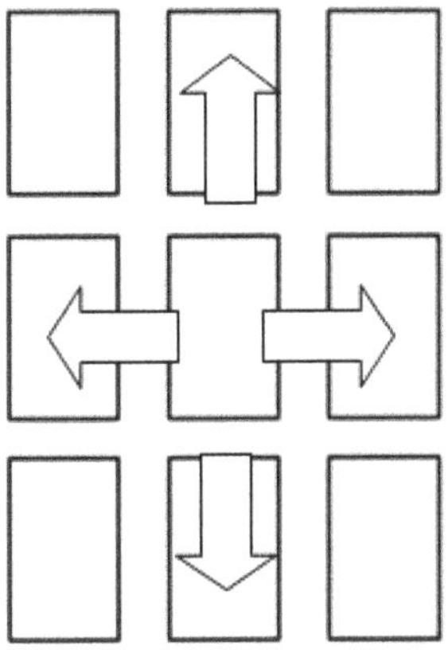

Figure 154: Avalanche from the Center Pile

It becomes very clear to the individual experiencing this simulation that a large number of cards can move with the introduction of a single card. This typically happens only after a number of cards have been introduced into the matrix, thereby increasing the criticality of the system. As criticality of the system reaches a critical mass, avalanches become the norm. These catastrophes or anomalies are therefore present in all systems, and the frequency of their occurrence is dependent upon the critical mass of the system. Remember those anomalies that occurred on your last project? They are a part of every project you will ever manage, and we must discover how to minimize the frequency of occurrence of those anomalies.

The 3X3 Sandpile model is very small, and therefore easily implemented manually. It can be thought to represent the behavior of nine simultaneous projects operating within a single organization. Upon investigation we discover that the edge effect experienced by the system is large, caused by the fact that eight of the nine cards form an edge of the matrix. Therefore a larger matrix might be more informative. However, a larger system rapidly becomes unwieldy for the user, and is best modeled on a computer.

It is clear that many grains of sand must be introduced into the system before any avalanches occur. As the critical mass increases, the avalanches become more frequent, until eventually they become the norm. (Note that the numbers recorded in **Figure 155** represent the number of avalanches, representing the movement of four grains of sand.)

What is of interest to the Project Manager is a perspective on the patterns occurring in a Chaotic system such as this. The frequency distribution of a 25X25 matrix is compiled in **Figure 155**.

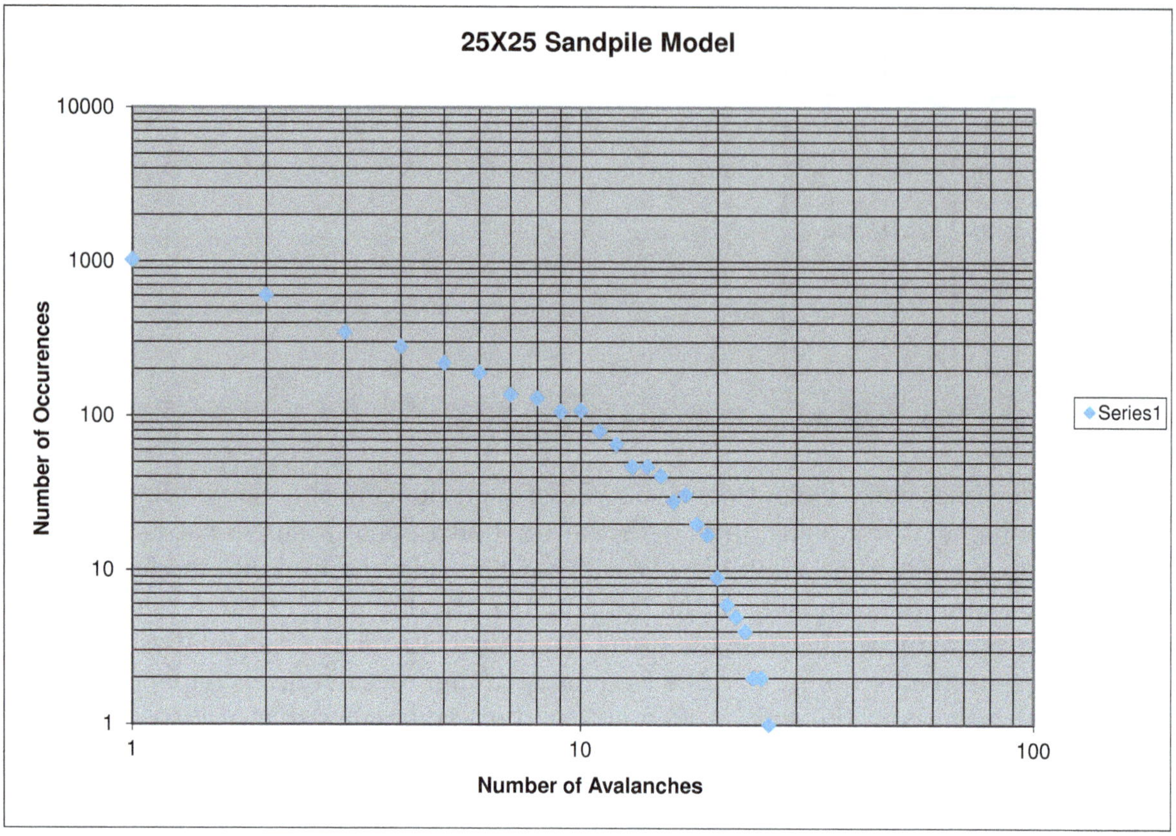

Figure 155: Frequency Distribution

Thus, the behavior of a Chaotic system follows a power law. Note that the curving downward of the line is caused by both edge effect and sample size. Larger data sets do not exhibit this phenomenon.

The project exists within a Chaotic system. There are numerous other projects within the system, all bumping into each other. The single project system will exhibit many small magnitude catastrophes throughout its duration, as well as fewer large catastrophes. These are the anomalies discussed earlier. They will exist in every project, and we must manage the project to account for their existence.

A Model of Chaos: Metered Traffic

A second model of Chaos is demonstrated in the behavior of metered traffic. What is noteworthy in this example is that a Chaotic problem is presented with a solution for handling this problem. The method of solution is applicable for the Project Manager, as well.

Metropolitan areas frequently experience traffic problems due to excessive volume. One method utilized to alleviate this problem is metering the flow of traffic into the system. This is typically accomplished on entrance ramps to freeway systems. Some have argued that this simply transfers the problem from freeways to surface streets. However, as we shall see, this is not the case.

The traffic system operates under simple set of rules. The traffic flow is analyzed, the difference between the actual traffic flow and the maximum possible flow is calculated, and the rate at which cars are metered into the system is set. The design of the system is focused on the maximization of traffic flow through the system. Obviously this system has all the elements necessary for Chaotic behavior.

The system was designed to maximize the flow of traffic, which is essentially maximizing the critical mass of the system. This would imply that the system would be more susceptible to catastrophe brought about by a small disturbance. Remember that the small disturbance in the Sandpile model was the introduction of a single grain of sand to the system. In the traffic case, the disturbances may be sourced from those elements already present in the system as well as those entering the system. As a tangible representation of an avalanche, imagine what occurs when a driver of one of those cars applies the brakes. The drivers immediately following also apply the brakes, and the disturbance avalanches throughout the system. In this case, a small tap on the brakes might impact ten thousand cars.

The solution presented by the traffic managers was to reduce the critical mass of the system. Instead of metering cars at a maximum capacity level, they instead chose to meter cars at a seventy-five percent level. This results in a buffer zone around each car, so that when brakes are applied it affects far fewer vehicles, perhaps ten cars. Thus the reduction in criticality of the system results in smaller, more manageable catastrophes. There is a counterintuitive conclusion to be drawn from this. *If cars are metered at a less than maximum capacity, more cars get through.* Although waiting time to enter the freeway is increased, overall trip time is reduced in the general case. Consistency is increased, and therefore the quality of the system is increased.

This conclusion was extended to the manufacturing factory floor, where Industrial Engineers reduced the rate of the bottleneck machine in the plant from maximum capacity down to seventy-five percent capacity. Again the result was counterintuitive. They were able to make more parts. *By reducing capacity, they were able to make more parts.* Chaos plays a very significant role in our lives. Lack of planning for Chaotic behavior can be a very costly oversight.

As a last note, one can discover many Chaotic models about us in the world. These models typically exhibit behavior that is more complex than is easily understood on a detailed level. The system level is more easily understood, such as in the case of the Windows operating system. The behavior is Chaotic, but yields an appropriate complex behavior for the system.

What about the Project Manager?

Dr. Fred Zimmerman (1997) most succinctly stated the role of the Project Manager when he noted that the Project Manager must embrace the fact that as a Project Manager, he or she will always be in trouble with someone, the question is just with whom. Likewise, Chaos exists on every project. Acceptance of this fact enables the Project Manager to accept the Chaotic behavior of the system as a necessary consequence of operating in an imperfect environment.

If we could insulate every project, so that it would never be impacted by other projects within the organization, we could significantly reduce or eliminate Chaos from our list of concerns. However, as Peters (1987) has pointed out, all white-collar work is essentially project oriented. This would imply that all Project Managers are faced with the likely occurrence of Chaos within their daily activities.

What can be done to manage a project within a Chaotic environment? The project manager must recognize that excess resources must be committed to the project in terms of management reserve, to account for the complex behavior of the organization. As stated in the traffic example, it was discovered that a committed reduction to seventy-five percent of the resources resulted in superior traffic flow. Thus the Project Manager must schedule the resources on the project to not exceed seventy-five percent utilization.

In project terms, this means that three days of effort might be required to complete a specific task. The duration utilized in the budget will reflect a three-day effort to accomplish the task. The schedule, however, will reflect a four-day completion cycle. This communicates that three days of work will be accomplished, but it will take four days to finish the work. Why is this the case?

Phone calls
Email
Voice mail
Conversations
Interruptions
Other projects
Other people
Anomalies
Others

The counterintuitive conclusion to be drawn from this discussion is as follows:

If you schedule projects for seventy-five percent capacity, you will get more work accomplished.

Project Managers will increase their success rate by reducing the rate at which they schedule projects. In many cases, organizations schedule personnel on projects at one hundred percent utilization, or worse yet one hundred twenty percent utilization. This is accomplished through a company culture that demands mandatory overtime of the employees. Project Managers must plan for Chaotic behavior on their projects by reducing the utilization of the resources to a more realistic level. Is seventy-five percent an appropriate level to compensate for these effects? That is for the Project Manager to determine in his or her specific organization and application.

It is clear that Chaos can be managed to account for all those things that occur during a project that cannot be predicted or incorporated into the project plan. It is time for Project Managers to become forward thinking, and understand that those anomalies that caused problems on their last projects are ever present in the system and will appear on their future projects.

Phase 5: Project Close-out

Details of Project Close-out are included in the section of this text dedicated to business rhythm and reporting. The final phase of the project is a critical one, and a checklist is provided that allows a practitioner to review the current state of his or her completed project and compare it to the original plan, allowing some prediction of future performance on projects based upon the accuracy of those estimates and execution. For further development, the reader is urged to consult one of the many Project Management textbooks available.

Program and Project Management in a Nutshell

The following elements are needed to successfully execute a program or a project: a design, test plan, program plan, and a schedule. If any one of those elements is missing, the program will be unsuccessful. The outcome will not necessarily deliver what the customer wants (no requirements), no end product (no design or poor design), not all of the tasks completed (no program plan or incomplete program plan), or the project not done on time (though schedule or an adequate schedule). Fundamentally, the above answers the questions of who, what, when, where, cost, and the risk of getting a product that works and is delivered within the cost budget.

Program Management/Project Management

There are five phases of Project Management:

(1) Project Conception and Initiation

An idea for a project will be carefully examined to determine whether or not it benefits the organization. During this phase, a decision-making table will identify if the project can realistically be completed.

(2) Project Definition and Planning

A project plan, project charter and/or project scope may be put in writing that outlines work to be performed. During this phase, a team should prioritize the project, calculate the budget and schedule, and determine what resources are needed. Two very useful documents for use in each phase of Project Management are the Integrated Master Plan (IMP) and the Integrated Master Schedule (IMS).

The Integrated Master Plan (IMP) and the Integrated Master Schedule
The Integrated Master Plan (IMP) and the Integrated Master Schedule (IMS) are used in the planning and scheduling of work efforts in complex systems development. The IMP is an event-driven strategy that documents the significant accomplishments and ties each accomplishment to a key program event. The primary purpose of the IMP and the supporting detailed schedules of the IMS is their use by the customer and Contractor for planning, executing and tracking the program's technical efforts, schedule status, cost status, and risk mitigation efforts.

The IMP is expanded to a time-based IMS to produce a networked and multi-layered schedule showing all detailed tasks required to accomplish the work effort contained in the IMP. The IMS flows directly from the IMP. The IMS is made up of "tasks" (SOW tasks) depicting the work effort needed to complete the criteria (the completion events defined in the IMP; milestones). It is a detailed, time-driven plan for program execution that helps to ensure the achievement of on-time delivery dates. It must show task (Statement Of Work elements - SOW) progress, interrelationships, and dependencies.

(3) Project Launch or Execution

Resource tasks are distributed and teams are informed of responsibilities.

(4) Project Performance and Control

Project Managers will compare project status and progress of the actual plan, as resources perform the scheduled work. During this phase, Project Managers may need to adjust schedules (IMS) or do what is necessary to keep the project on track.

(5) Project Close

An evaluation is necessary to highlight project success and/or learn from project history.

IMP/IMS Creation Process

The following figures are drawn using the Operational Activity view from the DoDAF Enterprise Architecture and the IDEF0 Modeling Language.

Data/Tools/ Process/Products (output of the process)

- Data we have
  - Physical architecture
  - Work Breakdown Structure (WBS)
  - Statement of Work (SOW)
  - RVTM
  - Constraints
- Tools
  - OV-5 (Excel, and Power point)
  - Microsoft Project
  - Microsoft Word

Figure 156: Developing an IMP and IMS

Data/Tools/ Process/Products (output of the process)

- Process outline description
  - Discover User needs - done
  - Discover requirements - done
  - Create functional architecture - done
  - Create Generic physical architecture - done
  - Create instantiated physical architecture – done
  - Allocate functions to instantiated physical architecture - done
  - Create Work Breakdown Structure (WBS) - done
  - Create Statement of Work (SOW) – done
  - Create IMP – to be done
  - Create IMS – to be done

Figure 157: Outputs of the IMP/IMS Process

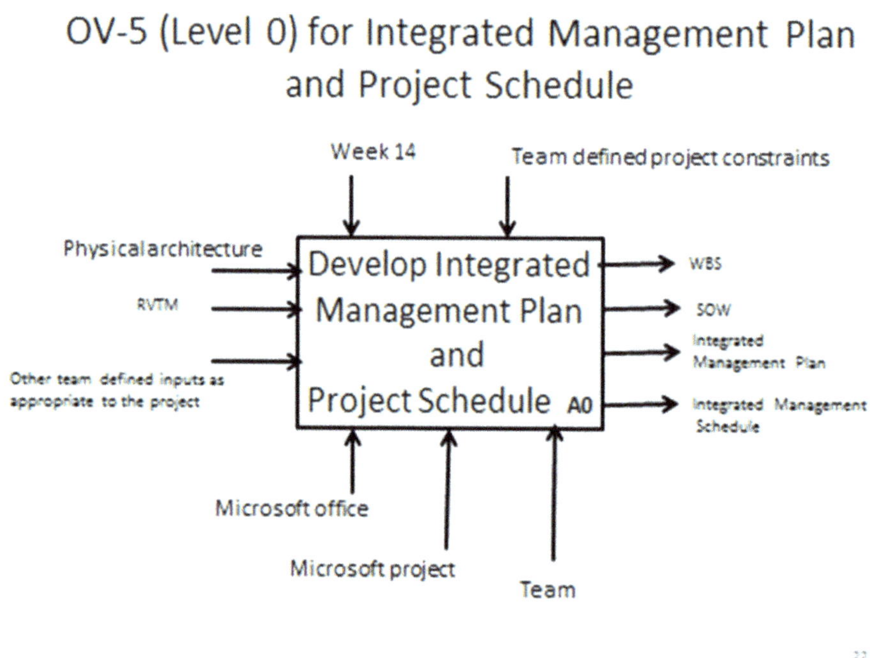

Figure 158: Using DoDAF to Develop a Plan

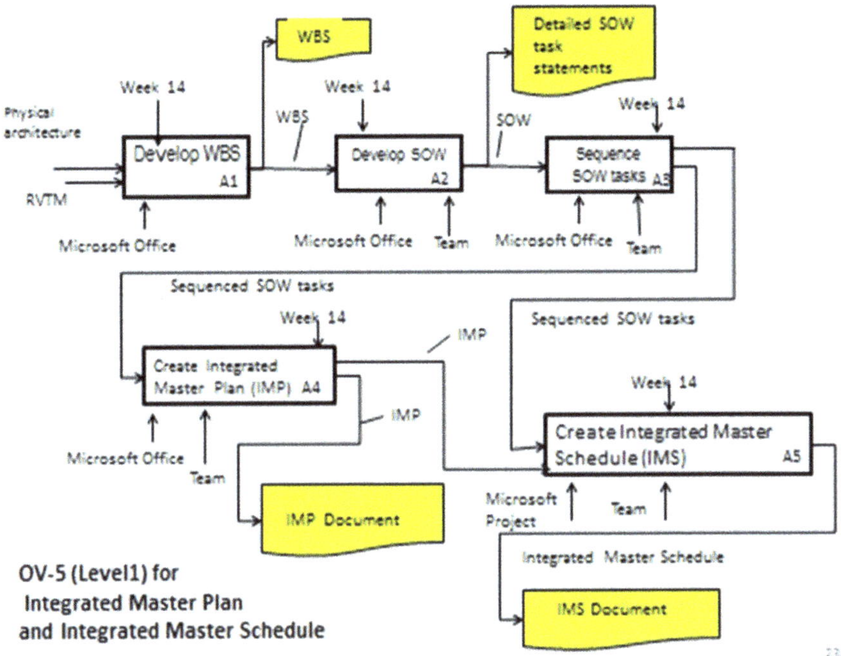

Figure 159: OV-5 for IMP/IMS Creation

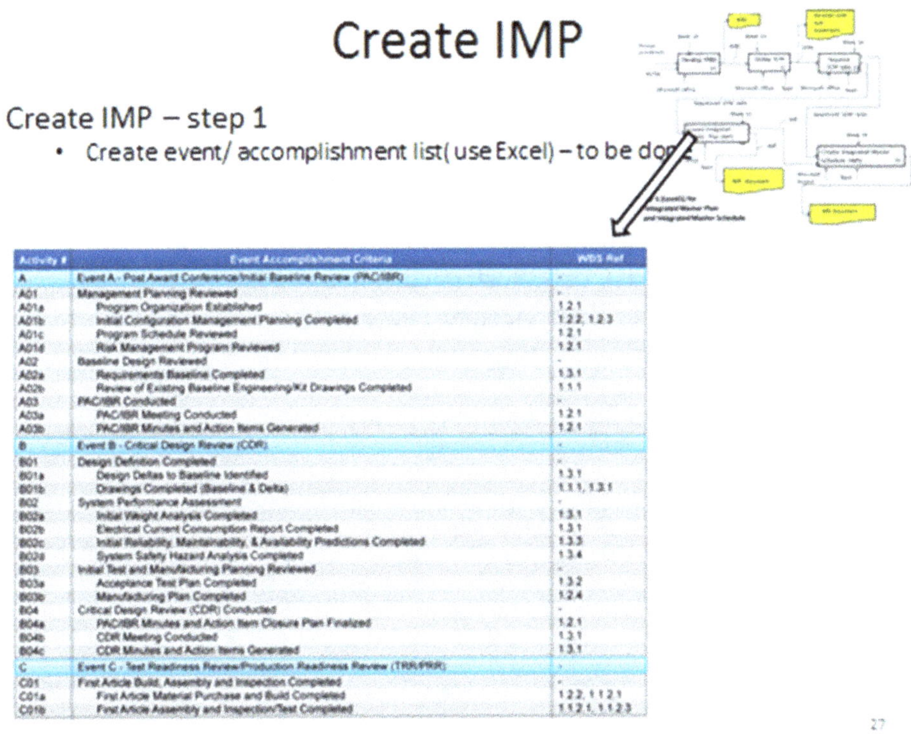

Activity #	Event Accomplishment Criteria	WBS Ref
A	Event A - Post Award Conference/Initial Baseline Review (PAC/IBR)	-
A01	Management Planning Reviewed	-
A01a	Program Organization Established	1.2.1
A01b	Initial Configuration Management Planning Completed	1.2.2, 1.2.3
A01c	Program Schedule Reviewed	1.2.1
A01d	Risk Management Program Reviewed	1.2.1
A02	Baseline Design Reviewed	-
A02a	Requirements Baseline Completed	1.3.1
A02b	Review of Existing Baseline Engineering/Kit Drawings Completed	1.1.1
A03	PAC/IBR Conducted	-
A03a	PAC/IBR Meeting Conducted	1.2.1
A03b	PAC/IBR Minutes and Action Items Generated	1.2.1
B	Event B - Critical Design Review (CDR)	-
B01	Design Definition Completed	-
B01a	Design Deltas to Baseline Identified	1.3.1
B01b	Drawings Completed (Baseline & Delta)	1.1.1, 1.3.1
B02	System Performance Assessment	-
B02a	Initial Weight Analysis Completed	1.3.1
B02b	Electrical Current Consumption Report Completed	1.3.1
B02c	Initial Reliability, Maintainability, & Availability Predictions Completed	1.3.3
B02d	System Safety Hazard Analysis Completed	1.3.4
B03	Initial Test and Manufacturing Planning Reviewed	-
B03a	Acceptance Test Plan Completed	1.3.2
B03b	Manufacturing Plan Completed	1.2.4
B04	Critical Design Review (CDR) Conducted	-
B04a	PAC/IBR Minutes and Action Item Closure Plan Finalized	1.2.1
B04b	CDR Meeting Conducted	1.3.1
B04c	CDR Minutes and Action Items Generated	1.3.1
C	Event C - Test Readiness Review/Production Readiness Review (TRR/PRR)	-
C01	First Article Build, Assembly and Inspection Completed	-
C01a	First Article Material Purchase and Build Completed	1.2.2, 1.1.2.1
C01b	First Article Assembly and Inspection/Test Completed	1.1.2.1, 1.1.2.3

Figure 160: Creating the IMP

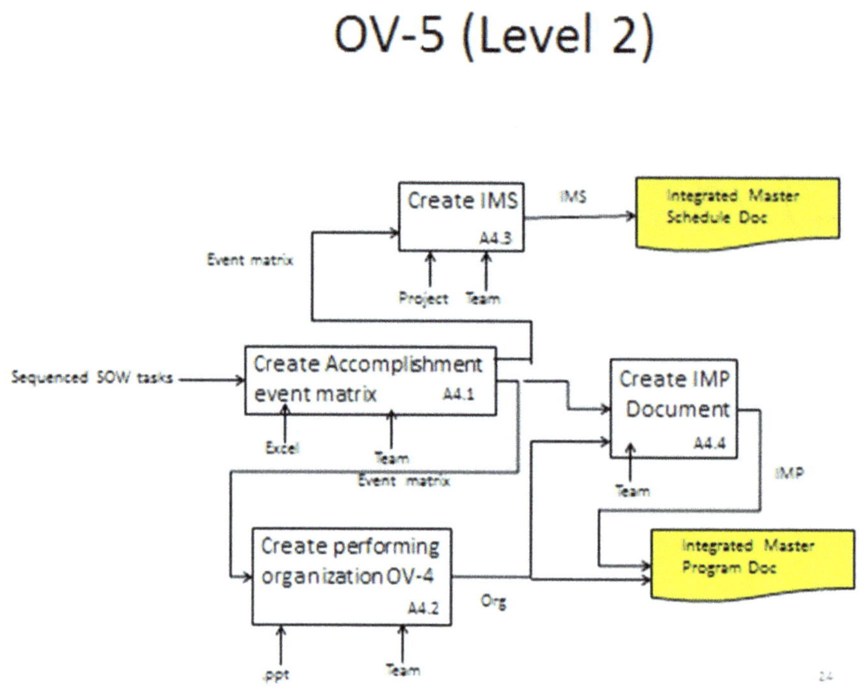

Figure 161: Drill Down of the OV-5

Create IMS

- In concert with the development of the IMP accomplishments and criteria matrix, begins the creation of a detailed project schedule
  - Inputs
    - IMP events (accomplishments and criteria matrix0
    - WBS
    - SOW
    - RVTM
- Output of effort
  - Project schedule

30

Figure 162: Inputs for the IMS

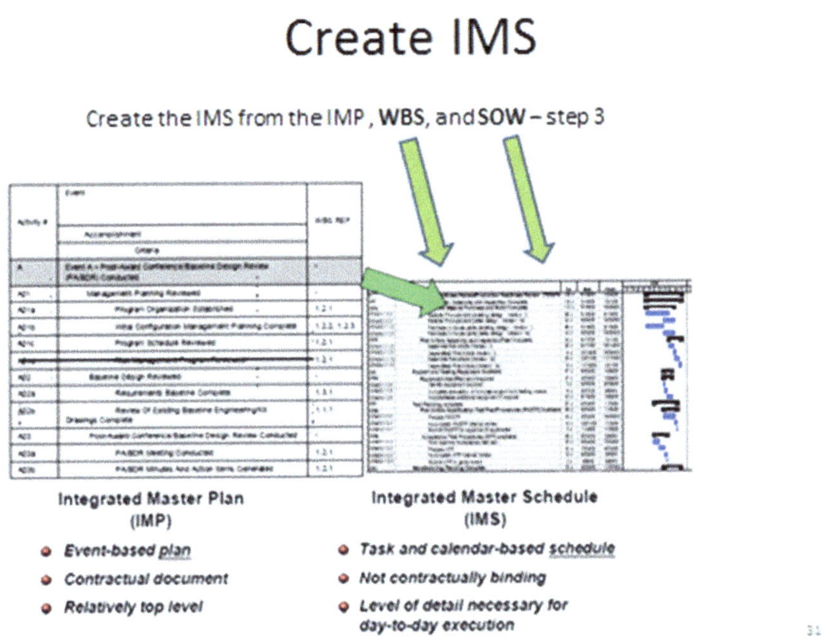

Figure 163: Creating the IMS

Create IMP Document

Document the IMP in " a Program Plan" Document

- Document outline

  Section 1- Introduction

  1. A short description of the program
  2. Assumptions and ground rules
  3. Event an action item dictionary
  4. Organization description

32

Figure 164: Creating a Program Plan

Create IMP Document

Section 2

1. Describe the IMP numbering system and list in table format program/project accomplishments and criteria.
    - Criteria can be either quantitative or qualitative must be measurable.
    - Provide summary, high-level, IMS schedule.

Section 3

1. IMP narratives
    - **Task** narratives may be used to describe the approach to performing tasks
    - **Process** narratives consist of concise summaries of the of key management and functional processes and how they relate to the development process and an overview of the efforts required to implement them,
    - **Risk** narratives describe the five or 10 top risks. The rest are typically described the risk identification and mitigation matrix.

Figure 165: IMP Documentation

Create IMP Document

Section 4 Glossary

1. Provide definitions for all terms and acronyms

Figure 166: IMP Documentation

IMP/IMS Creation Process Example – Wooden table

The following graphics prime example of how to create an IMP and IMS. The example is the design of a simple wooden table. We will show how the IMP is created, the relationship of the WBS, SOW, to the IMP and IMS.

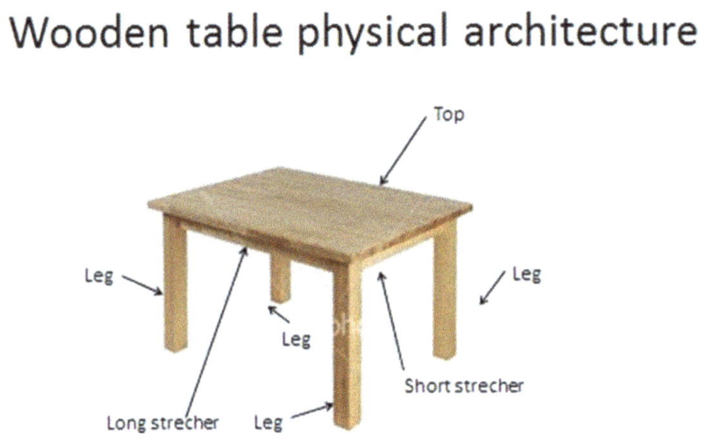

Figure 167: Wooden Table Physical Architecture

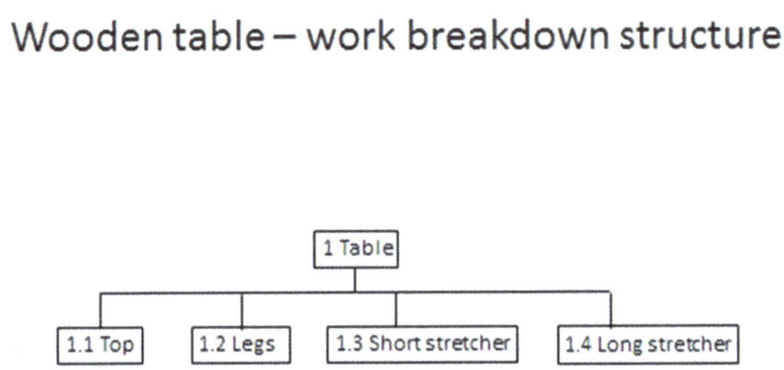

Figure 168: Wooden Table WBS

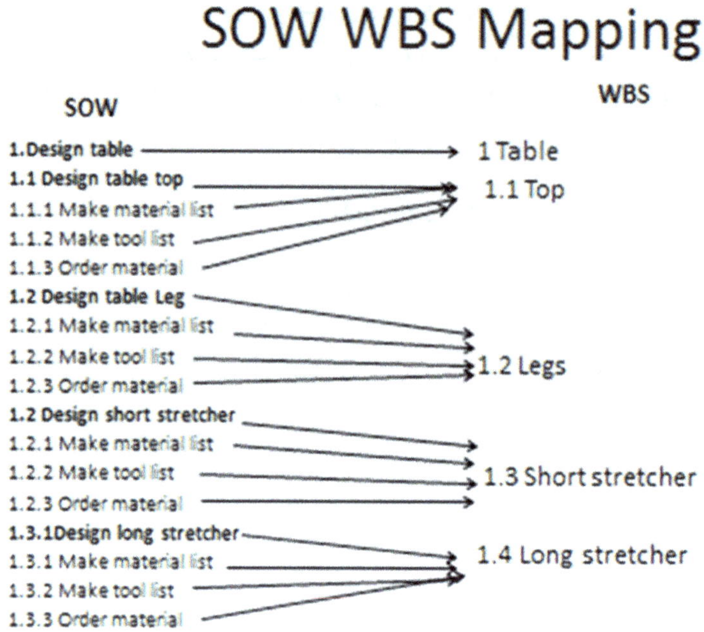

Figure 169: Mapping SOW to WBS

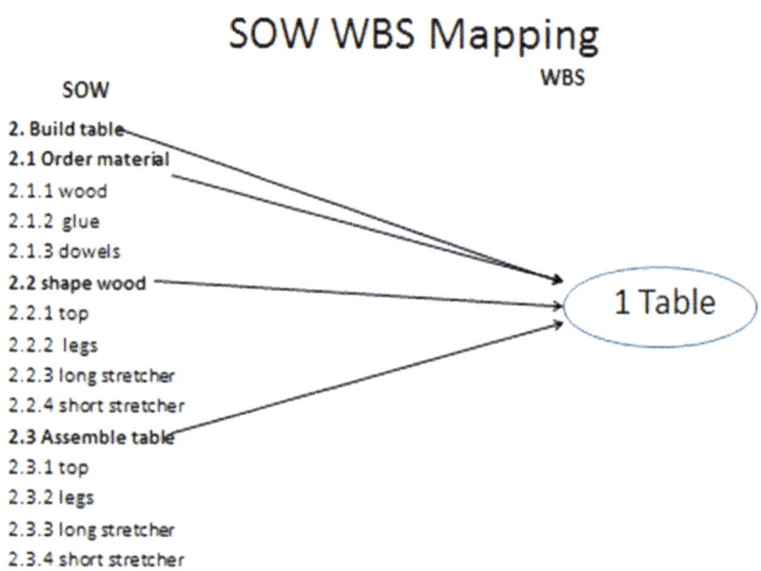

Figure 170: Mapping SOW to WBS

IMP (Activity Event list)

Activity #	Event	Event accomplishment criteria
A01	table designed	legs top, and stretchers design and drawings completed
A02	table parts shaped	
A03	table built	Table assembled
A04	Material arrives in shop	

Figure 171: IMP Development

IMS

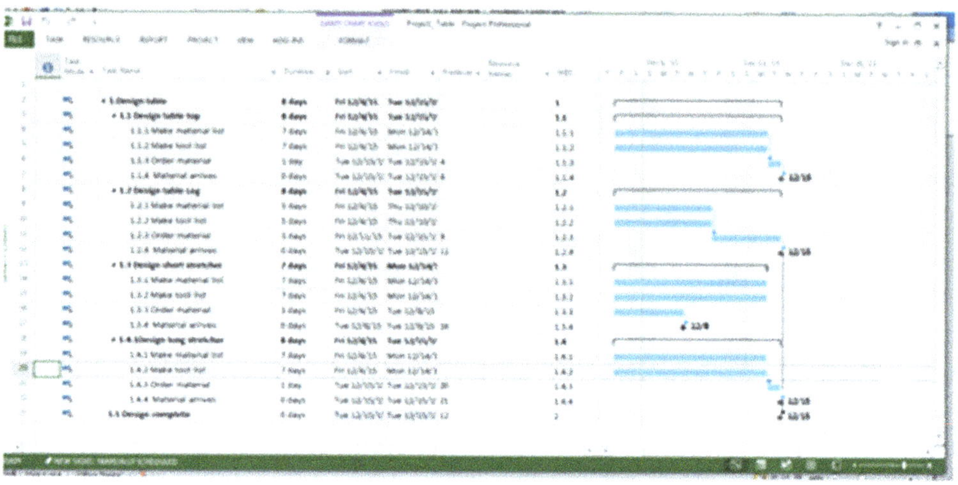

Figure 172: IMS Development

Chapter 13: The Proposal Effort

Systems Engineering is an extremely challenging subject for its instructors. We began the development of a Systems Engineering curriculum by focusing on the existing textbooks available within the domain space, and utilizing them as a backdrop for Systems Engineering programs at two major universities. Our fundamental intention was to develop the skills of Systems Engineers using proven methodologies in order to enable them to become effective contributors within their organizations.

During our first few years of educating Systems Engineers we found that we were well capable of imparting the subject matter, but it was extremely difficult to ascertain whether our students were building the skills necessary to become effective in the highly complex Systems Engineering domain. For this reason, we began investigating things like Bloom's Taxonomy in the hopes of better understanding the cognitive process and thereby maximizing the transfer of knowledge to our students.

Ultimately, we came to the conclusion that Systems Engineering is similar in domain expertise to something like Project Management. In essence, students can learn all of the necessary skills and approaches for effective Project Management in a classroom setting and from textbook documentation, but this does not guarantee that they will become effective Project Managers. Although this knowledge is critical for their development, there is no substitute for the hands-on approach. In fact, many highly effective Project Managers do not conform specifically to the methods and processes recommended within the Project Management doctrine; instead, they have created hybrid systems that enable them to complete their work effectively and efficiently. This implies that successful Project Managers have synthesized the available knowledge in the domain and have evaluated and developed their own appropriate approaches for the effective management of projects.

This implies that effective Project Managers are not born; they are developed. Additionally, they are not developed by the application of their skills on simple and successful projects. In many cases, their valuable skills are developed by experiences with highly complex and unsuccessful projects.

Previous textbooks offer specific, useful knowledge and approaches to Systems Engineering. However, the education of a Systems Engineer is not complete based upon the absorption of methodologies and processes as recommended in the Systems Engineering literature. In fact, the Systems Engineering educational process is not complete until its students have experienced the application of these principles in both successful and unsuccessful endeavors.

For these reasons, we have developed a highly experiential approach to teaching Systems Engineering, and present it in this text. Its method makes use of a challenging and difficult proposal process modeled on the real-world Systems Engineering domain, and the skills it teaches are directly applicable to a majority of the world's

Systems Engineering companies.
We believe that this approach is most effective at producing competent Systems Engineers as they exit their Engineering curricula upon completion of their degree programs.

The RFP, the SOO and the PBSS

The construction of this proposal project is relatively straightforward. In our systems curriculum, we have focused on the methodologies and processes during the first semester of Systems Engineering. The second semester of Systems Engineering includes a proposal effort, requiring the application of those skills obtained during the first semester of the curriculum. Strangely, we have seen highly effective students who have taken these courses in reverse order, thereby suggesting that some students benefit from the application of unlearned skills prior to being exposed to those skills in the classroom. It should be noted that those students without previous Systems Engineering classes were typically partnered with experienced Systems Engineering students for the proposal process.

During the first meeting of the second semester course, the students are split up into evenly distributed teams of five to fifteen students each, and are assigned the task of responding to a request for proposal. They are then given the opportunity to produce a response to the proposal in an effort to beat out the other student teams and be selected for the contract award at the end of the semester. We have on many occasions run this course with teams at widely separated universities in direct competition with one another.

After the initial team selection, documents are distributed to the teams representing the request for proposal, the statement of objectives, and the performance-based system specification. Examples of these documents are provided for the reader in the second volume of this text. It is recommended that a project be defined for the Systems Engineering course that will be in alignment with the student needs or organizational needs within the locale of the course. We have on several occasions written specific documents for inclusion of this approach within a single organization, with the project being of direct import to that organization.

Example proposals are included for the benefit of students taking the course. These proposals represent good approaches to the challenges posed to the students, but by no means represent the finest examples of generated proposals. We find that students benefit greatly from observing the typical magnitude and coherence of a submitted proposal, and the inclusion of these documents provide a baseline for expectations for students within the course.

The Proposal Schedule

While the students are involved in the generation of their response to proposal, the instructors act in the role of customer, contracting officer and Systems Engineering consultant for the project teams. Great care is taken to conduct the proposal in a professional atmosphere, requiring the instructors to maintain confidentiality amongst the project teams while imparting equivalent information and insight to each team in order to maintain fairness.

During the course of the semester, three major reviews of the project are scheduled. The first of these reviews occurs during the midpoint of the semester, and is known as the *blue team review*. Its function is to provide the students initial feedback on the technical approach and the coherence of their solutions with regard to the known user needs and requirements. In general, this review is treated as a preliminary design review, and the reviewers act as supporters and contributors to the development team. It is graded on the main subject areas required of the proposal, with grading provided in a red/ yellow/ green format. In the areas of concern, actionable recommendations are provided to each of the development teams. These reviews are conducted in private, and confidentiality is maintained between the teams.

At the three-quarters point of the semester, a second review is conducted with the project teams. This review is known as the *red team review*, and is similar to reviews conducted within organizations that determine if the proposal will be allowed to be submitted. This implies that the review team is performing a critical review of the work that has been accomplished, and is not acting in direct support of the proposal team. This review is quite effective at clarifying remaining work that must be completed by the project teams, and readies them for their final proposal presentations. These reviews are also conducted confidentially.

During the second to last class of the semester, the student teams offer their formal presentation of the proposals. They are required to follow all proposal rules and protocols, and are expected to present in a manner similar to professional presentations in industry. This includes appropriate presentation skills and professional dress. At the conclusion of their confidential presentations, they submit the written documentation required in the request for proposal for examination by the instructors. These reviews are also conducted confidentially, and comprise the final efforts for the course in the Systems Engineering curriculum.

During the subsequent week, the instructors review all submitted proposals, grading them on the specified criteria defined in the request for proposal. Each proposal is graded independently and the final grading of each instructor is combined in a final assessment of the effectiveness of the proposal responses. Ultimately, the instructors are free to select any proposal as the winner for the contract award, although they also have the freedom to select a prime contract awardee and directed awardees for specific areas of proposals that demonstrated excellence in their proposal response.

The final selection of the proposal winner is conducted during the last session of the class, and typically is an extremely dynamic and energy-filled affair. At the conclusion of this selection process, the students provide feedback on the course and the efforts conducted in the preparation of the proposal responses. The most common feedback provided is that on the final night of class the students feel the class was unbelievably difficult and challenging, and one they hope to never experience again in their lifetimes. Feedback received several weeks later often reflect a different sentiment; students at that point will often note that this was one of the most valuable courses they ever took in their Engineering curricula. The authors feel this is indicative of an effective learning process, and have found that in virtually every instance this approach has resulted in remarkable proposal efforts that are often of higher quality than those typically found in industry.

Summary and Conclusions

This text focuses on a systematic methodology that may be used by any Systems Engineer to approach complex and challenging problems in the systems development field. It provides a step-by-step process that is useful for the effective instruction of Systems Engineering within any University setting. This process includes the development of topical Systems Engineering skills based upon methods developed within the Systems Engineering domain, followed by a significant proposal development process, which will require the application of those skills in the preparation of a proposal response.

The authors have enjoyed significant success in the development of effective Systems Engineering students utilizing this process, and have successfully transitioned those students into the professional Systems Engineering domain in large defense companies. This is a significant measure of the success of this method of instruction, as the students brought into our organizations quickly proved to be highly effective contributors.

While it is true that we have developed what we consider to be an effective approach to the development of Systems Engineers, this approach is only as effective as the diligence of the students involved in the effort. It is imperative for the instructor to maintain the interest of the students and keep them engaged in the proposal process to the extent that they take the competition seriously. If that engagement is achieved, it is probable that the students will receive a practical grounding in the application of Systems Engineering methodologies and processes that will serve them well in the professional world.

References

Belassi, W., Tukel, O. (1996), A new framework for determining critical success/failure factors in projects, International Journal of Project Management

Breedveld, P (2003) Bond Graphs: Physical Systems Modeling I. University of Twente. Nederlands

Brown, A. (2012) Introduction to Science and the Scientific Method. Web Publication.

Buede, D.M. (2000) The Engineering Design of Systems. John Wiley and Sons, Inc. New York.

Cellier, F.E. (2007) Bond Graphs I, II, III. Web Presentation. Swiss Federal Institute of Technology

Cross, N. (2008) Engineering Design Methods: Strategies for Product Design. John Wiley and Sons. New York. New York.

De Bono, E (1970) Lateral Thinking: Creativity Step by Step, Harper & Row

De Bono, E (1999) Six Thinking Hats, Back Bay Books

Defense Science Board (1994) Report of the Defense Science Board Task Force on Acquiring Defense Software Commercially, The Undersecretary of Defense, Acquisition and Technology

Gonzales, L (2005) Deep Survival, W.W. Norton and Company, London, 295 pp.

Hoff, B (2002) The House on the Point, St. Martins Minotaur, New York, New York

Jones, C (1996) Patterns of Software Systems Failure and Success, International Computer Press

Kelley, J. (2001) The Art of Innovation. Doubleday, Random House, New York, NY

Lundin, S. et al. (2000) Fish! A Remarkable Way to Boost Morale and Improve Results, Hyperion

Luu, Andrew (2016) The CardCraft System. University of St. Thomas Masters Thesis. St. Paul, Minnesota.

Marchewka, J. (2003) Information Technology Project Management, Wiley

Monson, R. (2016) Engineering Leadership. Self Published. Saint Paul, Minnesota.

Monson R., Fried, A (2015) Effective Percolation Modeling. Flextech Conference Proceedings. Phoenix, AZ.

Murray, P (2003) Systems Thinking, NDIA Systems Engineering Conference, San Diego, CA

O'Connor, J, McDermott, I (1997) The Art of Systems Thinking, Thorsons, Hammersmith, London

Roedler, G.J., Jones, C (2005) Technical Measurement. INCOSE Document.

Rosenberg R., Karnopp, D. (1983) Introduction to Physical System Dynamics. McGraw-Hill Book Company. New York. New York.

Sherwood, Dennis (2002) Seeing the Forest for the Trees: A Manager's Guide to Applying Systems Thinking, Nicholas Brealey Publishing

Standish Group (1996) Chaos Report, Standish Group

Wakeman, L (2004) The Art of Seeing, photos.lois.co.uk

Website (September, 2007) http://www.qesnrecit.qc.ca/schools/olp/2mmsolve/cur98-4.htm

Website (September, 2007) http://suraj.lums.edu.pk/~cs564a05/Slides/CS564_15.pdf

Website (May, 2016) http://www.usabilitybok.org

Winograd, T (2006) Introduction to Human-Computer Interaction Design. Stanford University. California.

Wiseman, R. (2004) Did You Spot the Gorilla?: How to Recognise the Hidden Opportunities in Your Life, Arrow Books Ltd.

Appendix A – Verification and Validation Plan template

SONIQ
ACOUSTIC SYSTEMS

VERIFICATION AND VALIDATION PLAN

Project or System Name

P.O.C.: Project leader

Month, Year

Revision Sheet

Release No.	Date	Revision Description
Rev. 0	08/15/15	V&V Plan release

Verification and Validation Plan

NOTE TO AUTHOR: *Highlighted, italicized text throughout this template is provided solely as background information to assist you in creating this document. Please delete all such text, as well as the instructions in each section, prior to submitting this document.* ***ONLY YOUR PROJECT-SPECIFIC INFORMATION SHOULD APPEAR IN THE FINAL VERSION OF THIS DOCUMENT.***

The V&V Plan provides guidance for management and technical efforts throughout the test period. It establishes a comprehensive plan to communicate the nature and extent of testing necessary for a thorough evaluation of the system. This plan is used to coordinate the orderly scheduling of events by providing equipment specifications and organizational requirements, the test methodology to be employed, a list of the test materials to be delivered, and a schedule for user (tester) orientation and participation. Finally, it provides a written record of the required inputs, execution instructions, and expected results of the system test.

1.0 GENERAL INFORMATION

1.1 Purpose

Describe the purpose of the V&V Plan. (Why are we doing the project and the plan?)

1.2 Scope

Describe the scope of the V&V Plan as it relates to the project. (What is included in he plan? What is excluded from the plan?)

1.3 System Overview

Provide a brief system overview description as a point of reference for the remainder of the document. In addition, include the following:

- System name or title
- Operational status
  - Operational
  - Under development
  - Undergoing a major modification
- System environment and special conditions
- Include a graphical representation of the system

1.4 Project References

Provide a list of the references that were used in preparation of this document. *Examples of references are:*

- *Previously developed documents relating to the project*
- *Documentation concerning related projects*
- *Standard procedures documents*

1.5 Acronyms and Abbreviations

Provide a list of the acronyms and abbreviations used in this document and the meaning of each.

1.6 Points of Contact

1.6.1 Information

Provide a list of the points of organizational contact (POCs) that may be needed by the document user for informational and troubleshooting purposes. Include type of contact, contact name, department, telephone number, and e-mail address (if applicable). Points of contact may include but are not limited to helpdesk POC, development/maintenance POC, and operations POC.

1.6.2 Coordination

Provide a list of organizations that require coordination between the project and its specific support function (e.g., installation, manufacturing, service, training, etc.). Include a schedule for coordination activities.

2.0 TEST EVALUATION

2.1 Requirements Traceability Matrix

Prepare a functions/test matrix that lists all application functions on one axis and cross-reference them to all tests included in the test plan.

Req	Description	Inspect	Analyze	Test	Demonstrate	Test #
1	System shall operate on 440AC	X			X	T123
2	System shall operate saw blade at 2500 RPM				X	T456
3	System shall have a safety factor of 2.0 on bridge		X			N/A
4						
5						
6						

2.2 Test Evaluation Criteria

Decide the specific criteria that each segment of the system/subsystem must meet. Such criteria are described by the user of the system/subsystem and typically are a mix of functional and performance requirements, such as processing data within a certain time frame, producing a report, or responding to an online query within a certain amount of time.

2.3 User System Acceptance Criteria

Describe the minimum function and performance criteria that must be met for the system to be accepted as "fit for use" by the user or sponsoring organization.

3.0 TESTING SCHEDULE

3.1 Overall Test Schedule

Prepare a testing schedule to reflect the unit, integration, system acceptance, and release tests, as well as the time duration of each. This schedule should reflect the personnel involved in the test effort and the site location. In the test schedule, include the following information:

- Documentation review
- Test scripts
- Data preparation
- Test execution
- Output review
- System certification
- System release
- Return of test site to pretest condition

3.2 Security

Prepare a list of requirements necessary to ensure the integrity of the testing procedures, data, and site. Any special security considerations (e.g., passwords, classifications, security or monitoring software, etc) should be described in detail.

3.3 Testing Location

This section provides a description of testing locations. Each location should be under a separate section header, 3.3 - 3.x Locations may include Park facilities as well as customer sites and facilties. Identify the location at which the testing will be conducted, and the organizations participating in the test. List the tests to be performed at this location.

3.3.1 Milestone Chart

Provide a chart to depict the activities and events listed below. When preparing the chart, give consideration to all tests scheduled for this location. The activities and events will be presented in chronological order with supporting narrative, as necessary, and will depict, for example:

- The overall on-site test period by calendar date, and portions of the period assigned to major portions of the test.
- The pretest on-site period required for system test team orientation, familiarization, and for system debugging.
- The period assigned for the collection of database values, input values, and other operational data required for system test.
- The period assigned for user training, operator training, maintenance and control group training, and management orientation briefing.
- The period assigned for preparation, review, and approval of the test analysis report.

3.3.2 Equipment Requirements

Provide a chart or listing of the period of usage and quantity required of each item of equipment employed throughout the test period in which the system is to be tested. Any communications and test data reduction equipment will be included.

3.3.3 Software Requirements

Identify any software required in support of the testing when it is not a part of the system being tested. Include systems support, communications, and applications software, their recording and storage media, version number, and media type.

3.3.4 Personnel Requirements

Provide a listing of the personnel necessary to perform the test. For each of the personnel, this listing should provide the following information:

- Name, title, current organization
- Description of the required tasks to be performed
- Geographical location of the work to be performed
- Time required (dates needed)
- Whether the requirement is full time, part time, or as needed
- Any special skills required (i.e., programming language, machine familiarity)

3.3.5 Deliverable Materials

Itemize all materials that will be delivered as part of the system test, to include the quantity and full identification.

3.3.6 Testing Tools

Identify the testing tools to be used during the preparation for and execution of the test.

3.3.7 Site Supplied Materials

Describe any materials required to perform the test that need to be supplied at the test site. These materials could include desks, chairs, special equipment, office supplies, database and its media, as well as other input and their media.

4.0 TESTING CHARACTERISTICS

4.1 Testing Conditions

Indicate whether the testing will use the normal input and database or whether some special test input is to be used.

4.2 Extent of Testing

Indicate the extent of the testing to be employed. Where limited testing is to be employed, the test requirements will be presented either as a percentage of some well-defined total quantity or as a number of samples of discrete operating conditions or values. Also, indicate the rationale for adopting limited testing.

4.3 Data Recording

Indicate data recording requirements for the testing process, including data not normally recorded during system operation.

4.4 Testing Constraints

Indicate the anticipated limitations imposed on the testing because of system or test conditions (timing, interfaces, equipment, personnel).

4.5 Test Progression

In progressive or cumulative tests, include an explanation concerning the manner in which progression is made from one test to another so that the cycle or activity for each test is completely performed.

4.6 Test Evaluation

4.6.1 Test Data Criteria

Describe the rules by which test results will be evaluated.

4.6.1.1 Tolerance

Discuss the range over which a data output value or a system performance parameter can vary and still be considered acceptable.

4.6.2 Test Data Reduction

Describe the technique to be used for manipulation of the raw test data into a form suitable for evaluation, if applicable. *The available techniques may include:*

- *Manual collection and collation of system test output into test sequence order, followed by verification of the results.*
- *Automatic inspection of test results as obtained by data recording means using a test data reduction program followed by manual inspection of selected test results which do not lend themselves to complete reduction by automatic means.*
- *Automatic inspection of test results specifically recorded for manipulation by the test data reduction program. Test results as recorded, include all items of test significance. The test data reduction program contains an image of correct data output for an item by item comparison of data, and provides a summary of an evaluated test as output.*

5.0 TEST DESCRIPTION

This section provides a description of the tests. Each test should be under a separate section header, 5.1 - 5.x.

5.x Test Identifier

Provide a test name and identifier here for reference in the remainder of the section. Describe the test to be performed.

5.x.1 System Functions

Provide a detailed list of the system functions to be tested.

5.x.2 Test/Function Relationships

Provide a list of the tests that constitute the overall test activity. Include a test/function matrix summarizing the overall allocation of the system tests to the functions.

5.x.3 Means of Control

Indicate whether the test is to be controlled by manual, semiautomatic, or automatic means.

5.x.4 Test Data

Identify any security considerations in each of the following subsections.

5.x.5 Test Procedures

5.x.5.1 Procedures

Describe the step-by-step procedures to perform each test.

5.x.5.2 Setup

Describe or refer to standard operating procedures that describe the activities associated with setup of the facilities to conduct the test, including all routine machine activities.

5.x.5.3 Initialization

Itemize, in test sequence order, the activities associated with establishing the testing conditions, starting with the equipment in the setup condition. Initialization may include functions such as:

- Queuing of data input values for the test
- Queuing of support software
- Coordination of personnel actions associated with the test

5.x.5.4 Preparation

Describe, in sequence, any special operations such as:

- Inspection of test conditions
- Instructions for data recording
- Modifications of the data base
- Interim evaluation of test results

5.x.5.5 Termination

Itemize, in test sequence order, the activities associated with termination of the test, such as:

- Recording readouts and critical data from indicators for reference purposes
- Termination of operation of support software and test apparatus
- Collection of system and operator records of test results

Appendix B – Systems Engineering Management Plan template

SYSTEMS ENGINEERING MANAGEMENT PLAN (SEMP)

<Program Name>

[Contract Number]
[CDRL Number]

Issued: ***[Month day, year]***

Printed: ***[Month day, year]***

Recertified: ***[Month day, year or N/A]***

Supersedes: ***[Month day, year or N/A]***

Contact: ***[SE Lead name]***

Systems Engineering Manager: ______________ *[name]* ______ *[date]*

Technical Director: ______________ *[name]* ______ *[date]*

Program Manager ______________ *[name]* ______ *[date]*

SEMP Instructions

[A Systems Engineering Management Plan (SEMP) is required for each program/project. The SEMP must be signed by the Director/Manager of Systems Engineering within 60 days of contract award and Authorization to Proceed (ATP). This SEMP Template provides a model for preparation of a program/project SEMP.]

[This document provides the user with a standardized SEMP outline, and it contains a description of what is required in each paragraph. The italicized text, with the exception of Section and Paragraph titles, is a summary of what is required plus guidance about sources and content. The non-italicized text *provides standard material (boilerplate) that is useful as a starting point for developing the SEMP for a specific program. Items that are <italicized> such as this, are to be filled in by the author for a specific program/project. The SEMP author can tailor the contents of this outline for the specific contract or proposal that the SEMP is being written for, however the tailoring must be identified in Appendix C.]*

[The term "Systems Engineering" means the "An interdisciplinary approach encompassing the entire technical effort to evolve and verify an integrated and life-cycle balanced set of system people, products, and process solutions that satisfy customer mission needs. Systems Engineering encompasses:

a. The technical efforts related to the development, manufacturing, verification, deployment, operations, support, disposal of, and user training for, system products and processes;
b. The definition and management of the system configuration;
c. The translation of the system definition into work breakdown structures; and
d. Development of information for management decision-making.]

[SEMP Format]

[Application: Write the SEMP following this outline. The topics and outline provide comprehensive coverage. Each topic should be considered in the planning for a program, and then reviewed at the beginning of each subsequent phase. Some topics may only receive cursory consideration in early phases, with later expansion. (e.g. Producibility Analysis, System Implementation Transition)]

[Initial creation of a SEMP should require between 2-5 working days. Expansion of the SEMP to full planning during phase may take 2-4 weeks. Revision effort during later phases varies from 2 days to 3 weeks.]

Foreword

[Example with boilerplate text] This Carestream Systems Engineering Management Plan (SEMP), is submitted in accordance with policy. This document sets forth the methodologies, processes, organization roles and responsibilities of the *<Program Name>* Systems Engineering, Integration and Test (SEIT) effort for the *<Program Name>* System Development and Deployment (SDD) Program, and will be revised or updated, as specified by the Program Manager.

The following assumptions were identified when preparing this plan:

1. The *<Program Name>* Request for Proposal (RFP) and Statement of Work (SOW) will be updated to reflect the most recent status and scope of this contract, specifically the Contractor's roles and responsibilities with respect to this SEMP.
2. The Security Engineering team will support security accreditation. However, the contractor will not perform the security accreditation.
3. This plan will be updated periodically for comments received and contractor process improvements, and will reflect the most current technical details for specific technical activities to be performed by the Systems Engineering, Integration and Test (SEIT) Integrated Product Team (IPT).
4. The *<Program Name>* Requirements Document (ORD) will be used for informational data and is not a "test to" document.
5. The *<Program Name>* System Specification (SS) will be updated to capture all system level requirements that need to be verified and that this update will be reflected in an updated release(s) of the SS.

[The SEMP is to be considered a "living document" so that this document will always accurately reflect changes to the program and/or Systems Engineering approach in a timely manner throughout the program lifecycle. Revisions to the SEMP are to be controlled, communicated, and recorded in the below table. Waivers and Deviations are recorded in the second log and any item that is "To Be Determined" (TBD), "To Be Resolved" (TBR), or "To Be Supplied" (TBS) is recorded in the third log. Types of changes to the program that may necessitate revising the SEMP can include changes to milestone or delivery dates, introduction of a new system capability that would impact the top-level functional architecture, and the addition or removal of a key external interface to the system that would impact the system context diagram. The revision history should contain the Revision date and letter, and affected change pages.]

Record of Reviews and Change History

Date	*Review/Change*	*Authority*	*Remarks*

[This log is maintained to list all waivers and deviations that are applicable to the Technical Requirements Baseline.]

Waivers / Deviations Log

Date	*Waiver No.*	*Title*	**Disposition*

- Disposition is: Open, Closed or Superseded

[This log is maintained to denote items contained in this plan that need to be determined, resolved or specified prior to final submittal.]

TBD / TBR / TBS Log

Paragraph	*Item No.*	*TBD/TBR/TBS*	*Description of Item*

1 INTRODUCTION

1.1 Identification

[Provide a brief description of the purpose of the system to which this SEMP applies and a summary of the purpose and content of the SEMP. Base the description of the purpose of the system on the Customer's System Specification, if one exists, or, if not, on the Statement of Work (SOW). In addition, summarize the highlights of the SEMP. These will most likely need to be completed as one of the final steps after the body of the SEMP has been written and reviewed.]

This System Engineering Management Plan (SEMP) describes the engineering methodology for the *<Program Name>* Program Systems Development, Demonstration, Production, Deployment and Sustainment Phases. This SEMP applies to all program engineering activities, and defines the guidelines, processes, and controls will employ during the execution of the *<Program Name>*.

This SEMP is intended to: 1) provide technical direction to all engineering personnel and teammates performing Systems Engineering tasks; and, 2) describe the Systems Engineering, Integration and Test (SEIT) organizational roles and responsibilities within the program team structure. This SEMP is an internal compliance document under the auspices of the Technical Director/SEIT Manager. Program teammates and/or sub-contractor personnel will comply with processes as stated in the applicable subcontractor statement of work, and by supporting the methodologies described in this SEMP.

This Systems Engineering Management Plan (SEMP) has been prepared to describe the efforts required by Carestream to procure, develop, document, integrate and test (verify the integration) the operational systems for the *<Customer's Name>*. The SOW and System Specification have been used for the development of this SEMP.

The efforts described in this SEMP include the Systems Engineering interfaces to program management, engineering development, specialty engineering, installation, integration, test and integrated logistic support tasks. The *<Program Name>* system shall utilize the data, processes, and procedures already developed where practical. When applicable, Systems Engineering will deliver existing *<Program Name >* documentation to satisfy the *<Customer's Name>* requirements as specified in the System Specification.

The *<Program Name>* program will be implemented using an ***<incremental or spiral development>*** build approach. Each build will be integrated into a series of increasingly capable builds to mitigate integration risk and maximize the opportunity to detect and resolve problems early, thereby lowering overall program cost and schedule risks. The incremental build approach was adopted to mitigate the integration risks associated with a large and complex system, however it additionally offers the collateral benefit of providing end-users with early access to *<Program Name>* capabilities for evaluation, familiarization, and trial use. In Paragraph 1.2, System Description, Figure 1.2-1 illustrates a high-level, functional capabilities and interoperability view of the system integrated.

The *<Program Name>* Program *<name of the phase or phases>* Phase Program will be implemented in accordance with the plans, processes, and procedures as prescribed in the Program Plan (PP). The PP is the driving document for the development and implementation of the *<Program Name>* systems/subsystems, with respect to management and engineering management activities. This SEMP identifies and describes additional engineering processes and information that will be implemented by the *<Program Name>* program technical efforts. Refer to paragraph 1.5, Relationship to Other Program Plans and Documents, for more information concerning other program plans.

This SEMP has been designed to point to processes described in other program plans when appropriate, rather than repeating the information. During the life cycle of the program, this SEMP will be updated to incorporate customer direction, continuous process improvement initiatives, management directives and resulting engineering implementation improvements.

1.2 System Description

[Describe the system, using text and/or figures as appropriate. This system description should be used in all plans created for this project. Describe the scope of the system and provide the reader with a basic understanding of the program requirements, system capabilities, system architecture, factory test facility and training equipment. If there are teaming agreements, the charter of each team is delineated.]

[Example Text only] *The <Customer's Name> <Program Name> will modify eight (8) P-3C Update I aircraft with new mission system avionics and two (2) Update II.75 aircraft with new mission system avionics. The mission system avionics updates will be based on extensions to the US Navy's (USN) P-3C ASUW Improvement Program (AIP), Block Modification Upgrade Program (BMUP), and the FMS Case for the Royal Netherlands Navy (RNLN) Capabilities Upkeep Program (CUP) Program. The Scheduled Depot Level Maintenance (SDLM) will be performed on two (2) of the eight aircraft. The SDLM will be tailored from the US Navy standard. The <Customer's Name> P-3C system will use as a baseline the AIP mission system with selected components from the BMUP and the CUP mission systems and a number of Pakistan unique components. The configuration will have a fully capable Data Management System (DMS) consisting of a Mission Processor, a Video Display Controller (VDC), and a File Server. The supporting mission avionics will include a fully capable radar, Electronic Support Measures (ESM), acoustic, Electro Optical (EO)/Infrared (IR), communications, and Intercommunication System (ICS). In addition, a Survivability and Vulnerability (S&V) capability will be installed in each of the upgraded <Customer's Name> P-3C aircraft. Each aircraft will be fully populated prior to commencement of ground and flight functional tests so that all systems can be verified.*

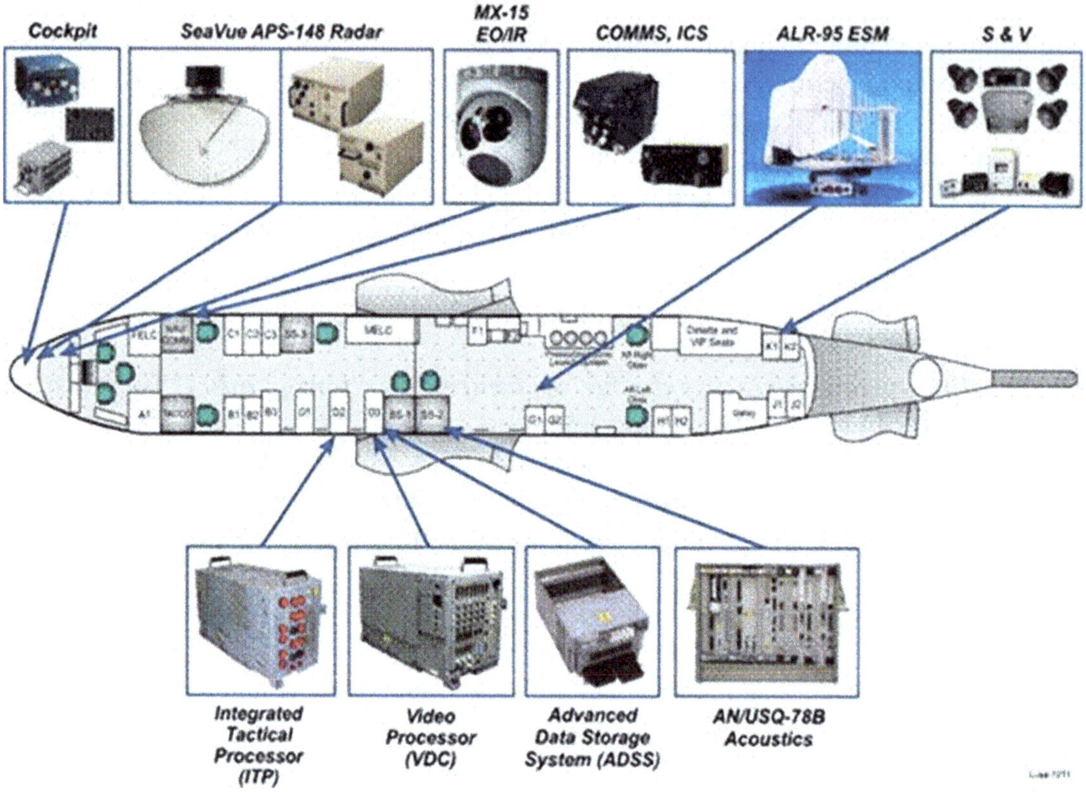

[Example]

Figure 1.3- <Customer's Name> Navy P3-C Mission Equipment Upgrades

1.3 Program Organization

[Provide lead in text here that describes the program's organization and provide a project organizational chart.]

The *<Program Name>* organizational structure depicted in Figure 1.3-1 uses the Integrated Product Team (IPT) approach. This organizational structure defines the areas of responsibilities for accomplishing the tasks as described in the *<Program Name>* SOW, a clear mapping to the Work Breakdown Structure (WBS) and accountability of each IPT for associated cost and schedule status. This IPT approach focuses on the design, development, integration and test of the hardware and software configuration items necessary to field the *<Program Name>* System.

- *The Program Director is responsible for the overall success of the <Program Name> program. This position focuses on the external interfaces with the customer and company executive management*
- *The Program Manager focuses on the internal daily management and execution of the cost, schedule, and technical performance of the program*
- *The Manager of Subcontracts is responsible for management policy and process and ensuring execution of that policy and process. In addition, the Manager of Subcontracts provides oversight of the subcontracts administration and materiel purchasing*

[Write to the specific program's organization]

The *<Program Name>* program organizational structure contains the following IPTs: Program Control; Systems Engineering; Integration and Test (SEIT); and Integrated Logistics Support (ILS). In addition, a Chief System Architect (CSA), and Technical Director (TD) are identified.

- *The Systems Engineering Manager is responsible for ensuring the functional to allocated requirement baseline will satisfy <Program Name> System-level capabilities*
- *The CSA is responsible for ensuring the top level architecture will satisfy <Program Name> System capabilities*

Refer to the *<Program Name>* Program Plan (PP) for more information concerning the details of the *<Program Name>* program organizational roles and responsibilities.

[Need to a program org chart for** <Program Name>* ***and paste below. This is an example Program Organization Chart only.]

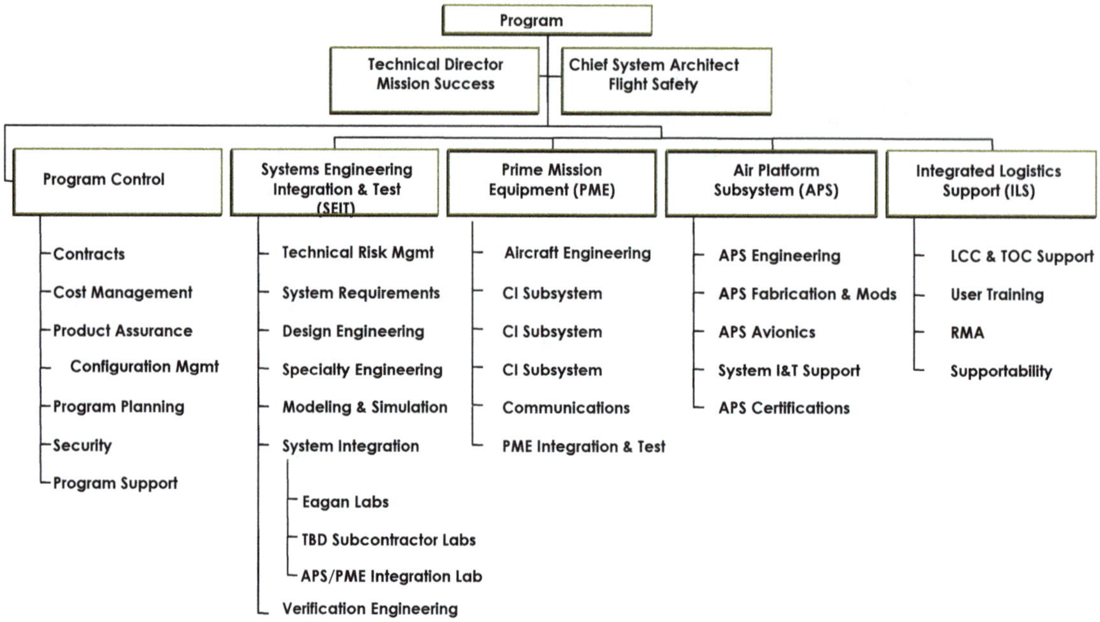

[Example Only]

Figure 3.2-1 <Program Name> Program Organization Chart

1.3.1 Program Control

[This is boilerplate text for this paragraph. The author must edit, as required, for the specific program this SEMP is addressing.]

The Program Control team is responsible for maintenance of the contract, maintenance of the program Integrated Master Plan/Integrated Master Schedule (IMP/IMS) and the overall cost baseline. The Program Control team will ensure cost and schedule performance is planned and reported as part of the Earned Value Management System (EVMS). The Program Control team provides the horizontal integration of the cost and schedule baselines across the program. Working with the Program Management Team (PMT), the Program Control team provides oversight of reproduction and deliverables. The Program Control team is also responsible for establishing and maintaining the Integrated Program Environment (IPE). The Program Control Manager is responsible for the following tasks:

- *Overall Business Operations for the <Program Name> Contract*
- *Management of the Program's Cost and Schedule Baseline*
- *Maintenance and Status of the IMP/IMS*
- *Status and Reporting of the EVMS*
- *Reproduction and Distribution of CDRLs*
- *Establishment and Maintenance of the Integrated Data Environment*

Please refer to Program Plan (PP), the Integrated Master Plan/Integrated Master Schedule (IMP/IMS), the Configuration Management Plan (CMP), and the Subcontract Management Plan (SMP) for more information concerning the details of <Program Name> Program Control roles and responsibilities.

1.3.2 Systems Engineering, Integration and Test (SEIT)

[This is boilerplate text for this paragraph. The author must edit, as required, for the specific program this SEMP is addressing.]

The SEIT Systems Engineering Manager is responsible for overall program technical implementation and provides the "horizontal integration" of the technical baseline across the program. The Systems Engineering Manager and IPT defines and enforces the program engineering process throughout the program lifecycle. The Systems Engineering Manager will coordinate with the PM Staff for technical guidance and program directives. In addition, the SE Manager is responsible for:

- *Program level design reviews and review minutes*
- *Technical reports*
- *Technical risk management*
- *Establishing the program's hardware, software and interface requirement baseline*
- *Decomposing functional requirements and establishing the allocated requirements baseline*
- *Establishing and maintaining the requirements management database*
- *Production of the program's Requirements Traceability Matrix (RTM)*
- *Coordination between IPTs in order to define functional capabilities and interface requirements at the configuration item level*
- *Performing trade studies and analyses*
- *Establishing the "design-to-baseline"*
- *Developing the system concept of operations document*
- *Defining the system level interface requirements in order to support interoperability capabilities with other ISR assets*
- *Performing hardware and software design engineering*
- *Simulation and modeling tasks in support of program engineering*
- *Performing engineering analyses and inspections in support of formal verification activities*
- *Providing engineering support to Functional Configuration Audits (FCAs), Physical Configuration Audits (PCAs) and Formal Qualification Reviews (FQRs)*
- *Security engineering*
- *Modeling the total system lifecycle costing*

1.3.3 Integration and Test

[This is boilerplate text for this paragraph. The author must edit, as required, for the specific program this SEMP is addressing.]

The SEIT Integration and Test Manager is responsible for test engineering, test operations, test reporting and test support. In addition, the Integration and Test IPT is responsible for test engineering, test operations, test reporting and test support. The SEIT Integration and Test Manager is responsible for:

- *Informal and formal system level integration activities*
- *Performing formal test planning*
- *Developing test requirements, including the Requirements Traceability and Verification Matrix (RTVM)*
- *Establishing the verification methods and test levels for all verification/validation activities to be accomplished, including aircraft engineering and flight certification requirements in accordance with applicable regulations and requirements for certification*
- *Developing formal test procedures and debug*
- *Performing formal test procedure validation*
- *Development and operation of the System Integration Labs (SILs)*
- *Conducting all Test Readiness Reviews (TRRs)*
- *Conducting formal tests and demonstrations, such as Development Test and Evaluation (DT&E) and Limited User Test (LUT)*
- *Performing post-test analysis*
- *Performing anomaly identification and tracking*
- *Preparing formal test reports*
- *Providing support to Functional Configuration Audits (FCAs), Physical Configuration Audits (PCAs) and Formal Qualification Reviews (FQRs)*

Please refer to the Master Test Plan (MTP), for more information concerning the details of *<Program Name>* Integration and Test roles and responsibilities.

1.3.4 Integrated Logistics Support

[This is boilerplate text for this paragraph. The author must edit, as required, for the specific program this SEMP is addressing.]

The Integrated Logistics Support (ILS) IPT is responsible for user training, system maintenance, depot requirements, reliability, maintainability and availability (RMA) engineering for the *<Program Name>* System. The ILS IPT exerts design influence to ensure ease maintainability and high reliability of *<Program Name>* hardware and software, and subsystems. The ILS IPT is responsible for ensuring *<Program Name>* human factors, environmental concerns, and health/safety concerns are satisfied. In addition, the ILS IPT is responsible for supporting the program's Functional Configuration Audits (FCAs), Physical Configuration Audits (PCAs), and Formal Qualification Reviews (FQRs). Reference: PLEN0100, PREN0200 and RFEN7580.

1.3.5 Chief System Architect (CSA)

[This is <u>boilerplate</u> text for this paragraph. The author must edit, as required, for the specific program this SEMP is addressing.]

The CSA is responsible for ensuring the evolved baseline architecture for the *<Program Name>* System established during the Concept Exploration (CE), Component Advanced Development (CAD) phases continues to support *<Program Name>* mission objectives and system requirements during the System Development and Demonstration (SDD), Low Rate Initial Production (LRIP) and Full Rate Production (FRP) contract phases. In addition, the focus of the CSA during the *<Program Name>* lifecycle phases is to evaluate and target technology insertion points in order to keep the <Program Name> up to date with the latest available technologies. The Chief System Architect (CSA) is responsible for managing the System Architecture IPT and ensuring the system level hardware and software architecture will satisfy *<Program Name>* System capabilities. Refer to the *<Program Name>* Program Plan (PP) for more information concerning the details of the *<Program Name>* program organizational roles and responsibilities.

1.3.6 Technical Director

[This is <u>boilerplate</u> text for this paragraph. The author must edit, as required, for the specific program this SEMP is addressing.]

The Technical Director (TD) provides technical oversight to all engineering activities. The TD is responsible for managing the technical, schedule, and cost performance baselines of the engineering team. The TD is expected to plan overall engineering activities in order to meet the program's technical, functional and performance requirements. The TD manages the execution of all engineering tasks and engineering contract deliverables. The TD supports all program Engineering, Operations, Risk Management, Quality, Program Control, and Finance processes. The TD is the primary technical interface with the customer concerning technical progress, engineering processes, technical plans, and engineering schedules.

1.4 Document Overview

This SEMP describes the *<Program Name>* overall SEIT responsibilities and tasks and provides: 1) a framework for SEIT activities to be performed during the lifecycle of the program; 2) a description of SEIT planning; and, 3) a description Systems Engineering methodologies, roles, responsibilities and relationships to other program engineering teams. The approach presented in this plan provides direction for implementing and documenting engineering activities for the Contractor Team. *<Program Name>* engineering activities described in this plan will be performed at the following locations: 1) *<Integration Lab Name and location(s) TBD>*; and, at the various subcontractor sites that are a part of the contactor's engineering team. A brief summary of the sections that are contained in this plan is listed in the following paragraphs.

Section 1 - Introduction

This section contains: the identification for the Systems Engineering Management Plan (SEMP); a summary description of the contractor's engineering efforts, and a brief overview of the structure of this SEMP and an overview of the contractor's *<Program Name>* program organization.

Section 2 - Applicable Documents

This section contains a listing of Government and Non-Government documents, specifications and plans that are referenced within this SEMP.

Section 3 – Systems Engineering Organization

This section contains a description of the contractor's SEIT team structure, team roles and responsibilities, and relationships to other contractor program plans and documents,

Section 4 – Systems Engineering Process

This section contains a description of the SEIT processes and methodologies to be implemented by the contractor's program engineering teams.

Section 5 – Documentation

This section contains a description of the minimum requirements for documentation developed by the contractor's engineering teams.

Appendix A – Terms and Definitions
Appendix B – Acronyms and Abbreviations
Appendix C –Guidance For Program Reviews

1.5 Relationship to Other Plans

The following plans, with the assigned *<Customer's Name and Program Name>* Contract Deliverable Requirement List (CDRL) sequence identifiers, will provide more specific details concerning the tasks and responsibilities associated with the SEIT engineering processes. The SEIT team will comply with items described in each of the other supporting program plans. In some cases, the responsibility for developing a plan is a SEIT responsibility, such as the Program Security Plan (PSP). Figure 1.5-1 depicts the *<Program Name>* program's plan hierarchy.

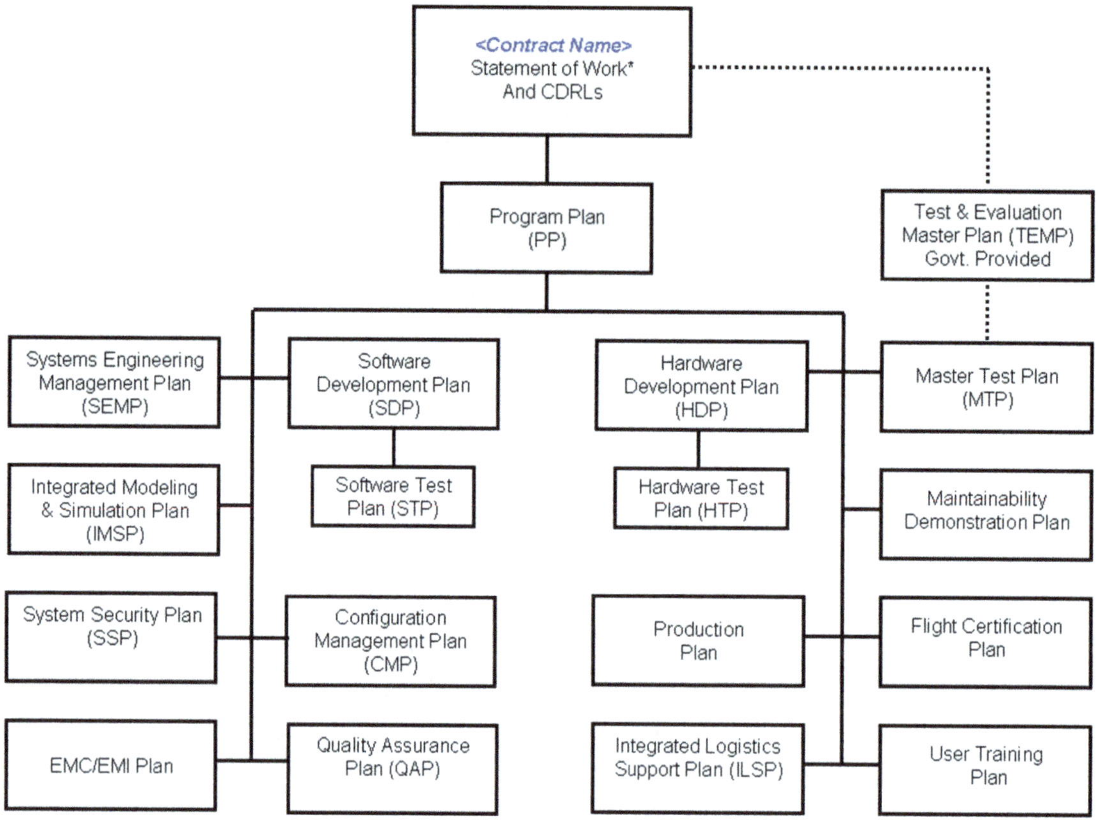

[Example]

Figure 1.5-1 <Program Name> Program Plan Hierarchy

1.5.1 Program Plan (PP)

The PP describes the *<Program Name>* program team's organization. Within this context, the Systems Engineering, Integration and Test (SEIT) team is assigned the charter of providing the overall technical "horizontal" leadership within the project's engineering team. The PP also provides details concerning overall project management functions and relationships, the project's approach to compliance with the Contract, SOW, LMC specific policies and procedures, the Contractor Work Breakdown Structure (CWBS), the Responsibility and Assignment Matrix (RAM) for cost account management, and specific duties and responsibilities of each project element within the overall project.

1.5.2 Engineering Project Plan (EPP)

The EPP contains the planning information created in accordance with the Project Planning [PREN0210] process and needed by all engineering technical disciplines involved for the project. The engineering project plan supports a common understanding of all assumptions, goals, commitments, and critical dependencies between groups, such as the Integrated Product Teams (IPTs) within the Integrated Program Environment (IPE). Together with other planning documents identified in Section 1.4, it comprises the complete set of planning information needed to manage and perform the *<Program Name>* Program.

1.5.3 Risk Management Plan (RMP)

The RMP describes the contractor's approach and processes for identification of risks, defining the scope and potential impact of the risks, and monitoring the management of the mitigation plans for the identified risks. The RMP also identifies opportunities for potential cost savings, schedule enhancements and quality improvements to design. The SEIT, specifically Systems Engineering, is responsible for preparing and implementing the risk/opportunity and management plan / procedures for the project's overall management team.

1.5.4 System Architecture Description Document (SADD)

The SADD provides a clear description of the overall system architecture and design, Concept of Operations (CONOPS) for the *<Program Name>*, and key assessments and assumptions of overall performance and requirements. The SADD is supplemented by a series of appendixes and supplemental documentation, including materials such as figures, reports, and System Architect[tm] files and architectural views. References: the Dept. of Defense Architecture Framework (DoDAF)

1.5.5 Software Development Plan (SDP)

The SDP describes the contractor's program's approach and processes for software design, development, integration and test. As part of the SEIT team, the Software IPT is responsible for the developing and implementing the SDP. The SEIT, specifically Systems Engineering, is responsible for ensuring software development processes and procedures are in compliance with the overall Systems Engineering process described within this SEMP and with the SDP.

1.5.6 Quality Assurance Plan (QAP)

The QAP describes the program's approach for monitoring all Systems Engineering and Development activities. The QAP also specifically addresses verification/evaluation tasks to be performed during formal testing (e.g., FQT, DT&E, OT&E).]

1.5.7 Test and Evaluation Master Plan (TEMP)

The test and evaluation management objectives and responsibilities for the *<Program Name>* program is planned in the TEMP. The TEMP defines the activities required for the project's integration and test program, and it describes entrance/exit criteria for all events leading to the successful completion of Milestone C. The Systems Engineering team will ensure that the project meets all technical objectives described in the TEMP and will focus on integration and test activities in order to meet these objectives.

1.5.8 Security Engineering Plan

The Security Engineering Plan describes the project's security engineering approach to establishing the requirements necessary to achieve any accreditation for the overall system and details concerning security controls and measures for individual configuration items. The Security Engineering Plan also includes the description of the project's security test objectives, security test planning, security test procedures and security test reporting.

1.5.9 Configuration Management Plan (CMP)

The CMP describes the contractor's approach to managing the product baselines as delivered for each Build or delivery to the customer. The CMP also provides a description of how project developed will be managed for configuration.

1.5.10 Master Test Plan (MTP)

The MTP is a comprehensive overview of the contractor's Integration and Test (IT) program. This SEMP defines specific processes for the engineering staff responsible for performing the integration and test functions. The MTP outlines the approach to the project's Integration and Test efforts, describes the organizational interfaces the project sustains, defines the test program objectives, and the entrance and exit criteria for major IT milestones and events.

1.5.11 Production Plan

The Production Plan and associated engineering analysis is a comprehensive overview of the project's approach to production engineering in the post-development phase. This SEMP defines specific processes for the engineering staff responsible for performing production-engineering analysis. The Production Plan outlines the approach to the project's engineering efforts to produce and supply the customer, describes the manufacturing efforts the project requires, defines the top-level project production objectives and activities and provides the specified project production requirements for the project's contract.

1.5.12 Deployment / Retrofit Plan

The Deployment and retrofit plan is included in the Integrated Logistics and Support Plan (ILSP) either directly or as an Appendix to the ILSP. The ILSP addresses the maintenance and maintainability demonstration approach for the program, which includes fault detection and isolation. This SEMP will describe the contractor's specific SEIT requirements for these approaches or cite specific plans that address these items.

1.5.13 Integrated Logistics Support Plan (ILSP)

The ILSP, prepared by the contractor's Logistics IPT is a comprehensive overview of the project's Integrated Logistics Support (ILS) program. This SEMP defines specific processes for the engineering staff responsible for performing the integrated logistics functions. The ILSP outlines the approach to the project's ILS efforts, describes the operational interface the project sustains, defines the logistics support program objectives and activities and provides the training and maintainability requirements for the *<Program Name>.*

2 APPLICABLE DOCUMENTS

This section presents a listing of the documents applicable to the *<Program Name>* SEMP implementation. The following documents, specifications, standards, and command media form a part of this document to the extent specified herein. In case of a conflict between these documents and the Statement of Work, the latter shall govern. Unless otherwise specified, all documents are considered as compliance documents. The dates associated with the listed documents were current at the time of publication of this SEMP.

2.1.1 Specifications and Plan

<Program Name> Statement of Objectives (SOO)
Date

<Program Name> Statement of Work (SOW)
Date

<Program Name> Operational Requirements Document (ORD)
Date

<Program Name> System Specification (SSS)
Date

<Program Name> Test and Evaluation Master Plan (TEMP)
Date

2.2 Other Documents

2.2.1 Standards

Electronic Industries Alliance (EIA) Standard 632
Processes for Engineering a System
Date

Electronic Industries Alliance (EIA) Standard 731
Systems Engineering Capability Model
Date

IEEE/EIA Standard 122207.0
Software Development
Date

3 Systems Engineering Management

3.1 Organizational Responsibilities

[Describe the organizational roles and responsibilities for the project here with respect to Systems Engineering. Typically, the Systems Engineering Manager directly reports to the project's Technical Director (TD) and is responsible for technical, cost and schedule status of Systems Engineering, integration and test activities within the project's management structure, as provided in the current project's organization chart depicted in a Figure in this section. The Systems Engineering team will direct and control engineering activities in order to accomplish the overall technical objectives of the project. The project's Systems Engineering team structure should also be provided in a Figure. Each element's roles and responsibilities with respect to the Project's Organizational structure are described in the following paragraphs. The Systems Engineering Manager delegates responsibility and allocates budgets to the appropriate management levels in the SEIT organization. The Systems Engineering Manager is

responsible for cost, schedule, and technical margins according to the project's Responsibility Assignment Matrix (RAM). The SEIT team provides support to Business Management and Program Planning in implementing the Integrated Master Plan/Integrated Master Schedule (IMP/IMS) and Earned Value Measurement System (EVMS). The SEIT organization works with the Subcontracts Administrator to ensure engineering subcontracts and inter-departmental work authorizations are executed properly.]

The *<Program Name>* Systems Engineering, Integration and Test (SEIT) concept provides the framework to ensure engineering discipline is maintained throughout the program lifecycle. Each Integrated Product Team (IPT), refer to paragraph 1.3, Program Organization, contributes to the overall engineering process described in Section 4.0, Systems Engineering Process.

The SEIT maintains focus on mission success, requirements satisfaction and best value for the customer. The SEIT establishes the technical plans, processes, design and development, and verification methods used to support *<Program Name>* development and deployment activities. The SEIT contains the engineering teams required to perform the implementation of the *<Program Name>* systems/subsystems. Together, the SEIT team is responsible for:

- *Mission Analysis*
- *Requirements Development*
- *Payload Design and Implementation*
- *Sub-system Integration*
- *System Level Integration*
- *Payload to Aircraft Integration*
- *Requirements Verification and Test Engineering*
- *User Training*
- *Aircraft Flight Certification*
- *Development System Maintenance*
- *Production Planning*
- *Deployment Readiness*
- *Sustaining Engineering / Operations and Maintenance (O&M)*

3.2 Systems Engineering Responsibilities

The SEIT Manager directly reports to the Program Management Team (PMT) and is responsible for technical, cost and schedule status of Systems Engineering, integration and test activities within the *<Program Name>* program's management structure, as depicted in Figure 1.3-1.The *<Program Name>* SEIT team will direct and control engineering activities in order to accomplish the overall technical objectives of the *<Program Name>* program. The *<Program Name>* SEIT team structure is depicted in Figure 3.2-1. Each element's roles and responsibilities with respect to the *<Program Name>* Program Organizational structure are described in the following paragraphs. The SEIT Manager delegates responsibility and allocates budgets to the appropriate management levels in the SEIT organization. The SEIT Manager is responsible for cost, schedule, and technical margins according to the *<Program Name>* Responsibility Assignment Matrix (RAM).

The SEIT team provides support to Business Management and Program Planning in implementing the IMP/IMS and Earned Value Measurement System (EVMS). The SEIT organization works with the Subcontracts Administrator to ensure engineering subcontracts and inter-departmental work authorizations are executed properly.

3.2.1 Objectives

[This section is Contract specific and the author must define the technical work that is to be performed under the contract's Statement of Work (SOW). Show the Contract Work Breakdown Structure (CWBS) at least to the third level, and describe the scope of each technical work package. Indicate the flow-down of the Contract Data Requirements List (CDRL), reference DD Form 1423 items, as applicable, to work packages (can be shown on the CWBS).]

3.2.2 Subcontractor Work Effort

[This section is Contract specific and the author must define the work efforts to be performed by subcontractors or teammates and their relationships between the various IPTs under contract. Identify the CWBS work packages containing this effort.]

3.2.3 Schedule

[This section is Contract specific and the author must provide the project's <u>top-level</u> (major milestones and tasks) schedule here as a Figure and with lead in text. Identify development phases and major milestones. This should be a network with CWBS elements, task descriptions with task dependencies and with the critical path (the path with the least float) identified, or a pointer to where the interested reader can find the program's Integrated Master Schedule (IMS) and Program Network depicting the critical path.]

3.2.4 Technical Organization

[This section is Program specific and the author must provide a technical project organization chart for engineering or Technical Operations here with lead in text. Also include diagrams of the related program organization and functional organizations. In creating this diagram, consider the principles of concurrent engineering. The author needs to create chart for <Program Name> and paste here. An example is provided below.]

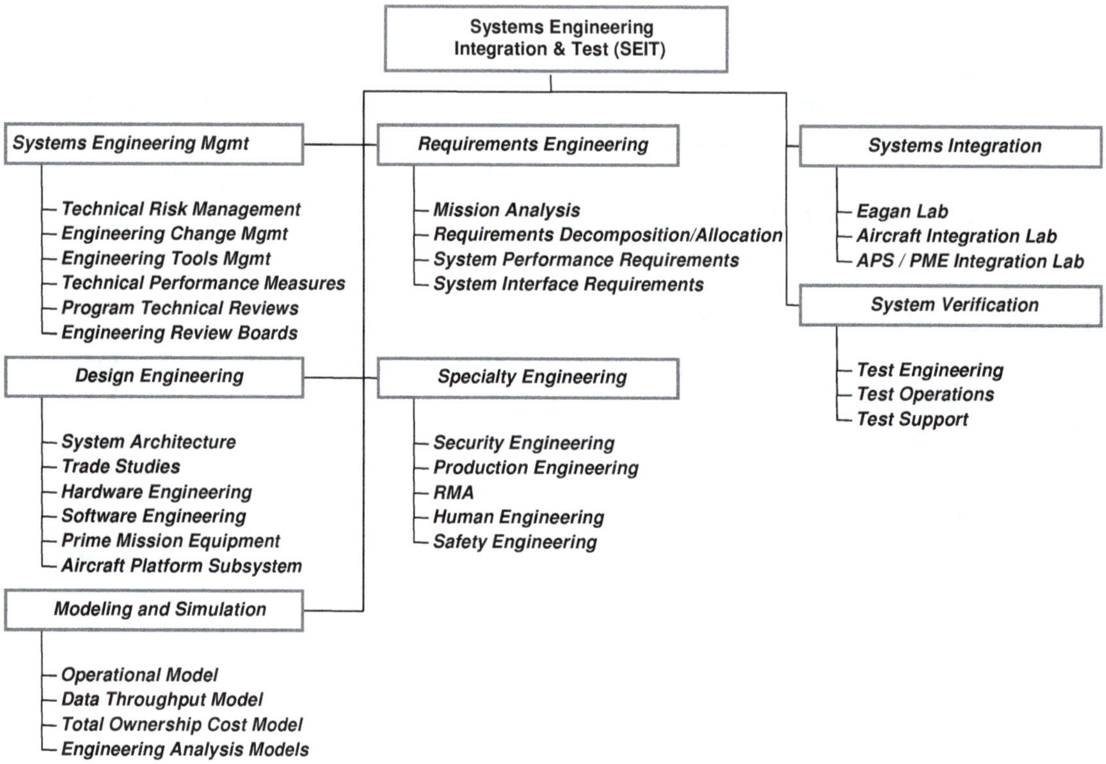

[Example Only]

Figure 3.2-1 <Program Name> SEIT Responsibilities

3.2.5 Technical Functions

[Provide a textual description of the responsibilities of each element in the project's technical organization. A cross-reference of responsibilities to the CWBS should be identified. Include subcontractor teams, or personnel, and include subcontractor-monitoring responsibilities.]

3.2.6 Program Interfaces

[Define the formal and informal technical interfaces to others. Consider customer and internal technical interface meetings, interface control working groups, and standards committees. Define how to control the interfaces - authority and responsibility.]

3.3 Technical Control

The following section describes the tasks and sub-tasks that must be accomplished in order for Systems Engineering to maintain technical "horizontal oversight" and technical control covering the entire lifecycle of the program. Requirements Management, Engineering Change Management, Technical Risk Management, Interface Control, Configuration Control and Technical Documents Control are discussed in the following paragraphs/sub-paragraphs.

3.3.1 Requirements Management

[Describe the project's approach to Requirements Management. Typically requirements management is the organized collection of data (inputs, as well as intermediate and final outputs) that provides the audit trail of decisions and rationale from initially stated needs and requirements to the current description of system people, products and process solutions. Ideally, it includes information such as functional and physical architectures, specification requirements, requirements traceability baselines, plans, critical issues, test results, models (e.g., finite element, simulation), automated tools, drawings, layouts, and the audit trail of changes to that information. The Requirements Management Database (e.g., DOORS, RTM, RDD, etc.) contains all the individual requirements for the system and all its subsystems to the lowest level. Each requirement is expressed as a "shall" statement. The RMD will maintain traceability of all requirements upward to original source material, and downward to functional allocations, and test procedure steps. This is commonly referred to as the Requirements Traceability Matrix (RTM). It provides building blocks for the System/Segment Specification (SSS), Hardware Configuration Item Specifications, Software Configuration Item Specifications, etc. The RMD can be maintained using the COTS software tool called Dynamic Object Oriented Requirements System (DOORS). The RMD will make it much easier to evaluate the impact of potential changes in system-level or component-level requirements. Requirements tracing will reveal which subsystems and lower level components may be impacted by a proposed change in the system-level requirements. Similarly, proposed changes to a lower-level requirements will have an impact on the ability to meet system level requirements, and that impact will be revealed by the RMD A description of the requirements management process and the tool chosen by the project to manage requirements is provided here.]

3.3.1.1 Requirements Decomposition and Allocation

The SEIT team is responsible for analyzing and defining the requirements baseline, which includes the development and decomposition process, and the implementation and control of this process across the *<Program Name>* program. The SEIT team will develop and maintain the functional and allocated requirements baseline for all *<Program Name>* components. The functional requirements baseline will be defined and derived using the provided *<Program Name>* Operational Requirements Document (ORD) and the *<Program Name>* System Specification (SS). The SEIT team will generate system level functional and performance requirement statements for each *<Program Name>*

The SEIT team will perform system requirements analyses for each configuration, including the assessment of mission requirements, requirements derivation, Configuration Item (CI) allocation, and traceability to verification documentation. The SEIT team will also evaluate technical requirements trades and present them to Program Management for review. Requirements documentation will be produced and maintained using *<name of tool to be used>* requirements database for traceability and accountability. The requirements database will be captured in the *<name of tool to be used>* requirements database tool and will provide the parent-child hierarchy of requirements and assigned requirement verification methods.

The SEIT team will establish an *<Program Name>* Specification Tree format and develop a system level specification tree for the *<Program Name>*. Figure 3.4.-1 depicts the *<Program Name>* specification hierarchy. The SEIT team will coordinate implementation at system/segment, subsystem and component levels. The SEIT team will also allocate *<Program Name>* system performance requirements to the subsystem and Configuration Item (CI) levels. This includes, but is not limited to, allocating *<Program Name>* size, weight, power and cooling (SWAP) budgets to subsystem levels (PME, APS, etc.).

[The SEMP author needs to create/obtain a specification tree for the program and paste it below.]

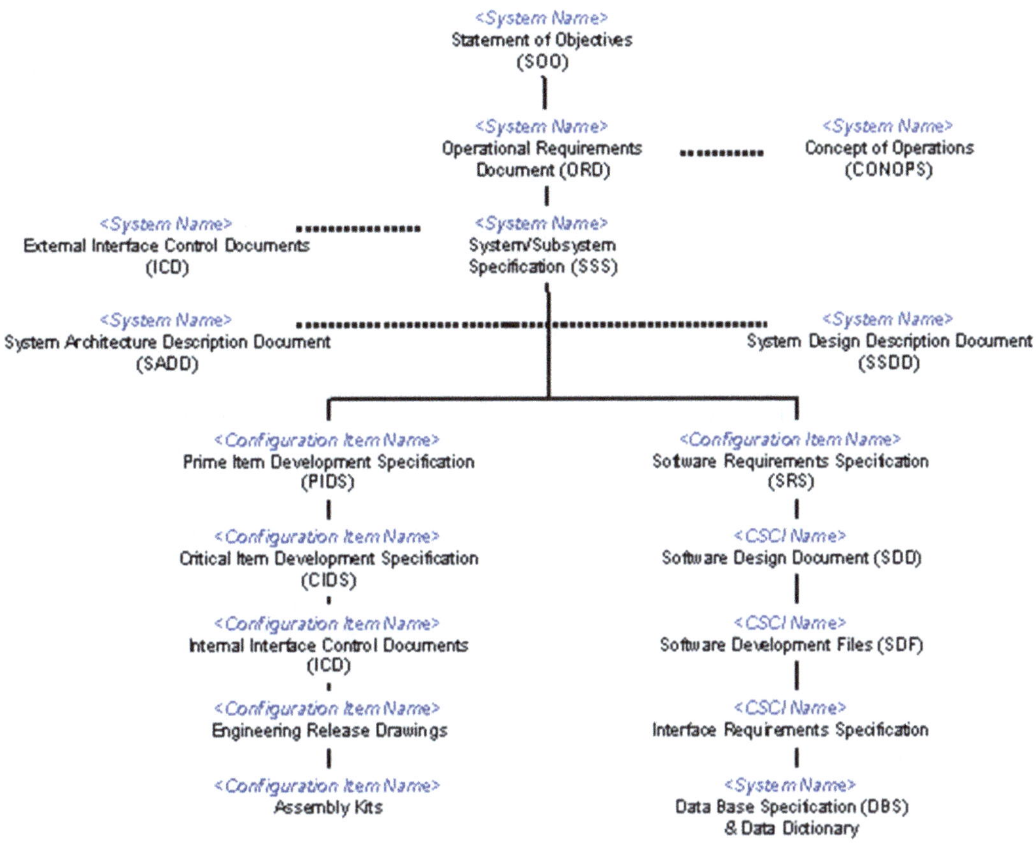

[Example Specification Tree]

Figure 3.4-1 <Program Name> Specification Tree

3.3.1.2 System Performance Requirements

The SEIT team will analyze and document the technical performance measures for the system test configurations and allocate performance margins to lower level subsystems and components. Design-to-Cost considerations will include technical performance margins, schedule margins, and cost margins. The SEIT team will track processes that estimate, monitor, record, and forecast time phased values of technical performance parameters during system all phases of the <Program Name> program. <Program Name> TPMs will include <Program Name> Key Performance Parameters (KPPs) and the SEIT team will status and report these TPMs at regular intervals per the <Program Name> IMP/IMS. See Paragraph 3.4.1 for more details concerning TPMs and the TPM process.

3.3.1.3 System Interface Requirements

The SEIT team is also responsible for identifying, developing, and coordinating <Program Name> interoperability interfaces. This is includes incorporating interface and interoperability requirements into applicable <Program Name> Interface Control Documents (ICDs). As a part of this task, the SEIT team will identify all external entities with which the <Program Name> will have an interface, and specify those requirements for the external interface for internal <Program Name> infrastructure capabilities. The SEIT team will develop and produce <Program Name> ICS, as applicable. ICDs will be produced at <Program Name> system level and will establish all inter-segment and subsystem interfaces. The SEIT team will provide the subsystem engineering teams with ICD technical direction and guidance. This effort includes internal and external interfaces for the <Program Name> system.

3.3.2 Engineering Change Management

[Describe the project's approach to engineering change management. Typically, the Systems Engineering team is responsible for managing and tracking Class I & II engineering changes for the project. The Systems Engineering team is normally responsible for establishing the Engineering Review Board (ERB) and will support the other program engineering boards as identified in the SEMP. The Configuration Management team and Program Control team is primarily responsible for describing, tracking and costing engineering changes received by the project. The Systems Engineering team is responsible for analyzing changes, identifying engineering change impacts, providing engineering change assessments, providing estimated cost, schedule and technical impacts to the Program Manager for review / approval. The ERB is the primary program engineering review board to identify and solicit change impacts from project engineering personnel and teams. Once the change impact assessments have been collected and estimated costs have been identified with specific implementation schedules, the Systems Engineering is responsible for forwarding those changes to the Program Manager. The Program Manager is responsible for determining whether a Change Control Board (CCB) is required to be scheduled for review of the assembled change package. Refer to the Configuration Management Plan (CMP, CDRL NNNN), for details concerning contract change management and pricing of Class I&II engineering changes.]

The SEIT team is responsible for managing and tracking Class I&II engineering changes for the <*Program Name*> program. The SEIT team is responsible for establishing the Engineering Review Board (ERB) and will support the other program engineering boards as identified in Section 4.3 of this SEMP. The Configuration Management team and Program Control team is primarily responsible for describing, tracking and costing engineering changes received by the <*Program Name*> program. The SEIT team is responsible for analyzing changes, identifying engineering change impacts, providing engineering change assessments, providing estimated cost, schedule and technical impacts to the Program Manager for review / approval. The ERB is the primary program engineering review board to identify and solicit change impacts from <*Program Name*> engineering personnel and teams. Once the change impact assessments have been collected and estimated costs have been identified with specific implementation schedules, the SEIT is responsible for forwarding those changes to the <*Program Name*> Program Manager. The <*Program Name*> Program Manager is responsible for determining whether a Change Control Board (CCB) is required to be scheduled for review of the assembled change package.

3.3.3 Technical Risk Management

[Describe the plans to manage and mitigate technical risks on the program. Identify risk levels, threshold criteria and decision authority. Identify methods to identify risk, analyze risk and track risks. The Systems Engineering team is responsible for identifying and reporting technical risks to the Program Manager. The Systems Engineering team will participate in the program's risk management process, and will support regularly scheduled meetings of the program's Risk Control Board (RCB). The Systems Engineering team will provide engineering inputs to, and updates for, technical risk mitigation plans. The Systems Engineering team will support the program's established risk management process as documented in the Risk Management Plan (RMP). The RMP will address the components, subsystems, and system architecture that pose the greatest technical, cost and schedule risks to the project. The Systems Engineering team will implement the objective set of criteria for development of risk mitigation plans, follow the risk numbering system, implement the program's established standard grading criteria, and provide the tracking requirements. The Systems Engineering team will ensure consistent use of the risk management process across engineering disciplines and will ensure technical risks are integrated into the risk process. In addition, the Systems Engineering team will support the program Watch List to maintain visibility of retired technical risks and those risks assessed as low. Refer to the Program Management Plan (PMP) and Risk Management Plan (RMP) for more information concerning the details of the project's risk management objectives and processes implemented.]

The SEIT team is responsible for supporting the overall Risk Management Process on the <*Program Name*> program and for identifying and reporting technical risks to the <*Program Name*> Program Manager. The SEIT will support the <*Program Name*> risk management process, and will support all risk management status meetings scheduled by the program. The SEIT team will provide engineering inputs to, and updates for technical risk mitigation plans presented in order to document the results in the Risk Review minutes.

The SEIT team will support the established risk/opportunity management process as documented in the Risk and Management Plan (RMP). The RMP will address the components, subsystems, and system architecture that pose the greatest technical, cost and schedule risks to the *<Program Name>* program.

The SEIT team will ensure consistent use of the risk/opportunity management process that has been established by the program and will ensure all technical risks are integrated into the risk/opportunity process. This process will include the tracking of system risks and their maturity as it applies to cost, schedule and performance for each risk item, and cross reference risks to the IMP/IMS in order to reduce risks to an acceptable level. In addition, the SEIT team will support the program's Watch List to maintain visibility of retired risks and those technical risks assessed as low. Refer to the *<Program Name>* Program Plan (PP) and Risk Management Plan (RMP) for more information concerning the details of the *<Program Name>* risk management objectives and systems implemented.

3.3.4 Interface Control

[Describe the plans to control interfaces within the system and between the system and other systems. Identify the interfaces to be controlled.]

3.3.5 Configuration Control

[Describe the plans to control baseline configuration of the system design. If a separate Configuration Management Plan exists, refer to it.]

3.3.6 Document Control

[Describe the plans to control the documents on the program. Consider master documents, copies, and distribution. Include documents created by the program team and those received from external sources.]

3.4 Performance Control

3.4.1 TPM Process

[Define the process to gather and calculate Technical Performance Measures (TPM). Define the frequency of measurement. Define the methods to disseminate the results, including management visibility.]

[Example and boilerplate]:
The Systems Engineering team is responsible for establishing the technical performance measures for the project. The project's system-level technical performance measurement (TPM) task can be logically decomposed into four primary classes and types of Technical Performance Measures:

1. The first class of TPMs is measures of Size, Weight and Power (SWAP) and indicators of the project's subsystem(s) technical data. These are Configuration Item (CI) metrics used to track key indicators of mission SWAP status that have bearing on engineering progress, and are indicators of engineering margins described in the system/subsystem specification baseline.
2. The second class of TPMs are those related to project specific mission performance in terms of either mission thread through-put measurements, or automated data processing equipment (ADPE) utilization, and loading measurements. These are generally related to the system-level performance requirements defined in the project's System/Segment Specification (SSS). The Systems Engineering team usually will measure or track subsystem level development metrics, typified by complexity, defect density, and software line of code (SLOC) production rate types of metrics.
3. The third class of TPMs is measures of Integration and Test processes and indicators of the project's technical status. These are I&T process metrics used to track key indicators of system status that have bearing on engineering progress, or are indicators of engineering's progress.
4. The fourth class of TPMs is qualitative and quantitative measures of the maturity and general trends for technologies that underpin the target system. These are

usually technology metrics used to calibrate technology evolution needed to sustain the end-user's mission requirements.

The Systems Engineering team will establish specific metrics, measurements, and tracking trends for the four classes of TPMs as indicators of overall engineering progress, technology leverage and risk opportunity, and mission performance, respectively. Figure 3.6.1-1 depicts the typical method for tracking TPMs.

The general approach to the TPMs will involve:

- *Projecting the TPM values and trend over time*
- *Taking periodic measurements and samples of the TPM*
- *Comparing the measurements and sample to the predictions*
- *Comparing the measurements and samples and projections to the requirements and margins between target goals and objectives*
- *Analyzing the differences*
- *Assessing the possible impact of the differences; and*
- *Refining the projections.*

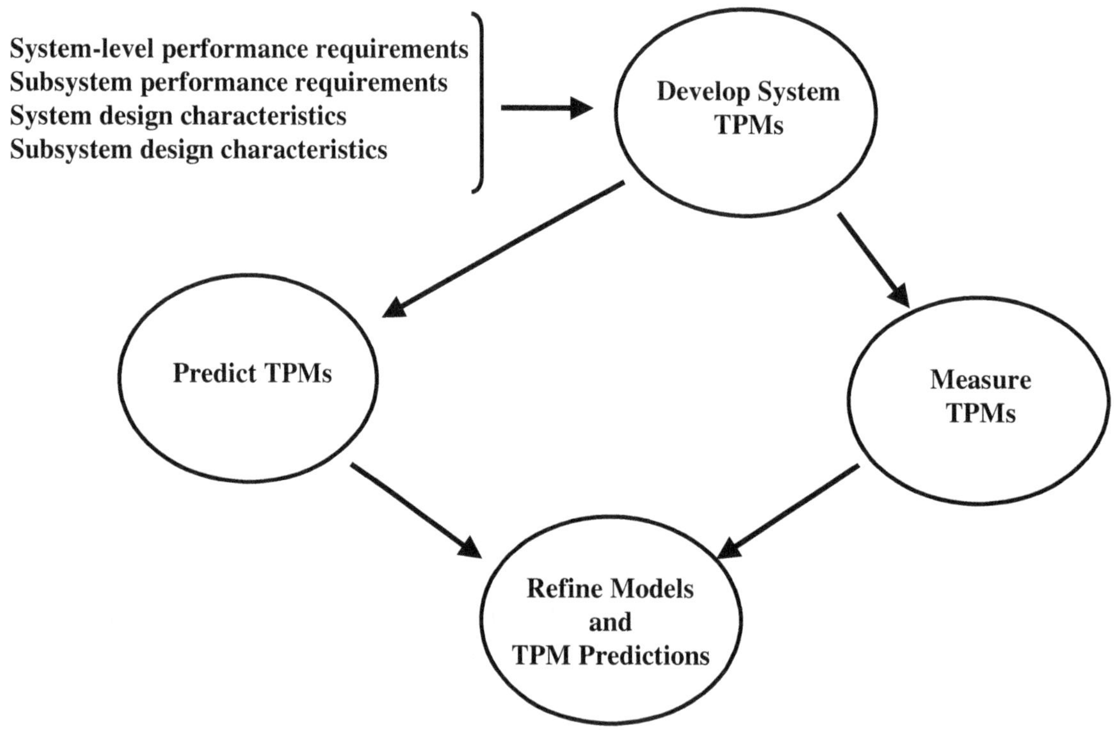

Figure 3.4-1 TPM Process

Each class of TPM motivates a slightly different approach to measurement and tracking, but in every case, Systems Engineering will maintain a database of TPMs. Key TPM status is reported at project reviews, particularly those that are suggestive of system risks.

Provide a description of the project's Technical Performance Measurement (TPM) reporting, which must provide a comprehensive and detailed itemization of TPM status, related analysis activities, and system implications.

3.4.2 Technical Performance Measures

[Define the TPMs to be calculated. Define the raw data required and sources. Define the calculation methods for each TPM.]

3.5 Program Phases

[This Section describes the program lifecycle phase as described in DoDI 5000.2 and applies primarily to Type I programs only. Refer to PRPM1000 and PRPM2000 for further clarification concerning program types and complexity factors. The following figure depicts the project phases, however the project contract shall take precedence over these requirements if there is any differences. These descriptions are to provide a framework for the Systems Engineering Management Plan (SEMP) and are for informational purposes only. Reference DoDI 5000.2]

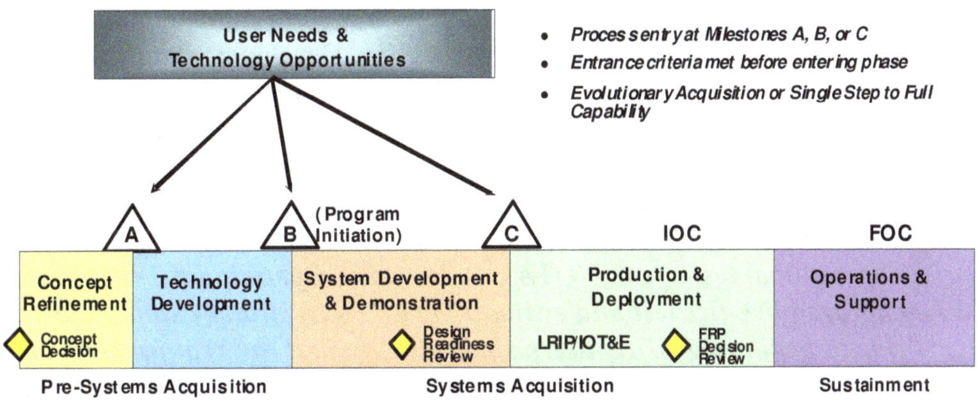

Figure 3.6-1, Defense Acquisition Management Framework.

3.5.1 Concept Exploration (CE) Phase

[Describe the reduction of technology risk and define the appropriate set of technologies to be integrated into a full system. Concept Exploration is a continuous technology discovery and development process reflecting close collaboration between the Science and Technology (S&T) community, the user, and the system developer. It is an iterative process designed to assess the viability of technologies while simultaneously refining user requirements. The project shall enter Technology Development at Milestone A when the Milestone Decision Authority (MDA) has approved the TDS. The tables in enclosure 3 identify all statutory and regulatory requirements applicable to Milestone A. This effort normally shall be funded only for the advanced development work. For business area capabilities, commercially available solutions shall be employed. (A toolkit of best practices is available at http://deskbook.dau.mil). A favorable Milestone A decision does not mean that a new acquisition program has been initiated. Reference DoDI 5000.2]

3.5.2 Component Advanced Development (CAD) Phase

[Describe an affordable increment of militarily useful capability, the technology for that increment has been demonstrated in a relevant environment, and a system can be developed for production within a short timeframe (normally less than five years); or when the MDA decides to terminate the effort. During the CAD phase, the user shall prepare the Capability Development Document (CDD) to support program initiation, refine the integrated architecture, and clarify how the program will lead to joint war fighting capability. The CDD builds on the ICD and provides the detailed operational performance parameters necessary to design the proposed system. A Milestone B decision follows the completion of CAD and then the SDD phase begins. Reference DoDI 5000.2]

3.5.3 System Design and Development (SDD) Phase

[Describe the development of the system or an increment of capability which would reduce integration and manufacturing risks (technology risk reduction occurs during Technology Development); ensure operational supportability with particular attention to reducing the logistics footprint; implement human systems integration (HSI); design for producibility; ensure affordability and the protection of critical program information (CPI) by implementing appropriate techniques such as anti-tamper; and demonstrate system integration, interoperability, safety, and utility. Development and demonstration are aided by the use of simulation-based acquisition and test and evaluation integrated into an efficient continuum and guided by a system acquisition strategy and test and evaluation master plan (TEMP). The independent planning of dedicated Initial Operational Test and Evaluation (IOT&E), as required by law, and Follow-on Operational Test and Evaluation (FOT&E), if required, shall be the responsibility of the appropriate operational test agency (OTA). A Director, Operational Test & Evaluation (DOT&E)-approved live-fire test and evaluation (LFT&E) strategy shall guide LFT&E activity. Prior to beginning SDD, users shall identify and the requirements authority shall approve a minimum set of key performance parameters (KPPs), included in the CDD that shall guide the efforts of this phase. These KPPs may be refined, with the approval of the requirements authority, as conditions warrant. Each set of KPPs shall only apply to the current increment of capability in development and demonstration (or to the entire system in a single step to full capability). At Milestone B, the PM shall prepare and the MDA shall approve an acquisition strategy to guide activity during SDD. The acquisition strategy shall include a TDS for the next technology spiral(s). Reference DoDI 5000.2]

3.5.4 Design Readiness Review (DRR)

[The Design Readiness Review during SDD provides an opportunity for mid-phase assessment of design maturity as evidenced by measures such as the number of subsystem and system design reviews successfully completed; the percentage of drawings completed; planned corrective actions to hardware/software deficiencies; adequate development testing; an assessment of environment, safety and occupational health risks; a completed failure modes and effects analysis; the identification of key system characteristics and critical manufacturing processes; an estimate of system reliability based on demonstrated reliability rates; etc. Successful completion of the Design

Readiness Review ends System Integration and continues the SDD phase into the System Demonstration effort. MDAs may, consistent with the intent of this paragraph, determine the form and content of the review.

3.5.5 Low Rate Initial Production (LRIP) Phase

[Provide a description of how the LRIP phase is to achieve an operational capability that satisfies mission needs. Operational test and evaluation shall determine the effectiveness and suitability of the system. The MDA shall make the decision to commit the Department of Defense to production at Milestone C. Milestone C authorizes entry into LRIP (for MDAPs and major systems), into production or procurement (for non-major systems that do not require LRIP) or into limited deployment in support of operational testing for MAIS programs or software-intensive systems with no production components. Reference DoDI 5000.2]

3.5.6 Full Rate Production (FRP) Phase

[An MDAP may not proceed beyond LRIP without approval of the MDA. The available knowledge to support this approval shall include demonstrated control of the manufacturing process and acceptable reliability, the collection of statistical process control data, and the demonstrated control and capability of other critical processes. The decision to continue beyond low-rate to full-rate production, or beyond limited deployment of AISs or software-intensive systems with no developmental hardware, shall require completion of IOT&E, submission of the Beyond LRIP Report for DOT&E Oversight Programs, and submission of the LFT&E Report (where applicable) to Congress, to the Secretary of Defense, and to the USD (AT&L). Continuation into full-rate production results from a successful Full-Rate Production Decision Review by the MDA (or person designated by the MDA). This effort delivers the fully funded quantity of systems and supporting materiel and services for the program or increment to the users. During this effort, units shall attain Initial Operational Capability. Reference DoDI 5000.2]

3.5.7 Deployment Phase

[The objective of this activity is the execution of a support program that meets operational support performance requirements and sustains the system in the most cost-effective manner over its total life cycle. When the system has reached the end of its useful life, it shall be disposed of in an appropriate manner. Operations and Support has two major efforts: Sustainment and Disposal. Reference DoDI 5000.2]

3.6 Formal Project Reviews

[Describe the project's formal reviews in this section. The SEIT team is responsible for the planning, coordination, preparation and presentation of Project System Level reviews. The SEIT team will coordinate across the program and engineering disciplines with respect to program reviews IAW the project's Statement of Work (SOW), and EIA Standard 632. Program reviews will be conducted in compliance with applicable Systems policies and procedures. The technical objectives to be accomplished by the Systems Engineering team for each project review are described in the following paragraphs.]

The SEIT team is responsible for the planning, coordination, preparation and presentation of <*Program Name*> System Level reviews. The SEIT team will coordinate across the program and engineering disciplines with respect to program reviews IAW the <*Program Name*> Statement of Work (SOW), and EIA Standard 632. Program reviews will be conducted in compliance with applicable policies. The technical objectives to be accomplished by the SEIT team for each program review are described in the following paragraphs. See Appendix D for more guidance information concerning program technical reviews.

3.6.1 Proof of Concept Review (POC)

[Reference the Concept Exploration Phase (CE) in paragraph 3.6.1 or the Contract's SOW and CDRLs for specific requirements that need to be documented in the SEMP.]

3.6.2 System Requirements Review (SRR)

The SEIT team will lead the SRR. The objective of the SRR is to ascertain the adequacy of the contractor's efforts in defining system requirements. It will be conducted when a significant portion of the system functional and performance requirements have been established.

3.6.3 System Design Review (SDR)

The SEIT team will lead the SDR. The SDR shall be conducted to evaluate the optimization, correlation, completeness, and risks associated with the allocated system level requirements. Also included is a summary review of the Systems Engineering process, which produced the allocated technical requirements and the engineering planning for the next phase of the contract's efforts. Basic manufacturing consideration will be reviewed and planning for production engineering in subsequent phases will be addressed. The SDR will be conducted when the system definition effort has proceeded to the point where system characteristics are defined and the Configuration Items (CIs) have been identified.

3.6.4 Software Specification Review (SSR)

The SEIT team will lead the SSR. The SRR shall be conducted when the Computer Software Configuration Item (CSCI) requirements and operational concepts have been initially finalized. The SRR is conducted when the CSCI requirements have been sufficiently defined to evaluate the contractor's responsiveness to and interpretation of the system or segment requirements. A successful SSR is predicated upon the contracting agency's determination that the Software Requirements Specification, the Interfaces and the CONOPS provide a satisfactory basis for proceeding into software preliminary design.

3.6.5 Preliminary Design Review (PDR)

The SEIT team will lead the PDR. The SEIT team will develop and coordinate the development of the System Level PDR across the project and engineering disciplines IAW the Statement of Work (SOW), and EIA Standard 632. The PDR will be conducted in compliance with applicable policies and procedures. The technical objectives to be accomplished by the SEIT team are as follows:

- Present a scaleable architecture, scaleable functions, modular design, and an open architecture design consistent with the:
    - Department of Defense Architecture Framework (DoDAF)
    - ***Example:*** Common Imagery Ground Surface System (CIGSS)
    - ***Example:*** Joint Technical Architecture (JTA)
    - ***Example:*** Joint Technical Architecture-Army (JTA-Army)
- Provide evidence that the preliminary design will meet the system KPPs, meet thresholds, and as many objective requirements as practical within the program's cost constraints.
- Present system capabilities for growth to meet other threshold and future objective mission and system requirements.

The SEIT team will be responsible for conduct the PDR, capturing the minutes and assigning owners to action items identified during the review. The SEIT team is responsible for ensuring all PDR action items are closed, or documenting any agreements that have been reached concerning resolution plans for closing action items identified.

3.6.6 Critical Design Review (CDR)

The SEIT team will lead the CDR. The SEIT team will develop and coordinate the development of the System Level CDR across the project and engineering disciplines IAW the Statement of Work (SOW), and EIA Standard 632. The CDR will be conducted in compliance with applicable policies and procedures. EIA Standard 731, Mil-Std-1521B will be used as guidance documents for this review. The technical objectives to be accomplished by the SEIT team are as follows:

- *Determine that the detail design of the configuration items under review satisfy the performance and engineering specialty requirements of the System/Segment Specification (SSS), Prime Item Development Specifications (PIDS), Critical Item Development Specifications (CIDS), and Computer Software Configuration Item (CSCI) specifications.*
- *Establish the detail design compatibility among the configuration items and other items of equipment, facilities, computer software and personnel,*
- *Assess configuration item risk areas (on a technical, cost, and schedule basis),*
- *Assess the results of the producibility analyses conducted on system hardware, and,*
- *Review the preliminary hardware product specifications. For CSCIs, this review will focus on the determination of the:*
    - Acceptability of the detailed design,
    - Performance and test characteristics of the design solution, and
    - The adequacy of the operation and support documentation.

The SEIT team will be responsible for conduct the CDR, capturing the minutes and assigning owners to action items identified during the review. SEIT is responsible for ensuring all CDR action items are closed, or documenting any agreements that have been reached concerning resolution plans for closing action items identified.

3.6.7 Test Readiness Review (TRR)

The SEIT team will lead the TRR. The SEIT team will develop and coordinate the development of the System Level TRR across the project and engineering disciplines IAW the Statement of Work (SOW), and EIA Standard 632. The TRR will be conducted in compliance with applicable Systems policies and procedures. The technical objectives to be accomplished by the SEIT team are as follows:

The TRR establishes that the necessary criteria for starting DT&E have been met and summarize the readiness of the system for test, largely based on the results of informal tests. The TRR also establishes that the DT&E test procedures are adequate and complete relative to the system requirements allocated to the build. The DT&E TRR is conducted at a minimum of 3 working days prior to the start of DT&E conduct.

The TRR will address the following:

- *Section 1 Introduction – a summary of the system under test at the CI level*
- *Section 2 Requirements Changes – a summary of any system requirements changes that affect build testing*
- *Section 3 Design Changes – a summary of any design changes, specifically changes to external or inter-segment designs since CDRs*
- *Section 4 Test Plan and TISs Changes*
- *Section 5 Test Procedure Changes*
- *Section 6 Informal test Results – summary of results from SIT dress-rehearsal*
- *Section 7 Test Resources – defines required test resources, including SIF resources*
- *Section 8 Test Limitations – identifies all known test limitations*
- *Section 9 System Problems – identifies open System Problem/Change Reports (SP/CRs), "make play" changes, and known discrepancies in the system under test baseline*
- *Section 10 Schedules*
- *Section 11 Documentation – identifies updates to documentation.*

The DT&E TRR immediately precedes the conduct of the formal requirements sell-off testing and is scheduled when reasonable confidence has been established that the integrated baseline has been completely debugged and that all system requirements have been established. The TRR involves the total engineering team, subsystem contractors, and Government resources, and establishes the product baseline for the DT&E baseline.

As a goal, the SEIT team will try to achieve the following level of build integration and test maturity before scheduling a the DT&E TRR:

(1) At least 90% of the SSS Requirements for the planned build have "passed" during the dry-run of the Test Procedures;

(2) No “show Stopper” problems are without a fix/work-around for the build;
(3) All project test resources are ready for the build testing; and,
(4) The Government has approved the build test plans, test information sheets and test procedures.

Shortfalls relative to these goals may not necessarily preclude the scheduling of the TRR, but will be used a guideline to prevent a false start of formal testing.

The SEIT team will conduct the DT&E TRR a minimum of 30 working days prior to the start of DT&E. The TRR will be structured consistent with Mil-Std 1521B, as tailored, for system level testing. At a summary level the DT&E TRR will establish that the necessary criteria for starting the DT&E have been met and will summarize the readiness of the system for test as follows:

Criteria

1. *Baseline test configuration with known change history to the IPT baselines is in-place;*
2. *Baseline configuration deficiencies are documented and closure plans identified;*
3. *DT&E Test Procedures are approved;*
4. *Test support requirements have been satisfied - all test resources are scheduled and available;*
5. *Safety requirements have been satisfied; and,*
6. *Limitations are identified.*

Status

1. *Configuration change status since IPT sub-system integration and test:*
    a. *Requirements or design changes, and*
    b. *Changes to approved test plans/procedures;*
2. *Test results status from DT&E Dry-run;*
3. *Requirement level Pass/Fail/Not-Tested status from DT&E Dry-Run;*
4. *Problem report status*

Upon Government approval of the TRR, DT&E will begin as soon as practical.

The SEIT team will be responsible for conduct the TRR, capturing the minutes and assigning owners to action items identified during the review. The SEIT is responsible for ensuring all TRR action items are closed, or documenting any agreements that have been reached concerning resolution plans for closing action items identified.

3.6.8 Functional Configuration Audit (FCA)

The Mission Success Manager is responsible for conducting the FCA with the Government auditors. The Contractor's team includes representatives from SEIT, CM and QA. The FCA is a formal audit to validate that the development of a Configuration Item (CI) has been satisfactorily been completed and that the CI has achieved both functional and performance requirements identified in the CI specification by the verification methods defined in the RTVM. In addition, the operational and support documentation shall be reviewed.

3.6.9 Physical Configuration Audit (PCA)

The Mission Success Manager is responsible for conducting the FCA with the Government auditors. The Contractor's team includes representatives from SEIT, CM and QA. The PCA is a formal review of the designated Configuration Item (CI) to verify the CI "as built" conforms to the technical documentation which defines the CI. This includes the engineering drawings and inspection of each hardware item to ensure the CI and documentation are consistent as defined in the RTVM.

3.6.10 Formal Qualification Review (FQR)

The Mission Success Manager is responsible for conducting the FQR with the Government auditors. The SEIT team is responsible for supporting the Government conducted FQR. The purpose of the FQR is be to verify that the actual performance of the configuration items of the system as determined through test, comply with the system specifications and interface control documents, and to identify the test report(s)/data which document results of qualification tests of the configuration items and the system as a whole. The Functional Configuration Audit (FCA) and Physical Configuration Audit (PCA) collect data necessary for FQR. FCA/PCA will be conducted at the discretion of the Government. The point of Government certification will be identified by the customer's project office and will depend upon the risk aspects of the deliverables, and project progress in successfully verifying the requirements of the configuration items and system specifications. For specific compliance requirements and guidance concerning the minimum topics to be discussed at the FQR, refer to the project SOW and reference MIL-STD-1521B. The Government will determine the format and content of this review.

3.6.11 Low Rate Initial Production Readiness Review (LRIPP)

The SEIT team will support the LRIPP. This review denotes the initial completion of Milestone C and is conducted by the Government during the Production and Deployment phase. The SEIT team is responsible for describing the program's technical approach to fulfilling the entire acquisition lifecycle of the project specific contract, if it is applicable.

3.6.12 Full Rate Production Readiness Review (FRPP)

The SEIT team will support the FRPP. This review denotes the final completion of Milestone C, the start of full rate production of the system, and is conducted by the Government during the Production and Deployment phase. The SEIT team is responsible for describing the program's technical approach to fulfilling the entire acquisition lifecycle of the project specific contract, if it is applicable.

4 SYSTEMS ENGINEERING PROCESS

[Provide a lead in paragraph concerning the compliance with Command Media and point to an Appendix that has the Compliance Matrix with the Command Media.]

The following description of the Systems Engineering process is based on current Industry Standards. The following paragraphs describe the major activities that must be performed by the SEIT team with concurrent engineering support from all of the IPTs within the program's organizational structure.

4.1 Process Description

[In the subsections that follow, describe the program-unique processes for reaching each major program milestone. Use standards as a reference, and do not restate the processes therein and provide a pointer to where that specific process is documented. Consider each activity to reach each major milestone documented in the SOW as specific tasking for engineering to perform . Specifically address the automated tools that will be used and their interconnection. The SEIT team will provide overall "horizontal" technical direction and guidance for the project as depicted in Figure 4.1-1. This includes establishing interaction with customer and contractor personnel that provide support to the program's engineering organization and specifically the SEIT team. The Systems Engineering Management Plan (SEMP) establishes the project's methodologies and processes, organizational roles and responsibilities for compliances. This SEMP template follows a typical Contract Data Requirements List (CDRL) defined in a Statement of Work (SOW), and specifies the basic inputs, processes, outputs, and tools to be used by a Systems Engineering staff, which includes the integration and test staff, during technical execution described in this Section. The project's Systems Engineering approach to development, integration, test, and demonstration is driven by the methodologies and approach described in DoDI 5000.2.

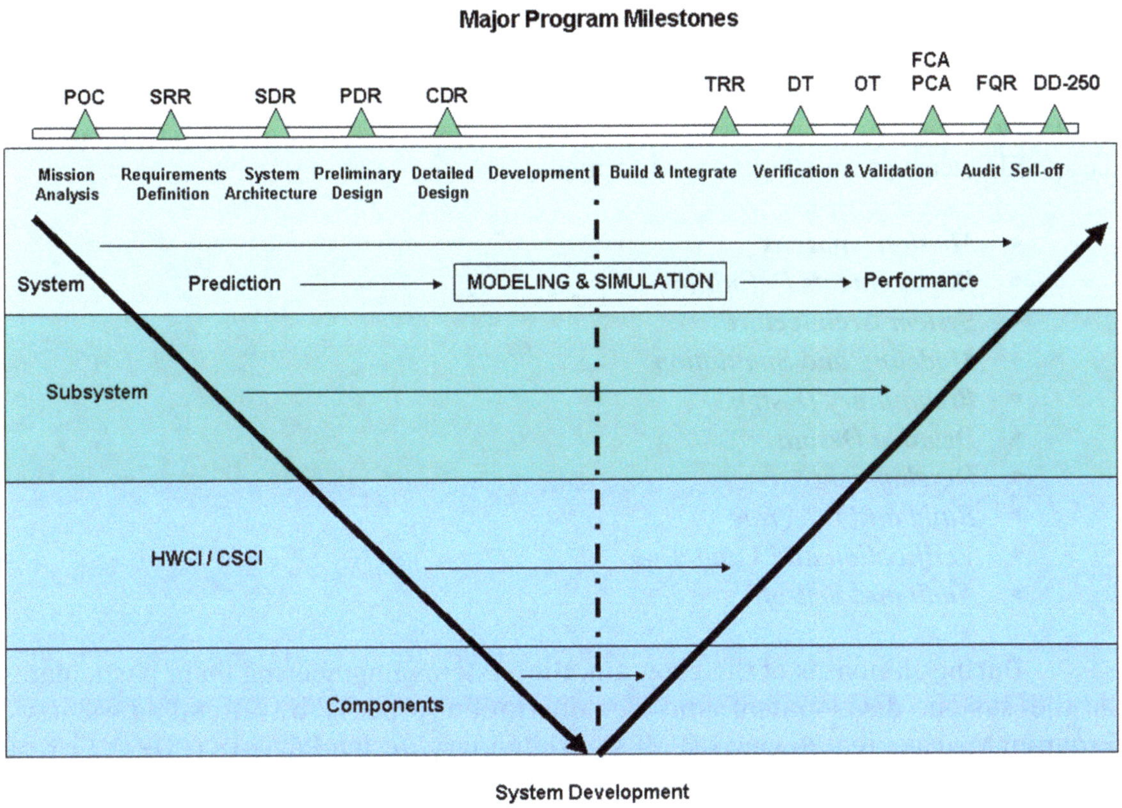

Figure 4.1-1, Horizontal Oversight of Systems Engineering

The Systems Engineering process flow consists of a series of technical reviews and audits and tasks that are performed during the lifecycle of a development program. Figure 4.1-1 illustrates the overall flow of a typical DoD, or Government program, and depicts the major milestones and activities that are usually performed. The following paragraphs summarize those program activities emphasizing Systems Engineering, Integration and Test (SEIT) team process tasks are performed:

- ***Mission Analysis***
- ***Requirements Definition***
- ***System Architecture***
- ***Modeling and Simulation***
- ***Preliminary Design***
- ***Detailed Design***
- ***Development***
- ***Build and Integrate***
- ***Verification and Validation***
- ***Audit and Sell-off***

During all phases of the program, the Systems Engineering team participates in the system development and demonstration process by attending reviews, Program Management Reviews (PMRs), Preliminary Design Reviews (PDRs), Critical Design Reviews (CDRs), Test Readiness Reviews (TRRs), and Formal Qualification Reviews (FQRs) as they occur. The following project's programmatic phases are described as sequential and it is the expectation that these activities are completed stepwise. However, many tasks described in the following paragraphs are repeated during the program lifecycle, and may be accomplished with some overlap during all phases of the project in order to achieve continuous process improvements, re-evaluate program costs, address dynamically changing program priorities, and most importantly to attain mission objectives and overall mission success.

4.1.1 Mission Analysis

The first task of the Systems Engineering process is the Mission Analysis task. Systems Engineering is responsible for the successful completion of this activity and associated detailed tasks. This process begins with a comprehensive analysis of the mission capabilities required from the system Capability Development Document (CDD), the acquisition contract's Statement of Objectives (SOO). The requirements are extracted and analyzed from the Operational Requirements Document (ORD) and the System/Segment Specification (SSS) baseline. Exit criteria from this phase is established when the Mission Analysis task has produced a functionally oriented Concept of Operations (CONOPS) that has also defined mission capabilities in terms of mission threads. These mission threads establish the functional and performance capabilities baseline of the system and are directly traced to the ORD and SSS.

- ***Inputs***
    - *Mission Capabilities*
    - *Statement of Objectives*
    - *Interoperability Directives*
    - *Operational Requirements Document (ORD)*
    - *System/Segment Specification (SSS)*
- ***Outputs***
    - *CONOPS*
    - *Mission Threads*
    - *System Requirements Traceability*
- ***Tools***
    - *List of tools required to support this task*

4.1.2 Requirements Definition

The second task of the Systems Engineering process is the System Requirements Analysis task. The Systems Engineering team is responsible for the successful completion of this task and associated detailed tasks. During this program phase, functional and performance system level requirements are analyzed to determine incremental build applicability and traceability to the System/Segment Specification (SSS), as well as determining the testability of the stated requirements. The requirements analysis for each delivery includes: 1) evaluation of the planned system level requirements, including quantitative performance requirements, to establish clear and concise meaning, 2) identification of duplicates or un-testable requirements; 3) establishing a verification approach and verification method for each requirement, and; 4) applying a binary pass and fail criteria to each requirement, including quantitative performance measures. During this phase, an updated allocated baseline of subsystem level requirements will be evaluated and documented using the Systems Engineering controlled requirements management tool (e.g., DOORS, RTM, RDD).

- ***Inputs***
  - *Mission Analysis*
  - *CONOPS*
  - *Mission objectives and requirements*
  - *System constraints (including loading and performance)*
  - *External interfaces required (Interoperability)*
  - *Legacy system requirements*
  - *RFP, SOW, System Specification and external interface definitions*
- ***Outputs***
  - *CONOPS*
  - *System Specification (Functional and Performance requirements)*
  - *Interface Requirements (External and Internal)*
  - *System Architecture loading and performance constraints*
  - *Security threat assessments*
  - *IRAD reuse strategy and technology insertion goals*
- ***Tools***
  - *Workstations and communications infrastructure*
  - *Office applications (MS Office products, etc.)*
  - *DOORS, RTM, RDD, etc.*
  - *A list of engineering tools that support this task that are program unique*

4.1.3 System Architecture

The third task of the Systems Engineering process is the System Architecture design task. Systems Engineering, working with the Chief Systems Architect (CSA) and other engineering integrated product teams, is responsible for the successful completion of this activity and associated detailed tasks. The primary purpose of the System Architecture design task is to examine alternative subsystems and common infrastructure solutions the can satisfy mission objectives, functional and performance requirements. The methodology used to develop the system architecture is Architecture Based Design (ABD). For additional details concerning the ABD methodology, refer to the USAF DoDAF (e.g., Operational Views, System Views, etc.) and IS&S EP 3.1.1. During this phase of the process the Systems Engineering team is dedicated to accomplishing the following tasks: 1) intermediate level program master schedule development; 2) incremental planning based on long lead acquisition configuration items (components for the system/segment and lower level components etc.); 3) trade studies; 4) modeling and simulation activities; 5) design to cost strategies and solutions; 6) make/buy evaluations; 7) technology insertion points; 8) specialty engineering tasks; 9) hardware / software design and development activities; and, 10) overall system level and integration tasks, including any required certification and safety qualification. The exit criteria for the this task is to baseline the system level architecture in order to meet mission objectives, demonstrate system level requirements traceability, and provide evidence to the customer that the overall system architecture is realistic, with respect to requirements, schedule and cost constraints. The planned review for determining satisfaction of this phase's exit criteria is the SDR and the PDR. The SEIT team will leverage data from the previous project phases in order to expeditiously satisfy these program milestones.

- ***Inputs***
  - *Mission Capabilities*
  - *Statement of Objectives*
  - *Interoperability Directives*
  - *Operational Requirements Document (ORD)*
  - *System/Segment Specification (SSS)*
- ***Processes***
  - *DODAF and Architecture Based Design (ABD)*
- ***Outputs***
  - *CONOPS*
  - *Mission Threads*
  - *System Requirements Traceability*
- ***Tools***
  - *System Architect^tm^*
  - *Visio^tm^*
  - *A list of engineering tools that support this task that are program unique*

4.1.4 Modeling and Simulation

The fourth task of the Systems Engineering process is Modeling and Simulation. It is the preferred method to demonstrate to the customer that the target system can support the functional and performance margins the system needs to have in order to satisfy mission objectives. This methodology serves two purposes – prediction and performance. Modeling and Simulation's goal is to reduce lifecycle costs and technical implementation. Modeling and Simulation is an extremely important task for Systems Engineering to accomplish. This task is very complex and analytical, usually requiring many engineers to develop the algorithms that drive the model and provide insight into the target system. Modeling and Simulation Inputs, Processes, Outputs and Tools:

- ***Inputs***
  - *Mission Analysis*
  - *CONOPS*
  - *Mission objectives and requirements, including Key Performance Parameters (KPP)*
  - *System constraints (including loading and performance)*
  - *External interfaces required (Interoperability)*
  - *Legacy system requirements*
  - *RFP, SOW, System Specification and external interface definitions*
- ***Processes***
  - *Use of an analytical tool which provides both static and dynamic performance predictions (No formal command media covers this topic)*
- ***Outputs***
  - *Analysis of the system's ability to meet Quantitative performance objectives and the margins associated with these objectives*
- ***Tools***
  - *HyPerformix Workbench*^{tm}
  - *Oralce Workbench*^{tm}
  - *High Level Architecture(HLA) Modeling Method*
  - *A list of engineering tools that support this task that are program unique*

4.1.5 Preliminary Design

The fifth task of the Systems Engineering process is the Subsystem Hardware / Software Preliminary Design activities. During this phase, Systems Engineering must provide overall "horizontal" technical oversight, final approval and direction concerning software development (CSCIs), hardware engineering (HWCIs), component level requirements allocation, incremental build planning, subsystem integration and verification activities.

- ***Inputs***
  - *Mission Capabilities*
  - *Statement of Objectives*
  - *Interoperability Directives*
  - *Operational Requirements Document (ORD)*
  - *System/Segment Specification (SSS)*
  - *System Requirements Analysis is underway*
  - *CONOPS (updated draft completed)*
  - *System Constraints (Design to Cost baseline established)*
  - *System loading and performance requirements defined*
  - *Performance Modeling analysis*
  - *System Security requirements defined*
  - *Functional to Allocated requirements baseline established*
  - *System Architecture plan established*
  - *TBD/TBR work-off plan*
- ***Processes***
  - *Architecture Based Design (Reference DoDAF)*
- ***Outputs***
  - *CONOPS*
  - *System/Segment Specification (SSS)*
    - *Functional and Performance requirements*
    - *Updated VCRM and RTM*
  - *Interface Requirements (External and Internal)*
  - *System Architecture loading and performance constraints*
  - *Security threat assessments*
  - *IRAD reuse strategy and technology insertion goals*
- ***Tools***
  - *Workstations and communications infrastructure*
  - *Office applications (MS Office products, etc.)*
  - *DOORS, RDD, RTM, etc.*
  - *A list of engineering tools that support this task that are program unique*

4.1.6 Detailed Design

The sixth task of the System Engineering process is the Detailed Design task. The Detailed Design task consists of two major activities. The first of these is the completion of the design of the software and hardware components using the design requirements as allocated to the components during the Preliminary Design Phase and approved by the customer at the Preliminary Design Review (PDR). The complete flow of data and execution control is usually specified using object-oriented design (OOD) for components to be implemented in and structured design using an automated system architecture tool for detailed components. The detailed design includes the logic to be employed and the algorithms to be implemented by each lower level component, such as software components and hardware components, in which they are embedded. Following the approval of the Critical Design Review (CDR), design is placed under internal configuration control with the project's Configuration Management (CM) team.

- ***Inputs***
  - *Mission Capabilities*
  - *Statement of Objectives*
  - *Interoperability Directives*
  - *Operational Requirements Document (ORD)*
  - *System/Segment Specification (SSS)*
    - *Functional and Performance requirements*
    - *Detailed requirement and derived test requirements*
  - *System Requirements Analysis is underway*
  - *CONOPS (updated draft completed)*
  - *System Constraints (Design to Cost baseline established)*
  - *System loading and performance requirements defined*
  - *Performance Modeling analysis*
  - *System Security requirements defined*
  - *Functional to Allocated requirements baseline established*
  - *System Architecture plan established*
  - *TBD/TBR work-off plan*
- ***Processes***
  - *Architecture Based Design (Reference DoDAF and ABD)*

- ***Outputs***
    - *CONOPS*
    - *System Specification (Functional and Performance requirements)*
        - *Functional and Performance requirements*
        - *Detailed requirement and derived test requirements*
        - *Updated RTM*
    - *Interface Requirements (External and Internal)*
    - *System Architecture loading and performance constraints updated*
    - *Security threat assessments*
    - *IRAD reuse strategy and technology insertion goals*
- ***Tools***
    - *Workstations and communications infrastructure*
    - *Office applications (MS Office products, etc.)*
    - *DOORS, RDD, RTM, etc.*
    - *A list of engineering tools that support this task that are program unique*

4.1.7 Build and Integrate

The seventh task of the Systems Engineering process is the Build and Integrate task. During this build and integrate phase of the program, Systems Engineering will provide technical oversight, final approval and direction concerning subsystem/segment integration, integration planning, component level requirements allocations, and verification planning / procedure development activities. The primary objectives of the build and integrate task are: 1) ensure that all requirements allocated to the delivery can be verified; 2) to establish a structured approach, ordering, and schedule for integrating the subsystem components; and 3) to establish complete requirements verification traceability to the Test Information Sheets (TISs) and detailed test procedures. The primary product of this program activity is the delivery's detailed test plan and test procedures. The detailed test plan describes the approach and test activities for the functionality to be provided by the subsystem engineering teams, or Integrated Product Teams (IPTs), with their planned deliveries. Delivery SSS and ICD requirements that have been assigned a verification method of demonstration or test are allocated to one or more Test Information Sheets and identified in the Requirements Traceability and Verification Matrix (RTVM). The SSS and ICD requirements that have been assigned a verification method of inspection or analysis are identified in the RTVM and assigned to the Systems Engineering team. Also during this phase, individual requirements are allocated to test procedures and test case outlines based upon a TIS and the allocations are loaded into the requirements database by using the project's Requirements Traceability tool (e.g., DOORS, RTM, RDD, etc.). Test procedures are elaborated to describe the inputs and outputs and define the proper execution sequence for the next task in the Systems Engineering process - verification and validation.

- ***Inputs***
  - *Mission Capabilities*
  - *Statement of Objectives*
  - *Interoperability Directives*
  - *Operational Requirements Document (ORD)*
  - *System/Segment Specification (SSS)*
    - *Functional and Performance requirements*
    - *Detailed requirement and derived test requirements*

- ***Outputs***
  - *CONOPS*
  - *Mission Threads*
  - *Updated derived test requirements and RTVM*
  - *System Requirements Traceability using the tool of choice*
- ***Tools***
  - *A list of engineering tools that support this task that are program unique*

4.1.8 Verification and Validation

The eighth task in the Systems Engineering process is the System/Subsystem Verification and Validation task. During this activity, Systems Engineering must provide overall "horizontal" technical oversight, final approval and direction concerning subsystem/segment verification (DT&E) and subsequent validation (OT&E). The primary responsibility for accomplishment of the Subsystem Integration resides within the integrated product development teams (IPTs), however it is the Integration and Test Team of the SEIT that is responsible for verification planning, procedure development and debug, dry-runs, test conduct and reporting. This includes support from the development IPTs and specialty engineering disciplines. The project's Integration and Test (I&T) team is responsible for integrating and testing the system/subsystems deliveries by installing the individual hardware/software configuration items (CIs) in the applicable System test facility and test environment. The SEIT team is responsible for developing and conducting formal TRRs. After the Verification and Validation task has been completed, the project is ready to conduct the final reviews and audits in order to support FCA/PCA and an FQR. The exit criterion for this phase is the successful verification of requirements allocated to the subsystems/segments for each test planned. The SEIT also supports the Government with Operational Testing in order to validate the mission objectives have been met by the system and the user needs and expectations have been also satisfied.

- ***Inputs***
  - *System/Segment Specification (SSS)*
    - *Functional and Performance requirements*
    - *Detailed requirement and derived test requirements*
    - *Interface requirements*
  - *Thread Based Scenarios (ABD)*
  - *System Test Procedures*
- ***Outputs***
  - *Test Results*
    - *Requirements Pass/Fail metrics*
  - *Problem Reports*
  - *Updated RTVM*
- ***Tools***
  - *A list of engineering tools that support this task that are program unique.*

4.1.9 Audits and Sell-off

The ninth and final task description of the Systems Engineering process is support to Government audits and reviews in order to achieve final sell-off of the system, otherwise known as a DD 250. The Functional Configuration Audit (FCA) determines that all hardware and software for the system have adequately satisfied the SSS requirements and the associated detailed requirements. The Physical Configuration Audit is accomplished to determine the correct quantities of Configuration Items (CI) and hardware have been delivered to the customer.

- ***Inputs***
    - *Mission Capabilities*
    - *Statement of Objectives*
    - *Interoperability Directives*
    - *Operational Requirements Document (ORD)*
    - *System/Segment Specification (SSS)*
        - *Functional and Performance requirements*
        - *Detailed requirement and derived test requirements*
        - *Updated RTVM*
- ***Outputs***
    - *CONOPS*
    - *Mission Threads*
    - *System Requirements Traceability*
    - *System Test Report*
    - *Analysis and Audit Data*
    - *Final RTVM*
- ***Tools***
    - *List of tools required to support this task]*

4.2 Related Processes

4.2.1 Electronics

[Define the specific electronic engineering development processes to be used. Define the types of system engineering support required.]

4.2.2 Software

[Define the specific software development processes to be used. Define the types of system engineering support required.]

4.2.3 Mechanical

[Define the specific mechanical engineering development processes to be used. Define the types of system engineering support required.]

4.2.4 Process Development

[Describe the methods by which program processes will be monitored and improved. Consider the impact on Systems Engineering and other processes.]

4.2.5 Additional Systems Engineering Activities

[This section is project specific. It contains a description of other areas not specifically included in previous sections but that are essential for proper planning and conduct of the Systems Engineering effort.]

4.2.5.1 Long Lead Items

[This section contains a description of the process by which long-lead items that affect the critical path of the program are defined/determined.]

4.2.5.2 Engineering Tools

[This section contains a description of the Systems Engineering tools that will be used on the program as well as the reliance on them and control of them:

- *a) Analysis tools*
- *b) Synthesis tools*
- *c) Control tools*
- *d) Reference tools*
- *e) Simulation tools*
- *f) Laboratory and other facility tools*

4.2.5.3 Requirements Management Tools

[Traceability of requirements will be maintained through use of a Requirements Management Database software tool. Systems Engineering will use some tool such as the Dynamic Object Oriented Requirements System (DOORS), RDD or RTM, etc. to manage the comprehensive requirements flow-down. "Parent to Child" relationships must be maintained and tracked with the Engineering Review Board (ERB) oversight chaired by the Technical Director (TD), or a delegated representative from System Engineering. The tool of choice must have features that include its interface capability

to modeling and analysis tools, ease of creating linkages, revision control, extensibility, and impact assessment. DOORS, RDD or RTM are the requirements management tools of choice for all programs/projects.]

4.2.5.4 Modeling and Simulation Tools

[There are some general technical areas where modeling and simulation is used in Systems Engineering, namely computer networking and communications. There are many modeling and simulation tools available for use by Systems Engineering. Examples of these modeling and simulation tools are Network II.5, SES Workbench, and Signal Processing Workbench. These tools require some effort on the user's part in the form of training or otherwise, to become proficient in their use.]

4.3 Trade Studies

4.3.1 Trade Study Process

[Describe the process to be used in performing trade studies. Consider trade matrix formats, depth of analysis, and types of information. Describe how to handle multi-dimensional interdependencies.]

4.3.2 Trade Study Tasks

[Identify the major system trade studies to be performed. Indicate the issues and constraints, including interdependencies.]

4.4 Requirements Allocation

[Describe the unique processes to be used to allocate requirements to system components. Use standards as a reference, and do not restate the processes therein.]

4.5 Analysis Methods and Tools

[Describe the mathematical, simulation, and/or prototyping methods to be used to optimize the design and measure its effectiveness. Define how these methods will be used to impact the design, in which phases. Identify the specific tools to be used. Describe how to handle multi-dimensional interdependencies.]

4.5.1 Design Analyses

[Identify the specific analyses, simulations, and/or prototypes that will be developed. Identify when each analysis will be performed, and by which personnel. Identify the issues and constraints on each, including interdependencies. Identify the results anticipated from each. Include technical and cost analyses. Consider the following possible types of analysis:

- *Performance Analysis (Throughput, Latency, Dynamic Range, Bandwidth, Memory, etc.)*
- *Logistic Support Analysis*
- *Life Cycle Cost*
- *Design to Cost*
- *Design for Manufacturability*
- *Design for Test*
- *Producibility]*

4.6 Documentation

[The SEIT team is responsible for producing the engineering CDRLs as called out by the Program SOW and as allocated in the CWBS and CWBS dictionary. Appendix E of this SEMP, provides the engineering team with typical engineering CDRLs to be produced on a Program.]

4.6.1 Specification Tree

[Show a hierarchical diagram of the specifications that will be created for the system and its elements. Identify the type of each specification. Indicate which specifications are internal and which must be delivered.]

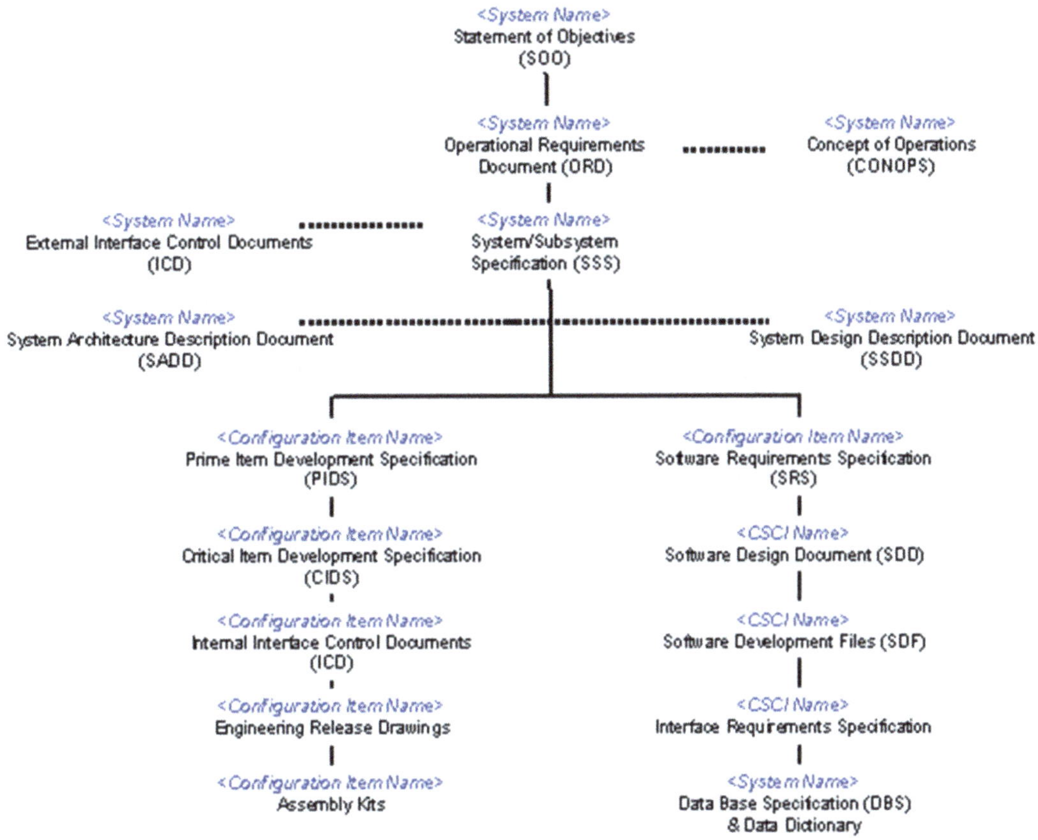

[Example Specification Tree]

Figure 4.6-1** <Program Name> **Specification Tree

4.6.2 Other Documents

[List other documents that will be created. Indicate which documents are internal and which must be delivered.]

4.6.3 Document Generation Methods

[Describe the methods to be used to generate documents. Identify the specific tools, and their required interconnections. Describe the review and sign-off process.]

5 SPECIALTY ENGINEERING

5.1 Control of Engineering Specialties

[Describe the integration and coordination of the various engineering specialties in each Systems Engineering phase. Indicate how Systems Engineering controls achieve the best mix of technical/performance values. Where the specialty programs overlap, define the responsibilities and authorities of each.]

5.2 Integrated Modeling and Simulation Plan (IMSP)

[Summarize the approach to Integrated Logistic Support (ILS). Refer to a separate ILS Plan, if available. Consider the aspects of Logistic Support Analysis, Provisioning, and Training. Indicate the information needed by ILS and its source and time of need. Indicate the expected results and when.]

5.3 Reliability/Maintainability/Availability Plan (RMA)

[Summarize the approach to Reliability/Maintainability/ Availability (RMA) analysis and control. Refer to a separate RMA Plan or plans, if available. Indicate the information needed by RMA and its source and time of need. Indicate the expected results and when.]

5.4 Safety Plan

[Summarize the approach to safety analysis and control. Refer to a separate Safety Plan, if available. Indicate the information needed by safety engineers and its source and time of need. Indicate the expected results and when.]

5.5 Human Factors Engineering Plan

[Summarize the approach to human factors analysis and control. Refer to a separate HFE Plan, if available. Indicate the information needed by human factors engineers and its source and time of need. Indicate the expected results and when.]

5.6 Security Plan

[Summarize the approach to security analysis and control. Refer to a separate Security Plan, if available. Indicate the information needed by security engineers and its source and time of need. Indicate the expected results and when.]

5.7 Integrated Logistics Support Plan (ILSP)

[Summarize the approach to Integrated Logistic Support (ILS). Refer to a separate ILS Plan, if available. Consider the aspects of Logistic Support Analysis, Provisioning, and Training. Indicate the information needed by ILS and its source and time of need. Indicate the expected results and when.]

5.8 EMI/EMC Plan

[Summarize the approach to electromagnetic effects analysis and control. Consider EMI, EMC, and TEMPEST as appropriate. Refer to a separate plan, if available. Indicate the information needed by electromagnetic effects engineers and its source and time of need. Indicate the expected results and when.]

5.9 Value Engineering Plan

[Summarize the approach to value engineering analysis and control. Refer to a separate plan, if available. Indicate the information needed by value engineers and its source and time of need. Indicate the expected results and when.]

5.10 (Other Plans)

[Describe other applicable specialty plans and programs such as: the Test & Evaluation Master Plan (TEMP); the Quality Assurance Plan (QAP); Nuclear Survivability Plan; and Parts Control Plan.]

6 SYSTEMS ENGINEERING DOCUMENTATION

6.1 General

The following paragraphs describe a minimum set of engineering documentation to be developed and maintained by project specific Systems Engineering IPTs. The project Statement of Work and customer provided Contract Data Requirements List (CDRL), DD Form 1423, and associated Data Item Description (DID), must be complied with by the project. Refer to the project Contract Work Breakdown Structure (CWBS) and CWBS Dictionary for more details concerning all required contract deliverables. Refer to the Program Management Plan (PMP) and the Responsibility Assignment Matrix (RAM) for the Cost Account Manager (CAM) responsible for preparation and delivery of documents to the customer. Not all documents listed in the following paragraphs are cited as contract deliverables, however they are required to support the overall engineering tasks of the project and to document specific project engineering tasks.

6.2 Concept of Operations (CONOPS)

The CONOPS will be developed and updated to contain the following information:

- ***Scope***
    - *Identification*
    - *System Overview*
    - *Document Overview*
- ***Referenced Documents***
- ***Current System or Situation***
    - *Background, Mission Objectives and Scope*
    - *Operational Policies and Constraints*
    - *Description of Current System or Situation*
    - *Users or Involved Personnel*
    - *Support Concept*
- ***Justification for and Nature of Changes***
    - *Justification for Change*
    - *Description of Needed Changes*
    - *Priorities of Changes*
    - *Changes Considered But Not Included*
    - *Assumptions and Constraints*
- ***Concept for New or Modified System***
    - *Background, Mission Objectives and Scope*
    - *Operational Policies and Constraints*
    - *Description of New System or Modified System*
    - *Users / Affected Personnel*
    - *Support Concept*
- ***Operational Scenarios***
- ***Summary of Impacts***
    - *Operational Impacts*
    - *Organizational Impacts*
    - *Impacts During Development*
- ***Analysis of the Proposed System***
    - *Summary of Advantages*
    - *Summary of Disadvantages / Limitations*
    - *Alternatives and Trade-offs Considered*
- ***Notes***
- ***Appendices***

6.3 System/Segment Specification (SSS)

The system functionality and associated requirements are typically documented in the System/Segment Specification (SSS) for a development program. The SSS defines the initial baseline of engineering requirements to which the system, subsystems and configuration items will eventually be designed and verified. These requirements define what has to be implemented to perform the mission. The SSS topics listed are intended as a guide, or trigger, for defining the project's system level and functional requirements baseline. Typically, a system specification will contain the following information:

- ***Scope***
  - *Identification*
  - *System Overview*
  - *Document Overview*
- ***Applicable Documents***
  - *Government Documents*
  - *Non-Government Documents*
- ***System Requirements***
  - ***Definition***
  - ***Characteristics***
    - *Performance Characteristics*
      - *States*
      - *Modes*
    - *System Capability Relationships*
    - *External Interface Requirements*
      - *External Interface Descriptions*
    - *Physical Characteristics*
    - *Reliability, Maintainability and Availability (RMA)*
    - *Environmental Conditions*
    - *Transportability*
    - *Flexibility and Expansion*
    - *Portability*

Section 6.3, SSS Outline, continued

-
  - ***Design and Construction***
    - *Materials*
    - *Electromagnetic Radiation*
    - *Nameplates and Product Markings*
    - *Workmanship*
    - *Interchangeability*
    - *Safety*
    - *Human Engineering*
    - *Nuclear Control*
    - *System Security*
    - *Government Furnished Property Usage*
    - *Computer Resources Reserve Capacity*
  - ***Documentation***
  - ***Logistics***
  - ***Personnel and Training***
  - ***Characteristics of Subordinate Elements***
    - *Segment Name(s) and Project Unique Identifier(s)*
  - *Precedence*
  - *Qualification*
  - *Standard Sample*
  - *Pre-Production Sample, Periodic Sample, Pilot or Pilot Lot*
- ***Quality Assurance Provisions***
  - *Responsibility for Inspection*
  - *Special Tests and Examinations*
  - *Requirements Cross Reference*
- ***Preparation for Delivery***
- ***Notes***
- ***Appendices***

6.4 Interface Control Document (ICD)

An ICD is used to document and control the interfaces between two systems and/or segments. An ICD should provide a common data reference for the two systems and/or segments. When there are more than two systems and/or segments, separate documents, as required, are prepared to identify all combinations of systems and/or segments. ICDs are to specify, in quantitative terms with tolerances, the mechanical and functional relations of the interfacing Configuration Items (CIs) to the level of detail necessary to permit detailed design. Functional interfaces shall specify the input, output requirements at the interface in term of voltages, pressures, etc. These functional interfaces must describe in great detail the functional relationships that are specified as requirements in the corresponding System Specification and HWCI and/or CSCI specifications. The basic functional interface described and contained in the System Specification need not be repeated, but referenced as a source requirement for the interface specified. The typical DoD content requirements have been listed below:

- ***Scope***
    - *Interface Description / Purpose*
    - *Specification Object Identifiers*
    - *Other Terminology*
- ***Applicable Documents***
    - *Government Documents*
    - *Non-Government Documents*
- ***Interface Requirements***
    - *Physical*
    - *Functional*
    - *Environmental*
    - *Safety*
- ***Quality Assurance Provisions***
- ***Notes***
- ***Appendices – Engineering Drawings, etc.***

6.5 Verification Cross Reference Matrix (VCRM)

The VCRM provides the traceability of requirements to the method of verification (i.e., inspection, analysis, demonstration and test), the test level at which the verification will take place (e.g., unit, subsystem, system, acceptance) and identifies those requirements which must be verified through specialty engineering (e.g., reliability, maintainability, safety). The information required in the VCRM and other reports concerning requirement traceability, are typically called out in the SOW for each program / customer. However, if the VCRM format and content are not cited in the Contract's SOW, then the following information will be included in the VCRM. The minimum content requirements for the VCRM have been listed below:

- ***VCRM Identification***
- ***Unique SSS Requirement Identifier / Number***
- ***SSS Paragraph***
- ***Requirement Text***
- ***Verification Method(s) Assigned***
- ***Test Level***
- ***Test Function(s) / TIS Identifier***

6.6 System/Segment Description Document (SSDD)

Traditionally the System Segment Design Document (SSDD) has been a formal text-based document with embedded graphics. In practice, the SSDD was rarely reviewed in detail, and the only exposure the customer had to the system design was via the System Design Review (SDR) briefing. A draft of the SSDD may be developed during the proposal phase of a program and submitted as a supporting document with the proposal. A preliminary version may also be developed during the Systems Requirement Analysis Phase. The final version of the SSDD is usually developed for the SDR and presented with briefing charts as a topic covered during the SDR presentation. The typical content requirements for an SSDD have been listed below:

- ***Scope***
    - *Identification*
    - *System Overview*
    - *Document Overview*
- ***Referenced Documents***
- ***Operational Concepts***
    - *Mission*
        - *User Needs*
        - *Primary Mission(s)*
        - *Secondary Mission(s)*
    - *Operational Environment*
    - *Support Environment*
        - *Support Concepts*
        - *Support Facilities*
        - *Supply*
        - *Government Agencies*
    - *Operational Scenarios*
- ***System Design***
    - *System Architecture*
        - *Functional Architecture*
        - *Physical Architecture*
    - *Hardware Configuration Item (HWCI) Identification*
        - *HWCI Name(s) and Purpose*
        - *SSS Requirement Allocation(s)*
        - *External Interfaces Addressed by the HWCI(s)*
        - *Design Constraint(s)*
    - *Computer Software Configuration Item (CSCI) Identification*
        - *CSCI Name(s) and Purpose*
        - *SSS Requirement Allocation(s)*
        - *External Interfaces Addressed by the CSCI(s)*
            - *Detailed Description of Each Interface*

Section 6.6, SSDD Outline, continued.

        - *Design Constraint(s)*
    - *Manual Operations Identification*
        - *Manual Operation(s) Name(s) and Purpose*
        - *SSS Requirement Allocation(s)*
        - *Design Constraint(s)*
    - *Internal Interfaces*
        - *HWCI to HWCI Interface Name(s)*
        - *HWCI to CSCI Interface Name(s)*
        - *CSCI to CSCI Interface Name(s)*
    - *External Interfaces*
        - *External Hardware Interface Name(s)*
        - *External Software Interface Name(s)*
    - *Specification Tree*
- ***Processing Resources***
    - *Processing Resource Name(s)*
    - *Memory Size*
    - *Word Size*
    - *Processing Speed*
    - *Character Set Standard*
    - *Instruction Set Architecture*
    - *Interrupt Capabilities*
    - *Direct Memory Access (DMA)*
    - *Channel Requirements*
    - *Auxiliary Storage*
    - *Growth Capabilities*
    - *Diagnostic Capabilities*
    - *Additional Computer Hardware Capabilities*
    - *Processing Resource Allocation*
- ***Systems Engineering Process***
    - *Technical Performance Measurement*
    - *Trade Studies*
    - *Analyses*
    - *Modeling and Simulation*
    - *Risk Management*
- ***Requirements Traceability***
- ***Notes***
- ***Appendices***

6.7 Risk Management Plan (RMP)

This section contains a description of the technical risk-management program to be covered in the RMP:

- ***Scope***
  - *Identification*
  - *System Overview*
  - *Document Overview*
- ***Applicable Documents***
  - *Government Documents*
  - *Non-Government Documents*
- ***Systems Engineering Risk Management Process***
  - *The approach and criteria for risk identification, prioritization, sensitivity assessment, handling, and risk impact integration into decision processes*
  - *The risks associated with the development, test and evaluation requirements*
  - *The plan to handle technical risk (e.g., additional prototyping, technology and integration verification, back-up development)*
- ***Risk Management Control***
  - *Risk control and monitoring measures including special verifications, technical performance measurement, and critical milestones A description of the method for relating TPMs and project identified risks to cost and schedule performance measurement and the relationship to the Contract Work Breakdown Structure (CWBS)*
- ***Systems Engineering Process Relative to Risk Management***
  - *Technical Performance Measurement*
  - *Trade Studies*
  - *Analyses*
  - *Modeling and Simulation*
- ***Requirements Traceability***
- ***Notes***
- ***Appendices***

Note that this section of the SEMP is only concerned with <u>technical risks</u>, not with cost, schedule or programmatic risks that have roots in other than technical issues.]

6.8 Trade Studies / Analyses

Trade Study reports are used to document the decision rationale for engineering choices that are made on the project, and for an auditable means to revisit decisions previously made without requiring a completely new trade study to be prepared. Trade Studies will include a variety of engineering techniques including complete or partial simulation of operations or maintenance. Trade Studies will be accomplished at different levels, and reporting data will vary dependent on the specific engineering trade considered. Trade Studies must include the results of system effectiveness, life cycle costs and associated risk analyses. Documented Trade Studies also support the capability to re-evaluate a previously considered engineering trade analysis covering cost, technical and schedule considerations. Trade Study content requirements for a typical project have been listed below:

- ***Introduction***
  - *Purpose of the Study*
  - *Summary Conclusion of the Study*
- ***Statement of the Problem***
  - *Description of the Issue(s) That Caused the Study*
    - *Identification of Functional and Technical Design Requirements*
    - *Identification of Possible Design Approaches and Design Characteristics*
  - *System Context*
- ***Evaluation Parameters***
  - *Technical Characteristics*
    - *Physical*
    - *Functional*
    - *Environmental*
    - *Safety*
  - *Programmatic Considerations*
    - *Cost*
    - *Schedule*
- ***Alternative Solutions Considered***
  - *Advantages*
  - *Disadvantages*
- ***Trade Study Report Index***
  - *Trades Required*
  - *Trades Completed*
  - *Trades To Be Completed*
- ***Appendices – Engineering Support Data***

6.9 Requirements Traceability Matrix (RTM)

The RTM provides the traceability of system level requirements to the allocated Configuration Item(s) (CI), the HWCI(s) and/or CSCI(s) specification, the method of verification (i.e., inspection, analysis, demonstration and test), the test level at which the verification will take place (e.g., unit, subsystem, system, acceptance) and identifies those requirements which must be verified through specialty engineering (e.g., reliability, maintainability, safety). The information required in the RTM and other reports concerning requirement traceability, are typically called out in the SOW for each program / customer. However, if the RTM format and content are not cited in the Contract's SOW, then the following information will be included in the RTM. The minimum content requirements for the RTM have been listed below:

- ***RTM Identification***
- ***Unique SSS Requirement Identifier / Number***
- ***SSS Paragraph***
- ***Requirement Text***
- ***Verification Method(s) Assigned***
- ***Test Level***
- ***Unique HWCI and/or CSCI Identifier***
- ***HWCI and/or CSCI Paragraph***
- ***HWCI and/or CSCI Requirement Text***
- ***Verification Method(s) Assigned***
- ***Test Event (e.g., FQT, DT&E, OT&E)***

6.10 Requirements Traceability and Verification Matrix (RTVM)

The RTVM provides the traceability of system level requirements to the allocated Configuration Item(s) (CI), the HWCI(s) and/or CSCI(s) specification, the method of verification (i.e., inspection, analysis, demonstration and test), the test level at which the verification will take place (e.g., unit, subsystem, system, acceptance) and identifies those requirements which must be verified through specialty engineering (e.g., reliability, maintainability, safety). In addition, the RTVM provides direct traceability of requirements to Test Information Sheets (TIS), Test Procedures/Test Cases/Test Steps and/or analysis reports and/or inspection reports. After any test level is completed, the RTVM is updated to provide audit information and becomes the project's master listing for all requirement verification activities. The following information will be included in the RTVM. The minimum content requirements for the RTVM have been listed below:

- ***RTVM Identification***
- ***Unique SSS Requirement Identifier / Number***
- ***SSS Paragraph***
- ***Requirement Text***
- ***Verification Method(s) Assigned***
- ***Test Level***
- ***Unique HWCI and/or CSCI Identifier***
- ***HWCI and/or CSCI Paragraph***
- ***HWCI and/or CSCI Requirement Text***
- ***Verification Method(s) Assigned***
- ***Test Event (e.g., FQT, DT&E, OT&E)***
- ***Test Information Sheet (TIS), Analysis Report and/or Inspection Report***
- ***Test Procedure/Test Case/Test Step***

After Test Conduct / Analysis / Inspection Completed

- ***Test Information Sheet (TIS), Analysis Report and/or Inspection Report***
- ***Test Procedure and Verification Step(s) / Paragraph(s)***
- ***Test Results (Pass, Fail, Not-tested)***
- ***Notes***
    - ***Waiver(s) / Deviation(s)***
    - ***Retest Scheduled Event***
    - ***Amplifying Data***

6.11 Hardware Configuration Item Specification (HWCI)

The HWCI Specification, or Prime Item Development Specification, or Critical Item Specification (PIDS/CIDS) is used to document and specify design requirements for hardware configuration items used on a project. The HWCI Specification contains a description of the materials used, allocated system level requirement, hardware configuration item fabrication, or construction. In addition, the HWCI Specification provides a description of the verification methods to be used to ensure each requirement has been met. Detailed hardware configuration item requirements are typically managed using a requirements management tool (e.g., DOORS, RTM or RDD, etc.) and requirements documents are normally published and maintained on-line using an Integrated Program Environment (IPE). The HWCI Specification specifies, in both quantitative terms with tolerances and in qualitative terms, the mechanical and functional relationships of the HWCI to the level of detail necessary to permit detailed design. Functional requirements shall specify both quantitative and qualitative, inputs and outputs required for the HWCI in terms of voltages, capacities, size, weight, power, etc. These specified capabilities should describe in sufficient detail the relationships that are traceable to the SSS and/or SSDD. The HWCI Specification provides the basis for detailed design implementation and verification. The basic functional capabilities described and contained in the SSS and/or SSDD, need not be repeated, but referenced as a source requirement for the functional capability specified. The minimum content requirements of the HWCI Specification have been listed below:

- ***Introduction***
  - *Scope*
  - *Classification*
- ***Applicable Documents***
  - *Government Documents*
  - *Non-Government Documents*
- ***Requirements***
  - *Qualification and First Article*
    - *Materials*
    - *Environmental Considerations*
    - *Recycled, Reclaimed and Recovered Materials*
    - *Design*
    - *Construction*
    - *Hardware*
    - *RMA*
    - *Transportability*
  - *Performance Characteristics*
    - *Energy Efficiency*
    - *Human Factors*

Section 6.12, HWCI Specification Outline, continued.

- *Safety*
- *Chemical and Physical Properties*
- *EMI Suppression*
- *Dimensions*
    - *Weight*
    - *Color*
    - *Finish*
    - *Identification Plate*
- *GFP*
- *Workmanship*

- ***Quality Assurance Provisions***
    - *Classification of Inspections*
    - *Inspection Conditions*
    - *Qualification Inspection*
    - *First Article Inspection*
    - *Conformance Inspection*
    - *Examination and Tests of Requirements*
- ***Packaging***
- ***Notes***
- ***Acronyms***
- ***Appendices***

6.12 Software Requirements Specification (SRS)

The SRS is used to document and specify design requirements for Computer Software Configuration Items (CSCI) used on a project. The SRS contains a description of the software coding language(s) used, how allocated system level requirements are to be achieved and how the CSCI is to be implemented. In addition, the SRS provides a description of the verification methods to be used to ensure each requirement has been met. Detailed CSCI requirements are typically managed using a requirements management tool (e.g., DOORS, RTM or RDD, etc.), and requirements documents are normally published and maintained on-line using an Integrated Program Environment (IPE). The SRS specifies, in both quantitative terms with tolerances and in qualitative terms, the functional relationships of the CSCI to the level of detail necessary to permit detailed design. Functional requirements shall specify, both quantitative and qualitative, inputs and outputs required for the CSCI in terms of dates throughput, data rates, data storage, display, processing capacity, etc. These specified capabilities should describe in sufficient detail the relationships that are traceable to the SSS and/or SSDD and provide the basis for detailed design implementation and verification. The basic functional capabilities described and contained in the SSS and/or SSDD, need not be repeated, but referenced as a source requirement for the functional capability specified. The minimum content requirements of the SRS have been listed below:

- ***Scope***
  - *Identification*
  - *System Overview*
  - *Document Overview*
- ***Referenced Documents***
  - *Government Documents*
  - *Non-Government Documents*
- ***Requirements***
  - *Required States and Modes*
  - *CSCI Capability Requirements*
    - *CSCI Capability (repeated as necessary)*
  - *External Interface Requirements*
    - *Interface Identification and Diagrams*
      - *Project Unique Interface Identifier (repeated as necessary)*
  - *Internal Interface*
  - *Internal Data*
  - *Adaptation*

Section 6.13, SRS Outline, continued

    - *Safety*
    - *Security*
    - *Environment*
    - *Computer Resource*
        - *Hardware*
        - *Software*
        - *Communications*
    - *Quality Factors*
    - *Design and Implementation Constraints*
    - *Personnel*
    - *Training*
    - *Logistics*
    - *Other*
    - *Packaging*
    - *Precedence and Criticality*
- ***Quality Assurance Provisions***
- ***Requirements Traceability***
- ***Notes***
- ***Appendices***

6.13 Master Test Plan (MTP)

The MTP is used to document the integration and test lifecycle on a project. The MTP contains a description of the mechanism by which the testing strategy is defined. The MTP includes a description of other test plans that are generated (e.g., build integration and test plans, acceptance test plans, qualification plans, etc.). The MTP details the various levels of testing to be accomplished in support of the project. The MTP defines the overall test approach, philosophy, strategy and test objectives for the project. Descriptions of test configurations, required resources, test environments, required tools, tests to be executed, levels of testing, test deliverables, and organizational responsibilities are provided in the MTP. In addition, the MTP identifies risk areas and planned contingencies were functions and/or requirements cannot be fully tested and the MTP established the acceptance criteria for each phase of the project's test program. The Quality Assurance Provisions typically described and contained in the SOW and SSS, need not be repeated, but are referenced as source paragraphs for the integration and test tasks to be performed. The minimum content requirements of the MTP have been listed below:

- ***Scope***
  - *Test Organization*
  - *Test Approach*
  - *Test Types*
    - *Developmental Tests*
    - *Preliminary Qualification Tests*
    - *Formal Qualification Tests*
    - *Human Factors*
    - *Acceptance Tests*
- ***Applicable Documents***
- ***Test Plan***
  - *Test Objectives*
  - *Test Item Configuration*
  - *Flight and Test Sequence Profiles*
  - *Test Schedules*
  - *Instrumentation*
  - *Data Processing and Reporting Procedures*
  - *Interface Identifications*
  - *Safety Considerations*
  - *Flow Diagram*
  - *Test Support Requirements*
- ***Requirements Traceabliltiy and Verification Matrix (RTVM)***
- ***Notes***
- ***Appendices***

6.14 Test Information Sheet (TIS)

Test Information Sheets (TIS) are developed to define a test approach for all allocated SSS and ICD requirements in a particular build that have been identified for verification either by test or demonstration verification methods. TISs are developed for functionally related groups of requirements, generally a system thread, to define the test information necessary to satisfy the individual Section 4.0, Quality Assurance Provisions, of the SSS and/or subordinate specifications allocated to formal test events. A TIS is written for each test, which may include several requirements for subsequent decomposition into test procedures and cases. The minimum content requirements for the TIS have been listed below:

- ***Scope***
- ***Applicable Documents***
- ***Test Information Sheet – Planning Information***
    - *Objective(s)*
    - *Specification Reference(s*
    - *Description of Test Item(s)*
    - *Configuration Under Test*
    - *Approach*
    - *Location, Schedule and Number of Tests*
    - *Support Equipment*
    - *Extent of Prime and Associate Contractor Participation*
    - *Extent of Government Participation*
    - *Instrumentation*
    - *Data Reduction and Analysis*
    - *Government Test Facilities*
- ***Notes***
- ***Requirements Allocation Matrix***

6.15 Test Procedures

Test Procedures are developed for informal and formal test activities. Test procedures will describe the input and output which will verify proper operation or performance of the system's functional capabilities. The procedures indicate the proper sequence of execution for the system and the expected results. For each input, there is an associated expected output, or set of outputs. Test procedures contain logical subsets of actions and responses to demonstrate end-to-end functions and sub-functions. Test procedures are developed in accordance with Test Information Sheets (TISs). Test procedures are submitted as appendices to the TISs. Test procedures are developed from the TIS; and from information contained in the source requirements document. The test procedures provide the verification steps of requirements allocated in the TIS. The minimum content requirements for the TIS based test procedure have been listed below:

- ***Introduction***
    - *Objective(s)*
    - *Test Item Description*
        - *Weight Changes During Flight, as applicable*
        - *Subsystem Descriptions*
        - *Design Features Descriptions*
        - *Anticipated Flight Restrictions and/or Subsystem Limitations*
        - *Identification and Location of non-production Equipments*
        - *Test Apparatus and Methods of Installation*
- ***Applicable Documents***
- ***Test Approach***
    - *Prerequisites*
    - *Postrequisites*
    - *Location, Schedule and Number of Tests*
    - *Extent of Prime and Associate Contractor Participation*
    - *Extent of Anticipated Government Participation*
    - *Instrumentation*
        - *Sensing and Recording Devices*
        - *Interface Wiring Diagrams (Between Sensors and Recording Devices)*
        - *Calibration Constants and Methods of Calibration*
        - *Accuracies Required*
        - *Check-Out Procedures*
        - *Operating Instructions*
    - *Test Phasing, Scheduling and Duration*

Section 6.16, Test Procedures, continued:

- ***Test Support Requirements***
- ***Test Operations***
    - *Test Case(s), Test Steps, Actions, Responses and Verifications*
- ***Post-Test Data Reduction and Analysis***
- ***Requirements Matrix***
- ***Notes***

6.16 Test Reports

The test report provides a summary of the testing activities that took place during scheduled test events, such as qualification testing, system testing and acceptance testing. The test report provides a mechanism for commenting on the overall readiness for use of the product, or system, and to document the results of the testing activity for historical purposes (e.g., FCA/PCA and FQR). The test report provides a chronology of what specific tests were conducted and by whom, when the tests were conducted, who witnessed the testing and a description of the test environment to include equipment configuration and baseline. All tools and/or simulators that were used during the testing are included in this report as well as any testing restrictions that were encountered. All testing incidents that occurred during the testing as well as documented deficiencies are described and listed. The minimum content requirements for the test report have been listed below:

- ***Introduction***
  - *High Level System Overview*
  - *Test Summary (who, what, when, where and why)*
- ***References***
  - *Specifications*
  - *Design Documents*
  - *Version Description Document (VDD)*
  - *Test Plans / Test Procedures*
  - *Test Logs*
- ***Test Environment***
  - *CM Baseline*
    - *System Description / Flight Article*
    - *Hardware Configuration Under Test*
    - *Software Version Description*
- ***Test Constraints***
  - *Known System Problems / Deficiencies*
  - *Test Limitations*
- ***Test Activities***
  - *Dates*
  - *Test Conducted*
  - *Deviations*
  - *Incidents*
  - *Discrepancies*
  - *Elapsed Time*
  - *Staff / Witnesses*
- ***Summary Assessment***
  - *Completeness*
  - *Limitations*
  - *Recommendations*
  - *Discrepancy Closure Plan*

6.17 Interface Requirements Specification (IRS)

The Interface Requirements Specification (IRS) specifies the requirements imposed on one or more systems, subsystems, hardware and software configuration items, manual operations, or other system components to achieve one or more interfaces among these entities. An IRS, by definition, can cover any number of interfaces. In addition, the IRS can be used to supplement the SSS, HWCI Specification(s) and/or SRS(s) as a basis for design and qualification testing of Systems/Segments and Configuration Item(s). The basic functional interface described and contained in the System/Segment Specification (SSS), need not be repeated, but referenced as a source requirement for the interface specified. The typical content requirements for an IRS have been listed below:

- ***Scope***
  - *Identification*
  - *System Overview*
  - *Document Overview*
- ***Applicable Documents***
  - *Government Documents*
  - *Non-Government Documents*
- ***Requirements***
  - *Interface Identification and Diagrams*
  - *Project Unique Interface Identifier, repeat as necessary*
    - *Interface Priority*
    - *Type of Interface*
    - *Data Characteristics*
    - *Data Element Assemblies*
    - *Communication Methods*
    - *Protocols*
    - *Other*
  - *Precedence and Criticality of Requirements*
- ***Quality Assurance Provisions***
  - *Inspection, Analysis, Test and Demonstration*
  - *Special Qualification Methods*
- ***Requirements Traceability***
- ***Notes***
- ***Appendices – Engineering Drawings, etc.***

6.18 Version Description Document (VDD)

The VDD contains an inventory of materials released and an inventory of software components contained in a compiled build. Changes installed since the last build, adaptation data, instructions for installation, possible problems with the build, and known errors are also described in the VDD. The primary purpose of the VDD is to document the software configuration under test. The VDD also provides the capability to recreate the test environment, if a retest is of the configuration under test is required at a later date. The typical content requirements for a VDD have been listed below:

- ***Scope***
  - *Identification*
  - *System Overview*
  - *Document Overview*
  - *Relationship to Other Plans*
- ***Referenced Documents***
- ***Version Description***
  - *Inventory of Items Released*
  - *Inventory of Software Contents*
  - *Changes Installed*
  - *Adaptation Data*
  - *Interface Compatibility*
  - *Related Documents*
  - *Summary of Change*
  - *Installation Instructions*
  - *Possible Problems and Known Errors*
- ***Notes***
- ***Appendices***

APPENDIX A TERMS AND DEFINITIONS

Activity - any step taken or function performed, both mental and physical, toward achieving some objective.

Allocated Baseline (ABL) - The initially approved documentation describing a subsystem's functional, performance, interoperability, and interface requirements that are allocated from those of the system or a higher level subsystem; interface requirements with interfacing subsystems; design constraints; derived requirements (functional and performance); and verification requirements and methods to demonstrate the achievement of those requirements and constraints. Generally, there is an allocated baseline for each subsystem to be developed.

Audit - An independent examination of a work product or set of work products to assess compliance with specifications, standards, contractual agreements, or other criteria. (Also, see Functional Configuration Audit and Physical Configuration Audit).

Automated Document Control Secure System (ADCS) - A system used in the securing/storing of firmware and hardware related design information and associated documentation.

Baseline - (1) an agreed-to description of a product, at a point in time, which serves as a basis for defining change, (2) an approved released document, or set of documents, each of a specific revision; the purpose of which is to provide a basis for managing change.

Change Management - The process of evaluating the impact of a requirement or design change on the system, analyzing the effects of a proposed change in terms of the system foundation architecture, performance, costs, and schedule criteria. Change management must be supported by configuration management to ensure that decisions to adopt a change in requirement, design, or implementation are reflected in system documentation, engineering drawings, or other representations of the system.

Change Request (CR) - A change request describes a problem encountered in the fabrication or utilization of an item. It is initiated by the operation that encounters the problem, and forwarded to the appropriate Engineering function, for problem analysis and problem correction.

Class I Change - Classification assigned when change has impact on contract scope or changes product/associated documentation to the extent it has impact on the procuring activity and/or requires their approval.

Class II Change - Classification assigned to changes that are not classified as Class I changes.

CMstat - A product definition and change management system, commercially available from the CMstat Corporation, and used for configuration management of engineering designs and related documentation.

Commercial and Government Entity (CAGE) Code - A five character alphanumeric code assigned to manufacturers and/or controllers of the design of items.

Commercial off the Shelf (COTS) - A product (hardware or software) sold, leased, or licensed to the general public in the course of normal business and used "as-is" or with minor modifications when acquired by the government.

Completeness - As it applies to requirements, a full consideration of all implications due to higher level requirements.

Component - A basic part of a system. Components may be personnel, hardware, software, facilities, data, material, services, and or techniques that satisfy one or more requirements in the lowest levels of the system architecture. Components may be subsystems and/or configuration items.

Computer Software Component (CSC) - A basic part of a computer software configuration item (CSCI). CSCs may be further decomposed into other CSCs. CSCs are also referred to as simply software components.

Computer Software Configuration Item (CSCI) - A software component (or aggregation of software components) that satisfies an end use function and is designated by the project (often with customer concurrence) for separate configuration management.

Concept of Operations (CONOPS) - An analysis that describes the intended basis for operation of an item. The CONOPS defines how an item is operated in the deployed environment.

Contract - In context of this document, this refers to legally binding contract requirements that bound the scope, responsibilities, liabilities and periods of performance associated with a Test requirement.

Configuration Baseline - The configuration information formally designated at a specific time during a system's or subsystem's life cycle. Configuration baselines, plus approved changes from those baselines, constitute the current configuration information.

Configuration Control - The element of configuration management concerning the systematic proposal, justification, evaluation, coordination, and disposition of proposed changes, and the implementation of all approved/released changes, in the configuration of a CI.

Configuration Control Board (CCB) - A board composed of technical of technical and administrative representatives who recommend approval or disapproval of proposed engineering changes, waivers, and deviations - from a CI's current approved configuration.

Configuration Documentation - The technical documentation that identifies and defines the items functional and physical characteristics.

Configuration Identification - The element of configuration management concerning the selection of CIs; the determination of the types of configuration documentation required for each CI; the issuance of numbers and other identifiers affixed to the CIs and to the technical documentation that defines the CI's configuration, including internal and external interfaces; the release of CIs and their associated configuration documentation; and the establishment of configuration baselines for CIs.

Configuration Item (CI) - An aggregation of software, hardware or both that is designated by the customer for separate configuration management.

Configuration Item Index (CII) - A product listing of controlled items within a product providing the "as-delivered" configuration report.

Configuration Item Number - A four-character identifier assigned to the genealogy (definition) of a product that is designated for configuration control. A three-digit suffix to the number may be used for each configuration of the product. The CI number is the base number for serialization of the product.

Configuration Management (CM) - A discipline applying technical and administrative direction and surveillance over configuration items.

Configuration Management Plan (CMP) - Document defining how CM will be implemented for a particular acquisition or program.

Configuration Status Accounting (CSA) - The recording, storing and accessing of information needed to manage configuration items effectively.

Configuration Status Log (CS Log) - A listing produced by the CSA system defining the as-designed configuration for all controlled items of a product serial.

Consistency - As it applies to requirements, an indication that upward traceability to higher level requirements is realized and that there is an absence of ambiguity, conflict, or error.

Constraint - (1) A restriction, limit or regulation. (2) A type of requirement that is not tradeable against other requirements.

Continuous Acquisition and Life Cycle Support (CALS) - Technology, standards, and procurement process related to the transition from a paper intensive activity to one operating with digital information. MIL-HDBK-59.

Contract Data Requirements List (CDRL) - Listing of the Data Items that are required to be delivered per contract.

Controlled Assembly (CA) - An item within a CI that is configuration verified and recorded in the as-delivered configuration report.

Control Revision List (CRL) - A detached listing produced by CMstat for each document defining the complete revision history of that document. The CRL is the controlling revision of the document and provides a complete cross-reference to the parts list (PL) revision, base/core document revision (e.g., drawing pictorial, listing, spec), and any other files/entities that represent product design information.

Critical Design Review (CDR) - A review conducted for each configuration item when detail design is essentially complete. (1) For hardware, this review determines that the detail design of the configuration item under review satisfies the performance and engineering specialty requirements of the HWCI development specifications; established the detail design compatibility among the configuration item and other items of equipment, facilities, computer software and personnel; assesses the configuration item risk areas; assesses the results of the producibility analyses conducted on system hardware; and reviews the preliminary hardware product specifications. (2) For software, this review focuses on the determination of the acceptability of the detailed design, performance, and test characteristics of the design solution, and on the adequacy of the operation and support documents. MIL-STD-1521

Customization - to adapt an organization process element to establish the project's defined process without changing what is required.

Data Item Description (DID) - A DID number is identified in Block 4 of CDRL DD Form 1423. The DID specifies content, format, preparation instructions and intended use of that particular CDRL.

Defined Process – A defined process is a "managed", stepwise process that is tailored from the organization's set of standard processes according to the organization's tailoring guidelines; has a maintained process description; and contributes work products, measures, and other process improvement information to the organization's process assets.

Derived Requirement - (1) A requirement that is further refined from a primary source requirement or a higher level derived requirement. (2) A requirement that results from choosing a specific implementation for a system element.

Detailed Design - The design solution resulting from detailed design activities (e.g., detail drawings, detail specifications, other design solution descriptions).

Developmental Configuration -The design and associated technical documentation that defines the evolving configuration of a configuration item during development.

Deviation - As it relates to configuration management, a written authorization, prior to manufacture, to depart from specific requirement(s).

Engineering Change Order (ECO) - An automated document within the CM state product definition system used to review and authorize implementation of changes to engineering designs and related documentation.

Engineering Change Proposal (ECP) - The documentation by which a proposed engineering change is described, justified, and submitted to the tasking activity for approval or disapproval.

Engineering Change Request (ECR) - An automated format within the CMstat product definition system for requesting changes or identifying problems to engineering designs and related documentation.

Engineering Information Release (EIR) - A document used for release or change of engineering designs and related documentation. This document has now been superseded by the ECO.

Engineering Management – The management of the engineering and technical effort required to transform a military requirement into an operational system. It includes the system engineering required to define the system performance parameters and preferred system configuration to satisfy the requirement, the planning and control of technical program tasks, integration of the engineering specialties, and the management of a totally integrated effort of design engineering, and production engineering to meet cost, technical performance and schedule objectives.

Engineering Review Board (ERB) – The ERB is chaired by the Program TD. The ERB charter is to analyze, plan and approve Class I and Class II changes and submit the approved changes to the CCB for approval when a designated cost threshold set by the Program Manager has been exceeded.

Engineering Specialties – Engineering specialties are those disciplines that support the design/development process by applying knowledge from a specific area to ensure system operability in the operational environment. These specialties include, but are not limited to, reliability, maintainability, and availability (RMA), logistics engineering areas as well as, security, configuration management, quality, EMC/EMI, human factors, standardization, manufacturing engineering, materials, and system safety.

Engineering Specialty Integration – The timely and appropriate intermeshing of engineering efforts and disciplines such as mission analysis, requirements engineering, reliability, maintainability, availability, logistics engineering, security engineering, hardware and software engineering, human factors, safety, value engineering, standardization, transportability, etc., to ensure their influence on system architecture and design.

Engineering Staff - The Engineering Staff includes the business segment directors and all other personnel who report directly to the Vice President (VP) of Engineering. Personnel from other organizations, e.g., Human Resources, support the VP of Engineering but are not direct members of the Engineering Staff.

Factory - The location (or locations) in which a product or system is developed and/or produced.

Firmware - A combination of software instructions/data that resides in a hardware device in a read only mode. The software cannot readily be modified under program control. DOD-STD-2167.

Formal Review - A review that is conducted in accordance with approved and established standards.

Functional Baseline (FBL) - The initially approved documentation describing a system's or configuration item's functional performance, interoperability, and interface requirements and the verification required to demonstrate the achievement of those specified requirements.

Functional Configuration Audit (FCA) - The formal examination of functional characteristics of a configuration item, prior to acceptance, to verify that the item has achieved the requirements specified in its functional and allocated configuration documentation.

Functional Requirement - A requirement that defines a function of the system under development.

Government Furnished Equipment (GFE) - Equipment supplied by the contracting agency to be used in the performance of the contract.

Government Furnished Item (GFI) - An item supplied by the contracting agency to be used in the performance of the contract.

Guidelines - Guidelines consist of information and recommendations, developed under the guidance of the applicable focus teams, that may be used to guide, train, and inform discipline practitioners in performing their assigned work tasks. Guidelines that are descriptive or step-oriented may exist as part of the organization's process assets and may be used as a starting point for developing organization-level or project-level procedures.

Hardware Configuration Item (HWCI) - A configuration item that is hardware.

Interchangeable Item - An item that possess such physical and functional characteristics as to be equivalent in performance, reliability, and maintainability, to another item of similar or identical purposes; and is capable of being exchanged for the other item without selection for fit or performance and without alteration of the items themselves or of adjoining items, except for adjustments.

Interface Control Document (ICD) - Control documentation which depicts physical and functional interfaces of related or co-functioning items.

Interface Control Working Group (ICWG) - A group made up of customer and contractor personnel established to control interface activity and resolve interface problems and documentation of interface agreements.

Integrated Product Definition System (IPDS) - A system (database) of storing and producing engineering definition of an item. IPDS is no longer accessible by the user to "input" design information and is primarily a conduit to the PDS database. The IPDS database is populated by CMstat.

Integrated Product Team (IPT) – A group of people with complementary skills and expertise who are committed to delivering specified work products in timely collaboration. An integrated team should include empowered representatives from organizations, disciplines, and functions that have a stake in the success of the work products. (Also known as an Integrated Team). CMMI v1.1

Interface Design Document (IDD) - Documentation that details the required interfaces. See ICD

Interface Requirements Specification (IRS) - Document that details the physical and functional interface requirements between related items.

Item - The entities that are implemented at the lowest level of the design hierarchy. An item may be a hardware component, a software unit, data (including training materials), or a manual procedure.

Life-Cycle Activity - A collection of tasks that is defined as a standard across life cycle models. The purpose is to foster consistent measurement to support estimation, tracking, and process improvement.

Life Cycle Model - A partitioning of the life of a product into phases and life cycle activities that guide the project from identifying customer needs through product retirement.

Line Replaceable Unit (LRU) - The lowest assembly (usually at the circuit card level) which can be replaced.

Managed Process - A performed process that is planned and executed in accordance with policy, employs skilled people having adequate resources to produce controlled outputs, involves relevant stakeholders, and is monitored, controlled and reviewed; and is evaluated for adherence to its process description.

Measure (n.) - A variable to which a value is assigned to represent one or more attributes. The term "measures" is used to refer collectively to base measures, derived measures, and indicators.

Measure (v.) - To ascertain a quantitative or qualitative value for an entity based on a standard of comparison.

Measurement - A set of operations having the object of determining a value of a measure

Media - Any machine readable, physical or electronic record containing a computer program/data.

Method - (1) A reasonably complete set of rules that establishes a precise and repeatable way of performing a task and arriving at a result. (2) An analytical technique/procedure for performing a task (how). Example requirements analysis methods are Structured Analysis, Functional Analysis, and Object Oriented Analysis.

Methods of Verification - The means by which a product (element or item) and/or system proves compliance. Specifically, these methods are test, analysis, demonstration, and inspection. Note: A specialized subset of analysis is referred to as qualification by similarity, when heritage and use are shown to satisfy qualification requirements. The definition of verification methods is as follows:

- *Inspection* - Inspection will provide physical non-destructive verification that the system meets requirements by visual examination or simple measurements of equipment, engineering drawings, flow diagrams and/or documents. Software inspection includes structured walkthroughs, on-line access of source code, and development tool outputs, and supporting documentation.
- *Analysis* - Analysis is the analytical validation that using mathematical representations, modeling and simulation, charts, graphs, diagrams, metrics, calculation, or data reduction meets a specification. Analysis includes conclusions drawn from test data, modeling based on system design and performance, and the extension of test-produced data to untested conditions.
- *Demonstration* - Demonstration is used to verify conformance of functional characteristics with specified requirements by actual manipulation, handling or operational functioning of software and equipment and/or terminal parameters. This includes content and accuracy of displays, comparison of system outputs with independently derived test cases, and prompt system recovery from induced failure conditions.
- *Test* - Tests will provide verification that a specification requirement is met by measuring, recording, and evaluating qualitative and quantitative data obtained during controlled exercises under all appropriate conditions using real and/or simulated stimulus. This includes verification of system performance, system functionality, and correct data distribution, for normal as well as stress conditions.

Metric - (an older term that is usually) synonymous with 'measure (n)'.

Model - A simplified representation of some aspect of the real world.

Modified COTS - COTS products that have been modified in some way**.** The modification may be by the contractor, the customer or by the original manufacturer at the direction of the contractor or customer. Ruggedized is COTS or modified COTS that is modified to meet specific service requirements.

Nomenclature - Alphanumeric designations (type designations) and descriptive titles (noun names) assigned by the government for standardized government identification of electronic equipment.

Non-developmental Item (NDI) - A generic term that covers products which satisfy the specified requirements with little or no development effort on the part of the contractor.

Non-Standard Parts Approval Request (NPAR) - DD Form 2052 or a form supplied by the customer, requesting to use a non-military part on a military contract.

Notice of Revision (NOR) - A document used to define revisions to specifications, drawings, associated lists, or other referenced documents which require revision after Engineering Change Proposal approval.

Operational Requirements Document (ORD) - A top-level set of requirements, usually developed by the Customer (as required by contract), for the purpose of defining critical mission needs for an item. Requirements derived from the ORD are typically captured in specifications, interface documents and CONOPS. (Reference: DoDI 5000.2).

Operational Scenario - A sequence of events expected during operation of system products. This includes the environmental conditions and usage rates as well as expected stimuli (inputs) and responses (outputs).

Organizational Measurements Database (OMD) - A database established to collect and make available data on the engineering processes and resulting engineering work products, particularly as they relate to the organization's standard engineering process. The OMD contains plan and actual project measurement's data, project profile/description data, and organizational specific data that can be used in proposal, process improvement, and trend analysis activities.

Organization's Standard Process - The operational definition of the basic process that guides the establishment of a common process across an organization. It describes the fundamental process elements that are expected to be included into any defined process. It also describes the relationships (e.g., ordering and interfaces) between these process elements.

Performance - A quantitative measure characterizing a physical or functional attribute relating to the execution of a mission or function. Performance attributes include quantity (how many or how much), quality (how well), coverage (how much area, how far), timeliness (how responsive, how frequent), and readiness (availability, mean time between failures). Performance is an attribute for all products, processes, and their associated personnel, including those for development, production, verification, deployment, operations, support, training, and disposal. Thus, supportability parameters, manufacturing process variability, reliability, and so forth, are performance measures.

Performance Requirement - How well the system products must perform a function along with the conditions under which the function is performed.

Physical Configuration Audit (PCA) - The formal examination of the "as-built" configuration of a configuration item against its technical documentation to establish or verify the configuration item's product baseline.

Point Release - A change to a software version release. A point release is a re-release, between version releases, of a computer software system or computer software configuration item. Typically used to correct problems in a test release or to correct high priority problems in a version release.

Point Release Number - A number used to identify a specific point release.

Preliminary Design Review (PDR) - Technical review of the basic design approach conducted to determine adequacy and risk resolution (technical, cost, and schedule basis). MIL-STD-1521

Problem/change report (P/CR) - Describes each problem detected on items under configuration control. The P/CR also describes corrective action taken and/or needed.

Procedure - Step by step instructions prepared by an organization that are used when performing test activities. Reference: IEEE-STD-610.

Process - The organization of people, tools, procedures, and standards into tasks to produce a specific result.

Process Architecture - A summary description that defines important characteristics of a set of process descriptions. For the organization standard process, it describes the ordering, interfaces, interdependencies, and other relationships between the process elements. It also describes the interfaces, dependencies, and other relationships to other external processes (e.g., program management, contract management, etc.).

Process Assets, Process Assets Library (PAL) - A collection of entities, maintained by an organization, for use by projects in developing, tailoring, maintaining, and implementing their engineering processes. The Engineering organization's process assets typically include:

- The organization's standard engineering process
- Descriptions of the engineering life cycles approved for use
- The guidelines and criteria for tailoring the organization's standard engineering process
- The organizational measurements database, and
- A library of engineering process-related documentation

It may also include information such as standards, procedures, software development plans, measurement plans, and process training materials. The PAL is comprised of both electronic and hardcopy information. Two major electronic components of the PAL include the organization's standard engineering process (currently located in the General Reference Server) and the organizational measurements database (a Microsoft Access implementation); both of the electronic databases are hosted on the PC network. The physical PAL is located at U1T23 and contains hardcopy information, index listings pointing to electronic information, and pointers to personnel and locations where additional hardcopy information is maintained.

Process Baseline - The current or operative set of processes and/or procedures that constitute the project's defined engineering process; once established, it may remain in effect throughout the life cycle of the project. It is a goal, but not a requirement, to update the process baseline for the project as continuous process improvement is performed on the organization's standard engineering process.

Process Description - A documented expression of a set of activities performed to achieve a given purpose that provides an operational definition of the major components of a process. The documentation specifies in a complete, precise, verifiable manner, the requirements, design, behavior, and other characteristics of a process. Process descriptions may be at the activity, project, or organization level.

Process Element - The fundamental unit of process description.

Product(s) - Products may be characterized as parts, items, elements, components, assemblies, subsystems, and systems. Various products may be assembled into a system. Typically the products are called out as the System/Unit(s) Under Test (S/UUT).

Product Baseline (PBL) - Approved documentation describing functional and physical characteristics of the configuration item.

Product Component - Any work product that must be engineered (requirements, design, implement, integrate) to achieve the intended use of the product throughout its life cycle.

Product Definition System (PDS) - A system of storing manufacturing definition information used to produce an item.

Production Readiness Review (PRR) - A review intended to determine the status of completion of the specific actions which must be satisfactorily accomplished prior to executing a production go-ahead decision. MIL-STD-1521

Project Deliverable - A work product to be delivered to the acquirer.

Project Library - The data, information, documentation, records, and databases that are accumulated during the development and the life cycle of a project; the contents of the project library are normally a mix of hardcopy and electronic information.

Project Measurements Database (PMD) - A database established by each project to collect its measures. This database must contain the information that satisfies both the customer's and the organization's measurement requirements.

Prototype - A model (physical, electronic, digital, analytical, etc.) of a product built or constructed for the purpose of: (a) assessing the feasibility of a new or unfamiliar technology, (b) assessing or mitigating technical risk, (c) validating requirements, (d) demonstrating critical features, (e) qualifying a product, (f) qualifying a process, (g) characterizing performance or product features, or (h) elucidating physical principles.

Qualification - (1) The formal act of determining that the Design Solution satisfies its specified requirements of the development baseline. (2) The formal act of determining that the requirements of the associated processes will be satisfied in time to meet the needs of the related end products and that the methods, tools and environment of each process will perform their intended functions.

Qualify - (1) Determine formally that a design solution satisfied its specified requirements of the development baseline. (2) Declare suitability and readiness for production, deployment, training, operations, support, and disposition.

Record ("the verb") - Set down in a manner that can be retrieved and viewed; the result may take many forms including, but not limited to, hand-written notes, hard-copy or electronic documents, and data recorded in computer-aided software engineering (CASE) and project management tools. [MIL-STD-498].

Release - Document or item is made available for use and is subject to configuration control.

Reliability - The probability that a system or component will perform its required functions under stated conditions for a specified period of time.

Retrofit - Incorporation of new design parts as a result of an approved change into an already accepted and/or operational item(s).

Requirements Traceability - The evidence of an association between a requirement and its parent requirement or between a requirement and its implementation.

Reusable Software - Software developed for the requirements of one application that can be used, in whole or in part, to satisfy the requirements of another application. MIL-STD-2167

Reuse - Items that have been developed by a specific project and then are used by other projects. The using project may, or may not, change the item.

Revision - An initial release, or re-release of a document. For software, the newer term "version" should be used.

Revision Number - A number used to identify a specific revision release.

Rework -Effort required to fix and retest any defect.

Requirements Traceability and Verification Matrix (RTVM) - A matrix that contains the requirements, the assigned verification method(s), verification criteria, and test procedure/test case allocations.

Simulation - A model of a product or its component(s) in operation. Simulations may be computer-based analogies to a product such as video games, or a practice role-play of an operation with or without tools (a simulation of a method or technique as in training classes).

Software Development Files (SDF) - A repository for the collection of material such as design, data, schedule, test requirements, test cases, test procedures and test results for development and support of software. MIL-STD-973.

Software Specification Review (SSR) - Review of the finalized CSCI requirements. MIL-STD-1521.

Software Trouble Report (STR) Form - See P/CR. Other commonly used terms, meaning the same thing, are SSTR, SP/CR, & TR.

Specification - A document that clearly and accurately describes requirements and other characteristics for a product and the procedures to be used to determine that the product satisfies these requirements.

Stakeholder - An individual or organization interested in the success of a product or system. Examples of stakeholder include customers, developers, engineering, management, manufacturing, users, etc.

Statement of Objectives (SOO) - High-level requirements document addressing the need for a new or upgraded system. Typically provided by DoD with a Request for Proposal (RFP). The SOO establishes the basis for developing a CONOPS. Typically evaluated during Operation Test (OT).

Statement of Work (SOW) - Describes the work to be performed, the documentation, equipment, tools and test equipment, facilities, and services to be supplied by the subcontractor in support of the deliverable items to the customer.

Subcontractor Data Requirement List (SDRL) - Same as CDRL, but is used when needed to separate/identify contractor data items from subcontractor data item requirements per DD Form 1423.

Success Criteria - Success criteria specify results that must be obtained to determine that a requirement has been satisfied as defined in the applicable test procedure verifying the requirement.

- *Pass/Fail* - This success criterion will be binary in nature. A verification method having a pass / fail criterion verifies a performance and / or design requirement as specified in the RTVM with no deviations or exceptions.

- *Tolerance/Limit* - This success criterion will be non-binary in nature. A verification method employing tolerance limit criteria will verify a performance and / or design requirement as specified in the RTVM when the results are specified to be within a pre-determined range.

Synthesis - The translation of input requirements (including performance, function, and interface) into possible solutions (resources and techniques) satisfying those inputs. Defines a physical architecture of people, product, and process solutions for a logical grouping of requirements (performance, function, and interface) and then designs those solutions.

System - An aggregation of end products and enabling products to achieve a given purpose. (Source: EIA-632).

System Architecture - A logical, physical structure that specifies interfaces and services provided by the components necessary to accomplish system functionality.

System Design Review (SDR) - The evaluation of the optimization, correlation, completeness, and risks associated with the technical requirements. MIL-STD-1521

Systems Engineering - An inter-disciplinary approach and means to enable the realization of successful systems.

System Engineering Process – The logical sequence of engineering activities and decisions which support the transformation of an operational need into a description of system performance parameters and a preferred system configuration.

System Requirements - characteristics of a system that identify the accomplishment levels needed to achieve specific objectives for a given set of conditions.

System Requirement Review (SRR) - Review to determine the adequacy of the efforts to define system requirements. Reference: MIL-STD-1521B.

System/Subsystem Specification (SSS) - Specification that is typically used for requirements sell-off during formal test activities, i.e., Development Test (DT), Qualification Testing and Acceptance Testing.

System Technical Requirements - A deconflicted set of stakeholder requirements stated in technical terms.

Tailor - To produce a process description by elaborating or adapting details of an incomplete process description. Specifically for LM MS2 TS, tailoring refers to developing a project's defined process from the organization standard process.

Task - (1) Well-defined unit of work in a defined process that provides a visible checkpoint into the status of the work product of the process. Tasks have readiness (entry) criteria and completion (exit) criteria. (2) a "formal" activity.

Technical Data Package (TDP) - The total set of engineering documents required to identify, build, test and maintain all of the items of a contract or product.

Technical Performance Measurement (TPM) – The continuing prediction and demonstration of the degree of anticipated or actual achievement of selected technical objectives. It includes an analysis of any differences among the "achievement to date", "current estimate", and the specification requirement. "Achievement to Date" is the value of a technical parameter estimated or measured in a particular test and/or analysis. "Current Estimate" is the predicted value of a technical parameter to be achieved by the end of the contract within existing resources.

Technical Performance Parameter - Technical Performance Parameters are a selected subset of the system's performance parameters used as the technical measures tracked in TPM. Technical Performance Parameters can be specification requirements; performance parameters such as measures of effectiveness and other key decision metrics used to guide and control progressive development; and design to cost requirements or goals.

Technical Program Planning and Control – The management of those design, development, tests and evaluations tasks required to progress from an operational need to the deployment and operation of the system by the user.

Technology - The tools, equipment, techniques, and methods that can be applied by people to accomplish a particular result.

Test Procedure Validation (Dry-runs) - This is the first-time execution of a step-by-step test procedure, or the execution of an automated test sequence against hardware, software, or both to make sure the procedure is technically and functionally accurate and adequate. Procedure Validation is also performed on an existing validated procedure upon a recognized need for re-validation. Validated procedures are also evaluated during the PCN process to determine if re-validation is required.

Test Readiness Review (TRR) - Review to determine if test procedures are complete and if the contractor is prepared for formal testing. MIL-STD-1521

Traceability - (1) The ability to trace the heritage and lineage of a requirement. (2) The ability to show upward compliance of derived requirements with higher-level parent requirements and downward completeness of derived requirements from higher level parent requirements.

Trade Study - An objective evaluation of alternative requirements, architectures, design approaches, or solutions using identical ground rules and criteria.

User Requirements Baseline - A configuration baseline of user requirements. This baseline is necessary to Systems Engineering domains in which there does not exist a single, clearly discernible, user or customer.

Value Engineering Change Proposal (VECP) - Proposed changes that will save the government money, require a contract change to implement, and will not degrade quality, reliability, performance, and maintainability.

Validation - (1) An activity that ensures that a set of technical requirements is consistent and complete with respect to parent requirements. (2) An activity that ensures that an end product stakeholder's true needs and expectations are met.

Verification - Verification is the process by which a product and/or system is shown to comply with specification requirements. To help distinguish "Validation" from "Verification" it is useful to consider the following:

- *Verification - "You built the product right"*
- *Validation - "You built the right product"*

In general, verification occurs prior to validation. That is, typically, we want to know if the product was "built right" prior to evaluating the product's performance in the user environment. The Validation Process, as in Verification and Validation (V&V), should not be confused with "Test Procedure Validation", or "Test Dry-runs", that occurs in the preparation phase of the test life cycle (consisting of plan, preparation, execute and assess).

Verification Criteria - Each requirement will be verified explicitly by comparison of expected results to actual results based on specified input values and / or conditions. The specific verification criteria applied to each requirement will be defined in the applicable test procedure or document verifying the requirement. The following success criteria will be applied to requirement verification acceptance activities.

Version - An initial release or re-release of a computer software configuration item, associated with a complete compilation or recompilation of the computer software configuration item.

Version Number - A number used to identify a specific software release. The number may be used to identify both major releases (version releases) as well as change releases (point releases) to the software; e.g., an "05.1a" version number identifies the fifth version release containing the "1a" set of changes

Version Description Document (VDD) - A document that accompanies and identifies a given version of a system or Computer Software Configuration Item.

Waiver - (1) For configuration management, a written authorization to accept an item, after manufacture, that departs from the specifications, but is considered usable "as is" or after repair by approved method. (2) For process management, a process waiver is the mechanism by which a project documents and obtains approval for the conscious and explicit elimination of a process requirement or substitution of a standard process element by one which does not satisfy the organization standard process when the standard is applicable.

Work Activity - A collection of work tasks spanning a fixed duration within a schedule of a project. Work activities may contain other work activities, as in a Work Breakdown Structure.

Work Breakdown Structure (WBS) - Product oriented family tree composed of Hardware, Software, Services, Data, and facilities, which relates items to each other and the end product. MIL-STD-881

Work In Process (WIP) - A proposed new version of a design or related document that is under consideration or review.

Work Instruction - Work instructions are synonymous with procedures.

Work Product - An artifact produced by a process. Any tangible item produced during a the process of developing or modifying a product. Examples of work products include the project plan, design document, source code, test plan, meeting minutes, schedule, budget, and problem report. A subset of the work products will be baselined and some will form the set of project deliverables. A key distinction between a work product and a product component is that a work product need not be engineered.

Work Task - The smallest unit of work subject to management accountability. A work task is small enough to allow adequate planning and control of a project, but large enough to avoid micro-management. The work to be accomplished is identified by a work package.

APPENDIX B ACRONYMS

The following acronym list provides the acronyms used in this SEMP template. For more information concerning engineering acronyms, the interested reader is referred to RFEN0050, Engineering Acronyms used by LM MS2 Engineering.

AC	Acceptance Criteria
ATP	Authorization to Proceed
BOM	Bill of Materials
CAD	Component Advanced Design
CAM	Cost Account Manager
CDR	Critical Design Review
CDRL	Contract Data Requirements List
CE	Concept Exploration
CI	Configuration Item
CIDS	Critical Item Development Specification
CM	Configuration Management
CMMI	Capability Maturity Model, Integrated
CMP	Configuration Management Plan
CONOPS	Concept of Operations
CSCI	Computer Software Configuration Item
CWBS	Contractor Work Breakdown Structure
DID	Data Item Description
DOD	Department of Defense
DODAF	Department of Defense Architecture Framework
DODI	Department of Defense Instruction
DT&E	Development Test and Evaluation
DTP	Detailed Test Procedure
ECP	Engineering Change Proposal
ERB	Engineering Review Board
FCA	Functional Configuration Audit
FQR	Formal Qualification Review
FRP	Full Rate Production
HDP	Hardware Development Plan
HDBK	Handbook
HTP	Hardware Test Plan
HWCI	Hardware Configuration Item

ICD	Interface Control Document
ILS	Integrated Logistics Support
ILSP	Integrated Logistics Support Plan
IMP/IMS	Integrated Master Plan/Integrated Master Schedule
IPT	Integrated Product Team
IRS	Interface Requirement Specification
IOT&E	Initial Operational Test & Evaluation
IPE	Integrated Program Environment
IPT	Integrated Product Team
LRIP	Low Rate Initial Production
M&S	Modeling and Simulation
MIL	Military
MTP	Master Test Plan
NDTI	Non-Destructive Test Instruction
ORD	Operational Requirements Document
PBR	Program Baseline Review
PCA	Physical Configuration Audit
PD	Program Directive
PDR	Preliminary Design Review
PIDS	Prime Item Development Specification
PM	Program Manager
PMP	Program Management Plan
QA	Quality Assurance
QAP	Quality Assurance Plan
QPR	Quantitative Performance Requirement
RAM	Responsibility Assignment Matrix
ROMP	Risk/Opportunity Management Plan
RTM	Requirements Traceability Matrix
RTVM	Requirements Traceability and Verification Matrix
SA	System Architect
SDD	System Development and Demonstration
SDP	Software Development Plan
SEIT	Systems Engineering, Integration and Test
SEMP	Systems Engineering Management Plan
SOO	Statement of Objectives
SOW	Statement of Work
SRS	Software Requirements Specification

SSDD	System/Segment Description Document
SSS	System/Segment Specification
STD	Standard
TPM	Technical Performance Measure
TRB	Test Review Board
TRR	Test Readiness Reviews
VC	Verification Criteria
VCRM	Verification Cross Reference Matrix
VDD	Version Description Document
VM	Verification Method
WBS	Work Breakdown Structure

APPENDIX C – GUIDANCE FOR PROGRAM REVIEWS

D.1 DESIGN REVIEW OBJECTIVES

The objectives of the Design Review are to:

1. *Ensure that specified design is sound and is traceable to system requirements;*
2. *Ensure that the design will meet the specified functional, performance and reliability requirements;*
3. *Ensure that the design satisfies DTC objectives;*
4. *Ensure that the risks are identified and that the attained status of the design justifies commitment of program resources to the next phase;*
5. *Ensure that engineering documentation is complete, accurate and technically adequate;*
6. *Ensure that the design is producible, testable, and maintainable;*
7. *Define action items necessary to resolve and design deficiencies found during the review;*
8. *Ensure that standard parts, standard release, design and environmental, health and safety requirements have been incorporated;*
9. *Report on the productivity of the engineering processes employed on the program via metrics as defined in the program Quantitative Management Plan;*
10. *Ensure that design artifacts are available, presentable, and in sufficient detail to assess the design.*

D.2 EIA 632 Table C.11

A) IDENTIFY TECHNICAL REVIEW OBJECTIVES AND REQUIREMENTS

Expected Outcomes - The following are identified and documented: (1) purpose and objectives of the review, (2) agenda requirements, (3) tasks to be completed at each required review, (4) entrance and exit requirements, (5) documentation requirements, (6) distribution requirements, (7) responsibilities of review participants

B) DETERMINE PROGRESS AGAINST EVENT-BASE PLAN

Expected Outcomes - The satisfaction of entrance requirements to the review are determined and documented

C) ESTABLISH TECHNICAL REVIEW BOARD, AGENDA AND SPEAKERS

Expected Outcomes - For each review, the following are established: (1) persons who will participate in the review, (2) chairperson, (3) secretary, (4) reviewers of the presentation, (5) agenda that meets review requirements and ensures that all required tasks are completed, and (6) members of the design team that will prepare the data package, prepare the presentation, prepare material for distribution at the review, make the presentation, answer questions, and accomplish tasks to close out action items

D) PREPARE TECHNICAL REVIEW PACKAGE AND PRESENTATION MATERIAL

Expected Outcomes - Comprehensive read-ahead material is prepared that includes sufficient information so that technical board members can understand the design and participate effectively in the review. Review team responsibilities, agendas, plans, and expectations from the review are defined and documented. A comprehensive set of presentation materials that describe the assigned design topics and that satisfy review objectives is prepared.

E) FACILITATE RESOLUTION OF EMERGING ISSUES

Expected Outcomes - Emerging issues identified and resolved prior to the review.

F) CONDUCT TECHNICAL REVIEW

Expected Outcomes - The following are assessed by the review: (1) maturity of system, or portion thereof, being engineered, (2) progress according to plans and requirements, (3) risks and variances in cost, schedule, and performance, and, (4) readiness to proceed with the next phase of development. Action items required to meet review objectives are generated, recorded, and assigned.

G) CLOSE-OUT REVIEW

Expected Outcomes - The following are completed for review close-out: (1) preparation and distribution of minutes that included purpose, time, place, attendees, decisions, action items, due date, and persons responsible for resolving action items, (2) resolution of action items, and (3) sign off by chairperson.

D.3 Guidance for PDR

This review shall be conducted for each configuration item or aggregate of configuration items to (1) evaluate the progress, technical adequacy, and risk resolution (on a technical, cost, and schedule basis) of the selected design approach, (2) determine its compatibility with performance and engineering specialty requirements of the Hardware Configuration Item (HWCI) development specification, (3) evaluate the degree of definition and assess the technical risk associated with the selected manufacturing methods/processes, and (4) establish the existence and compatibility of the physical and functional interfaces among the configuration item and other items of equipment, facilities, computer software, and personnel. For CSCIs, this review will focus on: (1) the evaluation of the progress, consistency, and technical adequacy of the selected top-level design and test approach, (2) compatibility between software requirements and preliminary design, and (3) on the preliminary version of the operation and support documents.

The PDR shall be a formal technical review of the basic design approach for a configuration item or for a functionally related group of configuration items. It shall be held after the hardware Development Specification(s), the Software Top Level Design Document (STLDD), the Software Test Plan (STP), the HWCI Test Plan, and preliminary versions of the Computer System Operator's Manual (CSOM), Software User's Manual (SUM), Computer System Diagnostic Manual (CSDM), and Computer Resources Integrated Support Document (CRISD) are available, but prior to the start of detailed design. For each configuration item the actions described below may be accomplished as a single event, or they may be spread over several events, depending on the nature and the extent of the development of the configuration item, and on provisions specified in the contract Statement of Work. A collective PDR for a group of configuration items, treating each configuration item individually, may be held when such an approach is advantageous to the contracting agency; such a collective PDR may also be spread over several events, as for a single configuration item. The overall technical program risks associated with each configuration item shall also be reviewed on a technical, cost, and schedule basis. For software, a technical understanding shall be reached on the validity and the degree of completeness of the STLDD, STP, and the preliminary versions of the CSOM, SUM, CSDM, and CRISD.

The contractor shall present the following for review by the contracting agency:

- HWCIs
- CSCIs
- Support Equipment
- Engineering Data
- Evaluation of Electrical, Mechanical, and Logical Designs
- Electromagnetic Compatibility
- Design Reliability
- Design Maintainability
- Human Factors
- System Safety
- Natural Environment
- Equipment and Part Standardization
- Value Engineering
- Transportability
- Test
- Maintenance and Maintenance Data (HWCIs)
- Spares and Government Furnished Property (GFP)
- Packaging/SDPE (Special Design Protective Equipment)
- Technical Manuals
- System Allocation Document
- Design Producibility and Manufacturing
- Post Review Action

D.4 Guidance for CDR

The CDR shall be conducted on each configuration item prior to fabrication/production/coding release to insure that the detail design solutions, as reflected in the Draft Hardware Product Specification, Software Detailed Design Document (SDDD), Data Base Design Document(s) (DBDD(s)), Interface Design Document(s) (IDD(s)), and engineering drawings satisfy requirements established by the hardware Development Specification and Software Top Level Design Document (STLDD). CDR shall be held after the Computer Software Operator's Manual (CSOM), Software User's Manual (SUM), Computer System Diagnostic Manual (CSDM), Software Programmer's Manual (SPM), and Firmware Support Manual (FSM) have been updated or newly released. For complex/large configuration items the CDR may be conducted on an incremental basis, i.e., progressive reviews are conducted versus a single CDR. The overall technical program risks associated with each configuration item shall also be reviewed on a technical (design and manufacturing), cost and schedule basis. For software, a technical understanding shall be reached on the validity and the degree of completeness of the SDDD, IDD(s), DBDD(s), STD, CRISD, SPM, and FSM, and preliminary versions of the CSOM, SUM, and CSDM.

The detail design as disclosed by the hardware Product Specification, drawings, schematics, mockups, etc., shall be reviewed against the HWCI Development Specification performance requirements. For other than facilities, the result of a successful CDR shall be the establishment of the design baseline for detailed fabrication/production planning i.e., the contractor is permitted to use the detail design as presented at CDR and reflected in the hardware Product Specification for planning for production and, if specifically authorized, for initial fabrication/production efforts.

CDR for a CSCI shall be a formal technical review of the CSCI detail design, including data base and interfaces. The CDR is normally accomplished for the purpose of establishing integrity of computer software design at the level of a Unit's logical design prior to coding and testing. CDR may be accomplished at a single review meeting or in increments during the development process corresponding to periods at which components or groups of components reach the completion of logical design. The primary product of the CDR is a formal identification of specific software documentation which will be released for coding and testing. By mutual agreement between the contractor and the contracting agency, CDRs may be scheduled concurrently for two or more CSCIs.

Since computer software development is an iterative process, the completion of a CDR for a CSCI is not necessarily sufficient for maintaining adequate visibility into the remaining development effort through testing.

Additional In-Progress Reviews may be scheduled post- CDR which address:

(a) Response to outstanding action items
(b) Modifications to design necessitated by approved ECPs or design/program errors
(c) Updating sizing and timing data

(d) Updated design information, as applicable
(e) Results obtained during in-house testing, including problems encountered and solutions implemented or proposed.

The contractor shall present the following for review by the contracting agency:

- HWCIs
- CSCIs
- Support Equipment
- Engineering Data
- Detailed Evaluation of Electrical, Mechanical, and Logical Designs
- Electromagnetic Compatibility
- Design Reliability
- Design Maintainability
- Human Factors
- System Safety
- Natural Environment
- Equipment and Parts Standardization
    - Equipment and Components
    - Parts
    - Assignment of Official Nomenclature
- Value Engineering
- Transportability
- Test
- Maintenance and Maintenance Data
- Spare Parts and Government Furnished Property (GFP)
- Packaging/SDPE
- System Allocation Document
- Design Producibility and Manufacturing
- Post Review Action

D.5 <u>Guidance for TRR</u>

The TRR shall be a formal review of the contractor's readiness to begin formal CSCI testing. It is conducted after software test procedures are available and CSC integration testing is complete. The purpose of TRR is for the contracting agency to determine whether the contractor is in fact ready to begin CSCI testing. A technical understanding shall be reached on the informal test results, and on the validity and the degree of completeness of the Computer System Operator's Manual (CSOM), Software User's Manual (SUM), and Computer System Diagnostic Manual (CSDM).

The contractor shall present the following for review:

- Requirements changes
- Design changes
- Software test plans and descriptions
- Software test procedures
- Software Unit test cases, procedures, and results
- Software test resources
- Test limitations
- Software problems
- Schedules
- Documentation Updates
- Post Review Action

After completing the TRR, the contractor shall publish and distribute copies of Review Minutes. The contracting agency officially acknowledges completion of a TRR. The accomplishment of the TRR shall be recorded on the configuration item Development Record by the contractor.

D.6 Guidance for FQR

The objective of the FQR shall be to verify that the actual performance of the configuration items of the system as determined through test comply with the hardware development specification(s), software requirements and interface requirements specifications, and to identify the test report(s)/data which document results of qualification tests of the configuration items. The point of Government certification will be determined by the contracting agency and will depend upon the nature of the program, risk aspects of the particular hardware and software, and contractor progress in successfully verifying the requirements of the configuration items. When feasible, the FQR shall be combined with the FCA at the end of configuration item/subsystem testing, prior to the PCA. If sufficient test results are not available at the FCA to ensure the configuration items will perform in their system environment, the FQR shall be conducted (post PCA) during System testing whenever the necessary tests have been successfully completed to enable certification of configuration items. For non-combined FCA/FQRs, traceability, correlation, and completeness of the FQR shall be maintained with the FCA and duplication of effort avoided.

The contractor shall present the following for review:

- FCA Requirements Satisfaction and Certification
- Audit System Qualification Records and Results
- Review FCA/PCA Minutes to Ensure Requirements Qualification
- Review Test results and Test Reports
- Provide Updates for the Configuration Item Development Record(s)
- Satisfaction of FCA General Requirements and Procedures
- Post Review Action

After completing the FQR, the contractor shall publish and distribute copies of Review Minutes. The contracting agency officially acknowledges completion of the FQR. The completion of the FQR shall be recorded on the configuration item Development Record by the contractor.

The Art of Systems Engineering
Volume 2

Anton Beck
Robert J. Monson, PhD

Introduction

The second volume of this text is provided as a supplementary inclusion for use by instructors choosing to teach systems engineering in a practitioner's environment. This text is constructed in two parts, the first being a typical assignment construct, which is provided to students of a systems design course during the initial class period. This assignment is provided in order to provide a skeleton of the technical needs and requirements for a major project effort.

This assignment is then utilized by student groups made up of 7 to 15 students, who developed a response to the request for proposal over the ensuing 14 weeks of the semester. The instructors act as contracting agents, customers, consultants, and evaluators of the finalized proposal.

In a typical semester, instructors will run a number of teams in competition during the development of this proposal, and during the final evening of class a don't select is performed based upon previously defined evaluation criteria. Included in this time is the initial assignment, A sample proposal document which is been produced by a team of students, and the evaluation technique utilize to provide down-select. The author's hope this will provide a useful guide for the development of a practitioner-oriented systems design course at the graduate level.

Project Assignments

OLICITATION, OFFER AND AWARD

1. THIS CONTRACT IS A RATED UNDER Euphoria (15 CFR	RATIN N/	PAG 1	O	PAGE 2

TRACT	3. SOLICITATION NGTF00010-09-R-	4. TYPE OF [SEALED BID [X NEGOTIATED	5. DATE 09Jan	6. REQUISITION/PURCHASE

ED Green Revolution	COD R0001	8. ADDRESS OFFER (If other than Item	COD
		TEL FAX	

n sealed bid solicitations "offer" and "offeror" mean "bid" and

SOLICITATION

electronic proposal copies submitted (one to Bob Monson and one to Tony Beck), oral presentation of proposal and copies due in class. The dues dates are to be determined. See Section L, Proposal Da TB

ON - LATE Submissions, will affect your

	A.	B. TELEPHONE (Include area *(NO COLLECT*	C. E-MAIL

11. TABLE OF

EC	DESCRIPTIO	PAGE(S	(X	SEC	DESCRIPTIO	PAGE(S
	PART I - THE				**PART II - CONTRACT**	
A	SOLICITATION/ CONTRACT	1	X	I	CONTRACT	8
B	SUPPLIES OR SERVICES AND PRICES/	2			**PART III - LIST OF DOCUMENTS, EXHIBITS AND OTHER**	
C	DESCRIPTION/ SPECS./ WORK	3	X	J	LIST OF	8-
D	PACKAGING AND	3			**PART IV - REPRESENTATIONS AND**	
E	INSPECTION AND	3	X	K	REPRESENTATIONS, CERTIFICATIONS OTHER STATEMENTS OF	9
F	DELIVERIES OR	3 -				
G	CONTRACT ADMINISTRATION	5 -	X	L	INSTRS., CONDS., AND NOTICES TO	16 -
H	SPECIAL CONTRACT	8	X	M	EVALUATION FACTORS FOR	2

OFFER (Must be fully completed by offeror)

Item 12 does not apply if the solicitation includes the provisions at 52.214-16, Minimum Bid Acceptance Period.

compliance with the above, the undersigned agrees, if this offer is accepted ______ calendar days (60 calendar days unless a different erted by the offeror) from the date for receipt of offers specified above, to furnish any or all items upon which prices are offered at the set opposite item, delivered at the designated point(s), within the time specified in the

SCOUNT FOR PROMPT				
KNOWLEDGMENT OF	AMENDMENT	DAT	AMENDMENT	DAT

AN ADDRES O OFFERO	COD	FACILIT	16. NAME AND TITLE OF PERSON AUTHORIZED SIGN OFFER (Type or
ELEPHONE (Include area	15C. CHECK IF REMITTANCE IS DIFFERENT FROM ABOVE - SUCH ADDRESS IN	17.	18. OFFER

AWARD (To be completed by NGRF)

CEPTED AS TO ITEMS	20.	21. ACCOUNTING AND	
THORITY FOR USING OTHER THAN FULL AND OPEN ☐) ☐)		23. SUBMIT INVOICES TO ADDRESS SHOWN (4 copies unless otherwise	ITE
MINISTERED BY (If other than Item COD		25. PAYMENT WILL BE MADE COD	
AME OF CONTRACTING OFFICER (Type or EMAIL		27. (Signature of Contracting	28. AWARD

RTANT - Award will be made on this

T13DGM00

33-

)

Request for Proposal (RFP)

PART I – THE SCHEDULE

Section B - Supplies or Services and Prices

ITEM NO	SUPPLIES/SERVICES	QUANTITY	UNIT	UNIT PRICE	AMOUNT
0001	System Development and Demonstration of the Breadfruit Enterprise Architecture Demonstration (BrEAD) and Systems Design Cost plus fixed fee	1	Lot		
				ESTIMATED COST	
				BASE FEE	________
0002	Technical, Financial and Administrative Data for Item 0001 (Exhibit A) Item 0002 is Not Separately Priced (NSP). The cost/price of Item 0002 is included in Item 0001. FOB: Destination	1	Lot		NSP
					________
				NET AMT	

Section C - Descriptions and Specifications

Item 0001 - The contractor shall perform in accordance with Attachment (1) Performance Based System Specification (PBSS) and (2), Statement of Objectives (SOO) for the New Green Revolution Foundation (NGRF) the System Development & Demonstration (SDD) Phase. Attachment (2) will be replaced with "Statement of Work (SOW) for the Breadfruit Enterprise Architecture Demonstration (BrEAD) and Systems Design SDD Phase" at contract award); and Attachment (3), Integrated Master Plan (inserted at time of award).

Item 0002 – The technical, financial, and administrative data to be furnished hereunder shall be in accordance with contractor defined format as approved by the NGRF contracting officer.

Section D – Packaging and Marking N/A

Section E - Inspection and Acceptance

Items 0001,– Inspection and acceptance of the supplies and services called for hereunder shall be performed at the Contractor's plant or such other facility as approved by the cognizant Procuring Contracting Officer (PCO)/Contract Administration Office (CAO). Acceptance of the Contractor's completed performance of each line item shall be acknowledged via NGRF form xxx, and signed by the Contractor and by the representative from the CAO.

Item 0002– Inspection and acceptance of the data to be furnished hereunder by the contractor shall be in accordance with contractor defined format as approved by the NGRF contracting officer.

INSPECTION AND ACCEPTANCE

(a) Inspection and acceptance of the supplies or services to be furnished hereunder shall be performed by the cognizant Contract Administration Office (CAO) at the contractor's or subcontractor's plant.

(b) Acceptance of all Contract Line Items/Subcontract Line Items (CLINs/SLINs) shall be made by signature of the accepting authority. Acceptance will only occur when the accepting authority is sure that inspections performed demonstrate compliance with contract requirements.

INSPECTION AND ACCEPTANCE OF TECHNICAL DATA AND INFORMATION

Inspection and acceptance of technical data and information will be performed by the Procuring Contracting Officer (PCO) or his duly authorized representative. Inspection of technical data and information will be performed by ensuring successful completion of the requirements set forth in the NGRF form xxx, Contract Data Requirements List (CDRL) and incorporation/resolution of NGRF review comments on the data items. Acceptance will be evidenced by execution of an unconditional NGRF form yyy, Material Inspection and Receiving Report, as appropriate, and/or upon receipt of a second endorsement acceptance by the PCO on the attachment to this contract entitled N/A. The attached form will not be used for high cost data such as drawings, specifications, and technical manuals.

Section F - Deliveries or Performance

Item 0001– The contract period of performance shall begin on the date of contract award (6/1/2009) and be completed by the date of Initial Operating Capability (IOC) in accordance with the "SDD Contract Top-Level Integrated Master Schedule (IMS)" (to be incorporated into the contract at time of award). The objective is to achieve IOC in Fiscal Year (FY) 09 (i.e. not later than 29 December2010) or earlier.

Items 0002,– The data to be furnished by the contractor shall be delivered in accordance with NGRF form xxx, Contract Data Requirements List (CDRL).

SHIPPING INSTRUCTIONS: To be inserted at time of award.

PLACE OF DELIVERY - NGRF FURNISHED MATERIAL

(a) The NGRF will furnish to the contractor for use in connection with this contract the following material at the time specified:

Material	Quantity	Date
None		

(b) The material will be delivered at the NGRF's expense to the location designated in the contractor's proposal for performance. Delivery includes delivery either directly to the factory/warehouse street location, or to a designated private siding if delivery is by rail.

(c) Only the material listed above in the quantity shown will be furnished by the NGRF. All other material required for performance of this contractor shall be furnished by the contractor. Such NGRF-furnished material shall be delivered at or near contractor's plant under NGRF bills of lading, free of expense to the contractor, on board the conveyance selected by the NGRF. When rail delivery is designated by the NGRF as a mode of transportation and shipment from a team track to the contractor's plant is necessary, the contractor agrees to arrange for prompt unloading of cars, pick-up and delivery of material to his plant free of expense to the NGRF.

TECHNICAL DATA AND INFORMATION

Technical Data and Information shall be delivered in accordance with the requirements of the Contract Data Requirements List, NGRF form xxx.

(a) The contractor shall concurrently deliver technical data and information per NGRF form xxx,

(b) Partial delivery of data is not acceptable unless specifically authorized on the NGRF form xxx, or unless approved in writing by the PCO.

(c) The NGRF review period provided on the NGRF form xxx for each item commences upon receipt of all required data .

Section G - Contract Administration Data

DESIGNATION OF CONTRACTING OFFICER'S REPRESENTATIVE (COR

(a) The Contracting Officer has designated [To be inserted at time of award] as the authorized Contracting Officer's Representative (COR) for this contract.

(b) The duties of the COR are limited to the following: [PCO should specify duties and responsibilities of the COR or reference Attachment – to be inserted at time of award]

CONTRACTOR'S AUTHORIZED CONTRACT COORDINATOR AND TECHNICAL LIAISON

(a) The contractor shall state below the name and telephone numbers of the contractor's employees responsible for coordination of contract functions/liaison with the Contracting Officer and/or Contract administrator, and providing technical assistance as required regarding product specifications, functionality, etc.

CONTRACT COORDINATOR:

NAME: ______________________________________

PHONE (BUS): ______________________________________

PHONE (AFTER HOURS): ___________________________

ALTERNATE:

NAME: ______________________________________

PHONE (BUS): ______________________________________

PHONE (AFTER HOURS): ___________________________

(b) The contractor shall notify the Contracting Officer and/or Contract Administrator in advance, in writing, of any changes in the above listed personnel.

DISCLOSURE, USE AND PROTECTION OF PROPRIETARY INFORMATION

(a) During the performance of this contract, the NGRF may use an independent services contractor (ISC), who is neither an agent nor employee of the NGRF. The ISC may be used to conduct reviews, evaluations, or independent verification and validations of technical documents submitted to the NGRF during performance.

(b) The use of an ISC is solely for the convenience of the NGRF. The ISC has no obligation to the prime contractor. The prime contractor is required to provide full cooperation, working facilities and access to the ISC for the purposes stated in paragraph (a) above.

(c) Since the ISC is neither an employee nor agent of the NGRF, any findings, recommendations, analyses, or conclusions of such a contractor are not those of the NGRF.

(d) The prime contractor acknowledges that the NGRF has the right to use ISCs as stated in paragraph (a) above. It is possible that under such an arrangement the ISC may require access to or the use of information (other than restricted cost or pricing data), which is proprietary to the prime contractor.

(e) To protect any such proprietary information from disclosure or use, and to establish the respective rights and duties of both the ISC and prime contractor, the prime contractor agrees to enter into a direct agreement with any ISC as the NGRF requires.

REIMBURSEMENT OF TRAVEL, PER DIEM, AND SPECIAL MATERIAL COSTS

(a) Area of Travel. Performance under this contract may require travel by contractor personnel. If travel, domestic or overseas, is required, the contractor is responsible for making all necessary arrangements for its personnel.

(b) Travel Policy. The NGRF will reimburse the contractor for allowable travel costs incurred by the contractor in performance of the contract.

(c) Travel. Travel and subsistence are authorized for travel beyond a fifty-mile radius of the contractor's office whenever a task assignment requires work to be accomplished at a temporary alternate worksite. No travel or subsistence shall be charged for work performed within a fifty-mile radius of the contractor's office. The contractor shall not be paid for travel or subsistence for contractor personnel who reside in the metropolitan area in which the tasks are being performed. Travel performed for personal convenience, in conjunction with personal recreation, or daily travel to and from work at the contractor's facility will not be reimbursed.

(1) For travel costs other than described in paragraph (c) above, the contractor shall be paid on the basis of actual amount paid to the extent that such travel is necessary for the performance of services under the contract and is authorized by the COR in writing.

(2) When transportation by privately owned conveyance is authorized, the contractor shall be paid on a mileage basis not to exceed the applicable NGRF transportation rate. Authorization for the use of privately owned conveyance shall be indicated in the basic contract. Distances traveled between points shall be shown on invoices as listed in standard highway mileage guides. Reimbursement will not exceed the mileage shown in the standard highway mileage guides.

(3) The contractor agrees, in the performance of necessary travel, to use the lowest cost mode commensurate with the requirements of the mission as set forth in the basic contract and in accordance with good traffic management principles. When it is necessary to use air or rail travel, the contractor agrees to use coach, tourist class, or similar accommodations to the extent consistent with the successful and economical accomplishment of the mission for which the travel is being performed.

(4) The contractor's invoices shall include receipts or other evidence substantiating actual costs incurred for authorized travel. In no event will such payments exceed the rates of common carriers.

(d) Vehicle and/or Truck Rentals. The contractor shall be reimbursed for actual rental/lease of special vehicles and/or trucks (i.e., of a type not normally used by the contractor in the conduct of its business) only if authorized in the basic contract or upon approval by the COR. Reimbursement of such rental shall be made based on actual amounts paid by the contractor. Use of rental/lease costs of vehicles and/or trucks that are of a type normally used by the contractor in the conduct of its business are not subject to reimbursement

(e) Car Rental. The contractor shall be reimbursed for car rental, exclusive of mileage charges, as authorized in the basic contract or upon approval by the COR, when the services are required to be performed beyond the normal commuting distance from the contractor's facilities. Car rental for a team on TDY at one site will be allowed for a minimum of four (4) persons per car, provided that such number or greater comprise the TDY team.

(f) Per Diem. The contractor shall not be paid for per diem for contractor personnel who reside in the metropolitan areas in which the tasks are being performed. Per Diem shall not be paid on services performed within a fifty-mile radius of the contractor's home office or the contractor's local office. Per Diem is authorized for contractor personnel beyond a fifty-mile radius of the contractor's home or local offices whenever a task assigned requires work to be done at a temporary alternate worksite. Per Diem shall be paid to the contractor only to the extent that overnight stay is necessary and authorized under this contract. The authorized per diem rate shall be the same as the prevailing per diem in the worksite locality. These rates will be based on rates contained in the FTR, JTR or SR. The applicable rate is authorized at a flat seventy-five (75%) percent on the day of departure from contractor's home or local office, and on the day of return. Reimbursement to the contractor for per diem shall be limited to actual payments to per diem defined herein. The contractor shall provide actual payments of per diem defined herein. The contractor shall provide supporting documentation for per diem expenses as evidence of actual payment.

AUTHORIZED CHANGES ONLY BY THE CONTRACTING OFFICER

(a) Except as specified in paragraph (b) below, no order, statement, or conduct of NGRF personnel who visit the contractor's facilities or in any other manner communicates with contractor personnel during the performance of this contract shall constitute a change under the "Changes" clause of this contract.

(b) The contractor shall not comply with any order, direction or request of NGRF personnel unless it is issued in writing and signed by the Contracting Officer, or is pursuant to specific authority otherwise included as a part of this contract.

(c) The Contracting Officer is the only person authorized to approve changes in any of the requirements of this contract and notwithstanding provisions contained elsewhere in this contract, the said authority remains solely the Contracting Officer's. In the event the contractor effects any change at the direction of any person other than the Contracting Officer, the change will be considered to have been made without authority and no adjustment will be made in the contract price to cover any increase in charges incurred as a result thereof. The address and telephone number of the Contracting Officer is

ENGINEERING CHANGES

(a) After contract award, the Contracting Officer may solicit, and the contractor is encouraged to propose independently, engineering changes to the equipment, software specifications or other requirements of this contract. These changes may be proposed for reasons of economy, improved performance, or to resolve increased data processing requirements. If the proposed changes are acceptable to both parties, the contractor shall submit a price change proposal to the NGRF for evaluation. Those proposed engineering changes that are acceptable to the NGRF will be processed as modifications to the contract.

(b) This applies only to those proposed changes identified by the contractor, as a proposal submitted pursuant to the provisions of this clause. As a minimum, the following information shall be submitted by the contractor with each proposal:

(1) A description of the difference between the existing contract requirement and the proposed change, and the comparative advantages and disadvantages of each.

(2) Itemized requirements of the contract that must be changed if the proposal is adopted, and the proposed revision to the contract for each such change.

(3) An estimate of the changes in performance costs, if any, that will result from adoption of the proposal.

(4) An evaluation of the effects the proposed change would have on collateral costs to the NGRF such as NGRF-furnished property costs, costs of related items, and costs of maintenance and operation.

(5) A statement of the time by which the change order adopting the proposal must be issued so as to obtain the maximum benefits of the changes during the remainder of this contract. Also, any effect on the contract completion time or delivery schedule shall be identified.

(c) Engineering change proposals submitted to the Contracting Officer shall be processed expeditiously. The NGRF shall not be liable for proposal preparation costs or any delay in acting upon any proposal submitted pursuant to this clause. The contractor has the right to withdraw, in whole or in part, any engineering change proposal not accepted by the NGRF within the period specified in the engineering change proposal. The decision of the Contracting Officer as to the acceptance of any such proposal under this contract shall be final and shall not be subject to the "Disputes" clause of the contract.

(d) The Contracting Officer may accept any engineering change proposal submitted pursuant to this clause by giving the contractor written notice thereof. This written notice may be given by issuance of a modification to this contract. Unless and until a modification is executed to incorporate an engineering change proposal under this contract, the contractor shall remain obligated to perform in accordance with the terms of the existing contract.

(e) If an engineering change proposal pursuant to this clause is accepted and applied to this contract, an equitable adjustment in the contract price and in any other affected provisions of this contract shall be made in accordance with the "Changes" clause.

(f) The contractor is requested to identify specifically any information contained in its engineering change proposal which it considers confidential and/or proprietary and which it prefers not to be disclosed to the public. The identification of information as confidential and/or proprietary is for information purposes only and shall not be binding on the NGRF to prevent disclosure of such information.

NGRF PROPERTY FOR THE PERFORMANCE OF THIS CONTRACT

(a) Definition. NGRF production and research property, as the term is used herein, shall consist of special tooling to which the NGRF has title or the right to acquire title.

(b) Authorization to Use NGRF Production and Research Property, Material, and Agency Peculiar Property Currently Covered by NGRF Contracts Without Rental Charge in Performing this Contract. NGRF production and research property, material, and agency peculiar property covered by the following listed NGRF contracts on the effective date of this contract is hereby authorized for use on a rent-free, non-interference basis in the performance of this contract and sub-contracts of any tier issued hereunder: Contract No(s): [To be inserted at time of award, as applicable.]

EARNED VALUE MANAGEMENT SYSTEM

(a) In the performance of this contract, the Contractor shall use an Earned Value Management System (EVMS) that has been recognized by the cognizant Administrative Contracting Officer (ACO).

(b) The NGRF may require integrated baseline reviews. Such reviews shall be scheduled as early as practicable and should be conducted within 180 calendar days after (1) contract award, (2) the exercise of significant contract options, or (3) the incorporation of major modifications. The objective of the integrated baseline review is for the NGRF and the Contractor to jointly assess areas, such as the Contractor's planning, to ensure complete coverage of the statement of work, logical scheduling of the work activities, adequate resourcing, and identification of inherent risks.

(c) Unless a waiver is granted by the ACO, Contractor-proposed EVMS changes require approval of the ACO prior to implementation. The ACO shall advise the Contractor of the acceptability of such changes within 30 calendar days after receipt of the notice of proposed changes from the Contractor. If the advance approval requirements are waived by the ACO, the Contractor shall disclose EVMS changes to the ACO at least 14 calendar days prior to the effective date of implementation.

(e) The Contractor agrees to provide access to all pertinent records and data requested by the ACO or duly authorized representative. Access is to permit NGRF surveillance to ensure that the EVMS complies, and continues to comply, with the criteria referenced in paragraph (a) of this clause.

(f) The Contractor shall require the following subcontractors to comply with the requirements of this clause:

(Contracting Officer to insert names of subcontractors selected for application of EVMS criteria in accordance with.)

To be inserted at time of award.

DISCLOSURE OF CONTRACT INFORMATION

(a) The Contractor shall not release to anyone outside the Contractor's organization any unclassified information (e.g., announcement of contract award), regardless of medium (e.g., film, tape, document), pertaining to any part of this contract or any program related to this contract, unless the Contracting Officer has given prior written approval.

(b) Requests for approval shall identify the specific information to be released, the medium to be used, and the purpose for the release. The Contractor shall submit its request to the Contracting Officer at least ten (10) days before the proposed date for release.

(c) The Contractor agrees to include a similar requirement in each subcontract under this contract. Subcontractors shall submit requests for authorization to release through the prime contractor to the Contracting Officer.

Section H – Special Contract Requirements – N/A

PART II – CONTRACT CLAUSES

Section I - Contract Clauses – N/A

PART III – LIST OF DOCUMENTS, EXHIBITS AND OTHER ATTACHMENTS

Section J - List of Documents, Exhibits and Other Attachments

Section J Table Of Contents

DOCUMENT TYPE	DESCRIPTION	PAGES	DATE
Exhibit A	NGRF form xxx, Contract Data Requirements List		
Attachment (1)	Breadfruit Enterprise Architecture Demonstration (BrEAD) and Systems Design Performance Based System Specification (PBSS)		
Attachment (2)	Statement of Work (SOW) for the NGRF Breadfruit Enterprise Architecture Demonstration (BrEAD) SDD Phase (Note: Statement of Objectives (SOO) for the Breadfruit Enterprise Architecture Demonstration (BrEAD) SDD Phase, as provided with the RFP will be replaced by the SOW at time of award.)		To be inserted at time of award; submitted by offerors as Annex D
Attachment (3)	Integrated Master Plan (IMP)		To be inserted at time of award; submitted by

			offerors as Annex E
Attachment (4)	SDD Contract Top-Level Integrated Master Schedule (IMS)		To be inserted at time of award; submitted by offerors as part of Annex F
Attachment (6)	Cost Plus Award Fee (CPAF) Plan for Breadfruit Enterprise Architecture Demonstration (BrEAD)	16	

PART IV – REPRESENTATIONS AND INSTRUCTIONS

Section K – Representations, Certifications and other statements of offerors

PLACE OF PERFORMANCE

(a) The offeror or respondent, in the performance of any contract resulting from this solicitation, [ ] intends, [ ] does not intend [check applicable block] to use one or more plants or facilities located at a different address from the address of the offeror or respondent as indicated in this proposal or response to request for information.

(b) If the offeror or respondent checks ``intends'' in paragraph (a) of this provision, it shall insert in the following spaces the required information:

Place of performance (street address, city, state, county, zip code)	Name and address of owner and operator of the plant or facility if other than offeror or respondent
______________________________	______________________________
______________________________	______________________________
______________________________	______________________________.

SMALL BUSINESS PROGRAM REPRESENTATIONS

(a)(1) The North American Industry Classification System (NAICS) code for this acquisition is 336411.

(2) The small business size standard is 1500 employees.

(3) The small business size standard for a concern which submits an offer in its own name, other than on a construction or service contract, but which proposes to furnish a product which it did not itself manufacture, is 500 employees.

(b) Representations. (1) The offeror represents as part of its offer that it [ ] is, [ ] is not a small business concern.

(2) [Complete only if the offeror represented itself as a small business concern in paragraph (b)(1) of this provision.] The offeror represents, for general statistical purposes, that it [ ] is, [ ] is not, a small disadvantaged business concern as defined in 13 CFR 124.1002.

(3) [Complete only if the offeror represented itself as a small business concern in paragraph (b)(1) of this provision.] The offeror represents as part of its offer that it [ ] is, [ ] is not a women-owned small business concern.

(4) Complete only if the offeror represented itself as a small business concern in paragraph (b)(1) of this provision. The offeror represents as part of its offer that it [ ] is, [ ] is not a veteran-owned small business concern.

(5) Complete only if the offeror represented itself as a veteran-owned small business concern in paragraph (b)(4) of this provision. The offeror represents as part of its offer that it [ ] is, [ ] is not a service-disabled veteran-owned small business concern.

(6) [Complete only if the offeror represented itself as a small business concern in paragraph (b)(1) of this provision.] The offeror represents, as part of its offer, that--

(i) It [ ] is, [ ] is not a HUBZone small business concern listed, on the date of this representation, on the List of Qualified HUBZone Small Business Concerns maintained by the Small Business Administration, and no material change in ownership and control, principal office, or HUBZone employee percentage has occurred since it was certified by the Small Business Administration in accordance with 13 CFR part 126; and

(ii) It [ ] is, [ ] is not a joint venture that complies with the requirements of 13 CFR part 126, and the representation in paragraph (b)(6)(i) of this provision is accurate for the HUBZone small business concern or concerns that are participating in the joint venture. [The offeror shall enter the name or names of the HUBZone small business concern or concerns that are participating in the joint venture:____________.] Each HUBZone small business concern participating in the joint venture shall submit a separate signed copy of the HUBZone representation.

(7) [Complete if offeror represented itself as disadvantaged in paragraph (b)(2) of this provision.] The offeror shall check the category in which its ownership falls:

[____] Black American.

[____] Hispanic American.

[____] Native American (American Indians, Eskimos, Aleuts, or Native Hawaiians).

[____] Asian-Pacific American (persons with origins from Burma, Thailand, Malaysia, Indonesia, Singapore, Brunei, Japan, China, Taiwan, Laos, Cambodia (Kampuchea), Vietnam, Korea, The Philippines, U.S. Trust Territory of the Pacific Islands (Republic of Palau), Republic of the Marshall Islands, Federated States of Micronesia, the Commonwealth of the Northern Mariana Islands, Guam, Samoa, Macao, Hong Kong, Fiji, Tonga, Kiribati, Tuvalu, or Nauru).

[____] Subcontinent Asian (Asian-Indian) American (persons with origins from India, Pakistan, Bangladesh, Sri Lanka, Bhutan, the Maldives Islands, or Nepal).

[____] Individual/concern, other than one of the preceding.

(c) Definitions. As used in this provision--

“Service-disabled veteran-owned small business concern”--

(1) Means a small business concern--

(i) Not less than 51 percent of which is owned by one or more service-disabled veterans or, in the case of any publicly owned business, not less than 51 percent of the stock of which is owned by one or more service-disabled veterans; and

(ii) The management and daily business operations of which are controlled by one or more service-disabled veterans or, in the case of a service-disabled veteran with permanent and severe disability, the spouse or permanent caregiver of such veteran.

(2) Service-disabled veteran means a veteran, as defined in 38 U.S.C. 101(2), with a disability that is service-connected, as defined in 38 U.S.C. 101(16).

"Small business concern," as used in this provision, means a concern, including its affiliates, that is independently owned and operated, not dominant in the field of operation in which it is bidding on NGRF contracts, and qualified as a small business under the criteria in 13 CFR Part 121 and the size standard in paragraph (a) of this provision.

“Veteran-owned small business concern” means a small business concern--

(1) Not less than 51 percent of which is owned by one or more veterans (as defined at 38 U.S.C. 101(2)) or, in the case of any publicly owned business, not less than 51 percent of the stock of which is owned by one or more veterans; and

(2) The management and daily business operations of which are controlled by one or more veterans.
"Women-owned small business concern," as used in this provision, means a small business concern--
(1) That is at least 51 percent owned by one or more women; or in the case of any publicly owned business, at least 51 percent of the stock of which is owned by one or more women; and
(2) Whose management and daily business operations are controlled by one or more women.
(d) Notice. (1) If this solicitation is for supplies and has been set aside, in whole or in part, for small business concerns, then the clause in this solicitation providing notice of the set-aside contains restrictions on the source of the end items to be furnished.
(2) Under 15 U.S.C. 645(d), any person who misrepresents a firm's status as a small, small disadvantaged, or women-owned small business concern in order to obtain a contract to be awarded under the preference programs established pursuant to section 8(a), 8(d), 9, or 15 of the Small Business Act or any other provision of Federal law that specifically references section 8(d) for a definition of program eligibility, shall--
(i) Be punished by imposition of fine, imprisonment, or both;
(ii) Be subject to administrative remedies, including suspension and debarment; and
(iii) Be ineligible for participation in programs conducted under the authority of the Act.

SMALL DISADVANTAGED BUSINESS STATUS
(a) General. This provision is used to assess an offeror's small disadvantaged business status for the purpose of obtaining a benefit on this solicitation. Status as a small business and status as a small disadvantaged business for general statistical purposes is covered by the provision at FAR 52.219-1, Small Business Program Representation.
(b) Representations. (1) General. The offeror represents, as part of its offer, that it is a small business under the size standard applicable to this acquisition; and either--
[ ] (i) It has received certification by the Small Business Administration as a small disadvantaged business concern consistent with 13 CFR 124, Subpart B; and
(A) No material change in disadvantaged ownership and control has occurred since its certification;
(B) Where the concern is owned by one or more disadvantaged individuals, the net worth of each individual upon whom the certification is based does not exceed $750,000 after taking into account the applicable exclusions set forth at 13 CFR 124.104(c)(2); and
(C) It is identified, on the date of its representation, as a certified small disadvantaged business concern in the database maintained by the Small Business Administration (PRO-Net); or
[ ] (ii) It has submitted a completed application to the Small Business Administration or a Private Certifier to be certified as a small disadvantaged business concern in accordance with 13 CFR 124, Subpart B, and a decision on that application is pending, and that no material change in disadvantaged ownership and control has occurred since its application was submitted.

(2) * For Joint Ventures. The offeror represents, as part of its offer, that it is a joint venture that complies with the requirements at 13 CFR 124.1002(f) and that the representation in paragraph (b)(1) of this provision is accurate for the small disadvantaged business concern that is participating in the joint venture. [The offeror shall enter the name of the small disadvantaged business concern that is participating in the joint venture:___________________.]
(c) Penalties and Remedies. Anyone who misrepresents any aspects of the disadvantaged status of a concern for the purposes of securing a contract or subcontract shall--
(1) Be punished by imposition of a fine, imprisonment, or both;
(2) Be subject to administrative remedies, including suspension and debarment; and
(3) Be ineligible for participation in programs conducted under the authority of the Small Business Act.

AFFIRMATIVE ACTION COMPLIANCE
The offeror represents that
(a) It [ ] has developed and has on file, [ ] has not developed and does not have on file, at each establishment, affirmative action programs required by the rules and regulations of the Secretary of Labor (41 CFR 60-1 and 60-2), or

(b) It [ ] has not previously had contracts subject to the written affirmative action programs requirement of the rules and regulations of the Secretary of Labor.

ROYALTY INFORMATION

(a) Cost or charges for royalties. When the response to this solicitation contains costs or charges for royalties totaling more than $250, the following information shall be included in the response relating to each separate item of royalty or license fee:

(1) Name and address of licensor.

(2) Date of license agreement.

(3) Patent numbers, patent application serial numbers, or other basis on which the royalty is payable.

(4) Brief description, including any part or model numbers of each contract item or component on which the royalty is payable.

(5) Percentage or dollar rate of royalty per unit.

(6) Unit price of contract item.

(7) Number of units.

(8) Total dollar amount of royalties.

(b) Copies of current licenses. In addition, if specifically requested by the Contracting Officer before execution of the contract, the offeror shall furnish a copy of the current license agreement and an identification of applicable claims of specific patents.

COST ACCOUNTING STANDARDS NOTICES AND CERTIFICATION

Note: This notice does not apply to small businesses or foreign NGRFs. This notice is in three parts, identified by Roman numerals I through III.

Offerors shall examine each part and provide the requested information in order to determine Cost Accounting Standards (CAS) requirements applicable to any resultant contract.

If the offeror is an educational institution, Part II does not apply unless the contemplated contract will be subject to full or modified CAS coverage pursuant to 48 CFR 9903.201-2(c)(5) or 9903.201-2(c)(6), respectively.

I. Disclosure Statement -- Cost Accounting Practices and Certification

(a) Any contract in excess of $500,000 resulting from this solicitation will be subject to the requirements of the Cost Accounting Standards Board (48 CFR Chapter 99), except for those contracts which are exempt as specified in 48 CFR 9903.201-1.

(b) Any offeror submitting a proposal which, if accepted, will result in a contract subject to the requirements of 48 CFR Chapter 99 must, as a condition of contracting, submit a Disclosure Statement as required by 48 CFR 9903.202. When required, the Disclosure Statement must be submitted as a part of the offeror's proposal under this solicitation unless the offeror has already submitted a Disclosure Statement disclosing the practices used in

connection with the pricing of this proposal. If an applicable Disclosure Statement has already been submitted, the offeror may satisfy the requirement for submission by providing the information requested in paragraph (c) of Part I of this provision.

Caution: In the absence of specific regulations or agreement, a practice disclosed in a Disclosure Statement shall not, by virtue of such disclosure, be deemed to be a proper, approved, or agreed-to practice for pricing proposals or accumulating and reporting contract performance cost data.

(c) Check the appropriate box below:

[ ]* (1) Certificate of Concurrent Submission of Disclosure Statement. The offeror hereby certifies that, as a part of the offer, copies of the Disclosure Statement have been submitted as follows:

(i) Original and one copy to the cognizant Administrative Contracting Officer (ACO) or cognizant Federal agency official authorized to act in that capacity (Federal official), as applicable; and

(ii) One copy to the cognizant Federal auditor.

(Disclosure must be on Form No. CASB DS-1 or CASB DS-2, as applicable. Forms may be obtained from the cognizant ACO or Federal official and/or from the loose-leaf version of the Federal Acquisition Regulation.)

Date of Disclosure Statement: ________________ Name and Address of Cognizant ACO or Federal Official Where Filed:

____

The offeror further certifies that the practices used in estimating costs in pricing this proposal are consistent with the cost accounting practices disclosed in the Disclosure Statement.

[ ]* (2) Certificate of Previously Submitted Disclosure Statement. The offeror hereby certifies that the required Disclosure Statement was filed as follows:

Date of Disclosure Statement: ________________ Name and Address of Cognizant ACO or Federal Official Where Filed:

____

The offeror further certifies that the practices used in estimating costs in pricing this proposal are consistent with the cost accounting practices disclosed in the applicable Disclosure Statement.

[ ]* (3) Certificate of Monetary Exemption. The offeror hereby certifies that the offeror, together with all divisions, subsidiaries, and affiliates under common control, did not receive net awards of negotiated prime contracts and subcontracts subject to CAS totaling $50 million or more in the cost accounting period immediately preceding the period in which this proposal was submitted. The offeror further certifies that if such status changes before an award resulting from this proposal, the offeror will advise the Contracting Officer immediately.

[ ]* (4) Certificate of Interim Exemption. The offeror hereby certifies that:

(i) the offeror first exceeded the monetary exemption for disclosure, as defined in (3) of this subsection, in the cost accounting period immediately preceding the period in which this offer was submitted and

(ii) in accordance with 48 CFR 9903.202-1, the offeror is not yet required to submit a Disclosure Statement.

The offeror further certifies that if an award resulting from this proposal has not been made within 90 days after the end of that period, the offeror will immediately submit a revised certificate to the Contracting Officer, in the form specified under subparagraph (c)(1) or (c)(2) of Part I of this provision, as appropriate, to verify submission of a completed Disclosure Statement.

Caution: Offerors currently required to disclose because they were awarded a CAS-covered prime contract or subcontract of $50 million or more in the current cost accounting period may not claim this exemption (4). Further, the exemption applies only in connection with proposals submitted before expiration of the 90-day period following the cost accounting period in which the monetary exemption was exceeded.

II. Cost Accounting Standards -- Eligibility for Modified Contract Coverage

If the offeror is eligible to use the modified provisions of 48 CFR 9903.201-2(b) and elects to do so, the offeror shall indicate by checking the box below. Checking the box below shall mean that the resultant contract is subject to the Disclosure and Consistency of Cost Accounting Practices clause in lieu of the Cost Accounting Standards clause.

[ ]* The offeror hereby claims an exemption from the Cost Accounting Standards clause under the provisions of 48 CFR 9903.201-2(b) and certifies that the offeror is eligible for use of the Disclosure and Consistency of Cost Accounting Practices clause because during the cost accounting period immediately preceding the period in which this proposal was submitted, the offeror received less than $50 million in awards of CAS-covered prime contracts and subcontracts. The offeror further certifies that if such status changes before an award resulting from this proposal, the offeror will advise the Contracting Officer immediately.

Caution: An offeror may not claim the above eligibility for modified contract coverage if this proposal is expected to result in the award of a CAS-covered contract of $50 million or more or if, during its current cost accounting period, the offeror has been awarded a single CAS-covered prime contract or subcontract of $50 million or more.

III. Additional Cost Accounting Standards Applicable to Existing Contracts

The offeror shall indicate below whether award of the contemplated contract would, in accordance with subparagraph (a)(3) of the Cost Accounting Standards clause, require a change in established cost accounting practices affecting existing contracts and subcontracts. [ ]* yes [ ]* no

PROPOSAL DISCLOSURE—COST ACCOUNTING PRACTICE CHANGES

The offeror shall check "yes" below if the contract award will result in a required or unilateral change in cost accounting practice, including unilateral changes requested to be desirable changes.

[ ] Yes [ ] No

If the offeror checked "Yes" above, the offeror shall--

(1) Prepare the price proposal in response to the solicitation using the changed practice for the period of performance for which the practice will be used; and

(2) Submit a description of the changed cost accounting practice to the Contracting Officer and the Cognizant Federal Agency Official as pricing support for the proposal.

NOTICE OF EARNED VALUE MANAGEMENT SYSTEM - POST AWARD IBR

(a) The offeror shall provide documentation that the Cognizant Federal Agency has determined that the proposed Earned Value Management System (EVMS) complies with the EVMS guidelines in ANSI/EIA Standard -748 (current version at time of solicitation).

tab (b) If the offeror proposes to use a system that has not been determined to be in compliance with the requirements of paragraph (a) of this provision, the offeror shall submit a comprehensive plan for compliance with the EVMS guidelines.

(1) The plan shall--

(i) Describe the EVMS the offeror intends to use in performance of the contracts;

(ii) Distinguish between the offeror's existing management system and modifications proposed to meet the guidelines;

(iii) Describe the management system and its application in terms of the EVMS guidelines;

(iv) Describe the proposed procedure for administration of the guidelines, as applied to subcontractors; and

(v) Provide documentation describing the process and results of any third-party or self-evaluation of the system's compliance with the EVMS guidelines.

(2) The offeror shall provide information and assistance as required by the Contracting Officer to support review of the plan.

(3) The NGRF will review and approve the offeror's plan for an EVMS before contract award.

(4) The offeror's EVMS plan must provide milestones that indicate when the offeror anticipates that the EVM system will be compliant with the ANSI/EIA Standard -748 guidelines.

(c) Offerors shall identify the major subcontractors, or major subcontracted effort if major subcontractors have not been selected, planned for application of the guidelines. The prime Contractor and the NGRF shall agree to subcontractors selected for application of the EVMS guidelines.

Section L - Instructions, Conditions and Notices to Bidders

TYPE OF CONTRACT

The NGRF contemplates award of a Cost Plus Award Fee (CPAF) contract, which includes Firm Fixed Price (FFP) line items, resulting from this solicitation.

IDENTIFICATION AND ASSERTION OF USE, RELEASE, OR DISCLOSURE RESTRICTIONS

(a) The terms used in this provision are defined in the following clause or clauses contained in this solicitation--

(1) If a successful offeror will be required to deliver technical data, the Rights in Technical Data--Noncommercial Items clause, or, if this solicitation contemplates a contract under the Small Business Innovative Research Program, the Rights in Noncommercial Technical Data and Computer Software--Small Business Innovative Research (SBIR) Program clause.

(2) If a successful offeror will not be required to deliver technical data, the Rights in Noncommercial Computer Software and Noncommercial Computer Software Documentation clause, or, if this solicitation contemplates a contract under the Small Business Innovative Research Program, the Rights in Noncommercial Technical Data and Computer Software--Small Business Innovative Research (SBIR) Program clause.

(b) The identification and assertion requirements in this provision apply only to technical data, including computer software documentation, or computer software to be delivered with other than unlimited rights. For contracts to be awarded under the Small Business Innovative Research Program, the notification and identification requirements do not apply to technical data or computer software that will be generated under the resulting contract. Notification and identification is not required for restrictions based solely on copyright.

(c) Offers submitted in response to this solicitation shall identify, to the extent known at the time an offer is submitted to the NGRF, the technical data or computer software that the Offeror, its subcontractors or suppliers, or potential subcontractors or suppliers, assert should be furnished to the NGRF with restrictions on use, release, or disclosure.

(d) The Offeror's assertions, including the assertions of its subcontractors or suppliers or potential subcontractors or suppliers, shall be submitted as an attachment to its offer in the following format, dated and signed by an official authorized to contractually obligate the Offeror:

Identification and Assertion of Restrictions on the NGRF's Use, Release, or Disclosure of Technical Data or Computer Software.

The Offeror asserts for itself, or the persons identified below, that the NGRF's rights to use, release, or disclose the following technical data or computer software should be restricted:

Technical Data or Computer Software to be Furnished With Restrictions*	Basis for Assertion**	Asserted Rights Category***	Name of Person Asserting Restrictions****
(LIST*****	(LIST)	(LIST)	(LIST)

* For technical data (other than computer software documentation) pertaining to items, components, or processes developed at private expense, identify both the deliverable technical data and each such item, component, or process. For computer software or computer software documentation identify the software or documentation.

** Generally, development at private expense, either exclusively or partially, is the only basis for asserting restrictions. For technical data, other than computer software documentation, development refers to development of the item, component, or process to which the data pertain. The NGRF's rights in computer software documentation generally may not be restricted. For computer software, development refers to the software. Indicate whether development was accomplished exclusively or partially at private expense. If development was not accomplished at private expense, or for computer software documentation, enter the specific basis for asserting restrictions.

*** Enter asserted rights category (e.g., NGRF purpose license rights from a prior contract, rights in SBIR data generated under another contract, limited, restricted, or NGRF purpose rights under this or a prior contract, or specially negotiated licenses).

**** Corporation, individual, or other person, as appropriate.

***** Enter "none" when all data or software will be submitted without restrictions.

Date ______________________________

Printed Name and Title ______________________________

____

Signature ______________________________

(End of identification and assertion)

(e) An offeror's failure to submit, complete, or sign the notification and identification required by paragraph (d) of this provision with its offer may render the offer ineligible for award.

(f) If the Offeror is awarded a contract, the assertions identified in paragraph (d) of this provision shall be listed in an attachment to that contract. Upon request by the Contracting Officer, the Offeror shall provide sufficient information to enable the Contracting Officer to evaluate any listed assertion.

TECHNICAL DATA OR COMPUTER SOFTWARE PREVIOUSLY DELIVERED TO THE NGRF

of all documents or other media incorporating technical data or computer software it intends to deliver under this contract with other than unlimited rights that are identical or substantially similar to documents or other media that the Offeror has produced for, delivered to, or is obligated to deliver to the NGRF under any contract or subcontract. The attachment shall identify--

(a) The contract number under which the data or software were produced;

(b) The contract number under which, and the name and address of the organization to whom, the data or software were most recently delivered or will be delivered; and

(c) Any limitations on the NGRF's rights to use or disclose the data or software, including, when applicable, identification of the earliest date the limitations expire.

NOTICE OF EARNED VALUE MANAGEMENT SYSTEM

(a) The offeror shall provide documentation that the cognizant Administrative Contracting Officer has recognized that the proposed Earned Value Management System (EVMS) complies with the NGRF EVMS criteria.

(b) If the offeror proposes to use a system that does not meet the requirements of paragraph (a) of this provision, the offeror shall submit a comprehensive plan for compliance with the EVMS criteria.

(1) The plan shall--

(i) Describe the EVMS the offeror intends to use in performance of the contract;

(ii) Distinguish between the offeror's existing management system and modifications proposed to meet the criteria;

(iii) Describe the management system and its application in terms of the govrnment EVMS criteria;

(iv) Describe the proposed procedure for administration of the criteria as applied to subcontractors; and

(v) Provide documentation describing the process and results of any third-party or self-evaluation of the system's compliance wEVMS criteria.

(2) The offeror shall provide information and assistance as required by the Contracting Officer to support review of the plan.

(3) The NGRF will review the offeror's plan for EVMS before contract award.

(c) Offerors shall identify the major subcontractors, or major subcontracted effort if major subcontractors have not been selected, planned for application of the criteria. The prime contractor and the NGRF shall agree to subcontractors selected for application of the EVMS criteria.

EXCLUSIVE TEAMING ARRANGEMENTS WHICH INHIBIT COMPETITION

Offerors who propose teaming arrangements on an exclusive basis will be evaluated to determine whether such teaming agreements inhibit competition. In order for the NGRF to evaluate whether the proposed agreements inhibit competition, offerors are required to (1) provide a copy of all teaming arrangements, and (2) explain why the teaming arrangements do not inhibit competition. The documentation must include, but is not limited to: structure of the teaming arrangement, responsibilities, and liabilities; financial responsibility; managerial responsibility and accountability; and applicable legal documents. The burden of proving that any exclusive teaming arrangement proposed does not restrict competition shall rest with the offeror. Offerors are advised that should the NGRF determine that any such proposed, exclusive teaming arrangement inhibits competition, (1) that determination may render the offeror's proposal ineligible for award, and (2) the Contracting Officer shall forward the matter to the appropriate authorities as prescribed by Federal Acquisition Regulation Part 3.3.

REALISM OF COST PROPOSALS

(a) An offeror's proposal is presumed to represent his best efforts to respond to the solicitation. Any inconsistency, whether real or apparent, between promised performance and cost should be explained in the proposal. For example, if the intended use of new and innovative production techniques is the basis for an abnormally low estimate, the nature of these techniques and their impact on cost should be explained. Additionally, if a corporate policy has been made to absorb a portion of the estimated cost, that should be stated in the proposal and the contract will include a clause which requires the offeror to absorb that portion of costs reflected in its cost proposal.

(b) Any significant inconsistency if unexplained, raises a fundamental issue of the offeror's understanding of the nature and scope of the work required and of his financial ability to perform the contract, and may be grounds for rejection of the proposal. The burden of proof as to cost credibility rests with the offeror.

LESS THAN FIVE PERCENT SMALL DISADVANTAGED BUSINESS SUBCONTRACTING GOAL

Offerors submitting Small and Small Disadvantaged Business Subcontracting Plans per FAR Clause 52.219-9, "Small Business Subcontracting Plan" which reflect a Small Disadvantaged Business (SDB) goal of less than five percent shall also provide, as a part of the subcontracting plan submission, those extenuating circumstances of why a five percent SDB goal cannot be proposed.

Section L - Instructions, Conditions and Notices to Bidders

Section L - PROPOSAL INSTRUCTIONS

PART A GENERAL INSTRUCTIONS

1.0 GENERAL

Offerors are required to submit sufficient information concerning the following areas to enable NGRF personnel to fully ascertain capabilities of the Offeror to perform the requirements. The proposal must be sufficient in detail and scope to permit evaluation and provide the evaluators a clear understanding of the Offeror's approach, expertise, experience and capability as required by the solicitation. All proposals must clearly and convincingly demonstrate that the Offeror has a thorough understanding of the requirements, associated risks and is able, willing and competent to devote the resources necessary to meet the requirements and has valid and practical solutions for all requirements and potential risk areas. Offerors shall respond to all requirements and are encouraged to respond to objectives to the greatest extent possible. Do not alter or rearrange the solicitation. The burden of proof as to technical capability, past performance, experience, and cost reasonableness rests with the Offeror. The Offerors are advised that the NGRF may incorporate into the final contract any part of the Offeror's proposal deemed beneficial to the NGRF. No Cost information (e.g. Dollar values, hourly rates) shall be included in any Volume except the Cost Volumes

In presenting material in the proposal, the Offeror is advised that quality of information is more important than quantity. Clarity, brevity, and logical organization should be emphasized during proposal preparation. It is the responsibility of the Offeror to present enough information to allow the various work efforts, support, and management approaches, as well as cost, to be meaningfully evaluated without discussions. Statements that the prospective Offeror understands, can or will comply with the specifications, and paraphrasing the requirements or parts thereof are considered inadequate and may render a rating of unsatisfactory.

The Offeror must include any data necessary to illustrate the adequacy of the various assumptions, approaches, and solutions to problems. Requested data may be satisfied by a range of substantiating data from design philosophy, analysis, and laboratory data. However, any information submitted should have a clear explanation as to where it came from and how it was derived. Charts or figures shall have sufficient detail (i.e., assumptions used, test conditions, data for test points shown) to permit NGRF evaluation and award without discussions. Failure to clearly and positively address a specific factor or subfactor may be considered a deficiency. There is no need to repeat information in more than one volume if an overlap exists; the detailed information must be included in the most logical place and summarized and referenced in other areas. Unnecessarily elaborate brochures or other presentation beyond that sufficient to present a complete and effective proposal is neither necessary nor desired.

Offerors are expected to comply with all requirements of the RFP. The NGRF will consider any exception or deviation to any term or condition of the RFP to be a deficiency, as defined in FAR 15.001.

Throughout these instructions, "Principal Subcontractors" are defined as a subcontractor who provides at least 10% of the proposed total cost/price (excluding the Offeror's profit/fee) for the contract for SDD.. Team Members are the entities that make up a joint venture or any other partnership or teaming arrangement that is formed for the purpose of responding to this solicitation. Critical team members are entities of the Offeror that perform a critical function in the performance of the resulting contract, whether it is technical or financial, and/or have important roles in any high or medium risk areas identified in the Offeror's proposal.
For the purpose of this acquisition, relevant is defined as something that has a logical connection with the matter under consideration. Such aspects of relevance include the type of effort (e.g., development, production, repair), the type of requirement (e.g., weapon systems, information systems, engineering services, programmed depot maintenance), service similarity, service complexity, contract type, contract dollar value, the division of the company that will do the work, and principal or critical subcontractors, teaming partners and/or joint ventures.

2.0 PROPOSAL FORMAT

Proposals must be formatted using Times New Roman 12 pt regular, italic, bold, or bold italic fonts (no reduction permitted), single-spaced with 1-inch margins all around, and formatted for/printed on standard 8.5 x 11 inch paper. All pages should be numbered with section and page numbers. When foldout pages are used they must not exceed 11 x 17 inches and will be counted as 1 page. Drawings may be provided separately and may be any size but should be folded to approximately 8.5 x 11 inch standard size and will count as 1 page. Graphs shall be presented in no smaller than a 10 pt font and should contain a grid, which allows values to be read directly from the graph to the same accuracy that a 10 x 10 to the ½ inch grid provides. Graphic resolution should be consistent with the purpose of the data presented. Figure callouts, tables, and titles must be presented in no smaller than a 10 pt font. Each Volume or Book shall be provided separately in a 3-ring binder. The binders shall be of an appropriate thickness for the number of pages it contains (e.g., a 30-page book shall not be placed in a 2-inch binder).

The Offeror will provide one complete copy of the proposal to the Procuring Contracting Officer (PCO) as electronic files fully compatible with Microsoft Office 2003, and for information not supported by MS Office products, with the latest Adobe Acrobat reader on a CD-ROM or flash drive. The Offeror will ensure that the Cost Volume is provided on a separate CD-ROM or flash drive. All CD-ROMs or flash drives must be labeled for content and the Offeror's name. If a discrepancy exists between the original paper copy of the proposal and the electronic copy, the paper copy will take precedence.

3.0 PROPOSAL CONTENT AND VOLUMES

The Offeror must present its proposal information in a manner that facilitates a one for one comparison between the information presented and this Proposal Instruction. Proposal information must be structured such that its paragraph number/letter is identical to the Proposal Instructions paragraph number to which it is responding, although the Offeror may add lower tier sub paragraphs. The Offeror must provide reasons it will not provide information for a particular paragraph. The proposal information instructions are structured by paragraph numbers where first, second, third and fourth parts correspond to the volume, book, section, and element, etc., in the Offeror's proposal.

Each volume of the proposal shall be submitted as one original and additional copies as specified in the table below. Submission of all volumes of the Original proposal to the address provided in section 4.0 will determine timeliness of receipt. Page recommendations for each volume, if any, are also specified in the table below as follows (title and table of contents pages do not count towards page limit):

Information submitted as an Annex to the proposal includes manuals, specifications, plans, procedures, and policies that exist as an official document of the company or facility, as well as other information requested in Part B Specific Instructions. Annexes do not have page limitations and do not count towards the suggested page count. The authorized Annexes are summarized in the table below.

Volume Number	Book Number	Volume Title	Recommended Pages	Copies Required
1		EXECUTIVE SUMMARY	5	1 Original/1
2		TECHNICAL		
	1	DESIGN APPROACH	50	1 Original/1Copies
		Annex A – Breadfruit Enterprise Architecture Demonstration (BrEAD) System Performance. Annex B – Detailed Reliability, Maintainability and Availability (RM&A) Predictions Technical Attachment 1 - Breadfruit Enterprise Architecture Demonstration (BrEAD) PBSS Tailorable Specification and Verification Matrix	As needed	1 Original/1 Copies
	2	PROGRAM AND SCHEDULE	20	1 Original/1Copies
		Annex C – CWBS Annex D – SOW Annex E – IMP Annex F – IMS Annex G – SEMP Annex H – Software Development Plan	As needed	1 Original/1 Copies
3		PAST PERFORMANCE Past Performance Attachment 1 –	N/A	
4		EXPERIENCE	N/A	

5		COST	As needed	1 Original/1 Copies

Each volume must contain the following information:

- Cover and title page
- Title of proposal and proposal number as applicable
- Offeror's name, address and POC
- RFP number
- Proposal volume/book number
- Copy number
- Table of Contents (The table of contents must provide sufficient detail to enable easy location of important elements)
- Use tabs and dividers

3.0 **PROPOSAL SUBMISSION**
Week 13 of the course.

PART B SPECIFIC INSTRUCTIONS

1.0 VOLUME 1 -EXECUTIVE SUMMARY

Note: This volume shall not contain any reference to cost or price aspects of the offer.

The purpose of the Executive Summary is to provide the evaluators with an overview of the Offeror's proposal. The Offeror must provide a concise abstract of the technical approach to be applied and should cogently define goals, technical approach, specific risks and expected outcomes. This volume should be divided into the following sections:

i. Offeror Summary Table: The Offeror shall complete the table below. The Offeror shall include all subcontractors and team members who will be involved with the performance or management of the program work and should list all sites where the work will be performed. If a teaming or subcontracting arrangement is proposed, identify the work share, distribution elements, and ratios that each Contractor will perform using the table below. Also provide a definition of the legal relationship between the entities if it is other than a Prime/subcontractor relationship.

Contractor Name (Indicate Prime, Team Member or Sub)	Place of Performance	Brief Work Description and/or Program Responsibility	Approximate % of Total Proposed Cost

ii. Technical

2.0 **VOLUME 2 - TECHNICAL**

This volume consists of two books – 1) Design Approach and 2) Program and Schedule. As previously stated, provide each book in a separate 3-ring binder.

All responses in this volume shall address the baseline Breadfruit Enterprise Architecture Demonstration (BrEAD) (CLIN 0001).

2.1 Book 1 – Design Approach - In this book, the Offeror shall describe the technical approach proposed by the Offeror to meet the Breadfruit Enterprise Architecture Demonstration (BrEAD) Performance Based System Specification (PBSS).

2.1.1 Overall Design Approach - Provide a description of the overall design approach and how it will meet the Breadfruit Enterprise Architecture Demonstration (BrEAD) PBSS requirements (note it is the responsibility of the Offereor to develop the baseline PBSS requirements). Provide specifics as instructed in the following sub-paragraphs.

2.1.1.1 Systems Overview - The Offeror shall provide an overview of the proposed Breadfruit Enterprise Architecture Demonstration (BrEAD) by describing the proposed concept and its ability to meet the requirements of the Breadfruit Enterprise Architecture Demonstration (BrEAD) PBSS. This description shall include the proposed Breadfruit Enterprise Architecture Demonstration (BrEAD) configuration Also, provide an overarching functional description including proposed systems, interfaces, and subsystems for Breadfruit Enterprise Architecture Demonstration (BrEAD) List all major components of Breadfruit Enterprise Architecture Demonstration (BrEAD). Describe the operational, Human Systems Integration (HSI) and logistics support concepts for the proposed solution in the form of a preliminary Day in the Life (DITL). Additionally provide an overview of the proposed Breadfruit Enterprise Architecture Demonstration (BrEAD) Systems Integration design approach and identify development challenges with proposed plans with regard to integrating all of the Breadfruit Enterprise Architecture Demonstration (BrEAD) elements including changes to existing designs. The Offeror shall identify and explain any assumptions, trade-offs and subsequent limitations and risks to the design approach.

2.1.1.4 External Systems – Provide a detailed description of any internal systems and/or any external system design dependences required to meet the requirements of the Breadfruit Enterprise Architecture Demonstration (BrEAD) PBSS.

2.1.1.5 Support System – Provide a detailed description of the design features that will drive the Supportability Approach including an overview of that Supportability approach encompassing the capability to meet the Breadfruit Enterprise Architecture Demonstration (BrEAD)PBSS and the SOO. Include a description of the approaches to Performance Based Logistics (PBL), Maintenance Planning, and incorporation of the tenets of Condition Base Maintenance Plus.

2.1.1.6 Hardware and Software Integration Overview – For the Breadfruit Enterprise Architecture Demonstration (BrEAD), provide a description of the approach for hardware and software integration and any development approach needed to integrate Breadfruit Enterprise Architecture Demonstration (BrEAD) elements. Address any inter-related software and hardware design/performance attributes. Provide a comprehensive block diagram in sufficient detail to show functional software and hardware elements, proposed systems, subsystems, and sensors including non-developmental item nomenclatures, identifications, sizes, and functionality of the proposed Breadfruit Enterprise Architecture Demonstration (BrEAD) components. Drawings, functional diagrams, analyses, and other material shall be used as needed to convey a clear understanding of the proposed design and implementation concepts. The Offeror shall provide a discussion on system/subsystem interfaces and their approach to integration challenges, such as, "interface boxes", Information Assurance/Program Protection, Security, and Anti-Tamper.

2.1.2 Performance Based Systems Specification (PBSS)- The Offeror shall Generate the Breadfruit Enterprise Architecture Demonstration (BrEAD) PBSS. Technical Attachment 1 (the PBSS) as proposed, may be incorporated into the contract as part of the Breadfruit Enterprise Architecture Demonstration (BrEAD) PBSS of the successful Offeror. This Attachment will also be used in the evaluation of the Offeror's proposed approach.

2.1.3 Open Systems Architecture - The Offeror shall describe the Breadfruit Enterprise Architecture Demonstration (BrEAD) open systems approach for using modular design standards, standards based interfaces, and widely-supported, consensus-based standards to achieve interoperability, evolvability, modularity, use of open standards, upgradeability, vendor independence, portability, reconfigurability , maintainability, technology insertion, reusability, removal of obsolete modules, and long-term supportability. Offeror shall provide the following as part of its proposal:

a. Evolvability – A detailed description of how the Offeror's proposed system will be able to survive changes due to technology advancement, operational need,

obsolescence, communication system replacement, additional functions, and data exchange requirements. Changes that can be accommodated by the proposed system and features that enhanced evolvability should address. Conversion of existing closed systems to an open architecture should also be addressed.
b. Portability – A detailed description of how the Offeror's proposed software will be able to survive different geographic locations operational environmental conditions.
c. Modularity – Describe in detail how the proposed system is partitioned into discrete, cohesive, and self-contained hardware and software units with well-defined interfaces that permit substitution of such units with similar components or products from alternative sources with minimum impact on existing units. A block diagram depicting hardware, key interfaces, hardware interfaces, a description of all employed interface standards, and software allocations to hardware should be provided.
d. Rationale for Modularization Choices – A detailed description of the Offeror's rationale for the modularization choices made to generate the design. At a minimum, the rationale shall explicitly address any tradeoffs performed, particularly those that compromise the modular and open nature of the system.
e. Employ a Layered Modular Architecture – A detailed description on how the proposed system architecture is layered, modular, and makes maximum use of Commercial-Off-the-Shelf/Non-developmental Item (COTS/NDI) hardware, operating systems, and middleware that utilize non-proprietary key Application Programming Interfaces (API) whenever practicable.
f. Maintainability – A description of how the Offeror's proposed system architecture will enhance the ease of modifications to software and hardware components to correct deficiencies in the deployed system. Provide a description of the hardware and software development environments, including tools, deficiency-tracking process, and documentation utilized for hardware/software modification and maintenance.
g. Technology Insertion and Refresh – A detailed description of how the Offeror's proposed system will allow for rapid and affordable technology insertion and refresh. For example, the Offeror should describe how the proposed system would allow incremental systems improvement through upgrades of individual hardware or software modules with newer modular components. At a minimum, the description shall address how the Offeror's architectural approach will support this requirement including how components from third party providers and reuse sources shall be included.
h. Reusability – Provide a detailed description of system, subsystem, and software components that the Offeror proposes to re-use.
i. Life-cycle Supportability – A detailed description of how the Offeror's proposed system mitigates technology obsolescence and dependence on proprietary or vendor-unique technology.
j. Traceability of System Requirements – A detailed description of the Offeror's approach for ensuring that all system requirements are accounted for through a demonstrated ability to trace each requirement to one or more modules.

Modules consist of components (one of the parts that make up a system and may be hardware and/or software), which are self-contained elements with well-defined, standards-based and published interfaces

.2 **Book 2 – Program and Schedule** - In this book, the Offeror shall describe its approach to meeting the overall Breadfruit Enterprise Architecture Demonstration (BrEAD)objectives To support the information requested in this book, provide as Annex C a Contract Work Breakdown Structure (CWBS). This CWBS should be used to support the development and evaluation of the SOW, IMP, and the IMS.

2.2.1 Proposed SOW and CDRLS – Propose, in Annex D, a detailed Contractor SOW that reflects all of the objectives stated in the NGRF provided SOO. The Offeror's proposed work statements in response to the SOO, coupled with the identification of the CDRL items for the appropriate task, will be the proposed Annex D SOW for this program. The SOW shall address all CDRLs, delineating the Offeror's proposed work task(s) required to develop each deliverable.

2.2.2 Integrated Master Plan – The Offeror shall provide an IMP as Annex E. The IMP will be incorporated in the contract. In general, the IMP shall provide the information as described below.

The Offeror shall submit an IMP that is structured as an event-based plan. Technical reviews applicable to the contracted event shall be included. The IMP shall include events, accomplishments that tie to these events and completion criteria for each accomplishment for the total contracted effort. The The Offeror shall describe the organizational structure of the proposed Contractor team with emphasis on IPT implementation; technical and budgetary purviews; and the approach to communication, including required NGRF participation and insight. Describe the relevant capabilities and commitment of the proposed Contractor team. Show what resources (e.g., staffing, facilities, communications, supportability, manufacturing capability, GFE) will be used and how they will be integrated in conducting the Breadfruit Enterprise Architecture Demonstration (BrEAD) program from SDD through transition to production and IOC. The Offeror shall provide a program plan that describes existing resources, practices and capabilities and the manner in which they will be applied to this program, identifying any new resources, capabilities or personnel, which do not currently exist or are assigned to other programs. The Offeror shall identify and describe organizational roles and responsibilities, key events and milestones, significant tasks, and success criteria for the Breadfruit Enterprise Architecture Demonstration (BrEAD), development, production and deployment. Success criteria may be presented in a tabular rather than narrative form.

2.2.3 Integrated Master Schedule – The Offeror shall demonstrate that the proposed approach will be able to meet the proposed IOC date. The Offeror shall provide an Integrated Master Schedule (IMS) in Annex F that delineates the planned schedule for all activities, events, milestones, and critical paths associated with all program. The IMS shall include the program schedule with technical tasks and activities necessary to complete the work effort scoped within the IMP. The Offeror shall use an estimated contract order award date of 1 Jan 09. The IMS shall be submitted in the Microsoft Project 2003. All additional schedule information shall be submitted in electronic format.

2.2.4 Systems Engineering (SE) Management - The Offeror shall provide their proposed Breadfruit Enterprise Architecture Demonstration (BrEAD) Systems Engineering Management Plan (SEMP) which shall be followed under this contract. The SEMP shall be included in the proposal as Annex G that will delineate all elements of the SE process and describe the integrated technical approach to the program. The SE process will ensure a progressive definition and clarification of the Breadfruit Enterprise Architecture Demonstration (BrEAD) system and subsystems design elements. The SE process will be a comprehensive, iterative, and recursive problem-solving process to allow for the definition and refinement of performance and functional requirements, define/refine the design of system elements to satisfy those requirements, and establish the system configuration. References to other documents provided in the proposal are acceptable. The Offeror's SEMP shall include:

- An executive summary, broad overview
- A description of the key technical and technical management processes. Provide Offeror's (and teammates, subcontractors, etc.) plans for continued process improvement.
- A description of the technical organization within the program Integrated Product Team (IPT) structure identifying roles and responsibilities, key personnel, and technical staffing requirements. Identify the primary participants within each IPT and the supporting participants to include the NGRF and subcontractors. Include a summary of the principle products of the IPTs. Include a description of technical working groups with roles, responsibilities, and proposed participants (e.g., Interface Working Group, Test Planning Working Group, and Technology Roadmap Working Group).
- Identity of the technical authority, stakeholders, and functional technical authorities on the program and the limit and scope of their responsibilities.
- A description of the relationships between Engineering and other Program Functional Organizations (e.g., Program Management, Test and Evaluation, Contract/Sub-contract Management, Financial Management, Manufacturing, Product Support, Risk Management)
- A description of the key SE positions, roles and responsibilities, and a detailed Systems Engineering organizational diagram.
- A description of how the Breadfruit Enterprise Architecture Demonstration (BrEAD) SE Management approach will be maintained and how the Systems Engineering Approach will be maintained throughout the Breadfruit Enterprise Architecture Demonstration (BrEAD)life-cycle phases
- A description of the automated tools and electronic documentation to be used to design, develop, integrate, test and support the Breadfruit Enterprise Architecture Demonstration (BrEAD).

- Flow down of technical and technical management plans and processes to the subcontractors or teammates, and how they participate in the processes.
- An event-based program plan (correlated and consistent with the IMP) for the efforts involved with the design, development, test, production, and sustainment, including planned block upgrades, technology insertion, etc.
- Planned event-driven technical reviews with entry and exit criteria and independent SME participation. The timing for the reviews shall be documented in the IMS. These reviews shall be conducted at the major system and total system level and shall include:
  - Systems Requirements Review
  - Systems Functional Review
  - Preliminary Design Review
  - Critical Design Review

- A description of the proposed set of program planning and specific plans such as TEMP, ISP, Software Development Plan, Product Support Plan, and Risk Management Plan, to ensure consistency and completeness.
- A description of the Technical Performance Measures (TPM) used to track the Breadfruit Enterprise Architecture Demonstration (BrEAD) key performance metrics
- Certification and qualification activities
- The process for trade studies and associated plans to support the design and development process
- A Configuration Management/Data Management (CM/DM) approach that addresses the entire system life cycle
- Process for advance notification of performance changes, diminishing manufacturing resources, part obsolescence and removal/replacement of assemblies and spares
- The intended Measures of Effectiveness (MOEs), how they interrelate, and criteria for the selection of Measures of Performance (MOPs) to support the evolving definition and verification of the system.
- The Human Systems Integration (HSI) process that will be implemented. The Offeror shall describe the coordination and management of requirements and trade-offs among manpower, personnel, training, Human Factors Engineering (HFE), safety, occupational health, personnel survivability, and habitability factors to achieve an effective, affordable and safe design. The Offeror shall describe the schedule and timing of HSI related tasks and analyses with respect to other program functions and tasks

2.2.5 Risk Management Plan - The Offeror shall identify all technical/performance, schedule, and cost yellow and red risks associated with their solution's cost, schedule and performance aspects as defined by the criteria in Table A. Describe program management risk assessment methodology and assumptions. Discuss the event or task, impact, and mitigation for each risk including an explanation on how the effect of mitigation plans will be measured. All risk mitigation plans shall be included in the IMS. Describe how these risk items relate to the critical path, system performance, schedule, and program cost. Discuss the external dependencies not under control of the Offeror that could have a serious impact on the ability to deliver the capabilities within the cost and schedule of the proposal. For each of these external dependencies, identify the responsible organization and discuss at least one possible work-around if the external dependency is late or does not meet the prescribed need. Describe the relationship between the IMP, IMS and the risk management program. Failure to provide a realistic risk assessment will be reflected in the NGRF's evaluation of the proposal as part of the cost and schedule realism assessment. Note: Cost risk will be evaluated under the cost factor.

Probability (Likelihood)					
5	G	Y	R	R	R
4	G	Y	Y	R	R
3	G	Y	Y	Y	R
2	G	G	G	Y	Y
1	G	G	G	G	Y
	1	2	3	4	5
	Consequence (Impact)				

Level	Likelihood
1	Not Likely
2	Low Likelihood
3	Likely
4	Highly Likely
5	Near Certainty

	TECHNICAL PERFORMANC	SCHEDUL	COS
1	Minimal or consequen to	Minimal or no ; mont	Minimal or impact;
2	Minor tion performance or can be tolerated; changes may be	No impact on critical able to meet key potential non critical schedule slip of 1 to 2	Budget increase for $1M-$25M; or prod ction cost of
3	Moderate performance supportabili workarounds are Moderate changes Require	Potential critical schedule slip of 2 to 3 potential impact to path, recoverable; impact to key	Budget increase SDD - ($25M - $125M) or production cost of $1M -$2
4	Significant performa or shortfall in workarounds may available ave consequenc changes may be	Potential – pat schedule slip of 4 to months; potential slip to critical path and is recoverable; potential to key	Budget increase SDD - ($126M - $225M) or production cost of $2M $5
5	Severe performance and cannot key threshold; no	Potential schedule ≥ months; critical unrecoverable; cannot key program	Budget increase >$225M or unit production cost of

TABLE A. Risk Identification Criteria

2.2.6 T&E - The Offeror shall provide a detailed description of their proposed T&E approach. Metrics that will be used to measure system test progress and effectiveness need to be included. The Offeror shall include a description of all required SIL assets, SIL connectivity (Interoperable/Information Exchange Requirement (IER) issues and how, and when, the SIL (or SILs) will be used during system integration and test to support the program major milestones. The Offeror shall describe in detail the capability of the SIL located at the Breadfruit Enterprise Architecture Demonstration (BrEAD) test site to perform Integrated Test (IT) and required Operational Test (OT).

The Offeror shall generate the Breadfruit Enterprise Architecture Demonstration (BrEAD) Verification Matrix, providing data depicting the methods of verification for each of the Breadfruit Enterprise Architecture Demonstration (BrEAD) PBSS requirements. .

2.2.7 Subcontract Management - Describe your strategy for subcontract management and control in the performance of this contract. In describing its strategy, the Offeror shall describe its approach and the complexity and variety of the work the subcontractors are to perform. Identify what portions of the contracts, or specific tasks, will be passed to the sub-contractors to ensure compliance with contract technical performance, cost, and schedule requirements. Also provide subcontractors/ subcontractor agreements to execute the requirements of the contract and its options.

2.2.8 Software Development - Offerors shall submit a draft version of their Software Development Plan (SDP) in accordance with the content defined in the Offeror's SOW as Annex H The SDP may be formatted as desired by the Offeror.

Section M- Evaluation factors for award

The evaluation factors for award are technical and management. The sum of the point scoring for both will be 100 points. The technical will be factor will be 70 of the 100 points. There are three technical sub factors: 1. Requirements (50 points), 2. Technical approach (25 points), and 3. System architecture (25 points). The management factor has two sub factors: Integrated Management Plan (50 points) and Integrated Management Schedule (50 points). The highest overall score will be the considered the best technical proposal.

– End Of Document -

Attachment 2

Statement of Objectives (SOO) For a Breadfruit Enterprise Architecture Demonstration (BrEAD) and Systems Design Phase

20 January, 2009
Version 2

1.0 PROGRAM SCOPE

The New Green Revolution Foundation (NGRF) a non-profit foundation has been doing research in several tropical island locations on expanding the used of breadfruit as a food source, and it believes that it could be a significant contribution to feeding a growing world population particularly for island nations. Currently breadfruit is an extremely perishable food source. It has a 48 hour shelf life after it is harvested from the breadfruit tree. The extremely short shelf life and lack of refrigeration on many of the islands has meant that the breadfruit has to be consumed locally in a very short time by the island populations. Preliminarily research results by the NGRF sponsored researchers has demonstrated that breadfruit can be processed into flour and other stable shelf life products using local technologies by the local population. The NGRF believes that a Breadfruit Enterprise Architecture Demonstration (BrEAD) needs to be developed to do further research and development to develop new breadfruit food products, to promote those products world wide, to facilitate the design of scalable breadfruit processing systems, to create a virtual collaborative knowledge sharing automation / communication system to share research and marketing information in real-time, and to seek funding from various sources, NGRFs and private organizations, to continue the development of breadfruit food products and breadfruit processing technology development.

2. PURPOSE

The purpose of the Breadfruit Enterprise Architecture Demonstration (BrEAD) System Development and Demonstration (SDD) acquisition phase is to: design, develop, fabricate, integrate, demonstrate, validate, test, and document the Breadfruit Enterprise Architecture (BrEAD) ensuring a scalable affordable, operationally supportable, tailorable breadfruit processing system, breadfruit product distribution system and a distributed virtual collaborative marketing and research communications system for future implementation.

There are currently ***x*** islands with breadfruit tree populations that are on average ***a***, with breadfruit that produce on average ***b*** breadfruit per tree and human populations available to harvest and process breadfruit that vary from ***y*** to ***z***. The proposed Enterprise Architecture and system architecture will be driven by and constrained by the program schedule and the NGRF Performance Based System Specification (PBSS) Attachment 1. This SOO is intended to provide enough information and detail to structure an executable program able to satisfy NGRF objectives. To this end, this SOO will specify critical program management, engineering and logistics processes considered necessary to meet these goals.

3.0 NGRF STRATEGIC ACQUISITION PHASE OBJECTIVES

Execution of the SDD Phase will demonstrate an operationally effective and suitable Breadfruit Enterprise Architecture Demonstration (BrEAD) The Breadfruit Enterprise Architecture Demonstration (BrEAD) at Initial Operating Capability IOC will be fully compliant with the PBSS requirements to ensure an operationally effective and suitable report from Operational Test and Evaluation (OT&E). The Breadfruit Enterprise Architecture Demonstration (BrEAD) will provide a 24 hour research data collection, data analysis, and data distribution capabilities for every technical and marketing research facility in the NGRF organization. It will be interoperable with local, state and federal agencies and private organizations as deemed necessary to promote breadfruit processing systems ,provide products world wide, and products standards world wide.

The Breadfruit Enterprise Architecture Demonstration (BrEAD) will provide an integrated System of Systems (SoS) that will consist of the following: (1) scalable breadfruit processing systems, (2) virtual world-wide integrated (technical and marketing) data sharing (3) world-wide breadfruit processing system user information, (4) local weather conditions at breadfruit processing sites (5) a suite of communications with Line-Of-Sight (LOS) and Beyond-Line-Of Sight (BLOS) capabilities; (6) Local Data Processing Equipments and Software (LDPES) and NGRF Center Data Processing Equipments and Software located at the St. Paul, Minnesota world-wide head quarters of NGRF.

The contractor will develop and submit in the proposal a statement of work that defines the efforts necessary to produce the Breadfruit Enterprise Architecture Demonstration (BrEAD).
The SOW will, at a minimum, address the following objectives:

1. Obtain a Breadfruit Enterprise Architecture Demonstration (BrEAD) that meets or exceeds the PBSS requirements;
2. Obtain a Breadfruit Enterprise Architecture Demonstration (BrEAD) that is economical to operate and maintain throughout its life cycle and ensure life cycle considerations are maximized throughout the design and integration process;
3. Obtain a Breadfruit Enterprise Architecture Demonstration (BrEAD) that is fully compliant with all applicable statutes and laws.
4. Obtain a Breadfruit Enterprise Architecture Demonstration (BrEAD) design that is implementable with Commercial-off-the-Shelf technology to support the program milestones and schedule;
5. Obtain a modular open systems architecture for hardware and software that supports efficient system upgrades and allows for interoperability with other systems. Obtain a design that maximizes the use of Commercial-Off-The-Shelf (COTS), Modified-COTS (MCOTS), NGRF-Off-The-Shelf (GOTS) items and Non-Developmental-Items;
6. Achieve effective integrated NGRF and Contractor product teams to execute the program. Integrated Product Teams (IPTs) will be the primary means used to gain insight and provide for timely exchange of information and data;
7. Employ a program management system based on the Integrated Master Plan/Integrated Master Schedule (IMP/IMS) that is able to provide accurate weekly cost, schedule and performance information to support Milestone C, IOC and throughout the life cycle of the program;

8. Employ an integrated NGRF/Contractor risk management system that actively identifies program risks and identifies mitigation steps for program risks;
9. Adequately protect Critical Program Information (CPI) throughout the program's lifecycle;
10. Employ an integrated, comprehensive Configuration Management (CM) system that maintains consistency of the Breadfruit Enterprise Architecture Demonstration (BrEAD) performance, compliance of functional and physical attributes with requirements, tracking of design and operational information and optimizes System Operational Effectiveness (SOE) throughout the life of the system;
11. Minimize the amount of NGRF Furnished items (NGRF Furnished Information, Equipment, Property) ;
12. Employ a comprehensive Systems Engineering Process In Accordance With (IAW), the INCOSE Handbook,
13. Employ an Architecture that ensures timely, external data exchanges from the NGRF Command Center Data Processing Equipments and Software (CDEPS) to a wide Community Of Interest (COI) ;
14. Support applicable COI forums promulgated via joint NGRF/Contractor IPT;
15. Employ a Breadfruit Enterprise Architecture Demonstration (BrEAD) Modeling and Simulation (M&S) strategy, which includes an integrated environment of accredited M&S tools, simulation and stimulation environment, data and resources that enables collaboration across functional disciplines;
16. Employ a Breadfruit Enterprise Architecture Demonstration (BrEAD) Human System Integration strategy to efficiently and safely operate, maintain and support a Breadfruit Enterprise Architecture Demonstration (BrEAD).
17. Obtain a Breadfruit Enterprise Architecture Demonstration (BrEAD) design that incorporates sound human engineering design to optimize the Breadfruit Enterprise Architecture Demonstration (BrEAD) and maintenance user interfaces considering workload, manpower, mission and task effectiveness, integration, information management, situational awareness and data fusion;
18. Employs a design maximizing the use of automation and decision aids for operators and maintainers to reduce workload, enhance coherence and information management and safety;
19. Develop a system that is safe to operate, maintain, and transport and does not introduce hazards;
20. Design and deliver a NGRF Breadfruit Enterprise Architecture Demonstration (BrEAD) System Integration Laboratory (SIL) that will be integrated with other NGRF research sites, and are able to simulate and stimulate the Breadfruit Enterprise Architecture Demonstration (BrEAD); and support subsystem/system test, integration, Validation & Verification (V&V), and system/software lifecycle sustainability;
21. Provide an integrated NGRF and Contractor test program, which demonstrates/validates the completeness of the design and development process, ensures design issues have been identified and solutions are available, and ensures compliance with Breadfruit Enterprise Architecture Demonstration (BrEAD) PBSS;

22. Develop Engineering Development Models (EDMs) to support developmental testing and the operational assessment of the Breadfruit Enterprise Architecture Demonstration (BrEAD);
23. Obtain a performance-based product support capability that:

    a. Provides a maintenance support concept based upon an effective Integrated Maintenance Concept (IMC);

24. Achieve an effective communication strategy that facilitates dialogue, ensures properly coordinated reviews, and audits, as well as the ability to assess progress towards meeting program requirements. At a minimum, the Contractor and appropriate subcontractors will support the following reviews, assessments and working groups and ensure appropriate entry and exit criteria are met.

    a. Integrated Baseline Reviews (IBR)
    b. Program Management Reviews
    c. System Requirements Reviews
    d. System Functional Reviews
    e. Preliminary Design Reviews/ Software Readiness Reviews
    f. Functional Configuration Audits (FCA)
    g. Physical Configuration Audit (PCA)
    h. Test Readiness Reviews
    i. Critical Design Reviews (CDR)
    j. Production Readiness Reviews (PRRs)
    k. System Verification Reviews
    l. Independent Expert Program Reviews
    m. Interface Control Working Groups
    n. Interoperable/Net-Centric Working Groups
    o. Human Systems Integration Working Groups
    p. Technical Interchange Meetings (TIMs)
    q. Test Plan Working Groups (TPWG) Meetings
    r. Technical Manual Management Team (TMMT) Meetings
    s. Independent Logistics Assessments (ILA)
    t. Initial Operational Capability Supportability Review (IOCSR)
    u. System Safety Working Groups
    v. Risk Management Reviews
    w. Software Audits
    x. Design Readiness Review (DRR)
    y. Operational Test and Readiness Review (OTRR)
    z. Build Release Reviews (BRR)
    aa. Software Qualification Reviews (SQR)
    bb. Software Verification and Validation Plan Reviews

cc. System Software Safety Working Groups
dd. Electromagnetic Environmental Effects (E3)/Spectrum Supportability (SS) working group(s).

1.1 **25. Obtain a System that is environmentally compliant with all applicable environmental**
1.2 **statutes and laws;**

26. Obtain a design which maximizes the use of standard manufacturing concepts and industrial
approved commercial best practices;

1.3

1.4 **3.1 <u>Detailed Objectives</u>**

1.5

1.6 **3.1.1 <u>Schedule:</u> It is an objective of the Breadfruit Enterprise Architecture Demonstration (BrEAD) program to achieve a January 20 FY10 IOC date or earlier. The Contractor will establish a program schedule that supports program milestones and allows adequate time for technical reviews, design, and test and evaluation to meet a January 20 FY10 IOC date.**

1.7

1.8 **3.1.2 <u>Open Systems Architecture (OSA):</u> The Breadfruit Enterprise Architecture Demonstration (BrEAD) will have Open Architecture designs and corresponding components. As part of this contract, the Contractor will be required to define, document, and follow an open systems approach for using modular design and standards-based interfaces.**

3.1.3 <u>Program Management:</u> The Contractor will use appropriate management techniques to support program control and gain program insight. The Contractor will develop and implement an Integrated Master Plan (IMP) and Integrated Master Schedule (IMS). The Contractor will develop and implement an IMS and utilize the Contractor Work Breakdown Structure (CWBS). The Contractor will provide in the IMP; the structure of events, accomplishments and criteria required to design, develop, integrate, produce, demonstrate, test the capabilities and functionalities, and support the design to provide an operationally effective and operationally suitable system to meet the Breadfruit Enterprise Architecture Demonstration (BrEAD) requirements. The IMS will be part of the Breadfruit Enterprise Architecture Demonstration (BrEAD) SDD contract. The Contractor will detail the required work, to include subcontractor efforts, in the IMP and IMS. Working in concert with the NGRF, the Contractor and subcontractors will consider the following areas in tailoring a program that supports the NGRF's objectives.

3.1.3.1 <u>Program Management Structure and System:</u> The NGRF management objective is to take full advantage of integrated NGRF and Contractor product teams to execute the program. IPTs will be the primary means used to gain insight and provide for timely exchange of information and data. The IPTs will consist of NGRF, Contractor and subcontractor personnel. The Contractor will propose a portfolio of metrics and Technical Performance Measures (TPMs) to support the continual assessment of program progress against defined plans and schedules. The program management system will be based on an IMP/IMS that is able to provide accurate and timely schedule and performance information to support IOC. A detailed program management system will be established, implemented and controlled as documented in the IMP/IMS and the Systems Engineering Master Plan (SEMP), including subcontractor management. The Contractor will maintain an IMP/IMS throughout the entire SDD contract and will provide access, licenses (if required) and training for IMP/IMS tools for the NGRF team. The system's detail will be such that the NGRF and Contractor can utilize schedule(s) for program planning, procurement planning, tracking, and execution of Breadfruit Enterprise Architecture Demonstration (BrEAD) SDD acquisition phase, decision/problem analysis, risk analysis and schedule reporting in accordance with the IMS. The IMS will contain the contract IMP events and milestones, accomplishments, criteria, required resources and activities from contract award to the completion of the contract in accordance with the IMS and IMP. IMS tasks will be directly traceable to control accounts included in the cost system, as collected by WBS and integrated with the Contract Performance Reports (CPRs). The Contractor will quantify the risk and provide optimistic, pessimistic, and most likely duration for each IMS activity and event. The Contractor will flow down the requirement of IMS to principal and/or critical subcontractors. It is the NGRF objective to minimize schedule risk during the program with respect to the IMP/IMS with particular emphasis on software development, SIL utilization, and test and evaluation including test asset allocation. The Contractor's schedule will have sufficient flexibility to accommodate for unforeseen technical issues without major impact to the IMP/IMS while maintaining a reasonable work and resource regimen. The Contractor will ensure schedule integration with its subcontractors and will ensure the validity of the subcontractor's schedule data including demonstrating effective methods for incorporating schedule data from subcontractors into the Contractor's IMS. The Contractor will participate in an IMS Schedule Risk Assessment (SRA) with the NGRF two weeks prior to the IBR. The Contractor will participate in a joint quarterly IMS-SRA following IBR continuing through the end of the contract period of performance with the NGRF in accordance with the IMS.

The Contractor will conduct quarterly Program Management Review (PMR) meetings at mutually agreed upon dates and locations. The PMR may be held in conjunction with the joint IMS-SRA review. During these reviews, the Contractor will present integrated cost, schedule, and technical performance status. Integrated Product Team leads or functional managers will include cost information in discussions of schedule status, technical performance, and risk, using earned value as an integrating tool. The following will be addressed: Cost/schedule trends, significant cost/schedule/technical variances, projected impacts, issues, quantified risk assessments, TPMs and corrective action plans. Additionally, the Contractor will present the status on personnel and staffing levels and compliance with the subcontractor management plan during the PMR.

3.1.3.2 Project Reporting:

3.1.3.2.1 Earned Value Management Systems (EVMS): The Contractor will establish, maintain, and use in the performance of this contract an integrated management system meeting the intent of the Industry Guidelines for Earned Value Management Systems (EVMS) ANSI/EIA-748-98 (reference DI-MGMT-81466 CPR; ANSI/EIA-748 (Latest Revision) and DI-MGMT-81650, IMS). The application of these concepts will provide for early indications of contract cost and schedule problems. Earned value assessments will correlate with technical achievement. A compliance review of the Contractor's EVMS will not be performed unless the NGRF program manager determines that it is necessary from IBR results or cost and schedule data quality assessments.

3.1.3.2.2 Cost and Software Data Reporting (CSDR): The Contractor will develop a CWBS and will systematically collect and report actual contract costs through Contractor Cost Data Reporting (CCDR) and Software Resources Data Reporting (SRDR). The Contractor will flow down the requirements of the CSDR to principal and critical subcontractors.

3.1.3.2.3 Contract Funds Status Reports (CFSRs): CFSR will be prepared IAW the Contractors reporting system as approved by the NGRF. The Contractor will flow down the requirements of CFSRs to principal and critical subcontractors.

3.1.3.2.4 Integrated Baseline Review: The Contractor will present its performance measurement baseline plan to the NGRF within six months of contract award, and subsequently, when required, following major changes to the baseline. The NGRF will verify during the IBR, and follow-on IBRs, when required, that the Contractor has established and maintains a reliable performance measurement baseline. The Contractor will ensure that the baseline includes the entire contract technical scope of work consistent with contract schedule requirements, and has adequate resources assigned. The Contractor will assure the NGRF that effective earned value methods are used to accurately assess contract cost, schedule, and technical performance. The Contractor will perform a self-assessment of the cost and schedule risk for the IBR. The IBR will be used to achieve a mutual understanding of the baseline plan, cost and schedule risk, and the underlying management processes used for planning and controlling the project. This review will be tailored to focus on risk areas identified by the NGRF.

3.1.3.3 Risk Management: The Contractor will implement a sound risk management system that facilitates risk identification and mitigates program risks. The risk management system will be integrated with program controls through the development of metrics to monitor program status, tied to the IMP/IMS. This system will be utilized to:

1) Perform timely and thorough assessment of cost, schedule and technical risk;
2) Identify and track all medium and high-risk areas;
3) Define, fund, and implement appropriate risk mitigation efforts and;
4) Provide for fall-back or alternate plans, including design off ramps in the event that the primary risk mitigation plans fail.

The risk assessment criteria for the program have been defined in Section L. After contract award, the NGRF and the Contractor will form, operate and co-chair an integrated Risk Management Board, based on the description above.

3.1.3.4 Data Management:

The NGRF intends to minimize data costs by acquiring only the minimum essential data to support effective management of the Breadfruit Enterprise Architecture Demonstration (BrEAD) Program. Data will be managed via the use of electronic media and an IDE to improve data transfer, NGRF access, efficiency, and affordability.

3.1.3.4.1 Integrated Digital Environment (IDE): Timely access to program data is of the utmost importance. The Contractor will use electronic media to improve data transfer efficiency and affordability. The IDE objective includes the utilization of a robust, cost-effective on-line digital data environment that allows the program acquisition and operational support activities, throughout the program's life-cycle, to digitally create, store, access, manipulate, share, and exchange programmatic and technical data. This data must be accessible to the Breadfruit Enterprise Architecture Demonstration (BrEAD) NGRF IPT at anytime from their workspaces (approximately 200 individuals). The Contractor will provide the maximum number of NGRF seats at no additional cost. The IDE will provide protection for program data in accordance with the Program Protection Plan (PPP). Strategy will include technical refresh provisions and any necessary accreditation activities. The Contractor will optimize the use of the IDE with its subcontractors and suppliers, as well as participating NGRF organizations. The Contractor will develop, update and maintain a Data Accession List (DAL) that identifies all data and documentation generated under the execution of the Breadfruit Enterprise Architecture Demonstration (BrEAD) effort.

A list of the data to be provided on the IDE will be attached to the IDE implementation plan.

IDE process controls, in addition to use of business tools (such as Associate Contractor nondisclosure agreements), will apply and will use system-level and data-level Metadata tags for identification and protection of digital form data (such as for classified, Export Control, Proprietary, or Distribution Statements).

The NGRF will require continued use of the IDE beyond the SDD program. The Contractor will provide the NGRF with strategies to ensure continued use of the IDE

3.1.3.4.2 Contract Data Requirements List (CDRLs): The RFP includes a list of CDRLs expected to be included in the contract. The CDRL list is as follows:

CDRL 0001 System Design Description
CDRL 0002 Integrated Program Plan (IMP)
CDRL 0003 Integrated Management Schedule (IMS)
CDRL 0004 Integrated Test Plan
CDRL 0005 Systems Engineering Management Plan (SEMP)
CDRL 0005 Software Development Plan
CDRL 0006 Validation and Verification Plan

CDRL 0007 Statement of Work
CDRL 0008 Up dated Performance Based Systems Specification (PBSS)
CDRL 0009 Configuration Management Plan
CDRL 00010 Data Management Plan
CDRL 00011 System Operational Effectiveness Plan
CDRL 00012 Quality Management Plan
CDRL 00013 Modeling and Simulation Plan
CDRL 00014 Human Engineering Task Analysis
CDRL 00015 Human Engineering Design Approach
CDRL 00016 Human Engineering Test Plan
CDRL 00017 System Safety Management Plan (SSMP)
CDRL 00018 Test and Evaluation Master Plan (TEMP)
CDRL 00019 Contractors Performance Based Logistics Planning (PBL) Strategy Report

The Contractor may request additions and/or deletions which will be subject to NGRF approval.

3.1.3.5 Configuration Management: The Contractor will identify and utilize a configuration management approach which establishes and maintains consistency of the Breadfruit Enterprise Architecture Demonstration (BrEAD) performance, compliance of functional and physical attributes with requirements, and tracking of design and operational information throughout system life. The approach will also consider regulating life-cycle costs. The Contractor, as part of the IDE, will employ an integrated change management system that identifies, prioritizes, tracks, allocates and manages program, system and software requirement changes via a Change Review Board. The approach will consider all elements of the Breadfruit Enterprise Architecture Demonstration (BrEAD) program including data rights, product support, training infrastructure and will address any peculiar software/hardware items. Specifically, the approach will address product support development and implementation within a PBL support strategy and will include Product Support Integrator requirements, vendor integration and COTS/MCOTS/GFE/NDI integration and management. The approach will consider the establishment of the functional, allocated and product baselines; the establishment of these baselines will be included in the IMP and identified in the IMS.

3.1.3.6 Specifications and Standards: If the use of a NGRF specification or standard is not specified, the NGRF encourages the use of generally accepted commercial and industry specifications and standards. Foreign specifications and standards may be used if existing equipment from non-U.S. suppliers is being proposed.

3.1.3.7 Systems Operational Effectiveness: The Contractor will develop, and submit to the NGRF for review and comment, a SOE Plan that will establish and document the Contractor's process for implementing SOE. The plan will detail the following:

a) Integration of SOE with the Systems Engineering process including the process for managing SOE in trade studies and trade space;
b) Integration of SOE with the supportability and logistics planning activities;
c) Specific measures for establishing appropriate SOE goals and evaluating attainment thereof, including test and evaluation and validation and inclusion in the SIL
d) Modeling and simulation tools and processes.

3.1.3.8 Quality Management System: The Contractor will develop and maintain a Quality Management System (QMS) that satisfies program objectives and meets the requirements of ANSI/ASQC-Q9001: 2000 or an equivalent QMS. The QMS procedures, planning, and all other documentation and data that comprise the QMS, will be made available to the NGRF for review. The NGRF may perform the inspections, verifications, and evaluations necessary to ascertain conformance to requirements and the adequacy of the implementing procedures. Third-party certification of the Contractor's QMS is not required. The Contractor will require subcontractors to maintain a QMS that achieves control of the quality, services and supplies provided.

3.1.4 Engineering

3.1.4.1 Systems Engineering: The Contractor will develop and update as necessary throughout the SDD phase a comprehensive SEMP that describes the Contractor's methodology for fully integrating the NGRF's requirements and PBSS. The SEMP will demonstrate an integrated systems engineering approach towards development of the requirements, configuration baseline definitions, system design, system integration PBL product support requirements, Open Architecture implementation, system architecture quality and concordance, Human Systems Integration (HSI) and logistics resources (maintenance personnel, support equipment and repair parts) that cumulatively provide the capabilities specified in the PBSS. Additionally, the Contractor will address how production capabilities and processes will be incorporated into the systems engineering approach. In particular, the Contractor will focus on a design engineering producibility program identifying and implementing the design and fabrication/assembly processes which will result in the system achieving its production performance, cost and schedule requirements. The Contractor will generate TPMs, and track and report the status of the TPMs. The Contractor will utilize a requirements traceability tool, (such as TRACE, Dynamic Object Oriented Requirements System (DOORS) or a compatible automated tool), throughout the design, development, production and verification phases of the program. The Contractor will provide sufficient access, licenses (if required) and training for the requirements traceability tool to the NGRF team. The traceability tool will include all levels of the system and subsystem requirements, derived requirements and links to the capabilities documents, and all testing requirements.

The Contractor will ensure HSI will be a fundamental driver for the design by implementing a strategy to coordinate and manage issues across the program concerning manpower, personnel, training, Human Factors Engineering (HFE), safety, occupational health hazards, personnel survivability, and habitability factors.

3.1.4.2 Modeling and Simulation (M&S): The Contractor will employ integrated M&S tools as appropriate in the Breadfruit Enterprise Architecture Demonstration (BrEAD) development and testing process as a means of optimizing the system design and demonstrating compliance with the PBSS. A key aspect of the Breadfruit Enterprise Architecture Demonstration (BrEAD) M&S strategy is an integrated environment of accredited M&S tools, data, and resources that enables collaboration across functional disciplines. M&S will integrate both simulation and stimulation capabilities. Development and utilization of this environment will place emphasis on open sharing of common tools/data and avoidance of proprietary components. All M&S tools will be certified prior to testing. The Contractor will provide a detailed description of their proposed approach to the development and application of M&S assets in support of T&E, training, and other SDD development/engineering activities.

Modeling and simulation, including SIL M&S that includes Human-In-the-Loop (HIL), will leverage the Human Engineering Simulation Concept Report and human engineering related task analysis deliverables to characterize human interface functionality, as appropriate.

3.1.4.3 Human Factors Engineering (HFE): The Contractor will use human-centered HFE design principles and procedures, for all Breadfruit Enterprise Architecture Demonstration (BrEAD) systems, sub-systems and components to optimize operator and maintainer task performance, minimize operator and maintainer error, and minimize any safety, personnel health hazards, or environmental impacts. The Contractor will prepare and submit a Human Engineering Program Plan (HEPP) to the NGRF for approval.

The Contractor will provide a detailed Human Engineering Task Analysis which includes analysis of all functions and tasks allocated to Breadfruit Enterprise Architecture Demonstration (BrEAD) operators and maintainers, to include identification of critical tasks and operation in degraded modes. Information on each task will be provided to a level sufficient to identify operator and maintainer problem areas that can adversely affect mission accomplishment and to propose and evaluate corrective actions, including design solutions.

The Contractor will evaluate and test the Breadfruit Enterprise Architecture Demonstration (BrEAD) user-interface design, including implementation of controls and displays as specified in applicable design and requirements documents. All design decisions/validation related to Breadfruit Enterprise Architecture Demonstration (BrEAD) workstation configuration and layout, operator performance, controls and displays including information presentation and display management, decision aiding, adaptive system features and automation will be supported by Human Interface Laboratory (HIL) simulations. The Contractor will prepare and submit to the NGRF for approval a Human Engineering Simulation

Concept/Report. Human Engineering activities will include demonstration and validation of Breadfruit Enterprise Architecture Demonstration (BrEAD) requirements for human performance, workload, personnel survivability, safety, and habitability. In addition to HIL simulations as required, the Contractor will perform evaluations and testing in functional operator workstations or mockups, and/or through computer models to verify overall system design.

The Contractor will use working groups, including NGRF working groups to assist in ensuring human engineering criteria as well as overall Breadfruit Enterprise Architecture Demonstration (BrEAD) human systems related performance requirements are met. The Contractor will develop and submit a Human Engineering Test Plan (HETP) and a Human Engineering Test Report (HETR).

3.1.4.4 Software: The Contractor will define a software development approach appropriate for the computer software effort to be performed under this solicitation. This approach will be documented in a NGRF approved Software Development Plan (SDP). The Contractor will follow this SDP for all computer software to be developed or maintained under this effort. The SDP will be updated as changes occur as a result of process improvement reviews and resubmitted for NGRF approval.

The SDP will contain the information defined in IEEE/EIA Std. 12207.1 section 5.2.1 (generic content) and the Plans or Procedures in Table 1 of IEEE/EIA Std. 12207.1. In all cases the level of detail will be sufficient to define all software development processes, activities and tasks to be conducted. Information provided must include, at a minimum, specific standards, methods, tools, actions, strategies, and responsibilities associated with development and qualification.

The Breadfruit Enterprise Architecture Demonstration (BrEAD) software will utilize a High Order Language (HOL), which follows ANSI, IEEE. The use of assembly language or low-level code will be restricted to processing time-constrained and memory-constrained functions. Standards will be established and utilized for embedding comments in source code per best industry practices. The comment standards for banners, headers, and special comments will be described in Contractor-approved standards or an equivalent methodology.

3.1.4.5 System Safety: The Contractor will ensure the Breadfruit Enterprise Architecture Demonstration (BrEAD) design is safe and efficient and conforms to current commercial standards for safety and human engineering. This effort will address hardware, software, and procedural safety and will eliminate or control hazards to personnel, the Breadfruit Enterprise Architecture Demonstration (BrEAD), other assets, and the environment to acceptable levels during all phases of the Breadfruit Enterprise Architecture Demonstration (BrEAD) lifecycle. To ensure a safe system design, the Contractor will develop and submit a System Safety Management Plan (SSMP) for NGRF approval to indicate how the Contractor and their subcontractors plan to manage and implement a system safety program. The Contractor will develop and submit a System Safety Program Plan (SSPP) for NGRF approval that details how the Contractor plans to implement the Breadfruit Enterprise Architecture Demonstration (BrEAD) SSMP. Hazards will be identified in various hazard analyses (such as a Preliminary Hazard Analysis [PHA], Subsystem Hazard Analysis [SSHA], System Hazard Analysis [SHA], and Operating and Support Hazard Analysis [O&SHA]) performed at various times during the design process commensurate with the maturity of the system design, with residual hazards/risks documented in Safety Assessment Reports prior to system test/fielding. The Breadfruit Enterprise Architecture Demonstration (BrEAD) will ensure against degradation or negation of safety features during operations, maintenance, storage, and shipping. The Breadfruit Enterprise Architecture Demonstration (BrEAD) will have fail-safe features with adequate redundancy in both hardware and software and be capable of being rendered safe during emergency or abnormal situations.

3.1.4.5.1 Software Safety: The objective is to ensure that the Breadfruit Enterprise Architecture Demonstration (BrEAD) software is safe. The Breadfruit Enterprise Architecture Demonstration (BrEAD) software will prevent or mitigate single point failures that could disable the entire system. Any software mitigation that is required to deal with hazards will be clearly identified in all software requirements documents and tools, architectural documentation or models as well as software design documentation and design tools. Hazards that could result from operator errors or operator input of improper navigation, control, payload operation, or environmental data will be mitigated by software monitoring and checks of operator inputs. The Breadfruit Enterprise Architecture Demonstration (BrEAD) software will perform appropriate safety checks to assure that dynamic updates of data do not adversely affect the health or safety of the Breadfruit Enterprise Architecture Demonstration (BrEAD).

3.1.4.5.2 Critical Safety Items: Breadfruit Enterprise Architecture Demonstration (BrEAD) Critical Safety Items (CSIs) and fatigue/fracture critical parts will be identified early in the design process through Failure Modes Effects and Criticality Analysis (FMECA) and tracked to ensure proper item management throughout the system lifecycle.

3.1.4.6 System Integration Laboratory (SIL): The Contractor will develop, deliver, install and maintain at a
facility hardware, firmware, and software components for a NGRF Integration Test Facility to support the Breadfruit Enterprise Architecture Demonstration (BrEAD) Integrated Test (IT) and Operational Test (OT) program. The SIL will be used to verify and validate systems integration prior to deployment around the world.

3.1.4.7 Department of Defense Architecture Framework (DODAF): The Contractor will develop and deliver the Breadfruit Enterprise Architecture Demonstration (BrEAD) products in compliance with the current DODAF Volume I and II.

3.1.4.8 Design Decision Documentation. Utilize and maintain a Decision Database to document system-wide design decisions including decisions about the system's behavioral design and decisions affecting the selection and design of systems components and/or subsystems. Include the tradeoffs and other assessments, the decisions made, their justification, the actions taken including revised plans or changes to the baselines or functional architecture proposed or made as a result of the decisions. The decision database will be maintained over the life of the contract and will be readily available to and directly accessible by the NGRF. The Contractor will prepare and submit to the NGRF an archived image of the data in the Decision Database at the time of the close out of each technical review and audit.

3.1.5 Test & Evaluation (T&E): The Contractor, in conjunction with the NGRF, will develop and execute a detailed integrated NGRF and Contractor test program to demonstrate the completeness of the design and development process, to ensure design problems have been identified and solutions are available, to determine compliance with the Breadfruit Enterprise Architecture Demonstration (BrEAD) PBSS. Test planning will cover the Contractor testing, developmental testing, and operational testing.

The Integrated Test Team (ITT) will be responsible for developing the Integrated Test Plan (ITP), approved by the cognizant authority per the Test and Evaluation Master Plan (TEMP). The ITT should explore the most efficient use of existing and planned T&E resources including, but not limited to, systems integration labs, simulation and analysis tools, test ranges, telemetry and data gathering capabilities. A test V&V table will be provided to the NGRF to validate the PBSS compliance methodology. The NGRF will assess the test requirements, work with the Contractor if refinement is necessary and adjudicate any requests for test requirement removal or modification.

3.1.5.1 Problem Reporting

The NGRF will have insight into the Contractor's problem reporting database when system integration testing begins in the laboratory. The Contractor will ensure that full access to the problem reporting database is provided to the NGRF. The NGRF and Contractor will form, operate and co-chair a review board to prioritize problems for resolution.

3.1.6 Supportability: Breadfruit Enterprise Architecture Demonstration (BrEAD) will be designed to minimize scheduled and unscheduled maintenance, minimize the number of maintenance personnel, maximize SOE. The Contractor will determine and/or develop the necessary tools, support equipment, Logistics Management Information (LMI), technical documentation and skills to perform maintenance. The Contractor's supportability analysis and product development will reflect the following considerations:

a) The Breadfruit Enterprise Architecture Demonstration (BrEAD) will be supported using a performance based logistics concept. The performance based support system may be fully organic, fully commercial or a combination of both. The supply chain will maximize SOE by using a sense-

and-respond process for identifying material requirements and will provide operating unit flexibility and self-sufficiency to mobilize, deploy and sustain;

b) The Breadfruit Enterprise Architecture Demonstration (BrEAD) will include a PHM subsystem that will facilitate reliability, maintainability and supportability requirements;

c) Minimize Total Ownership Cost;

3.1.6.1 Performance Based Logistics Planning: PBL is an enabling factor for achieving a high level of SOE. Therefore, the NGRF intends to support the Breadfruit Enterprise Architecture Demonstration (BrEAD) through the application of a system-unique, PBL strategy. The Contractor will design, and report to the NGRF, a conceptual PBL strategy for use in supporting the Breadfruit Enterprise Architecture Demonstration (BrEAD). The strategy will optimize the Breadfruit Enterprise Architecture Demonstration (BrEAD) design for life cycle support and provide for continuous improvement in reliability, system readiness, TOC and product support processes. The NGRF may comment upon and approve the concept for use as a design-to support strategy or select portions of the proposed strategy. The Contractor will then utilize the approved strategy to influence the design and testing of the Breadfruit Enterprise Architecture Demonstration (BrEAD) subsystems and development of supportability products and data. The PBL strategy will comply with all relevant public laws, regulations and service preferences. The Contractors PBL Strategy Report will discuss the following at a minimum:

a) The Contractor's process for evaluating SOE impacts of the strategy;
b) The Contractor's process for influencing UAS design for enhanced supportability using the strategy, SDOE;
c) Assumptions used in analyzing and selecting a strategy;
d) NGRF actions necessary for implementing the strategy such as requests for waivers to policy or regulations;
e) Proposed performance metrics and incentives that may be used to achieve high levels of supportability and cost control;
f) Test and Evaluation and verification steps the Contractor will use to validate the strategy.

3.1.6.2 Maintenance Planning: The Contractor will incorporate processes for maintenance planning to reduce life-cycle cost, maintenance man-hours and system down time. The Contractor will develop a Logistic Management Information (LMI) database with appropriate LMI levels assigned, using a standardized LMI database software program capable of producing common output reports such as a Maintenance Plan, FMECA, Task Narrative, Provisioning Technical Documentation (PTD), Support Equipment Recommendation Data (SERD) and Training Requirements.

3.1.6.2.1 Integrated Maintenance Concept (IMC): The Contractor will maximize integration to a depth that maximizes SOE and cost effectiveness. In addition, the

Contractor will implement maintenance planning concepts that take full advantage of SDOE and SOE to reduce life cycle costs.

3.1.6.3 Technical Manuals/Technical Data: The Contractor will prepare Breadfruit Enterprise Architecture Demonstration (BrEAD) technical manuals in the Interactive Electronic Technical Manuals (IETMs) format, and in accordance with a NGRF approved specification. Technical data will be accurate and in sufficient detail to support all planned levels of maintenance and operation. The technical data will include systems/functional description and theory of operation, maintenance and operating procedures, troubleshooting and testing procedures, fault isolation, Illustrated Parts Breakdown (IPB), Maintenance Requirement Cards (MRC), periodic maintenance and phased maintenance requirements, preventive maintenance, and corrective maintenance authorized for each level of maintenance. The Contractor will provide validated/verified technical data required for training, operations, testing, and maintenance. All updated publications and associated source data, XML files, schemas, Document Type Definition (DTD), business rules exchange, style sheets, and Format Output Specification Instance (FOSI) will be considered NGRF property at the conclusion of the contract.

3.1.6.4 Reliability, Maintainability, Supportability (RMS) Prognostics and Health Management (PHM) Engineering: The Contractor will maximize the inherent RMS and PHM capabilities of the Breadfruit Enterprise Architecture Demonstration (BrEAD) in a cost effective manner that will minimize O&S costs utilizing all RMS/PHM PBSS objectives as follows:

1. RMS/BIT/PHM Program Plan
2. RMS/BIT Organization and Structure
3. Schedule for RMS/BIT/PHM Tasks and Milestones
4. Reliability Modeling, Allocation and Predictions (e.g. RELEX ™, RAPTOR)
5. Maintainability Allocations, Analyses and Predictions
6. BIT and Testability Allocations, Analyses and Predictions
7. PHM Allocations, Analyses and Predictions
8. Failure Reporting, Analysis, and Corrective Action System (FRACAS)
9. Failure Review Board (FRB)
10. Fault Tree Analysis (FTA)
11. Failure Mode Effect and Criticality Analysis (FMECA)
12. Level of Repair Analysis (LORA)
13. Reliability Development/Growth Process and Tracking
14. Maintainability Demonstrations
15. BIT and Testability Demonstrations
16. Manufacturing Screening
17. RMS/BIT Testing and Tracking
18. PHM Program Plan
19. Reliability and BIT Performance Maturation
20. Parts Control/Application Program
21. Parts Derating Program

22. RMS/BIT Lessons Learned Incorporation
23. CBM+
24. Reliability Centered Maintenance (RCM) Analysis Plan
25. Maintenance Task Analysis (MTA)
26. Supportability Analysis
27. Maintenance Concept/Planning

The objective of PHM is to increase the survivability, reliability, maintainability, supportability, readiness and affordability of the Breadfruit Enterprise Architecture Demonstration (BrEAD) over the life cycle of the system. Specific sub systems and failure modes to be monitored by the PHM system will be determined by study of the system FMECAs and failure mode risk index, and a return on investment analysis of the benefit of versus the cost to implement a PHM system monitoring algorithm. Proposed failure modes to be monitored will be judged on effectiveness and maturity of the proposed monitoring algorithm relative to current and/or similar systems, and the potential benefit of new algorithms. The PHM system will allow for both hardware and software growth in order to address emerging failure modes over the life cycle of the program.

3.1.6.5 Unique Identification (UID): The Contractor will establish an effective serialized management program on all components and end items that are repairable, life-limited or time-controlled, configuration tracked items, and/or assigned logs and/or records. The Contractor will also be responsible for configuration management and associated data.

3.1.6.6 Support Equipment: The Contractor will ensure systems engineering processes are used to effectively trace supportability requirements to the Breadfruit Enterprise Architecture Demonstration (BrEAD) functional and physical architecture. The Contractor will develop and submit to the NGRF for review and comments, a Support Equipment Program Plan (SEPP). The SEPP will detail the Contractor's process for managing its SE development activities. At a minimum, the SEPP will detail how the Contractor plans to integrate SE engineering with the Systems Engineering process (unless adequately addressed in the SEMP), test and evaluation methods and LFp control and reduction efforts.

3.1.7 Training

3.1.7.1 Objective: The objective of the Breadfruit Enterprise Architecture Demonstration (BrEAD) training development program is to provide a Training System (TS) capable of producing operators and maintenance personnel in sufficient numbers to perform meet Integrated Test (IT) and OPEVAL requirements. The delivered TS will provide Cadre and Initial training system IOC date. The TS will include all hardware, software, courseware, technical and training documentation, and manuals necessary to train instructors, operators, and support personnel to full mission-ready proficiency. All documentation, Interactive Multi-media Instruction

(IMI), instructional material, and training devices will maintain concurrency with Breadfruit Enterprise Architecture Demonstration (BrEAD) configuration.

3.1.8 Environment, Safety and Occupational Health (ESOH): The NGRF's objective is to reduce risks and Total Ownership Cost (TOC) by incorporating ESOH objectives into the systems engineering process, integrating elements of Contractor ESOH programs into the NGRF's ESOH programs, and ensuring continuous compliance with Federal, State, and local ESOH regulations. The Contractor will draft and submit to the NGRF, a Hazardous Material Management Program (HMMP) Plan. The HMMP Plan will detail the Contractor's management processes and controls, and outline requirements for periodic HMMP reports.

3.1.9 Transition to Production: The Contractor will maintain a management organization with the ability to coordinate all phases of the Breadfruit Enterprise Architecture Demonstration (BrEAD) transition to production while minimizing program risk.

3.1.10 Materials and Processes: The NGRF's objective is for all materials and processes considered for application in the Breadfruit Enterprise Architecture Demonstration (BrEAD) to be subjected to rigorous evaluation in a well-defined and documented design development test program. The principle objectives of the testing are to establish material system performance in the defined operational environments; identify, characterize, and optimize associated stable processes; verify methods used in the evaluation of materials, and establish the design. Design properties (in the appropriate chemical, thermal, and climatic environments) will be established during development testing to support transition and application of the material systems and processes into the Breadfruit Enterprise Architecture Demonstration (BrEAD). Environmentally conditioned tests will be performed at the development test level to meet design conditions. Whenever materials are proposed for which only a limited amount of data is available, the NGRF will be provided with sufficient background data so that a determination of the suitability of the material can be made. All testing and data used in establishing material system performance will be documented and submitted to the NGRF, including any data obtained from previous design development test programs.

3.1.10.1 Materials Producibility: The Contractor will develop and implement processes and procedures required to establish material systems maturity, repeatability, and integrity.

3.1.10.2 Material Repairability: Repair concepts and procedures will be developed to fully restore the structural, electrical or other applicable functional properties. Repair concepts and procedures will be demonstrated and validated. Use of previously qualified materials, processes, concepts, and procedures will be substantiated. Any studies or tests performed to develop and validate any repair materials, processes, concepts, and procedures will be documented in an Advanced Material Repair Development Program Plan. Results of developmental efforts will be documented in an Advanced Material Repair Development Report. Any existing data used to substantiate repair materials, processes, concepts, and procedures will also be documented.

3.1.11 Manufacturing: The Contractor will ensure implementation of processes and systems that consider manufacturing, quality and design functions in achieving a balanced product design which meets cost, schedule and performance requirements with acceptable risk. Appropriate practices for implementation should include production cost modeling; identification of key characteristics and processes; variability reduction; electronic simulation of the manufacturing environment; Cost As an Independent Variable (CAIV); cost/performance trade studies; manufacturing capability assessments/PRRs; product and process validation; and key supplier relationships.

3.1.11.1 Production Quality and Manufacturing Efficiency: The Contractor will implement processes and systems to assure program affordability through product quality and manufacturing efficiency. The following elements may be considered as appropriate practices for implementation: product improvement initiatives; variability reduction on product and process; manufacturing process control and continuous improvement; key supplier relationships.

The Contractor will identify drawings and process documents critical characteristics or critical processes. These critical characteristics will be inspectable and/or testable during the manufacturing/ fabrication/assembly processes. - END OF DOCUMENT -

Performance Based System Spec (PBSS)

Attachment 1 - Breadfruit Enterprise Architecture Demonstration (BrEAD) and Systems Design Performance Based Systems Specification (PBSS)

The PBSS is provided to the contractor as an initial performance based set of requirements, is general in nature and requires more detail. It is the expectation of the NGRF that the contractor will provide additional paragraphs, and more detail so that the contractors design team is able to a provide systems design and deliver a system that meets the objectives of the SOO. The updated PBSS requirements (CDRL 0008) are deliverable as part of the proposal.

1. The BrEAD shall be comprised of the Breadfruit Enterprise Architecture Demonstration (BrEAD) that forms the organizational structure that will: be an advocate for breadfruit as a component of the next green revolution, design, market, collect data, distribute and provide logistics support for breadfruit processing systems and the System Design (SD) for breadfruit processing equipment.

2. The Breadfruit Enterprise Architecture Demonstration (BrEAD) shall provide an automated research, marketing, and user data collection capability for NGRF head quarters in St. Paul, Minnesota that will provide 24 hour per day data from researchers, marketers and users world wide.
3. The BrEAD shall provide organizational liaison to the Community of Interest (COI) and automated electronic connectivity for real time data interchange. The community of interest is comprised of the NGRF, government and private organizations that have or could have an interest in breadfruit products.
4. The initial baseline breadfruit systems design shall be based on the characterization of any prototype existing breadfruit processing systems. The characterization shall include, but not be limited to; operator usability, reliability, maintainability and manufacturability. The findings shall be documented in a requirements document and a product description.
5. The breadfruit processing system shall be operable in geographic regions that do not have an electrical power infrastructure and in geographic regions that have an electrical power infrastructure with minimal equipment differences.
6. The design of the breadfruit processing system shall be modular, such that any specific geographic requirements can be accommodated without breadfruit processing system redesign.
7. The breadfruit processing system human machine interface shall be intuitive; that is the human operation of the system shall require minimal documentation- ideally none.
8. Any design modifications that are indicated from operational data gathered from users of breadfruit processing systems in use in the field shall be analyzed and considered for incorporation as design modifications to future the breadfruit processing system designs.
9. The breadfruit processing system shall be easily repairable with a minimum of tools in the field
10. The breadfruit processing system shall have a failure rate of one failure per 10, 000 hours.
11. The mean time to repair the breadfruit processing system shall be 15 minutes.
12. The breadfruit processing system shall extendible. Extendibility is the capability of the breadfruit processing system to produce more than one food product or food product component.
13. The breadfruit processing system shall be scalable ; the range of scalability to be considered is from family use to industrial scale use that could provide large quantities of breadfruit products. The range anticipated is from a few ponds per day per family to several tons per day for an industrial scale facility. The specific values shall be determined and documented in the proposal.
14. The BrEAD shall provide weather data gathered through out the world where breadfruit is used. It shall provide hourly temperature, pressure,

humidity, and rain fall data to the Central Data Collection and Analysis Center of the NGRF located in St. Paul, Minnesota.

15. The BrEAD architecture shall have distributed local weather sensors, and data collection systems and shall be interoperable with the Central Data Collection and Analysis Center. The user data collection human machine interface at the data collection facilities shall be intuitive to the user.
16. The breadfruit processing system shall operate from 40 degrees Fahrenheit to 120 degrees Fahrenheit.
17. The Systems Integration Laboratory (SIL) shall be used to evaluate candidate architectures for the BrEAD , the breadfruit processing systems, and the intuitive human interface capability desired.
18. The breadfruit processing shall operate in 100% relative humidity.

Evaluation Method

The evaluation method is typically encoded into a spreadsheet, in order to make calculations automatic. This is crucial, as it provides the opportunity for the reviewer to evaluate each independent portion of the proposal purely on merit, and does not support continuous monitoring of the proposal score, which can result in a biased evaluation of the teams.

This methodology has frequently resulted in surprising results for the reviewers, and is highly recommended.

Semmar Proposal evaluation - 05/1/2012 - evaluator Tony Beck

Section M - Evaluation factors for award: The evaluation factors for award are technical and management. The sum of the point scoring for both will be 100 points. The technical will be factor will be 70 of the 100 points. There are three technical sub factors: 1. Requirements (50 points), 2. Technical approach (25 points), and 3. System architecture (25 points). The management factor has two sub factors: Integrated Management Plan (50 points) and Integrated Management Schedule (50 points). The highest overall score will be the considered the best technical proposal. T1=Requirements (50 points),T2 =Technical approach (25 points), T3=System architecture (25 points), M1=Integrated Management Plan (50 points) , M2= Integrated Management Schedule (50 points). Objective # = bidder offering above and beyond program requirements. Values in score cells range from 0 to 10.

Index #	Sub factors	Subfactor Evaluation Descriptions (Sec L) &Check List Questions	score	Comments
1	L	Proposal Format		The proposal was difficult to evaluate because the proposer did not follow the outline in section L!! Instead they invented their own outline
2	L	Proposal Content & Volumes		The proposal content was godd to excellent. The diviation form the outline in section L of the RFP reduced the score of the proposal.
		Executive Summary Vol 1		Good executive summary, overall well constructed and compelling arguments.
		TECHNICAL Volume 2		
		DESIGN APPROACH - Book		
		2.1 Design Approach	3.0	An overall description of the design approach is not provided ; instead a glossary is provided. What should have been a 10, because they put
3	T1	2.1.1 Overall Design Approach	8.0	No text provided; the paragraph used as a header.Infrmation here was provided in the proposers own outline and hard to find. . So instaed of
4	T2	2.1.1.1 Systems Overview	10.0	Good description provided in the day in the life; and the aothour used the putline difined in Section L of the proposal.
5	T2	2.1.1.4 External Systems	8.0	The interfaces are indentified , but the impacts on the enterprise system architecture are not described; nor are the impacts on the systems
6	T2	2.1.1.5 Support System	10.0	Well done, and you followed the outline in section L.
6	T2	2.1.1.6 Hardware and Software Integration Overview	9.0	Adequate description
8	T1	2.1.2 Performance Based Systems Specification (PBSS)	5.0	Only a refference given to technical attachment; no discussion of the signficance or key requirments.
9	T3	2.1.3 Open Systems Architecture	10.0	Good write up
10	M1	2.2.1 Proposed SOW and CDRLS	9.0	This would have been a 10 if they had followed the outline in section L.
11	M2	2.2.2 Integrated Master Plan	9.0	This would have been a 10 if they had followed the outline in section L.
12	M2	2.2.3 Integrated Master Schedule	7.0	No critical pathdescription; and this would have been and 8 if they had followed the outline in section L.
13	M1	2.2.4 Systems Engineering (SE) Management	8.0	There was no summary schedule described; and this would have been and 9 if they had followed the outline in section L.
14	M1	2.2.5 Risk Management Plan	9.0	Would have been a 10 if you had folowed the outline in section L.
15	M1	2.2.6 T&E	6.0	Acceptable, and would have been a 7 if you had followed the outline in section L.
16	M1	2.2.7 Subcontract Management	9.0	Would have been a 10 if you had folowed the outline in section L.
17	M1	2.2.8 Software Development	3.0	No mention of the software development plan; however there was a discussion in a section 3.2.9 software and hardware Overveiw that did
18	T3	Annex A – Automated Bridge Toll and Health Monitoring	10.0	Good
19	T3	Annex B – Detailed Reliability, Maintainability and Availability	10.0	Good
20	T1	Technical Attachment 1 - Automated Bridge Toll	7.0	Good PBSS, no RVTM
21	M1	Annex C – CWBS	7.0	no definition of the work, only titles
22	M1	Annex D – SOW	5.0	No detailed stat ment of work
23	M1	Annex E – IMP	10.0	Good IMP
24	M2	Annex F – IMS	8.0	No narrative expanantions.
25	M1	Annex G – SEMP	10.0	Excellent SEMP
26	M2	Annex H – Software Development Plan	10.0	Excellent SDP

Figure 173: Section scoring of evaluation spreadsheet

Team	T1	33.20	M1	40.50		T1+T2+T3	56.86
None	T2	23.13	M2	42.50		M1+M2	24.90
	T3	24.90				Total Score	81.76

Figure 174: Evaluation summary block of spreadsheet

Proposal 1: BRAINS Proposal

BRAINS Proposal
for the
Breadfruit Enterprise Architecture Demonstration
(BrEAD)

RFP #: E00010-07-R-0001

Volume # 1: Executive Summary
Proposal #: BRN3425

Prepared for:
New Green Revolution Foundation (NGRF)
2134 University Ave, Saint Paul, MN 55206

Prepared by:
Bright Robust Architect for the Institute of New Systems (BRAINS)
917 Washington Ave SE, Minneapolis, MN 55414
POC: Danielle Eastberg 952-994-5393

Volume 1 – Executive Summary

1.1. Purpose

BRAINS wrote this proposal with the intention to raise sufficient funds to design the breadfruit transformation process from natural fruit into flour, and the Breadfruit Enterprise Architecture Demonstration (BrEAD).

1.2. Scope

This project proposal aims to relief millions of people from hunger by creating a long lasting product, and easier to be shipped to geographically remote populations. BRAINS identified all the activities, resources, risks, estimated cost, and assumptions needed to implement the breadfruit processing system and the enterprise architecture at the Republic of the Marshall Islands (RMI).

The breadfruit processing system shall be a reliable, scalable, and friendly design that prolongs the shelf life of such fruit by supporting its transformation into flour, and by incorporating innovative green ideas and low energy consumption machinery, for small and large scale systems.

The enterprise architecture shall be capable of supporting the production of breadfruit flour as well as capable of taking care of legal, marketing and sales, and training activities to complement the business core activity.

BRAINS created this reliable project proposal successfully goes beyond the expectations by significantly improving the quality of life of millions of farmers by bringing electricity to their villages, thus creating new social, economical, and cultural opportunities to future generations, families and communities.

1.3. Stakeholders

BRAINS recognizes that BrEAD and the breadfruit processing system will mark the beginning of a new life for millions of people, and given the world-wide dimensionality of the project, BRAINS concentrated all its efforts into one particular nation, the RMI. These are the main stakeholders grouped by (geographical) categories we have identified, and to some extent these are replicated across the world.

Minneapolis/St. Paul:

- NGRF
- BRAINS
- Legal Department (BRAINS)
- US Government
- Breadfruit Institute (Hawaii)
- Local population
- Consumers
- Land Owners (chiefs, workers, and their families)
- Environmental organizations
- Local Packaging companies
- Local transportation companies
- Local RMI Government

1.4. User Needs

BRAINS prides itself in getting to know customers individually and indentifying their specific needs to maximize the system design. This is a summary of the main user needs.

- ***NGRF***
    - Better understanding of the activities, resources, time line, and others related to the research, design, implementation, production, testing, and results of the project
- ***Local population***
    - Guarantee that waste disposal, and chemicals used at the production system will not impact health, environmental, or in general their quality of life at the island
    - Benefits from participating on the processing system; new learning opportunities, training, cultural events, etc.
    - Equal opportunity to participate (to be hired)
- ***Consumers***
    - Long lasting breadfruit-based product
    - Good quality and low cost breadfruit-based product
    - New recipes and training to incorporate new product in daily diet
- ***Land Owners (chiefs, workers, and their families)***
    - Sign a lease and define the terms upon the land will be used by the members of the family
    - Monetary incentive in exchange of the breadfruit, maintenance and use of the land, and family members that will labor the land
    - Machinery, location, tools, and resources to transform the breadfruit into flour
    - Boats, cars, trucks to transport breadfruit if necessary
    - Training and support for breadfruit processing systems
    - Continue communication with BrEAD's members, feedback, and marketing initiatives
- ***Environmental organizations***
    - Guarantee of environmental preservation measures
    - Chemical wastes, and garbage are properly collected and treated
    - Pesticides used to prevent tree and fruits damage are adequate
    - Breadfruit trees are not in danger of life extinction because of proposed processing system
- ***Transportation employees***
    - Logistic information as where to pick up breadfruits and where to deliver it
    - Trucks, cars or boats to transport the breadfruit
    - Other tools and resources needed for management purposes, and for safely transporting natural breadfruit or flour
    - Monetary incentive in exchange of their work
- ***US and Local RMI Governments***

- Guarantee that legal local laws regarding taxes, health insurance, environmental, transportation, exportation/importation regulations, and labor laws, are respected by BRAINS.
- ***Breadfruit Institute (Hawaii)***
    - Resources available in-site to conduct experiments and testing laboratories
    - Data information (results) regarding current experiments in-site
    - Feedback from local farmers regarding daily operations, new approaches, breadfruit description, etc
    - Continuous bi-direction communication between BRAINS and chiefs (land owner) if needed
- ***Legal Department (BRAINS)***
    - Better understanding negotiation standards and rules with local farmers owners of the land
    - Legal requirements to export/import materials to/from RMI
- ***Local packaging companies***
    - Logistic information
        - Location of collection points
        - Flour quantities and packaging presentation design
        - Location of selling points to deliver the packages of flour

1.5. Main Operational Nodes

In order to promote the use and consumption of breadfruit, and to promote and encourage people to adhere to versatile fruit flour in their daily diet, BRAINS focused all its efforts on the Republic of the Marshall Islands (RMI) to provide a local (but flexible) reliable process to transform breadfruit into flour. The motivation behind the initiative to select a specific country is to learn about as much as possible about this location thus building a real proposal controlling as many variables as possible.

The RMI is a considerable large set of small islands and atolls where the breadfruit naturally grows, and the locals daily consume it. Currently, they consume the breadfruit on a daily basis, but there is no such thing as breadfruit flour process. Local population has currently no mechanical neither electrical process already established for neither a small scale (family size), nor large size (villages or more than one island). Even more, all the breadfruit consumed is only for family consumption, and no surplus is been generated for sale or economical benefit. BRAINS proposed a novel process to produce a long-term breadfruit product: flour.

Flour is a very familiar ingredient in the civilized cultures. Wheat flour is the main ingredient used for a wide range of daily products like bread, cakes, pancakes, biscuits, etc. It is also an important ingredient in many other recipes. It is certainly a highly consumed ingredient around the world because of it nutritional values and many applications. This is an enormous advantage for people that daily uses breadfruit, and cannot afford or cannot consume regular flour because of their geographical location. BRAINS completed a research phase to have a better understanding of the Marshall Islands. More of this information can be found at Attachment 2 - Tech. Volume. The main operational nodes regarding geographical location are **Hawaii, Minneapolis/St Paul**, and the **Republic of the Marshall Islands**.

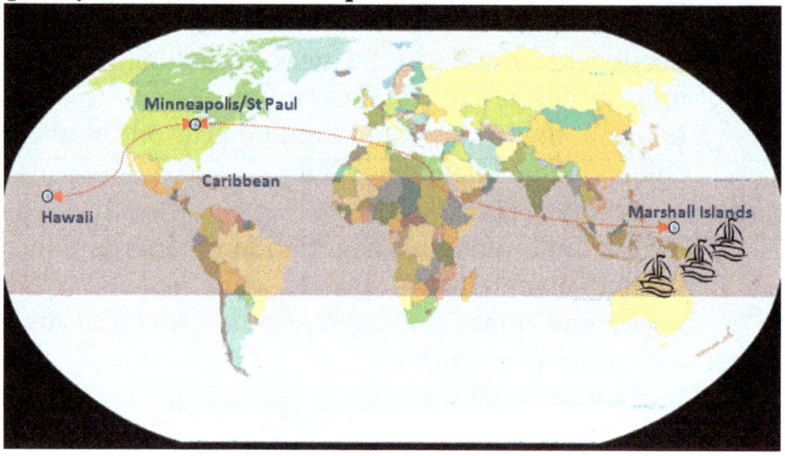

Figure 175. From left to right: Hawaii, Minneapolis/St Paul, and the Marshall Islands

1.6. Program

BRAINS's strategy is to provide a **reliable** process and robust architecture that will consistently provide breadfruit to hungry people. Therefore BRAINS provides simple, non-cultural shocking solutions to communities with low resources.

1.6.1. Abstract of Technical Approach

BRAINS proposes to lease the land from the local farmers by contacting the chief (owner of the land) of the local tribes, to collect breadfruit and transform it into flour. Once breadfruit is gathered together, it will be transported by the local employees to the processing units (factories) where it will be processed.. Once the flour is produced it will be shipped to local companies for packaging, and redistributed to small local stores for locals to buy it for a small price.

1.6.2. Breadfruit Processing System

1.6.2.1. Gathering of breadfruit

The process starts with a collection phase when local farmers who own and work on the land (rented under a predefined lease) will collect and grab the breadfruit from the trees and transport it back to the place where the breadfruit will be processed. To determine if a breadfruit is ready or not we described physical characteristics of immature, mature, and ripe breadfruit.

1.6.2.2. Delivery of breadfruit

Natural breadfruit will be collected in large amounts (relatively to the scale of the production site) before processing it. Locals will gather breadfruit from trees, and BRAINS will provide transportation from a pre-defined collection point to the processing factories.

1.6.2.3. Processing of breadfruit

This is the main transformation phase where the breadfruit after been collected will be transformed into flour. For each scale (small and large) BRAINS identified reliable and non-cultural shocking, low-maintenance machinery. The process itself has multiple stages like cleaning, peeling, slicing, blanching, drying, grinding, and screening (BRAINS provide a very detailed operational procedure for each). The main final product is breadfruit flour. BRAINS identified an innovative energy supply unit with multiple community benefits, and low-cost low-maintenance wind turbine. Refer to the Technical Volume for more information.

1.6.2.4. Shipping of breadfruit flour

Once the breadfruit is transformed into flour it will be shipped to a local company to package it.

1.6.2.5. Packaging of breadfruit flour

On this step, the breadfruit flour is packaged on plastic bags for distribution.

1.6.3. Enterprise Architecture

It is mainly composed of professional employees, offices, and office and communication resources to support the back-office activities such as legal, financial, marketing, logistics, etc. needed to guarantee the breadfruit processing system daily's production. They contact contractors, land owners, and manufacturers (among others), and reach out local population through marketing initiatives to promote the breadfruit flour.

1.7. External Systems

BRAINS identified a set of internal and external subsystems to guarantee the requirements of the BrEAD PBSS. Internal subsystems are all phases regarding processing breadfruit such as collection, cleaning, peeling, slicing, drying, grinding, shipping, packaging, and distributing. Some are located in Minneapolis for such as IT solutions, and others are located at the RMI companies for shipping and packaging. All interfaces are explicitly defined and smoothly interrelated.

Subcontractor	(Location) Specialty	Task Work Order
RMI Private Businesses	(RMI) Shipping	Provide inter-atoll shipping of materials
Matson Navigation	(RMI) Shipping	Provide international shipping of end-product
Majuro Packaging Facility	(RMI) Packaging	Package flour for distribution
Ralik Store	(RMI) Distribution, retail and wholesale	Provide the capability to distribute the end-product to the end user and distribution node
Parallel Track Engineering Company Pvt. Ltd.	(RMI) Storage and material handling solutions	Provide storage and material handling solutions

Techfluent	(Twin Cities - MN, US) Information Technology	Provide IT Service Management and solutions

Table 1 – Offeror's Summary Table

1.8. Open Systems Architecture

BRAINS identified non-proprietary commercial off the shelf artifacts to be used at the breadfruit flour production sites to make systems easily upgradeable and modular. All resources can be easily found, replaced, tailored, or upgraded if needed.

1.9. Qualifications

Bright Robust Architects for the Institute of New Systems (BRAINS) has been a leading global systems engineering firm for 38 years. BRAINS engage in research, design, development, and integration of infrastructure systems and services. BRAINS is headquartered out of Minneapolis, MN-USA, and currently employs 10,200 people worldwide. We distinguish ourselves by our passion for originality and initiative.

Contributors:

Volume Editor: Gabriela Farias

Executive Summary: Gabriela Farias

References:

About Breadfruit:

[1] http://www.capetrib.com.au/breadfruit.htm
[2] http://www.kew.org/ksheets/fruits.html
[3] http://www.agroforestry.net/tti/A.altilis-breadfruit.pdf.
[4] http://www.wisegeek.com/what-is-breadfruit.htm
[5] http://archives.starbulletin.com/1997/07/30/features/request.html
[6] http://www.hort.purdue.edu/newcrop/morton/breadfruit.html#Season
[7] http://www.agribusinessonline.com/crops/postbrdf.asp
[8] http://postharvest.ucdavis.edu/produce/producefacts/Fruit/breadfruit.shtml

About Flour:

[9] http://en.wikipedia.org/wiki/Flour
[10] http://www.recipezaar.com/library/getentry.zsp?id=64
[11] http://whatscookingamerica.net/Bread/FlourTypes.htm
[12] http://www.bigoven.com/whatis.aspx?id=Flour
[13] http://foodsubs.com/Flour.html
[14] http://cat.inist.fr/?aModele=afficheN&cpsidt=1315902
[15] http://www.shipton-mill.com/Other-Flour-Classification.html

About Republic of the Marshall Islands:

[16] http://www.rmiembassyus.org/
[17] http://en.wikipedia.org/wiki/Marshall_Islands
[18] http://www.visitmarshallislands.com/
[19] http://www.imf.org/external/country/mhl/index.htm
[20] http://www.adb.org/marshallislands/default.asp
[21] http://www.prel.org/aboutprel/organization/pacific-service-region/rmi.aspx
[22] https://www.cia.gov/library/publications/the-world-factbook/geos/rm.html
[23] http://www.state.gov/r/pa/ei/bgn/26551.htm
[24]
http://ec.europa.eu/development/geographical/regionscountries/countries/country_profile.cfm?cid=rm&type=short&lng=en
[25] http://dic.academic.ru/dic.nsf/enwiki/9554225
[26] http://www.everyculture.com/Ma-Ni/Marshall-Islands.html
[27] http://www.privateislandsonline.com/marshallislands_info.htm

About Hawaii – Breadfruit Institute:

[27] http://en.wikipedia.org/wiki/Breadfruit
[28] http://www.ntbg.org/breadfruit/

BRAINS Proposal
for the
Breadfruit Enterprise Architecture Demonstration
(BrEAD)

RFP #: E00010-07-R-0001

Volume # 2: Technical
Book 1: Design Approach
Copy #: 1 of 2

Proposal #: BRN3425

Prepared for:
New Green Revolution Foundation (NGRF)
2134 University Ave, Saint Paul, MN 55206

Prepared by:
Bright Robust Architect for the Institute of New Systems (BRAINS)
917 Washington Ave SE, Minneapolis, MN 55414
POC: Danielle Eastberg 952-994-5393

Table of Figures

2. Volume 2 - Technical Volume

2.1. Design Approach

The breadfruit has the tremendous potential to reduce and alleviate the hunger of millions of citizens around the world. A scalable process definition is needed to nourish a bigger population, as well as an Enterprise definition to support its operation around the world. BRAINS will demonstrate a flexible, robust, and friendly breadfruit processing system and a mature Enterprise architecture that will meet and exceed the requirements. Our strategy is to focus on making our system reliable. We chose reliability as a strategy so that we could build a strong foundation that works efficiently and can therefore accommodate future expansions. We plan to work closely with the NGRF board of directors throughout our process to ensure we are collaborating and helping to create their vision.

2.1.1. Overall Design Approach

BRAINS design approach follows the steps of the V- diagram shown Figure 176. To begin the proposal the system users were determined. The users are categorized according to their geographical location including the Marshall Islands, Minneapolis/St. Paul, and Hawaii (OV-1 for more information, and the executive summary). Next the user needs were determined and a very in-depth research was done to understand the local population of the Republic Marshall Islands. The requirements to meet those user needs were derived and merged them with the given NGRF requirements. The requirements were used to generate a functional analysis. Using the functional analysis, the subsystem architecture was generated. From there, a physical architecture was created. Some of these steps were duplicated to elaborate the breadfruit processing itself and the enterprise architecture definition.

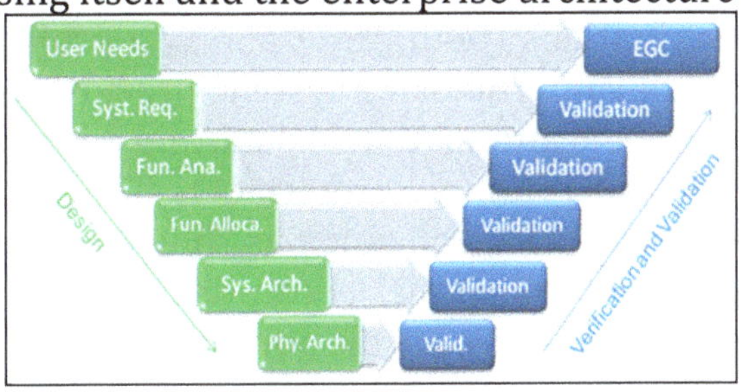

Figure 176: V – Diagram

2.1.1.1. Systems Overview

2.1.1.1.1. Breadfruit Enterprise Architecture Configuration

BRAINS will establish an enterprise architecture that will implement at minimum all the requirements in NGRF's request for proposal (NGRF). Utilizing the techniques of the DoD Architecture Framework (DoDAF), BRAINS created the top level OV-5 diagram shown on Figure 177 for the Breadfruit Enterprise Architecture Demonstration (BrEAD).

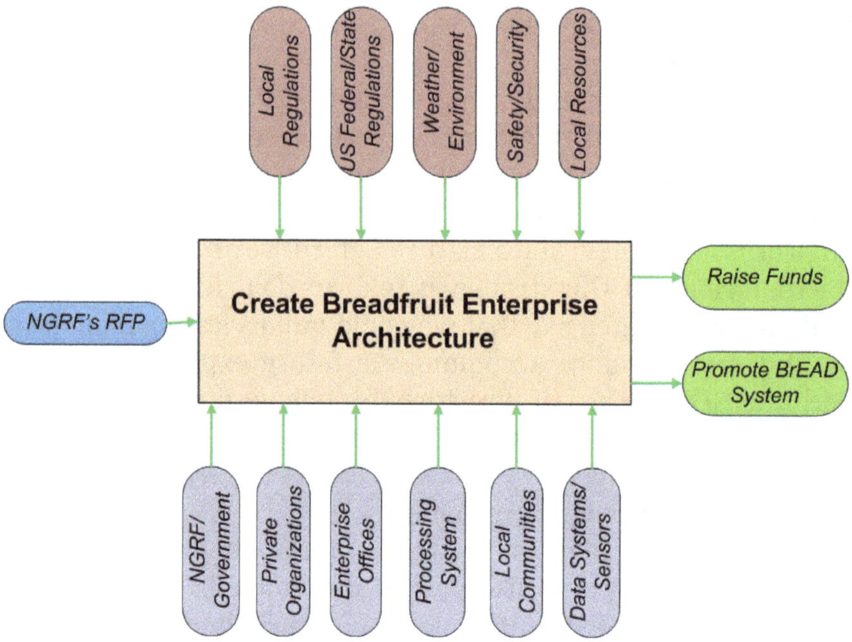

Figure 177: Operational View Top Level Diagram

Upon receiving NGRF's request for proposal, BRAINS first identified the people and elements/equipments that will be necessary for the establishment of BrEAD. These items are depicted on the bottom section of the diagram. Regarding people/elements, BRAINS will contact and establish relationship with NGRF, all applicable governments (local, state, and federal), local communities, as well as the appropriate private and Non-governmental Organizations (both profit and non-profit). In addition, BRAINS will develop scalable breadfruit processing system; procure systems for the transfer, collection, sharing, storing, and automation of data; and open convenient local offices that will administer field operations. Furthermore, BRAINS has identified all the external factors that the enterprise will abide as shown on the upper section of the diagram. In doing its operations, the enterprise will observe the applicable laws/regulations that are put forward by government entities at all levels. The enterprise will also conduct extensive study on the weather/environmental conditions, any safety/security concerns of each location, and the relevant local resources that are available such as transportation, communication systems, electricity, etc. Once realized, these implementations will lead to the enterprise's goal of promoting the breadfruit system worldwide and raising funds as shown by the output of the diagram.

In addition to the top level operational activities model described above, BRAINS developed detailed OV-5 diagram in order to map all the enterprise functions and insure specific activities are well defined as displayed on Figure 178. Again this drill-down diagram is developed based on the DoDAF architecture. After conducting thorough review of all the requirements presented in the NGRF provided Statement of Objectives (SOO) and the Systems Design Performance Based Specification (PBSS),

BRAINS will perform the following activities in sequence or simultaneously as feasible:

- **Indentify Locations**: BRAINS will consult with NGRF, private/NGO organizations, and both US and local governments/communities when selecting suitable location to implement the BrEAD. BRAINS will also review all the external factors that will impact in the selection process such as weather/environment considerations, compliance with local/US regulations, safety/security, and the local resources that will ease the service of the enterprise to the local communities. As described in the OV-1, BRAINS identified Marshall Islands for the first implementation of the breadfruit processing systems. Obviously the location selection will be based on the abundance of breadfruit in the area of consideration, and the user needs in order to make sure we're first operating in the regions where help is most needed.
- **Design breadfruit processing system**: With consultation of NGRF and other organizations, BRAINS will design an efficient and reliable processing system that will:
    - Comply with all NGRF, US, and local requirements.
    - Operate in different weather/environmental conditions with minimal or no modification.
    - Operate with/without local resources available such as power.
    - Be intuitive and well-suited with local cultures.
    - Be capable of processing breadfruit into stable shelf life product of flour
    - Have low maintainability and failure rate as required by NGRF.
    - Be scalable from industrial level to household scale
    - Be modular so that it's applicable to all breadfruit growing areas with minimal changes.

    BRAINS created an Integrated Product Team (IPT) within the enterprise architecture organization that will develop, design, and implement the breadfruit processing system. This IPT organization is described in program planning volume of the proposal.

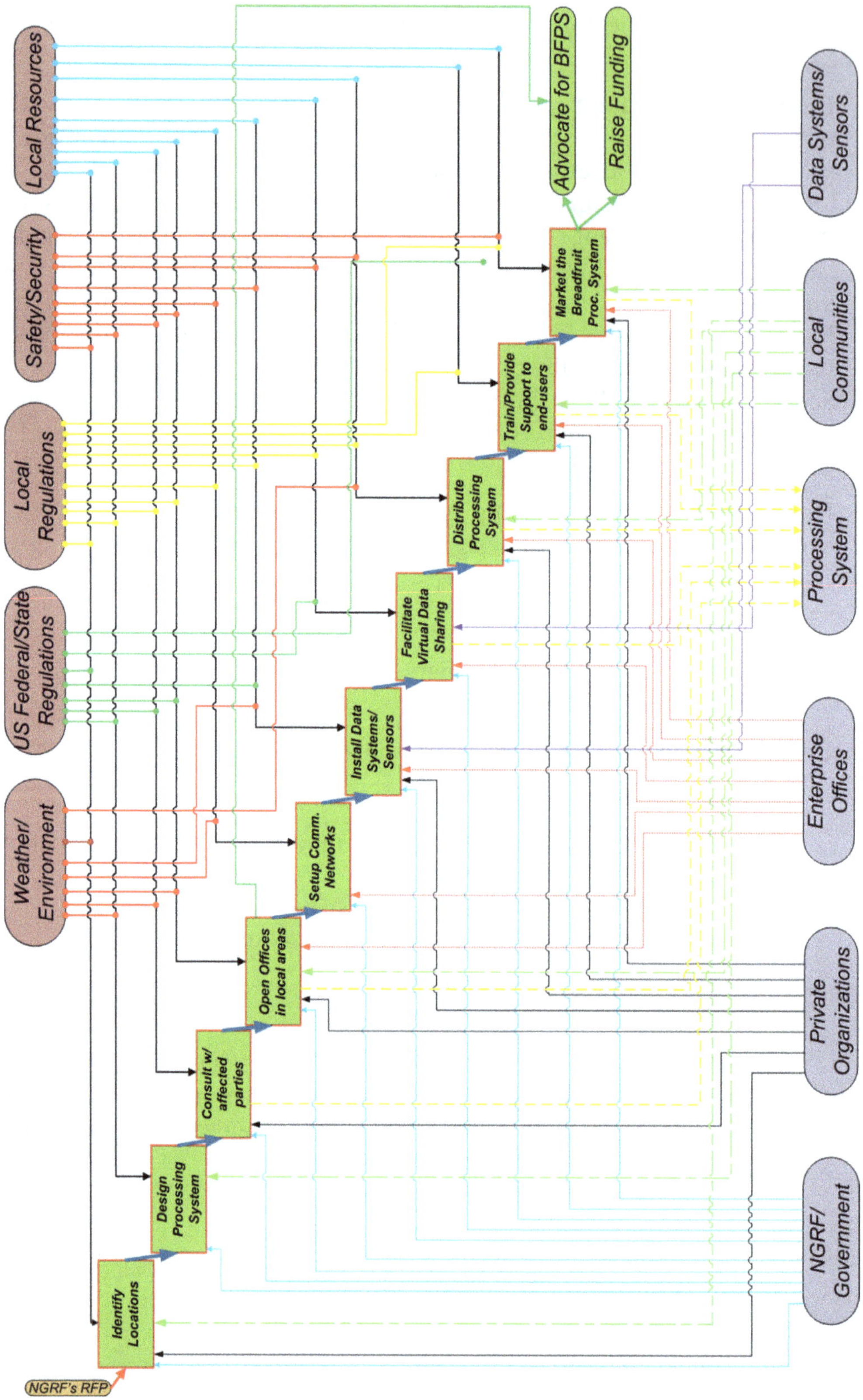

Figure 178: Enterprise Drill Down Operational View

- **Consult with affected parties**: While reviewing all the external factors, BRAINS will constant discuss about its operational activities with all the stakeholders such as NGRF, government institutions, local communities, as well as the Breadfruit Institute in Hawaii. BRAINS will also consult with private and non-governmental organization that may have stake in the project such local labor unions, consumers, distributors, environmental/wild life groups, transportation (ground/overseas) companies, , IT providers, Legal Department/FDA, Machinery manufacturers, etc.
- **Open local offices**: After a location is identified and all stakeholders are consulted, BRAINS will proceed to form local administrative offices that will:
    - Advocate, promote and market the breadfruit processing system locally.
    - Bridge the gap between NGRF headquarters in St. Paul, MN and its local operations.
    - Regionally administer and execute all other daily enterprise activities.
    - Be embassy for NGRF diplomatic regional relations.
    - Maintain and manage enterprise properties.
    - Locally collect and provide real time data to NGRF Central Data Collection and Analysis Center in St. Paul, MN.
    - Solve any regional labor, supplier, and/or sub-contractor issues.

  The enterprise local office organization will consist of the Vice-President and General Manager and his/her operational, functional, technical, financial, and legal subordinates. The Vice-President and General Manager will report directly to head of the enterprise architecture in St. Paul.
- **Set up internal networks and data systems/sensors**: The enterprise will install all the necessary amenities and resources that are required for effective operations of the local offices. Internal information sharing/storage networks will be created such as SharePoint, and Shared Drives. These networks and all other email and application servers will be hosted on the Breadfruit Enterprise Architecture Information network (BEAI) located in Minneapolis, MN. These networks will enable an easy and automated virtual data sharing and communication between local offices, NGRF, and other stakeholders. The enterprise will also install the Wireless Weather Station sensor (TE821W) in the breadfruit processing field to collect non-satellite weather data. In addition, to monitor the condition and health of the breadfruit processing equipments.The data from these sensors will be continuously sent to the NGRF Central Data Collection and Analysis Center in St. Paul, MN for analysis.

Other activities of the Breadfruit Enterprise Architecture include facilitating virtual data sharing of all the stakeholders and distributing the processing systems to the various local field operations. The enterprise will also train and provide support to all the enterprise workforce, clients, and end-users of the system. In addition, while observing all the external factors and utilizing its people and elements, the enterprise will promote and market the breadfruit products in both local and global communities.

2.1.1.1.2.Breadfruit Processing Systems Configuration

The Breadfruit processing system is a sequence of organized stages that will guarantee the transformation of breadfruit into flour. To successfully achieve this result, the process itself is first introduced as a black box (main orange box labeled as "Process Breadfruit" on Figure 179). Each yellow balloon on the left, right, bottom, and top of the box, are the main triggers, expected outcomes, materials, and external factors respectively. They represent the people, elements, or any other component that will be needed throughout the processing system.

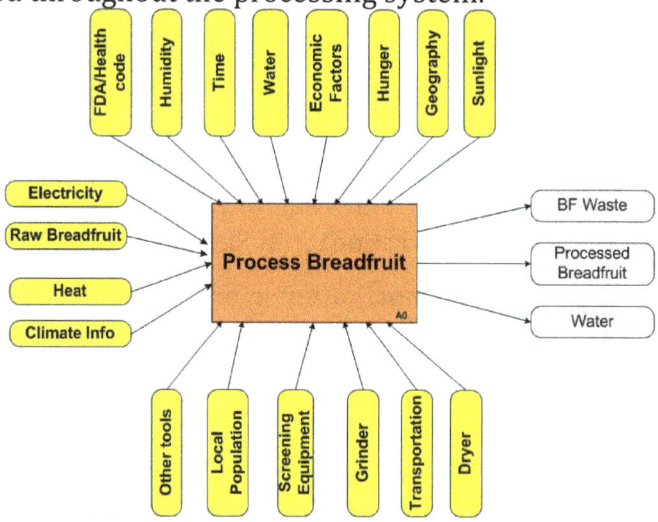

Figure 179: OV-5 Breadfruit Processing System

As mentioned, the processing itself has multiple stages: cleaning, peeling, slicing, drying, and grinding. Figure 180 shows the OV-5 drill down. Each step will be explained in the following sections.

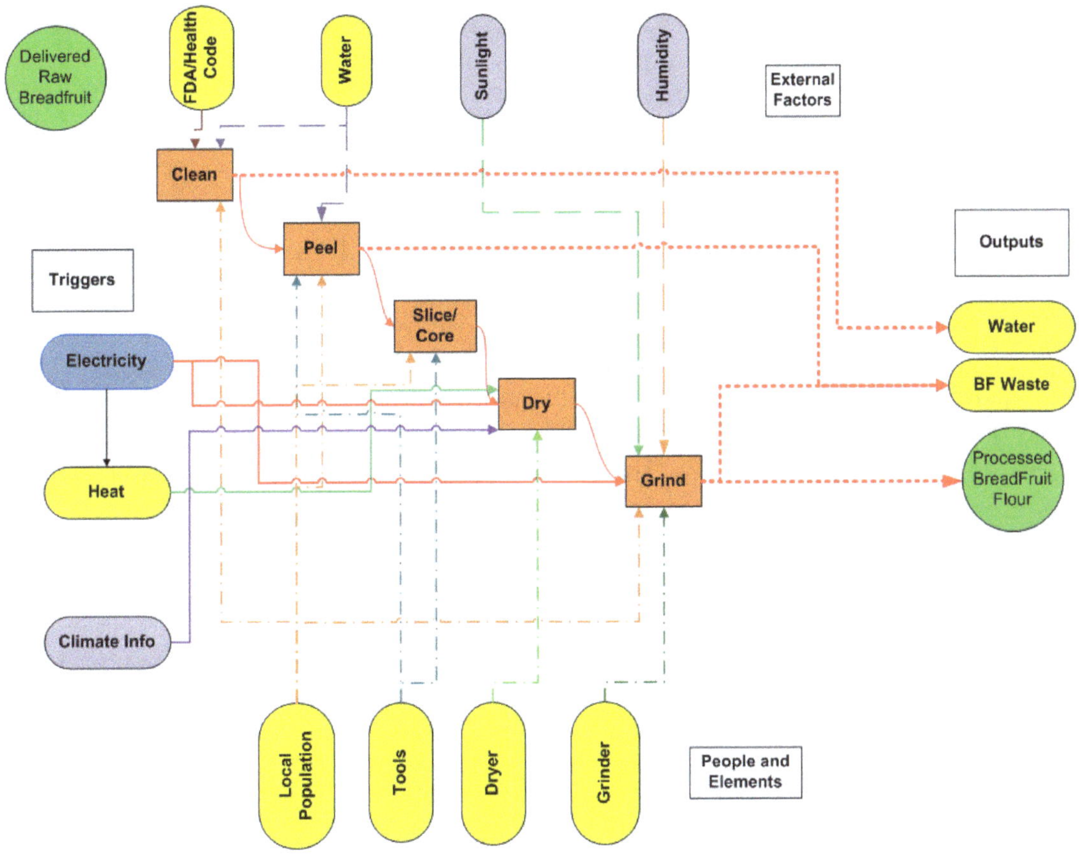

Figure 180: OV-5 Drill down Breadfruit Processing System

2.1.1.1.2.1 Collection of Breadfruit

BRAINS identified these three types of breadfruit at the Marshall Islands (MRI).

Artocarpus Altilis	Artocarpus Mariannensis	Artocarpus Camansi

Figure 181: Breadfruit types in MRI

More than 20 names are known for different cultivated varieties in the Marshall Islands (see Figure 6). These include: petaaktak, bukdoļ, mākinono, mejwaan, mejenwe, and metete. Petaaktak and Bukdoļ are seedless. Mejwaan is a seeded breadfruit variety with five lobed leaves. Metete is also seeded, with narrower, more sharply pointed leaves. The smooth-skinned cultivars normally bear fruit earlier, taste sweeter, and have firmer flesh. The rough-skinned types bear fruit later, may taste bitter, have looser flesh, and occasionally produce seeds. At the RMI local habitants find these different breadfruit sizes and weight:

- Maximum weight: 3 kg (6.5 lbs)
- Minimum weight: 1.2 kg (2.5 lbs)
- Large: 2.0 to 3.0 kg (4.4 to 6.5 lbs)
- Small: 1.2 to 2.0 kg (2.5 to 4.4 lbs)

It is worth emphasizing the fact that the proximity of the processing center to the source of supply for fresh raw materials presents major advantages; some are as follows:

- Possibility to pick at the best suitable moment.
- Reduction of losses by handling/transportation.
- Minimizes raw material transport costs.
- Possibility to use simpler/cheaper receptacles for raw material transport.

The major causes of breadfruit deterioration include the following:

- Growth and activity of micro-organisms.
- Activities of the natural food enzymes.
- Insects, parasites and rodents.
- Temperature, both heat and cold.
- Moisture and dryness.
- Air and in particular oxygen.
- Light.
- Time

BRAINS identified six strategic collection points around the Marshall Islands to collect breadfruit as shown on Figure 1817. Research indicates that most of the local farmers are not on the main islands. Farmers and their families are on the remaining islands, where they harvest breadfruit on daily basis as well as coconuts, bananas and fresh fish to directly prepare and consume. These collection points (orange stars in Figure 181) where local tribal chiefs will exchange collected breadfruit for a small incentive or for flour as agreed. Given the long distances between the islands BRAINS propose to have multiple collection points that will considerable reduce the distance to be travelled by boats. BRAINS also preferred not to use the main to cities as centralized collection points because we assume the breadfruit trees are mostly located on remote islands (or even desolated).

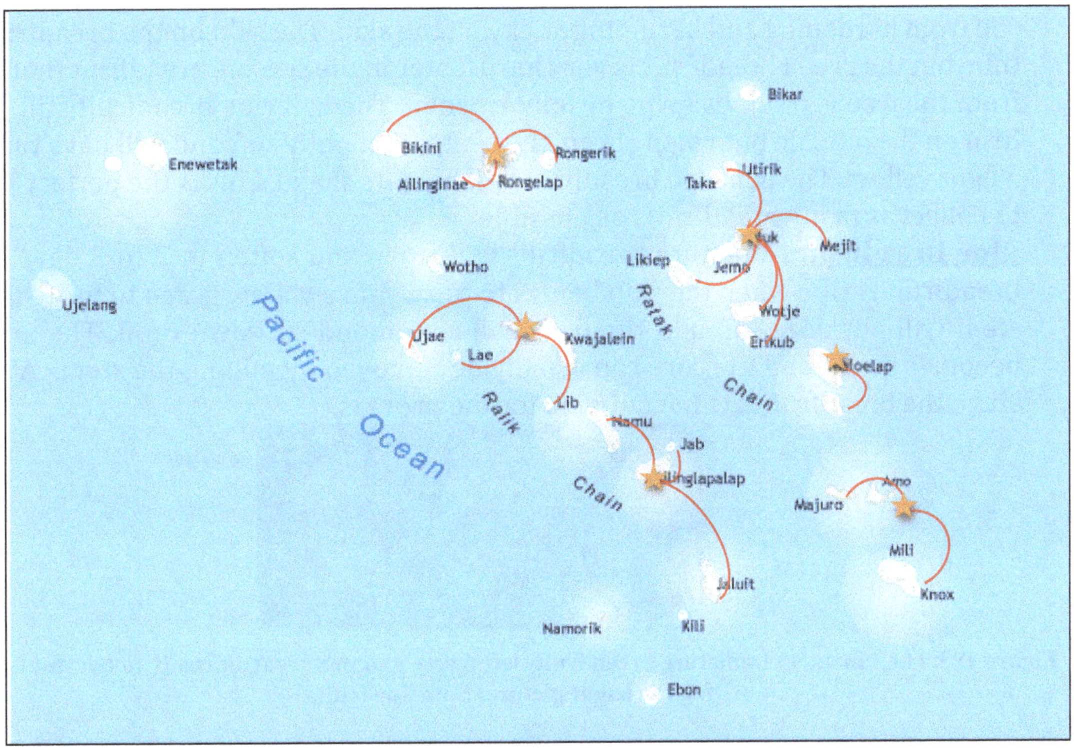

Figure 182: Breadfruit collection points

Fruit reception at the collection points is performed mainly for following purposes:

- checking of sanitary and freshness status
- control of varieties and fruit wholeness
- evaluation maturity degree
- collection of data about quantities received in connection to the source of supply: outside growers/farmers, own farm (large scale)

BRAINS identified the most important breadfruit physical characteristics to easily determine the maturity level of the breadfruit:

Immature Breadfruit: The skin of the immature breadfruit is very bumpy and the fruit is bright green. When picked, the immature breadfruit drips with white milky sap. To avoid getting the sticky latex stuck on the skin, immediately drop the breadfruit into a container of water and allow soaking for 10 to 15 minutes. Be sure to rinse the container when the breadfruit is removed to prevent the sap from sticking on the bottom. Remove any of the white sap from the skin as soon as possible or it will stick to the skin even after it is cooked. Pick an immature breadfruit when it is about 3 to 4 inches in diameter, tennis ball to base ball sized. This is not the best stage for collection.

Mature Breadfruit: The breadfruit is mature when the skin yellows and the milky sap begin to ooze though the skin. The bottom of the breadfruit may begin to brown but the stem will have white sap when picked. The mature breadfruit is hard all the way to the core. Soak the breadfruit as soon as it is picked to keep the sap from hardening and becoming sticky on the skin. The skin on the breadfruit is thin, but the green breadfruit is very hard. Later in the season, breadfruit that falls from the tree is often half ripe or fully ripened. The half ripe breadfruit still feels firm on the outside but when sliced, the center is slightly soft and will have turned a light yellow. The half ripe breadfruit will hold its shape. This is the perfect stage to collect as many breadfruits as possible.

Ripe Breadfruit: A mature breadfruit will ripen and soften in 2 to 3 days. The breadfruit is ripe when the skin beings to turn from a yellow green to brown. The stem will not ooze sap and the fruit will soften and sit down on it. The aroma becomes sweet and the core can usually be removed by pulling the stem. At this stage the breadfruit it is not suitable for the process.

Figure 183: (Left picture) Immature green fruit (left fruit), and mature fruit ready to harvest (right fruit). (Right picture) Soft ripe fruit.

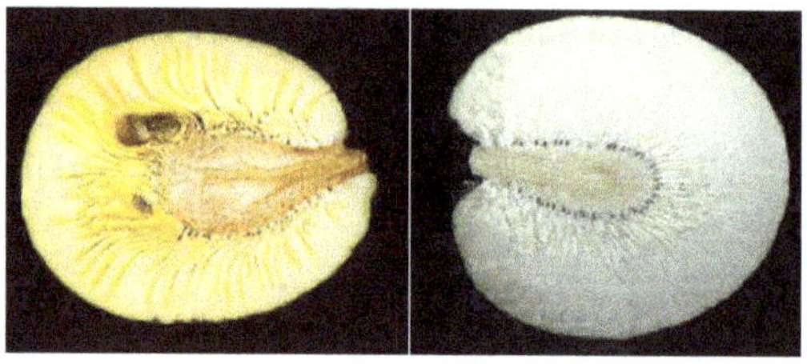

Figure 184: Mature breadfruit – Yellow flesh (left), and mature breadfruit- White flesh.

Regarding the task to be done at the collection points, BRAINS identified that good quality breadfruits are mature-green, firm, with intact stem, and free from defects (such as blemishes, sunscald, cracking, bruising, and insect damage) and decay. Uniformity of shape, size, and weight is also important as quality factors.

Some specific desirable characteristics to look for at the ***collection points*** are:

- No wounds from harvesting or handling, including punctures, scratches or bruises
- No insect damage
- Stem green and intact
- No yellowing of the peel or stem

Fruit reception at the collection points is performed mainly for following purposes:

- Checking of sanitary and freshness status
- Control of varieties and fruit wholeness
- Evaluation maturity degree
- Collection of data about quantities received in connection to the source of supply: outside growers/farmers, own farm (large scale).

Variety control is needed in order to identify that the fruit belongs to an accepted variety as not all are suitable for different technological processes.
Fruit maturity degree is significant as industrial maturity is required for some processing/preservation methods while for others there is the need for an edible maturity when the fruit has full taste and flavor. Some useful checks/control to be performed at reception are summarized in Table 1.

Checks at each delivery/raw material lot	Color Texture Taste Flavor Appearance Refractometric extract Number per kg Variety Sanitary evaluation
Checks at each ten lots (for the same raw material)	Density Water content: oven method Total sugars, reducing sugars Total acidity
Audits - every six months - on five different lots	Ascorbic acid Mineral substances Tannic substances Pectic substances

Table 1: Raw Material Control - Fresh breadfruit at Reception

Finally, the collection of breadfruit has to be as short as possible in order to avoid flavor losses, texture modification, weight losses and other deterioration that can take place over this period.
Some basic rules for this step are as follows:

- Keep products in the shade, without any possible direct contact with sunlight
- Avoid dust as much as possible
- Avoid excessive heat
- Avoid any possible contamination
- Store in a place protected from possible attack by rodents, insects, etc.

Cold storage is always highly preferred to ambient temperature. For this reason a very good manufacturing practice is to use a cool room for each processing centre; this is very useful for small and medium processing units as well. If not available, we recommend to keep the breadfruit inside a room, under the roof or outside under a tree avoiding direct sunlight.

2.1.1.1.2.2 Breadfruit Processing: Cleaning Phase

Harvested fruits are washed to remove soil, micro-organisms and pesticide residues. Fruit washing is a mandatory processing step; it would be wise to eliminate spoiled fruit before washing in order to avoid the pollution of washing tools and/or equipment and the contamination of fruit during washing. Washing efficiency can be gauged by the total number of micro-organisms present on fruit surface before and after washing - best result are when there is a six fold reduction. The water from the final wash should be free from moulds and yeast; a small quantity of bacteria is acceptable. Washing will be done before the fruit is cut in order to avoid losing high nutritive value soluble substances (vitamins, minerals, sugars, etc.).

2.1.1.1.2.3 Breadfruit Processing: Peeling Phase

Peeling is the first step in the process. BRAINS recommended a manual process for this step due to reliability, availability, and capacity needed. A manual process is reliable since it's independent of any other sources such as electricity or system breakdown. The tool that needs to do the task could be found almost everywhere, mainly in the kitchen. The skill and training requires for the job is minimum. Personnel could learn the process or pick up the procedure by watching or even discussing. Since, the company output is small; a manual process is the best approach as money needs to be used in more complex areas of the process. The knife shown in Figure 185 will be used for small scale.

Figure 185: Knife for manual breadfruit peeling

Advantages: (as compared to automatic process)

1) It's faster and low cost
2) A lot more reliable
3) It's easy to find (availability)

Specifications

1. **Input(s) :**
    a. **People:** This method is intended mainly for family use. A person can peel 15 breadfruits in an hour and no training is needed.
    b. **Material:** Kitchen knife as shown in Figure 185.
    c. **Breadfruit:** Raw good quality breadfruit
    d. **Time:** The average time for using the knife is approximately
    4 minutes/breadfruit = 31.5kg/hour, = 252kg/day
    e. **Weather conditions:** Peeling can be done either inside/outside household. However, it's preferably done inside a household during a rainy day for sanitary purposes.
    f. **Location:** When not in use, the knife can be used for other purposes.

g. **Assumption:** Good quality breadfruit

2. **Output(s):**
   a. **Waste:** The peelings can be used to feed livestocks.
   b. **Breadfruit:** The peeled breadfruit

3. **Procedure:**
   1) Pick up the breadfruit and using one hand to hold the breadfruit
   2) With the other hand, slowly slide the breadfruit with the knife away from your body to peel up the skin.
   3) Rotate the breadfruit to an unpeel area, and frequently rinse with water. See Figure 186
   4) Continue step 2 until the breadfruit is peeled.
   5) Pass it to the next step.

Figure 186: Latex substance released when the peel is removed

   6) General recommendations: To make it easier to peel, cut a thin slice off each end slice down the side of the fruit carefully, curving the slice toward the bottom. Turn the breadfruit after each slice of skin is removed. Continue around the breadfruit until peeled. The breadfruit flesh turns brown as soon as it is cut. Have a container of water handy to rinse the breadfruit as it is cut. Soak any pieces that will not use immediately in water. This will also reduce the sticky sap the may build up on the knife. Once the pieces have been soaked for a few minutes, they will no longer turn brown when exposed to air and can be stored in a sealed in a plastic bag and refrigerated until needed. The mature breadfruit will become half ripe in 1 to 2 days.

4. **General considerations:**
   a. Cautious: The user should remind them self to be more cautious when using the knife.
   b. Accident: A person can accidently drop the breadfruit and destroy it.

5. **Maintenance**
   a. Daily**:** A knife should be clean and sharp on a daily base or as needed.
   b. If some damage the knife, there shall be spare knife that can be used as backup.

For the large scale peeling, BRAINS proposes using the peeler knife shown on Figure 187. The difference between the large scale and small scale is the large scale worker has the option to choose the peeler knife or the knife. The advantages are very similar with only few differences.

Figure 187: Large Scale Peeling Knife

Advantages: (as compared to automatic process)

1) It's faster and low cost
2) A lot more reliable
3) It's easy to find (availability)
4) More choices (peeler knife or knife)

Specifications

1. **Input(s) :**
   a. **People:** 1 person per factory per day since a person can peel 4 minutes/breadfruit and each breadfruit weighs 2.1kg. Therefore for an 8 hours shift, a person can peel 252kg/day. Our goal per factory is 240kg/day. Again, since peeling is manual process, no training is needed. Might be helpful if there is a tutorial or a demo.
   b. **Material:** a kitchen knife or the peeler knife of Figure 187. It depends on what the worker is comfortable with.
   c. **Breadfruit:** Raw good quality breadfruit
   d. **Time:** The average time for using the knife or the peeler is approximately

      4 minutes/breadfruit = 31.5kg/hour, = 252kg/day.
   e. **Weather Condition:** Since this will be a factory size process, we want to use the lean manufacture. Therefore, production will be in the factory as rainy, windy, and cloudy days should not affect our performance.
   f. **Location:** When not in use, the knife and peeler knife can be used for other purposes.
   g. **Assumption:** Good quality breadfruit
2. **Output(s):**
   a. **Waste:** The peelings can be used to feed livestocks or burn as a form of energy to feed into other processes.
   b. **Breadfruit:** The peeled breadfruit
3. **Procedure:**
   1) Pick up the breadfruit and using one hand to hold the breadfruit
   2) With the other hand, slowly slide the breadfruit with the knife away from your body to peel up the skin.
   3) Rotate the breadfruit to an unpeel area.
   4) Continue step 2 until the breadfruit is peeled.
   5) Pass it to the next step.
   6) General recommendations while peeling: To make it easier to peel, cut a thin slice off each end slice down the side of the fruit carefully, curving the slice toward the bottom. Turn the breadfruit after each slice of skin

is removed. Continue around the breadfruit until peeled. The breadfruit flesh turns brown as soon as it is cut. Have a container of water handy to rinse the breadfruit as it is cut. Soak any pieces that will not use immediately in water. This will also reduce the sticky sap the may build up on the knife. Once the pieces have been soaked for a few minutes, they will no longer turn brown when exposed to air and can be stored in a sealed in a plastic bag and refrigerated until needed.

4. **General considerations:**
   a. **Cautious:** There should be alert signs around the factory for the large scale. The sign should remind the worker to be more cautious when using the knife or the peeler knife.
   b. **Accident:** A worker can accidently drop the breadfruit and destroy it.
5. **Maintenance**
   a. Daily: A knife should be clean and sharp on a daily base or as needed. If some damage the knife or peeler knife, there shall be spare that can be used as backup.

2.1.1.1.2.4 Breadfruit Processing: Slicing Phase

The breadfruit slicing process will be an extension of the peeling process, in that it will be done manually. The same people who peel the breadfruit will perform the slicing operation. After washing the breadfruit once again to remove any fragments of the peel, the breadfruit can be sliced using a traditional kitchen knife. The first cut should expose the core of the breadfruit, so that any seeds and extra material can be removed. This part of the process can be performed with any kind of spoon or shell. Subsequently, the slicing should produce small chunks or thin slices of breadfruit, as this will expedite the drying process.

Specifications:

1. **Input(s)**
   a. **People:** The speed of this process is limited by the bottleneck in the dryer, and so one person should be able to slice 15kg of breadfruit in an hour, which amounts to seven whole breadfruits. At this stage, the breadfruit is still saturated with water.
   b. **Material:**
      a. A kitchen knife or a peeler knife from previous phase). It depends on what the worker is comfortable with.
      b. Water to wash the breadfruit while peeling it
   c. **Breadfruit:** Peeled raw good quality breadfruit
   d. **Time:** Each breadfruit should take approximately 8 minutes to peel and slice.
   e. **Weather Condition:** Since this will be a factory size process, we want to use the lean manufacture. Therefore, production will be in the factory as rainy, windy, and cloudy days should not affect our performance.
   f. **Location:** This part of the process will take place in the same location as the other steps, which would be inside a building.

g. **Assumption:**
    a. Good quality breadfruit
    b. The natives already know how to slice other vegetables and fruits, making this an easy extension of their skill.

2. **Output(s)**
    a. The waste products are the water from washing the fruit, and the waste seeds and material from the core of the fruit. The usable chunks or slices are ready to be placed into the dryer, where the water content will evaporate.
3. **Procedure**
    1) Obtain peeled breadfruit.
    2) Take knife and cut breadfruit in half.
    3) Remove seeds and core material from breadfruit.
    4) Cut remaining breadfruit into 1/8" thick slices or chunks of 2" cube size.
    5) Place in dryer when finished
4. **General Considerations**
    a. The main point of safety is to position the breadfruit so that any cutting motion will not injure the hand.
    b. It is also prudent to prevent any breadfruit pieces from falling on the floor, as this would render them as waste product, and thus reduce the final output of breadfruit flour.
5. **Maintenance**
    a. The only maintenance needed is to maintain the sharpness of the blade that is used. Having replacement knives will be useful in case the blade breaks. Investing in a blade sharpener might be expensive.

2.1.1.1.2.5 Breadfruit Processing: Drying Phase

Drying is most widely used method to remove water from foods for two reasons:

a. to prevent microorganisms and hence preserve the food, and
b. To reduce the weight and bulk of food for cheaper transport/storage/grinding.

If correct drying method is used, the nutritional quality, color, flavor and texture of rehydrated foods are slightly less than fresh food but, for most people, this has only minor nutritional significance as dried foods form one component in the diet. However, with incorrectly applied method, there is a chance of greater loss in nutritional and eating qualities and more seriously, a risk of microbial spoilage and food poisoning. Therefore it becomes very important to choose a correct drying method according to the food type. 'Hot Air drying' is most commonly used method for food drying. For effective drying, air should be hot, dry and moving. The dryness of air is termed 'humidity' - the lower the humidity, the drier the air. If a new type of drier is to be introduced, or if a new type of food is to be dried, it is necessary to carry out some experiments to find the rate of drying. The results obtained can then be used to find the time that the food should spend in the drier before the moisture content reduces to the desired amount. To find the rate of drying, Food is weighed, placed in the drier for some definite time interval. It is then removed, reweighed and replaced. This is continued until the weight of the food no longer changes.

Generally, the drying process will follow the characteristics of Figure 188. It shows two distinct phases of drying - the 'constant rate drying' and 'falling rate drying'.

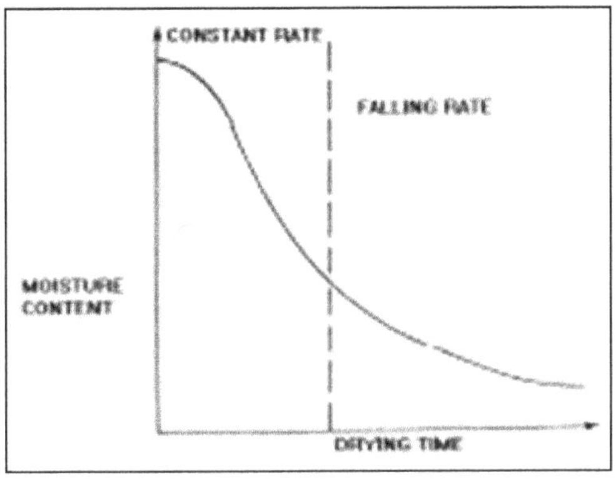

Figure 188: Drying of Food

In the constant rate, the surface of the food remains wet and there are chances of contamination by moulds and bacteria. In the falling rate the surface is dry and the risk of spoilage is much smaller. The food should therefore be dried to a weight that corresponds to the end of the constant rate period as quickly as possible

Drying Rate (weight/time) = (**Initial Weight - Final Weight**) / **Time interval**

As far as Breadfruit is concerned, our proposal focuses on Marshal Island, where sunlight is available with no scarcity at all, and obviously that was our first choice when we started exploring various possible methods for drying of Breadfruits. There are definitely some advanced techniques by using photo cells, big and wide solar panels etc. which can offer higher drying rates, but considering the lifestyle of people on Marshal Island as well as their education level, we dropped the idea of using method which requires operator to have some minimum technical knowledge while working on advanced technology equipments. Then we started searching for technique which will be easy to understand for any illiterate personnel regardless of scalability of process, whether for family use or for industrial use. Finally we selected two methods, a 'Solar Tent Dryer' for family use and 'Electric Tray Dryer' for industrial use.

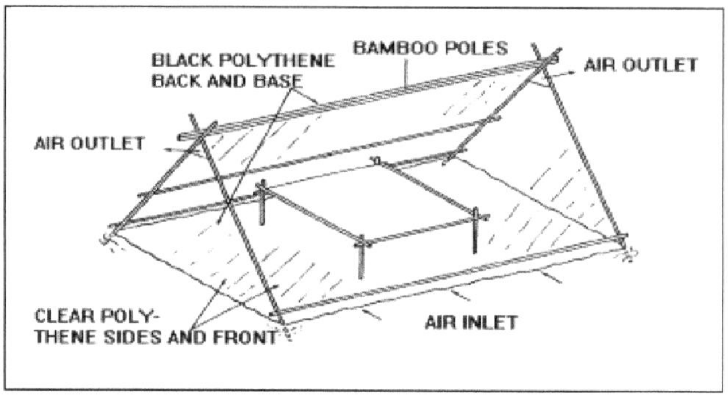

Figure 189: Solar Tent Dryer for small scale

As shown in Figure 189, an inexpensive and simple to build 'Solar tent Dryer'. It consists of a frame of wooden bamboo poles covered with plastic sheet. Black plastic should be used on the wall facing away from the sun. The transparent polythene shall be used for other two sides and that will be useful for the inspection. The food to be dried is placed on a rack above the ground. There is a small air gap at the bottom for air circulation. This dryer can accommodate maximum 10 Pounds of Breadfruit slices at a time

Advantages: (as compared to normal sun drying)

1. Tent dryers provide protection against rain, insects and dust
2. Faster Drying Rate (0.25 kg/hr)
3. Tent dryers can also be taken down and stored when not in use.

Specifications:

1. **Input(s) :**
   a. **People:** This method is intended mainly for family use. So once a dryer is built, even a single person can put the breadfruits for drying. If he could get help from other family members that would be plus.
   b. **Material:** (i) Wooden bamboo poles (ii) Polythene sheet (black, transparent); Length of black sheet = approx. (3 X length of bamboo poles), Length of Transparent sheet = approx. (2 X length of bamboo poles), Plywood board of length = length of bamboo poles, nylon rope
   c. **Breadfruit:** for faster drying, it is recommended to put very thin breadfruit slices. The fresh fruit shall be peeled, then sliced and then spread evenly throughout the dryer area inside tent
   d. **Time:** Since we have not carried out any experiments with breadfruit, we assumed that breadfruit slices will have 30% moisture remaining in it after peeling and slicing.
   Drying rate for solar tent dryer = 0.25 kg/hour
   10 Pounds (4.55kg) will have 1.36 kg moisture in it
   Drying time (hour) = Moisture content in Breadfruit (kg) / Drying Rate (kg/hour)
   = 1.36 (kg)/ 0.25 (kg/hour)
   = 5.45 Hours = **5 Hours 27 Minutes for 10 Pounds of Breadfruit**
   On a bright sunny day it's possible to dry 20 Pounds

e. **Weather conditions:** For effective drying, this method shall be used with bright sunlight only, if weather gets cloudy or it starts raining, then no need to rush out and bring the food in, the polythene sheets will protect the fruit slices from rain water.
f. **Location:** When not in use, this dryer shall be disassembled and kept inside some shade, just to avoid fading of black polythene sheets. It is recommended not to keep any weights over it. The material (bamboo poles, polythene sheets etc.) shall not be used anywhere else to avoid risk of food contamination.

2. **Output(s):**
   a. Waste: No major waste except some small breadfruit particles (something similar to bread crumbs which remain in toaster)
   b. Breadfruit: the dry slices (looking like potato chips) with almost 0% moisture
3. **Procedure:**
   1) Clean the polythene sheet with moist cloth
   2) Peel, cut and slice the Breadfruit.
   3) Spread the breadfruit slices evenly along the entire inside area of tent (on black polythene sheet)
   4) Pull tent in a position where black polythene sheet gets maximum sunlight
   5) Take Breadfruit slices out after 5.5-6 hours
4. **General considerations:** For longer life of Dryer
   a. Bring the tent out only when bright sunlight is available
   b. Avoid using this dryer if the wind is blowing very badly
   c. Always clean the black polythene spread sheet before use
5. **Maintenance**
   a. **Daily:** No daily maintenance is needed
   b. If some damage happens to polythene sheets, then patch up the damaged part
   c. Dismantle the complete assembly when it is not in use.
6. **Technology Support:** Practical Action, The Schumacher Centre for Technology and Development will provide the Technology. It's a charity and their aim is to bring life changing effects in lives of poor people across the world with simple technology solutions. Once the system is built, the maintenance or any adjustments if needed can be carried out locally.

For large scale, BRAINS proposes, the 1m cubicle structure with Electric Fan Heater (2kW) fitted at the bottom shown in Figure 190. The trays containing Breadfruit slices will be arranged offset to each other to ensure zigzag circulation of hot air. This dryer can accommodate 12 Pounds of Breadfruit slices at a time and has a drying rate of 5kg/hour.

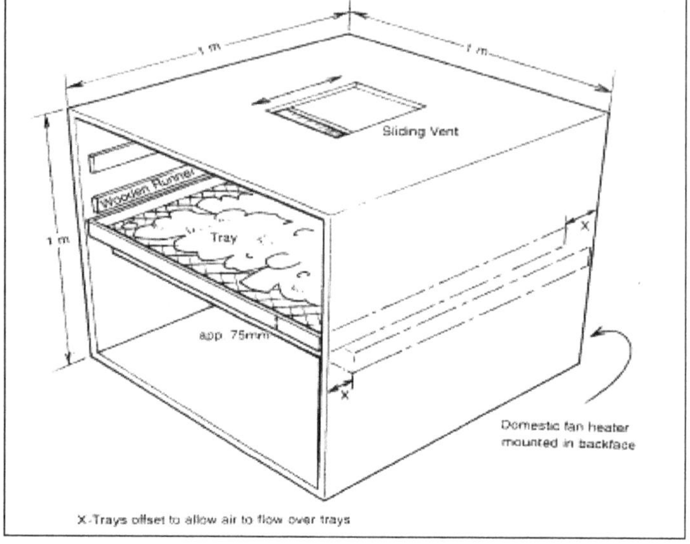

Figure 190: Electric Tray Dryer for large scale

Specifications:

1. **Input(s):**
   a. **People:** 1 person/dryer. All trays can be taken out at a time since they are placed in a frame which fits in a dryer. So once the operator keeps 12 Pounds for drying, he can refill the seconds set of trays with breadfruit slices till drying is in progress. So as soon as the 1st batch of trays removed from dryer, operator will have other batch ready to put in dryer.
   b. **Materials:** This is an Electric Dryer. It requires 2kW single phase Domestic Electric Fan Heater and Electric Power Supply feed with minimum voltage regulation.
   c. **Breadfruit:** for faster drying, it is recommended to put very thin breadfruit slices. The fresh fruit shall be peeled, then sliced and then spread evenly throughout the tray area.
   d. **Time:** As stated above, we assumed that breadfruit slices will have 30% moisture remaining in it after peeling and slicing.
      Drying rate for Electric Tray Dryer = 5 kg/hour
      12 Pounds (5.45kg) will have 1.63 kg moisture in it
      Drying time (hour) = Moisture content in Breadfruit (kg) / Drying Rate (kg/hour)
      = 1.63 (kg)/ 5 (kg/hour) = 0.32 Hours
      = approx.**20 Minutes for 12 Pounds of Breadfruit**
      i.e. 36 Pounds/hour so in a day it's possible to process 250-300 Pounds
   e. **Weather conditions:** Since the Dryer will be placed inside factory and Electrically operated, it is totally independent of outside weather conditions
   f. **Location:** The Dryer shall be placed near Electrical Power Socket. The layout of factory shall be planned in such a way that, the

breadfruit slices shall be transferred quickly in drying area after peeling and slicing.

2. **Output(s)**
   a. **Waste:** No major waste except some small breadfruit particles (something similar to bread crumbs which remain in toaster
   b. **Breadfruit:** the dry slices (looking like potato chips) with almost 0% moisture
3. **Procedure**
   1) Clean all the trays to remove previous day wastes. The trays are to be cleaned only once in a day, either at beginning or at ends of shift.
   2) Peel, cut and slice the Breadfruit.
   3) Spread the breadfruit slices evenly inside each tray
   4) Turn the Power ON
   5) Take Breadfruit slices out after 20 minutes
4. **General considerations:** For General Safety
   a. Never operate the dryer without safety gloves.
   b. Never touch Electric Heater when plugged in.
5. **Maintenance**
   a. **Daily:** Clean all the trays to remove previous day wastes. The trays are to be cleaned only once in a day, either at beginning or end of shift.
   b. If Heater is not working, just check for any loose Electrical connections, Check if Power Supply is working or not. If everything looks perfect then call the electrician.
6. **Technology Support:** Practical Action, The Schumacher Centre for Technology and Development will provide the Technology. It's a charity and their aim is to bring life changing effects in lives of poor people across the world with simple technology solutions. Once the system is built, the maintenance or any adjustments if needed can be carried out by local electrician.

2.1.1.1.2.6 Breadfruit Processing: Grinding Phase

The purpose of the grinder is to take dried breadfruit pieces and crush them to fine flour. Flour can be packaged and stored for longer periods of time and is the most versatile product breadfruit can make, allowing customers to make a variety of foods. The BRAINS architecture consists of six plants to handle the yearly harvest. Each plant will handle 1/6th the total breadfruit processing capacity of the Marshall Islands. The capacity is estimated at 252,000 kg/year. Using a conservative estimate of 100 working days, the grinder must process 240 kg/day. Using 8 hr working days there is a 30kg/hr rate required for the grinder to handle.

After researching many flour grinders, BRAINS is proposing two options to meet manufacture requirements. A large scale grinder would handle our factory processing – large scale operations plants. Then a popular and effective hand operated grinder for household operations. Through the combination of several processing facilities each, with a manufacture capacity of 240kg/day, and the ability for customers to have their own breadfruit grinder will allow flour production for individual families to well over a ton of flour production.

As an answer to the home grinding demand BRAINS has chosen to provide the CTI (Compatible Technologies International) Omega VI Grinder. It is a hand driven burr mill that will produce fine flour from a variety of foods including dried breadfruit. It has been designed specifically for use in developing countries to provide means for families to use and sell their home product. This small scale grinder is in use in Africa, Central America, the Caribbean, Asia, and North America. A picture of the grinder is below in Figure 191.

Specifications

1. **Input(s):** The input to the hand grinder is dried crushed breadfruit or other grains or nuts. The grinder is manually or electrically operated. For manual operation either a person to push the foot pedals or rotate the handle is required. For the motor either gas or electricity is required depending on the motor. There are many gas and electric motors available on the market that will suit this purpose.
2. **Output(s):** The grinder will provide about 15kg/hr of product for hand cranked, and 150kg/hr with a motor. It will output fine flour for dried breadfruit.
3. **Procedure:** The unit may be operated by hand crank, pedal crank, or a motor operating at six times the hand-crank rate (1.5 HP motor recommended).
4. **General Considerations:** The small scale grinder is very easy to use and self-explanatory. Cleaning can be accomplished by whipping down the unit.
5. **Maintenance:** There is little to break in this mechanical device and no maintenance is required.

Figure 191: Small Scale Home-Use Grinder

The breadfruit grinding phase requirement is to reduce dried breadfruit slices to fine powder (flour). For the large scale grinder, the purpose is to provide the factories with enough grinding speed and capacity to handle the amount of breadfruit that is harvested described above. The flour will be stored in drums for shipping to customers. BRAINS found the Pleasant Hill Meadow Steel Burr Grinder to be the most effective for grinding breadfruit into flower. The company is well established offering a host of grinders from large to home use for many different applications.

Specifications

1. **Input(s):** Many grinders are only suitable for grains; however this type of grinder can handle many types of foods including wet or oily foods. The grinder provides up to over 200kg/hr throughput rate which is very effective for its small size. The unit is compact and small requiring only a small mounted 5 HP motor. The motor can operate electricity of 1 phase 240V or 3 phase 480V, which required about 15.5 amps at 240V 1 phase. The grinder has a large bucket on top for the conveyor belts to drop the dried breadfruit pieces into. The grinder will accept anything that fits in down into the mill inlet. The mill inlet has dimension of 2.5" x 5". It will accept a large range of products from beans, grains, spices, coconut, and more. Once the breadfruit is dumped into the mill inlet there is an internal auger that forces the product into the burrs and grinds it. This allows the dried breadfruit pieces to be large and odd shaped which is expected from the peeling and slicing process.

   The grinder should have at least one person watching the grinder at all times. The conveyer belt will be manually dropping the product into the grinder but it is important for an operator to watch for any malfunctions, jams, or other problems where the shut off switch should be pressed. In addition, the grinder has an adjustment for the fineness of the ground breadfruit. The grinder is not meant to be stored or used outside. Since it will be processing food for consumption, it must be kept in a clean indoor environment free from dirt and other pollutants.
2. **Output(s):** The only output of the grinder is fine ground breadfruit flour. The burr mill grinds up the breadfruit as fine as desired and outputs through the chute on the side. The only other output would be heat from the motor which should be well ventilated and kept clear of any blockages or other equipment.
3. **Procedure:** Operation the grinder is very simple. Once installed - mounted securely on a strong surface capable of handling vibration and electrical power is connected - simply press the On button and start dropping food into the hopper. Monitor flour output granularity and if desired, turn adjustment knob to change grinding fineness. To turn off press the Off button.
4. **General Considerations:** Hands and non-food product should be kept out of the hopper while on as it may cause injury or damage the unit. Monitor the hopper in case of jams the grinder should be powered down, the jam removed, and continue operation.
5. **Maintenance:** The grinder does not require maintenance of the mill except cleaning and the occasional replacement of the mill burrs, which takes only 10 minutes. The mill body is high strength steel with a non-toxic finish for long life and low corrosion. The hopper and grinder should be cleaned daily to prevent breadfruit in the grinder from developing mold and contaminating new processed breadfruit. Figure 192 shows a picture of the grinder.

Figure 192: Large Scale Breadfruit Grinder

2.1.1.1.2.7 Breadfruit Processing: Factory Layout

The factory is designed to create a streamlined process that will produce breadfruit flour with ease. The input to the factory is the collected breadfruit, clean water, and electricity by virtue of the wind turbines. The breadfruit would be handled manually by two individuals each processing at 15 kilograms per hour. The output of peeled and cored breadfruit chunks would be sent to the pair of dryers that also process breadfruit at the same rate. There will be one grinder per factory that can process in excess of 30 kilograms of dried breadfruit per hour. The resulting breadfruit flour will be stored in a drum that can hold an entire day's output of flour. Aside from the flour, the used water from washing the breadfruit in the initial stage will be the outputs of the process.

2.1.1.1.2.8 Breadfruit Processing: Outputs

The production process outputs breadfruit flour suitable for cookies. The breadfruit was processed into flour and used to supplement wheat flour in different percentages or completely. Different types of flour are suited to different items and all flours are different therefore we suggest not to switch from one type to another since it could easily ruin the recipe. To achieve success in baking, it is important to know what the right flour is for the job. Wheat cake flour is already being used for baking cookies and cakes thus we identified specific characteristics relevant to the production of wheat flour so that it can benefit the breadfruit flour production.

The cake flour is milled from a variety of soft wheat, which is lighter, contains lower levels of gluten, and results in baked items that are tender and less dense. Breadfruit is gluten-free, thus the breadfruit flour should have similar characteristics to the wheat cake flour, regarding gluten levels. Flour that is bleached naturally as it ages is labeled "unbleached," while chemically treated flour is labeled "bleached." Bleached flour has less protein than unbleached. Bleached is best for pie crusts, cookies, quick breads, pancakes and waffles. The advantage of using bleached flour rather than unbleached is that the chemicals used for bleaching (usually chlorine, which evaporates after it is added to the flour) act as a preservative so that the flour will not develop an off flavor or spoil after a short period. The chemicals also prevent dough from becoming discolored and provide more consistent results when baking, however the chemicals affect the gluten strength of the flour, therefore bread makers often prefer unbleached flour.

Flour must be kept cool and dry. All flours, even white flour, have a limited shelf life. Millers recommend that flours be stored for no more than 6 months. The main change that occurs is the oxidation of oils when flour is exposed to air and natural oils cause this flour to turn rancid quickly at room temperature. Flour should be stored, covered, in a cook and dry area. This prevents the flour from absorbing moisture and odors and from attracting insects and rodents. It is better not to mix new flour with old if you are not using the flour regularly. Do not store flour near soap powder, onions or other foods and products with strong odors. If a refrigerator is available store the flour in it during hot weather. If freezer space is available, flour can be repackaged in airtight, moisture-proof containers, labeled and placed in the freezer at 0 degrees F. If flour is stored like this, it will keep well for several years. Keep flour in the refrigerator the year around.

Breadfruit Waste and Water: Breadfruit waste is classified as the peelings and the core of the breadfruit, the dirty water used while cleaning and peeling, and any not useful piece of machine that needed to be replaced. BRAINS identified multiple destinations for each; the peelings and the core of the breadfruit can either be placed on a plastic bag and be disposed, or if no garbage collection system is available BRAINS suggests to local workers collecting all the peelings and cores and burn it. Water with waste should be disposed as well since it is no longer suitable for human consumption. This liquid can be disposed back to the sand since it does not contain any pieces of breadfruit in it. BRAINS strongly recommend not disposing this water back into the Ocean. Finally any piece of any machine that was replaced can be either keep as a backup if it is in good conditions, or can be disposed to be collected by the garbage collection system, or as another option it can be sell.

2.1.1.1.2.9 Renewable Energy Power

Most of the communities in third world countries do not have access to basic electricity. This is especially true in rural areas where breadfruit usually grows. After extensive study of the needs of local communities where we plan to implement our breadfruit processing system (BPS), BRAINS is proposing a renewable energy system that solves the essential problems of these communities. Since the global initiative of NGRF is to reach out all the communities in breadfruit growing areas with or without power.

Our solution is based on the Poor Man's wind turbine by Richard Lewis. As its name suggest, this wind turbine is designed essentially for the poor people that do not have the means to purchase expensive parts or the skills to build complex wind turbine. At the same time, it can generate sufficient and scalable power for household use and small industrials as will be shown in the following sections.

With the concept of Poor Man's wind turbine, BRAINS would like to accomplish two main objectives that directly complement NGRF's mission of eliminating hunger in breadfruit growing communities.

1. To create cheap and dependable green source of power that will help us implement reliable and efficient breadfruit processing system that will produce high levels of breadfruit products in the local communities.
2. To use the wind turbine as an educational tool to teach third world communities about renewable energy and how they can use it for variety of applications that will improve their standard of living.

Thus, this wind turbine concept does not only address the immediate problem of breadfruit processing but it has wider implications for the development of the communities. BRAINS will empower the local people by showing them basic techniques of building wind turbines using available local resources. Once the locals are able to produce their own energy without depending on outside help, BRAINS believes they will be able to use it for other applications such as basic lighting, pumping water, etc.

As mentioned, the poor man's wind turbine is designed knowing the resources that are available to these communities. The most attractive feature is that it can be mostly constructed from scrap parts that might be readily available to the users. This is very inexpensive way to generate power.

BRAINS plans to build four of these wind turbines at each of the breadfruit processing site. Each wind turbine will be designed to generate 3kW of direct power at all times. This provides up to 12kW of power. This would be enough power for the processing systems (dryer and grinder) and basic lighting. Any time more power is needed, additional wind turbine can be added. For detailed analysis and parts procurement of the proposed wind turbine, see Annex A.

2.1.1.1.2.9.1 Back-up Power systems

BRAINS proposed a back-up battery system that will be used in the case of wind turbine loss. These batteries will be charged through the wind turbine when it's not in use (preferable non-work hours). In addition to the batteries, each breadfruit processing site will have a back-up generator for use in rare situations of both wind turbine and battery power loss. Complete details of these systems are given in Annex A.

2.1.1.1.2.10 Estimated Production Analysis

Study of Daily Quantities Production of Breadfruit at Marshall Islands

- Facts about the Breadfruit:

- Breadfruit tree height 26 m (85ft)
- Single tree produces 50 to 150 fruits per year, but it is estimated 25 fruits per tree are collected
- Main season May to July or September
- Minimum weight 1.2 kg (2.5 lbs) , and maximum weight 3 kg (6.5 lbs) , average of 2.1 kg (4.63 lbs)
- Season from May - September (*5 months total: May, June, July, August, and September*)
- Total of 24 islands in the RMI

-- Assumptions:

- We assumed an average number of *50 breadfruit trees* on any island
- All these trees belong to the same family's land (lease)
- Given expected high demand of breadfruit, *100 fruits per tree* instead of 25
- *Six (6) breadfruit processing plants*
- One processing plant every four islands

-- Summary:

- 50 trees x 24 islands x 100 breadfruits/tree = 120,000 breadfruits per season *(to collect on 150 days)*
- 120,000 x 2.1 Kg = 252,000 kg breadfruit/season (555,564 lbs) approx
- Daily processing: approx. 240 kg breadfruit/daily per processing plants (529,109 lbs)
- *Total daily for all processing plants: approx. 1680 kg (3,703 lbs)*

Through scalability, the scope of this project aims to nourish as many of the local population within the Marshall Islands as possible (on July 2008 the population has grown up to 89,000 people approximately). We estimate a large and considerable nearby communities and villages will be directly benefited from this production.

Study of Estimated Nourished Population

- Facts about breadfruit process into flour:

-Initially the fruit contains 75% water, 25% solids

- After drying, the weight of solids has not changed (only water is removed) and the moisture content is reduced to 8%

- Assumptions:

-Daily production ***1680 kg*** (3,703 lbs) at RMI

- ***1260 Kg of water***
- ***420 kg of solids***

- After drying, the solids content is (100 – 8) = 92 %

- Summary:

- If 420 Kg solids is 92%, then the total weight of product = 100 / 92 * 420 = ***456,52 Kg (1006 lbs) approx. in one day***

-Therefore 1680 Kg – 456,52 Kg = ***1223,48 Kg (lbs) approx. of water is removed!!!!***

- What will locals be able to prepare with this amount of flour ?

- 456 Kg equals to 3356 cups of flour!
- We can daily produce 6716 pancakes!
- .Or we can daily produce 1342 cakes!

2.1.1.4. External Systems

In this section, we describe the external systems that will be part of the Breadfruit Enterprise Architecture Demonstration (BrEAD).

2.1.1.4.1.Enterprise Communication and Data Networks

As required by NGRF, BRAINS has to provide an automated system to enable 24/7 real time data sharing among all the stakeholders of the Breadfruit Enterprise Architecture Demonstration (BrEAD). Some of the stakeholders are:

- The researchers of BrEAD and Breadfruit Institute in Hawaii and elsewhere.
- Private organizations - both profit and non-profit that have interests in BrEAD such as Compatible Technology International (CTI).
- The world wide end-users
- The main community of Interest (COI), NGRF and applicable governments.

Some of the data that will be shared among the stakeholders are weather/environmental information of the breadfruit processing sites received from real time satellite weather feed (Blue Skies Unlimited Weather Views Satellite) and the site's wireless weather station; the stakeholders will also share the breadfruit marketing information that will be gathered by the enterprise architecture. In addition other data will be collected and shared at processing plants such the number of breadfruits processed each day, end of day inventory, daily amount of shipped product etc.

After studying and making the necessary trade analysis for different options, BRAINS has determined the best way to implement this automated system is to outsource to an experienced IT service provider. BRAINS has concluded this is the best option both in terms of cost and efficiency.

BRAINS sent a bid to number of IT providers, and selected Techfluent to be the IT services provider for BrEAD. Some of the reasons why BRAINS has chosen Techfluent over other competitors are its location here in the twin cities, its unparalleled experience in IT services, and the best service it provides with an unbeatable price. Through this automated system, Techfluent will host the application/help desk services and provide secure and reliable network that will be at peak performance at all times. The specific services will be described in detail in the following sections.

Figure 193 shows a pictorial view of how the data will be collected and shared among all the stakeholders. As depicted on the figure, the center of the system is Techfluent and there is two way communications among all the stakeholders. This is a state of the art system that will allow the BrEAD stakeholders to share data easily, promptly and efficiently.

2.1.1.4.1.1. Local Offices Communication Networks

The local offices will only need internet connection to share data and have access to the system. For the Marshall Islands where the BrEAD will be delivered first, the local office will be located in Majuro, the main commercial city, and will subscribe to high speed internet through the Marshall Islands National Telecommunications Authority (NTAMAR) which is the sole internet provide of the country. The NTAMAR provides high speed internet connections of up to 512 kilo bit per second which is relatively fast compared to other third world countries. The email services and SharePoint sites of the BrEAD local offices employees will be part of Techfluent services. The NTAMAR will also provide telephone (local and long distance), mobile (local and long distance), and facsimile services to the BrEAD local offices. As advertized on their website [8], the NTAMAR offers these services as a bundle to individuals, businesses, and organizations in most of the country including Majuro. These communication systems (internet, email, telephone, mobile, and facsimile) will enable the local offices to stay connected and share information with the rest of the stakeholders. Since the local offices provide the most important information about the state of the breadfruit processing system, the BRAINS chief information technologist will specifically oversee the local office communication infrastructure and implement the best solutions.

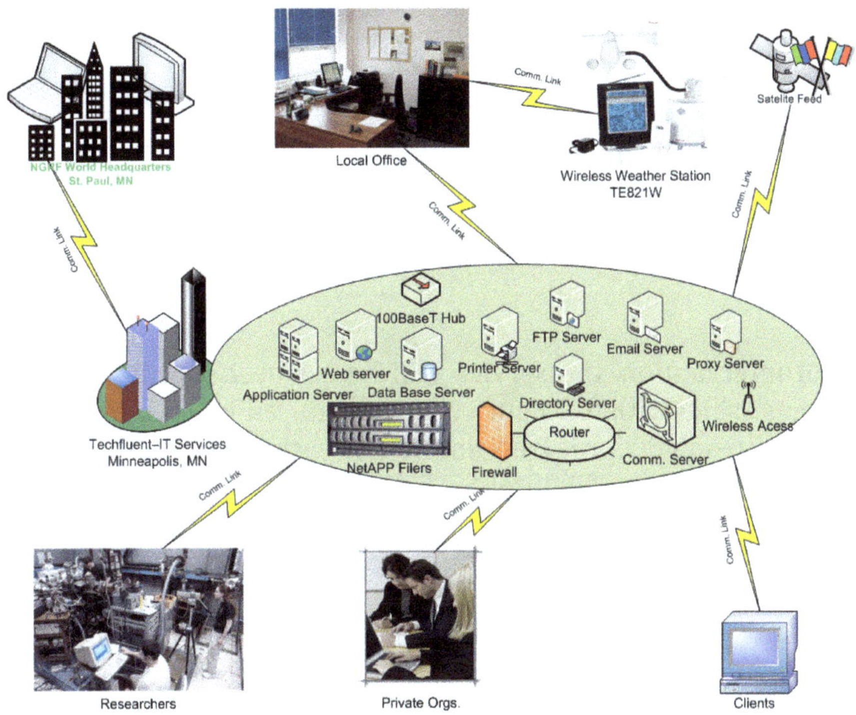

Figure 193: Pictorial view of the BrEAD data sharing system among the communities of interest.

Using the standard IT outsourcing template provided by the IT Service Management Community (ITSM) - see reference [7], we now describe the services provided by Techfluent to its customer, BrEAD in detail. BRAINS modified the template to make it more intuitive and easy to understand for its customer, NGRF. This method of illustration is widely used for IT outsourcing for describing provider and customer relations. It gives a clear picture of all the business services for which Techfluent is responsible for and their specifications. Please note the following descriptions only provide top-level overview of the services and are not the only requirements agreed by Techfluent and BRAINS.

2.1.1.4.1.2. Email Services

This service delivers electronic messages to and from all BrEAD Employees.
The standard features of this service are:

- Sends and receives email messages, both internally and via the Internet.
- Provides a calendar feature with scheduling capability.
- Provides an easily accessible Global address list of email addresses and "white page" information.
- Allows for up to 1Gigabit of mailbox storage space by default.
- Sends and receives file attachments (such as Word documents, Excel spreadsheets, etc).
- Provides the ability to create "distribution groups" holding a number of email addresses for easy distribution.
- Deletes Email and calendar items over 90 days old automatically.
- Provides access to mailbox via a browser (Outlook Web Access).
- Allows for shared messages, contacts, and calendar items through Public Folders.
- Provides virus and spam protection.

Techfluent will deliver these service via BrEAD desktops/laptops, blackberries and other handheld devices. The service will be available 24 hours 7 days per week including holidays.

In the case of interruption, the service will be restored within 1 hour for low to medium impacts while within 2 hours will be restored when more severe incident occurs. Help with resolving problems and issues with Email service can be obtained via Techfluent Service Desk (1-800-456-5655).

2.1.1.4.1.3. Instant Messaging Service

This service tracks presence of BrEAD employees on the network and allows them to send quick text messages.

The standard features of this service are:

- Provides real-time collaboration tool that allows employees to initiate quick chats and/or attend virtual meetings with other BrEAD Instant Messaging users.
- Provides the ability to see who is currently online (presence information).
- Provides ability to share documents and use an electronic whiteboard.
- Maintains lists of high-frequency contacts.
- Encrypts chat messages.

Techfluent will deliver this service via BrEAD desktops and laptops. The service will be available 24 hours 7 days per week including holidays.

In the case of interruption, the service will be restored within 1 hour for low to medium impacts while within 2 hours will be restored when more severe incident occurs. Help with resolving problems and issues with this service can be obtained via Techfluent Service Desk (1-800-456-5655).

2.1.1.4.1.4. Knowledge Sharing/SharePoint Service

This service allows teams of people to share information specific to that project team or department (e.g., discussion threads, calendars and documents) via easy-to use websites.
The standard features of this service are:

- Allows information sharing among designated members.
- Stores and retrieves shared documents.
- Secures and manages access to individual collaboration websites.
- Provides ability to specify and modify delivered web page content.
- Provides ability to search across collaboration websites for information.
- Provides up to 10Gigabit of space per site by default.
- Allows secure 2-way delivery of e-mail text and attachments; creates a secure "tunnel" between partner mail servers.
- Provides ability to choose whether all messages are encrypted automatically or only specific messages

Techfluent will deliver this service via BrEAD desktops and laptops. The service will be available 24 hours 7 days per week including holidays.
In the case of interruption, the service will be restored within 1 hour for low to medium impacts while within 2 hours will be restored when more severe incident occurs. Help with resolving problems and issues with Knowledge Sharing/SharePoint Service can be obtained via Techfluent Service Desk (1-800-456-5655).

2.1.1.4.1.5. Website Service

This service hosts and supports the design, upgrades and maintenance of main BrEAD public website (BrEAD.ORG).

The standard features of this service are:

- Provides state of the art website for the advocacy of breadfruit based on BRAINS and NGRF requirements.
- Accommodates large number of users at any given time.
- Accessible through all the common web browsers (Internet Explorer, Netscape, Mozilla Firefox).
- Provides customized live weather feed from the Blue Skies Unlimited Weather Views satellite from all the breadfruit processing sites on the main page.
- Provides upgrades 900 GB web space.
- Handles at least 80, 000 GB traffic.
- Provides up to 99.9% server uptime.

The service will be available 24 hours 7 days per week including holidays. In the case of interruption, the service will be restored within 1 hour for low to medium impacts while within 2 hours will be restored when more severe incident occurs. Help for resolving problems and issues with the website can be obtained via Techfluent Service Desk (1-800-456-5655).

The IT providers will develop a website that will advertise and market breadfruit. The website will describe the availability of breadfruit, its uses, recipes, nutritional information and much more. The website will allow people, organizations, and governments to realize the benefit in producing local breadfruit in developing countries, or undeveloped countries, as an efficient and low cost means of providing food. Governments spend billions of dollars on foreign food aid that is throwing a fish to a hungry person, but the breadfruit marketing and website will show how developing countries can fish themselves and have a simple infrastructure put in place to provide food for many. The website will help generate interest from world hunger advocacy groups and help generate subsidies to put the factory in place. The initial requirements provided to the IT firm for the website are provided below as the features and content that will be available on the website.

- Recipe sharing.
- Provide information on how to process the breadfruit.
- Nutritional information on breadfruit.
- Biography on the benefits of the architecture for the local community.
- Provide information on the different varieties of breadfruit and where it's found.
- Section describing our mission and purpose.
- Shelf-life information for breadfruit.
- Where to purchase breadfruit.
- Contain the BRAINS production architecture showing how breadfruit is harvested and delivered as flour.
- Description of the processing system.
- Explain how efficiency in producing breadfruit reduces environmental pollutants.
- Provide a user message board allowing forum discussion of breadfruit.
- Allow donations to be submitted online to provide cash or certain products to needed developing areas with breadfruit.
- Capability for managing website content and updating news pages.
- Contain breadfruit harvest and production numbers for current and past seasons as provided from the NGRF marketing database.

2.1.1.4.1.6. Blue Skies Satellite Live Weather Feed

Techfluent will customize BrEAD website to deliver view of weather conditions for the processing sites. This is a fully customizable weather station that will be integrated in to the website. This is one way of collecting local weather data and making available to the stakeholders. Shows a sample of how the weather view will look like on BrEAD's website.

Figure 194: Sample weather view on BrEAD's website

2.1.1.4.1.7. Honeywell Full Wireless Weather Station

In addition to the live weather feed from the Blue Skies Satellite, BRAINS has identified Honeywell's full weather station device (TE821W) for site weather collection. This device is specifically designed for weather sensing of remote areas that have no communication systems. It uses small long lasting batteries and it senses all the weather conditions of the site such as temperature, humidity, rainfall, wind speeds, etc. BRAINS proposes this device to aid local office to collect more precise weather data of the breadfruit processing sites that will be used in SIL analysis.

2.1.1.5. Support Systems

In this section, we describe the approaches and the principle that BRAINS will implement in order to meet the requirements of the Enterprise Architecture and the Statement of Objectives (SOO).

2.1.1.5.1. Integrated Logistics Support

This section provides logistics concepts and logistics planning process necessary to execute integrated logistics support for the BrEAD system such as managing the flow of goods, information and other resources such as personnel and vendors in order to meet the requirements of the stakeholders. This involves the integration of transportation, material-handling, packaging, warehousing and inventory. This section also describes all of the necessary aspects required to ensure the effective and economical support of the system, subsystems, network, and associated components.

2.1.1.5.1.1. Supply Chain Management

The BrEAD system consists of a network of interconnected businesses to provide the product and service packages necessary to meet all the requirements. The supply chain management spans all movement and storage of raw materials, work-in-process inventory, and finished goods from point-of-origin to point-of-consumption. In order to better understand and satisfy the customers' needs, a number of organizations and businesses are involved in the BrEAD system logistics operations. This strategy will improve efficiency while reducing management control of daily logistics operations. Less control and more supply chain partners will promote trust and collaboration among partners, thus improving product visibility and mission success.

2.1.1.5.1.2. Logistics Strategy

The logistic and maintenance support infrastructure is a major element of the BrEAD system and this is properly addressed during the system design process. The areas of system support that are considered throughout the system engineering design process are the logistics associated with initial acquisition, production, and distribution of the system and its elements to user operational sites, and the subsequent sustaining maintenance and support of the system.

The BrEAD system logistics strategy addresses the number, location and missions of suppliers, product facilities, distribution centers, warehouses, cross-docks and customers early in the system development. Early in the planning phase, logistics engineers, logistics support analysts, and supply chain management experts will define the strategy and coordination across the supply chain. The goal of this strategy is to gain the most efficient and effective performance of the system.

Figure 195 depicts the BrEAD system supply chain model. This model shows the flow of resources in the overall supply chain network. It starts with the local farmers supplying the breadfruit. The purchased breadfruit will then be delivered to the collection points. There will be six collection points throughout RMI. At each collection point, breadfruit will be checked and sorted for quality. From the collection point, breadfruit is delivered to the processing facilities to be processed into flour. The processed flour is then transported to the packaging facility to be packaged. The warehouse facility will keep track of inventory, and control the distribution of the finished product. The BrEAD logistics strategy optimizes a steady flow of material through a network of transport links and storage node.

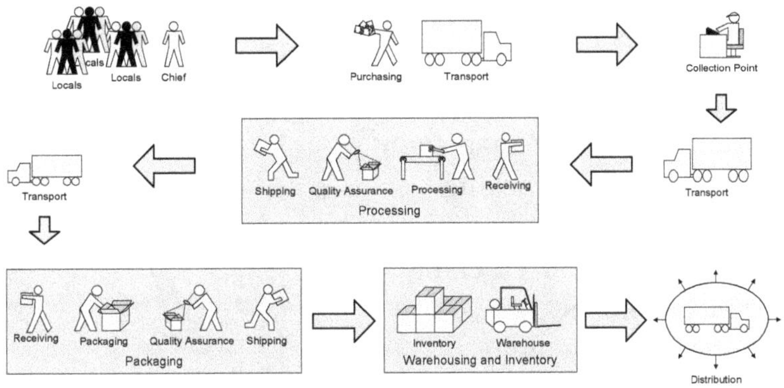

Figure 195: BrEAD Supply Chain Model

There are two focuses in business logistics: internal focus and external focus. Internal focus involves inbound logistics such as physical supply and internal logistics such as manufacturing, while external focus involves outbound logistics such as physical distribution.

The activities involve in the physical supply logistics are:

- Forecasting
- Supplier selection
- Procurement/purchasing
- Inventory management
- Transportation
- Information flow

The activities involve in the manufacturing logistics are:

- Production planning
- Procurement/purchasing
- Materials handling
- Inventory management
- Packaging/transportation
- Information flow

The activities involve in the physical distribution logistics are:

- Material processing
- Inventory management
- Transportation
- Product installation
- Customer service
- Information flow

2.1.1.5.1.3. Logistics Management Information System

A Logistics Management Information System (LMIS) has been developed to support the BrEAD system and provide an infrastructure that integrates all logistics management information. The LMIS manages the following functions:

- Configuration and Data Management
- Material Ordering and Tracking
- Repairable Management
- Repairable Ordering and Tracking
- Inventory Management
- Transportation Management
- Maintenance Management and Analysis
- Trouble and Problem Reporting
- Status Reports

2.1.1.5.1.4. Vendors

Given BRAINS system design configuration, the production process commences, involving a prime manufacturer and a number of different suppliers. Table 2 lists several suppliers of materials and services to successfully implement BRAINS integrated logistics support plan.

Table 2: Vendor List

MATERIALS/SERVICES	VENDORS
Raw materials	
Breadfruit	Local Farmers/tribal chiefs
Production materials	
Dehydration material (syrup, sugar, salt)	ED & F Man, Fortune Century Overseas Co. Ltd, Salt Products Company
Product Packaging (paper flour bag)	INDEVCO and Haicheng Sitong Plastic Weave Co. Ltd.
Shipping Container	Arena Products Inc.
Replacement/spare parts	
Grinder parts	Prater-Sterling, Pleasant Hill Grain Company
Turbine parts	Ginlong, GE Energy, Vestas Wind Systems
Generator parts	Generac Power Systems
Powder packaging machine parts	Guangzhou All-bloom Machinery Co. Ltd, Guangzhoujuwa Packing Machinery Co. Ltd
Equipments	
Tooling/test equipment	Brijport Brian Machine Tool, Factory Maintenance Services Ltd
Safety equipment	Mangla Plastic Industries, Jayco Safety Products Pvt. Ltd.
Training equipment	Various Suppliers
Storage and material handling solutions	Parallel Track Engineering Company Pvt. Ltd.
Other materials	
Fuel for company vehicles/transport	Marshalls Energy Company (MEC)

MATERIALS/SERVICES	VENDORS
Shipping/Transport	
Inter-atoll shipping	RMI private businesses and government-sponsored services
International shipping	Matson Navigation, PM&O Lines
Air Transportation	Air Marshall Islands (AMI)
Warehouse	
Private warehouse	Owned and operated by Majuro Packaging Facility
Public warehouse (general merchandise)	RMI private businesses
Distributor/store	
Retail and Wholesale	Ralik Store
Telecommunications	
Telephone service	National Telecommunications Authority (NTA)
Cellular service	National Telecommunications Authority (NTA)
Internet access	National Telecommunications Authority (NTA)
Information Technology	
IT Service Management and Solutions	Techfluent

2.1.1.5.1.5. Personnel

Given the personnel requirements for the BrEAD system, the next step is to identify available resources and those individuals who can be employed to operate and maintain the system. Selected personnel are evaluated relative to current skills as compared to the necessary skills for the system. The differences dictate the needs for training. Table 3 lists the necessary personnel to successfully run and maintain the processing facilities.

Each processing plant will have one Plant Engineer which has the primary role of ensuring the overall success of the facility. There is one Production Engineer assigned for each facility whose primary responsibility is to maintain the steady and efficient flow of materials in and out of the production process. There are two Quality Assurance Engineer assign at each site to ensure products and services are designed and produced to meet or exceed customer requirements. One person per facility is in charge of inventory management of raw materials, work-in-progress, and finished goods. Procurement personnel are in charge of the acquisition of goods and/or services involve in all phases of production. Simple procurement involves repeat purchasing, and complex procurement involves finding better solution and/or long term partnership with suppliers. There are two procurement personnel per each facility. Operational Analyst provides advance analytical support by identifying, capturing and manipulating key data, analyzing data, preparing and distributing management reports to help management make their decisions. Each Operational Analyst supports two facilities.

Table 3: Personnel

Research and Development	
Position	**No.**
Project Management	1
Production Management	1
Logistics and Support Management	1
Market Analyst	1
System Engineer	1
Processing Operators	
Position	**No.**
Fruit Sorter/Grader	18
Cleaning Station	12
Peeling Station	18
Slicing Station	12
Blanching Station	18
Drying Station	12
Grinding Station	12
Screening Station	24
Packaging Station	24

Production/Operation	
Position	**No.**
Plant Engineer	6
Production Engineer	6
Quality Assurance Engineer	12
Inventory Management	6
Procurement Personnel	12
Operational Analyst	3
Maintenance Crew	24
Operator Training Instructor	3
Marketing and Sales	3
Transportation Management	6
Receiving/Handling/Support Crew	24
Janitorial Crew	18
Customer Service	6
Processing Operator	150

*Sourcing for six factories

Given the requirements for the system, it is assumed that all available resources personnel currently possess the necessary qualifications for entry into positions requiring a basic skill level. On the basis of these assumptions, it is necessary to develop a training program that will:

- Train entry-level personnel in the fundamentals of system operation for basic skill level requirements
- Train entry-level personnel in the performance of operator and maintenance functions to satisfy the intermediate skill level requirements
- Train entry-level personnel in the performance of operator and maintenance functions to satisfy the high skill level requirements for the system

2.1.1.5.1.6. Training

Training will be accomplished through formal structured programs and/or on-the-job training depending on the skill level required. Training requirements include both the training of new personnel and replacement personnel.

2.1.1.5.1.7. Obsolescence Management

The strategy for BrEAD obsolescence management is based on a proactive approach that identifies obsolescence issues as early in the process as possible. This ensures the optimum window of availability and support, and maximizes the time frame to implement a resolution to reduce risk. The same proactive approach is continued to be applied during the production and sustainment phases.

2.1.1.5.2. Logistics Support System

The logistic support concept addresses all logistics support activities from the development phase through production, verification, integration, and operations phases. The primary objective of the logistic support effort is to effectively control resources allocations while ensuring the system meets the operational requirements. Meeting these objectives requires the development of a support infrastructure that:

- Provides maintenance support for each facility
- Procures and provides spare parts and supplies
- Control inventory and security of assets, including packaging, handling, storage, and transportation
- Supports development and sustainment of training of operators, maintainers, supervisors, and network administrators
- Identifies logistics support alternatives during the operational phase

Logistics reviews and inspections will be conducted throughout the integrated logistics support process. A Performance Based Logistics (PBL) will be conducted at each system node to identify strength and weaknesses in the system, and validate system configuration and processes and ensure its effectiveness. At each review, audit and/or inspection a comment/resolution document is generated. When a discrepancy is identified, it is documented for resolution.

2.1.1.5.2.1. Maintenance and Test Logistics

This section establishes the framework in which organizational, intermediate (regional) and depot maintenance will be accomplished for the BrEAD system. It provides details on the maintenance support process for the system and associated equipment. The maintenance and support concept is developed during the conceptual design phase, and evolves from the definition of system operational requirements.

Corrective and preventive maintenance will be performed on the system itself at the site where the system is used, intermediate shop near the operational site, and//or at a depot. The following describes the basic criteria and differences between maintenance levels:

1. Organizational Maintenance
    - On-site corrective and preventive maintenance
    - Low skill level personnel
        i. Visual inspection
        ii. Operational checkout

iii. Minor servicing
iv. External adjustments
v. Removal and replacement of some components

2. Intermediate Maintenance
    - Test and support equipment
    - Medium skill level personnel
        i. Detailed inspection and system checkout
        ii. Major servicing
        iii. Major equipment repair and modifications
        iv. Complicated adjustments
        v. Limited calibration
3. Depot Maintenance
    - Detailed maintenance, overhaul, calibration
    - High skill level personnel
        i. Complicated facility adjustments
        ii. Complex equipment repairs and modification
        iii. Overhaul and rebuild
        iv. Detailed calibration

The BrEAD system has been designed to be supported using common tools and test equipment whenever possible. Common tools and test equipment are classified as items that are common and standardized, and easily and readily procured. Should a calibration requirement be identified, maintenance of tools and equipment will be in accordance with the specifications and applicable calibration regulations identified by the Original Equipment Manufacturer (OEM). Commercial-Off-The-Shelf (COTS) products requiring calibration will be calibrated in accordance with the prescribed vendor equipment calibration schedule. All maintenance performed will be documented and reported. This report is completed by site administrators and/or maintenance personnel.

2.1.1.5.2.2. System Maintenance and Repair Policy

Repair policy is initially developed and established during the conceptual design phase, and subsequently updated as the design progresses and the results of the level-of-repair and supportability analyses become available. The following maintenance and repair policy will be provided at different maintenance level:

1. Organizational Maintenance
    a. Unscheduled maintenance: the faulty unit is removed and replaced. It is sent to the intermediate maintenance for repair
    b. Scheduled maintenance: the unit scheduled for maintenance is removed, replaced, and sent to intermediate maintenance shop at 6-month intervals
    c. Support Factors
        i. Support and test equipment: built-in self-test
        ii. Supply: spare units

        iii. Personnel skill level: basic
2. Intermediate Maintenance
    a. Unscheduled maintenance: accomplish fault isolation to unit and repair through piece part replacement. Discard faulty parts.
    b. Scheduled maintenance: disassemble unit and remove module for calibration. Then replace module and checkout unit.
    c. Supply Factors
        i. Support and test equipment: unit test set, module and circuit board tester
        ii. Supply: spare units, modules and piece parts
        iii. Personnel skill level: intermediate
3. Depot Maintenance
    a. Unscheduled maintenance
        i. Overhaul and repair: accomplish fault isolation to piece part and repair
        ii. Calibration lab: accomplish repair and/or calibration of module
        iii. Discard faulty modules
    b. Support Factors
        i. Support and test equipment: standard items
        ii. Supply: modules and piece parts
        iii. Personnel skill level: advance

2.1.1.5.2.3. Data Repository

The BrEAD data repository is located at St Paul/Minneapolis headquarters. This location is the central repository for all technical documentation and data collected on the BrEAD system to be used by researchers, marketers, and users worldwide. Data Management maintained the masters of all technical documents and data collected. The purpose of the data repository is to maintain masters of all technical documents and data collected in one location.

All documentation and data is restored electronically whenever possible. Automated electronic connectivity for real time data interchange is provided worldwide. The site documentation will consist of necessary technical manuals, procedures documentation, and support planning documentation.

2.1.1.5.2.4. Warehouse and Support Facility

The BrEAD warehouse and support facility coordinates the flow of materials, functions as the central distribution point for processed and packaged end-product, and provides storage for spare and repair parts. The warehouse and support facility provides the following:

- Material and Inventory Management
- Material Shipment and Notification
- Spares/Repair Parts Management

- Repairable Management

2.1.1.6. Hardware and Software Integration Overview

The following is a description of the approach for hardware and software integration and any development approach need to integrate necessary elements. Hardware is separated into two types. One is network hardware, and another is desktop and laptop computers, and mobile devices. Network functionality is to communicate between different main operational nodes (according to OV-1), store data, and secure data communication. Desktop and laptop computers provide an interface between all the users to enter, update, remove, share and read data between the participants remotely and located. Two types of software are used. One is to handle network system, another to enable the communication between different mobile devices (including laptops and desktop computers) and the network system. This software is provided by Techfluent (as BRAINS identified it as the IT provider).

- All mobile devices, laptops and desktop computers are connected through ISP to enable communication via a local internet network. This will be provided by Techfluent
- Each mobile device, laptops and desktop have install corporate software clients such as an email client application, a Instant Messaging client application, an internet explorer such as Internet Explorer or Firefox, and direct access to SharePoint
- Some of the information will be encrypted with security purposes

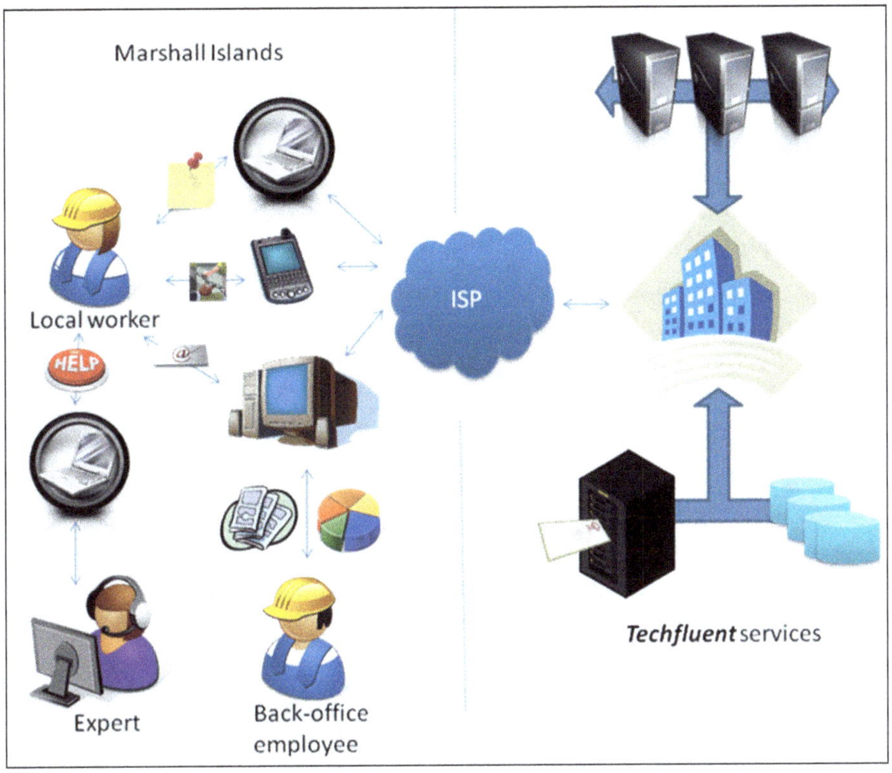

Figure 196: Hardware and Software Integration Overview

Some examples of data exchange using **E-mail**:

- Schedule meetings when needed to organize or disseminate top-bottom action plan
- Communicate and programmed deadlines
- Broadcast general information relevant to all geographically remote locations such as organizational changes, new definition of procedures, new available machinery, etc.
- Motivating news regarding breadfruit flour increasing consumption around the world
- BrEAD's PNL and financial prognostics
- New procedures information dissemination

Some examples of documents shared using **SharePoint**:

- Breadfruit daily production
- Weather daily conditions
- Wind turbines performance and efficiency
- New founds regarding processing methodologies, new approaches, experimentation results, feedback
- Recipes
- Social and community response to the implementation of small and large scale production sites
- Financial improvement after processing factories maturation
- Difficulties or challenges encountered while locally implementing the proposal made by BRAINS, as well as modifications made to it

Some examples of information published on **Website**:

- Provide real-time local weather conditions
- Warning messages and information about the proximity of natural disasters
- All type of information and statistics relevant to local and global production, and distribution
- Nutritional values and new recipes
- Contests and marketing initiatives to be locally implemented

Some examples of information communicated using **Instant Messaging**:

- Provide an online help desk or technical support to remote locations to support the daily production and relieve technical daily challenges
- Replace phone lines and satellite communication to quickly reach out local and remote located professionals, experts, or similar when needed

2.1.2. Performance Based System Specifications (PBSS)

System Requirements

The requirements for the project were developed from the source requirements in the Performance Based Systems Specification (PBSS) and Statement of Objectives (SOO). The SOO provides guidance toward achieving the PBSS requirements. From the PBSS, the requirements that make up the BRAINS proposal are a derivation. BRAINS has chosen to use DOORS® database program tool to enter, track, trace, and output system requirements.

Requirements Derivation

The figure below shows to traceability and requirements approach taken. The PBSS requirements were entered into doors and then decomposed into another database. Since many of the PBSS requirements are compound requirements (asking for many things in one sentence or requirement) the BRAINS team separated out each individual requirement from the full PBSS text to several more concise statements. The decomposition allowed BRAINS to further derive the requirements according to the architecture we are designing. Many requirements will be flown down to subcontractors where work is outsourced or used internally for BRAINS to ensure we meet our design and NGRF requirements. The derived requirements will be furnished to the NGRF via CDRL 0008 Updated Performance Based Systems Specification (PBSS) for approval.

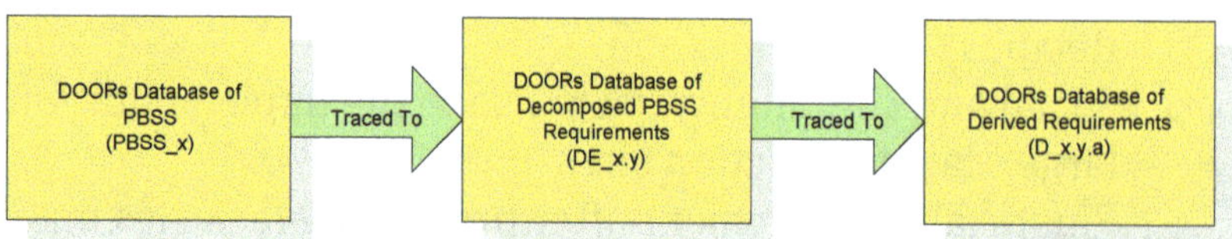

In order to create unique requirement ID's that would provide traceability, an identifier for each requirement the following scheme was developed: PBSS requirement ID's were in sequential order by the identifier PBSS_x, where x identifies the number 1 through 18 of PBSS requirements from the NGRF. The decomposed PBSS requirements are identified by DE_x.y, where x is the PBSS number and y is the index of the decomposed PBSS. For example if there are five decomposed requirements, y will have values 1-5. See Technical Attachment 1 for a complete listing of the derived and updated PBSS requirements with traceability and verification methods.

Requirements Allocation

Requirements allocation was done by BRAINS in order to ensure tracking of system design data is properly delivered in the CDRL deliveries. In this regard a traceability matrix was developed to ensure the proper CDRL contained the contract data. In some cases where there is not an appropriate CDRL, the requirement is traced to test procedures or reports. The requirements traceability discussed in the previous section will allow BRAINS to ensure each subsystem and component is properly receiving the flow down of high level system requirements. See Technical Attachment 1 for the table of requirements and traceability.

Verification and Validation

The verification and validation of requirements will be a combination of validation through manual test methods (test procedures) performed in the SIL, analysis of design or test data reports, demonstration of the product, and inspection of system components. The verification method of how requirements will be validated is provided in the requirements table in the requirements section. The diagram below provides a simple summary of the verification and validation process. The derived requirements serve as the source flowing down to inspection, analysis, test, or demonstration methods of verification/validation. For each type of requirement, the verification/validation process will be a different one. For analysis for example, all the documentation required to show the requirement is met will be provided. For inspection, a physical inspection will take place with a Quality Assurance representative of NGRF to verify BRAINS inspection records. For test verification methods, the test plan will be responsible for including procedures that verify the test. Any demonstration requirements will be for system operations such as showing the factory successfully outputs the flour at its specified rate.

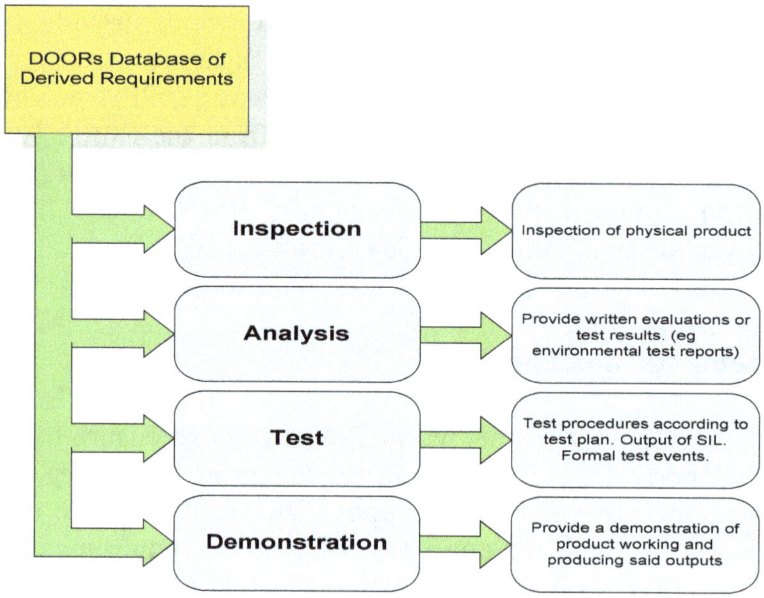

Figure 197: Verification and Validation Process

2.1.3. Open Systems Architecture

BRAINS developed all the designs for the Breadfruit Enterprise Architecture Demonstration (BrEAD) and the Breadfruit Processing Systems (BPS) in accordance with the customer's specifications of the open systems architecture. BRAINS' demonstrated excellence in the fields of systems engineering and systems integration allowed the use of commercial off the shelf (COTS) components that are proven to work for the entire design of the BrEAD and BPS. BRAINS's wide vendor base also allowed flexibility in selecting open system architecture standards and interfaces that are widely supported. To better describe the conformity of BRAINS's proposal to Open System Architecture, we divide the BrEAD into three logical segments, namely:

1. Information Technology platforms.
2. BrEAD's System Integration Laboratory (SIL).
3. Breadfruit processing systems.

These architectures are described in the following sections.

2.1.3.1 Evolvability

The software and hardware used in all the three segments of BrEAD above are completely open architecture and do not contain any propriety systems. For the Information Technology, Techfluent which will be BrEAD's IT service host and provider uses the most widely used network and application solutions. Specifically BrEAD's software and hardware items such as operating systems, email/web servers, storage devices, firewalls/security devices, and database applications are all based on well defined standards that will be easy to replace with minimal down time. They are the popular brands of Microsoft, Cisco Systems, Dell, hp, and Symantec. As agreed by Techfluent, none of BrEAD's assets will have closed systems that will need the conversion to an open architecture. Techfluent has license agreements with its vendors. Any time that a more useful and newer product comes out, such as new Microsoft Office product, Techfluent will purchase the license and upgrade the systems. This will cause minimal or no interruption to the network flow. As evident, these systems also use the communication and data exchange standards such as TCP/IP and other file formats that will not need extra requirements. For more information on the services provided by Techfluent, see the Enterprise Communication and Data Network section of the proposal.

The software and hardware items of the BrEAD System Integration Lab (SIL) will be primarily used for environmental, qualification, modeling and simulation of the breadfruit processing systems. In terms of their evolvability, these items are very similar to the platforms of the IT services. Again, there are no propriety or closed systems in the lab. They consist of Pro-E model softwares, Microsoft Office products, temperature sensors/chambers, and other systems that are widely open architecture and will not have any evolvability issues. BrEAD will have license agreements with its vendors and make the necessary upgrades and/or changes as necessary.

Lastly, the breadfruit processing systems mainly contain the breadfruit peeler, slicer, dryer and grinder. As explained in their respective sections, these equipments are fully hand or hardware controlled and do not contain any interfaces that need to be accommodated. Since these equipments are discrete systems and do not feed any data to other systems, they can be replaced anytime that a more advanced system is identified without any impact to the overall operation. In addition, the other site specific systems of Poor Man's wind turbine and Honeywell's full weather station are also discrete and open architecture in their respective applications. The small weather station is portable wireless full weather sensor that accepts USB standards for data transmission. Based on Honeywell's specification of the device, they will keep up any technology changes of the USB interfaces and will provide any necessary upgrades to its customers. The turbine and its control and energy saving components are also hand or hardware operated that will not be affected by technology evolvability.

2.1.3.2 Portability

BRAINS doesn't propose any new software development for BrEAD. All the software that is used for the development of the architecture is commercial off the shelf (COTS) items that are proven to work in all the environments and conditions that BrEAD will operate. The local offices and breadfruit processing sites will only have internationally standardized software and hardware systems such as Microsoft Office and IBM hardware platforms that are established to work in any geographic location.

2.1.3.3 Modularity

As briefly mentioned above, all systems and components of BrEAD are separated into discrete and independent units that have well defined interfaces which will allow easy substitution with no or minimum impact on the existing systems. For BrEAD's

information technology, Techfluent uses the popular IBM copper/fiber network switches that will allow the follow of data to a back-up unit while a hardware or software component is being replaced or worked on. This guarantees no down time of the service. The diagram of Figure 198 shows top-level diagram of Techfluent's modular system interfaces. This modularity that Techfluent offers was one of the main reasons why BRAINS selected as the IT provider over other competitors. For instance, if the email server needs to be replaced, the network switch will be toggled to a temporary email server that will continue the service until the main server is restored. As described in the local office communication and network section, the local offices also use the enterpriser's servers, SharePoint, and websites that are hosted here in the Twin cities through Techfluent's modular systems. This eliminates the need to have separate information technology infrastructure for each local office. Since these local offices will be located in the commercial cities of the breadfruit growing countries, BRAINS has identified there are internet providers for these areas.
\

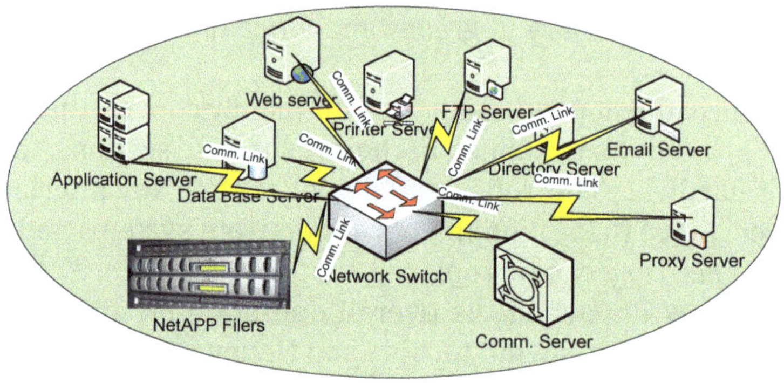

Figure 198: Techfluent Modular systems

The System Integration Lab (SIL) and the breadfruit processing systems are all physically separated units that have no dependence each other.

2.1.3.4 Rationale for Modularization Choices

Before generating any designs, BRAINS has done extensive evaluation on the needs of the end-users. Since BrEAD will mainly operate in third world countries which are well behind on technology, BRAINS considered these user constraints for all system designs. The processing systems are selected based on their performance, how intuitive and user/culturally friendly they are. Also equally important, the main tradeoffs that BRAINS made are the modularity and open system architecture aspects. Since we expect these systems to be deployed in geographically different locations, portability and modularity were key factors in the selection. As mentioned before, Techfluent has offered the highest conformity to open systems architecture over its competitors.

2.1.3.5 Employ a Layered Modular Architecture

There are no hardware or software items that will be developed for either the BrEAD or the BPS. All the systems in this proposal are fully commercial-Off-the-Shelf/Non-developmental Item (COTS/NDI) and non-proprietary components.

2.1.3.6 Maintainability

BRAINS is not proposing any new software or hardware development. Thus there won't be any proprietary software that needs to be maintained. Techfluent will sustain all BrEAD's Information Technology infrastructures. For the processing systems and the SIL, there will be periodic scheduled maintenances. There will be basic tools and a trained technician at each processing site and SIL that will solve any minor fixes as there will be spare parts for replacements. BRAINS has selected these systems in accordance with NGRF's requirements of failure rate and mean time to repair. Refer to Annex B for more information.

2.1.3.7 Technology Insertion and Refresh

Since the BRAINS's proposed processing systems are all discrete and independent systems, any specific component can be replaced and/or refreshed without affecting other systems.

2.1.3.8 Reusability

When a wind turbine is dissembled, many of the parts can be reused as they have long warrantee from the manufacturers. Also other SIL software and hardware items may be reused for other applications as needed.

2.1.3.9 Life-cycle Supportability

There are no proprietary or vendor-unique technologies that will be used in either BrEAD or BPS. All the proposed components have wide base of vendors that can accommodate for any obsolete items.

2.1.3.10 Traceability of System Requirements

The requirements for the project were developed from the source requirements in the Performance Based Systems Specification (PBSS) and Statement of Objectives (SOO). The PBSS requirements were entered into doors and then decomposed into another database. As discussed previously, many requirements will be flown down to subcontractors where work is outsourced or used internally for BRAINS to ensure we meet our design and NGRF requirements. The traceability was done using a system of requirements ID's that track parent to child requirements from the highest level PBSS requirement all the way down to the lowest level component or system that needs that requirement. In this fashion, every system, subsystem, and component will have the appropriate requirements traced to it.

2.1.4 References

1. Richard Lewis, *Poor Man's Guide to Wind Power and Battery Systems.*
2. Michael Schmidt, Strategic Energy Institute, *Georgia Institute of Technology*, www.energy.gatech.edu
3. National Lightning Safety Institute (NLSI), *Lightning Hazard Reduction at Wind Farms*, http://www.lightningsafety.com/nlsi_lhm/wind1.html
4. Brian McNiff, *Wind Turbine Lightning Protection Project*, National Renewable Energy Laboratory (NREL), http://www.nrel.gov/docs/fy02osti/31115.pdf
5. Renewable Energy Research Laboratory, University of Massachusetts at Amherst, *Wind Power: Siting in Communities*, http://www.ceere.org/rerl/about_wind/RERL_Fact_Sheet_4_Siting.pdf
6. Blue Skies Unlimited Weather Views, *customisable Weather Station*, http://www.liveweatherfeed.com/weathersystem/1weatherstation.cfm
7. ITSM Community – *IT Service Management Catalog template*, http://www.itsmcommunity.org/downloads/Service_Catalog_Example.pdf
8. Marshall Islands National Telecommunications Authority (NTAMAR), http://www.ntamar.net.
9. Browning, T. (1988). Sources of Schedule Risk in Complex System Development. *Proceedings of the Eighth Annual International Symposium of INCOSE*. Vancouver.
10. Netzer, A. (n.d.). Integrated Master Plan (IMP) and Integrated Master Schedule (IMS). Retrieved April 14, 2009, from www.fas.org/spp/military/program/imint/98b030.pdf.
11. http://importfood.com/howtoslicemango.html
12. http://cooking101.abccook.com/tools-and-techniques/knives.html
13. http://www.weibull.com/
14. http://www.weibull.com/LifeDataWeb/competing_failure_modes.htm
15. RAM Guide, *Office of the Undersecretary of Defense*, Washington, DC, August 3, 2005
16. http://www.floridaits.com/SEMP/
17. http://www.fhwa.dot.gov/cadiv/segb/views/document/sections/Section8/8_4_2.htm
18. General Dynamics C4 Systems template, Bloomington, MN
19. System engineering management, Blanchard, Benjamin S. Knovel (Firm) 3rd ed. Hoboken, NJ : John Wiley c2004
20. Human factors and ergonomics in manufacturing. *Institute of Industrial Engineers (1981-)*, Human Factors and Ergonomics Society, International Ergonomics Association, [New York, N.Y.] : John Wiley & Sons, Inc. c1997-
21. SDP, http://sepo.spawar.navy.mil/SDP_Template.doc
22. http://www.wholesalesolar.com/back-up-power.html
23. http://www.generac.com/Default.aspx

24. http://www.pmhut.com/wp-content/uploads/2008/01/risk_management.pdf
25. Handbook of Industrial Drying, Edited by Arun S. Mujumdar
26. Practical Actions, Small Scale Drying Technologies; https://practicalaction.org/

S.R.Lab Instruments; http://www.srlabinstruments.com/tray_dryer.htm

Annex A – Breadfruit Enterprise Architecture Demonstration (BrEAD) System Performance

A.1 Measures of Performance

The success of the processing system will be measured on its output. The flour output will differentiate our system from other related systems.

A.2 Disaster Recovery Strategy

The processing system shall have a backup generator to provide power if the wind turbine is damaged or unusable. The backup system will be available at all times and automatically switch on when main power failure is detected. The processing system shall have spare materials available to construct additional wind turbines.

A.3 Assumptions

Only a select few species of breadfruit will be considered for this system. This will allow the process system to focus on the specific tools required to transform the breadfruit.

The citizens of the Marshall Islands will be a welcoming and accommodating culture who will support our minimally-invasive processing system.

The citizens of the Marshall Islands will be organized for the collection of the breadfruit from the source to the collection point, where the processing system will be.

A.4 Metrics for measurement

Breadfruit gathered: 1680 kg per day

Factory metrics

Peelers output: 15 kg/hr
Dryer output: 15 kg/hr
Grinder output: 30 kg/hr => 240 kg/day
Factory output (total with 6 factories) 1680 kg/day

Metrics for Grinder

Large Scale Meadows Stone Burr Mill Grinder [1] Phase Voltage Motor Grinding Speed Flour Output	 Single 120 875 rpm 45 kg/hour
Small Scale CTI Omega VI Grinder Power Grinding speed Flour Output	 Hand Crank/Manual 60 rpm 15 kg/hour

Metrics for Dryer

Large Scale Dryer Power Required Capacity Drying Rate Drying time	 2 kW 12 pounds (5.45 kg) 5 kg/hr 20 minutes
Small Scale Dryer Power Required Capacity Drying Rate Drying Time	 0 kW 10 pounds (4.55 kg) 0.25 kg/hr 5.46 hours

References

1. Meadows Mills Company, http://www.meadowsmills.com/sbm8.htm

2. Compatible Technology International, Omega VI Grinder: http://www.compatibletechnology.org/whatwedo/devices/BM_Omega6brochure.pdf

Wind Turbine Performance

In order to generate power from a wind turbine, there has to be reliable wind speed. BRAINS has invested extensive research to study the climate of the tropical and sub-tropical countries where breadfruit usually grows. The main observation that we found in our research is that these areas have mostly dependable and consistent wind speeds. Table 1 on the following page summarizes the monthly average wind speeds of some of the countries that breadfruit heavily grows. The four countries listed on the table (Marshall Islands, Jamaica, Kiribati, and Haiti) represent good sample of the major breadfruit areas. The lowest monthly average wind speed is above 9 mph. As will be shown in the following pages, 9 mph is sufficient speed for our application.

	COUNTRY			
	WIND SPEED (MPH)			
MONTH	**MARSHAL ISLAND**	**JAMAICA**	**KIRIBATI**	**HAITI**
JANUARY	16.80	19.89	11.19	11.19
FEBRAURY	16.30	17.40	12.43	10.57
MARCH	16.30	18.65	12.43	11.19
APRIL	15.00	18.65	11.19	9.32
MAY	13.90	18.65	10.57	11.19
JUNE	13.10	19.89	10.57	13.67
JULY	11.20	20.51	10.57	12.43
AUGUST	9.80	18.65	11.19	12.43
SEPTEMBER	9.10	16.16	11.19	10.57
OCTOBER	10.00	16.16	11.19	9.32

NOVEMBER	12.20	16.16	10.57	9.32
DECEMBER	16.50	10.57	11.19	9.32

Table 1: Average wind speeds of some sample tropical countryIn the Pacific islands and many other islands, the months of March through May are usually the harvest time for breadfruit. Thus our processing system will have the highest demand for power in this period. Coincidently, the wind speeds are high in these months according to Table 1. This makes the wind turbine concept an ideal solution to the problem. In addition to the average monthly wind speeds shown on Table 1, the daily wind speed average on a typical spring season day is roughly as shown on Figure 2 below. In tropical and sub-tropical areas wind speeds are usually strongest in the mornings and evenings and slowest in the middle of the day when the sun is strongest. This sample is taken in the month of March in the Marshall Islands. This again emphasizes how reliable wind speeds are through out the day.

	Afternoon	Evening	Night	Morning
	More sun than clouds. Warm.	Passing clouds. Warm.	Broken clouds. Mild.	More sun than clouds. Warm.
Temp	29°C	28°C	28°C	28°C
Wind Speed/Dir	37 km/h / NE	35 km/h / ENE	32 km/h / ENE	31 km/h / ENE
Humidity	72%	72%	70%	70%
Dew Point	23°C	22°C	21°C	22°C
Comfort Level	33°C	32°C	30°C	31°C
Visibility	13.7 km	13.8 km	15.2 km	15.2 km
6hr. Precip. Total				
6hr. Precip. Probability	30%	0%	30%	30%

Figure 199: Daily wind speeds in the Marshall Islands

To give a sense of how much power we can generate from the Poor Man's wind turbine if we use the average daily wind speed shown on figure 2, an optimal 14 feet blade diameter poor man's turbine would put-out about 52 kilowatt-hours on the day shown. This is significantly large power that will be sufficient for all our breadfruit processing needs. For this calculation, we used the formula below [1] which is basic thumb rule for measuring wind turbine outputs.

Power = (.00477) (radius^2)(wind speed^3) (1)

Where the radius is measured in feet and wind speed is in miles/hour. As the formula indicates, the power can be significantly increased with an increase of either the rotor radius and/or the wind speed. Nevertheless, a larger rotor will also increase the weight and the cost of the wind turbine. Naturally wind speeds fluctuate throughout the day, month, and year even though it's relatively stable and predictable in the tropical/sub-tropical areas where breadfruit is grown. In our data, the lowest wind speed is about 9 mph which corresponds to 3kwh/day.

In order to avoid any power interruption due to low wind speeds, we will always have reserve wind turbines that will power-up whenever the active turbines start producing less power than needed to compensate for the lost power. In addition to these back-up wind turbines, we will also have another layer of back-up power in the form of batteries. The batteries will be charged by the wind turbines and will be used in a rare case where we lose all the turbines to natural disasters or in the unlikely case where momentarily there is no wind in the area.

Part list of the Wind Turbine

Figure 3 below is a general wind turbine schematic with the labels of the main components of the system. In general, the main design parameters of the wind turbine are rotor diameter, height, rotor blade design, and generator type [2]. These design parameters are location specific and each location will have different design based on the particular environment. The turbine parts are selected with an optimal design in mind.

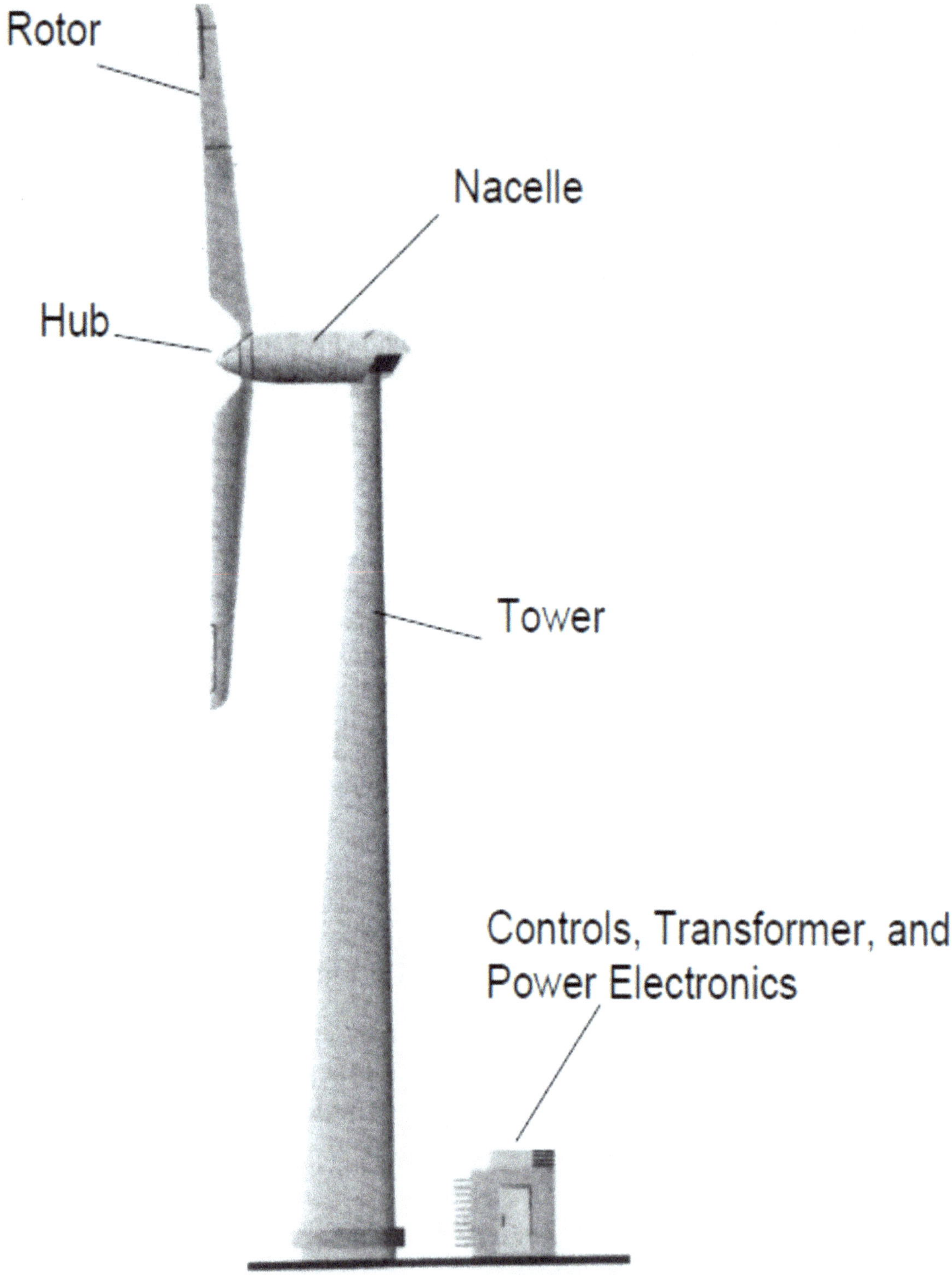

Figure 200: Wind turbine with main sections labeled (courtesy of GaTech)

The main parts that the Poor Man's wind turbine will be constructed are the generator, furling system, steel pipes for the tower, and PVC pipes for the rotor blades as summarized on the table below.

QUANTITY	PART NUMBER / DESCRIPTION
1	P56MD019 - Permanent Magnet DC Motor
1	307702 - Architectural Solid Bronze Hinge for Furling System
2	J-M 531111 - Steel Pipe Tower
1	098179194361 - PW EAGLE - PVC pipe
1	Gearbox
1	Pulley
1	Cement
1	Paint
1	Nuts
1 1	1N4001 - Bridge Rectifier LM341 – Voltage Regulator

Table 2: Poor Man's wind turbine parts list

For our generator, we will use Imperial Electric's Permanent Magnet DC Motor (Model: P56MD019). These motors are ideal for the application since they have large shaft size (3/4 inch) and strong bearings. These motors are relatively cheap – only $35/each and readily available in large quantities. This motor is particularly rated at 1150 RPM with 5.5 Horsepower. Alternatively, car alternators that can be found as scrap parts in junk yards can be used as a generator for household use scale.

The wind turbine tower will be constructed from regular steel pipe that are readily available in hardware stores or construction sites as scrap parts. In our wind turbine, we will use J-M Manufacturing Company's model # 531111. These pipes are 20 feet sections with 1.5 inches inside diameter. They can withstand any weight of the wind turbine up to 250 lb which is more than the maximum weight of any wind turbine that we'll build. The local people can always find these pipes in construction areas if they wish to build their own wind turbine and can not afford to buy. However, J-M Manufacturing sells these pipes for only $3.49/each which is very cheap as only two of these pipes are necessary to construct the turbine.

The rotor blades will be built from PW Eagle's PVC pipes (model # 098179194361). These pipes hold up well and do not bend much to the wind due to their thickness [1]. PW Eagle will supply these pipes to BRAINS for about $37/each. Each wind turbine will have three blades and there's only one PVC pipe required to make the three blades. Alternatively, for household wind turbine, these pipes can usually be found in construction areas and junk yards as remains. Thus if someone in unable to buy one, they can always try to find them for free.

Other parts of the turbine such as nuts and bolts, paint, gearbox, pulley, and cement are readily available in local stores with large quantities. Thus, to stimulate and help local businesses, we will buy these parts from the local hardware stores. Lastly, the alternating current voltage generated by the wind turbine will be rectified through basic diode bridge rectifier such as 1N4001. This rectifier is available in the electronic parts carriers such as Newark. A quantity of 25 pieces sells less than $10. To prevent the over charging of the batteries by the wind turbines, BRAINS will use the popular LM341 voltage regulators of the National Semiconductor which are also available in the electronic parts carrier of Avnet.

Support and Logistics

BRAINS will create an easy to follow step by step manual for constructing such a wind turbine based on Richard Lewis's Poor Man's Wind Turbine design [1]. BRAINS technicians and wind turbine engineers will construct all turbines for the breadfruit processing systems. BRAINS will also provide material and logistical support to local communities and families that want to construct wind turbines for their applications. In doing this, BRAINS will partner with the US military which has a lot of bases and stake in the Marshall Islands. BRAINS is hopeful that this partnership will be successful in advocating this renewable energy initiative.

Lightning Hazard protection

The tropical and sub-tropical areas where breadfruit grows such as Pacific Islands are known to have seasonal and occasional lightning. According to the National Lightning Safety Institute (NLSI), lightning may have the following effects on wind turbines [3]:

- Potential rise of the wind power plant
- Voltage drop across cable shields
- Heating
- Melting
- Magnetic induction
- Repeated H-field impulses

In return, these effects may damage some of the components of the wind turbine such as the rotor, and bearings [3].

To mitigate these lightning risks, BRAINS will fully adapt the Wind Turbine Lightning Protection techniques of the National Renewable Energy Laboratory (NREL) when constructing the turbines [4]. The NREL provides the best methods in lightning hazard prevention such as shielding specific components, current surge protections, grounding, and specific rotor designs to minimize lightning hazard.

Wind turbine Spacing

Since BRAINS plans to have six wind turbines for each breadfruit processing plant, another design consideration is the spacing between individual turbines. According to the Renewable Research Lab at the University of Massachusetts [5], "spacing turbines too tightly leads to reduced performance and increased maintenance due to higher turbulence at the downwind turbines." Thus, it becomes essential that the six turbines are separated appropriately so that the risks are mitigated. Reference [5] and other wind turbine literatures suggest that a conservative spacing of the turbines is 2 - 10 rotor diameters (RD). Thus, for instance, 14 feet rotor diameter turbines should be spaced approximately 50 feet apart depending on the location. BRAINS will use these literatures to determine the exact spacing of the six turbines once specific location is identified and the environmental and wind characteristics are taken into account.

In summary, as shown above, an optimal Poor Man's wind turbine can be constructed with parts cost of less than $150. Another feature of this wind turbine is that it is easy to build with simple manual instructions that will be provided by BRAINS. In addition, this turbine is scalable depending on how much power is needed. A typical family in third world country that doesn't have power hungry furniture in their house can build smaller version with less resources into its construction.

In order to increase efficiency and generate maximum power from each wind turbine, BRAINS will adapt the power optimization techniques that are put forward by the Strategic Energy Institute at Georgia Tech. Depending on the wind characteristics of the location, fixed or variable speed turbine will be chosen. Then optimization steps will be followed to insure maximum efficiency. Figure 4 on the next page shows flow chart model of the process. More details on the design and optimization of the Poor Man's wind turbine can be found on the references provided.

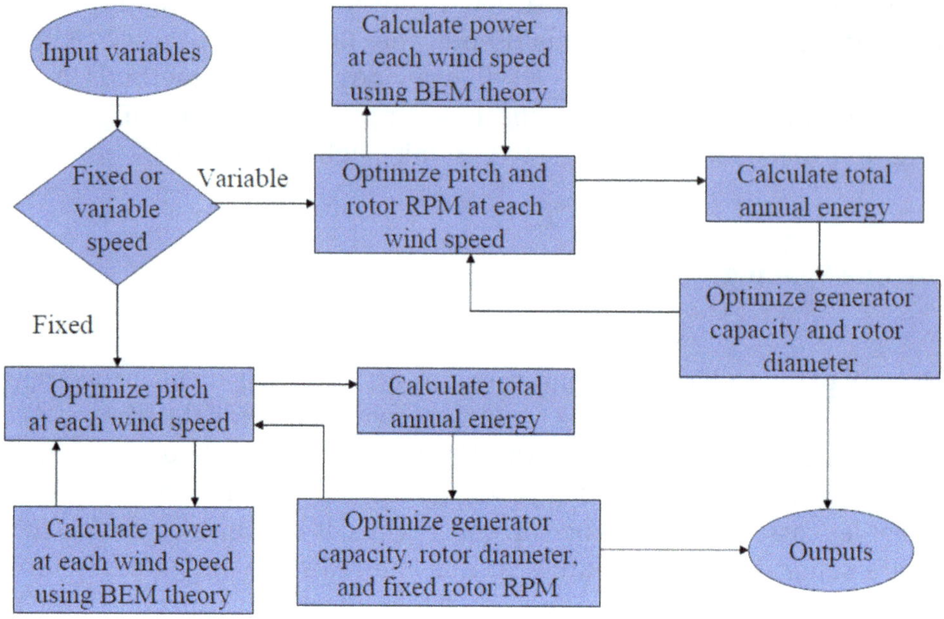

Figure 201 wind turbine optimization model (courtesy of GaTech)

REFERENCES:

1. Richard Lewis, *Poor Man's Guide to Wind Power and Battery Systems.*
2. Michael Schmidt, Strategic Energy Institute, *Georgia Institute of Technology,* www.energy.gatech.edu
3. National Lightning Safety Institute (NLSI), *Lightning Hazard Reduction at Wind Farms,* http://www.lightningsafety.com/nlsi_lhm/wind1.html
4. Brian McNiff, *Wind Turbine Lightning Protection Project,* National Renewable Energy Laboratory (NREL), http://www.nrel.gov/docs/fy02osti/31115.pdf
5. Renewable Energy Research Laboratory, University of Massachusetts at Amherst, *Wind Power: Siting in Communities,* http://www.ceere.org/rerl/about_wind/RERL_Fact_Sheet_4_Siting.pdf

Back up Power System Performance

Wind power is the primary source of power. The size of wind turbine installed in the Marshall Island ranges from 1 kW to 20 kW. In order to run the breadfruit processing system (from peeling to drying) about 5kw is required. When the wind power speed is less than expected and also when unexpected power outages happens because of , for instance, typhoons, back up power is needed to replace the primary power grid and run the breadfruit processing system. Therefore, batteries and generators are used as a backup power source.

Back Up Generators

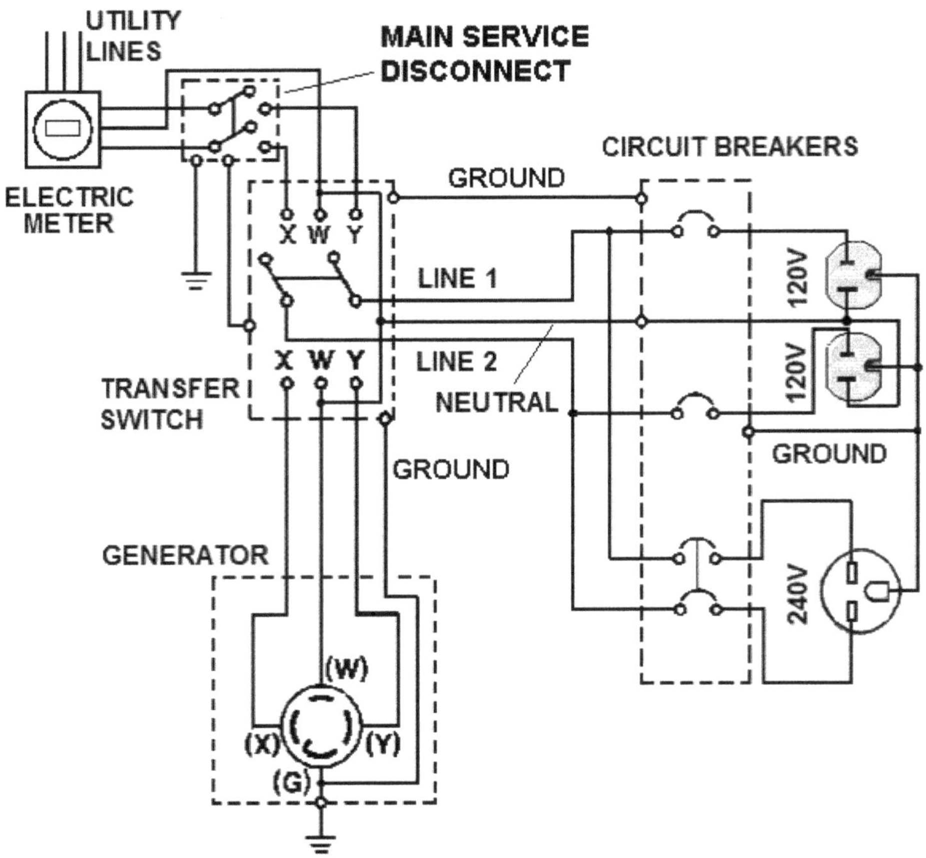

As it can be described in the above diagram, the main grid (wind turbine) is connected to Transfer Switches and the transfer switches in turn is connected to standby generator. An **automatic transfer switch** includes a control circuit that senses mains voltage. When utility power is interrupted, it starts up the generator, disconnects the load from the wind power turbine and connects it to the generator.

It then continues to monitor mains voltage, and when it is restored, switches your household wiring from the generator back to the utility. When the generator is disconnected it normally goes into a cool-down process and then automatically shuts down.

The Guardian 5kw air cooled standby generator with automatic transfer switches (ATS).

Battery System Backup

Besides, for short energy use, we use basic backup power package. This power backup package is prewired the inverter AC and DC sides with circuit protection in a single enclosure, with the inverter mounted on a swing door on the front. This compact assembly saves space, on-site time and component shopping.

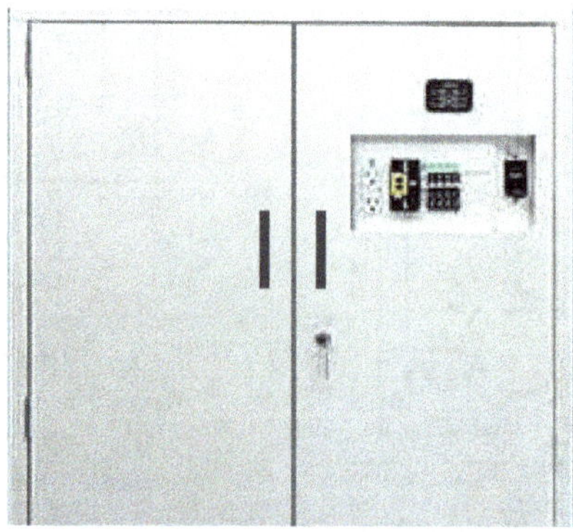

Above: 9.6 kWh Basic Backup Power Package

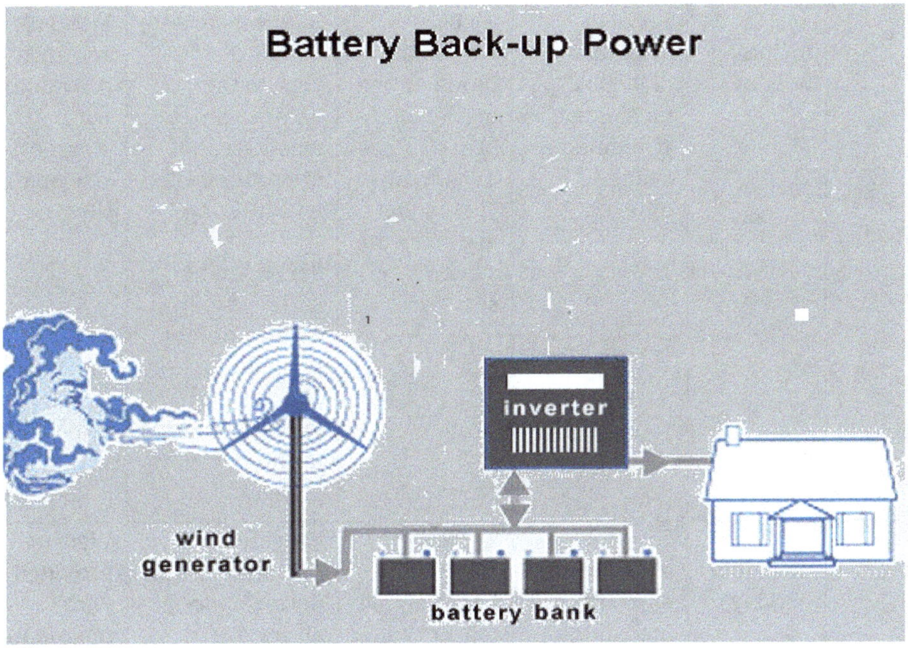

Battery Bank System

Major components in Basic Backup Power Package:

1. Inverters: Converts DC power in batteries to AC. It contains AC transfer relay that senses grid power. If grid power (Wind power) drops out, inverters switch in and continue to supply the load circuits. Here the size of the inverters is 6KW.
2. Batteries: Lead acid storage batteries typically 1.2 kilowatt-hour of storage each.

   These battery banks are automatically charged by grid power when it is available. If the grid goes down, power is drawn out of the batteries to run the essential appliances.

The basic power package will be located near the wind power generation in a situation where they are vented to the outdoors.

Techfluent IT Services Performance Summary

The following table provides a quick snapshot summary of the service performance that will be provided by Techfluent to BrEAD. Again, in preparing this summary, BRAINS used the standard IT outsourcing template provided by the IT Service Management Community (ITSM). See reference [1].

Service	Service Hours	Availability	Cost	Service Speed	Quantity	Risk Mitigation Plan

Email	24x7 including holidays	Service availability of 99.9% including scheduled change outages.	To be negotiated based on the market and other Techfluent clients that are using the same service.	Requests to add, delete, or change the name of an email account are processed within 1 day. Service to be restored within 1 hour for low to medium outages and within 2 hours for more severe outages.	Mailbox account provided with storage size up to 1 GB	100% of Email and calendar items deleted after 90 days. 100% of Email and calendar items are backed up every 24 hours.
Website	24x7 including holidays	Service availability of 99.9% including scheduled change outages.	To be negotiated based on the market and other Techfluent clients that are using the same service.	Provides maximum peek performance at all service hours. Service to be restored within 1 hour for low to medium outages and within 2 hours for more severe outages.	Website provided with upgradable 900 GB space and 80,000 GB traffic.	100% of sites backed up every 24 hours.
Instant Messaging	24x7 including holidays	Service availability of 99.9% including scheduled change outages.	To be negotiated based on the market and other Techfluent clients that are using the same service.	Requests for client software are processed within 1 –2 days (self service installation available immediately via website download). Service to be restored within 1 hour for low to medium outages and within 2 hours for more severe outages.	Unlimited	None

SharePoint	24x7 including holidays	Service availability of 99.9% including scheduled change outages.	To be negotiated based on the market and other Techfluent clients that are using the same service.	Requests to add, delete, or change the name of SharePoint are processed within 1 - 2 days. Service to be restored within 1 hour for low to medium outages and within 2 hours for more severe outages.	Up to 10 GB of storage space per site.	100% of sites backed up every 24 hours.

REFERENCES:

1. ITSM Community – IT Service Management Catalog template, http://www.itsmcommunity.org/downloads/Service_Catalog_Example.pdf
2. Marshall Islands National Telecommunications Authority (NTAMAR), http://www.ntamar.net.

Annex B – Detailed Reliability, Maintainability and Availability (RM&A) Predictions

1.9 Scope

The primary objective of Breadfruit Enterprise Architecture Demonstration (BREAD) is to provide quality flour for a low cost that able to satisfies the demand of the local population and also to promote worldwide. Furthermore, BrEAD likes to facilitate the design to a scalable breadfruit processing systems, to create a virtual collaborative knowledge sharing automation and communication system to share research and marketing information in real-time. This cannot be done if we don't have a firm RMA processes in place. This annex derives and supports our objective. It addresses reliability, maintainability, and availability (RMA) as essential elements of product capability. It focuses on what can be done to achieve satisfactory levels of RMA and how to assess RMA. This section introduces RMA, what it is, why it is important, and predictions for the RMA processes.

1.10 RMA Defined

RMA refers to three related characteristics of the BrEAD and its operational support: reliability, maintainability, and availability.

1.10.1 Reliability

Reliability is the probability of an item to perform a required function under stated conditions for a specific period of time. Reliability is further divided into product reliability and logistics reliability.

1.10.2 Maintainability

Maintainability is the ability of an item to be retained in, or restored to, a specific condition when maintenance is performed by personnel having specified skill levels, using prescribed procedures and resources, at each prescribed level of maintenance and repair.

1.10.3 Availability

Availability is the measure of the degree to which an item is in an operable state and can be committed at the start of a task when the task is called for at an unknown (random) point in time. Availability as measured by the user is a function of how often failures occur and corrective maintenance is required, how often preventative maintenance is performed, how quickly indicated failures can be isolated and repaired, how quickly preventative maintenance tasks can be performed, and how long logistics support delays contribute to down time.

1.10.4 Factors Affecting RMA

Many factors are important to RMA: BrEAD systems subsystems design; COT systems and subsystems quality; manufacturing quality; the environment in which the BrEAD those systems are transported, handled, stored, and operated; the design and development of the BreAD support systems; the level of training and skills of the people operating and maintaining the BREAD system; the availability of material required to repair the BREAD system; and the diagnostic aids and tools available to them. All these factors must be understood to achieve a system with a desired level of RMA. During pre-systems acquisition, the most important activity is to understand the users' needs and constraints. During BREAD system development, the most important RMA activity is to identify potential failure mechanisms and to make design changes to remove them. During production, the most important RMA activity is to ensure quality in manufacturing so that the inherent RMA qualities of the design are not degraded. Finally, in operations and support, the most important RMA activity is to monitor performance in order to facilitate retention of RMA capability, to enable improvements in design (if there is to be a new design increment), or of the BREAD support system (including the support concept, spare parts storage, etc.).

1.11 Importance of RMA

Achieving specified levels of RMA for the BREAD system is important for many reasons; specifically the affect RMA has on readiness, system safety, mission success, total ownership cost, and logistics footprint.

1.11.1 Readiness

Readiness is the state of preparedness of systems to meet a task, based on adequate and trained personnel, material condition, support systems, numbers of units available, etc. Poor RMA will cause readiness to fall below needed levels or increase the cost of achieving them. Effective diagnostics helps assure both system/task readiness and efficient repair/return to ready status.

System Safety

Inadequate reliability or false failure indications of components deemed Critical Safety Items (CSI) may directly jeopardize the safety of the user(s) of that component's system and result in a loss of life. The ability to safely complete a task is the direct result of the ability of the CSI associated with the BREAD system reliably performing to design intent.

1.11.2 Total Ownership Cost

The concept of Total Ownership Cost (TOC) is an attempt to capture the true cost of design, development, ownership and support of BREAD BREAD system. At the individual program level, TOC is synonymous with the life cycle cost of the BREAD system. To the extent that this system can be designed to be more reliable (fewer failures) and more maintainable (fewer resources needed) with no exorbitant increase in the cost of the BREAD subsystem components or spares, the TOC for this system will be lower.

1.11.3 Logistics Footprint

The logistics footprint of the BREAD system consists of the number of logistics personnel and the material needed in a given theater of operations. Improved RMA reduces the size of the logistics footprint related to the number of required spares, maintenance personnel, and support equipment as well as the force size needed to successfully accomplish a task.

Reliability

The Contractor's BREAD design uses system design features and components with proven operational experience. BREAD incorporates operationally proven subsystems. The proposed system requires minimal Non-Recurring Engineering (NRE) to adjust to the environment. Each system or subsystem consists of existing systems placed in similar environments and will easily meet the reliability requirements. The system has the ability to function in a wide range of environments, altitudes, and communications regimes. Each system has a path for future growth. Our system features and capabilities result in a solution that is innovative, flexible and low risk.
The proposed design uses existing Commercial Off the Shelf (COTS) processing infrastructure, and hardware components that have high Mean Time Between Failures (MTBF). The Contractor's approach integrates COTS products that provide a redundant network without requiring software modification or special network configuration. The total solution has system reliability performance that surpasses the Performance Based System Specification (PBSS) requirement.

2.1 Background

To identify reliability matrix, Table B.1 will be used for the BrEAD System.

Table B.1: Reliability Parameters

Parameter	Description
Failure Rate (λ)	The total number of failures within an item population, divided by the total time expended by that population,

	during a particular measurement interval under stated conditions.
Hazard Rate	Instantaneous failure rate. At any point in the life of an item, the incremental change in the number of failures per associated incremental change in time.
Mean Time Between Failure (MTBF)	A basic measure of reliability for repairable fielded systems. The average time between all system maintenance actions. Maintenance actions may be for repair or preventive purposes. (RAC Toolkit) An alternative definition: The time (i.e. operating hours, flight hours) between the need for maintenance actions to restore a system to fully operational condition, including confirmation that no fault exists (a No Defect maintenance action) This parameter provides the frequency of the need for maintenance and complements the labor hour parameter to project maintenance workload. This parameter is also used to identify unscheduled maintenance (MTBUMA) and Scheduled maintenance (MTBSMA)
Mean Time Between Repair (MTBR)	A basic measure of reliability for repairable fielded systems. The average time between all system maintenance actions requiring removal and replacement or in-situ repairs of a box or subsystem.
Mean Time Between Critical Failure (MTBCF)	A measure of system reliability that includes the effects of any fault tolerance that may exist. The average time between failures that cause a loss of a system function defined as "critical" by the customer. (RAC Toolkit)
Mean Time To Failure (MTTF)	A basic measure of reliability for non-repairable systems. Average failure free operating time, during a particular measurement period under stated conditions.

2.2 Ground Rules and Assumptions

The reliability analysis ground rules and assumptions have lead to a conservative reliability prediction. The following ground rules provide a framework for the reliability analysis.

- A serial MTBF prediction for each subsystem is sufficient for the reliability analysis.
- All MTBF numbers are evaluated using the Ground Benign environment.

2.2.1 Predictions

The bottom-up BREAD reliability prediction reveals that our system meets the MTBF requirements. The vendor supplied reliability data is a mix of predicted and field data. The BREAD reliability prediction for the system is based on the vendor data.

2.3 To calculate the reliability for the system:

Since an overall system reliability prediction can be made by looking at the reliabilities of the components that make up the whole system or product. We will examine the methods of performing such calculations. The reliability-wise configuration of components must be determined beforehand. For this reason, we will first look at different component/subsystem configurations, also known as structural properties. The components will be assumed to be statistically independent, which implies that failure information for one component provides no information about or does not affect other components in the system. There are two main types configuration. They are serial and parallel. Below is the description for each type.

2.3.1 Series Systems

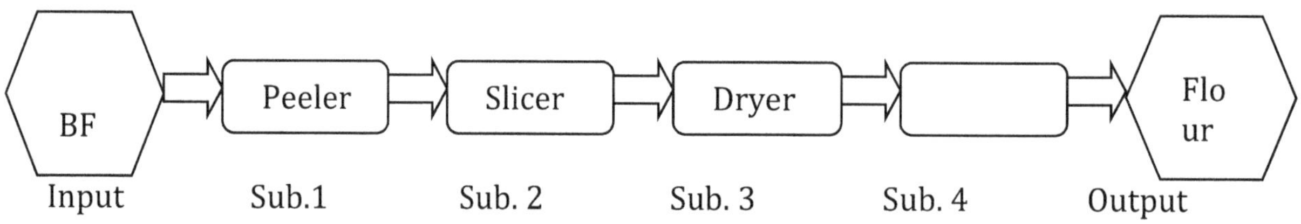

Figure B.1: Series Systems

In a series configuration, a failure of any component results in failure for the entire system. In most cases when considering complete systems at their basic subsystem level, it is found that these are arranged reliability-wise in a series configuration. For example, in our processing procedures, there are consist of four basic subsystems: the Peeling, slicing, drying, and grinding. These are reliability-wise in series and a failure of any of these subsystems will cause a system failure. In other words, all of the units in a series system must succeed for the system to succeed.
The reliability of the system is the probability that peeler subsystem succeeds slicers subsystem succeeds and all of the other subsystems in the system succeed. So, all subsystems must succeed for the system to succeed. The reliability of the system is then given by:

$$R_s = P(X_1 \cap X_2 \cap \ldots \cap X_n) = P(X_1)P(X_2 \mid X_1)P(X_3 \mid X_1X_2)\ldots P(X_n \mid X_1X_2\ldots X_{n-1})$$

$$R_s = P(X_1 \cap X_2 \cap \ldots \cap X_n) = P(X_1)P(X_2 \mid X_1)P(X_3 \mid X_1X_2)\ldots P(X_n \mid X_1X_2\ldots X_{n-1}) \quad (1)$$

Where:
RS = reliability of the system.
Xi = event of unit i being operational.
P(Xi) = probability that unit i is operational.

In the case where the failure of a component affects the failure rates of other components (i.e. the life distribution characteristics of the other components change when one fails), then the conditional probabilities in Eqn. (1) must be considered. However, in the case of independent components, Eqn. (1) becomes:

$$R_S = P(X_1)P(X_2)\ldots P(X_n)$$ $$R_S = P(X_1)P(X_2)\ldots P(X_n)$$

Or:

$$R_S = \prod_{i=1}^{n} P(X_i)$$ $$R_S = \prod_{i=1}^{n} P(X_i)$$

Or, in terms of individual component reliability:

$$R_S = \prod_{i=1}^{n} R_i$$ $$R_S = \prod_{i=1}^{n} R_i$$ (2)

2.3.2 Parallel Systems

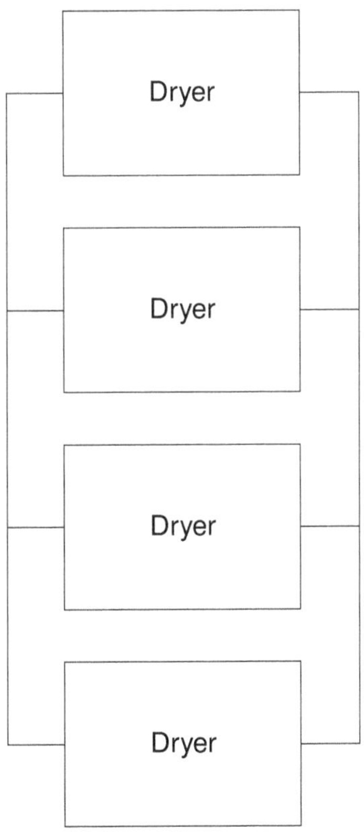

Figure B.2: parallel system.

In a parallel system as show above in Figure B.2 at least one of the dryers must succeed for the system to succeed. The probability of failure, or unreliability, for a system with n statistically independent parallel components is the probability that dryer 1 fails and dryer 2 fails and all of the other dryers in the system fail. So in a parallel system, all n dryer must fail for the system to fail. Put another way, if dryer 1 succeeds or dryer 2 succeeds or any of the n units succeeds, then the system succeeds and produce as expected output. However, if dryer 1 and 2 suceed, 3 and 4 fail, then the output product is 50% of the expected product. Again, the unreliability of the system is then given by:

$$Q_s = P(X_1 \cap X_2 \cap \ldots \cap X_n) = P(X_1)P(X_2 \mid X_1)P(X_3 \mid X_1X_2)\ldots P(X_n \mid X_1X_2\ldots X_{n-1})$$

$$Q_s = P(X_1 \cap X_2 \cap \ldots \cap X_n) = P(X_1)P(X_2 \mid X_1)P(X_3 \mid X_1X_2)\ldots P(X_n \mid X_1X_2\ldots X_{n-1}) \quad (3)$$

Where:

Qs = unreliability of the system.

Xi = event of failure of unit i.

P(Xi) = probability of failure of unit i.

In the case where the failure of a component affects the failure rates of other components, then the conditional probabilities in Eqn. (3) must be considered. However, in the case of independent components, Eqn. (3) becomes:

$$Q_s = P(X_1)P(X_2)\ldots P(X_n)$$ $$Q_s = P(X_1)P(X_2)\ldots P(X_n)$$

Or:

$$Q_s = \prod_{i=1}^{n} P(X_i)$$ $$Q_s = \prod_{i=1}^{n} P(X_i)$$

Or, in terms of component unreliability:

$$Q_s = \prod_{i=1}^{n} Q_i$$ $$Q_s = \prod_{i=1}^{n} Q_i \quad (4)$$

Observe the contrast with the series system, in which the system reliability was the product of the component reliabilities; whereas the parallel system has the overall system unreliability as the product of the component unreliability.

The reliability of the parallel system is then given by:

$$R_s = 1 - Q_s = 1 - (Q_1 \cdot Q_2 \cdot \ldots \cdot Q_n) = 1 - [(1 - R_1) \cdot (1 - R_2) \cdot \ldots \cdot (1 - R_n)] = 1 - \prod_{i=1}^{n}(1 - R_i)$$

2.3.3 Calculations and Results

The process system is a series configuration. The large scale process has sub-systems that setup in a parallel configuration as show in Figure B.3. Small scale process doesn't have specific parallel subsystems. A small scale process could execute each subsystem as a parallel task. For example, a person could peel all the breadfruits before slice, dry, and grind. The configuration for the small scale system is shown in Figure B.4 for non-parallel subsystems.

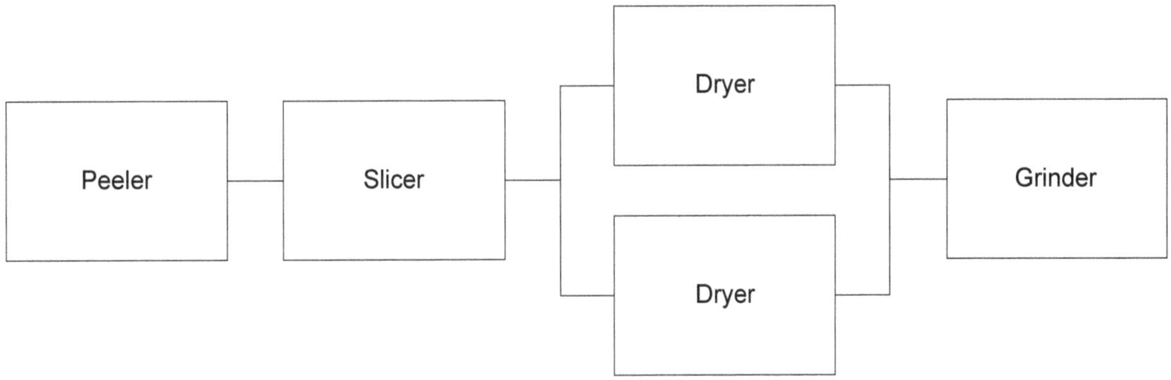

Figure B.3: Large Scale Process System

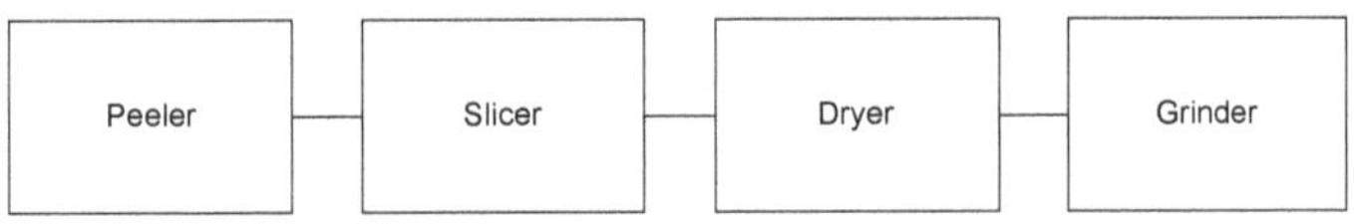

Figure B.4: Small Scale System

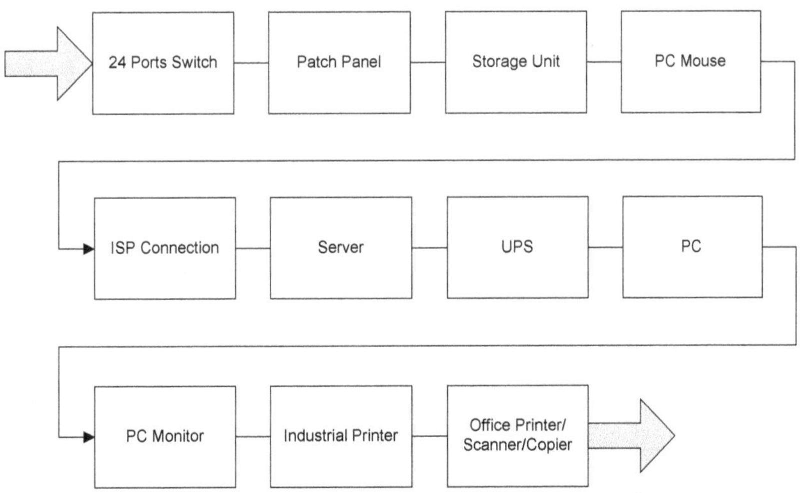

Figure B.5: System Integration Lab

From the figures above, the calculation for each system is shown below.

Small Scale Process:

Rsystem = Rpeeler * Rslicer * Rdryer * Rgrinder

Rsystem = .999 * .999 * .995 * .995

Rsystem = 98.8%

Large Scale Process:

Rsystem = Rpeeler * Rslicer * (1-(1-Rdryer) *(1-Rdryer)) * Rgrinder

Rsystem = .999 * .999 * (1-(1-.995)*(1-.995)) * .995

Rsystem = 99.3%

Small Scale Process:

Rsystem = Rswitch*Rpatch*Rstorage*Rmouse*Risp*Rserver*Rups*Rpc*Rmon*RI.print*RP.S.C

Rsystem = .913*.8*.8*.75*.75*.8*.5*.75*.75*.778*.5

Rsystem = 2.88

To calculate the MTBF for the system:

ExtendedF.R = Quantity*FailureRate

X=ExtendedF.R

MTBF = 1000000/X

Table B.2: Small Scale Series MTBF

Description (Small)	Required Quantity	Total Quantity	MTBF	F.R.	Extended F.R.
Peeler	1	3	3333300	.1	.3
Slicer	1	3	3333300	.1	.3
Dryer	1	1	2000000	.5	.5
Grinder	1	1	2000000	.5	.5
			Total Failure Rate		1.6
			MTBF		2916600

Table B.3: Large Scale Serial MTBF

Description (Large)	Required Quantity	Total Quantity	MTBF	F.R.	Extended F.R.
Peeler	1	2	5000000	.1	.2
Slicer	1	2	5000000	.1	.2
Dryer	2	2	1000000	.5	1
Grinder	1	1	2000000	.5	.5
			Total Failure Rate		1.9
			MTBF		6842100

Table B.3 shows that 2 dryers are needed for full operation capacity; the calculation result doesn't take into account that if one dryer fails, the output capacity is 50% the full operation capacity.

Table B.4: SIL Serial MTBF

Description (SIL)	Required Quantity	Total Quantity	MTBF	F.R.	Extended F.R.
24-port Network Switch	2	2	115000	8.7	17.4
Patch Panel	2	2	50000	20.0	40.0
Storage Unit	2	2	50000	20.0	40.0
PC Mouse	5	10	40000	25.0	250.0
ISP Connection	5	10	40000	25.0	250.0
Server	2	2	50000	20.0	40.0
UPS	2	2	20000	50.0	100.0
PC	5	10	40000	25.0	250.0
PC Monitor	5	10	40000	25.0	250.0
Industrial Printer	2	5	45000	22.2	111.1
Office Printer/Scanner/Copier	1	1	20000	50.0	50.0
			Total Failure Rate		1398.5
			MTBF		364.68

Maintainability

Many different parameters are used for maintainability. They include quantitative measures such as mean time to repair (MTTR), max time to repair (Mmax), and maintenance ratio (MR). Table 5 lists some of these quantitative measures that are mainly concerned with time. Maintainability is also a function of finding failures; therefore diagnostics is important and is characterized with metrics such as built-in-test (BIT) effectiveness, fault detection, isolation and false alarm rates. Maintainability is also concerned with economical considerations and ease of maintenance. The ease of maintenance is indirectly indicated, or measured, by accessibility, accuracy of diagnostics, level of standardization, and human factors-related considerations. Features of the design, such as the level and accuracy of embedded diagnostics instrumentation and prognostics, can increase the maintainability of the BrEAD Processing system. Table B.5 identifies several maintainability metrics that will be used for the BrEAD system.

3.1 Background

Table B.5: Quantitative Measures of Maintainability

Parameter	Description
Mean Time to Repair (MTTR)	The total number of failures within an item population, divided by the total time expended by

Also called Mean Corrective Maintenance Time (Mct)	that population, during a particular measurement interval under stated conditions.
Maximum Active Corrective Maintenance Time (Mmax)	That value of maintenance downtime below which one can expect a specified percent of all corrective maintenance actions to be completed. Must be stated at a given percentile point, usually the 90th or 95th. Primarily related to the lognormal distribution.
Mean Preventive Maintenance Time (Mpt')	A composite value representing the arithmetic average of the maintenance cycle times for the individual preventive maintenance actions (periodic inspection, calibration, scheduled replacement, etc.) for a system.
Median Active Corrective Maintenance Time (Mct)	That value of corrective maintenance time that divides all downtime values for corrective maintenance such that 50% are equal to or less than the median and 50% are equal to or greater than the median.
Mean Active Maintenance Time (Mct')	The mean or average elapsed time needed to perform maintenance (preventive and corrective), excluding logistic and administrative delays.
Mean Time to Restore System (MTTRS)	For highly redundant systems, the mean or average time to switch to a redundant backup unit.
Mean Downtime (MDT)	The mean or average time that a system is not operational due to repair or preventative maintenance. Includes logistics and administrative delays.
Maintenance Ratio (MR)	A measure of the total maintenance labor burden required to maintain an item. It is expressed as the cumulative number of labor hours of maintenance expended in direct labor during a given period divided by the cumulative number of life units during the same period.
Percent BIT Fault Detection (Pfd)	The ratio of the number of faults detected by the system BIT to the total number of faults experienced by the system, expressed as a percent.
Percent BIT Fault Isolation (Pfi)	The ratio of detected faults that was unambiguously isolated to a single replaceable unit or other rule identified in the procurement specification (i.e. to a group of 3 or less replaceable units).
Percent False Alarms (Pfa)	The ratio of detected (indicated) failures to the total indicated failures plus verified failures, expressed as a percent.

3.2 Prediction

The BREAD system incorporates maintainability design solutions from industry best practices including knowledge obtained from previous experience installing similar type equipment. For BREAD a high reliability factor with low maintenance solutions has been chosen for the process system and subsystems.

- Simplicity – Our design incorporates easy accessibility and servicing for assembly racks in Central. As some of our tool could be found or made easily with any scrap parts
- Smart design – Standardized mounting brackets ensure easy of repair of field equipment
- Logistics – Common parts have been selected for use to lower the amount of inventory required on hand.
- Supportability – Built-in test and fault isolation are designed into the system.

3.2.1 Ground Rules and Assumptions

Our maintainability analysis uses several ground rules and assumptions that lead to a conservative maintainability prediction. The following ground rules provide a framework for the maintainability analysis.

- Maintainability predictions and Line Replaceable Unit (LRU) replacement are acceptable at the specified level in the tables shown in Section.
- All tools are available at the maintainer.
- Adequate support/help either with COTs product or BRAINS is available at the time of repair.
- All replacement parts have the same Size, Weight, and Power requirements.

3.3 Maintainability Predictions

The Mean Time to Repair (MTTR) is used to measure the Maintainability of the system. Maintainability predictions below are calculated based on a Fault Detection (FD), Fault Isolation (FI), Remove, and Replace Times. These times are predicted for each piece of equipment. Our maintenance approach for the system shows compliance when calculated with the availability requirements as shown in the Availability Section

The calculation of the MTTR is as follows:

X is any part

$$X = FD + FI + Rmove + Rplace$$

$$MTTR = Average\,(X1 \ldots\ldots Xn)$$

Our design solution and features resulted in predicted maintenance time at IOC that surpasses the availability requirement. Please reference Table B.6, Table B.7, and Table B.8 below for the predicted Central, SIL, and Node repair times (respectively).

Table B.6: PROCESS LARGE SCALE MTTR Prediction

Description (Central)	Total Quantity	FD/FI	Remove	Replace	Total (Mins)	Total (Hrs)
Peeler	6	1	5	5	11	.183
Slicer	6	1	5	5	11	.183
Dryer	12	5	15	30	50	.833
Grinder	6	10	20	20	50	.833
					MTTR	.508

Table B.7: SIL MTTR Prediction

Description (SIL)	Total Quantity	FD/FI	Remove	Replace	Total (Mins)	Total (Hrs)
24-port Network Switch	2	5	10	10	25	0.42
Storage Unit	2	5	20	20	45	0.75
Server	2	20	15	40	75	1.25
UPS	2	10	30	30	70	1.17
PC	10	5	5	5	15	0.25
Patch Panel	6	5	5	5	15	0.25
PC Mouse	10	5	5	5	15	0.25
ISP Connection	10	20	10	5	35	0.58
PC Monitor	10	5	5	5	15	0.25
Industrial Printer	5	5	20	20	45	0.75
Office Printer/Scanner/Copier	1	5	10	10	25	0.42
					MTTR	0.57

Table B.8: OTHER MTTR Prediction

Description (Node)	Total Quantity	FD/FI	Remove	Replace	Total (Mins)	Total (Hrs)
Wind Turbine	40	10	30	38	78	1.30
Equipment Dull	10	10	20	30	60	1.00
Network Connection	1	10	10	20	40	0.67
Server Hardware	25	10	40	40	90	1.50
Environment	1	10	10	10	30	0.5
					MTTR	0.99

Availability

The BREAD system design incorporates the optimal design solution in order to exceed the Operational Availability (Ao) requirements. By having a high availability for the system we are best able to meet the requirement of being operational 8 hours a day for 150 days in a season. To obtain the high availability, the system has to meet these requirements:

- MTBF has to be high
- MTTR has to be low

- Design of subsystems has to support MTBF and MTTR

4.1 Ground Rules and Assumptions

The BREAD system has several ground rules and assumptions that help guide the design. The following are parts of the process:

The tools required for repairing the system are standard and not special test equipment is required.
The information obtained from vendors is realistic and the equipment has been sufficiently fielded to prove the numbers.
Spare parts are available as needed.

4.2 Availability Predictions

The Ao is used to measure the effective operational (uptime) time of the system. The Ao calculation for the system (subsystems) is based on the Reliability and Maintainability predictions.

The equation for the Ao is:

Ao = MTBF/(MTBF + MTTR)

To calculate the overall system availability the equation is:

Ao = Ao1 * Ao2 ......Ao3

When accounting for redundancy use:

Ao = 1-(1-Ao)N)

The BREAD design solution resulted in predicted Ao time at Initial Operational Capability (IOC) that surpasses the Availability requirement.

Summary

The BREAD system design meets the Reliability, Maintainability, and Availability requirements as specified in the RVTM. The RMA evaluation of the system has shown that the design uses equipment with high MTBFs. The design and use of equipment that is easily repaired or replaced helps to drive the Operational Availability requirements of the system. By designing and using parts as prescribed throughout this document we are able to drive a high Operational Availability of at least 99.9% exceeding the stated requirement.**Technical Attachment 1 – Breadfruit Enterprise Architecture Demonstration (BrEAD) PBSS Tailorable Specification and Verification Matrix**

Requirements Allocation

Requirements allocation was done by BRAINS in order to ensure tracking of system design data is properly delivered in the CDRL deliveries. In this regard a traceability matrix was developed to ensure the proper CDRL contained the contract data. In some cases where there is not an appropriate CDRL, the requirement is traced to test procedures or reports.

ID	Text	Type	CDRL
PBSS_1	1. The BrEAD shall be comprised of the Breadfruit Enterprise Architecture Demonstration (BrEAD) that forms the organizational structure that will: be an advocate for breadfruit as a component of the next green revolution, design, market, collect data, distribute and provide logistics support for breadfruit processing systems and the System Design (SD) for breadfruit processing equipment.	Requirement	0001, 0008, 0011, 0013, 0017, 0019
PBSS_2	2. The Breadfruit Enterprise Architecture Demonstration (BrEAD) shall provide an automated research, marketing, and user data collection capability for NGRF head quarters in St. Paul, Minnesota that will provide 24 hour per day data from researchers, marketers and users world wide.	Requirement	0001, 0005, 0008, 0011, 0019
PBSS_3	3. The BrEAD shall provide organizational liaison to the Community of Interest (COI) and automated electronic connectivity for real time data interchange. The community of interest is comprised of the NGRF, government and private organizations that have or could have an interest in breadfruit products.	Requirement	0002, 0005, 0008, 0019
PBSS_4	4. The initial baseline breadfruit systems design shall be based on the characterization of any prototype existing breadfruit processing systems. The characterization shall include, but not be limited to; operator usability, reliability, maintainability and manufacturability. The findings shall be documented in a requirements document and a product description.	Requirement	0001, 0005, 0008, 0011, 0013, 0016, 0018, 0019
PBSS_5	5. The breadfruit processing system shall be operable in geographic regions that do not have an electrical power infrastructure and in geographic regions that have an electrical power infrastructure with minimal equipment differences.	Requirement	0001, 0006, 0008, 0015, 0017, 0018
PBSS_6	6. The design of the breadfruit processing system shall be modular, such that any specific geographic requirements can be accommodated without breadfruit processing system redesign.	Requirement	0001, 0006, 0008, 0015, 0017, 0018
PBSS_7	7. The breadfruit processing system human machine interface shall be intuitive; that is the human operation of the system shall require minimal documentation- ideally none.	Requirement	0008, 0013, 0014, 0015, 0016
PBSS_8	8. Any design modifications that are indicated from operational data gathered from users of breadfruit processing systems in use in the field shall be analyzed and considered for incorporation as design modifications to future the breadfruit processing system designs.	Requirement	0004, 0008, 0012, 0013, 0014, 0015, 0016

PBSS_9	9. The breadfruit processing system shall be easily repairable with a minimum of tools in the field	Requirement	0008, 0013, 0014, 0015, 0016
PBSS_10	10. The breadfruit processing system shall have a failure rate of one failure per 10, 000 hours.	Requirement	0008, 0011, 0012
PBSS_11	11. The mean time to repair the breadfruit processing system shall be 15 minutes.	Requirement	0008, 0013, 0014, 0015, 0016
PBSS_12	12. The breadfruit processing system shall extendible. Extendibility is the capability of the breadfruit processing system to produce more than one food product or food product component.	Requirement	0001, 0008, 0011, 0018
PBSS_13	13. The breadfruit processing system shall be scalable ; the range of scalability to be considered is from family use to industrial scale use that could provide large quantities of breadfruit products. The range anticipated is from a few ponds per day per family to several tons per day for an industrial scale facility. The specific values shall be determined and documented in the proposal.	Requirement	0001, 0008, 0015, 0016, 0018
PBSS_14	14. The BrEAD shall provide weather data gathered through out the world where breadfruit is used. It shall provide hourly temperature, pressure, humidity, and rain fall data to the Central Data Collection and Analysis Center of the NGRF located in St. Paul, Minnesota.	Requirement	0001, 0008, 0005, 0008,
PBSS_15	15. The BrEAD architecture shall have distributed local weather sensors, and data collection systems and shall be interoperable with the Central Data Collection and Analysis Center. The user data collection human machine interface at the data collection facilities shall be intuitive to the user.	Requirement	0001, 0005, 0008, 0010, 0011, 0015, 0016, 0018
PBSS_16	16. The breadfruit processing system shall operate from 40 degrees Fahrenheit to 120 degrees Fahrenheit.	Requirement	0018
PBSS_17	17. The Systems Integration Laboratory (SIL) shall be used to evaluate candidate architectures for the BrEAD , the breadfruit processing systems, and the intuitive human interface capability desired.	Requirement	0014, 0015, 0016, 0018
PBSS_18	18. The breadfruit processing shall operate in 100% relative humidity.	Requirement	0012, 0018

The list of derived requirements below shows the element ID, the derived requirement, and the verification method. The verification method is identified by the codes I, A, D, T, which stands for Inspection, Analysis, Demonstration, and Test respectively. The Verification and Validation section describes how BRAINS will use these to satisfy the updated PBSS requirements.

ID	Derived	Verification Method
H1	**Breadfruit Advocate and System Design**	
D1.1.1	The BrEAD shall conduct advertising campaigns promoting Breadfruit uses, health benefits, and environmentally friendly processing and production as compared to other food products.	I

D1.1.2	The BrEAD shall establish a website containing information on breadfruit.	I
D1.1.2.1	The BrEAD website shall provide recipe sharing.	I
D1.1.2.2	The BrEAD website shall provide information on how to process the breadfruit.	I
D1.1.2.3	The BrEAD website shall provide nutritional information on breadfruit.	I
D1.1.2.4	The BrEAD website shall provide a biography on the benefits of the architecture for the local community.	I
D1.1.2.5	The BrEAD website shall provide a description of the processing system.	I
D1.1.2.6	The BrEAD website shall provide information on the different varieties of breadfruit and where its found.	I
D1.1.2.7	The BrEAD website shall contain a "General Information" section describing our mission and purpose.	I
D1.1.2.8	The BrEAD website shall contain shelf-life information for breadfruit.	I
D1.1.2.9	The BrEAD website shall contain information on where to purchase breadfruit.	I
D1.1.2.10	The BrEAD website shall contain the BRAINS production architecture showing how breadfruit is harvested and delivered as flour.	I
D1.1.2.11	The BrEAD website shall explain how efficiency in producing breadfruit reduces environmental pollutants.	I
D1.1.2.12	The BrEAD website shall provide a user message board allowing forum discussion of breadfruit.	I
D1.1.2.13	The BrEAD website shall allow donations to be submitted online to provide cash or certain products to needed developing areas with breadfruit.	I
D1.1.2.14	The BrEAD website shall provide the capability for managing website content and updating news pages	I
D1.1.2.15	The BrEAD website shall contain breadfruit harvest and production numbers for current and past seasons as provided from the NGRF marketing database.	I
D1.2.1	BrEAD shall design a product that processes breadfruit into flour	D
D1.2.1.1	The processing system shall be established in 6 strategically located process facilities	A
D1.2.1.1.1	The processing facility shall have a receiving area for breadfruit deliveries by truck	A
D1.2.1.1.2	The processing facility shall have 3 phase electrical power at 240VAC	A
D1.2.1.1.3	The processing facility shall have single phase electrical power at 120VAC	A
D1.2.1.1.4	The processing facility shall have windmill power generation	A,D
D1.2.1.1.4.1	The wind turbines shall provide 3kW per day minimum of power	A
D1.2.1.1.5	The processing facilitiy shall use battery power for wind generated electrical storage	A,D
D1.2.1.1.6	The processing facility shall provide backup power capable of servicing all equipment	A,D
D1.2.1.1.7	The processing facility shall use a diesel powered generator for its source of backup power	A,D
D1.2.1.1.8	The processing facility shall have a backup generator	I
D1.2.1.1.8.1	The backup generator shall start on detection of main power failure	D
D1.2.1.1.8.2	The backup generator shall provide power to meet the facilities full operational needs without interuption	A,D
D1.2.1.1.9	The processing facility shall provide sufficient space for the breadfruit storage, processing, and breadfruit flour storage	A,I

D1.2.1.1.10	The processing facility shall provide work tables for workers to peel and slice breadfruit	I
D1.2.1.1.11	The processing facility shall provide storage for sliced breadfruit that is air sealable	I
D1.2.1.1.12	The processing facility shall provide means for drying ovens to be installed	A,I
D1.2.1.1.13	The processing facility shall have two conveyer belts for dried breadfruit to be placed on going to the grinder hopper.	A,I
D1.2.1.1.14	The processing facility shall have a grinder capable of grinding dried breadfruit into flour	D
D1.3.1	BrEAD shall conduct advertising campaigns that inform potential consumers and buyers of the breadfruit processing system the economic benefits of the system.	A
D1.4.1	BrEAD shall collect relevant data to support BrEAD Enterprise operations and requirements worldwide	A
D1.4.1.1	BrEAD shall collect demand data for service locations by comparing sold product to inventories.	A
D1.4.1.2	BrEAD shall provide current production rates to BrEAD on a daily basis.	A,T
D1.4.1.3	BrEAD shall provide projected future production rates that meet projected future demand and satisfy negative inventories to BrEAD on a daily basis.	A
D1.4.1.4	BrEAD shall provide communications links from local processin plants to local offices providing real time data on the processign plant	T
D1.4.1.5	BrEAD shall provide communications links from local offices that transmit the data from processing plants to the NGRF servers	T
D1.5.1	BrEAD shall provide a distribution plan that will allow delivery of the breadfruit processing system to customers.	A
D1.5.1.1	The distribution plan will identify locations of facilities	A
D1.5.1.2	The distribution plan will identify methods to distribute breadfruit flour	A
D1.5.1.3	The distribution plan will identify how the breadfruit small scale grinders will be sold to locals	A
D1.6.1	The BrEAD shall have an inventory tracking tool to coordinate shipping and receiving logistics.	T
D1.6.1.1	The BrEAD inventory tracking tool should be automated to recognize low inventory and reorder when necessary.	T
D1.7.1	BrEAD shall document engineering design for breadfruit processing equipment.	A
H2	**Marketing**	
D2.1.1	The BrEAD shall establish an automated information sharing system.	A
D2.1.1.1	The BrEAD information sharing system shall document marketing strategies and consumer response.	A

D2.1.1.2	The BrEAD information sharing system shall be linked to the Breadfruit Institute's database.	T
D2.1.1.3	The BrEAD information sharing system shall be accessible 24 hours per day via a secured website.	T
D2.1.1.4	The BrEAD marketing team shall promote BRAINS architecture as a green company	A
D2.1.1.5	The BrEAD marketing team shall promote breadfruit via local print advertisement	D
D2.1.1.6	The BrEAD marketing team shall promote breadfruit via local television advertisement	D
D2.1.1.7	The BrEAD marketing team shall promote breadfruit via internet advertisement	D
H3	**Organizational Liaisons**	
D3.1.1	The BrEAD shall establish a committee containing a Community of Interest representative, a government representative, a BrEAD representative, and an NGRF representative for each region of interest.	N/A
D3.1.1.1	The BrEAD shall establish a real-time forum to enable communications between committee members.	D
D3.1.1.2	The BrEAD shall store data collected at processing plants in a web based database that will allow a Community of Interest representative to securely access and view breadfruit processing statistics.	T
D3.1.1.3	The database shall store the following information: the number of breadfruits processed each day, end of day inventory, daily amount of shipped product.	T
H4	**Design Reuse**	
D4.1.1	The BrEAD shall leverage existing processing knowledge by researching current techniques.	A
D4.1.1.1	The BrEAD shall provide a white paper of current uses, processing techniques, operator usability, reliability, maintainability, and manufacturability.	A
D4.1.1.2	The BrEAD shall provide a trade study of commercially available equipment to be used in our processing system.	A
H5	**Power**	
D5.1.1	The breadfruit processing system shall be operable without electrical infrastructure provisions.	T
D5.1.1.1	The breadfruit home use processing system shall be operable manually.	D
D5.1.1.2	The breadfruit processing system shall have a power supply and regulator that allows use of at least the following power sources: generator, solar panels, wind turbines, or water turbines.	A
D5.1.1.2.1	The BrEAD shall utilize wind power through the use of windmill turbines to generate enough electrical power to sustain factory operation.	A,T

D5.1.1.3	The BrEAD shall utilize a 5kW generator to provide uninterupted power to the factory.	T
D5.1.2	The breadfruit processing system shall be design to use existing power infrastructures.	D
D5.1.2.1	The breadfruit processing system shall run off of 110VAC power.	A
D5.1.2.2	The breadfruit processing system shall be designed to run off 240VAC power.	A
D5.1.2.3	The breadfruit processing system shall be capable of using wind generated power if no power infrastructure exists.	A,T
H6	**Modularity**	
D6.1.1	The breadfruit processing subsystems shall be modular.	A
D6.1.2	The breadfruit processing components shall use modular and Commercial Off the Shelf (COTS) components when available.	A
D6.1.3	The BrEAD processing system shall be operable in any geographical region.	A
D6.1.3.1	The BrEAD home use processing equipment shall be operable in any geographical region.	A
D6.1.3.2	The BrEAD large scale production facility shall be operable in any geographical region.	A
H7	**Human Engineering**	
D7.1.1	The BrEAD system shall be capable of operation with minimal training.	A
D7.1.2	The Breadfruit system shall maximize graphical instructions and descriptions.	I
D7.1.3	The BrEAD shall provide training on operating the breadfruit system.	A
D7.1.4	A Human Machine Interface (HMI) certified consultant shall be involved in the design of the BrEAD process.	A
D8.1.1	An Human Machine Interface (HMI) certified consultant shall be used to analyze user operation for potential design changes to the BrEAD process.	A
D8.1.2	The BrEAD shall maintain a collection of spare parts for common equipment used in processing.	A
D8.1.3	The BrEAD shall provide technical support for on-site repairs.	D
D9.1.1	The BrEAD user manuals shall contain a troubleshooting section for isolating and repairing problems.	A
D9.1.1.1	The BrEAD shall provide a user manual for operating and maintaining the system.	A
D9.1.1.2	The BrEAD shall provide maintenance information for the Breadfruit system.	A
D9.2.1	The breadfruit processing system shall require less than 10 tools in the field to repair	A
H8	**Reliability/Maintainability**	
D10.1.1	The BrEAD processing system shall have a failure rate of one failure per 10,000 hours.	A
D10.1.1.1	The BrEAD component trade study shall include all Mean Time Between Failure (MTBF) data from COTS vendors.	A
D10.1.1.2	The SIL shall provide prototype testing and MTBF analysis for processing component selection to the BrEAD.	A

D10.1.2	The BrEAD shall provide data on the reliability and life cycle of the BrEAD system.	I
D11.1.1	The Mean Time to Repair (MTTR) for the BrEAD Processing system shall be 15 minutes.	A
D11.1.1.1	The BrEAD component trade study shall include all MTTR data from COTS vendors.	A
D11.1.1.2	The SIL shall provide prototype testing and MTTR analysis for processing component selection to the BrEAD.	A
H9	**Scalability/Extendibility**	
D12.1.1	The BrEAD processing system shall be extendible.	A
D12.1.1.1	The BrEAD factory subsystem shall be modular and capable of outputting various breadfruit products with minimal system changes.	A
D12.1.1.2	The breadfruit processing system shall have an adjustable grinder that allows production of flour or coarse breadfruit.	T,D
D13.1.1	The breadfruit processing system shall be scalable.	A
D13.1.1.1	The BrEAD will provide design solutions for the various levels of scalability.	A
D13.1.1.2	The BrEAD system shall be scalable by changing components while maintaining the original process.	T
D13.1.2	The industrial scale solution shall provide 370 pounds of breadfruit flour per hour of operation.	T,D
D13.1.3	The family use scale solution shall provide 23 pounds per hour of breadfruit flour when operated by hand.	T,D
H10	**Weather Systems**	
D14.1.1	Each BrEAD processing center shall contain a weather reporting unit that collects local weather data on an hourly basis.	T
D14.2.1	Each weather reporting unit will report temperature, pressure, humidity, and rain fall data to the BrEAD Central Data Collection and Analysis Center via hourly uplink.	T
D14.2.1.1	The weather station shall report temperature, pressure, humidity, and rain fall data to the local plant computers.	T
D14.2.1.2	The local plant computers shall forward environmental data from the weather station to the central data center servers via web link hourly.	T
D15.1.1	The Breadfruit processing centers shall have a set of sensors that read temperature, pressure, humidity, and rain fall.	T
D15.2.1	The weather reporting unit of each breadfruit processing center shall provide the weather data via IP to the Central Data Collection and Analysis Center.	T
D15.3.1	The BrEAD weather data collection at the collection facilities shall be fully automated such that no manual interaction is required to send data to the CDCAC.	T,D
H11	**Environmental**	

D16.1.1	The BrEAD processing system shall operate from 40 degrees Fahrenheit to 120 degrees Fahrenheit.	T
D16.1.1.1	The grinder shall operate in ambient temperatures from 40 degrees Fahrenheit to 120 degrees Fahrenheit.	T
D16.1.1.2	The drying ovens shall operate in ambient temperatures from 40 degrees Fahrenheit to 120 degrees Fahrenheit.	T
D16.1.1.3	The conveyer belts shall operate in ambient temperatures from 40 degrees Fahrenheit to 120 degrees Fahrenheit.	T
D16.1.1.4	The drying ovens shall operate in ambient temperatures from 40 degrees Fahrenheit to 120 degrees Fahrenheit.	T
D16.1.1.5	The storage containers shall maintain less than 100 degrees Fahrenheit in ambient temperatures from 40 degrees Fahrenheit to 120 degrees Fahrenheit.	T
H12	**Test and Integration**	
D17.1.1	The Systems Integration Lab (SIL) shall provide data on process improvement recommendations to BrEAD.	A
D17.1.2	The breadfruit institute shall provide research and design data to the BrEAD.	A
D17.1.3	The BrEAD shall provide process feedback to the BrEAD.	A
D17.1.4	The BrEAD shall collect and analyze design feedback from the SIL, the breadfruit institute, and BrEAD.	A
D17.1.5	The BrEAD shall incorporate design feedback and provide future design recommendations to the SIL, the breadfruit institute, and BrEAD.	A
D17.1.6	The BrEAD shall coordinate the design improvements among the SIL, the breadfruit institute, and BrEAD.	A
D17.1.7	The SIL shall utilize certified HMI experts for the design and development of the BrEAD processing system.	A
D17.1.8	The SIL shall be used to evaluate candidate architectures for the BrEAD.	A
D17.1.9	The SIL shall be used to evaluate the BrEAD processes and provide process design recommendations.	A
D17.1.10	The SIL shall support bread fruit processing simulation software	T,D
D17.1.10.1	The processing software shall be able to simulate the BrEAD architecture components	T
D17.1.10.2	The processing software shall be able to support the modeling of real and theoretical data	T
D17.1.11	The SIL shall support structural software for the test of the component of the processing system	D
D17.1.11.1	BRAINS shall be able to generate recommendations for improvement	A
D17.1.11.2	The structural software shall be able to support the modeling of real and theoretical data	D
D17.1.12	The SIL shall support dryer modeling software for the test of the drying process	T
D17.1.12.1	The dryer modeling software shall be able to generate to support recommendations for dyer improvement	T
D17.1.12.2	The dryer modeling software shall be able to support the modeling of real and theoretical data	T
D17.1.13	The SIL shall support subsystem/system test	D
D17.1.14	The SIL shall support integration test	D
D17.1.15	The SIL shall support verification and validation	D
D17.1.16	The SIL shall support system/software lifecycle sustainability	A

D17.1.17	Changes and Improvements to the processing system shall be managed by the SIL	A
D17.1.18	The BrEAD shall coordinate the design improvements among the SIL, the breadfruit institute, and BrEAD	A
D17.1.19	The SIL shall provide prototype testing and MTTR analysis for processing component selection to the BrEAD	T
D17.1.20	The SIL shall provide for extracting data from test points at internal boundaries between subsystems	T
D17.1.21	The SIL shall provide for extracting data from the normal system outputs	T
D17.1.22	The SIL shall be designed to be flexible enough to accommodate system updates	A,D
D18.1.1	The breadfruit processing system shall operate in 100% relative humidity.	T
D18.1.1.1	The grinder shall operate in 100% relative humidity.	T
D18.1.1.2	The drying ovens shall operate in 100% relative humidity.	T
D18.1.1.3	The conveyer belt assemblies shall operate in 100% relative humidity.	T
D18.1.1.4	The storage containers shall maintain less than 70% humidity in 100% relative humidity environments.	T
D18.1.1.5	The generator shall operate in 100% relative humidity	T

Attachment 2 - OV-1

This section describes a high-level operational overview of the main operational nodes. These operational nodes will be described since each one participates and successfully will support the breadfruit process to meet all the requirements. This section describes the participants, their geographical location, main responsibilities, communication processes with other entities and stakeholders, etc. This section also describes how the team members of this group decided to focus most of our energy and effort to come up with the most down to earth proposal; our strategy focuses on a specific geographical area specifically the Marshall Islands, Minneapolis/St Paul, and Hawaii. Thus a formal description of each one is provided with figures, facts, and detailed descriptions. Finally, this section explains how this system is flexible to meet the requirements at any remote geographical location with a detailed explanation from a top-bottom and vice-versa perspectives.

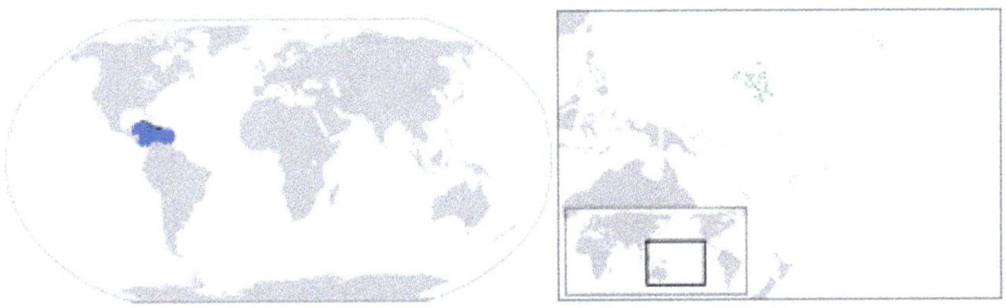

Figure 1- From left to right: (a) Caribbean (blue area) and (b) Marshall Islands inside the rectangle

Breadfruit is now grown in close to 90 tropical countries around the globe (See Figure 1) and holds great promise for alleviating hunger in the tropics (see Figure 2). Breadfruit is cultivated on most Pacific islands, with the exception of New Zealand and Easter Island. In the late 1700s several seedless varieties were introduced to Jamaica and St. Vincent from Tahiti, and a Tongan variety was introduced to Martinique and Cayenne via Mauritius. These Polynesian varieties were then spread throughout the Caribbean and to Central and South America, Africa, India, Southeast Asia, Madagascar, the Maldives, the Seychelles, Indonesia, Sri Lanka, and northern Australia. Breadfruit is also found in south Florida. Our proposal understands the importance of building a flexible enterprise and breadfruit process models. We identified the Tropic of Cancer as the zone where most of these countries are geographically located.

Figure 2 – East Pacific countries harvesting breadfruit. Blue sign at the right indicates relative location of the Marshall Islands to Australia and Asia.

A. Republic of Marshall Islands

Describing the RMI geographically, political, socially and cultural provided us with a better understanding of this country policies, life style, who are the main participants and key actors to guarantee the success of our project proposal. It is unimaginable how much learned, and how this learning process opened our eyes to reality. The bottom line is that we identified much more in detailed stakeholders, their needs (user needs), and from there on we increased the list of requirements, and finally created a robust and flexible system architecture that will satisfy their needs.

On this section you will learn about the geographical location and distribution of the set of islands, the local population, official languages and dialects, the climate, agriculture, land owners and how to do business, taxes, labor laws, available infrastructure, exporting policies and benefits from political relationships with the US Government.

1. Geographical location

The Marshall Islands consists of 29 atolls and 5 isolated islands (see figure 3) with a total area of 181 km2 or 69.8 sq mi. Close to 10 of these islands are uninhabited; Majuro is the capital city (it is the largest city of the Islands), and where two-thirds of the nation's population lives between here and Ebeye. Majuro Atoll is the site of the heaviest population, and the nation's capital. Ebeye Island (in Kwajalein Atoll), is the second major population center, and only other urban area.

The distances between the main islands are:

- Majuro Atoll is located approximately 225 nautical miles southeast of Kwajalein
- Maloelap Atoll is located about 226 nautical miles east of Kwajalein
- Wotje Atoll is located about 164 nautical miles east of Kwajalein
- Jaluit Atoll is located 243 nautical miles south of Kwajalein
- Mille Atoll is located 95 nautical miles south of Majuro
- Enewetak Atoll lies 330 nautical miles northeast of Kwajalein

The outer islands are sparsely populated due to lack of employment opportunities and economic development; however life on the outer atolls is generally still fairly traditional. Given the placements of the Marshall Islands and the location of the islands relative to one another, the entire republic must be considered a "hard to reach" area. Travel amongst these areas is only available via cost prohibitive air travel or time consuming open ocean ship transport. The only uniform mode of communication between the urban centers and outlying areas is by short wave radio. The geographical distribution of the Islands motivated us to use this scenario as our main focus of attention. Its geographical complexity allowed us to understand how local habitants can process the breadfruit, think about how they will be able to distribute it, and from the point of view of the Enterprise, we indentified and analyzed multiple aspects (population density, transportation, time distribution, culture, territory distribution and lands, etc) highly relevant to the success of this proposal. This proposal even though it is focused on the Marshall Islands, this country was only an example and almost a simulation environment for the purpose of our own understanding and objective's setting. This proposal is flexible enough to reach other countries around the globe as well (detailed information is provided at each section as alternative solutions, suggestions, or comments).

2. National population

This nation of roughly 62,000 persons (see Figure 4) is located north of Nauru and Kiribati, east of the Federated States of Micronesia, and south of the U.S. territory of Wake Island, to which it lays claim. It is a relatively young population since roughly 25,000 people is under 34 years old, and the biggest population percentage is infants under 4 years old.

BRAINS assume that most of them are completely capable to perform any of the activities related to the breadfruit processing, such as fruit collection, peeling, carrying a few of them for relative short distances, etc.

BRAINS also predict that most of the local farmers will adopt this process to include in their daily diets the breadfruit and to feed themselves and their own infant(s), especially in those islands where there is not much commerce and other food providers thus aliments rather than fish, coconuts, and similar locally grown products.

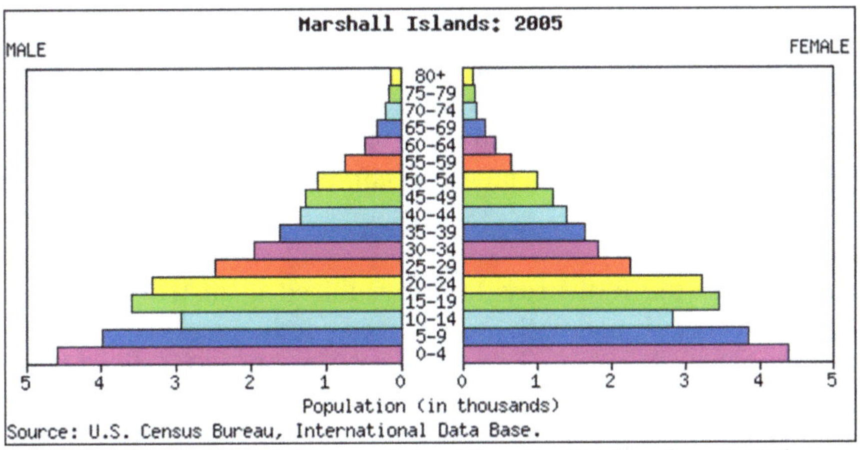

Figure 4- Population of the Marshall Islands - 2005

3. Official languages

Although English is an official language and is spoken widely, though not fluently, Marshallese is used by the government. Two major dialects have emerged, one in the Ralik chain and one in the Ratak chain of atolls.

4. Climate and agricultural production

The climate is hot and humid, with a wet season from May to November. Occasionally the Islands suffer from typhoons, prolonged droughts, and flooding in the capital city of Majuro and other urban centers as well. Typhoons in recent years have destroyed suitable breadfruit trees for consumption and canoe construction on many of the atolls.

Agricultural production is concentrated on small farms, and the most important commercial crops are coconuts, breadfruit, tomatoes, and melons. The interior of the islands, often covered with dense jungle-like growth, is usually planted with breadfruit trees as well as hardy beach shrubs that are exposed to salty spray from the ocean surf. The dominating vegetation of these coral islands is made up of the coconut palm (Cocos nucifera) followed by pandanus, breadfruit (Artocarpus artilis), Piosonia grandis, Messerschmidia argentea, Morinda citrifolia and Premna obtusifolia. Shrubs that form an outstanding part of the vegetation are Scaevola sericia and Pemphis acidia (see Figure 3). Small-scale industry is limited to handicrafts and fish processing (fishing has been critical to the economy of this island nation). On the other hand, tourism remains the best hope for future added income. In general, the islands have few natural resources, and imports far exceed exports.

5. Political Issues

The government of the Marshall Islands operates under a mixed parliamentary-presidential system, and elections are held every four years. In 1779, the Government of the Marshall Islands was officially established and the country became self-governing. However in 1986, the Compact of Free Association with the United States entered into force, granting the Republic of the Marshall Islands its sovereignty. The Compact provided US defense of the islands in change for continued US military use at one of the islands (it must be noted that one third of the available land mass in the Marshall Islands is no longer habitable due to irradiation from U.S. nuclear testing). However, the economy of the MSI is heavily dependent on payments made on the Compact of Free Association. In 2007 the Marshall Islands joined the International Labor Organization thus its labor laws will comply with international benchmarks.

6. Infrastructure

The urban areas of the RMI provide the required infrastructure for international business. Majuro is equipped with modernized roads, access to telecommunications, reliable energy services, shipping facilities and an international airport with regular flights from Honolulu. Due to the widely dispersed atolls, developing infrastructure in the outer islands remains challenging.

Summary of Infrastructure	
Telecommunications	• Telephone, cellular, Internet and long distance services
Energy	• Power service available on Majuro, Ebeye, Jaluit and Kili • Diesel fuel supplied on Majuro
Air Transportation	• Continental Airlines and Aloha Airlines from Honolulu to RMI • Air Marshall Islands for inter-island transport
Shipping	• International shipping - Matson Navigation and PM&O Lines • Inter-island shipping – Government sponsored service and private companies • Commercial docks in Majuro and Kwajalein atolls • Deep water ports and freight container facilities in Majuro • Over 50 private mooring locations
Roads	• 37 miles of roads on Majuro • 7 miles of roads on Ebeye (including causeway to Gugeegue)

Source: Interviews, NTA Annual Report, MEC Annual Report, Marshall Islands Embassy

Table 1 – Summary of Infrastructure at the RMI

- ***Telecommunications:*** The National Telecommunication Authority (NTA) provides local telephone service, cellular service, Internet access and long distance service to commercial and residential customers in the Marshall Islands. Established in 1987 to privatize telecommunications, the NTA is an independent entity with public shareholders and a Board of Directors. The government is currently the majority shareholder. International long distance rates from the RMI are $1.25 and $1.00 per minute during peak and non-peak periods, respectively.
- ***Energy:*** The Marshalls Energy Company (MEC), a government entity, provides power on Majuro and Jaluit Atolls and supplies diesel fuel to fishing vessels and government boats. This combination of revenue streams allows MEC to offer domestic power at affordable rates, without government subsidy. MEC's success is attributed to effective management, customer orientation and commitment to affordable service. Local governments on Kwajalein and Kili Atolls sponsor power services to Ebeye and Kili islands. MEC and MRD recently initiated the use of solar panels to power the outer islands. RMI and the European Community signed a new Country Strategy Paper in October 2007 for the period of 2008–13 and with a total budget of € 5.8 M. Given the difficulty of delivering secure energy access to the outer islands, promotion of renewable energy is the best technical and economic option to ensure secure energy supplies.
- ***Air Transportation:*** Three airlines currently provide international service. Active airstrips on Majuro and Kwajalein are capable of landing commercial jet aircraft, including 737s. Airstrips on Enewetok and Jaluit have the potential for larger aircraft landings as well. Continental Air Micronesia operates flights from Honolulu to Majuro and Kwajalein with continuing service to the Federated States of Micronesia and Guam three times per week. Aloha Airlines flies to Majuro and Kwajalein from Honolulu once a week. Air Nauru initiated service.

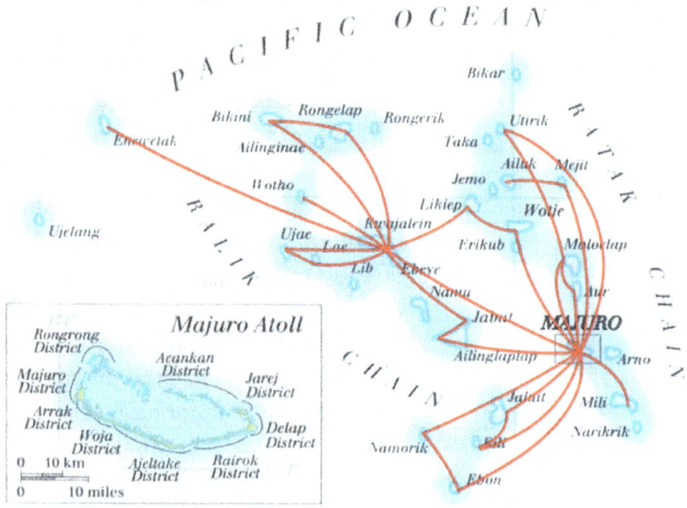

Figure 5 – Airline Route Map at RMI

7. Land Owners and Land Leases

The land in the RMI is divided into land parcels called watos. Watos may extend across an island from lagoon side to ocean side, or they may be located entirely in the interior of the island. Land ownership is customary and is restricted to natural-born Marshallese citizens. Non-Marshallese may lease land on a long-term basis, but are precluded by law from land ownership.

Land lease costs in the RMI range from a low of $2,000 per acre for residential land to $4,000 per acre for commercial purposes. Land can be leased in perpetuity; however, the mortgage on the land lease cannot exceed 50 years. In addition, mortgages can only be held against the lease on the land, not against the title to the land itself. Long-term land leases have become popular in Majuro, and a lease that allows the United States Army to use large segments of Kwajalein Atoll provides income for chiefs and land-holders of Kwajalein. A land lease, is not legally binding anywhere in the Marshall Islands without the signatures of all three title holders - iroij, alap and senior rijerbal. Foreigners cannot own islands in the Marshall Islands, but long leases are quite common. Leaseholds are available for up to 50 years and are renewable. Lease rates are negotiable with the respective land owners.
Regarding the ownership and use of land there are three primary classes of land ownership as determined by the three social classes in the Marshalls: iroij (male) or leroij (for female) (chief); alap (owner or elder); and rijerbal (worker or commoner). The Iroij have ultimate control of such things as land tenure, resource use and distribution, and dispute settlement. The Alap supervises the maintenance of lands and daily activities. The Rijerbal are responsible for all daily work on the land including cleaning, farming, and construction activities. For example: suppose a leroij (female chief) has two children (a boy and a girl) when she dies. Assuming that the leroij has no brothers or sisters, her title and all of her land rights pass to her two children. However, as the eldest, the brother inherits the title of iroij for the land, but both the brother and the sister inherit the rights of the iroij/leroij to the land. As the

titled iroij, the brother has the right to make decisions on behalf of everyone else who may also hold iroij/leroij rights on that land – in this case, his sister. When the sister dies, her leroij rights pass to her children, but the iroij title for the land remains with her brother as he is the eldest surviving member of the family. When the brother dies, considering that he was the sole survivor of that generation, the title of iroij goes to his sister's eldest child.

Major decisions affecting the use of land require the consent of the alap and senior rijerbal as well.

The agricultural division will assist local farming and livestock production through fiscal incentives, facilitating delivery of produce to the US Army in Kwajalein, to the outer islands, and to the neighboring countries on the national airline, Air Marshall Islands, as well as private run ships. In addition, chips from breadfruit, has great potential for finding a ready niche market in Hawaii, the U.S. mainland and the Pacific Rim countries.

8. Exporting, doing business and Investments at the Marshall Islands

For a foreign investor to do business in the Marshall Islands as an individual, a partnership, a joint venture, or a corporation (domestic or foreign), the investor must first obtain a Foreign Investment Business License. Here are some other important conditions prepared by the local government:

- The National Government imposes import taxes on goods and materials imported into the Marshall Islands.
- With respect to their employees, businesses are required to withhold and pay to the National Government a tax imposed on the employee's wages and salaries.
- Local governments do impose sales taxes:
    - In Majuro, the sales tax on most items is 4 percent of the selling price.
    - In Ebeye, the sales tax on most items is 2 percent of the FOB price.
- Under the Immigration and Emigration Act 1986, all non-citizens entering the Marshall Islands, with certain exceptions (e.g., diplomats and government workers), must have valid passports and entry permits (visas).

According to the 2007 edition of "Doing Business," prepared by the World Bank's private sector development department, declared the Marshall Islands to be the world's "Best Performer" for its ease and low expense in hiring and firing employees. But the study gave the Marshall Islands extremely-low ratings for its protection of investors and contract enforcement.

While the RMI encourages foreign investment, the following sectors are reserved for Marshallese citizens:

Business Opportunities Reserved for Marshallese Citizens	
Small scale agriculture for local markets	Motor vehicle rental
Small scale mariculture for local markets	Retail shops with quarterly turnover of $1,000
Bakeries and pastry shops	Laundromat and dry cleaning
Motor garages and fuel filling stations	Tailor and sewing shops
Land taxi operations	Video rental
Handicraft shops	Delicatessen or food take-out

Table 2– Summary table: Business Opportunities Reserved for Marshallese Citizens

BRAINS agrees that the small scale agriculture for local markets will be respected all through this proposal.

All articles wholly grown, made or produced in the Republic of the Marshall Islands have duty-free access into the United States except for the following categories of products: watches, clocks, buttons, textile and apparel articles, footwear, handbags, luggage, flat good, work gloves, and leather wearing apparel. Articles exported from the RMI qualify for this duty-free treatment if sum of the cost or value of the materials produced in the RMI, and the direct costs of processing operations performed in the RMI are not less than 35% of the appraised value of the merchandise at the time of its importation into the United States. In other words only 35% of the customs value of the U.S. import must be contributed in the Marshall Islands. As much as 15% of the value of the product may be contributed to this 35% added-value requirement when materials produced in the customs territory of the United States are used.

BRAINS also identified some important organizations helping in the process of beginning a new business:

- Trade and Investment Services Division, Ministry of Resource and Development: Its objectives are to facilitate and promote foreign and domestic investment in the Marshall Islands.
- Chamber of Commerce, Marshall Islands: It consists of members of around 50 local businesses, which employ approximately 2,000 Marshallese employees. Its purpose is to develop, encourage, promote, and protect the commercial, professional, financial, and general business interest of the Marshall Islands.
- Ministry of Resources and Development: It provides assistance in: product and market development, Investment and business development, and Energy services.
- Small Business Development Center: Provides training and technical assistance to small and medium sizes businesses in Business planning, Marketing, Taxation, Record Keeping, and other related small business topics. Also facilitates Micro-Credit Loan and advices small businesses on loan requirements.
- Bank of Guam, Bank of the Marshall Islands, and the Ministry of Foreign Affairs, among others.

9. Human Resources at the RMI

Males typically perform activities associated with the sea and sky (fishing, canoe building, gathering drinking coconuts, capturing birds) while females dominate activities on the land (digging arrowroot or gathering pandanus fronds).
With respect to employees, businesses are also required to make quarterly social security contributions.

Businesses are required to make quarterly health insurance contributions.

With a minimum wage of $2.00 per hour and an unemployment rate approaching 30%, the RMI offers a large pool of affordable and readily available labor. Education and training initiatives are expected to improve in the coming years to increase the skilled labor base. The College of the Marshall Islands (CMI) provides vocational training services for the local market and recently introduced a successful services training program to boost the tourism industry. CMI plans to initiate similar programs in marine resources and business administration.

BRAINS suggest that our employees (if any required at the RMI) will be relocated on either of the two main islands (Majuro or Kwajalein). On either one, the employees will benefit from electricity, telecommunications, Internet, food, commerce, transportation between the islands, and much other services and products.

Figure 6 – Majuro Atoll Tourist Map

10. Labor laws

Most of the labor laws deal directly with the hiring of non-citizen labor and indirectly with the working conditions of citizen labor. However, there are minimum wage laws and exemptions to the minimum wage laws that deal directly with labor in general. The following is a list of pertinent labor laws: The Protection of Resident Workers Act, which sets forth the procedures that must be followed before importing non-citizen workers (including prior advertisement of the job in the Marshall Islands and employment of qualified residents);

The Protection of Resident Workers (Amendment) Act of 1990, which among other things requires the repatriation after two years of employment of non-resident workers (other than United States citizens);

The Nonresident Worker (Fee) Act 1987, which requires employers to make quarterly contributions into a training fund for Marshallese in the amount of 25 cents per hour for work done by aliens (other than citizens of the United States and Palau);

The Nonresident Worker's Health Certificate Act, which requires that non-citizen workers and family members entering the Marshall Islands obtain and maintain a certificate of freedom from communicable diseases, including AIDS;

The Minimum Wage Act of 1986 as amended in 1995 which sets the minimum wage at $2.00 per hour, in non-export oriented industries; and

The Minimum Wage (Amendment) Act 1989, which exempts from application of the minimum wage act:

- non-citizen employees who are employed by private sector employers that have been authorized by the National Government to invest or to do business in the Marshall Islands, and;
- Marshallese trainees and apprentices. For more information on the Marshall Islands Labor Laws, contact the Division of Labor, Ministry of Foreign Affairs and Trade, PO Box 1349, Majuro, MH 96960.

Next we will describe the main operations performed in Hawaii.

B. Hawaii

Hawaii is a strategic point in this proposal because of the Breadfruit Institute and their research activities and projects. The Breadfruit Institute was created by the National Tropical Botanical Garden in 2002 to increase focus on breadfruit as a highly nutritional answer to global food shortages.

The mission of the Breadfruit Institute is to promote the conservation and use of breadfruit for food and reforestation. The Institute is taking a leading role in the conservation of breadfruit diversity and ethno botanical research documenting traditional uses and cultural practices involving breadfruit. Thousands of species have been gathered from throughout the tropical world, through hundreds of field expeditions by staff and through collaborations with other institutions and researchers. Its living collections include the largest assemblages of native Hawaiian plant species and of breadfruit cultivars in existence.

The First International Symposium on Breadfruit Research and Development in 2007 gathered multiple and diverse of entities like the Breadfruit Institute, the Secretariat of the Pacific Community, and the Technical Centre for Agricultural and Rural Cooperation, as well as researches and growers from both the public and private sector in the African, Caribbean and Pacific (ACP) region, to open up opportunities for greater use of breadfruit diversity through sharing information with the aim to increase the awareness of the importance of the breadfruit in the food systems of the ACP.

In the Pacific, there has been also some progress lately in trying to address this: there have been workshops on breadfruit processing and the Pacific Agricultural Plant Genetic Resources Network (PAPGREN) has developed a regional approach to genetic resources conservation.

Secretariat of the Pacific Community (SPC)
CTA Technical Centre for Agricultural and Rural Cooperation ACP-EU
National Tropical Botanical Garden (NTBG) Breadfruit Institute (BFI)

Figure 7- Participants of the First International Symposium on Breadfruit Research and Development (2007)

BRAINS truly believe this proposal will help them (see Figure 7) with data collection, statistics, and unknown local perspectives about the breadfruit to further increase their information resources and continue with biology and social commitment. This international-scope proposal will help them to:

- Gather relevant information regarding research and development on breadfruit
- Define through experimentation global strategies for sustainable and rational conservation and use of breadfruit genetic resources
- Identify, develop, and identify strategies in place for improving post-harvest handling, and promoting food product development
- Collaborate and network with geographically remote locations in breadfruit research and development

Finally we described the operational importance of Minneapolis/St Paul node.

C. Minneapolis/St Paul

As it was requested the headquarters will be located at Minneapolis/St Paul cities. After a wide range of options, BRAINS selected an IT service provider Techfluent located in the Minneapolis/St Paul cities that will outsource all our requirements regarding communication, infrastructure, data transfer, etc between the RMI, Hawaii, and the Headquarters.

Attachment 3 - SIL

System Integration Laboratory (SIL)

As its name implies, the System Integration Laboratory SIL has the objectives of testing, assembling and integrating all the engineered components of the new system into an effectively operating whole, and demonstrating that the system meets all of its operational requirements.
The BrEAD's SIL goal is to identify malfunction at any level of the system (component, subsystem) and provide recommendation to fix it. Then qualify the enterprise architecture, the processing systems and all the human interfaces prior to the implementation and deployment.
In order to carry out successfully the integration operations, the SIL shall support the following functions:

- Modeling & Simulation
- Testing

1 - Modeling & Simulation (M&S)

To be successful in Modeling and simulation operation, the SIL will first of all define the problem to be studied, including a written statement of the problem-solving objective before any attempt to conceptualize the model. Then identify, specify, and gather data in support of the model; and capture the conceptualized model using the constructs of a simulation language. The SIL will establish that the model executes as intended and that the desired accuracy or correspondence exists between the model and the system. The SIL pay a particular attention to the simulation output analysis to draw the correct inferences and make recommendations for problem resolutions or supply supportive evidential information for the decisions resulting from the simulation.
The SIL shall use only certified and accredited simulation and Modeling tools after a careful study of the main features such as

- Input
- Processing
- Output
- Environment
- Vendor
- Cost

2 - Testing

For the testing operations the SIL will follow this systematic approach:

- System component/element test
- Subsystem integration
- Total system integration

A - System component/element test

All the elements of the system shall be tested according to simplified test model below

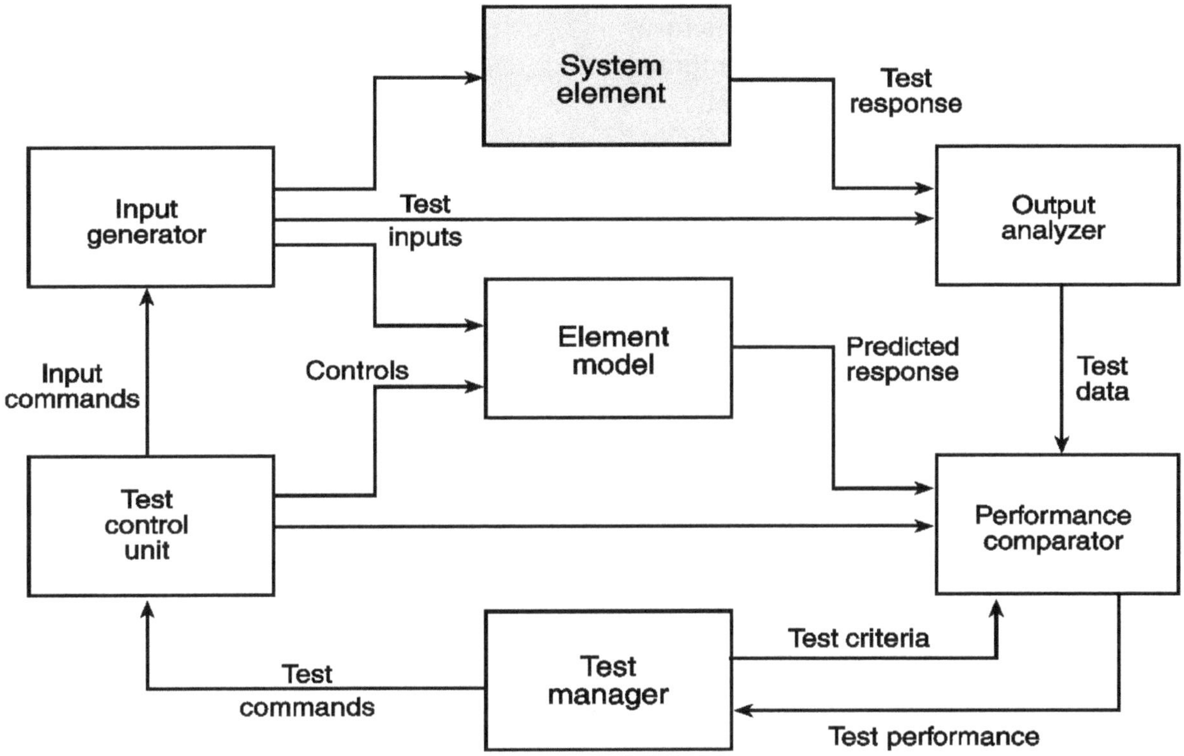

Fig SIL 1 - System element test configuration.

It is important to note that the input generator must convert test commands into exact replicas, functionally environmentally, operationally and physically, of the inputs that the component is expected to receive. These may be a sequence of typical inputs covering the range expected under operational conditions. The input signals in the same or simulated form are also fed to the element model.
The output analyzer converts any outputs that are not already in terms of quantitative physical measures into such form and the comparison and assessment is performed in real time whenever practicable to enable rapid diagnosis of the source of deviations from expected results.

For the BrEAD system architecture and the processing system, we choose to maximize the use of Commercial-Off-The-Shelf subsystem that is the use of components is minimized to reduce risk of failure and increase the availability of the system. However, as our system has open system architecture, any time a component needs to be changed or upgraded, SIL shall evaluate its performance according to the model.

B – Subsystem Integration test

At this stage, the parts are systematically aggregated, and the assembly is tested to reveal and correct any faulty interfaces or component functions
The subsystem integration should begin with components that have only external inputs, either from the system environment or from other subsystems.
Examples of such components include:

- Subsystem support structures
- Signal or data input components (e.g., external control transducers)
- Subsystem power supplies

The subsystem integration should follow the model below

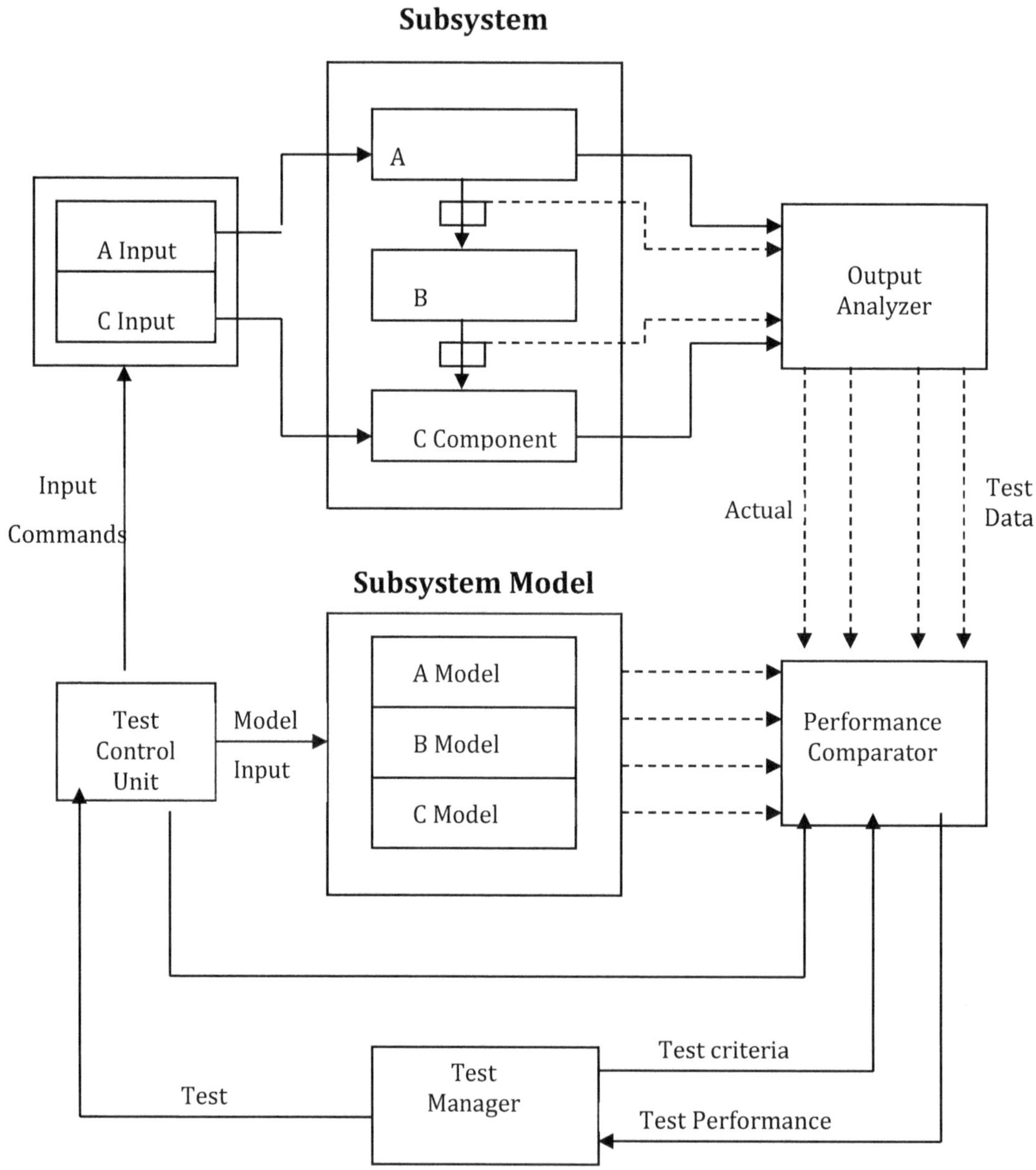

Fig SIL 2 – Subsystem Test Configuration

The configuration above should be assembled as follows:

1. Start with A, which has no internal inputs. Test A's outputs.
2. Add B and test its output. If faulty, check if input from A is correct.
3. Add C and test its output

Such integration sequence does not require the construction of input generators to provide internal functions, and should rapidly converge on the source of a faulty component or interface.
The methodology of the above figure works in the great majority of cases but must be, of course, carefully reviewed in the light of any special circumstances.
In the light of the described approach, all the subsystem of the BrEAD systems architecture and the processing system shall be tested by the SIL before the implementation.
Here is the list of the subsystems:

- Peeling subsystem
- Grinder subsystem
- Drying subsystem
- Power subsystem
- Data collection subsystem
- Data sharing system

C - Total system integration

SIL shall test the total system integration prior to its use. The integration of the whole system from its subsystems is based on the same general principles as those governing the integration of individual subsystems. However, the scale is relatively great and complex, thus more detail planning and direction are needed.

D - SIL facility and equipments

A specially designed facility is required to support the integration and test of systems and their major subsystems. The same facility will be use for the assembly and integration of total systems.
All the equipments of the SIL must meet exacting standards:

1. Accuracy

The inputs and measurements should be several times more precise than the tolerances on the system element inputs and responses. There must be calibration standards available for ensuring that the test equipment is in proper adjustment.

2. Reliability

The test equipment must be highly reliable to minimize test discrepancies due to test equipment errors. It should be either equipped with self-test monitors or subjected to frequent checks.

3. Flexibility

To minimize costs where possible, test equipment should be designed to serve several purposes, although not at the expense of accuracy or reliability. It is frequently possible to use some of the equipment designed for component tests also for integration testing

We recommend that such a facility should be gradually built up and equipped during system development to serve as a test bed for risk reduction testing, and may be assembled in part from subsystem test facilities according to the following requirements:

- The SIL shall support bread fruit processing simulation software
    - The processing software shall be able to generate recommendations for improvement
    - The processing software shall be able to support the modeling of real and theoretical data
- The SIL shall support structural software for the test of the component of the processing system
    - The structural software shall be able to generate recommendations for improvement
    - The structural software shall be able to support the modeling of real and theoretical data
- The processing software shall be able to generate recommendations for improvement
- The SIL shall support dryer modeling software for the test of the drying process
    - The dryer modeling software shall be able to generate recommendations for improvement
    - The dryer modeling software shall be able to support the modeling of real and theoretical data
- The SIL shall support subsystem/system test
- The SIL shall support integration test
- The SIL shall support verification and validation
- The SIL shall support system/software lifecycle sustainability
- Changes and Improvements to the processing system shall be managed by the SIL
- The BrEAD shall coordinate the design improvements among the SIL, the breadfruit institute, and Natural Breadfruit
- The SIL shall provide prototype testing and MTTR analysis for processing component selection to the BrEAD.
- The SIL shall provide for extracting data from test points at internal boundaries between subsystems
- The SIL shall provide for extracting data from the normal system outputs
- The SIL shall be designed to be flexible enough to accommodate system updates

From these requirements, we can derive the necessary equipments for the SIL:

Quantity Description

Equipment		Software
Quantity	Description	Microsoft office Microsoft Visio Pro-E Advanced Structural Dynamics Pro-E Mechanism Matlab Matix
1	Thermotron Temperature Chamber	
1	Thermotron Humidity Chamber	
1	24-port Network Switch	

1	Patch Panel	MathCAD C++ Test
2	Storage Unit	
1	Industrial Printer	
1	UPS	
3	PC	
3	PC Monitor	
3	PC Keyboard	
3	PC Mouse	
1	Office Printer/Scanner/Copier	
1	Server	
2	ISP Connection	

Acronyms List:

Term	Definition
BF Capture	Breadfruit Capture Team
BrEAD	Breadfruit Enterprise Architecture Demonstration
BRAINS	Bright Robust Architects of the Institute of New Systems
COTS	Commercial Off-The-Shelf
CPT	Cross product Team
DFM	Design for Manufacturing
DOD	Department of Defense
DOORS	Dynamic Object Oriented Requirements System
EIPT	Enterprise Integration Product Team
FIT	Fault-Isolation Test
FMEA	Failure, Modes, and Effects Analysis
FoS	Family of Systems
GB	Giga-Byte
I&T	Integration and Test
I/O	Input/Output
ILS	Integrated Logistic Support
IPT	Integrated product team
MOE	Measure of effectiveness
MOP	Measure of performance
NGRF	New Green Revolution Foundation
NQMS	NGRF Quality Management System
OSA	Open Systems Architecture
PC	Personal Computer
RAM-T	Reliability, Availability, Maintainability - Testability
SDD	System Development and Demonstration
SDP	Software Development Plan
SE	Systems Engineering
SEMP	Systems Engineering Management Plan
SEP	Systems Engineering Process
SIL	System Integration Lab
SMP	Subcontract Management Plan
SOW	Statement of Work
SRR	System Requirements Review
SRS	Software Requirements Specification
SSD	Systems and Services Development
SSDD	System/Subsystem Design Description

Term	Definition
WBS	Work Breakdown Structure

Contributors:

Volume Editors: Abdi Hashi, Leland Smith

Overall Design Approach: Abdi Hashi, Gabriella Farias

Systems Overview: Breadfruit Capture Team

External Systems: Abdi Hashi

Support System: Will Tamayo, Abdi Hashi

Hardware and Software Integration Overview: Gabriela Farias

PBSS: Leland Smith

Open System Architecture: Abdi Hashi

Annex A – BrEAD System Performance: Aditya Srinath, Abdi Hashi, Mesfin Woldo

Annex B – Detailed Reliability, Maintainability and Availability Prediction: Khuong Vu

Technical Attachment 1 - Breadfruit Enterprise Architecture Demonstration (BrEAD) PBSS Tailorable Specification and Verification Matrix: Leland Smith

Technical Attachment 2 – OV-1: Gabriella Farias

Technical Attachment 3 - SIL: Edem Kavi

BRAINS Proposal
for the
Breadfruit Enterprise Architecture Demonstration
(BrEAD)

RFP #: E00010-07-R-0001

Volume # 2: Technical
Book 2: Program & Schedule
Copy #1 of 2:

Proposal #: BRN3425

Prepared for:
New Green Revolution Foundation (NGRF)
2134 University Ave, Saint Paul, MN 55206

Prepared by:
Bright Robust Architect for the Institute of New Systems (BRAINS)
917 Washington Ave SE, Minneapolis, MN 55414
POC: Danielle Eastberg 952-994-5393

Table of Contents

2.2. Program and Schedule

The BRAINS program plan was written to ensure 100% compliance with the objectives of the BrEAD. The scope of the plan includes the proposed Statement of Work (SOW), our program management structure, the Integrated Master Plan (IMP), the Integrated Master Schedule (IMS), the Systems Engineering Management Plan

(SEMP), the Risk Management Plan (RMP), and the Software Development Plan (SDP). In addition, a Contract Work Breakdown Structure (CWBS) has been provided in Annex C to support the proposed SOW, IMP, and IMS.

The goal of the BrEAD is to eventually reach out and feed a million people with the underutilized Breadfruit. The program plan has specified how this will be accomplished while meeting all contract requirements and being on the contract schedule. The outcomes of our plan are as follows:

- Feed hungry people by providing them with a breadfruit product
- Advocate breadfruit for the New Green Revolution Foundation (NGRF)
- Design a scalable system to process breadfruit into a storable product
- Set up logistics to deliver and distribute breadfruit
- Set up our first implementation in the Marshall Islands
- Meet all contract requirements and launch the BrEAD by January 10th, 2010

BRAINS has met or exceeded all of the requirements specified in the customer Performance Based Systems Specification (PBSS), with the exception of the following requirement in which we were granted customer relief:

> *The Breadfruit Enterprise Architecture Demonstration (BrEAD) shall provide an automated research, marketing, and user data collection capability for NGRF headquarters in St. Paul, Minnesota that will provide 24 hour per day data from researchers, marketers and users worldwide.*

The "24 hour per day" portion of the requirement was relaxed and we designed our system accordingly.

The BRAINS approach to meeting the BrEAD objectives are addressed in the following areas:

- The resources and events used to implement the BrEAD
- The planned schedule for all activities to meet the customer Initial Operating Capability (IOC) date
- Systems Engineering (SE) process to manage the complexity of the BrEAD
- Technical, schedule, and contract risk considerations and mitigation strategies
- Subcontracting decisions and terms
- Detailed dates of events, milestones and the critical path
- Cost analysis

2.2.1. Proposed SOW and CDRLS

The BRAINS Group seeks to create a processing system capable of providing food security to those in need. In addition, an enterprise architecture will be designed to enable, empower and support all the activities related to the breadfruit process in order to support NGRF. By using western technology, BRAINS will create a virtual collaborative knowledge by sharing automation and communication systems to consolidate research and marketing information in real-time.

2.2.2. Integrated Master Plan

The following Integrated Master Plan (IMP) details BRAINS major events, accomplishments, and the criteria for completion for each event. This plan will outline BRAINS resources and capabilities and how they will be utilized in the BrEAD infrastructure. This includes staffing, testing facilities, and organizational roles. This section will also describe our success criteria for completing planned events.

We have created the events of the IMP is such a manner that it will drive the tasks of the Integrated Master Schedule (IMS). The IMP is also linked to the risk mitigation plan. The IMP will be structured as follows:

Identifier #	Event
	Significant Accomplishment
	Accomplishment Criteria
	Task

The Integrated Master Plan is located in Annex E.

2.2.2.4. BRAINS Organizational Structure

BRAINS is a global systems engineering firm. Our organization has a wealth of resources through our Finance, Sales & Marketing, Research & Development, Human Resources, Legal, Engineering, Services, and Operations departments. Our Breadfruit Capture team, shown below in Figure 5, has a strong background in systems consulting and we bring years of experience in carrying out business propositions. We propose to not only develop and produce a system for the BrEAD, but to also run and maintain it under our Operations department after it launches. More details on the capabilities of BRAINS can be found in our company profile located in Attachment 7.

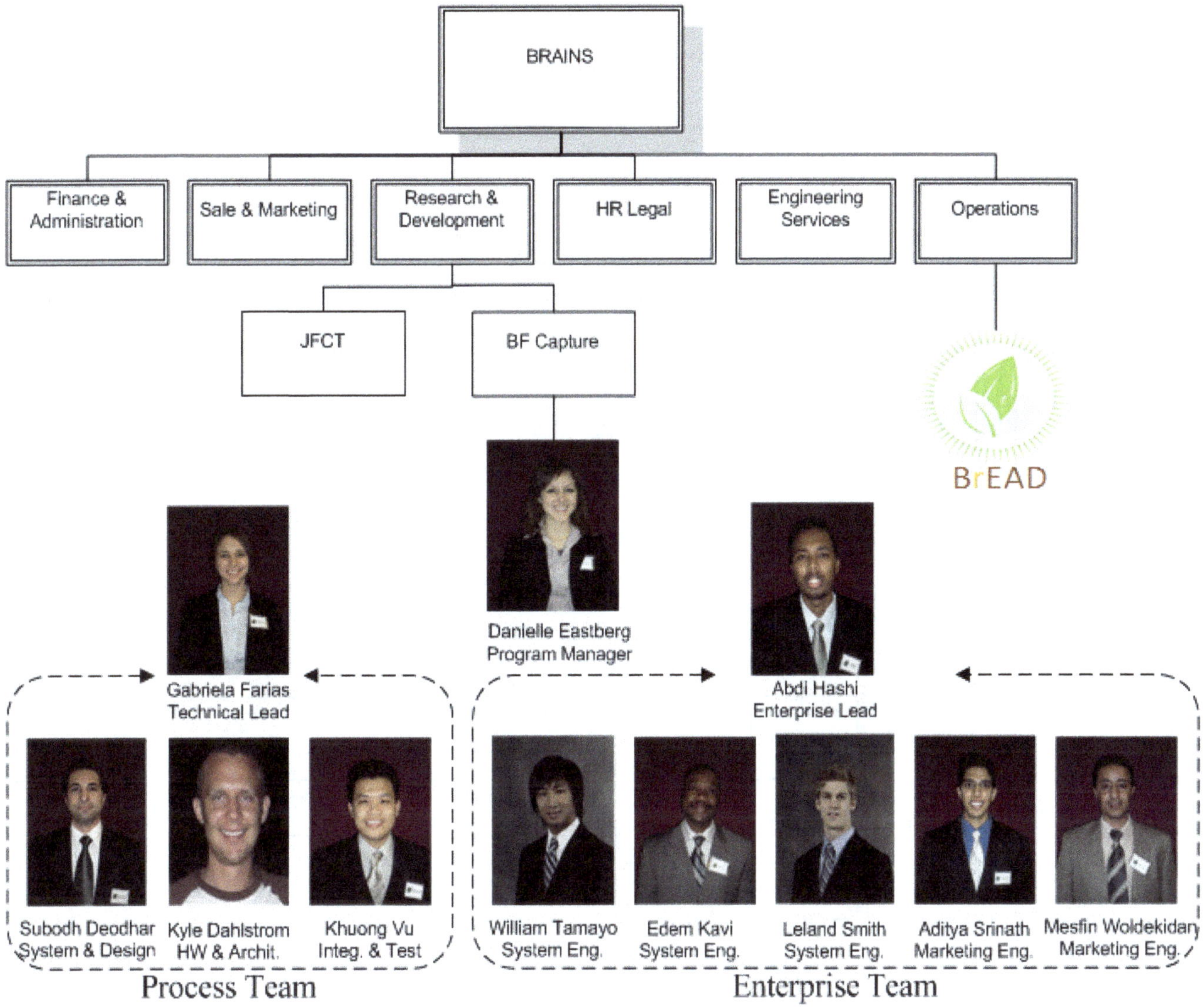

Figure 202. BF Capture Team Organizational Relationship Chart

The Program OV-4 organization relationship chart (Figure 6) below shows the overall Program and Systems Engineering Management structure to support the BrEAD implementation. Some of the major sections of this approach are the Program Management Office (PMO), Cross Product Team (CPT), Enterprise Integrate Product Team (EIPT), Process Integrate Product Team (PIPT), and System Integration Lab (SIL).

These Teams mainly work on the design, integration, and marketing of the products. Furthermore, during the operation phase, they shall provide an automated research, marketing, and user data collection capability such as hourly temperature, pressure, humidity and rain fall data for NGRF headquarters located in St. Paul, Minnesota on a daily basis.

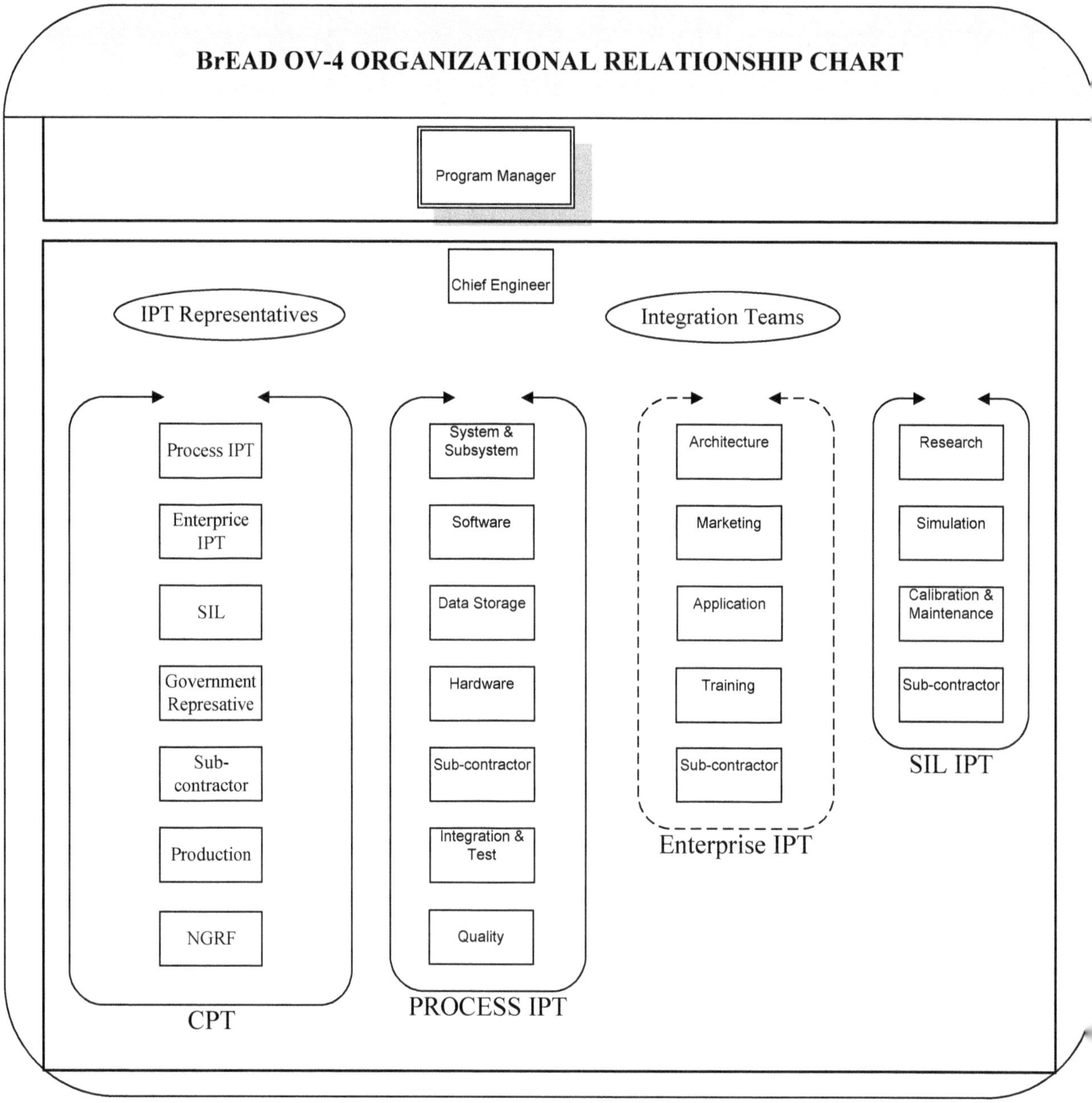

Figure 203. Program OV-4 Organizational Relationship Chart

2.2.2.2 Technical & Budgetary Purviews

The technical responsibilities of the BrEAD fall under our technical leads.Gabriela Farias is responsible for overseeing the processing functionality and Abdi Hashi is responsible for overseeing the enterprise architecture. Our project manager, Danielle Eastberg, is also acting as the Cost Account Manager (CAM) and will be responsible for the budgetary oversight if awarded.

2.2.2.3 Communication Approach

Included in the IMP are monthly working group meetings with the NGRF Board of Directors and the appropriate NGRF Points Of Contact (POCs). This will be held by teleconference and will ensure that NGRF is a part of the BrEAD process and their vision is being carried out.

In addition, we have established a data systems communication plan in order to meet the requirements of the RFP and distribute our information. The details of this plan are located in Volume 2: Technical Approach, Book 1.

2.2.2.4 Resources

We have developed relationships with existing suppliers, packagers, and contractors within the Marshall Islands. More details are in Technical Volume 1, Book 1 of this proposal. Using these resources we will bring the BrEAD program from System Development & Demonstration (SDD) to production. Our Operations department will then request to take on full operating responsibilities.

2.2.2.5 Success Criteria

Each significant accomplishment in the IMP will be measured for progress using the status completion tool in Microsoft Project 2003. We will be able to gage the maturity level of all accomplishments and tasks at any time throughout the program. These measures will be monitored through weekly status reports.

We will definitively measure a completed accomplishment by establishing that all the accomplishment criteria and tasks below it have been completed. All appropriated reviews will have been conducted, forms approved and signed, and documentation delivered.

2.2.2.6. System Node Operational View 2

Operational node connectivity

This section discusses the operational connectivity between all the stakeholders of the system and also the relationship they have with each order.

In the first step, BRAINS identifies all the organization that could have any input in the enterprise architecture of the BrEAD system. The first ones is of course, the New Green Revolution Foundation (NGRF) and the users of the system, the private organizations who could share the same breadfruit interest, the federal and state governments, the Breadfruit Institute, the local office in Marshal Islands and finally the contractors and vendors.

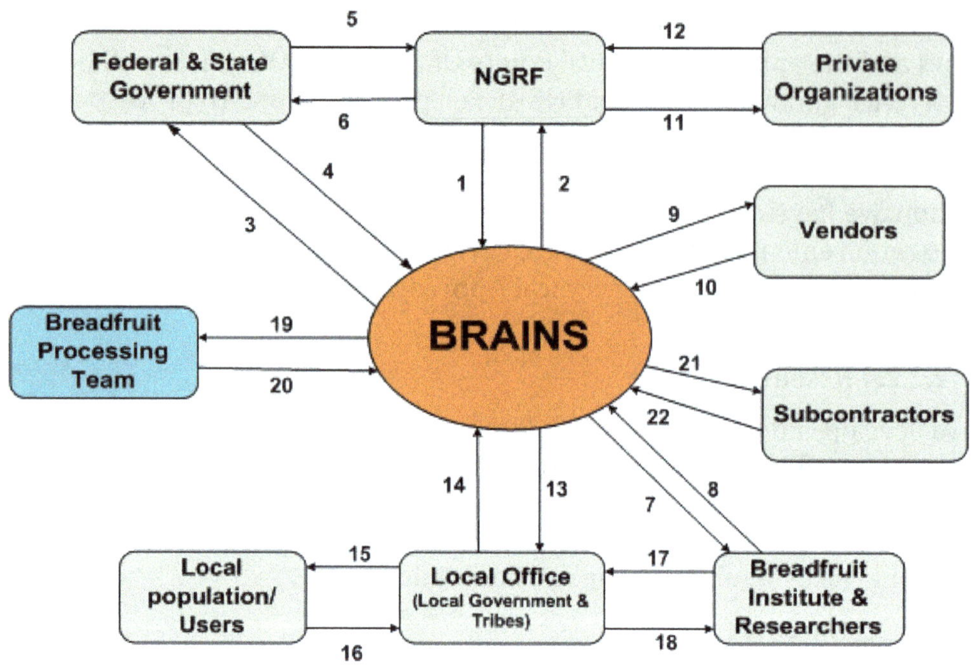

Figure 204. Operational Node connectivity (OV-2) Enterprise architecture

During the next step BRAINS organizes the relationship between all the actors for better flow of information needed for the design of the system.

- NGRF sends the PBSS and the request for proposal to BRAINS through a direct relation and after the completion of the proposal, the same relation type will be use to send back all the applicable document of the system created. In addition BRAINS decide to contact all the private organizations trough him. This approach is intended to facilitate our relationship with NGRF and have the possibility to involve them in the fund raising process.
- Though the rows requirements contain some information on the Federal and State laws, BRAINS initiates a direct contact with these institutions to make sure all the regulations are capture in the system.
- As the BrEAD system will be implemented in mainly in the islands and other locations that have different regulations and culture, BRAINS decides to start by opening a local office in Marshall Islands to capture all the local regulations and the user details. That local office will take into account all the specificity of the users. Such option will help to adjust the system if necessary before spreading out on the field.
- The Breadfruit institute is one of valuable information resource for the breadfruit so BRAINS decides to have a direct contact with that organization. This direct relation will not prevent the local office to request and receive information directly from the same institution too.
- The vendors and subcontractors will provide directly all the needed and requested information to BRAINS.

- Finally one important aspect of BrEAD, the system in creation is the processing system. BRAINS will send trough the breadfruit processing node all the requirements relating to the processing system and will receive in the same way the feedback upon its design.

The same analysis methodology is used to identify all the stakeholders of the processing system and also their relationship. The figure 2 below gives the details.

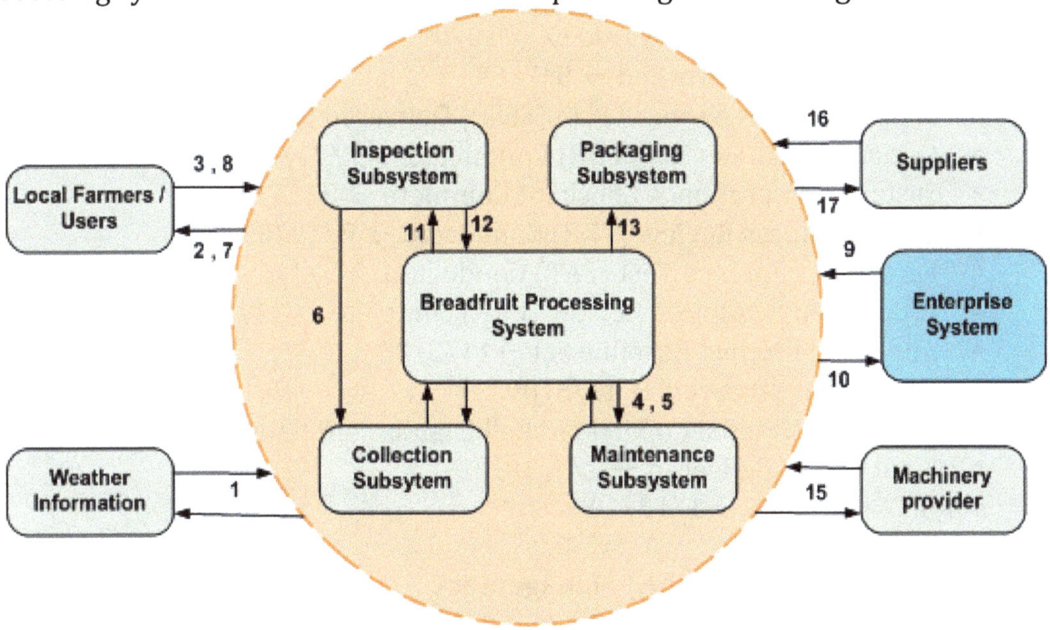

Figure 205. Operational Node connectivity (OV-2) Processing System

One thing worth mentioning over the processing system is the identification of four internal subsystems namely: Inspection, Collection, Packaging and Maintenance.

All the information exchanged through the needline are detailed in the information matrices (OV3).

2.2.3. Integrated Master Schedule

BRAINS is submitting an Integrated Master Schedule (IMS) in Annex F to detail our approach to meeting the proposed Initial Operating Capability (IOC) date of January 10th, 2010. We are submitting our IMS in the form of an embedded Microsoft Project 2003 Gantt chart (.mpp). The IMS includes the planned schedule for all activities, events, milestones, and the critical path that will ensure a complete BrEAD roll-out by the contract IOC date. The schedule as set allows ample time for design, reviews, testing, verification, validation, and construction of the BrEAD enterprise. The IMS was written in correlation with IMP to include the technical tasks that need to take place after contract award.

The following events and milestone have been identified with zero slack time, and therefore define our critical path for project completion:

- Product Baseline Description Developed → 1/22/09
- Initial CDRLS Completed & Delivered to NGRF → 3/26/09
- DoD Architecture Framework Delivered to NGRF → 4/16/09
- Preliminary Design Review (PDR) Conducted → 3/29/09
- Contract Proposal Presented → 4/30/09
- Written Proposal Delivered → 4/30/09
- NGRF Contract Awarded → 5/7/09
- Supply Support Services Plan (SSSP) Delivered to NGRF → 7/3/09
- Critical Design Review (CDR) Conducted → 7/21/09
- Systems Integration & Checkout Complete → 8/13/09
- Test Readiness Review (TRR) Conducted → 8/19/09
- Factory Acceptance Test (FAT) Conducted → 9/1/09
- Production Readiness Review (PRR) complete → 11/11/09
- Approved Product Baseline Set →11/3/09
- Equipment Procured → 11/5/09
- BrEAD Facilities set up in Marshall Islands complete → 12/28/09
- Final CDRLS delivered → 1/5/10
- BrEAD Launch → 1/10/10

2.2.4. Systems Engineering (SE) Management

As a program management tool, the Systems Engineering Management Plan (SEMP) provides a framework for organizing and implementing the technical and management support for successful program execution. This SEMP satisfies CDRL 0005 Systems Engineering Management Plan (SEMP) and is applicable for the lifecycle of the Breadfruit Enterprise Architecture Demonstration (BrEAD) SEMP.

2.2.5. Risk Management Plan

A well structured Risk Management Plan is necessary for any successful project. Risk Analysis is definitely not a onetime task. It's an ongoing process throughout the lifetime of project, but the best time to perform risk analysis is just prior to beginning of project.
The proposed Risk Management Plan has been discussed first followed by its application to BrEAD Project are following steps involved in assessing as well as managing risks and all of them shall be followed for effective risk management.

1. Risk Identification
2. Categorize Risks
3. Risk Impact Assessment
4. Risk Management
5. Risk Monitoring and Controlling

Figure 206 shows, how different stages are interrelated and information shall be exchanged for effective risk management

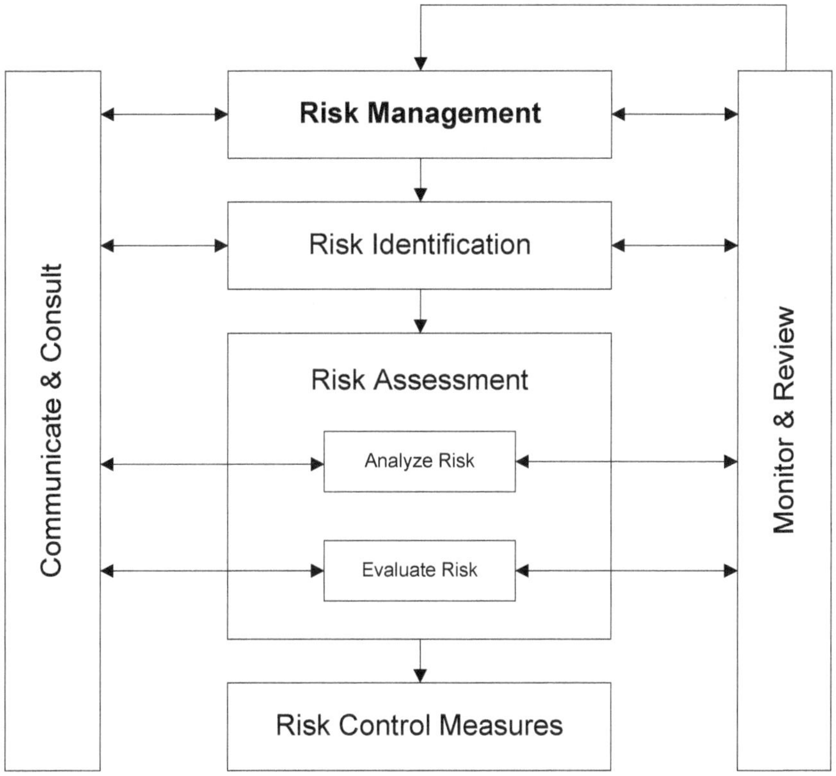

Figure 206. Risk Management Plan

2.2.5.6. Risk Identification:

There are numerous ways to identify risks. The risks which are listed down after brainstorming about all possible occurrences shall be used as preliminary stage document, but systematically it can be done in following way

- The task for which the team has no expertise shall be identified first. The duration and cost estimates for these tasks are more likely to be inaccurate
- Duration and cost estimates which are aggressive shall be double checked. Team members who worked on such estimates shall be consulted about their confidence in their estimates. This activity shall be carried out especially for tasks on critical path.
- Situations where the team has limited number of resources shall be identified.
- If there are chances of having several predecessors, then it could result in considerable delay. Such tasks shall be listed down.
- The list of identified risks shall be reviewed by some experienced Project Manager for comments and suggestions.

2.2.5.7. Categorize Risks:

All the identified risks in first step shall be categorized into different categories like Technical Risks, Schedule Risks, Contract Terms and Conditions Risks etc.

2.2.5.8. Risk Impact Assessment:

For each identified risk event, the likelihood of occurrence and its impact on project objectives shall be assessed. This information shall be used to prioritize the risk using established threshold criteria. The risk matrix shall be used for this purpose. The likelihood and impact classification for developing Risk Matrix shall be according to tables shown below. It will help to decide the level of impact for three different types of risks namely technical, schedule and cost risk. It has prescribed limits from levels of impact 1 to 5. The second table defines the levels for deciding the likelihood of event from 1 as lowest to 5 as highest. The risk matrix shall be prepared to identify risks in Red, Yellow and Green zones. The risks which are appearing in Red Zone shall be listed down as "Hot List" Items and they shall always be kept on top priority for risk assessment as well as reviewing. The second priority shall be given to items in Yellow Zone and those items shall be listed down as "Watch List" items. The items appearing in Green Zone are the items with lowest likelihood of occurrence with lowest impact on project, so they shall be given last priority.

Error! Reference source not found.

2.2.5.9. Risk Management: The risk can be managed in following ways

a. **Avoidance:** The plan can be changed or more resources can be used or schedule can be changed in order to avoid the risk. More in depth planning can sometimes help to avoid occurrence of risk.

b. **Transference:** If avoidance seems impossible, then the risk can be transferred to third party e.g. by buying insurance. Another way sometimes adopted is to enter into fixed price contract.

c. **Mitigation:** Mitigation seeks to reduce the probability and consequences of risk event to acceptable threshold by taking actions ahead of time, thereby decreasing the likelihood of problem.

d. **Acceptance:** The final way to manage risk is to respond to the risk item with contingency plan. The contingency plan should include any work that must be done ahead of time to make the contingency successful.

2.2.5.10. Risk Monitoring and Controlling:

a. **Monitoring:** Identified Risks shall be reviewed periodically and ensured that mitigation plan has already been set. Priority shall be given to HIGH risk items. Such review sessions shall be included in project schedule from initial stages. The purpose of review session is to

- Ensure that all of the requirements of the Risk Management Plan are being implemented.
- Assess currently defined risks
- Evaluate effectiveness of actions taken
- Decide Status of actions to be taken
- Validate previous risk assessment (likelihood and impact)
- Validate previous assumptions
- Identify new risks
- Track Risk Response
- Complete any necessary Communications

b. **Controlling:** The purpose of this stage is to

- Validate mitigation strategies and alternatives
- Take corrective action when actual events occur
- Assess impact on the project of actions taken ($$, time, resources)
- Identify new risks resulting from risk mitigation actions
- Ensure the Project Plan (including the Risk Management Plan) is maintained
- Ensure change control addresses risks associated with the proposed change
- Revise the Risk Assessment Questionnaire and other risk management documents to capture results of mitigation actions.
- Revise Risk Response Plan
- Complete any necessary Communications

Risk Management for BrEAD:

1) **Risk Identification:** BRAINS have used all the techniques listed above for identifying risks. The risks are identified and categorized as shown

   i. **Technical Risks:**

   All the risks associated with technology have been listed. The identified risks are mostly associated with Processing Stages, since the technology used there is very simple and has more chances of failure. The overall system consideration brings out risks related with IT. The mitigation plan for all these risks has been discussed in the next section.

Technical Risks

Sr.No.	Risk	Description

1	Weather Station Failure	Weather station will provide current weather conditions and forecast data to BF processing system. The system may become unavailable because of any miscellaneous problem (failure of battery, sensor, recorder etc.)
2	Transportation System Failure	Collected Breadfruits must be transported immediately for Processing. Transportation is also essential for moving processed breadfruit containers to packaging unit. The transportation system may become unavailable for Planned Outage (for maintenance of vehicles) or Forced Outage (for any misc. reason)
3	Failure of Dryer/ Grinder	The entire process is interlinked by these two equipments. The failure of one will result in downtime for the complete system.
4	Wind Power Failure	Since most of the islands are not connected to Electrical Power Grid, the wind Power is the only power source for processing and lighting. The power supply system may become unavailable because of very high or very low wind speeds or any other miscellaneous reason
5	Communication System Failure	There are various communication links as indicated in Data System Design diagram. For nodes located in metros/ developed regions, communication link failure is not a major issue (as various other alternatives are available) but for location(s) on Marshal Island it could create problems.
6	Data Storage System Failure	All Data (weather, processing, maintenance feedback etc.) will be stored on operating systems/storage media in headquarters. Storage media/ operating system may crash.
7	Unavailability of specialized Maintenance personnel	Specialized maintenance personnel/crew will take care of all types of major technology failures. The member/crew may be unavailable because of maintenance work at other location/other misc. issues.
8	Inspection Fails	Inspection at the collection point or at the factory does not pass
9	Natural Disaster	Natural Disaster such as Hurricanes or Typhoons that wipe out the breadfruit crop. Or damage the processing facility.

ii. Schedule Risks

Sr.No.	Risk	Description
1	Product & Process Novelty – Uncertainty with Breadfruit	Since BRAINS has never handled any project relatec to Breadfruit, we have to collect a lot of informatio and do some foundation work before moving ahea with further process designs, and that might tak considerable time

2	Resource availability	There are all chances that, the resources needed for system setup may not be easily available. The delay in getting any resources like land, electricity, processing equipments etc. will affect schedule
3	System Design Approval Delays	The BRAINS has to take approval from NGRF as well as related Government agencies before proceeding with system setup, and that might take time
4	Iteration Scope and Duration	There are chances that some of the activities not listed in scope of work will get added at later stage after contract award. If it happens then time needed to fulfill those requirements may cause delay
5	Dependencies on supplier teammates	BRAINS will have to interact with number of product as well as machinery suppliers, for material needs and sometimes for assistance. They might have some other priorities, and that can take time
6	Lead time for materials	The processing machinery as well as material and any other miscellaneous material which has to be ordered from places outside Marshall Island might take long time for delivery
7	Interrelations between Processing and Enterprise teams	Processing and Enterprise Teams have to exchange lot of information for carrying out activities at respective ends, most of the activities depends on inputs provided by each other, this communication might take time
8	Testing Requirements	As per NGRF's requirements, testing and evaluation has to be carried out before actual system setup. For processing equipments which are to be transported to Marshall Island from outside are to be tested by simulating weather conditions similar to Island. It might take considerable time
9	Decision/Mgmt Availability	If authorized approving authority is unavailable, then further activities might have to be kept on hold

iii. Cost Risks:

Sr.No.	Risk	Description
1	Revision in Machinery Prices	Prices of Processing equipments quoted by suppliers might change at the time of actual ordering
2	Revision in Transportation Cost	Since most of the material is to be transported to Marshall Island from there might be a considerable increase in transportation cost because of unanticipated miscellaneous material which is needed for SDD

3	Unavailability of low cost material for wind turbines	BRAINS have anticipated that, low cost parts will get easily available for constructing wind turbines. If it doesn't happen, then we might have to opt for standard branded parts which are way costlier.
4	Any incidental charges which are not anticipated at Proposal Stage	BRAINS might have to pay any legal processing fees or even penalties during SDD phase

2) Risk Impact Assessment:

With the help of definitions in Risk tables, the Risk matrices are developed for all the three categories identified above. The risk matrices are as shown below

a. **Technical Risks:** The various risks identified earlier have been marked in risk matrix according to their probability of occurrence and resulting consequences. The risks which might result in delayed production are the ones on priority and listed in "Hot"

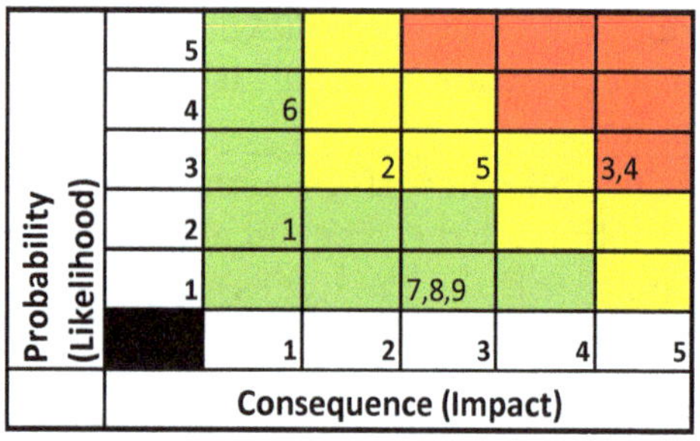

Risk which may cause considerable production delay has max. impact

"Hot List" Items	"Watch" Items
3) failure of Dryer/ Grinder 4) Wind Power Failure	2) Transportation System Failure 5) Communication System Failure
"New" Items	**"Retired" Items**
	1) Weather Station Failure 6) Data Storage System Failure 7) Unavailability of Specialized Maintenance Personnel 8) Inspection Fails 9) Natural Disaster

b. **Schedule Risks:** With same criteria, schedule risks have been classified and shown in risk matrix as below

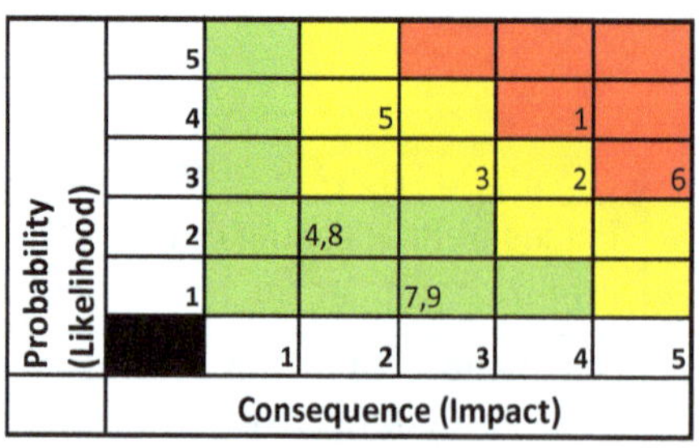

Risk which may cause considerable production delay has max. impact

"Hot List" Items	"Watch" Items
1)Product & Process Novelty 6) Lead time for materials	2) Resouce Availability 3) System Design Approval Delays 5) Dependencies on supplier teammates
"New" Items	**"Retired" Items**
8) Testing Requirements	4) Iteration Scope and Duration 7) Interrelations between Processing & Enterprise 9) Decision/Mgmt Availability

c. **Cost Risks:** With the same criteria, cost risks have been classified and shown in risk matrix as below. Since most of the processing equipments are low cost equipments, any revision in prices will definitely not cause price impact greater than 1 million. Any unanticipated charges will also be in a range so as to be classified as "retired" items.

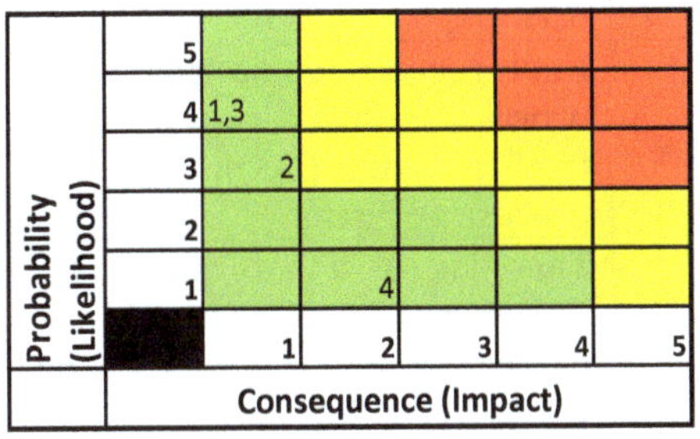

Risk which may cause considerable production delay has max. impact

"Hot List" Items	"Watch" Items
"New" Items	**"Retired" Items**
	1) Revision in Machinery Prices 2) Revision in Transportation Cost 3) Unavilability of low cost material for wind turbines 4) Any incidental charges not anticipated at Proposal Stage

3) **Risk Management:** Risk Management Plan for various risks is as under

   a. **Technical Risks:**

   Since most of the technical risks are associated with processing side, there is a valid mitigation plan and technique for each and every risk. Most of the identified techniques are 'reduction' techniques. These are actually some preventive or corrective actions which are to be followed in order to reduce the impact of risk. The mitigation plan for various identified technical risks has been listed below.

Risk Management Plan for Technical Risks

Sr.No.	Risk	Mitigation type	Mitigation Description
1	Weather Station Failure	Reduction	Use of weather data from <www.weather.com> till system restores
2	Transportation System Failure	Reduction	Planned Outage: Local farmers shall be informed in advance about rescheduling their drop-off times. Forced Outage: Rental transportation system shall be used
3	Failure of Dryer/ Grinder	Reduction	1) Specialized Maintenance personnel shall be contacted immediately 2)Collection Points shall be informed about taking necessary actions (for sending and receiving breadfruits for processing)
4	Wind Power Failure	Reduction	1) Use of Backup Power Generator 2) Further BF processing shall be kept on hold 3) same set of actions taken for Dryer/Grinder Failure risk mitigation shall be followed
5	Communication System Failure	Reduction	1) Ongoing processing activities shall be completed and all new processing activities shall be delayed till data communication exchange (demand, weather info etc.) restores 2) Nearest island shall be contacted for reporting communication link failure problem to Headquarter 3) Possibility of having a satellite phone shall be checked (?)
6	Data Storage System Failure	Reduction	Data back-up shall be taken on weekly/monthly basis on high-quality storage media

7	Unavailability of specialized Maintenance personnel	Avoidance	Operators on processing equipments shall be trained to rectify minor routine faults. They shall be provided with a checklist to find out the cause, solution and expected repair time if major fault occurs.
8	Inspection Fails	Acceptance	Have regular monitoring of our machines and processes. Have a plan in place to clean out old food product (decontaminate) and start the process up again.
9	Natural Disaster	Transference/ Acceptance	The Risk of a natural disaster can either be transferred by purchasing insurance or accepted by dealing with the consequences if the disaster were to occur.

b. **Schedule Risks:**

The risk management technique for all schedule risks is 'Acceptance'. The technique of having contingency plan if the problem occurs shall be used. In some cases, it could be addition of resources while in some cases, that includes some work which must be done ahead of time to make the contingency successful. It is ensured that, the additional resources which are to be considered are readily available.

c. **Cost Risks:**

As mentioned for schedule risk, 'Acceptance' will be the strategy for dealing with Cost Risks. In some cases, it would not be possible to have additional resources; in that case, the extra buffer added in cost estimate will take care of increased cost.

4) Risk Monitoring and Controlling:

Identified Risks have been reviewed and mitigation plan has already been set for all the risks. The project schedule has been structured in order to review these risks and their current status periodically. The review session will focus mainly on list of preventive actions and their status. The risks which may result in production delay are in red zone and at top of our priority for review meetings. All the project team members have been instructed to bring out newly identified risks for review meeting. The risks and necessary actions which are to be communicated to higher management will be finalized after review meeting only. The Controlling phase deals with deciding and approving action plan for the risks which have already been occurred. That action plan could be allocation of additional budget or resources in order to avoid 'side effects' resulting from that risk. In some cases, new risk could result from the mitigation plan; such risks will be discussed. BRAINS has expert and experienced project team for BrEAD Project, which includes members who have track record of successful completion of many similar projects in past. This risk management plan has been developed by project team after careful planning and discussions.

2.2.6. T&E

2.2.6.6. SIL (System Integration Lab):

SIL and its functions have been mentioned in detail under different section which plays a vital role in Testing & Evaluation phase. It deals with modeling and simulation based performance evaluation of all engineered components of system, has capability of building the mathematical model of all the dynamic systems, and is intelligent enough to provide recommendations about further improvements in system. Once all subsystems are analyzed by SIL, the T&E will have backup of simulated full system run which will be handy for actual testing. SIL might not simulate each and every operating condition like weather, varying wind speeds etc. though it will apply some correction factors. In those cases, they will post remark for T&E team to double check system functionality under those conditions.

2.2.6.7. Sub-system Testing:

Each of the subsystems like communication, transportation, dryer etc. shall be tested on three different levels

2.2.6.8. Preliminary Inspection:

Each sub-system shall be checked thoroughly to ensure that, it has been prepared and developed as per given specifications. The manufacturer's catalogues supplied with the product shall be studied thoroughly. Inspection shall be carried out to verify dimensions, material, color, accessories etc. to check whether they match the specification or not. If the manufacturers are required to provide any Test Certificates, then those certificates shall be checked and verified before moving into actual operational Tests

2.2.6.9. Operational Test:

The operational check shall be carried out to verify and validate output data/product for the specific input. All the sub-system specifications provided in manufacturer's catalogue shall be verified. e.g.: output processing rate for dryer, bits transfer rate for data storage system etc. These experiments shall be carried out with actual input. The test results shall be observed for minimum, maximum and average values of input signal.

2.2.6.10. Fault Protection Devices:

The inbuilt or externally provided protection devices for any abnormal input/output conditions shall be checked by simulating that abnormal input/output. e.g.: thermostat provided for dryer shall switch off the power supply to dryer, when the temperature exceeds the maximum allowed limit

2.2.6.11. System Safety Issues:

During operational check, the system shall be inspected to identify any safety issues which might occur during operation. The precautions and instructions for the operator shall be listed down in the safety instructions manual for the operators.

2.2.6.12. Environmental Test Chamber:

Once all the operational checks are completed, then attention shall be paid towards environmental testing. The weather on Marshal Island is very hot and humid. Since all the processing, communication equipments are to be transported from outside, it's necessary that, those equipments shall be tested and their operating capability shall be verified under similar weather conditions. If equipment is to be transferred from place like Twin Cities, then Environmental Test Chamber (Thermotron) shall be used. Thermotron's temperature and humidity test chambers are able to provide rapid product temperature change rates and use varying levels of relative humidity to locate design problems prior to shipping of products, improving product quality and reliability. Any previously unidentified system output like operating noise or waste product disposal, which might result environmental pollution (air, noise) must be noted down in operator's manual along with action plan for its minimization

2.2.6.13. Feedback Evaluation:

The test results, comments marked by Testing Team, any observed discrepancies shall be analyzed and listed down in Feedback Report. If the observed variations are in acceptable tolerance band, then the sub-system shall be transferred for Full System Testing but if not then manufacturer shall be contacted to make necessary modifications in their system. In some cases if the required result producing capability appears to be out of standard range of equipments, then instead of going for any non-standard solution at higher cost, management shall be contacted to change the manufacturer. The feedback report must be verified by Project Manager, before moving into Full System Testing.

2.2.6.14. Full System Testing:

All the subsystems shall be arranged in exactly the same way as planned in layout and used in same sequence as per process description. The specific quantity of breadfruits shall be gathered and then output quantity and processing time shall be calculated. These breadfruits shall then be transferred through all the processing stages till the flour is obtained. The calculated time shall be verified against the actual time, If they are approximately the same, and also if the breadfruit flour clears all the product quality tests, then the processing system shall be reported Okay. Similarly communication, data storage system, their integration with weather station, onsite maintenance crew etc. shall be actually simulated and tested.

2.2.7. Subcontract Management

To reduce costs and gain access to specialized skill sets, BRAINS has taken the opportunity to benefit from the products and services of small businesses and large corporations. BRAINS has developed a strategy to manage and control subcontract activities. BRAINS has also developed detailed process for the selection and management of subcontracted partners to ensure that the performance of program exceeds the expectations of all the stakeholders.

2.2.7.6. Subcontract Management Plan

Subcontract Management Plan (SMP) will be established for each major subcontractor to assure successful program execution. The SMP describes the process used by BRAINS to select qualified subcontractors and manage them efficiently. The SMP sets the process to manage and integrate project activities. The SMP shall be updated annually or as required to reflect all the changes in subcontractor management functions and processes.

The SMP will list required documentation which includes the Statement of Work (SOW), Subcontract Task Work Order (TWO), and other documentation such as status reporting and invoice procedures. The SMP provides guidance for subcontract management activities such as methods and processes; roles and responsibilities; and management oversight to minimize program impacts due to cost, schedule, quality and technical performance. The SMP specifies the necessary coordination of quality assurance (QA) and configuration management (CM), and applies the appropriate control to the subcontractor. The SMP also specifies the process for change control.

2.2.7.7. Subcontracting Processes

BRAINS approach to managing subcontractors is based on the following:

- Clearly defining and establishing effective channels of communications
- Clearly defining, in the Statement of Work, each subcontractor responsibilities for the deliverables and services outlined in its areas of responsibility
- All constraint imposed on the subcontractor, including scheduling and budget constraints, will be clearly define in the Statement of Work
- Identifying risk areas at the beginning of TWO and implementing risk mitigation procedures to minimize the risk
- Clearly identifying subcontractor's quality requirements, including the requirement to allow independent quality inspection of materials and processes
- Using lessons learned to maximize the potential for success

2.2.7.8. Coordination Between Prime Contractor and Subcontractors

Each subcontractor organization will have one representative to interface with the prime contractor for contractual matters. The prime contractor will identify a single point of contact (POC) for contractual matters. On a day-to-day operation, all subcontractor personnel are free to interact with any prime contractor personnel as necessary to fulfill the subcontract.

2.2.7.9. Subcontract Control

The subcontractor's performance is continually monitored and assessed against agreed standards and established performance metrics for the subcontracted activities. The subcontract manager conducts reviews as needed to ensure timely and effective completion of subcontract activities. Corrective actions must be implemented for any unperformed or unsatisfactory work in accordance with the established standards. During regular quality reviews, any unsatisfactory work will be documented, reported, and the path forward will be determined to correct any discrepancy or non-compliance.

A subcontract TWO must be formalized when issuing a requirement to a major subcontractor. The TWO defines the scope of work, required level of performance, schedule milestones, performance metrics, and any unique requirements.

2.2.7.10. Proposed Subcontractors

BRAINS has carefully selected a team of core subcontractors and vendors whose capabilities can meet and/or exceed the requirements issued in the TWO needed by the BrEAD system. Table 1 lists the selected subcontractors and their responsibilities.

Subcontractor	Specialty	Task Work Order
RMI private businesses	Shipping	Provide inter-atoll shipping of materials
Matson Navigation	Shipping	Provide international shipping of end-product
Majuro Packaging Facility	Packaging	Package flour for distribution
Ralik Store	Distribution, retail and wholesale	Provide the capability to distribute the end-product to the end user and distribution node
Parallel Track Engineering Company Pvt. Ltd.	Storage and material handling solutions	Provide storage and material handling solutions
Techfluent	Information Technology	Provide IT Service Management and Solutions

2.2.8. Software Development

This Software Development Plan (SDP) establishes the software development approach, methodologies, tools, and procedures to be used during the analysis, design, development, testing, integration, deployment, and maintenance of the software for BrEAD and the processing system. The Software Development Plan (SDP) describes a developer's plans for conducting a software development effort. The term "software development" in this document is meant to include new development, modification, reuse, reengineering, maintenance, and all other activities resulting in software products. This SDP is a dynamic document and shall be updated on a periodic basis to reflect organizational changes, lessons learned, new tools, and advances in methodologies.

The SDP will not be directly applicable to the BrEAD engineer, since typical BrEAD activities do not include software development. However, the SDP should be a requirement for the subcontractors responsible for developing and submitting the SDP document for a software development effort. This is located in Annex H.

References

DoD Architecture Framework
Version 1.5, Volume II: Product Descriptions

System engineering management
Blanchard, Benjamin S. Knovel (Firm)
3rd ed. Hoboken, NJ : John Wiley c2004

Human factors and ergonomics in manufacturing
Institute of Industrial Engineers (1981-); Human Factors and Ergonomics Society.; International Ergonomics Association.
[New York, N.Y.] : John Wiley & Sons, Inc. c1997-

http://sepo.spawar.navy.mil/SDP_Template.doc.

Browning, T. (1988). Sources of Schedule Risk in Complex System Development. Proceedings of the Eighth Annual International Symposium of INCOSE. Vancouver.

Netzer, A. (n.d.). Integrated Master Plan (IMP) and Integrated Master Schedule (IMS). Retrieved April 14, 2009, from www.fas.org/spp/military/program/imint/98b030.pdf.

Annex C –CWBS:

Top level WBS

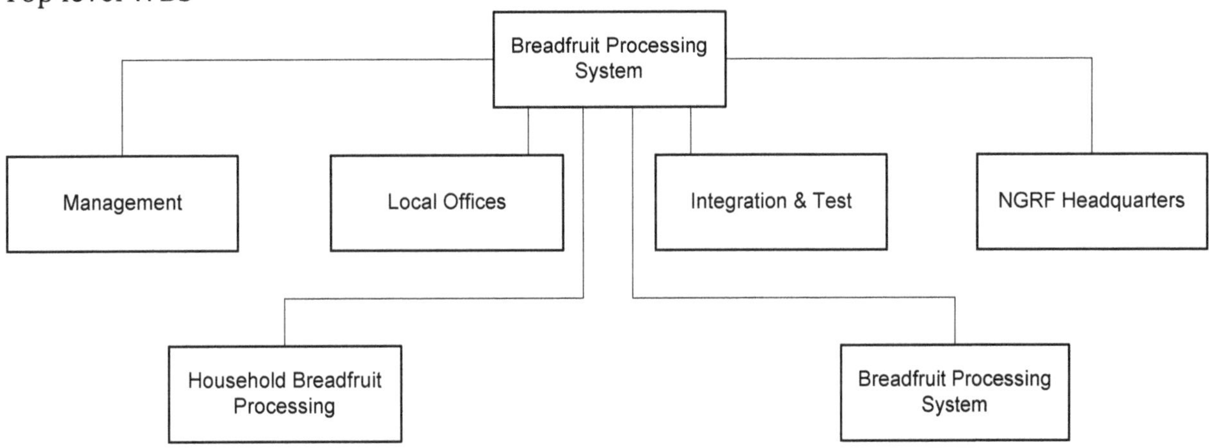

Level 2: Management WBS

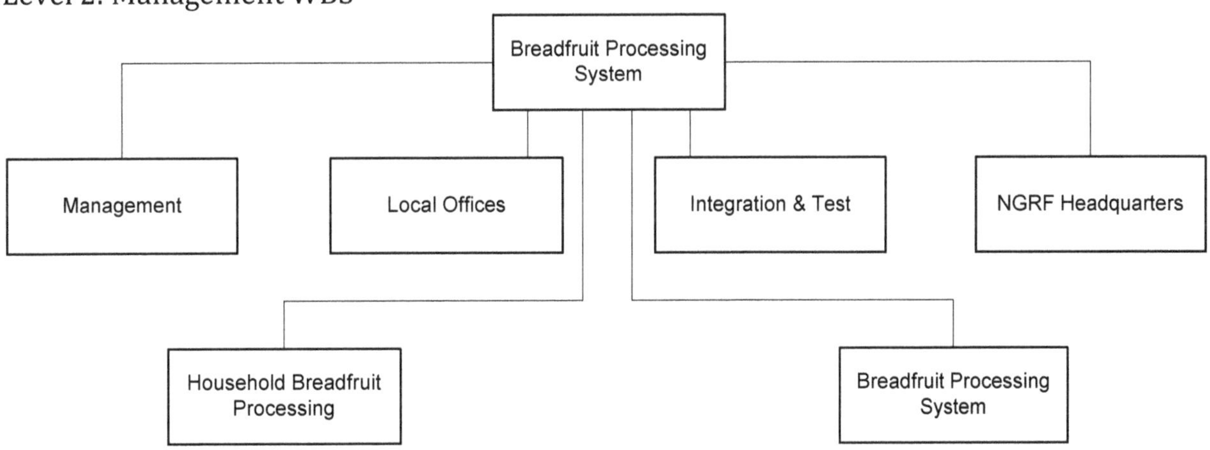

Level 2: Local Offices WBS

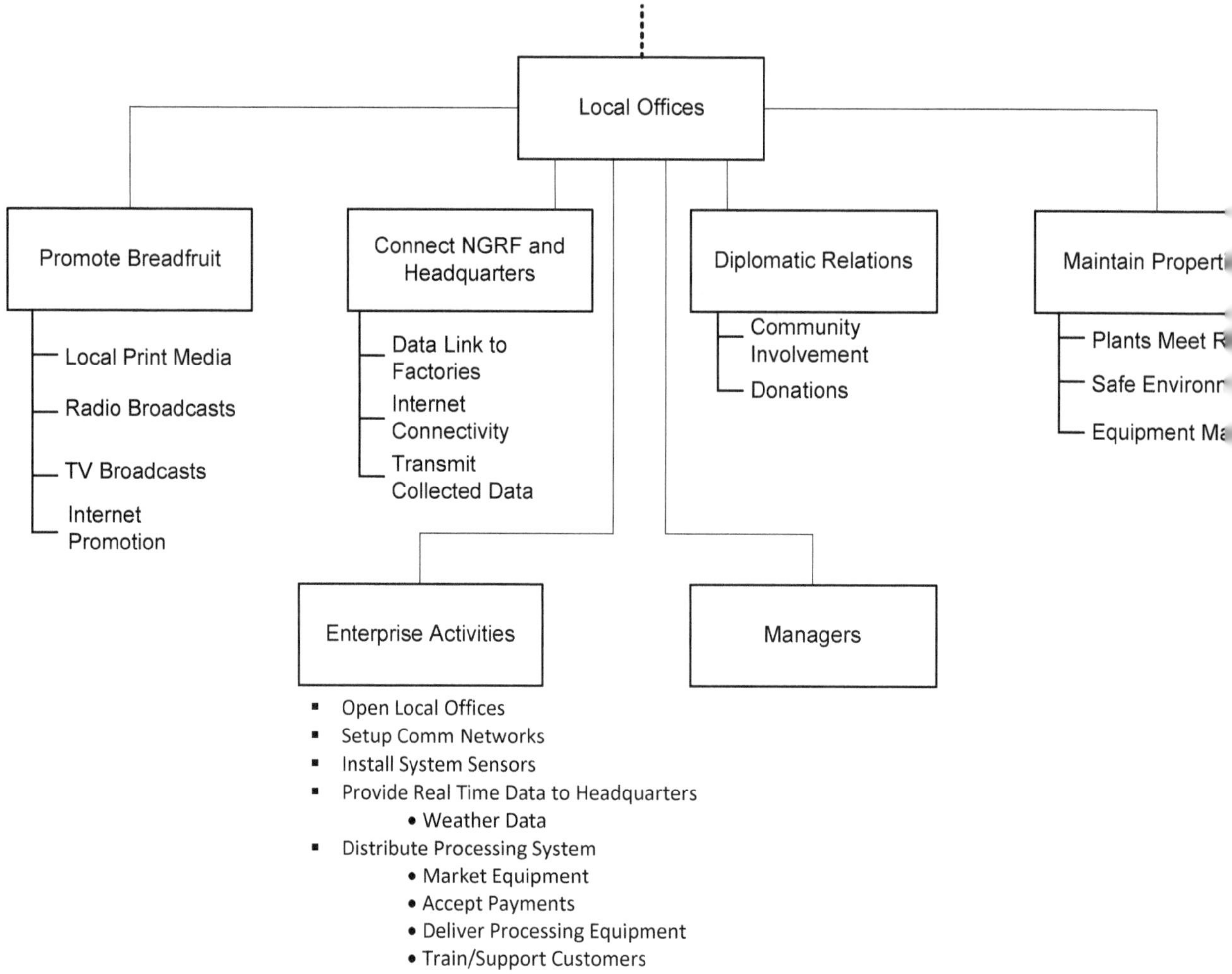

Level 3 (Under Local Offices): Managers WBS

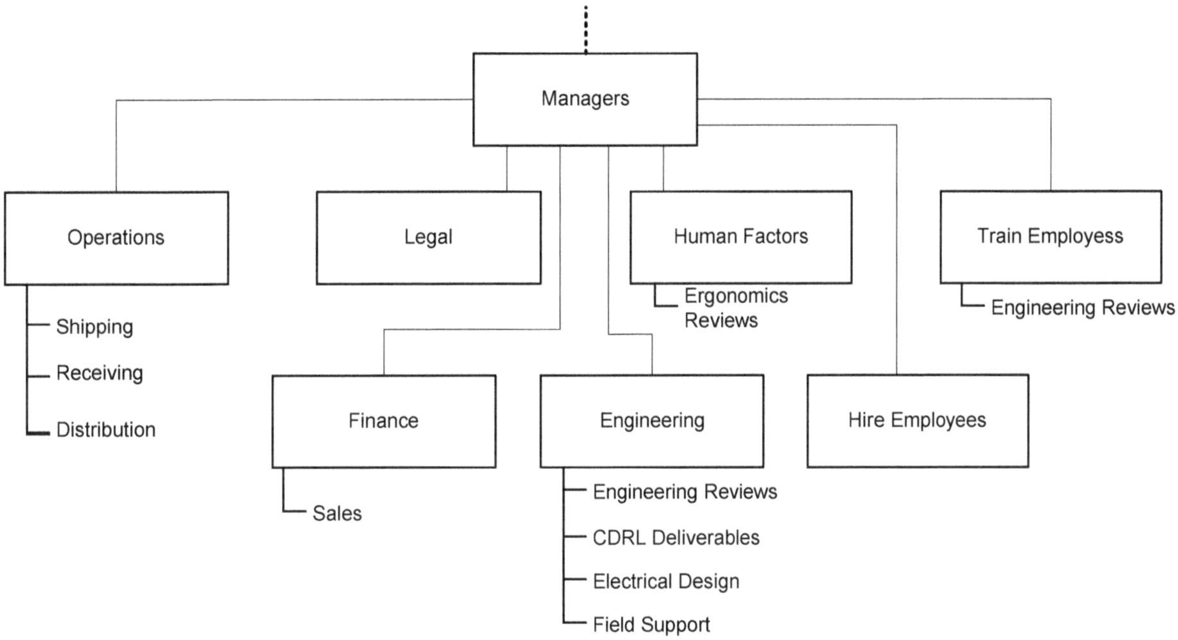

Level 2: Integration and Test

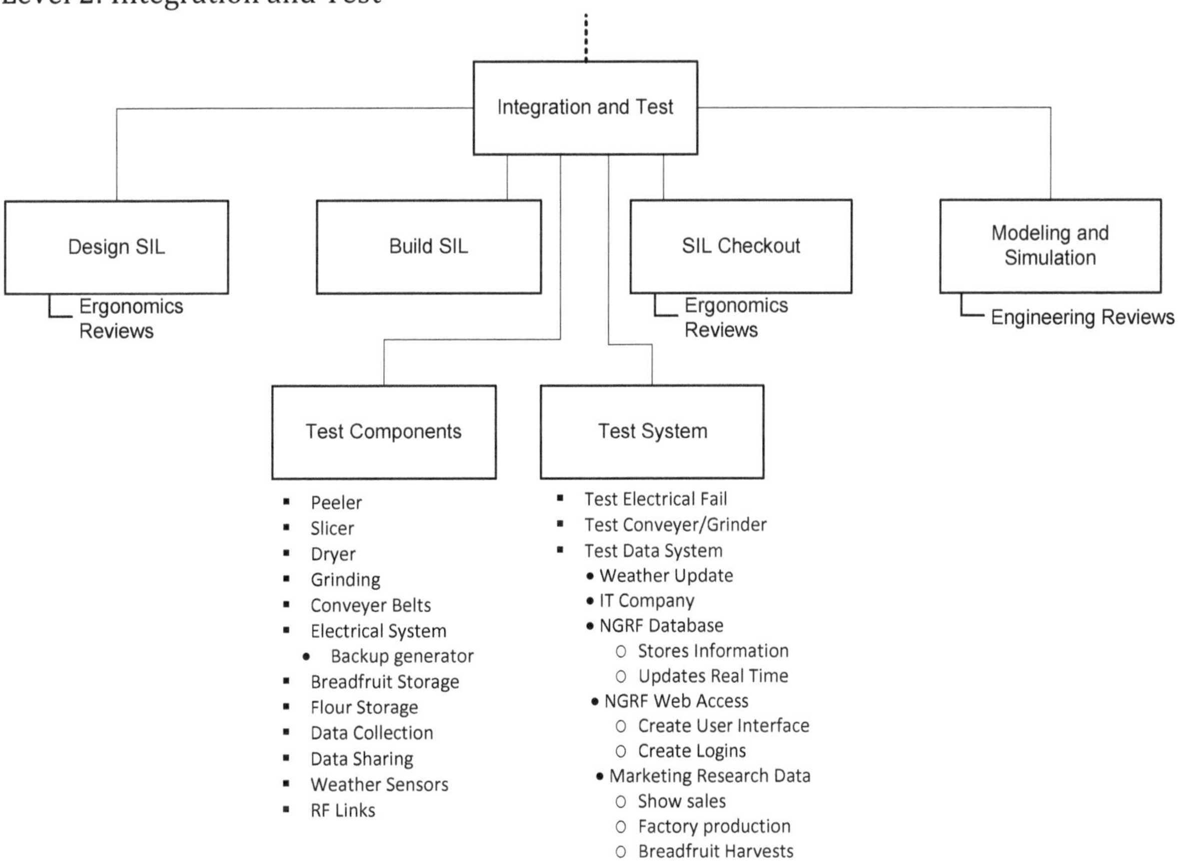

Level 2: NGRF Headquarters

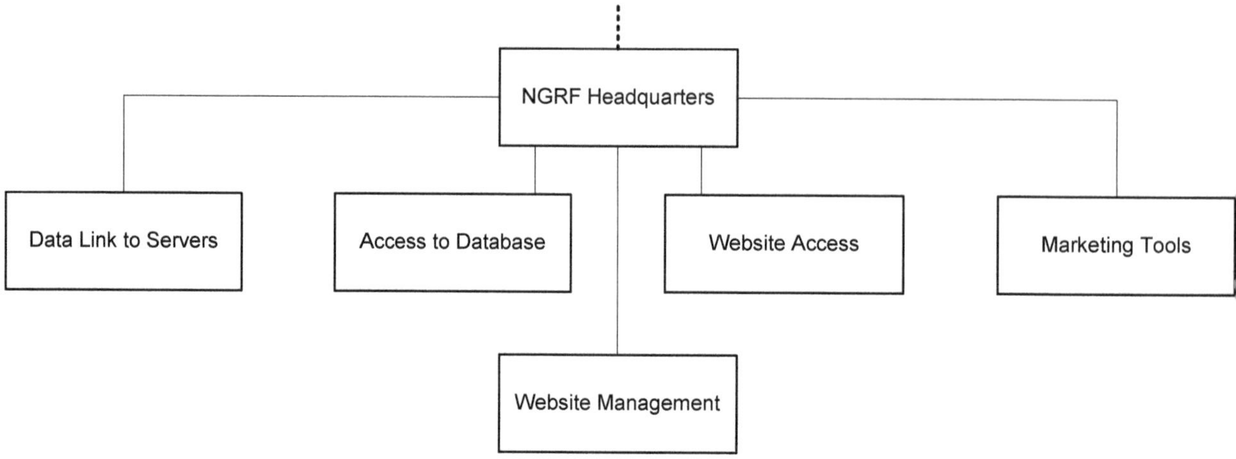

Level 2: Household Breadfruit Processor

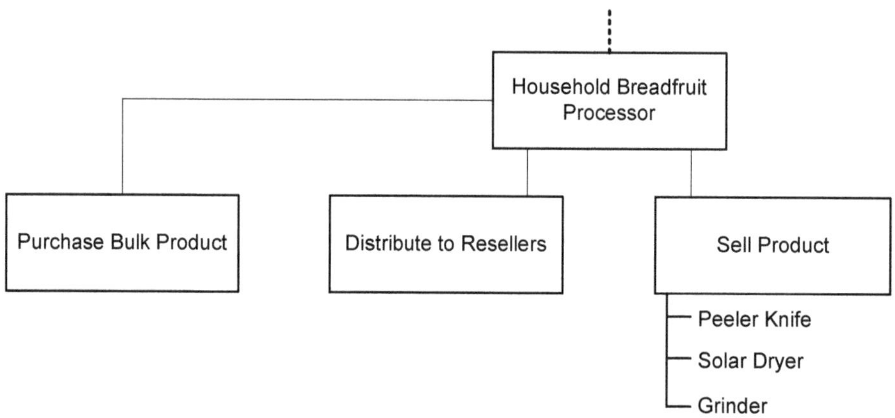

Level 2: Processing Facility

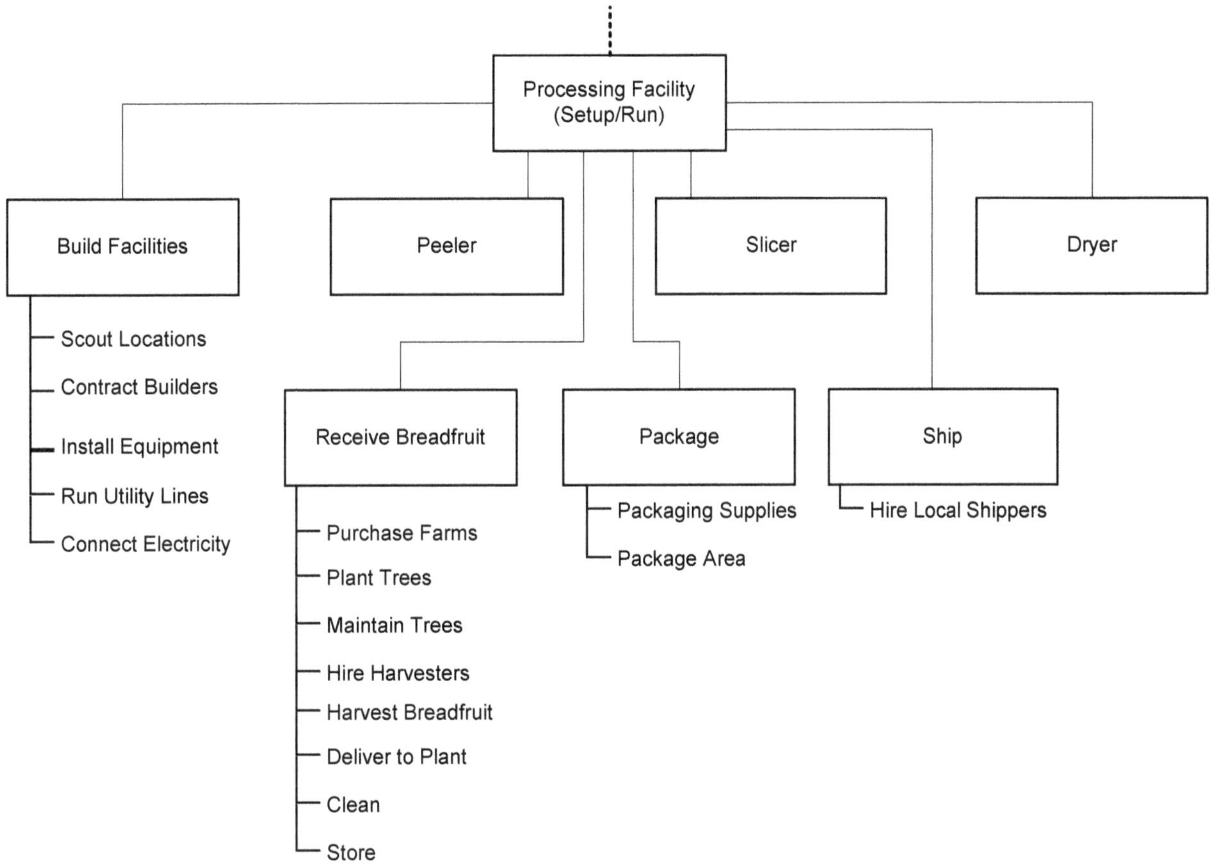

Level 2: Processing Facility

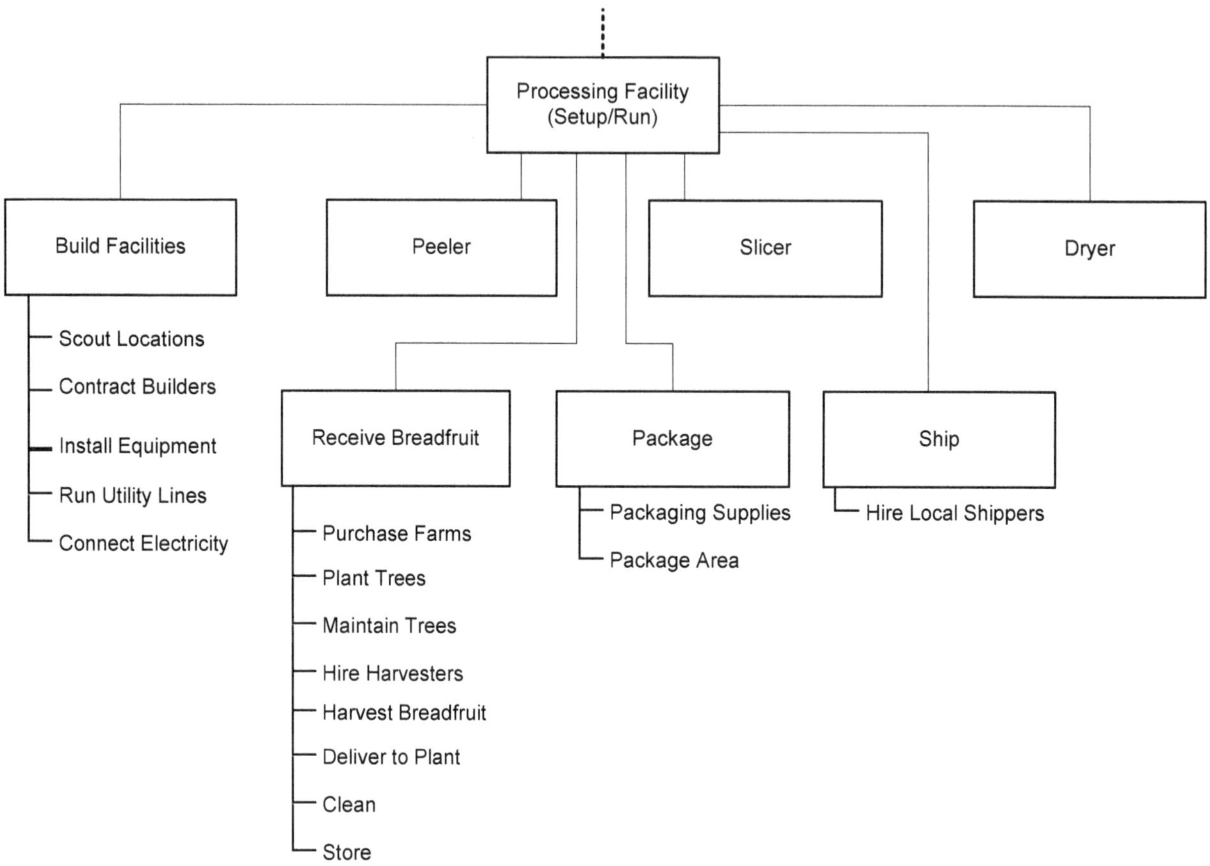

Breadfruit Processing System

- Management
    - Form Liaisons
        - Politicians
        - Local Communities
        - Interest Groups
    - Monitor Program Schedule
    - Milestone Management
        - Customer Reviews
    - CM Process
        - Engineering Reviews
    - Status Meetings
    - Procurement
        - Equipment Purchase
- Local Offices
    - Promote Breadfruit

        - Local print media
        - Radio broadcasts
        - TV broadcasts
        - Internet Promotion
    - Connect to NGRF and Headquarters
        - Data Link to Factories
        - Internet Connectivity
        - Transmit Collected Data
    - Diplomatic Relations
        - Community Involvement
        - Donations
    - Maintain Properties
        - Plants Meet Regulations
        - Safe Environment
        - Equipment Maintenance
    - Enterprise Activities
        - Open Local Offices
        - Setup Comm Networks
        - Install System Sensors
        - Provide Real Time Data to Headquarters
            - Weather Data
        - Distribute Processing System
            - Market Equipment
            - Accept Payments
            - Deliver Processing Equipment
            - Train/Support Customers
    - Managers
        - Operations
            - Shipping
            - Receiving
            - Distribution
        - Engineering
            - Design Documentation
            - CDRL Deliverables
            - Electrical Design
            - Field Support
        - Legal
        - Finance
            - Sales

- Human Factors
  - Ergonomics Reviews
- Hire Employees
- Train Employees

- Integration and Test
  - Design SIL
  - Build SIL
  - SIL Checkout
  - Modeling and Simulation
  - Test Components
    - Peeler
    - Slicer
    - Dryer
    - Grinding
    - Conveyer Belts
    - Electrical System
      - Backup generator
    - Breadfruit Storage
    - Flour Storage
    - Data Collection
    - Data Sharing
    - Weather Sensors
    - RF Links
  - Test System
    - Test Electrical Fail
    - Test Conveyer/Grinder
    - Test Data System
      - Weather Update
      - IT Company
      - NGRF Database
        - Stores Information
        - Updates Real Time
      - NGRF Web Access
        - Create User Interface
        - Create Logins
      - Marketing Research Data
        - Show sales
        - Factory production
        - Breadfruit Harvests

- NGRF Headquarters
    - Data Link to Servers
    - Access to Database
    - Website Access
    - Marketing Tools
    - Website Management
- Household Breadfruit Processor
    - Purchase Bulk Product
    - Distribute to Reseller
    - Sell Product
        - Peeler knife
        - Solar Dryer
        - Grinder
- Processing Facility
    - Build Facilities
        - Scout Locations
        - Contract Builders
        - Install Equipment
        - Run Utility lines
        - Connect Electricity
    - Receive Breadfruit
        - Purchase Farms
        - Plant Trees
        - Maintain Trees
        - Hire Harvesters
        - Harvest Breadfruit
        - Deliver to Plant
        - Clean
        - Store
    - Peeler
    - Slicer
    - Dryer
    - Grinder
    - Conveyer Belts
    - Backup generator
    - Breadfruit Storage
    - Flour Storage
    - Data Collection
    - Package
        - Packaging Supplies

        - Workstations
    - Ship
        - Hire Local Shippers

Annex D - SOW

1. Introduction
    a. Background
    b. Purpose
    c. Program Scope
2. Requirements
    a. Program Management
        i. DoDAF
        ii. Program Planning
        iii. Program Schedule
        iv. Contract Work Breakdown Structure
        v. Reporting
            1. Earned Value Management System
            2. Cost and Data Reporting
            3. Contract Funds Status Reports
        vi. Technical Performance Measures
        vii. Technical Reviews
            1. Integrated Baseline Review
            2. Program Management Reviews
            3. System Requirements Reviews
            4. System Functional Reviews
            5. Preliminary Design Reviews
            6. Functional Configuration Audits
            7. Physical Configuration Audit
            8. Test Readiness Reviews
            9. Critical Design Reviews
            10. Production Readiness Reviews
            11. System Verification Reviews
            12. Independent Expert Program Reviews
            13. Interface Control Working Groups
            14. Interoperable/Net-Centric Working Groups
            15. Human Systems Integration Working Groups
            16. Technical Interchange Meetings
            17. Test Plan Working Groups
            18. Technical Manual Management Team Meetings
            19. Independent Logistics ssessments
            20. Initial Operational Capability Supportability Review
            21. Systme Safety Working Groups
            22. Risk Managemnt Reviews
            23. Software Audits
            24. Design Readiness Review
            25. Operational Test and Readiness Review
            26. Build Release Reviews
            27. Software Qualification Reviews
            28. Software Verification and Validation Plan Reviews
            29. System Software Safety Working Groups

30. Electromagnetic Environmental Effects/Spectrum Supportability working groups

b. Systems Development

i. Systems Architecture

ii. Systems Engineering

iii. Industrial Engineering

3. Data Management

a. IT outsourced

4. Quality Management
5. Verification and Validation
6. Contract Data Requirements List

D.1 Introduction

D.1.1 Background

The BRAINS Group seeks to create a processing system capable of providing food security to those in need. In addition, an enterprise architecture will be designed to enable, empower and support all the activities related to the breadfruit process in order to make a profitable business. By using western technology, BRAINS will create a virtual collaborative knowledge by sharing automation and communication systems to consolidate research and marketing information in real-time.

The major problem with breadfruit is its shelf life in natural form. If left unprocessed, a typical breadfruit will begin to decay after 48 hours. Therefore, BRAINS will develop a processing system to prolong the usable life of the breadfruit. By using current technology to expedite the process, the breadfruit will be transformed into an economically viable and nutritious product that will help millions of hungry individuals.

D.1.2 Purpose

BRAINS wrote this proposal with the intention to raise sufficient funds to design the breadfruit transformation process system, and the Breadfruit Enterprise Architecture Demonstration (BrEAD).

D.1.3 Scope

This Statement of Work (SOW) defines the effort required for the design, development, fabrication, integration, demonstration, validation, test and documentation of the Breadfruit Enterprise Architecture Demonstration (BrEAD). It includes the associated program management, human engineering, and logistic support planning requirements.

D.2 Requirements

D.2.1 Program Management

BRAINS shall address the following program management objectives per the SOO:

1. BRAINS (BRAINS) shall provide a breadfruit processing system that is fully compliant with all applicable statutes and laws of the nation it is implemented in.
2. BRAINS shall provide an enterprise architecture that empowers the natives who operate the machines while providing the necessary support infrastructure.
3. BRAINS shall provide a system that is environmentally compliant with all applicable environmental statutes and laws

BRAINS shall use appropriate management techniques to support program control and gain program insight. They shall develop and implement an Integrated Master Plan (IMP) and Integrated Master Schedule (IMS). The contractor shall develop and implement the IMS and utilize the Contractor Work Breakdown Structure (CWBS) that separates tasks. BRAINS shall provide in the IMP: the structure of events, accomplishments and criteria required to design, develop, integrate, produce, demonstrate, test the capabilities and functionalities, and support the design to provide an operationally effective and operationally suitable system to meet the requirements. BRAINS shall also detail the required work, to include subcontractor efforts, in the IMP and IMS.

D.2.1.1 DODAF

The contractor shall develop and deliver integrated architecture products in compliance with the current DODAF version 1.5.

D.2.1.2 Program Planning

The contractor shall address the following program planning and control objectives per the SOO:

1. BRAINS shall provide a processing system and enterprise architecture that meets or exceeds the PBSS requirements
2. BRAINS shall employ a program management system based on the IMP/IMS that is able to provide accurate weekly cost, schedule and performance information to support the defined milestones of the project.
3. BRAINS shall employ an integrated NGRF/BRAINS risk management system that actively identifies program risks and identifies mitigation steps for said risks.

D.2.1.3 Program Schedule

D.2.1.3.1 Performance Based System Specifications

BRAINS will create a Performance Based Systems Specifications as an initial set of performance based requirements. BRAINS shall provide additional paragraphs and more detail to the NGRF to provide a system design and deliver a system that meets the objectives of the SOO. (CDRL 0009)

D.2.1.3.2 Integrated Master Plan

BRAINS will develop an Integrated Master Plan (IMP) outlining the structure of events, accomplishments and criteria required to design, develop, integrate, produce, demonstrate, test the capabilities and functionalities, and support the design to provide an operationally effective and operationally suitable system to meet the requirements that the NGRF has posed. The IMP shall address these events including subcontractor effort throughout the entire contracted effort. (CDRL 0002)

D.2.1.3.3 Integrated Master Schedule

BRAINS shall develop an Integrated Master Schedule (IMS) containing the contract IMP events and milestones, accomplishments, criteria, required resources and activities from contract award to the completion of the contract utilizing Microsoft Project 2007 software based on an estimated contract award date of 1-10-2010. The IMS must support program milestones and allows adequate time for technical reviews, design, and test and evaluation to meet the initial operating date of 1-10-2010.

Tasks on the IMS will be closely integrated with the Work Breakdown Structure (WBS) and the Contract Performance Reports (CPRs). BRAINS shall quantify each risk and provide the optimistic, pessimistic, and most likely duration for each IMS activity and event. The NGRF shall minimize schedule risk during the program with respect to the IMP/IMS with particular emphasis on software development, SIL utilization, and test and evaluation including test asset allocation. The contractor's schedule shall have sufficient flexibility to accommodate unforeseen technical issues without major impact to the IMP/IMS while maintaining a reasonable work and resource regimen. BRAINS shall ensure schedule integration with its subcontractors and the validity of the subcontractor's schedule data including demonstrating effective methods for incorporating schedule data from subcontractors into the contractor's IMS. BRAINS shall participate in an IMS Schedule Risk Assessment (SRA) with the NGRF two weeks prior to the IBR. BRAINS shall include technical reviews applicable to the contract and provide access, licenses (if required) and training for IMS tools for the NGRF team. (CDRL 0003)

D.2.4 Contract Work Breakdown Structure
BRAINS shall provide a Contractor Work Breakdown Structure (CWBS) in accordance with the Integrated Master Schedule (IMS). BRAINS will provide in the CWBS the structure of events, accomplishments and criteria required to design, develop, integrate, produce, demonstrate, test and support the design to provide an operationally effective and operationally suitable system to meet the Breadfruit Processing System.

D.2.5 Reporting
D.2.5.1 Earned Value Management System
BRAINS shall establish, maintain, and use an integrated management system meeting the intent of the Industry Guidelines for Earned Value Management System (EVMS) ANSI/EIA-748-98. The application of these concepts will provide for early indications of contract cost and schedule problems. A compliance review of the contractor's EVMS shall not be performed unless the Government program manager determines that it is necessary from Integrated Baseline Review (IBR) results.

Reference:
DI-MGMT-81466 CPR
ANSI/EIA-748 (Latest Revision)
DI-MGMT-81650
CDRL 0003 Integrated Master Schedule (IMS)
Integrated Baseline Review (IBR) results

D.2.5.2 Cost and Data Reporting
BRAINS shall collect and report actual contract costs through Contractor Cost Data Reporting (CCDR) and Software Resources Data Reporting (SRDR). BRAINS shall provide requirements of the CSDR to principal and critical subcontractors only.

Reference:
Contractor Work Breakdown Schedule (CWBS)

D.2.5.3 Contract Funds Status Reports

BRAINS shall provide Contract Funds Status Reports (CFSRs) as approved by the NGRF. The contractor shall provide requirements of the CSDR to principal and critical subcontractors only.

D.2.5.4 Technical Performance Measures

BRAINS shall utilize the Contractor Work Breakdown Structure (CWBS) and the contractor's program organizational structure in a manner that permits costs and schedule performance measurement by elements of either or both structures as needed. In addition to the CWBS, the contractor shall provide a portfolio of metrics and Technical Performance Measures (TPMs) to continually assess the program progress against defined plans and schedules.

Reference:

CDRL 0002 Integrated Master Plan (IMP)
CDRL 0003 Integrated Master Schedule (IMS)
CDRL 00018 Software Program Measurement Plan (SPMP)

D.3 Technical Reviews

D.3.1.3.1 Integrated Baseline Reviews (IBR)

BRAINS shall conduct Integrated Baseline Reviews (IBR) to present its performance measurement baseline plan to the NGRF within six months of contract award, and subsequently, when required, following major changes to the baseline. The NGRF shall verify during the IBR, and follow-on IBRs, when required, that BRAINS has established and maintains a reliable performance measurement baseline. BRAINS shall ensure that the baseline includes the entire contract technical scope of work consistent with contract schedule requirements, and has adequate resources assigned. BRAINS shall assure the NGRF that effective earned value methods are used to accurately assess contract cost, schedule, and technical performance. BRAINS shall perform a self-assessment of the cost and schedule risk for the IBR. The IBR shall be used to achieve a mutual understanding of the baseline plan, cost and schedule risk, and the underlying management processes used for planning and controlling the project. This review shall be tailored to focus on risk areas identified by the NGRF.

D.3.1.3.2 Program Management Reviews

BRAINS shall conduct quarterly Program Management Review (PMR) meetings at mutually agreed upon dates and locations. The PMR may be held in conjunction with the joint IMS-SRA review. During these reviews, BRAINS shall present integrated cost, schedule, and technical performance status. Integrated Product Team leads or functional managers shall include cost information in discussions of schedule status, technical performance, and risk using earned value as an integrating tool. The following shall be addressed: Cost/schedule trends, significant cost/schedule/technical variances, projected impacts, issues, quantified risk assessments, Total Productive Maintenance (TPMs) and corrective action plans. Additionally, BRAINS shall present the status on personnel and staffing levels and compliance with the subcontractor management plan during the PMR.

D.3.1.3.3 System Requirements Reviews

BRAINS shall conduct a System Requirements Review (SRR) to ensure that the system under review can proceed into the System Development and Demonstration phase, and that all system requirements and performance requirements are defined and are consistent with cost (program budget), schedule (program schedule), risk, and other system constraints. This review shall assess the system requirements as captured in the system specification, and ensure that the system requirements are consistent with the preferred system solution as well as available technologies. Of critical importance to this review is an understanding of the program technical risk inherent in the system specification and in the systems engineering plan. Determining an acceptable level of risk is key to a successful review.

BRAINS shall repeat the SRR after the start of System Development and Demonstration to clarify BRAINS's understanding of redefined or new user requirements.

D.3.1.3.4 System Functional Reviews

BRAINS shall conduct a System Functional Review (SFR) to ensure that the system under review can proceed into preliminary design and that all system requirements and functional performance requirements are defined and are consistent with cost (program budget), schedule (program schedule), risk, and other system constraints. This review shall assess the system functional requirements as captured in system specifications (functional baseline), and ensure that all required system performance is fully decomposed and defined in the functional baseline.

System performance may be decomposed and traced to lower-level subsystem functionality that may define hardware and software requirements. The SFR shall determine whether the systems functional definition is fully decomposed to a low level, and whether the IPT is prepared to start preliminary design.

The SFR shall determine whether the system's lower-level performance requirements are fully defined and consistent with the mature system concept, and whether lower-level systems requirements trace to top-level system performance. A successful SFR shall be predicated upon the IPT's determination that the system performance requirements, lower level performance requirements, and plans for design and development form a satisfactory basis for proceeding into preliminary design.

D.3.1.3.5 Preliminary Design Reviews/Software Readiness Reviews

BRAINS shall conduct a Preliminary Design Review (PDR)/Software Readiness Review
(SRR) to ensure that the system under review can proceed into detailed design, and can meet the stated performance requirements within cost (program budget), schedule (program schedule), risk, and other system constraints. This review shall assess the system preliminary design as captured in performance specifications for

each configuration item in the system (allocated baseline), and ensure that each function in the functional baseline has been allocated to one or more system configuration items. Configuration items may consist of hardware and software elements and include such items as airframes, avionics, weapons, crew systems, engines, trainers/training, etc.

The program manager shall conduct a PDR/SRR for each subsystem or configuration item, leading to an overall system PDR/SRR. When individual reviews have been conducted, the emphasis of the overall system PDR should focus on configuration item functional and physical interface design, as well as overall system design requirements. The PDR/SRR determines whether the hardware, human, and software preliminary designs are complete, and whether the IPT is prepared to start detailed design and test procedure development.

The PDR/SRR evaluates the set of subsystem requirements to determine whether they correctly and completely implement all system requirements allocated to the subsystem. The PDR/SRR also determines whether subsystem requirements trace with the system design. At this review the IPT shall review the results of peer reviews of requirements and preliminary design documentation. A successful review is predicated on the IPTs determination that the subsystem requirements, subsystem preliminary design, results of peer reviews, and plans for development and testing form a satisfactory basis for proceeding into detailed design and test procedure development.

D.3.1.3.6 Functional Configuration Audits

BRAINS shall conduct a Functional Configuration Audit (FCA) to verify that the configuration item has achieved the performance specified in its functional, or allocated, configuration identification documents, and that it's operational and support documents are complete and satisfactory. FCA is the formal examination of the "as-tested" characteristics of a configuration item where the focus is on design and performance suitability. FCA provides a systematic comparison of requirements with the results of tests, analyses, or inspections.
Normally, test/analysis data is collected from a prototype or pre-production article. When a prototype of pre-production article is not produced, test/analysis data from the first article can be used. The FCA demonstrates System Development and Demonstration is sufficiently mature for entrance into Low Rate Initial Production.

D.3.1.3.7 Physical Configuration Audit

BRAINS shall conduct a Physical Configuration Audit (PCA) to review the engineering drawings, technical documentation/data management, hardware and software. The PCA is the formal examination of the "as built" physical configuration or "as coded" total system software conforms to the design and construction process or technical documentation that defines it. It is performed during the Production and Deployment Phase and focuses on production suitability. The PCA formalizes the Product Baseline, including specifications and the technical data package so that future changes can be made only through formal configuration change control procedures.

D.3.1.3.8 Test Readiness Reviews

BRAINS shall conduct a Test Readiness Review (TRR) to ensure that the subsystem or system under review is ready to proceed into formal test. The TRR assesses test objectives, test methods and procedures, scope of tests, and safety and confirms that required test resources have been properly identified and coordinated to support planned tests. The TRR verifies the traceability of planned tests to program requirements and user needs. The TRR determines the completeness of test procedures and their compliance with test plans and descriptions. The TRR assesses the system under review for development maturity, cost/ schedule effectiveness, and risk to determine readiness to proceed to formal testing. In addition to adequate planning and management, to be effective the program manager should follow-up with the outcomes of the TRR.

Readiness to convene a TRR shall be predicated on the program manager's and Test and Evaluation Working-level Integrated Product Team's determination that preliminary testing, functional testing, and pre-qualification testing results form a satisfactory basis for proceeding with a TRR and subsequent initiation of formal, system-level Developmental Test.

D.3.1.3.9 Critical Design Reviews

BRAINS shall conduct a Critical Design Review (CDR) to ensure that the system under review can proceed into system fabrication, demonstration, and test; and can meet the stated performance requirements within cost (program budget), schedule (program schedule), risk, and other system constraints. This review shall assess the system final design as captured in product specifications for each configuration item in the system (product baseline), and ensure that each product in the product baseline has been captured in the detailed design documentation. Product specifications for hardware enable the fabrication of configuration items, and may include production drawings. Product specifications for software (e.g., Software Design Documents) enable coding of a Computer Software Configuration Item. Configuration items may consist of hardware and software elements, and include items such as trainers/training. The subsystem detailed designs shall be evaluated to determine whether they correctly and completely implement all system requirements allocated to the subsystem, and whether the traceability of final subsystem requirements to final system detail design is maintained. At this review, the Integrated Product Team shall also review the results of peer reviews on requirements and final detail design documentation, and ensure that the latest estimates of cost (development, production, and support) are consistent with the detail design. A successful review is predicated on the Integrated Product Team's determination that the subsystem requirements, subsystem detail design, results of peer reviews, and plans for testing form a satisfactory basis for proceeding into system fabrication, demonstration and test.

D.3.1.3.10 Production Readiness Reviews

BRAINS shall conduct a Production Readiness Review (PRR) to determine if the design is ready for production and if the producer has accomplished adequate production planning. The review examines risk; it determines if production or production preparations incur unacceptable risks that might breach thresholds of schedule, performance, cost, or other established criteria.

The review shall evaluate the full, production-configured system to determine if it correctly and completely implements all system requirements. The review shall determine whether the traceability of final system requirements to the final production system is maintained.

At this review, the Integrated Product Team shall review the readiness of the manufacturing processes, the Quality Management System, and the production planning (i.e., facilities, tooling and test equipment capacity, personnel development and certification, process documentation, inventory management, supplier management, etc.). A successful review is predicated on the Integrated Product Team's determination that the system requirements are fully met in the final production configuration, and that production capability forms a satisfactory basis for proceeding into Low-Rate Initial Production and Full-Rate Production.

D.3.1.3.11 System Verification Reviews

BRAINS shall conduct a System Verification Review (SVR) to ensure that the system under review can proceed into Low-Rate Initial Production and Full-Rate Production within cost (program budget), schedule (program schedule), risk, and other system constraints. This review shall be an audit trail from the Critical Design Review. It assesses the system final product, as evidenced in its production configuration, and determines if it meets the functional requirements (derived from the Capability Development Document and draft Capability Production Document) documented in the Functional, Allocated, and Product Baselines. The SVR establishes and verifies final product performance.

D.3.1.3.12 Independent Expert Programs

BRAINS shall conduct Independent Expert Program Review (IEPR) with an independent team of subject matter experts for the Systems Requirements Review, Systems Functional Review, and Critical Design Review. Independent experts shall review programs and report on technology and development risk, cost, schedule, design, development, project management processes and the application of systems and software engineering best practices. The team shall report their findings directly to the program manager.

Independent Expert Program Reviews shall be a mechanism for defining the health of programs as well as identifying get well strategies for sick programs. This presentation shall define IEPRs, contrast IEPRs with other assessment and review techniques, discuss the benefits, costs, and risks of IEPRs, and identify NGRF sources for IEPR services.

D.3.1.3.13 Interface Control Working Groups

BRAINS shall assemble an Interface Control Working Group (ICWG) comprised of appropriate technical representatives from the interfacing activities and other interested participating organizations. The ICWG serves as a forum to develop and provide interface requirements, as well as, focus on interface detail definition and issues. ICWG members include representatives from other government agencies and offices, contractors, and other industry organizations. The ICWG shall coordinate meetings and documents to the public.

D.3.1.3.14 Interoperable/Net-Centric Working Groups

BRAINS shall assemble an Interoperable/Net-Centric Working Group to design with interfaces (both internal and external) to promote ease of evolutionary growth through replacement of hardware, middleware, and software components, and to satisfy requirements for interoperability with existing and projected multi-service capabilities without redesign of entire systems or large portions thereof;

It shall be interoperable with local, state and federal agencies as deemed necessary to the timely collection of tolls, notifications of toll violations, accidents, vehicles on law enforcement alert lists, bridge surface conditions, and local weather conditions as a minimum.

D.3.1.3.15 Human Systems Integration Working Groups

BRAINS shall assemble a Human Systems Integration (HSI) Working Group consisting of interdisciplinary technical and management processes for integrating human considerations within and across all system elements; an essential enabler to systems engineering practice. Human Systems Integration shall be embedded in Systems Engineering practices, leading to the efficient delivery of effective systems.

BRAINS shall ensure HSI shall be a fundamental driver for the design by implementing a strategy to coordinate and manage issues across the program concerning manpower, personnel, training, Human Factors Engineering (HFE), safety, occupational health hazards, personnel survivability, and habitability factors.

D.3.1.3.16 Technical Interchange Meetings

BRAINS shall conduct and support Technical Interchange Meetings (TIMs), in conjunction with IPT reviews and working group meetings, to facilitate identification and resolution of engineering issues. TIMs shall address Operator Machine Interface (OMI), hardware, software, and installation engineering, and demonstration planning. TIMs shall be held as required and may be combined with the other contract meetings. Technical documentation required for each TIM shall be based on the purpose of the meeting and shall be specified in an agenda.

D.3.1.3.17 Test Plan Working Group Meetings

BRAINS shall establish a Test Plan Working Group (TPWG) to identify test processes and procedures based on industry best practices and lessons learned from the previous service center migrations. The TPWG shall establish a robust testing environment to ensure the successful migration of test mechanisms into the consolidated environment. The Test Plan shall detail three iterations forming into an Operational Readiness Test (ORT) phase, an Operational Readiness Demonstration (ORD) phase, and the testing criteria that must be satisfied during the move. Test scripts and test scenarios, which are reviewed by the TPWG, shall ensure that unique and/or site specific testing criteria are identified and incorporated into the overall test approach.

D.3.1.3.18 Technical Manual Management Team Meetings

BRAINS shall conduct Technical Manual Management Team (TMMT) Meetings to prepare BRAINS technical manuals in the Interactive Electronic Technical Manuals (IETMs) format, and in accordance with a NGRF approved specification. Technical data shall be accurate and in sufficient detail to support all planned levels of maintenance and operation. The technical data shall include systems/functional description and theory of operation, maintenance and operating procedures, troubleshooting and testing procedures, fault isolation, Illustrated Parts Breakdown (IPB), Maintenance Requirement Cards (MRC), periodic maintenance and phased maintenance requirements, preventive maintenance, and corrective maintenance authorized for each level of maintenance. BRAINS shall provide validated/verified technical data required for training, operations, testing, and maintenance. All updated publications and associated source data, XML files, schemas, Document Type Definition (DTD), business rules exchange, style sheets, and Format Output Specification Instance (FOSI) shall be considered NGRF property at the conclusion of the contract.

D.3.1.3.19 Independent Logistics Assessments

BRAINS shall conduct Independent Logistics Assessments (ILA) independent of systems developers to help ensure an impartial evaluation of program status. The ILA process, although viewed and implemented differently by each service, provides the program manager, program logisticians, and ultimately, the Milestone Decision Authority (MDA) with a confidence in the Integrated Logistics Support (ILS) planning and implementation. Assessments of logistics support programs prior to system fielding also help ensure timely awareness of potential deficiencies requiring attention and corrective action.

The ILA shall make certain that the system in acquisition can be supported effectively both during the acquisition process and throughout its service life. The ILA shall assess the product support management processes needed to achieve the performance objectives outlined in applicable capabilities documents. In addition to assessing product support planning for the sustainment elements, planning documents should be reviewed to ensure that they project effective product support strategies. Product support planning and implementation processes must demonstrate sufficient life-cycle management planning to promote effective program management and execution of the activities necessary to acquire and subsequently sustain the product successfully.

D.3.1.3.20 Initial Operational Capability Supportability Review

BRAINS shall conduct and Initial Operational Capability Supportability Review (IOCSR) to ensure systems and equipment are fully supportable at IOC levels and affordable over their life cycle. The purpose of the IOCSR shall be to positively impact supportability programs through augmented management attention, realignment of funds, or other available means, and to communicate the ILS posture of systems & equipment to our customers. The IOCSR process shall provide quality and timely information to decision authorities regarding ILS support. To accomplish the objectives, specific functions must be accomplished and certain products and support services must be provided.

D.3.1.3.21 System Safety Working Groups

BRAINS shall assemble System Safety Working Groups optimize the acquisition process from development to disposal. The NGRF Lead System Safety Engineer, appointed by the Program Manager (PM), shall be the primary safety point of contact for all aspects of the system. He or she shall develop a system safety management approach for the acquisition program and documents the approach in the NGRF's System Safety Management Plan (SSMP). The NGRF Lead System Safety Engineer also shall ensure BRAINS has a System Safety Program Plan (SSPP) for development of the system. To successfully carry out the system safety program for a given acquisition program, the NGRF Lead System Safety Engineer shall establish a System Safety Working Group (SSWG) made up of NGRF and contractor representatives. The SSWG shall be responsible for implementing system safety program requirements outlined in the SSMP and SSPP.

D.3.1.3.22 Risk Management Reviews

BRAINS shall conduct Risk Management Reviews as early as possible and throughout the program addressing timely and thorough assessment of cost, schedule and technical risk; identifying and tracking all medium and high-risk areas; defining, funding, and implementing appropriate risk mitigation efforts and; providing for fall-back or alternate plans, including design off ramps in the event that the primary risk mitigation plans fail.

D.3.1.3.23 Software Audits

BRAINS shall conduct Software Audits in which one or more auditors who are not members of the software development organization conduct an independent examination of a software product, software process, or set of software processes to assess compliance with specifications, standards, contractual agreements, or other criteria. Software audits shall be distinct from software peer reviews and software management reviews in that they are conducted by personnel external to, and independent of, the software development organization, and are concerned with compliance of products or processes, rather than with their technical content, technical quality, or managerial implications.

D.3.1.3.24 Design Readiness Reviews

BRAINS shall conduct a Design Readiness Review (DRR) to provide an opportunity for mid-phase assessment of design maturity as evidenced by measures such as the number of subsystem and system design reviews successfully completed; the percentage of drawings completed; planned corrective actions to hardware/software deficiencies; adequate development testing; an assessment of environment, safety and occupational health risks; a completed failuremodes and effects analysis; the identification of key system characteristics and critical manufacturing processes; an estimate of system reliability based on demonstrated reliability rates; etc. Successful completion of the Design Readiness Review ends System Integration and continues into the System Demonstration effort.

D.3.1.3.25 Operational Test and Readiness Reviews

BRAINS shall conduct Operational Test and Readiness Reviews (OTRR) as a multi-disciplined product and process assessment to ensure that the "production configuration" system can proceed into Initial Operational Test and Evaluation with a high probability of successfully completing the operational testing. Successful performance during operational test generally indicates that the system is suitable and effective for service introduction. The Full Rate
Production Decision may hinge on this successful determination. The understanding of available system performance to meet the Capability Production Document is important to the OTRR. The OTRR is complete when the Service Acquisition Executive evaluates and determines materiel system readiness for Initial Operational Test and Evaluation.

D.3.1.3.26 Build Release Reviews

BRAINS shall conduct Build Release Reviews (BRR) as a system-level review, conducted at the completion of subsystem integration testing to determine whether subsystem integration testing is complete and to verify the system is ready to start customer system integration testing.
At this review, the customer reviews the results of subsystem integration testing and supplier system integration testing and supplier system integration and testing. Emphasis is placed on understanding problems identified by the supplier in the delivered system software. A successful review is predicated on the customer's determination that the delivered system software forms a satisfactory basis for proceeding into customer system integration testing. The Program
Manager shall serve as the approval authority.

The Program Manager (or designee) shall be responsible for attending the review, affirming the completion of BRR checklist items, and approving satisfactory completion of the review. The test teams review BRR deliverables and attend the review to ensure that they and their teams understand: 1) what is in the build (e.g., Requirements, etc), and 2) known problems with the
Design Agent's implementation. By BRR, all facilities changes shall be incorporated, verified and ready for use. Modifications should be scheduled (with parts ordered and a delivery date set, risk mitigation plans in place, etc.), with a goal of modifications implemented.

D.3.1.3.27 Software Qualification Reviews

BRAINS shall conduct Software Qualification Reviews (SQR) to uniquely identify each software requirement in the software requirements document(s), and evaluate against the requirements traceability matrix that tracks each requirement through the Software Design Document(s), to the Software Qualification Test Plan and into the Software Qualification Test (SQT) Procedure, showing that all software requirements are implemented and verified. The requirements traceability matrix shall indicate the test in the SQT Test Procedure verifying each requirement, and the SQT Test Procedure shall indicate the test and step that verifies the requirement. If any identified requirements cannot be verified using the final software, lower level assemblies shall be tested or code shall be inspected. The Organization's Software Quality Assurance representative shall verify and record the results.

D.3.1.3.28 Software Verification and Validation Plan Reviews

BRAINS shall conduct Software Verification and Validation (V&V) Plan Reviews to cover all V&V activities to be performed during all phases of the life cycle. The major V&V activity during this phase shall be to develop a concept of how the system is to be reviewed and tested. Test concepts shall involve simple generation of test cases by a user representative or may require the development of elaborate simulators and test data generators. Without an adequate V&V concept and plan, the cost, schedule, and complexity of the project may be poorly estimated due to the lack of adequate test capabilities and data.

D.3.1.3.29 Software System Safety Working Groups

BRAINS shall assemble Software System Safety Working Groups (SwSSWG) chaired by the software safety point of contact. The SwSSWG shall be the primary means of communication between the PM and the developer's software safety program. The SwSSWG schedule shall include periodic meetings, meetings prior to major program milestones, and splinter meetings as required to fully addressing the software safety issues. The SwSSWG is an IPT that is responsible for oversight and management of the software safety program. Membership of the IPT shall include:

- The safety principals [safety point of contact, safety PM(s)],
- System safety engineering,
- Software safety engineering,
- Systems Engineering,
- Software Engineering,
- Software Testing,
- Independent Verification and Validation (IV&V),
- Software Quality Assurance,
- Software Configuration Management, and
- Human Factors Engineering.

D.3.1.3.30 Electromagnetic Environmental Effects /Spectrum Supportability Working Groups

BRAINS shall assemble Electromagnetic Environmental Effect (E3)/Spectrum

Supportability (SS) Working Groups to review E3 control requirements established early in the acquisition process and to ensure that Department of Defense (DOD) equipment, subsystems, and systems are designed to be self-compatible and operate compatibly in the operational electromagnetic environment. E3 control applies to the electromagnetic interactions of both spectrum-dependent and non-spectrum dependent objects within the operational environment. E3 is the impact of the electromagnetic environment upon the operational capability of military forces, equipment, systems, and platforms. It encompasses all electromagnetic disciplines, including compatibility, interference; vulnerability, pulse; electrostatic discharge; hazards of radiation to personnel, ordnance, and volatile materials; and natural phenomena effects, of lightning and precipitation static.

BRAINS shall determine as to whether the electromagnetic spectrum necessary to support the operation of spectrum-dependent equipment or system during its expected lifecycle is, or shall be, available (that is, from system development, through developmental and operational testing, to actual operation in the electromagnetic environment.) The assessment of equipment or system as having "spectrum supportability is based upon, as a minimum, receipt of equipment spectrum certification, reasonable assurance of the availability of sufficient frequencies for operation, and consideration of electromagnetic compatibility (EMC).

D.4 Systems Development

D.4.1 Systems Architecture

BRAINS shall develop Open Architecture designs and corresponding components. As part of this contract, BRAINS shall be required to define, document, and follow an open systems approach for using modular design and standards-based interfaces. BRAINS shall develop, maintain, and use an Open System Management Plan (OSMP) demonstrating compliance during
all design reviews. As part of an OSMP, BRAINS shall identify to the NGRF all Commercial-Off-the-Shelf/Non-development Item (COTS/NDI) components, their functionality, and provide copies of license agreements related to the use of these components for NGRF approval prior to use.

To achieve an Open System Architecture (OSA), it is an objective of the BRAINS process system to use an open systems approach to:

1) Facilitate development of a self-sustainable architecture that allows for affordable operability;

2) Ensure that the system design is sufficiently flexible and robust to accommodate evolving technology and requirements;

3) Facilitate integration with other systems and use of commercial products from multiple sources both in the initial design and in future enhancements;

4) Enable technology insertion as currently available commercial products mature and new commercial products become available in the future;

5) Design for cost-effective integration, control of, and communication with future payloads;

6) Allow continued access to technologies and products supported by many suppliers (a broad industrial base which does not restrict available sources to the detriment of competition);

7) Design with OSA interfaces (both internal and external) to promote ease of evolutionary growth through replacement of hardware, middleware, and software components, and to satisfy requirements for interoperability with existing and projected multi-service capabilities without redesign of entire systems or large portions thereof;

8) Enable supportability, portability, distribution of processing, scalability, fault tolerance, reduction in life-cycle cost, share resource management, obsolescence and replacement or upgrade of system hardware, middleware and software elements and components;

9) Mitigate the risks associated with technology obsolescence and the ability to replace obsolete components during the SDD and Production phases of the program;

10) Address the migration of proprietary or closed system equipment or interfaces in the proposed system;

11) Allow for the insertion of hardware, middleware and software with minimal impact to physical elements, components, and functions;

BRAINS shall address how they shall provide data and information to the NGRF that may be used to support third-party development and delivery of competitive alternatives for subsystems on an ongoing basis.

D.4.2 Systems Engineering

BRAINS shall address the following systems engineering objectives per the SOO:

1) BRAINS shall employ a comprehensive Systems Engineering Process in accordance with the International Council on Systems Engineering (INCOSE) Handbook

2) BRAINS shall employ a Service Oriented Architecture that embodies the concepts of ensuring timely, external data exchanges from the data center to any member of the enterprise architecture

3) BRAINS shall employ a testing strategy, which includes an integrated environment of accredited M&S tools, simulation and stimulation environment, data and resources that enables collaboration across functional disciplines.

4) BRAINS shall design and deliver an enterprise architecture system (BrEAD) that will create an independent, self-sustaining network; it will support subsystem/system test, integration, V&V, and system/software lifecycle sustainability.

5) BRAINS shall provide a design which maximizes the use of standard manufacturing concepts and industrial approved commercial best practices, along with use of commercially available products.

BRAINS shall develop and update as necessary throughout the SDD phase a comprehensive Systems Engineering Management Plan (SEMP) that describes BRAINS's methodology for fully integrating the NGRF's requirements and PBSS. The SEMP shall demonstrate an integrated systems engineering approach towards development of the requirements, configuration baseline definitions, system design, system integration Problem Based Learning (PBL) product support requirements, Open Architecture implementation, system architecture quality and concordance, logistics resources (maintenance personnel, support equipment and repair parts) that cumulatively provide the capabilities specified in the PBSS. Additionally, BRAINS shall address how production capabilities and processes shall be incorporated into the systems engineering approach. In particular, BRAINS shall focus on a process design engineering reducibility program identifying and implementing the design and fabrication/assembly processes which shall result in the system achieving its production performance, cost and schedule requirements. BRAINS shall generate TPMs, and track and report the status of the TPMs. BRAINS shall utilize a requirements traceability tool, (such as TRACE, Dynamic Object Oriented Requirements System (DOORS) or a compatible automated tool), throughout the design, development, production and verification phases of the program. BRAINS shall provide sufficient access, licenses (if required) and training for the requirements traceability tool to the NGRF team. The traceability tool shall include all levels of the system and subsystem requirements, derived requirements and links to the capabilities documents and all testing requirements.

BRAINS shall provide a System Design Document in narrative form using non-technical terms. It shall provide a high-level system architecture diagram showing a subsystem breakout of the system. The high-level system architecture or subsystem diagrams shall show interfaces to external systems and supply a high-level context diagram for the system and subsystems. The allocation of the functional requirements into this design document shall refer to the requirements traceability matrix. (CDRL 0001)

D.5 Data Management

Because many I.T. services are required, BRAINS has selected a company that specializes in data and server management. This company will be the sole provider of technical service to the entire BrEAD system.

BRAINS shall provide a Data Management Plan (DMP) establishing an overall plan for the data management requirements for a specific project. This DMP will be provided through the data company that BRAINS has selected. The DMP shall provide the necessary management and control of the contractually identified data items. The prime functions include administration of contract record keeping, data copying control, data quality control, acquisition/administration of supplier data, storage and retrieval systems, maintenance and control of supplier-developed information and Purchaser-furnished information, identification and handling of property rights-in-data, pricing data, and planning, scheduling, and delivery of data. (CDRL 00011)

D.5.1 I.T. Outsourced

BRAINS will outsource its data services to Techfluent for the I.T. infrastructure and data support. Their experience in data operations and equipment maintenance and upgrades will be crucial in setting up the system quickly in our location of interest.

D.6 Quality Management

BRAINS shall develop and maintain a Quality Management System (QMS) that satisfies program objectives and meets the requirements of ANSI/ASQC-Q9001: 2000 or an equivalent QMS. The QMS procedures, planning, and all other documentation and data that comprise the QMS, shall be made available to the NGRF for review. The NGRF may perform the inspections, verifications, and evaluations necessary to ascertain conformance to requirements and the adequacy of the implementing procedures. Third-party certification of BRAINS's QMS is not required. BRAINS shall require subcontractors to maintain a QMS that achieves control of the quality, services and supplies provided.

BRAINS shall provide a Quality Management Plan documenting that the delivered products satisfy contractual agreements, meet or exceed quality standards. The level of detail contained in the Quality Management Plan should be consistent with the complexity, size, intended use, mission criticality, and cost of failure of the system development effort. (CDRL 00013)

D.7 Verification and Validation

BRAINS shall provide a Verification and Validation Plan describing the testing strategies that shall be used throughout the life-cycle phases. This plan shall include descriptions of contractor, NGRF, and appropriate independent assessments required by the project.

D.8 Contract Data Requirements List

- CDRL 0001 System Design Description

- CDRL 0002 Integrated Program Plan (IMP)
- CDRL 0003 Integrated Management Schedule (IMS)
- CDRL 0004 Integrated Test Plan
- CDRL 0005 Systems Engineering Management Plan (SEMP)
- CDRL 0005 Software Development Plan
- CDRL 0006 Validation and Verification Plan
- CDRL 0007 Statement of Work
- CDRL 0008 Updated Performance Based Systems Specification (PBSS)
- CDRL 0009 Configuration Management Plan
- CDRL 00010 Data Management Plan
- CDRL 00011 System Operation Effectiveness Plan
- CDRL 00012 Quality Management Plan
- CDRL 00013 Modeling and Simulation Plan
- CDRL 00014 Human Engineering Task Analysis
- CDRL 00015 Human Engineering Design Apporach
- CDRL 00016 Human Engineering Test Plan
- CDRL 00017 System Safety Management Plan (SSMP)
- CDRL 00018 Test and Evaluation Master Plan (TEMP)
- CDRL 00019 Contractors Performance Based Logistics Planning (PBL) Strategy Report

D.9 Risk Management

BRAINS will implement a sound risk management system that eases risk identification and mitigates program risks. The risk management system shall be integrated with program controls through the development of metrics to monitor program status, and tied to the IMP/IMS. This system shall be utilized to:

1) Perform timely and thorough assessment of cost, schedule and technical risk;
2) Identify and track all medium and high-risk areas;
3) Define, fund, and implement appropriate risk mitigation efforts and;
4) Provide for fall-back or alternate plans, including design off ramps in the event that the primary risk mitigation plans fail.

D.10 Integrated Data Environment

Timely access to program data is of the utmost importance. BRAINS shall use electronic media to improve data transfer efficiency and affordability. The Integrated Data Environment (IDE) objective includes the utilization of a robust, cost-effective on-line digital data environment that allows the program acquisition and operational support activities, throughout the program's life-cycle, to digitally create, store, access, manipulate, share, and exchange programmatic and technical data. This data must be accessible to the BRAINS Integrated Product Team (IPT) at anytime from their workspaces (approximately 200 individuals). BRAINS shall provide the maximum number of NGRF seats at no additional cost. The IDE shall provide protection for program data in accordance with the Program Protection Plan (PPP). Strategy shall include technical refresh provisions and any necessary accreditation activities. BRAINS shall optimize the use of the IDE with its subcontractors and suppliers, as well as participating NGRF organizations. BRAINS shall develop, update and maintain a Data Accession List (DAL) that identifies all data and documentation generated under the execution of the BrEAD effort.

BRAINS shall define how the IDE shall be used to facilitate the integration of the IMP, IMS, and WBS to provide the data necessary for the NGRF to assess the concept's ability to meet the requirements of the PBSS (verification matrix) and provide the necessary flight clearances for the flight test program. The IMS shall include sufficient time for NGRF approval of data at significant events.

A list of the data to be provided on the IDE shall be attached to the IDE implementation plan.

IDE process controls, in addition to use of business tools (such as Associate contractor nondisclosure agreements), shall apply and shall use system-level and data-level Metadata tags for identification and protection of digital form data (such as for classified, Export Control, Proprietary, or Distribution Statements). The NGRF shall require continued use of the IDE beyond the SDD program. BRAINS shall provide the NGRF with strategies to ensure continued use of the IDE.

D.11 Design Decision Documentation

Utilize and maintain a Decision Database to document system-wide design decisions including decisions about the system's behavioral design and decisions affecting the selection and design of systems components and/or subsystems. Include the tradeoffs and other assessments, the decisions made, their justification, the actions taken including revised plans or changes to the baselines or functional architecture proposed or made as a result of the decisions. The decision database shall be maintained over the life of the contract and shall be readily available to and directly accessible by the NGRF. BRAINS shall prepare and submit to the NGRF an archived image of the data in the Decision Database at the time of the close out of each technical review and audit.

D.12 Problem Reporting

The NGRF shall have insight into BRAINS's problem reporting database when system integration testing begins in the laboratory. BRAINS shall ensure that full access to the problem-reporting database is provided to the NGRF. The NGRF and contractor shall form, operate and co-chair a review board to prioritize problems for resolution.

D.13 Unique Identification (UID)

BRAINS shall establish an effective serialized management program on all components

and end items that are repairable, life-limited or time-controlled, configuration tracked items,
and/or assigned logs and/or records. BRAINS shall also be responsible for configuration management and associated data.

D.14 Transition to Production

BRAINS shall maintain a management organization with the ability to coordinate all phases of the BrEAD transition to production while minimizing program risk.

Annex E – IMP

Event #	Event
	Significant Accomplishment
	Accomplishment Criteria
	Task
Event 1	Event 1 – Pre-contract Engineering Plan for Start-up Activities
	Pre-contract Start-up Activities Complete
	Internal kick-off meeting complete
	Internal Staffing for BF Capture team Approved
	Capture team proposal schedule made
	Pre-contract Engineering Plan written
	Traceability Matrix Complete
	BrEAD PBSS Spec & Verification Matrix Complete
	Contract Work Breakdown Structure (CWBS) Complete
	Statement of Work Complete
	Integrated Master Plan (IMP) Complete
	Integrated Master Schedule (IMS) Complete
	Systems Engineering Management Plan (SEMP) Complete
	Software Development Plan Complete
	Risk Management Plan Complete
Event 2	Event 2 - System Definition Complete
	Product Baseline Description Developed
	Initial CDRLS Completed
	Initial CDRLS Delivered to NGRF
	Status Report 2 delivered to NGRF
	System Function Review (SFR) prepared
	Requirements Update Complete
Event 3	Event 3 - Preliminary Design Complete
	COTS Component Analysis Complete
	Component selection and acquisition
	Electrical/Power Systems Engineering Complete
	Product Definition Complete
	Open Architecture Profile Developed
	DoD Architecture Framework Documentation Developed
	Preliminary Design CDRLS Delivered to NGRF
	Open Architecture Profile delivered to NGRF
	DoD Architecture Framework Delivered to NGRF
	Preliminary Design Review Preparation and Conduct
	PDR Technical Package Prepared
	PDR Conducted

Event 4	Event 4 - Proposal Presented & Delivered
	Oral Presentation Developed
	Written Proposal Developed
	Executive Summary (Vol I) Developed
	Technical Volume (Vol II) Developed
	Contract Proposal Presented
	Written Proposal Delivered
Event 5	Event 5 - BrEAD Critical Design Complete
	NGRF Contract Awarded
	Major Reviews and Milestones
	Post Award Meeting with NGRF
	Integrated Baseline Review (IBR) Complete
	Critical Design Review Complete
	Test Readiness Review Complete
	Formal Qualification Complete
	Production Readiness Review (PRR) Complete
	Program Planning Revised
	Integrated Master Schedule (IMS) Updated with Current Progress
	Engineering Critical Design Revised
	DoD Architecture Framework Documentation Revised
	System Safety Program Plan (SSPP) Complete
	Develop System Safety Program Plan
	Review by Gov't of Marshall Islands
	Deliver SSPP to NGRF
	Initial Marketing Plan Complete
	Marketing Plan Developed
	Maintenance Task Analysis Report Complete
	Maintenance Task Analysis Report
	Task Analysis Interviews Conducted in Marshall Islands
	Surveys Distributed& Collected in Marshall Islands
	Supply Support Services Plan (SSSP) Complete
	SSSP Developed
	SSSP Reviewed by NGRF
	SSSP Reviewed by Suppliers
	SSSP Delivered to NGRF
	Initial System/Subsystem Design Description (SDDD) Complete
	Develop SDDD
	Requirements Allocation Matrix Updated
	Complete Peer Review
	Update SDDD
	Submit Initial SDDD to NGRF

	Life Cycle Cost Report Delivered to NGRF
	Work Load, Tasking, & Modeling Plan Developed
	Reliability & Maintainability Plan (RMA) Developed
	Critical Design Status Reports & CDRLS Delivered to NGRF
	Critical Design Review Complete
	Develop CDR Agenda
	CDR Internal Technical Review
	CDR Conducted
Event 6	Event 6 - BrEAD Test Readiness & Functional Testing Complete
	Test Readiness Review (TRR) Complete
	Test & Certification Plan Developed
	Review by NGRF of Test & Certification Plan
	Factory Acceptance Test (FAT) Developed
	Software Development Complete
	Received Code from Suppliers
	Systems Integration & Checkout Complete
	Develop Test Procedures
	Define Test Scenarios
	Define Analysis Required
	Writing Procedures
	Factory Acceptance Testing (FAT) Complete
	FAT TRR Complete
	TRR Technical Documents Prepared
	Conduct TRR
	FAT Conduct Complete
	Perform FAT Test Procedure
	Document Findings
	Analyze Findings for Impact
	Develop FAT Test Report
Event 7	Event 7 - BrEAD Qualification Testing Complete
	Training Management & System Familiarization Complete
	Initial Planned Maintenance System (PMS) Complete
	Develop PMS
	Submit PMS to NGRF
	Functional/Physical Configuration Audit Complete
	Functional/Physical Configuration Audit Conducted
	Installation of Systems Integration Laboratory (SIL)
	Testing, Modeling, Simulation conducted at SIL
	Qualification Testing for Factory Complete
	Qualification Testing for Grinder Conducted
	Qualification Testing for Dryer Conducted

	Qualification Testing for BreadFruit Flour Conducted
	Qualification Test for Wind Turbines Conducted
	Maintenance Tests Complete
	Maintainability Demo Conducted
Event 8	Event 8 - Production Readiness
	Engineering Planning Revised
	System/Subsystem Specification Revised
	System/Subsystem Design Description Revised
	Technical Report Complete
	Develop Technical Report
	Deliver Technical Report to NGRF
	Marketing Plan Complete
	Initial Marketing Plan revised
	Marketing Plan Delivered to NGRF
	BrEAD Facilities set up in Marshall Islands
	Production Readiness Review (PRR) Complete
	Equipment Tools & Test Equipment List Complete
	Reliability Prediction & Supporting Data Complete
	Spares List Complete
	PRR Complete
	PRR Agenda delivered to NGRF
	PRR Technical Package Prepared & Delivered to NGRF
	Conduct PRR
	Approved Product Baseline Set
	Product Baseline Description Document Delivered to NGRF
	Training Conference Conducted
	Final Life Cycle Cost Complete
	Initial Operating Capability (IOC) Complete
	Initial Operating Capability (IOC) Review
Event 9	Event 9 - Contract Complete
	Equipment Procured
	Purchase orders for slicing & peeling equipment, dryers, grinders, wind turbines
	Procure Slicing equipment
	Procure Peeling equipment
	Procure Wind Turbine Equipment
	Procure Drying Equipment
	Procure Grinding Equipment
	Procure storage bins
	Recurring Reviews Conducted
	Program Management Reviews
	Conduct PMR1

	Conduct PMR2
	Conduct PMR3
	Conduct PMR4
	NGRF Board of Directors Working Group Meeting
	Conduct Working Group Meeting 1
	Conduct Working Group Meeting 2
	Conduct Working Group Meeting 3
	Conduct Working Group Meeting 4
	Conduct Working Group Meeting 5
	Conduct Working Group Meeting 6
	Conduct Working Group Meeting 7
	Conduct Working Group Meeting 8
	Conduct Working Group Meeting 9
	Final CRDLs Delivered
	BrEAD Launch

Annex F – IMS

BrEAD IMS.mpp

ID	Task Name	Duration	Start	Finish
1	Breadfruit Enterprise Architecture Demonstration (BrEAD)	252 days	Thu 1/22/09	Sun 1/10/10
2	Event 1 - Precontract Start-up Activities Complete	30 days	Thu 1/22/09	Wed 3/4/09
3	RFP Delivered by NGRF	1 day	Thu 1/22/09	Thu 1/22/09
4	Program Management Processes Complete	1 day	Thu 1/22/09	Thu 1/22/09
5	Conduct Internal Kick-off Meeting	1 day	Thu 1/22/09	Thu 1/22/09
6	Engineering Planning for Start-up Activities Complete	30 days	Thu 1/22/09	Wed 3/4/09
24	Event 2 - System Definition Through Preliminary Design Comple	78 days	Thu 1/22/09	Mon 5/11/09
25	Product Baseline Description Document Developed	1 day	Thu 1/22/09	Thu 1/22/09
26	Initial CDRLS Completed & Delivered to NGRF	30 days	Mon 2/2/09	Thu 3/26/09
27	Status Report 1 delivered to NGRF	1 day	Mon 3/23/09	Mon 3/23/09
28	System Functional Review (SFR) prepared	1 day	Mon 3/23/09	Mon 3/23/09
29	Requirements Update Complete	1 day	Mon 3/23/09	Mon 3/23/09
30	Engineering Planning for Preliminary Design Complete	78 days	Thu 1/22/09	Mon 5/11/09
31	COTS Component Analysis Complete	45 days	Thu 1/22/09	Wed 3/25/09
32	Component Selection and Acquisition	45 days	Thu 1/22/09	Wed 3/25/09
33	Electrical/Power Systems Engineering Complete	30 days	Thu 1/22/09	Wed 3/4/09
34	Product Definition Complete	45 days	Thu 1/22/09	Wed 3/25/09
35	Open Architecture Profile Developed	5 days	Mon 5/4/09	Fri 5/8/09
36	DoD Architecture Framework Documentation Developed	60 days	Thu 1/22/09	Wed 4/15/09
37	Preliminary Design CDRLS Delivered to NGRF	18 days	Thu 4/16/09	Mon 5/11/09
38	Open Architecture Profile delivered to NGRF	1 day	Mon 5/11/09	Mon 5/11/09
39	DoD Architecture Framework Delivered to NGRF	1 day	Thu 4/16/09	Thu 4/16/09
40	Preliminary Design Review Preparation and Conduct	45 days	Mon 2/9/09	Fri 4/10/09
41	PDR Technical Package Prepared	45 days	Mon 2/9/09	Fri 4/10/09
42	PDR Conducted	0 days	Sun 3/29/09	Sun 3/29/09
43	Event 3 - Proposal Presented & Delivered	49 days	Mon 2/23/09	Thu 4/30/09
44	Oral Presentation Developed	5 days	Mon 3/9/09	Fri 3/13/09
45	Written Proposal Developed	45 days	Mon 2/23/09	Fri 4/24/09
46	Executive Summary (Vol I) Developed	10 days	Mon 3/9/09	Fri 3/20/09
47	Technical Volume (Vol II) Developed	45 days	Mon 2/23/09	Fri 4/24/09

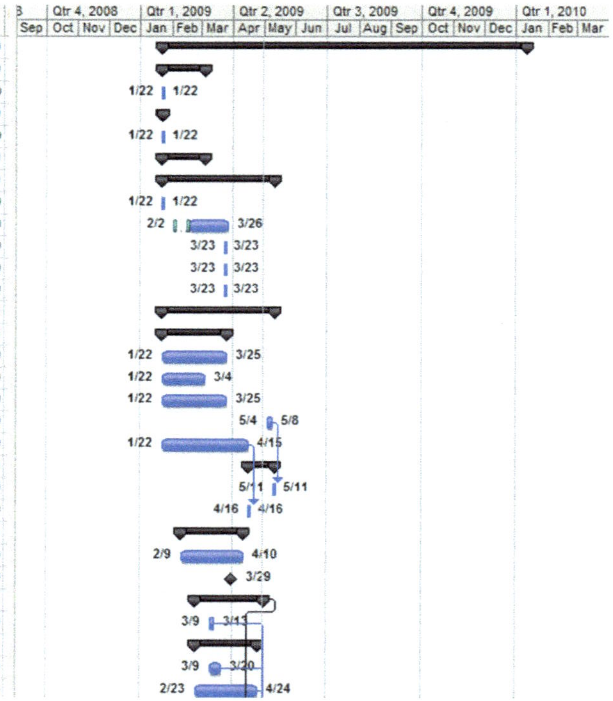

ID	Task Name	Duration	Start	Finish
48	Contract Proposal Presented	1 day	Thu 4/30/09	Thu 4/30/09
49	Written Proposal Delivered	1 day	Thu 4/30/09	Thu 4/30/09
50	⊟ **Event 4 - BrEAD Critical Design Complete**	**62 days**	**Fri 5/1/09**	**Mon 7/27/09**
51	NGRF Contract Awarded	1 day	Fri 5/1/09	Fri 5/1/09
52	⊟ **Major Reviews and Milestones**	**57 days**	**Fri 5/1/09**	**Tue 7/21/09**
53	Post Award Meeting with NGRF	1 day	Fri 5/1/09	Fri 5/1/09
54	Integrated Baseline Review (IBR) Complete	0 days	Sun 5/3/09	Sun 5/3/09
55	Critical Design Review Complete	0 days	Tue 7/21/09	Tue 7/21/09
56	Test Readiness Review Complete	0 days	Sun 6/14/09	Sun 6/14/09
57	Formal Qualification Complete	0 days	Sun 6/28/09	Sun 6/28/09
58	Production Readiness Review (PRR) Complete	0 days	Sun 7/12/09	Sun 7/12/09
59	⊟ **Program Planning Revised**	**1 day**	**Tue 7/21/09**	**Tue 7/21/09**
60	Integrated Master Schedule (IMS) Updated with Current Progres	1 day	Tue 7/21/09	Tue 7/21/09
61	⊟ **Engineering Critical Design Revised**	**62 days**	**Fri 5/1/09**	**Mon 7/27/09**
62	DoD Architecture Framework Documentation Revised	3 days	Fri 5/1/09	Tue 5/5/09
63	⊟ **System Safety Program Plan (SSPP) Complete**	**16 days**	**Mon 5/4/09**	**Mon 5/25/09**
64	Develop System Safety Program Plan	10 days	Mon 5/4/09	Fri 5/15/09
65	Review by Gov't of Marshall Islands	5 days	Mon 5/18/09	Fri 5/22/09
66	Deliver SSPP to NGRF	1 day	Mon 5/25/09	Mon 5/25/09
67	⊞ **Initial Marketing Plan Complete**	**15 days**	**Mon 6/29/09**	**Fri 7/17/09**
69	⊟ **Maintenaince Task Analysis Report Complete**	**23 days**	**Fri 5/1/09**	**Tue 6/2/09**
70	Maintenaince Task Analysis Report	12 days	Mon 5/18/09	Tue 6/2/09
71	Task Analysis Interviews Conducted in Marshall Islands	5 days	Mon 5/11/09	Fri 5/15/09
72	Surveys Distributed& Collected in Marshall Islands	20 days	Fri 5/1/09	Thu 5/28/09
73	⊟ **Supply Support Services Plan (SSSP) Complete**	**46 days**	**Fri 5/1/09**	**Fri 7/3/09**
74	SSSP Developed	30 days	Fri 5/1/09	Thu 6/11/09
75	SSSP Reviewed by NGRF	5 days	Fri 6/12/09	Thu 6/18/09
76	SSSP Reviewed by Suppliers	10 days	Fri 6/19/09	Thu 7/2/09

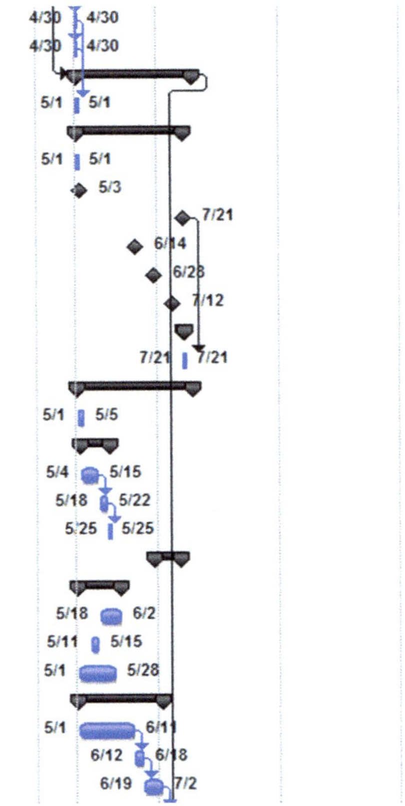

ID		Task Name	Duration	Start	Finish
78		⊞ Initial System/Subsystem Design Description (SDDD) Co	62 days	Fri 5/1/09	Mon 7/27/09
87		Critical Design Status Reports & CDRLS Delivered to NGRF	1 day	Fri 5/1/09	Fri 5/1/09
88		⊟ Critical Design Review Complete	7 days	Mon 7/13/09	Tue 7/21/09
89		Develop CDR Agenda	1 day	Mon 7/13/09	Mon 7/13/09
90		CDR Internal Technical Review	5 days	Tue 7/14/09	Mon 7/20/09
91		CDR Conducted	1 day	Tue 7/21/09	Tue 7/21/09
92		⊟ Event 5 - BrEAD Test Readiness & Functional Testing Complete	40 days	Tue 7/28/09	Mon 9/21/09
93		⊟ Test Readiness Review (TRR) Complete	40 days	Tue 7/28/09	Mon 9/21/09
94		⊟ Test & Certification Plan Developed	6 days	Tue 7/28/09	Tue 8/4/09
95		Review by NGRF of Test & Certification Plan	1 day	Tue 7/28/09	Tue 7/28/09
96		Factory Acceptance Test (FAT) Developed	5 days	Wed 7/29/09	Tue 8/4/09
97		⊟ Software Development Complete	40 days	Tue 7/28/09	Mon 9/21/09
98		Received Code from Suppliers	40 days	Tue 7/28/09	Mon 9/21/09
99		⊟ Systems Integration & Checkout Complete	13 days	Tue 7/28/09	Thu 8/13/09
100		⊟ Develop Test Procedures	13 days	Tue 7/28/09	Thu 8/13/09
101		Define Test Scenarios	2 days	Tue 7/28/09	Wed 7/29/09
102		Define Analysis Required	4 days	Thu 7/30/09	Tue 8/4/09
103		Writing Procedures	7 days	Wed 8/5/09	Thu 8/13/09
104		⊟ Factory Acceptance Testing (FAT) Complete	13 days	Fri 8/14/09	Tue 9/1/09
105		⊟ FAT TRR Complete	4 days	Fri 8/14/09	Wed 8/19/09
106		TRR Technical Documents Prepared	3 days	Fri 8/14/09	Tue 8/18/09
107		Conduct TRR	1 day	Wed 8/19/09	Wed 8/19/09
108		⊟ FAT Conduct Complete	9 days	Thu 8/20/09	Tue 9/1/09
109		Perform FAT Test Procedure	4 days	Thu 8/20/09	Tue 8/25/09
110		Document Findings	1 day	Wed 8/26/09	Wed 8/26/09
111		Analyze Findings for Impact	2 days	Thu 8/27/09	Fri 8/28/09
112		Develop FAT Test Report	2 days	Mon 8/31/09	Tue 9/1/09
113		⊟ Event 6 - BrEAD Qualification Testing & Production Complete	70 days	Tue 9/22/09	Mon 12/28/09

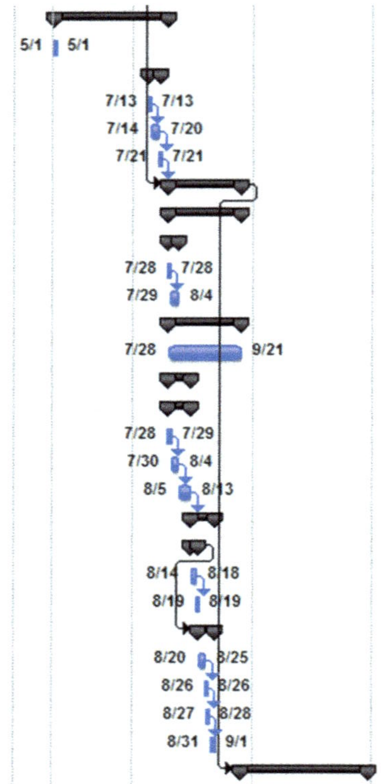

ID	Task Name	Duration	Start	Finish
123	**Training Management & System Familiarization Complete**	**33 days**	**Tue 9/22/09**	**Thu 11/5/09**
138	BrEAD Facilities set up in Marshall Islands	70 days	Tue 9/22/09	Mon 12/28/09
139	**Production Readiness Review (PRR) Complete**	**7 days**	**Tue 11/3/09**	**Wed 11/11/09**
140	Equipment Tools & Test Equipment List Complete	1 day	Tue 11/3/09	Tue 11/3/09
141	Reliability Prediction & Supporting Data Complete	5 days	Tue 11/3/09	Mon 11/9/09
142	Spares List Complete	1 day	Tue 11/3/09	Tue 11/3/09
143	**PRR Complete**	**7 days**	**Tue 11/3/09**	**Wed 11/11/09**
144	PRR Agenda delivered to NGRF	1 day	Tue 11/3/09	Tue 11/3/09
145	PRR Technical Package Prepared & Delivered to NGRF	5 days	Wed 11/4/09	Tue 11/10/09
146	Conduct PRR	1 day	Wed 11/11/09	Wed 11/11/09
147	**Approved Product Baseline Set**	**1 day**	**Tue 11/3/09**	**Tue 11/3/09**
148	Product Baseline Description Document Delivered to NGRF	1 day	Tue 11/3/09	Tue 11/3/09
149	Training Conference Conducted	5 days	Tue 11/3/09	Mon 11/9/09
150	Final Life Cycle Cost Complete	3 days	Tue 11/3/09	Thu 11/5/09
151	**Event 7 - Contract Complete**	**181 days**	**Fri 5/1/09**	**Sun 1/10/10**
152	**Equipment Procured**	**181 days**	**Fri 5/1/09**	**Sun 1/10/10**
153	**Purchase orders for slicing & peeling equipment, dryer**	**1 day**	**Thu 11/5/09**	**Thu 11/5/09**
154	Procure Slicing equipment	1 day	Thu 11/5/09	Thu 11/5/09
155	Procure Peeling equipment	1 day	Thu 11/5/09	Thu 11/5/09
156	Procure Wind Turbine Equipment	1 day	Thu 11/5/09	Thu 11/5/09
157	Procure Drying Equipment	1 day	Thu 11/5/09	Thu 11/5/09
158	Procure Grinding Equipment	1 day	Thu 11/5/09	Thu 11/5/09
159	Procure storage bins	1 day	Thu 11/5/09	Thu 11/5/09
160	**Recurring Reviews Conducted**	**177 days**	**Fri 5/1/09**	**Tue 1/5/10**
161	**Program Management Reviews**	**175 days**	**Fri 5/1/09**	**Fri 1/1/10**
162	Conduct PMR1	0 days	Fri 5/1/09	Fri 5/1/09
163	Conduct PMR2	0 days	Sat 8/1/09	Sat 8/1/09
164	Conduct PMR3	0 days	Sun 11/1/09	Sun 11/1/09
165	Conduct PMR4	0 days	Fri 1/1/10	Fri 1/1/10

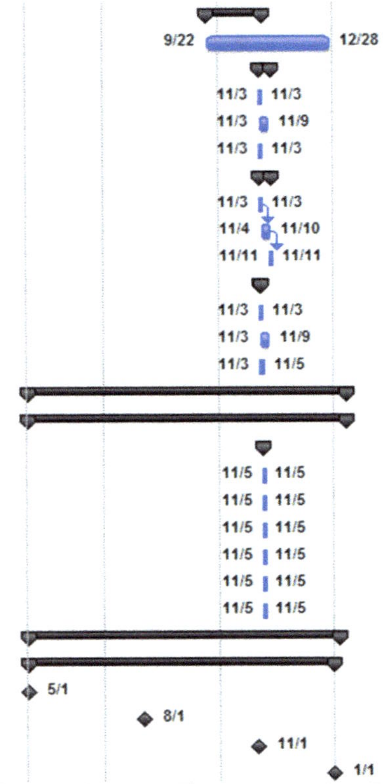

ID		Task Name	Duration	Start	Finish
166		**NGRF Board of Directors Working Group Meeting**	**175 days**	**Fri 5/1/09**	**Fri 1/1/10**
167		Conduct Working Group Meeting 1	0 days	Fri 5/1/09	Fri 5/1/09
168		Conduct Working Group Meeting 2	0 days	Mon 6/1/09	Mon 6/1/09
169		Conduct Working Group Meeting 3	0 days	Wed 7/1/09	Wed 7/1/09
170		Conduct Working Group Meeting 4	0 days	Sat 8/1/09	Sat 8/1/09
171		Conduct Working Group Meeting 5	0 days	Tue 9/1/09	Tue 9/1/09
172		Conduct Working Group Meeting 6	0 days	Thu 10/1/09	Thu 10/1/09
173		Conduct Working Group Meeting 7	0 days	Sun 11/1/09	Sun 11/1/09
174		Conduct Working Group Meeting 8	0 days	Tue 12/1/09	Tue 12/1/09
175		Conduct Working Group Meeting 9	0 days	Fri 1/1/10	Fri 1/1/10
176		Final CRDLs Delivered	0 days	Tue 1/5/10	Tue 1/5/10
177		BrEAD Launch	0 days	Sun 1/10/10	Sun 1/10/10

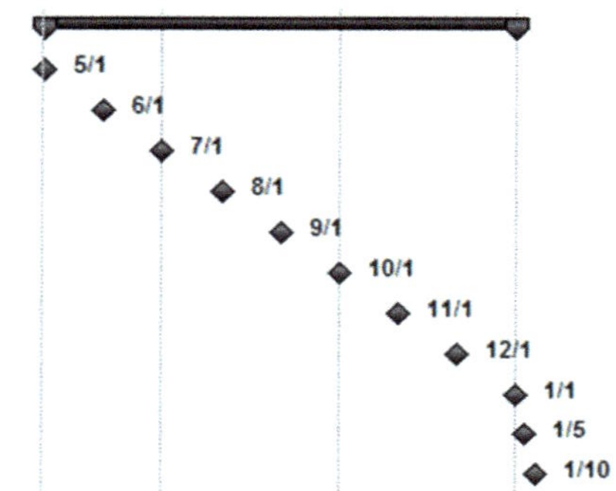

Introduction

System Engineering Management Plan (SEMP)

As a program management tool, the Systems Engineering Management Plan (SEMP) provides a framework for organizing and implementing the technical and management support for successful program execution. This SEMP satisfies CDRL 0005 Systems Engineering Management Plan (SEMP) and is applicable for the lifecycle of the Breadfruit Enterprise Architecture Demonstration (BrEAD) SEMP.

BrEAD SEMP

This SEMP is followed by the BRAINS team and describes the BrEAD program development execution and management from concept through initial build and evaluation phases. This SEMP identifies and describes program-specific management procedures and techniques. It also identifies pertinent plans and schedules, including technical milestones and reviews. The BRAINS team participates in workshops with the IPTs to describe the roles and responsibilities in order to better integrate the plans, processes and tools described in this SEMP with the IPTs.

Scope and Purpose

This document describes the activities, processes, and tools used by Bright Robust Architects for the Institute of New Systems (BRAINS) to develop a Breadfruit Processing System and Breadfruit Enterprise Architecture Demonstration (BrEAD) for the National Green Revolution

Foundation (NGRF).

The New Green Revolution Foundation (NGRF) is a non-profit foundation and has been doing research in the Marshall Islands on expanding the use of breadfruit as food source to feed the growing world population, particularly the Marshall Island. The purpose of the NGRF is to write a proposal to raise sufficient funds to develop the Enterprise Architecture Demonstration (BrEAD) to do further research and development to develop new breadfruit food products, to promote those products worldwide, to facilitate the design of scalable breadfruit processing systems, to create a virtual collaborative knowledge sharing automation and communication system to share research and marketing information in real-time, and to seek funding from various sources, to continue the development of breadfruit food products and breadfruit processing technology development.

This SEMP identifies program-specific development processes and technical management procedures BRAINS will use during the BrEAD System Development and Demonstration (SDD) phase. This document describes how organization-specific procedures apply to the BrEAD program and also captures additional plans and process tailoring specific to BrEAD.

Document Overview

1. Introduction – BrEAD SEMP purpose, applicability, and contents.

2. Applicable Documents - Identifies Government and Non-Government documents referenced here. Where possible, this document refers the reader to additional documents containing the desired information rather than duplicating that information here.

3. Program Description - Describes BREAD, tools and facilities, program schedules, deliverable descriptions, and plans and process integration.

4. Technical Program Planning and Control - Describes overall development team structure and responsibilities. It also describes program controls, risk management, technical performance measurement, technical reviews and audits, configuration management, and subcontracting monitoring and control.

5. System Design and Development Process - Describes BREAD design development approach including the plan for requirements definition and analysis, architecture, software, and hardware design, and integration and test. It also describes plans and processes for technical reviews and subcontractor management.

6. Engineering Support and Specialty Integration - Describes specialty considerations including reliability and maintainability, survivability/vulnerability, Operability/Human Engineering, safety engineering, integrated logistics support, design-to-cost, and producibility.

Appendices – Appendix A includes a glossary of acronyms and terms.

Applicable Documents
The locations of documents referenced in this SEMP are listed below in Sections 2.1 and 2.2.
Government Documents
The following documents of the issue in effect on the issue date of this document form part of this document to the extent specified herein.

- Non-Government Documents
- NGRF
- Exhibit A NGRF Form XXX Contract Data Requirements List
- Breadfruit Enterprise Architecture Demonstration (BrEAD) – Performance Based Specification (PBSS).
- Statement of Work (SOW) for the NGRF Breadfruit Enterprise Architecture Demonstration (BrEAD) SDD phase
- Integrated Master Plan (IMP)
- SDD Contract Top-Level Integrated Master Schedule (IMS)

- Cost Plus Award Fee (CPAF) Plan for Breadfruit Enterprise Architecture Demonstration (BrEAD)
- Statement of Objectives (SOO) for a Breadfruit Enterprise Architecture Demonstration (BrEAD) and Systems Design Phase.
- Solicitation, Offer and Award RFP E00010-07-R-0001
- BREAD Repository
- Design-to-Cost Plan
- EMI/EMP/EMC Control Plan
- Facilities Management Plan
- Engineering Tools List
- Staffing
- Trade Study Results
- Hardness Assurance Maintenance Surveillance Plan
- Hardware Development Plan (HDP)
- IPT Roles and Responsibilities
- Deliverables
- Peer Review Plan
- Software Development Plan (SDP)
- Survivability Program Plan
- Program Training Plan
- Trade Studies
- BREAD Process Framework
- The procedures listed below comprise the BREAD Process Framework.
- BRAINS
- User Requirements Definition
- System Requirements Definition
- System Design
- Detailed Specifications
- System High Level Design
- System Detailed Design
- System Implementation
- Software/Hardware Test
- System Integration
- System Verification
- Customer Acceptance Test
- System Design concepts Definition
- High Level Design
- Detailed Level Design
- Implementation
- Test
- System Requirements Analysis and Design
- System Requirements Analysis
- Preliminary Design
- Detailed Design

- Implementation
- SW Integration and Test
- SYS Integration and Test
- Verification
- Formal Test
- Marketing and Customer
- Wind Turbine
- Peeler
- Grinder
- Slicer
- Dryer
- Generator
- Outsourcing
- Manufacture Facilities/SIL
- Verify System
- Define Hardware Requirements
- Develop Preliminary Hardware Design
- Develop Detailed Hardware Design.
- Fabricate and Assemble Prototype Hardware
- Integrate Hardware
- Verify Hardware.
- Simulation

Precedence
If the requirements in this document or the documents referenced herein conflict with contractually higher-level documents, the order of precedence is:

The contract: The contract overrides requirements in contractually specified documents.
The contractually specified documents to the extent specified in the contract. The contractually specified documents override requirements in this SEMP.

This SEMP, which overrides requirements in the documents referenced herein, to the extent specified.

Conflicts shall be referred in writing to the procuring activity for resolution through BRAINS Contracts.

BrEAD Program Description

System Description
The Breadfruit Enterprise Architecture Demonstration (BrEAD) program has been initiated to design, develop, produce, and deploy a new system that will fulfill the requirements as specified in Performance Based Specification (PBSS). The BrEAD

utilizes the breadfruit to provide a storable food product and provide a scalable design for a system that will disseminate the product to millions of people worldwide. The BrEAD also uses the profit money collected from the breadfruit's products to expand the company in better and more advanced processing equipments.

The proposed process for the Breadfruit Enterprise Architecture Demonstration is divided into the following subsystems as show in the OV-4 below: power, peeler, slicer, grinder, dryer, blancher, and backup power.

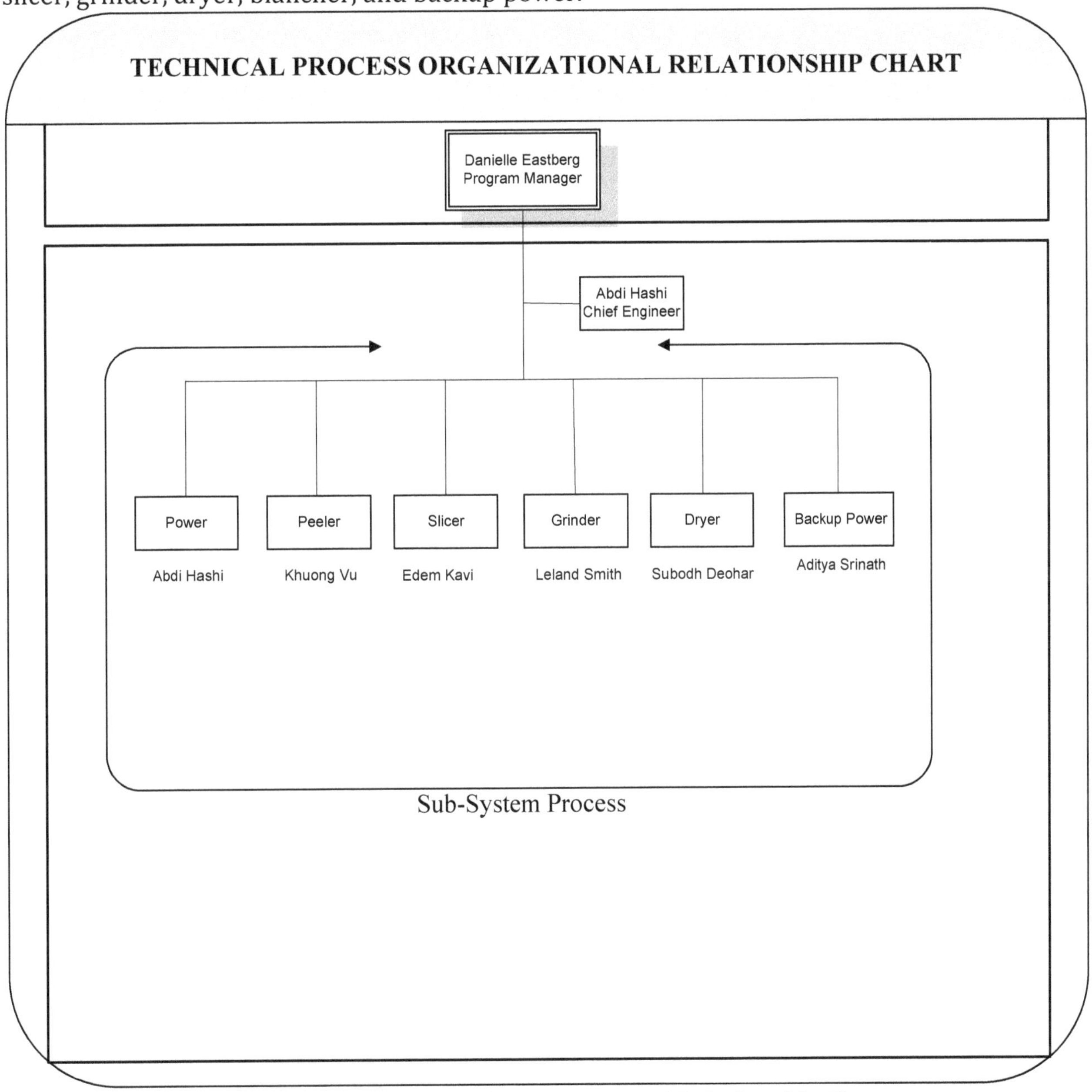

Figure 207. OV-4 Process Technical Organizational Relationship Chart

Open System Architecture

The BrEAD BF Capture team has designed the BrEAD system to meet and exceed the Open System Architecture (OSA) requirements. The processing design architecture is able to react quickly without delaying other the system downtime when it comes to maintenance and repair. Additionally, since all of the sub-systems depend on the power sub-system as the main source, a mechanism is designed with the backup power kicks in case there is no power. The OSA will use services provided by subcontractors and COTS products. This will make BrEAD easily upgradeable, reduce sensitivity to failure and make system alterations and maintenance user friendly.

System Constraints

Most BrEAD components or sub-system will be derived from commercial, modified commercial, or government off-the-shelf technologies and products. The BrEAD system will leverage industry standard interfaces and products that have a product roadmap to facilitate low-cost, low-risk technology insertion and system growth.

System constraints include the following:

A Reuse Evaluation Analysis Report will be generated for reused components or sub-system.

The system, or portions of it, will be offline for scheduled maintenance

All subsystems will be maintenance/calibrate twice a year to prevent downtime.

Manual operation is used only when main and backup powers are no longer available.

Spare parts are available for backup to keep the plant operates efficiency.

Systems Engineering tools and facilities

Tools

The BRAINS team is using or will use the following tools as shown in Table 4 on the BREAD program.

Collaboration Architecture	Mechanical Design	Electrical Design	System/Software Design	
Collaboration ProjectLink ProductView		Analog Circuit Analysis MathCAD PSpice	System/Software Estimation	Specialties and Support Seismograph
Risk Management Active Risk Manager Technical Performance Measures TechView	Structural Analysis Pro-Eng Advanced Structural Dynamics Pro-Eng Mechanism ANSYS	Digital Circuit Analysis Hyperlynx (Mentor Graphics)	Requirements Management DOORS	Decision Analysis

Collaboration Architecture	Mechanical Design	Electrical Design	System/Software Design	
Product Development	Thermal Analysis	Analog ASIC Design	Modeling and Simulation Matlab/MATIXx	
Information Management	EMI Analysis	Digital ASIC Design Mentor Graphics	Configuration Management Clearcase/Clearcase Multisite	
	Printed Circuit Design PSpice	FPGA/PLD Design ModelSim	Change Management ClearQuest	
Desktop Office Suite Microsoft Office	Mechanical Design Solid Model Design EDS-Unigraphics Pro/Engineer	Test and Measurement LabView	Software Implementation	
Project Management Microsoft Project Visio2007	Other Design Design-to-Cost Price Systems		Testing and Verification C++ Test WIM Tester	
	Visualization 2-D Viewing Adobe Acrobat 3-D Viewing EDS-VisViewPro			

Table 4. BrEAD Tools

Engineering tools that have customer guideline will follow those guidelines. The tools identified with customer guidelines are:

Dynamic Object Oriented Requirements System (DOORS) for linking requirements traceability.

Engineering tools used in executing the BrEAD program are managed and controlled as described in the following sections. In addition, version level and history information for all tools used in the engineering environment is maintained in the BREAD Repository. This program data repository stores program-related information using ProjectLink. Data related to individual tool versions, Service Releases, and patch levels (and other pertinent information) is maintained at the facility as well. This repository provides an interface for the customer.

Application Server-Based Tools

Many program and BF Capture's tools use a client/server relationship where a small client software application resides on the user's local desktop, while the main application software resides on a system application server. When the user opens the client software on their desktop, this accesses the tool-specific software from the application server. This approach forces all users to access the same version of the tool, and, since upgrades or enhancements are installed on the server, all users have immediate access to the new capability. This way administrators can better manage tool use and version control across the program.

Desktop Tool Installations

Engineering tools that need to reside on the desktop will be controlled through server-based release areas. The sys admin will create a release area for each tool, as directed by the program environment lead. As users request tool installations for program use, the appropriate software is downloaded from the release area to the user's desktop. The release area ensures that all program users receive the correct software, and gives system administrators a central point to implement patches and upgrades.

Facilities

Facilities that will be used during BREAD system development and implementation are described in the Facilities Management Plan.

Schedules

The top-level program schedule is contained in CDRL 0003 Integrated Master Schedule. On a weekly basis, Integrated Master Plan (IMP) / Integrated Master Schedule (IMS) starts, stops, and earned value data are uploaded for integration with the customer overall program schedule. The top level program plan is contained in CDRL 0002.

Deliverables

Table 5, below, lists BREAD program deliverables by the Contract Line Items (CLI) identified.

CLIN	Qty/UM	Description
001	1 Lot	Sub-systems deliver
001	1 Lot	Process Demonstration
001	1 Lot	Sub-system manual Deliver
001	1 Lot	Training according to the SOW

Table 5. Deliverable Items

Specifications

The current BREAD system specifications will be maintained as modules in the Telelogic DOORS requirements management system, providing traceability up to the NGRF requirements management system.

Plans and Process Integration

The BRAINS technical program is based on best-in-class International Standards Organization (ISO) and Capability Maturity Model Integrated (CMMI) certified processes. Work products created solely by BRAINS will follow the ISO and CMMI certified program management process under Program Management Process, Engineering Quality Assurance, and Process Improvement. The common processes identified for the BREAD program are listed in Table 6 below.

Common Processes	Implementation
Risk Management	Tailored program Risk Mgt Plan
System Integration	BrEAD Process
Preliminary Design	BrEAD Process
Decision Analysis Resolution	BrEAD Process
Configuration Management	BrEAD Process
User Requirements Definition	BrEAD Process
System Requirements Definition	BrEAD Process
System Architecture Methodologies	BrEAD Process
Detailed Specifications	BrEAD Process
Peer Review	BrEAD Process
Project Monitoring and Control	BrEAD Process
Project Repository Structure	BrEAD Process
Requirements Management	BrEAD Process

Table 6. Common Processes

Using these processes the required work products relevant to this SEMP will be developed. Where applicable, plans and processes are referenced rather than reproduced here in full. Plans are generally referred to throughout this SEMP where the relevant subject matter is discussed.
The Software Development Plan (SDP) describes the software methodologies and software management used for software development, including software generation and reuse for systems end items, and the COTS software product integration and management.

The Hardware Development Plan (HDP) plans, methodologies, and management are for hardware development including COTS hardware product integration and management.

- Security Considerations
- Information Protection
- BREAD security considerations are to be determined.
- System Security Engineering
- Technical Program Planning and Control
- Organization and Responsibilities

This section of the SEMP describes how the Contractor develops its plan on delivering the BrEAD to the NGRF. It includes how the Contractor is organized, the policies and procedures for Systems Engineering; include organizational responsibilities and control of subcontracted engineering.

Breadfruit (BF) Capture Team

The BrEAD program is developed by a group of 11 talented individuals that are carefully selected by the BRAINS PMO. These individuals named themselves the Breadfruit (BF) Capture Team (Figure 2). The team structures as follow: 1 program manager, 2 leads, and 8 engineers. Each individual is selected based on their specialties and background experiences as shown below.

Danielle Eastberg

Name: Danielle Eastberg
Title: Program Manager
Age: 45
Background: BS in Computer Engineering - University of Minnesota (1987)
MS in System Engineering - University of St. Thomas (1990)
PHD in Electrical Engineering – University of Minnesota (1995)
Specialty: Very persuasive and a powerful speaker

Gabriel Farias

Name: Gabriel Farias
Title: Technical Lead Engineer
Age: 38
Background: BS in Computer Science - University of Minnesota (1994)
MS in Software Engineering - University of St. Minnesota (1998)
Specialty: Everything technical

Abdi Hashi

Name: Abdi Hashi
Title: Enterprise Lead Engineer
Age: 48
Background: BS in Electrical Engineer - University of Minnesota (1985)
MS in System Engineering - University of St. Minnesota (1988)
Master in Business Adminstration – Univerisy of St. Thomas (1995)
Specialty: Marketing

Subodh Deodhar

Name: Subodh Deodhar
Title: System & Design Engineer
Age: 50
Background: BS in Electrical Engineering - University of Minnesota (1983)
MS in Electrical Engineering - University of St. Thomas (1986)

Specialty: Subsystem design

Kyle Dahlstrom
Name: Kyle Dahlstrom
Title: Hardware & Architecture Engineer
Age: 45
Background: BS in Electrical Engineering - University of Minnesota (1987)
MS in System Engineering - University of Mineesota (1992)
Specialty: Architectecture Modeling

Khuong Vu
Name: Khuong Vu
Title: Integration & Test Engineer
Age: 28
Background: BS in Computer Engineering - University of Minnesota (2004)
BS in Electrical Engineering - University of Minnesota (2004)
MS in Electrical Engineering - University of St. Minnesota
(2006)
Specialty: Test integration and Design

William Tamayo
Name: William Tamayo
Title: System Engineer
Age: 42
Background: BS in Computer Science - University of Minnesota (1990)
MS in System Engineering - University of St. Thomas (1994)
Specialty: System Design

Edem Kavi
Name: Edem Kavi
Title: System Engineer
Age: 39
Background: BS in Electrical Engineering - University of Minnesota (1994)
MS in System Engineering - University of St. Thomas (1998)
Specialty: System Integration

Leland Smith
Name: Leland Smith
Title: System Engineer
Age: 32

Background: BS in Computer Science - University of Minnesota (2000)
MS in Software Engineering - University of St. Thomas (2003)
Specialty: System Architecture

Aditya Srinath
Name: Aditya Srinath
Title: Marketing Engineer
Age: 34
Background: BS in Mechanical Engineering - University of Minnesota (1999)
Master of Business management - University of St. Thomas (2002)
Specialty: International market study

Mesfin Woldekidan
Name: Mesfin Woldenkidan
Title: Marketing Engineer
Age: 38
Background: BS in Electical Engineer - University of Minnesota (1994)
MS in Electrical Engineering - University of St. Thomas (1998)
Specialty: Marketing, mainly in a foreign country

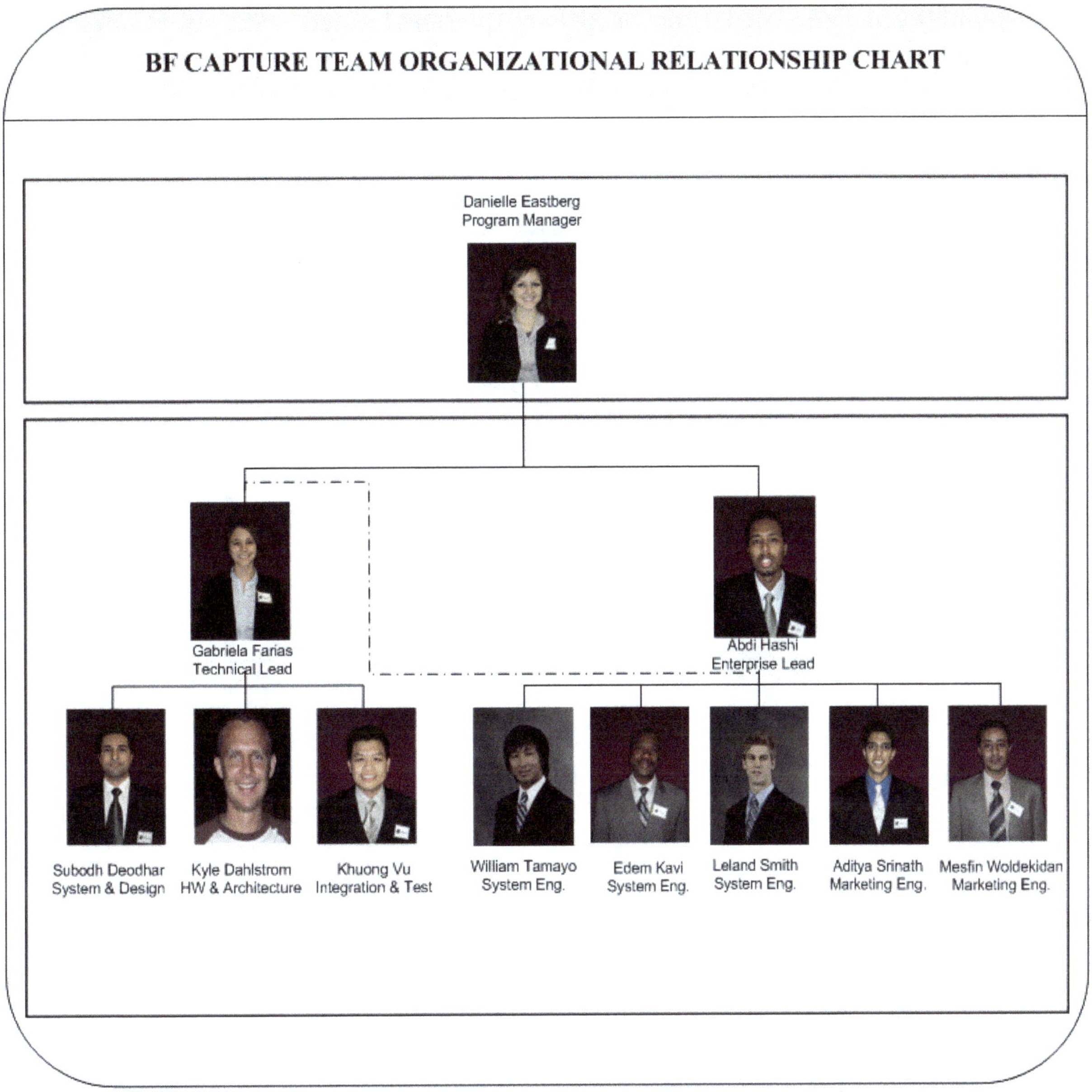

Figure 208. BF Capture Team Organizational Relationship Chart

Breadfruit Enterprise Architecture Demonstration (BrEAD)

The Program OV-4 organization relationship chart (Figure 3) shows the overall Program and Systems Engineering Management structure, to support the BrEAD implementation. Some of the major sections of this approach are the Program Management Office (PMO), Cross Product Team (CPT), Enterprise Integrate Product Team (EIPT), Process Integrate Product Team (PIPT), and System Integration Lab (SIL).

These Teams mainly work on the design, integration, and marketing of the products. Furthermore, during the operation phase, they shall provide an automated research, marketing, and user data collection capability such as hourly temperature, pressure, humidity and rain fall data for NGRF head quarters located in St. Paul, Minnesota on a daily base.

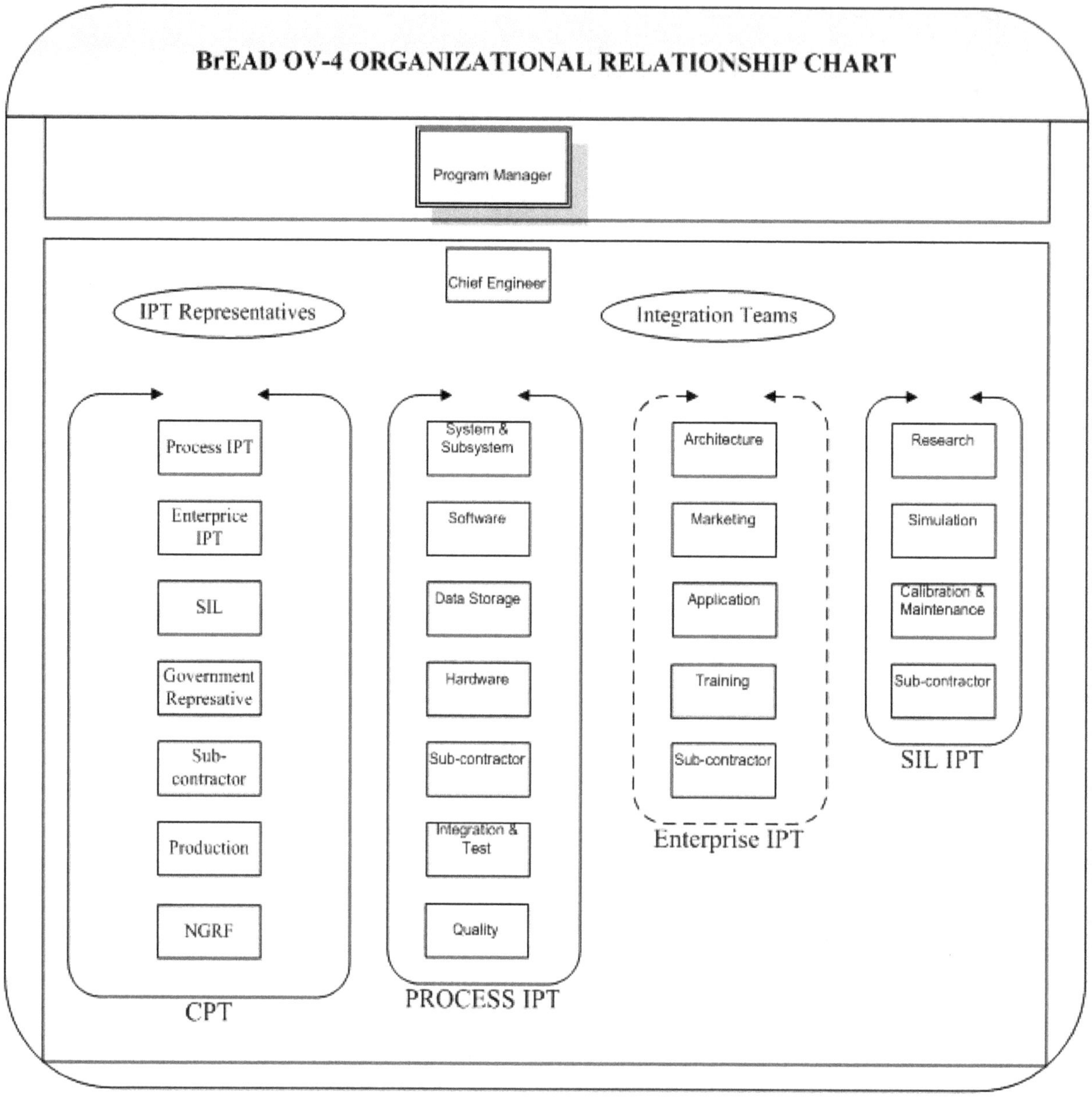

Figure 209. Program OV-4 Organizational Relationship Chart

Program Management Office (PMO)

The PMO is responsibility for overall contract execution, schedule and cost. The project is led by the Program Manager (PM), whom will also serve as the lead government liaison throughout the program implementation. The PMO will maintain a portfolio of metrics and Technical Performance Measures (TPMs) to support the continual assessment of program progress against the defined plans and schedules. The schedule status, technical performance, and risk will be continually monitored and updated using earned value as an integrating tool.

System Engineering and Integration Test (SEIT)
The SEIT is responsibility for systems engineering management. The SEIT is led by the Chief Engineer (CE), whom will serve as the primary contact for the government CE throughout the program implementation. SEIT membership includes the IPTs (EIPT and PIPT) captains and CPT functional representatives, as well as representatives from the following government functions; Engineering Services, Research and Development. A representative from Sub-Contractor firm will also be included as deemed necessary during the required program stages.

Integrated Product Teams (IPTs)
Each IPT is responsibility for product development and delivery. The IPT's are multi-disciplinary groups used to speed communication and remove functional silos. The group consists primarily of Engineers different expertise. Representatives from the Contract and Sub-Contractors are involved within each IPT based on the IPT's demand. The government representative doesn't are not involve with any IPT as it will be discussed in the CPT. There are three IPTs for this program; EIPT, PIPT and SIL. Each IPT captain is a member of the SEIT and reports to the CE. IPTs will be the primary means used to gain insight and provide for timely exchange of information and data.

Cross Product Team (CPT)
Because the NGRF is a non-profit organization, most of the money for the BrEAD's development comes from the government or donation from large organizations. The purpose of the CPT is to support activities as required across the IPT's. The CPT's bring specialized expertise to help the IPT's to meet their specification, schedule and budget. This mean an engineer might be in both PIPT and EIPT as his main responsibilities reside in the PIPT. The CPT contains captain and lead in each IPT, sub-contractors, and government representatives. The SEIT is different that the SEIT as the SEIT is lead by the Chief Engineer. The captain of CPT is an IPT lead and reports to the Chief Engineer as well as other IPT lead.

System Constraints
Most BREAD components will be derived from commercial or government off-the-shelf technologies and products. The BREAD system will leverage industry standard interfaces and products that have a product roadmap to facilitate low-cost, low-risk technology insertion and system growth.
System constraints include the following:
All device drivers are small hardware-intensive embedded software. Most will be 'commercial off the shelf' (COTS), with some development and/or enhancements. Where possible, no separate software documentation will be generated; i.e., it will be part of the corresponding hardware documentation.
No software will be taken from government inventory

A Reuse Evaluation Analysis Report will be generated for reused components. For software reuse, refer to the Software Development Plan (SDP) for details on the report.

Technical Review Board

The Technical Review Board is an internal technical review authority for the BREAD program. This group provides technical oversight and approval. The scope of their charter includes technical strategy, internal approval of the technical baselines, make/buy and trade studies oversight, technical risks, technical performance measures, and technical aspects of external communications and proposals.

Work Breakdown Structure (WBS)

The WBS (Work Breakdown Structure) is provided as a separate program deliverable. Each System development specification is allocated to a WBS element. Each WBS element is composed of discrete Work Packages and Planning Packages. Work Packages are composed of tasks (milestones) scheduled in the program Integrated Master Schedule. Work authorization, cost reporting, and integration to scheduling are described in the Earned Value Management System Plan.

Risk Management

The approach to be followed for identifying and managing program risks is described in Risk Management Plan.

System Test Planning

BrEAD system test planning will be managed within the Integration, Simulation, and Test IPT. This IPT will be responsible for all BrEAD requirements verification, System Integration Labs (SILs), special test equipment for BrEAD subsystems, System-level testing, and System integration activities, test software, acceptance test development, and Formal Qualification.

Integration, Simulation, and Test IPT Lead has overall BrEAD integration and test program responsibility. Integration, Simulation, and Test IPT responsibilities include requirements verification, Design Verification, and OSA Qualification. Integration, Simulation, and Test IPT responsibilities include overseeing and managing all system integration activities, system requirements verification, test software development, system design verification and acceptance test development. Integration and Test Planning will be executed according to BRAINS System Integration and System Test Plan. An Integrated Test Plan (CDRL 0004) will capture the entire BrEAD program test and evaluation process. It will become the Test and Evaluation Master Plan (CDRL 00021) for BrEAD and will delineate all test documentation required for the BrEAD verification program. In addition to BRAINS process-specific documents, it will also include BrEAD program documentation: the Validation and Verification Plan (CDRL007), and the Acceptance Test Requirements Plan.

Decision and Control Process

See Section 5.2.2 for a discussion of the BREAD program decision and control process.

Technical Performance Measurement (TPM)

TPMs will be selected to meet one or more of the following objectives:
Support the BrEAD system-level Measures of Effectiveness (MOEs), Measures of Performance

(MOPs) and Key Performance Parameters (KPPs)

Monitor areas of potential technical, schedule, or cost risk. Monitor progress on requirements that need improvement in the state-of-the-art in order to achieve compliance.
TPMs will only be defined and measured for the Prototype design, not for the sub-systems.
Technical Performance Measurement will follow BRAINS phases User Requirements Definition and System Requirements Definition.

Planning

Measurement methods for each TPM will be defined prior to reporting TPMs at each major system design review:
SRR (System Requirements Review)
PDR (Preliminary Design Review)
CDR (Critical Design Review)
TRR (Test Readiness Review)

Reporting

Technical Performance Measures are reviewed internally on a monthly basis. Detailed status is reported on a monthly basis in the Cross Domain Group (CDG) meeting. In addition, summary TPMs are reported at the monthly Program Management reviews.

Quality Management

Quality assurance requirements, management processes and plans are contained in the Quality

Configuration and Data Management

Program configuration and data management are described in the BREAD Configuration/Data

Management Plan.

Subcontract Management Monitoring and Control
Subcontract Management Plans (SMPs) will be established for each major subcontractor. Each SMP includes information regarding the Subcontract Management Team charter, membership, responsibilities and accountability, and an explanation of the processes, procedures, and tools to be used in subcontractor management. The SMP will list documentation required in the subcontract relationship, which includes the Subcontract, the Statement of Work (including deliverables, acceptance criteria and schedule), Data Items (DIs), Security

Requirements, and other documentation such as specifications and reference documents.

The SMP specifies the performance management and subcontractor monitoring plan, including schedule, quality, technical performance, and predictive measures to be used. The plan documents program and risk management plans, including the process for corrective action where required. The plan specifies formal and informal communications plans, including formal Program Reviews, Technical Interchange Meetings, Risk Management Meetings, and regularly scheduled status reports. The process for change controls is part of the SMP.
The BRAINS Subcontract Program Manager creates SMPs for respective subcontracts and posts them in the Program Repository. The Subcontract Program Manager maintains the SMPs as needed.

System Design and Development
To support early sub-system development activities, the BrEAD program will deliver several different method based on the sub-system. A manual process is recommended for the early state of the project. These will consist primarily of Commercial Off-the Shelf (COTS) components or existing methods or systems. There will be modification of existing systems for specific needs such as the wind turbine for power consumption and other subsystems for the process system.
Deliveries of the real subsystems or final subsystems are later as it needs to pass the simulation in the SIL. Subsystem deliveries occur after system verification testing and Prototype systems are delivered after system verification testing and environmental testing. References to the tailored organizational processes are provided within this document. All program processes and procedures will be made available to team members through the project repository.
To support timely subsystems delivery, all design and development processes will run concurrently. For example, one process will produce the wind turbine subsystem and the other will produce the grinder subsystems. Each subsystem has to fulfill its requirements as well as the system level requirements. The subsystems will fulfill the physical and environmental requirements as specifies in the SOO and PBSS. In addition, different versions of the design revision may have different functional and/or performance requirements, based on availability of suitable COTS components and system development activity needs. Engineering Processes used for BREAD system design and development, also known as phases, are defined in the following sections. Process details do not appear in this document but are available on a secure, password controlled, server. These processes represent a merging of BRAINS tailored organizational processes specific to BREAD.

System Requirements Development
System Requirements Analysis:
System requirements analysis will follow BRAINS phases System Requirements Definition, System Design Concepts, Detailed Specification, System Requirements Analysis/Design, and Requirements Analysis. The Telelogic DOORS tool will be used for requirements management. These phases and others referenced, describe the

methods, procedures, and tools for analysis of missions/operations and environments; identification of functional and performance requirements for development, manufacturing, deployment, operations, support, training and disposal; and determination of constraints. They also document the process for defining the functional and performance requirements for quality factors such as producibility, testability, disreputability, survivability, reliability and maintainability, product support, and for the engineering specialty areas.

Functional Analysis and Architecture Development

System functional analysis and architecture development will follow BRAINS phases System Requirements, System Design Concepts Definition, High Level Design, and System Architecture Methodologies. System functional analysis and architecture development will comply with the applicable sections of the Joint Technical Architecture. The BF Capture team will decompose and allocate the requirements in the Procurement Control Drawing (PCD) to establish a functional baseline. Functions shall be analyzed by decomposing higher-level functions identified through requirements analysis into lower-level functions. Performance requirements associated with the higher level shall be allocated to lower functions. The resulting description describes the product in logical terms and in terms of the required performance. System/subsystem functional requirements shall be documented with Functional Flow Block Diagrams and/or Unified Modeling Language (UML) models, Time Line Analyses, and Requirements Allocation Sheets. The Telelogic DOORS requirements management tool will verify that the functional architecture derived from the functional analysis is upward- and downward-traceable to the validated requirements baseline.

System Physical Architecture Synthesis

System physical architecture development will follow BRAINS phase System Design and System Architecture Methodologies.

Validation during Requirements Development

During the system requirements development process, modeling and simulation will be used as applicable in conjunction with trade studies to support development of a consistent, reasonable, and verifiable set of requirements. The Telelogic DOORS tool will validate that the requirements baseline established from requirements analysis is both upward-and-downward-traceable to customer requirements, program and enterprise constraints and external constraints.

The Subsystem Design Description presents architecture views.

System Analysis

Complete BREAD process system analysis will require knowledge from many engineering disciplines, including the specialty disciplines discussed in later section. The Systems Engineering and Architecture IPT will coordinate these specialty engineers and initiate studies from the specialty areas. The results will be returned to the Systems Engineering and Architecture IPT for integration into overall system analysis. Details of the interfaces between the Systems Engineering and Architecture IPT and the engineering specialty areas are discussed in later section. Quality factors and engineering specialties requirements will be flowed down from and managed in DOORS, the same as other IPT disciplines. Meetings occur regularly between these groups and the other IPTs and NGRF and government Representatives. TPM measures and control mechanisms will follow the guidelines.

Effectiveness Analysis
Analyzing a system like BrEAD begins with defining Measures of Effectiveness (MOEs). MOEs focus on the most important needs as seen from the user's viewpoint. MOEs are defined independently of any specific system implementation.

Once MOEs are identified, Measures of Performance (MOPs) can be derived from them. MOPs are defined from the supplier's viewpoint and focus on system technical characteristics. System-level Key Performance Parameters (KPPs) are derived from MOPs.

When system architecture is defined, Technical Performance Measures (TPMs) are identified for the elements of that architecture. Some TPMs are defined to support the MOPs and KPPs, and through them, the MOEs. Others are defined to ensure proper management attention to key performance requirements, perceived risk areas, and to ensure that planned progress is made on those requirements where advances in the state of the art are needed to achieve compliance; and to identify areas of over achievement where tradeoff opportunities may exist.
Trade Studies

Trade studies will follow Decision Analysis and Resolution Procedures.
Trade study results will be maintained in the BREAD Repository. The subscription and notification facilities of the ProjectLink tool will be used to ensure that relevant parties are informed of the results

Preliminary Design
Hardware and Software Preliminary Design
Hardware and Software Preliminary Design will follow the BRAINS phase Preliminary Design.

Detailed Design
Hardware and Software Detailed Design
Hardware and Software Detailed Design will follow BRAINS Phase Detailed Design.
Implementation

Hardware Design Release
Hardware Design release for build will follow BRAINS Phase Detailed Design.
Hardware Build of Engineering Development Unit
Hardware build of Engineering Development Units will follow BRAINS Phase Implementation and TCP Activity E-020-040: Fabricate and Assemble Prototype Hardware.

Software Implementation and Unit Testing
Software Implementation and Unit testing will follow BRAINS Phases Implementation, Software/Hardware Test, and TCP Activity E-030-040: Implement Software Elements.

Validation during Implementation
During implementation, the primary validation tool will be unit testing by the designers to ensure that individual units meet requirements allocated to them.

Requirements Management
Requirements management will follow BRAINS Phase User Requirements Definition and System Requirements Definition. The Telelogic DOORS requirements management tool will verify that the functional architecture from functional analysis is both upward- and downward-traceable to the validated requirements baseline.

All BRAINS requirements will be managed in DOORS and mirrored to the NGRF DOORS database, providing a common tool set and traceability at both the NGRF and IPTs. The architecture development will be linked to requirements in DOORS which have traceability to the NGRF DOORS BRAINS PCD. A mirror copy of NGRF DOORS PCD will be used until BRAIN CI specifications are in formal NGRF configuration control (after CDR). Released approved specification DOORS modules will be uploaded for the NGRF to link to the NGRF's common DOORS database.

Requirements changes will be discussed and coordinated in several venues: 1) The BRAINS community of associate contractors discuss potential changes and develop a sense of whether changes are in or out of scope; 2) The Technical Review Board will also discuss potential changes and make change of scope recommendations through the program office to the NGRF; 3) The NGRF boards (for example the Engineering Review Board, Interface Control Board, and Program Decision Board) will be supported. These boards also discuss potential changes.
Technical Reviews

Peer Reviews
Peer Reviews for the BREAD program will follow document Doc123456789, BRAINS Engineering Review Handbook. The Peer Review Plan lists BREAD items requiring Peer Reviews.

Phase Kickoff Reviews
Reviews are held with the program team to help management determine if the program is executing consistently with its documented program plans. They focus on the entire program and are conducted at the completion of lifecycle phases. These reviews will be conducted according to Engineering Kickoff Phase Packages and Engineering Status Review. The following reviews will be held for BREAD. Target completion dates are relative to BREAD program events.

Kick-off	14 days after contract award
Planning Commitment	60 days after contract award
System Requirements/Design	14 days before SRR
Preliminary Design	14 days before PDR
Critical Design	14 days before CDR
Integration Readiness	14 days before End Item Verification (EIV)
Test Readiness	0 days after EIV
Production Readiness	14 days before Production Readiness Review (PRR) #2
Closeout	End of Contract

Customer Reviews
BREAD Customer Reviews will follow Operating Standard – Design and Development Process and the Engineering Review Handbook. Specific customer reviews are shown in the Integrated Master Schedule and are consistent with the SOW.

Engineering Support and Specialty Integration
The design, development and production of the BREAD program require integration across many engineering and program disciplines. This section addresses BREAD approach to integrating engineering support and specialty disciplines during design and development. Integrating Engineering Support and Specialties will follow BRAINS Support Engineering Process Flows as applied within the program IPT structure. ES&S activities are the responsibility of the Systems Engineering and Architecture IPT and the Sustainment IPT as indicated in the following sections.

Reliability, Availability, Maintainability - Testability Engineering
The BREAD Reliability, Availability, Maintainability - Testability (RAM-T) engineering program will drive the BREAD design to ensure the BrEAD meets or exceeds the BrEAD RAM-T requirements. The BREAD RAM-T engineering process will be detailed in Reliability, Availalility, Maintainability, and Testability Annex B.

Maintainability Engineering
Maintainability Engineering is addressed as part of RAM-T Engineering.

Human Factors
Human Factors Engineering (HFE) defines the integration of human characteristics into system definition, design, development, and evaluation to optimize human-machine performance under operational conditions. The HFE goal maximizes the ability of an individual or crew to operate and maintain a system at required levels by eliminating design-induced difficulty and error. Human factors engineers work with systems engineers to design and evaluate human-system interfaces to ensure they are compatible with the capabilities and limitations of the potential government-user population.

HFE is conducted during all phases of system development, including requirements specification, design, and testing and evaluation. During requirements specification,

HFE evaluates predecessor systems and operator tasks; analyzes needs; analyzes and allocates functions; and analyzes tasks and associated workload. In the design phase, HFE evaluates alternative designs using equipment mock-ups and software Prototypes; performs software usability testing; refines analysis of tasks and workload; and uses modeling tools such as human figure models to evaluate crew station and workplace design and operator procedures. For testing and evaluation, HFE confirms that the design meets HFE specification requirements; measures operator task performance; and identifies any undesirable design or procedural features. The Sustainment IPT is responsible for HFE.

Integrated Logistic Support (ILS) Engineering

The purpose of ILS engineering is to address logistics considerations early in the system acquisition life cycle throughout the SDD phase by designing supportability capabilities into the ICS. From early ILS planning to Post Production Support, the BREAD program team's staff of logistics engineers, logistics support analysts, and supply chain management experts will achieve the best balance between Life Cycle Cost (LCC) and operational availability (Ao). The staff is comprised of individuals with training and experience across all ILS disciplines, including RAM-T engineering, HFE, System Safety, Maintenance planning, Provisioning, Supply Support, and Performance Based Logistics (PBL). Some ILS disciplines and logistics engineering activities Staff logisticians are experienced in performing all ILS-related MIL-STD tasks and requirements.

The Sustainment IPT is responsible for RAM-T, ILS, and PBL. Areas expected to be covered in the Sustainment IPT are:

- PBL and Logistics Planning
- Logistic Support Analysis (LSA)
- Supportability Analysis
- Life Cycle Cost Analysis
- Affordability and Total Ownership Cost Support
- Level of Repair Analysis/Repair Level Analysis (LORA/RLA)
- Use Studies
- Trade Analysis
- Audits and Reviews
- Maintenance Planning
- Maintenance Task Analysis
- Provisioning Data
- Manpower, Personnel and Training Requirements Analysis
- Packaging, Handling, Storage, and Transportability
- Supply Chain Modeling
- Demand Forecasting
- Analytic Model Development
- System Supportability Forecasting
- System Requirements Forecasting

The BREAD Team focuses on issues that affect system support, which includes elements such as test and support equipment, supply support, personnel and training, transportation and materials handling and computer resources. All are necessary for implementing material flow and distribution functions, as well as sustaining the life cycle support of the system throughout its period of use.

Advanced planning ensures the optimum balance between the system in operation and its related support requirements. The need to increase overall productivity in a resource constrained environment has placed extra emphasis on all aspects of the system product life cycle with Performance-Based Logistics assuming a major role. Provisioning and Sparing

The Sustainment Engineer is also responsible for a comprehensive Supportability Analysis in accordance with the BREAD Supportability Strategy to achieve logistic design influence and identify supportability issues including a provisioning and supply support assessment.
An availability analysis will be part of the Reliability, Availability, and Maintainability-Testability (RAM-T) Case Management Plan. Under the requirement for supportability test and evaluation, the BREAD program team will conduct a Logistics Demonstration Readiness Review. Prior to the readiness review, the team will identify the overall approach, activities, procedures, and equipment required for the logistics demonstration. The team will provide all test equipment, parts, tools, consumables, and other resources required to conduct the demonstrations.

The Repair Parts and Special Tools List (RPSTL) technical data will be part of the Training group and will be created and maintained for access to provisioning data. RPSTL data will provide parts data needed for implementation of Logistic Decision Support and Battle Damage and Repair (BDAR) programs including BDAR kit requirements. The BREAD program team will do the following:

Review damage mode analyses to help engineering identify combat design criteria for servicing, cleaning, scheduled adjustments/calibrations, inspections, scheduled component replacement/retirements, component wear, and damage limits
Support the engineering development system to interface with the NGRF as well as the government representative to support spare parts requisitioning.

Training Plan:

External Training
The BrEAD Training Plan will be created to include information from Training Job/Task Analysis and Training Support Packages. The BrEAD STP will provide a high-level yet detailed blueprint for organizing training for individual and collective training tasks performed in an instructor-led, computer-based, embedded, or simulator training environment.

The training required for the development team in executing the BrEAD program will be detailed in the Program Training Plan. Subsystems and equipment training is required for field engineer who has to support the product and its life cycle.

Internal Training

Internal training is provided to individuals based on their specific role on the BrEAD program. From a functional organization stand point, a role-based training matrix will be used to identify the specific processes, tools, and methods training to enable each individual to perform their functional role on the program. This internal training is managed and controlled within the functional organization.

In addition, program specific training is provided to augment the functional role based training. This includes program orientation and training specific to each IPT. The program specific training is managed and controlled within each IPT.

System Safety Engineering

Safety Engineering is responsible for eliminating or mitigating potential hazards to personnel safety and health, and to the environment. Safety Engineering is involved in review and analysis of product design in all phases of the design process (requirements development, product design, and product test). Safety draws from and contributes to all engineering and IPT groups to ensure the final product is safe, both functionally and for materials used in construction, operation, and maintenance of the final product. The Sustainment engineer is responsible for Safety Engineering.

Survivability/Vulnerability Engineering

BRAINS will prepare and maintain a Survivability Program Plan and update it as required to reflect compliance with changes made to the system specification. The Survivability Program Plan will address system survival to nuclear weapon-generated hostile environments. Most relevant to electronic system design are the initial nuclear radiation (INR) environments comprised of transient radiation effects (TRE) from gamma dose rate, neutron fluence, and ionizing dose. BRAINS will perform element and sub-element compliance assessments to ensure compliance with the system specification. BRAINS will provide a Hardness Assurance Maintenance Surveillance Plan and Hardness and Survivability data to verify the level of system hardness and ensure that it is retained throughout the system life cycle. BRAINS will use the NGRF Integrated Survivability Plan if there is one as a guide.

BRAINS will perform requirement analysis, allocation and design assessments of system element survivability requirements. Design assessments will determine the degree to which system element requirements have been achieved within the element/sub-elements. Results of the analyses and assessments will be documented and be made available to the NGRF for review.

BRAINS will provide inputs to the NGRF survivability analyses. Inputs will be provided for integration into NGRF documents.

BRAINS will participate in the Integrated Survivability Working Group and will form a BRAINS survivability working group.

BRAINS will develop a verification approach/concept for element survivability requirements, including system verification logic networks and verification objectives. BRAINS will perform verification analyses and assessments, as well as support verification test and analysis planning, execution and evaluation.

Value Engineering

Value Engineering (VE) is the systematic application of recognized techniques by a multi-disciplined team to identify product or service function, establish its worth, generate alternatives and provide the functions needed to fulfill the program's purpose of the program at the lowest lifecycle cost without sacrificing program safety, quality, or environmental attributes.

VE includes cost reductions, schedule reductions, environmental requirements modification, and operational procedures that can be improved through functional cost evaluation.

VE will be incorporated into activities described in Section 6.10 for Design to Cost using Cost as an Independent Variable strategy. The following areas will be considered during the CAIV (Cost as an Independent Variable) studies:

- Identify and evaluate function, cost and worth
- Develop new and unusual alternatives for required functions
- Determine the best and lowest life-cycle cost alternatives
- Develop fully supported recommendations
- Report to management the study results

All IPTs are responsible for identifying VE opportunities. The PM and Systems Engineering and Architecture IPTs will implement, monitor, and report VE activities to the customer.

Design-to-Cost Engineering

BRAINS will implement the BrEAD Life Cycle Cost Containment Strategy. The program will incorporate principles of Cost as an Independent Variable (CAIV) to balance cost, performance, and schedule and provide Total Ownership Cost estimates to track progress toward requirements.

Strategy will include implementing CAIV and its impacts. Cost will be treated as an independent variable with the same status as function and performance. The strategy will discuss unit cost, development cost, lifecycle cost, total ownership costs, and their interrelationships as they apply to system development. Primary focus will be cost balance, control and impact to strategy and design.

Producibility Engineering

Producibility Engineering, designing for ease of manufacture and test, will follow BRAINS' Handbook." The key aspects of the BRAINS Product Design strategies are discussed within the BRAINS Enterprise Production Plan.

Product Design for Manufacturing (DFM) Strategies

Up-front DFM goals will be communicated and agreed upon between the systems, design, and production groups. The design will be analyzed at successive steps during development. The following is a short list of examples:

- Minimize total number and unique number of parts.
- Breakdown design into modular sub-assemblies.
- Reduce or simplify part orientation requirements.
- Maximize the use of automation over manual assembly operations.

Testability Goals and Objectives

Up-front testability goals will be communicated and agreed upon between design, manufacturing, and reliability groups. The design will be reviewed against these goals at appropriate design phases. The following is a non-inclusive list of factors evaluated during design reviews.

- Circuit cards are In-Circuit Testable (ICT)
- Circuit card is compatible with ICT test fixture
- Connector and mounting keying to prevent accidental damage
- Tolerances consistent with Automated Test Equipment (ATE) capability
- Modules are functionally testable
- Modules can be independently tested.
- Module interface compatible with ATE
- Redundant circuits independently testable
- Minimum operator adjustments required
- Diagnostic/Fault Isolation capability built into functional code or diagnostic code available for fault isolation
- Tolerances consistent with ATE capability
- Product Testability Analyses

Product testability requires high test detection and fault isolation capabilities. The BF Capture team goal is to meet or exceed NGRF requirements for fault detection and isolation. These goals support ease of initial production and ongoing maintenance. The design will be analyzed at appropriate times during the project. During initial design, the parts selection and schematic diagram will be analyzed for fault detection. Failures will be categorized by built-in test (BIT) detection, ICT detection, and undetected. Undetected failures will be evaluated and corrective action taken as appropriate. This provides rapid feedback to the design team as they incorporate fault detection and isolation into the initial design.

Design Reviews
Peer reviews, engineering status reviews and customer reviews will occur at appropriate BRAINS development phases, as described in section 5.9. These reviews focus on certain design aspects.

CCA Reviews
Each Circuit Card Assembly (CCA) will be reviewed at discrete points during development.
Sub-Assembly Reviews
Sub-assemblies consist of groups of CCAs or wired chassis assemblies or power supply sub-assemblies. These assemblies will go through discrete reviews.
Miscellaneous Cables/Accessories
Miscellaneous cables and accessories will also be reviewed. These may be reviewed individually or along with another review. The following breakdown is an example of attributes evaluated during this review:

- Use of standard parts
- Use of standard tools and processes
- Subsystems Reviews

Each process subsystem will go through two design reviews. The following breakdown is an example of attributes evaluated during these reviews:

Solid Model
Review any complex steps performed at design level
Verify by using the simulation software in the System Integration Lab (SIL)

Review design issues

- Producibility Review after first build.
- Build Time (actual and estimated future).
- Build Defects (actual and estimated future).
- Design Score (actual and estimated future).

Long Lead Items
Long lead items are addressed within the BRAINS Enterprise Production Plan. This document discusses the process for determining and managing long lead times affecting the critical path of the program. Updates to the Enterprise Production plan, and any related inputs to the IMP/IMS are planned prior to PDR, CDR, and LRIP.

EMC/EMP/EMI
The BREAD design will meet EMC/EMP/EMI requirements and permit the overall system to satisfactorily perform its stated mission in the operating environment.

BRAINS will establish an EMI/EMP/EMC Control Program and Control Plan to facilitate an EMI/EMP-compliant design. EMI/EMP control will be important throughout the design phase. EMI/EMP design compatibility will be tested for compliance in accordance with its test methods.
To ensure compliance, BRAINS starts at the component-selection level, and then checks PWB designs against EMI/Signal Integrity guidelines. Power supplies, connectors, and internal wiring are reviewed for emission/susceptibility concerns. The metal enclosure is the last link in this chain and one of the most critical. The enclosure has no open apertures for airflow; all access panels are sealed with conductive environmental gaskets and attached with screws spaced to limit RF leakage. Metal-shielded external connector backshells should be used to terminate all shielded I/O cables to ensure signal cables do not become antennas for high-frequency EMI. Enclosures of this type have been previously qualified to requirements that equal or exceed radiation levels specified for this application.
Electromagnetic Interference Control

Establishing an EMC/EMP/EMI Control Plan during the early stages of system design will help meet BREAD requirements. EMI is controlled by a multi-step process to achieve a cost-effective design solution. Source suppression is always the preferred method; however, a practical design requires a balanced approach to achieve containment at the board level, at the subassembly enclosure level, at the interface wiring level, and at the power input level as well as at any other interfaces.

Test Approach
EMI/EMC Qualification testing will be performed to a customer-approved written test procedure. The procedure will detail all tests required to meet the requirements

TEMPEST Control
During development, a TEMPEST Control Plan will describe design details and disciplines employed to meet requirements imposed on various products. This plan details additional processes used to design the enclosure and circuitry to reduce and control compromising signal emanations from this unit. This plan also addresses the RED/BLACK signal interface issues, circuit isolation, circuit separation, and the enclosure's electrical shielding including minimizing the number of openings, gasketing materials to be used on cover panels, shielding the display window opening, and optimizing connector shielding. BRAINS has developed and will use EMC/TEMPEST checklists during development and for all design reviews. Checklists cover the mechanical and electrical features needed to ensure shielding, bonding, grounding circuit layout, cabling, and filtering concerns addressed during design.
Environmental Stress Screening

An Environmental Stress Screening (ESS) plan will be developed as part of the Enterprise Production Plan. The ESS plan will include functional/performance verification and environmental verification for BREAD hardware. The Verification Cross Reference Matrix will delineate how each requirement will be verified (Inspection, Analysis, Demonstration, Test), and at what test phase the verification will take place (Design Verification, Qualification Test, Manufacturing Test). Acceptance Test Procedure (ATP) will be developed and conducted on BREAD hardware at their respective test stations.

Annex H - Software Development Plan

1 - Introduction

1.1 Identification

This Software Development Plan (SDP) establishes the software development approach, methodologies, tools, and procedures to be used during the analysis, design, development, testing, integration, deployment, and maintenance of the software for BrEAD and the processing system. The Software Development Plan (SDP) describes a developer's plans for conducting a software development effort. The term "software development" in this document is meant to include new development, modification, reuse, reengineering, maintenance, and all other activities resulting in software products. This SDP is a dynamic document and shall be updated on a periodic basis to reflect organizational changes, lessons learned, new tools, and advances in methodologies.

The SDP will not be directly applicable to the BrEAD engineer, since typical BrEAD activities do not include software development. However, the SDP should be a requirement for the subcontractors responsible for developing and submitting the SDP document for a software development effort

Two types of Configuration Items (CI) have been defined for the BREAD software: Computer Software Configuration Items (CSCI) consisting of developed software CIs that include Commercial Off-The-Shelf (COTS) and their associated files.

1.2 Overview

The BrEAD will provide weather data gathered trough out the world where breadfruit is used and will share hourly temperature, pressure, humidity, rain fall data with the Central Data Collection and Analysis Center of the NGRF located in St Paul, Minneapolis.

Therefore the development of the BrEAD system involves the use and the upgrade of the two types of software mentioned previously.

This SDP was developed in conformance with MIL-STD-498. It is structured in sections following the format and content provisions of Data Item Description (DID)Data Item Description (DID) DI-IPSC-81427. Each section identifies tailoring applied to the structure and instructions for content defined in the DID.

1.3 Reference Documents

The documents listed below were either used to create this document or are referenced in it:

Software Development and Documentation, MIL-STD-498

Technical Reviews and Audits for Systems, Equipment, and Computer Software, MIL-STD-1521

Software Development Plan, Data Item Description DI-IPSC-81427

Integrated program plan, CDRL0002

Software Development Plan, CDRL0006

Configuration Management Plan, CDRL00010

Quality management plan, CDRL00013
Validation and verification plan, CDRL0007
Integrated test plan, CDRL0004
Institute of Electrical and Electronics Engineers (IEEE)/Electronic Industries Association (EIA) 12207 Series, IEEE and EIA, March 1998
SSC San Diego PAL, Reference g
Etc.

2 - Overview of Software Development Planning

The BrEAD will apply an Incremental (Preplanned Product Improvement) strategy to develop and evolve the functional capabilities of the BrEAD architecture Software. The BrEAD architecture Software requirements are controlled by the SEMP. All issues concerning cost, schedule, and incremental build content must be negotiated with the IMP.

Table 7: Key Features of Program Strategies

Program Strategy	Define All Requirements First?	Multiple Development Cycles?	Field Interim Software?
Grand Design	Yes	No	No
Incremental (Preplanned Product Improvement)	Yes	Yes	YES
Evolutionary	No	Yes	Yes

2.1 Requirements and Development

The MIL-STD 498 Incremental Life Cycle Model and the DID for the SDP have been used to guide the content and format developing the SDP. Overall, IEEE/EIA 12207 Supporting Processes and Organizational Process have been applied in to the management and control of the BrEAD software Project The integration of the MIL-STD 498's Incremental Life Cycle Model into the context of IEEE/EIA 12207 disciplines is illustrated in Figure 1. Section 5 contains the specifics of the BrEAD software implementation of the processes in Figure 1.

The allocation of functional requirements for each build will be based on the IMP and documented in the Build Plan for each build. The Build Plan will also include build specific acceptance criteria and installation/user training requirements. The process will integrate reusable software from existing sources with newly-developed software. Software design and coding will be performed by the Software Development Group using an object oriented design approach and generate class and object process interaction diagrams. Artifacts and evidence of results of software development activities will be deposited in Software Development Files (SDFs)Software Development Files (SDFs) and Software Engineering Notebooks (SENs).Software Engineering Notebooks (SENs) These artifacts, along with pertinent project references will be deposited and maintained in a Software Development Library (SDL)Software Development Library (SDL) and made available to support management reviews, metrics calculations, quality audits, product evaluations, and preparation of product deliverables.

Following integration of reusable and new software units by the Software Test and Evaluation Group, software Formal Qualification Testing (FQT)Formal Qualification Testing (FQT) will be performed in accordance with processes defined in Section 5. A Test Readiness Review (TRR) will be conducted to verify readiness for system level testing. The System Test Organization will prepare a Test Plan (TP) for a system level FQT and execute test cases defined in the implementing Test

Description (TD). They will generate Problem/Change Reports (P/CRs) to describe software errors uncovered during test results analyses. The System Test Organization will be supported by the BrEAD software providing analysis and repair as required. The separation of organizational entities and their testing responsibilities is illustrated in Figure 2.
Software Configuration Management (SCM),Software Configuration Management (SCM), Software Quality Assurance (SQA)Software Quality Assurance (SQA), and Software Product Evaluation, Corrective Action, and preparation for software delivery will follow detailed processes described in Section 5 of this SDP.

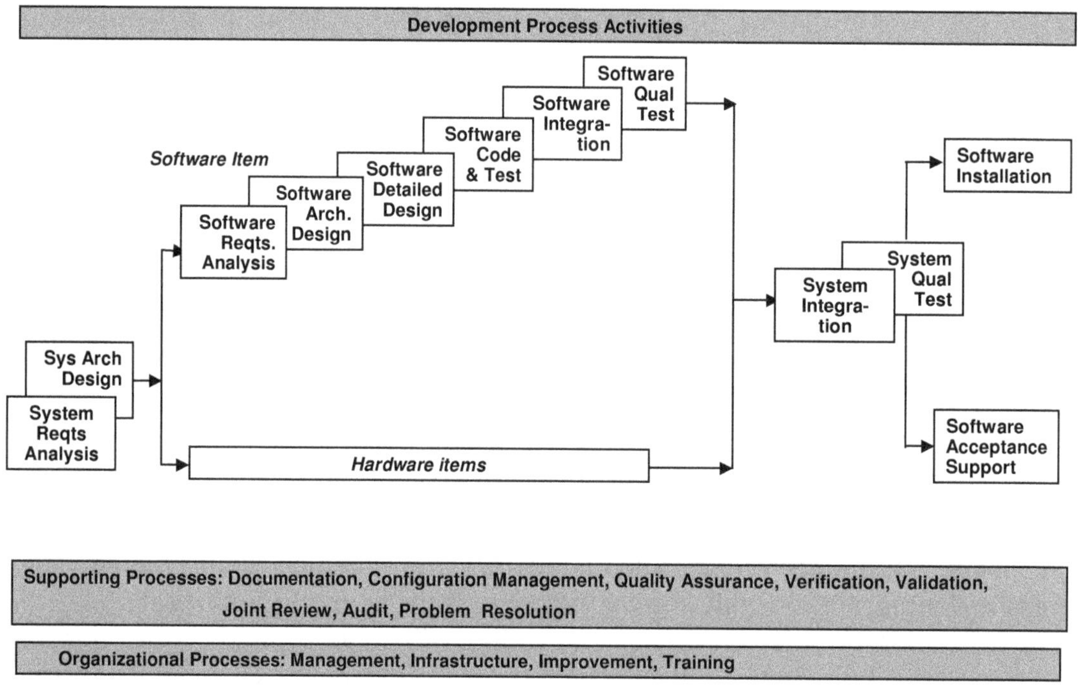

2.2 Project Documentation

In order to produce the project documentation the contractor will use Microsoft office (i.e., Microsoft Word, Microsoft Excel, etc.)

2.33 Training Requirements

This section defines the software development training requirements for the project software engineers and that will be described in the following sections.

3 – General Software Development Activities

3.1 Development Process

BrEAD architecture software development will conform to MIL-STD-498. The development approach will be to apply SEMP and technical work processes to the project

3.2 Development Methods

BrEAD architecture software development will apply the following general methods:

The project will follow the defined processes documented in Section 5 to conduct software requirements analysis and manage the Software Requirements Specification (SRS). Express software requirements in language that addresses a single performance objective per statement and promotes measurable verification. Construct a software architecture that will consist of reusable software components and components to be developed. Allocate software requirements to one or more components of that architecture. Use an automated data base tool to capture, cross-reference, trace, and document requirements.

The Software Development Group and the Software Test and Evaluation Group collaborate during software requirements analysis and then each proceed with separate activities that are intended to ensure the software complies with the detailed software requirement specification. These activities for the Software Development Group and the Software Test and Evaluation Group are defined in Table 2. The activities for each group will be addressed separately in Section 5.

Table 8: Development Activities Group Allocation

Phase / Responsible Group	Software Requirements Analysis	Software Design		Software Imp & Unit Test	Software Integration & Test	CSCI Qualification Test
		Architectural	Detailed			
Software Development Group	Detailed Requirement Specification	Architectural Modeling	Detailed Design	Unit Dev & Test	Unit Rework	CSCI Rework
Software Test and Evaluation Group		Test Planning	Test Case Design	Dev Test Procedures	Incrementally Execute Tests	Perform Qual Test

The project will only design and develop software to meet requirements that cannot be satisfied by reusable software. New software will be based on principles of object-oriented design and exploit object-oriented features associated with the selected high-level language and development environment. New software design will be defined at top-level and detailed design stages of development in the SDD. Software design will promote ease of future growth, specifically new application interfaces. Top-level design will be expressed in a graphical form. Detailed design will also be expressed in a graphical form depicting classes, relationships, operations, and attributes.

The project will adhere to the standards required by the SDP for design, coding, and test methods for new software.

The project will reuse software for requirements that can be satisfied by selected Non-Development Item (NDI)Non-Development Item (NDI) software.

The project will modify (as necessary), unit test, integrate, and document reused software following the same processes used for new software. While reused code will not be expected to conform to a single coding standard, changed source code must be supplemented with sufficient new comments and standard code headers to meet commenting provisions of the coding standard and to promote understandability

3.3 Product Standards

BrEAD architecture software development will comply with applicable directions contained in the documents listed in Table 3. These documents impose standards that are applicable to software requirements, design, coding, testing, and data.

Table 9: Standards and Specifications Applicable to Software Development

Document	Description
MIL-STD-498	Software Development and Documentation, 5 December 1995
MIL-STD-961D	DoD Standard Practice Defense Specifications
MIL-STD-973	Configuration Management
Etc.	

4 – Detailed Software Development Activities

4.1 Project Planning and Oversight

This SDP shall be maintained and modified to reflect the current plans, policies, processes, resources, and standards affecting the BrEAD architecture. It shall be the Software Project Manager's responsibility to keep abreast of industry technology changes and programmatic direction from the Program Manager that would require modification of this plan.

.4.2 Software Development Planning

The plan for all software development shall employ software engineering 'best' practices in verification and validation, configuration management, peer reviews, project tracking and oversight, and software quality assurance. Microsoft Project will be used to develop and maintain the BrEAD architecture master plan and schedule. In addition, each build will be subject to the planning process described in Figure 3 and the build's Microsoft Project plan will be rolled up into the master project plan contained in Section 7 of this document. The build project plans will be made available to all participants. The Software Project Manager will use weekly project-wide meetings to maintain the status of the software project and to resolve any conflicts or changes that might occur. The Software Project Manager will employ a database to record action item assignments, status, and resolutions.

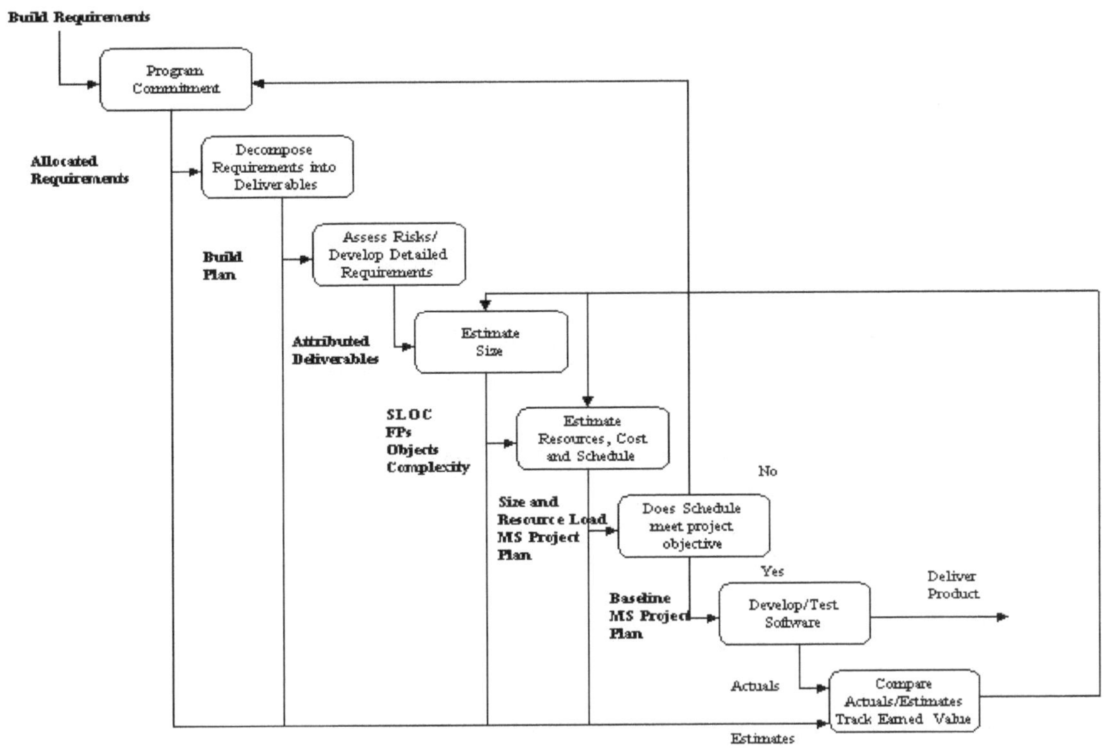

Figure 210: Project Planning and Oversight Process

4.3 CSCI Test Planning

CSCI Test Planning consists of the two levels of testing listed below:
CSCI Integration Testing
CSCI Qualification Testing.
Planning for and conducting the individual CSCI component testing is the responsibility of the Software Development Manager and the Software Test and Evaluation Manager. Test procedures will be devised for each functional component of a CSCI. The intent is to verify that the CSCI's functional components individually meet their SRS and Interface Requirements Specification (IRS)/Interface Design Document (IDD)Interface Design Document (IDD) requirements and that the CSCI taken as a whole meets the performance requirements of the SRS for that CSCI. Paragraph 5.9 documents the processes for software test planning; test case construction; test procedure development, conduct, results analysis, and reporting.

System Test Planning

The intent of system test planning is to validate that the system meets its performance requirements. It will be the responsibility of the System Test Manager to direct the development of system test plans and procedures based on the System/Subsystem Specification (SSS), and conduct the system tests. Paragraph 5.11 documents the processes for system test planning, test case construction, test procedure development, test conduct, results analysis, test reporting and participation of the Software Test and Evaluation Group.

Software Installation Planning
Software installation is at the direction of the Project Manager and will be performed according to the procedures referenced in paragraph 5.11 of this SDP and as addressed in the Build Plan for each deliverable build.

Software Transition Planning
The BrEAD architecture will remain at the CDEPS for Life Cycle Maintenance (LCM); therefore, there is no need to identify a transition process

4.4 Establishing a Software Development Environment

These paragraphs describe the approach to establish and maintain the physical resources used to develop and deliver the BrEAD architecture software.

A Software Test Environment
The Software Test Environment (STE) for the BrEAD architecture will be housed at the SIL facilities. Many STE components will be the same as those used in the SEE.

B Software Development Library
The Software Development Library (SDL) is the master library of all programs, documents, reports, manuals, specifications, reference documents, and correspondence associated with the BrEAD architecture. The SCM Manager functions as the Software Librarian.
The SCM Manager controls the Software Development Files (SDFs), the records/minutes of formal reviews, the SEN series, and the Change Control Documents. The SDF is the control and tracking document for software components during all phases of the software development process. The Change Control Documents include the Problem/Change Report (P/CR) and the Review and Response (R&R)Review and Response (R&R) Form.
The SCM Manager will produce the BrEAD architecture software Master Document Status Summary and the Change Control Document Status Summary. For administrative purposes, the SDL is subdivided into three distinct libraries: Document, Program, and Correspondence & Reference. BrEAD architecture Software Development Library is located at CDEPS. BrEAD architecture Software Manager and SCM Manager will maintain control over all their software support items.

C. Document Library

The Document Library, located in BW HQ, consists of all specifications, manuals, documents, the SEN series, and other reports directly related to BrEAD architecture software. The associated P/CRs, SDFs, Document Review and Response (R&R)

Forms, and the records/minutes of the formal reviews and other formal project meetings are included.

Documents approved for baseline by the BrEAD architecture Software Project Manager will be given to the SCM Manager for Configuration Control (CC).Configuration Control (CC). The SCM Manager will be responsible for the preparation of the cover and title pages with appropriate signatures and the List of Effective Pages. Both machine (magnetic) and hard (paper) copy will be maintained in the Document Library. Upon receipt of changes to baseline documents, the Software Project Manager will ensure that the change is reviewed for completeness, accuracy, and form. When discrepancies are found, the change will be returned to the change originator for correction or resolution of the discrepancy. Upon approval of changes by the Software Project Manager, the SCM Manager will make changes to the baseline document and print a hard copy.

Clearly labeled drafts and preliminary iterations of BrEAD architecture software documents may be placed in the Document Library at the discretion of the Software Project Manager until such documents are readied to be base lined. These documents may include technical, operations, maintenance, and specification manuals. Once base lined, all preliminary or draft issues will be disposed of as directed by the Software Project Manager. All updates and subsequent issues of documents must be maintained in the Document Library. Only current editions will be filed.

Document R&R Forms are maintained in the Document Library. A copy of Software Quality Assurance (SQA) Quality Assurance (QA) review comments will be held in a file until responses are received and signed. The signed R&R forms will then be filed and may not be removed from the library. Working copies may be requested from the SCM Manager. All Test Reports will be maintained in the Document Library. The originals will not be removed from the library. Copies may be requested from the SCM Manager.

Software Program Library

The library consists of program disks and listings of programs that are under development and are being maintained in the SDL until delivered as a program package. Programs or program changes under development will be maintained by the Software Development Group. Once ready for base lining, program modules are entered into the Program Library only by the SCM Manager. These procedures are subject to monitoring by the SQA Group's activities.

Correspondence & Reference Library.

Only correspondence and items approved by the BrEAD architecture Software Project Manager are accepted by the SCM Manager for inclusion in the Correspondence Library. Correspondence will be identified by originator, originator serial number, date of origination, and subject matter.

Software Development Files

SDFs will provide visibility into the development status of the individual CSCIs. The Software Development Group will maintain SDFs for each Software Unit (SU) or collection of SUs that make up a functional component of a CSCI. SDFs will be the principal working logs for assigned programmers. The Software Development Group will maintain SDFs primarily on electronic media and periodically ensure the completeness, consistency, and accuracy of SDFs with respect to specifications, design, and user manuals.

The Software Development Group will organize SDFs to contain the following information:

Introduction - Statement of purpose and objectives, lists the contents of the SDF.

Requirements - SU allocated requirements with a cross reference to the parent CSCI requirements and pertinent programmer notes.

Design Data - Schedules, Status, and the design in the design depiction method selected for the CSCI.

Source Code - Contains listings, by directory, of source code files.

Test Plan and Procedures - Current unit test plan and procedures.

Test Reports - Unit test results and reports.

Review and Audit Comments - Record of reviews and sign-off signatures resulting from reviews and informal audits

Software Development File Approach

At the beginning of the software planning phase, an SDF will be created for each CSCI functional component.

SDFs will be maintained by the Software Configuration Control Manager and will be periodically audited by the Software Development Manager and SQA Group. With the exception of metrics, each programmer is responsible for submitting all material and information required for preparation and maintenance of the SDFs.

Software Development File Format

In the front of each SDF is a task tracking sheet that contains schedules and actual dates of development milestones for the software component. The SDF contains sections applicable to each phase of the development of the software component. Each section of the SDF begins with a header that identifies the phase, software component, and the contract. Each section contains metrics data, schedules, and phase-specific data such as interfaces or specification paragraph references. The SCM Manager will maintain the SDF database with input from responsible engineers.

Software Development File Measurements

The collection of measurements is performed in each software development phase. All defect containment information resulting from peer reviews and unit testing will be recorded in the SDFs by the Software Configuration Control Manager. This database will be the basis for metrics estimation analysis at the end of each development phase and at project end.

Non-Deliverable Software

Non-deliverable software shall consist of all the software that is not specifically required to be delivered, but is used in the design, development, manufacture, inspection, or test of deliverable products. It may include but not be limited to the items listed below:
Software used to design or support the design of a deliverable product which may include databases, design analysis, modeling, simulation, digitizers, and graphic plotters.
Software used for in-process testing of deliverable products, or to generate input data or acceptance criteria data for the test program.

5 - System Requirements Analysis
The Software Configuration Control Group will process requests for clarification, change, and waiver/deviation in accordance with SCM procedures defined in Software Configuration Management Plan (SCMP). The Software Project Manager will convene a Local Software Configuration Control Board (LCCB)Software Configuration Control Board (SCCB). to direct revision of base lined documents, review changes for the SSS, and submit approved change requests as ECPs to System Configuration Control Board (SCCB) for final approval.

a Analysis of User Input
The Software Project Manager, acting as Chair of the LCCB, will assign candidate Problem/Change Requests (P/CRs) to the Software Development Group for analysis of their impact on requirements documents. The Software Development Group will process requests and prepare recommendations for the LCCB as follows:
Requests for clarification will be evaluated to determine if one or more requirements need to be reworded. If a clarification requires a change to a requirement, it will be processed as a request for change.
Requests and rationale for change will be evaluated with respect to the status of software development to determine the impact on schedule, effort, and cost for software requirements analysis, design, implementation, integration, and testing.
Requests and supporting rationale for waiver/deviation of requirements will be evaluated with respect to their impact on the overall processing integrity of BrEAD architecture software.

b Operational Concept
The LCCB will analyze requests for clarification, change, or waiver/deviation that impact the OCD, and/or SSS and prepare an impact assessment. They will evaluate constraints imposed by such factors as interfacing systems, system architecture, NDI software selections, and hardware capabilities and forward them to the sponsor for action. Changes impacting the OCD and/or SSS will be forwarded as ECPs with supporting material to the SCCB for consideration.

c System Requirements
The SCCB in the Program Manager's office will control documents constituting the system requirements baseline; respond to requests for clarification, correction, or

waivers/deviations; analyze impacts; revise the OCD and SSS; record system requirements measures; and manage the system requirements change process.

6 - System Design

System wide design and system architectural design decisions will be based on the requirements that have been derived from the SOO (Statement of Objectives).

7 - Software Requirements Analysis

The software requirements for BrEAD architecture are allocated from the SSS by the Systems Engineering Group. The Program Manager's SCCB is the final authority for approving the allocation of requirements from the SSS to the individual builds as well as the system as a whole. The Software Project Manager will apply the following process to develop, document, and manage allocated software requirements for the BrEAD architecture

7.1 Software Requirements Development Process

Purpose

The purpose of the Software Requirements Change process is to control the software requirements baseline; respond to requests for clarification, correction, or waivers; revise the SRS; and manage the change process. When the SCCB directs changes to the SRS via an approved ECP, the Software Requirements Team will prepare revised change pages in accordance with the Software Requirements Change Process. Changes affecting the SSS will also be submitted to the SCCB where the System Requirements Change process will take precedence.

Roles and Responsibilities

The roles and responsibilities for this process are listed below:
Software Requirements Team manages the requirements data base.
Software Development Group performs analysis to identify algorithms, high level data flow, interfaces and logical functions.
SQA Group verifies process is performed.
IV&V Group validates traceability.
SCM Group performs configuration control of artifacts.

Entry Criteria

The entry criteria for this process are listed below:
Software Requirements Team, Software Development Group, SCM Group, SQA Group, and IV&V Group processes defined and staff trained
System level requirements specification approved by system level CCB

Inputs

The inputs for this process are listed below:
Approved System level specification, requirements allocated to software identified
Approved SCPs, ECPs
Accumulated PC/Rs

Process Activity
The steps needed to complete this process are listed below:
Perform initial database entry and attribution of the system requirements allocated to software.
Perform detailed analysis and update of requirements in database to ensure completeness, testability, consistency. Perform detailed data flow, control flow, and algorithm analysis. Document the qualification method for each of the software requirements and enter into a test planning database and as an attribute for each requirement in the database. Define acceptance criteria for baseline revision (i.e., % targeted requirements passing test). Document in a Build Plan the content (i.e., requirements, PC/Rs) of current increment and/or revision and associated acceptance criteria. Produce draft of the SRS. Develop a traceability matrix between the SSS and the SRS.

Members of the SQA Group and IV&V Group provide analysis and Design Review (DR) comments on requirements and SRS draft.

Perform Formal Inspection to ensure that the SRS meets the system requirements allocated to software. Update SRS based on the sum of the accumulated analysis and comments.
Obtain approval of the SRS by the SCCB.

Negotiate Build Plan with sponsor, gaining agreement on content, schedule, costs, and acceptance criteria. Publish current SRS and Build Plan.

Outputs
The outputs of this process are listed below:
SRS signed
Requirements attributed and quantified in a database
Build Plan with documented acceptance criteria
Test requirements database initialized.

Exit Criteria
The exit criteria for this process are listed below:

SCCB approved SRS
A commitment has been reached with the sponsor on the content of the current increment and/or baseline revision including the acceptance criteria for the current baseline.
Process Measurements

The following data shall be collected:

Earned Value Schedule

Event	% Complete

Initial entry and attribution of build requirements into database	10
SRS Level requirements data flow, interfaces, and associated algorithms defined	40
Requirements allocated to build cataloguing and attribution complete in RM database.	55
SRS trace to SSS complete	65
Start Formal Inspection of SRS draft	75
Finish Formal Inspection and finalize SRS	90
SCCB Baseline Requirements	100

Track total requirements allocated to build
Track % requirements allocated to build that are fully attributed
Track % build allocated requirements testable (e.g., requirements that can be validated by explicit test)
Track actual vs planned staff hours expended
Track actual vs planned costs expended

7.2 Software Design

The software design approach for the BrEAD architecture will be Object-Oriented Design (OOD). The methodology, definition of terms, and notation for new software design will be based on the Booch Method for OOD. This section describes the software design process and the artifacts produced by this process.

Software design will occur incrementally. A top-level design for BrEAD software will be developed during the architectural design phase. During the detailed design phase, the software necessary to meet the capabilities assigned to BrEAD software incremental builds will undergo detailed design to a level sufficient for implementation in support of targeted capabilities.
A software architectural model of BrEAD architecture will be placed under developmental configuration management following the top-level design review. Updates and refinements to the model will occur incrementally during detailed design. An updated model will be placed under developmental configuration management following each successful formal inspection.
The design step produces a data design, an updated architectural design, and a procedural design. The data design transforms the information model created during analysis into the data structures that will be required to implement the software. The procedural design transforms the functional model into a procedural description of the software.

7.2.1 CSCI-Wide Design Decisions

CSCI-wide software design decisions for the BrEAD architecture system software will be a continuous effort. The Software Project Manager will direct the conduct of design analyses and trade-offs through Software Design Team meetings involving all concerned. The Software Design Team will conduct domain analyses, evaluate CSCI-wide design issues, and formulate decisions for new development, reuse of existing components, and hardware/software architectures.

The system-wide design principles for BrEAD architecture are listed below:
Establish an initial capability for BrEAD architecture software upon which to build future enhancements
Use COTS open system architecture hardware and system software components
Maximize reuse of existing software application packages.

7.2.2 CSCI Architectural Modeling and Specification Process

Purpose

The Architectural Design step transforms the model created during requirements analysis into the discrete software components that will be required to implement the system software. Detailed design transforms the software component architectural model into an implementation description of the software consisting of Software Units (SUs), their internal data, algorithms, interfaces, and control flow. The software architectural model of the project will be placed under developmental configuration management following the Formal Inspection of the architectural design. Updates and refinements to the model will occur incrementally during detailed design and the updated model will be placed under developmental configuration management following completion of the Formal Inspection of the detailed design.

Roles and Responsibilities

The roles and responsibilities for this process are listed below:
Software Design Team performs analysis to create the functional model of the software system
SQA Group verifies process is performed
IV&V Group validates traceability
SCM Group performs configuration control of artifacts.

Entry Criteria

The entry criteria for this process are listed below:
Formal modeling methodology selected Engineering has been trained in the application of the modeling discipline.

Inputs

The inputs for this process are listed below:
Approved SRS
Build Plan with documented acceptance criteria.

Process Activity

The steps needed to complete this process are listed below:
Review available documentation on requirements, assumptions, and constraints.
Formulate and evaluate architectural alternatives. Select architectural model.
Define software components and internal software units, their interactions, data flow, and functional processes.

Allocate known requirements to the architectural components. Identify and select COTS open system architecture hardware and system software components. Maximize reuse of existing software application packages.

Design and document the initial Graphical User Interface (GUI) and position actions.

Document architectural design in formal design method (i.e., IDEF0, Inter-action Diagrams, etc).

Conduct technical review to ensure architectural design meets the requirements allocated to software. Repeat process several times until architectural design meets approval.

Develop initial database scheme to support the system software.
Conduct a technical review of the GUI screen designs to ensure they satisfy system level operational needs and conform to accepted Human Computer Interface (HCI) standards.

Develop a traceability matrix between the components of the architectural design and the SRS.
The SQA Group and IV&V Group provide analysis and comment on architectural model and GUI design.

Conduct Formal Inspection of architectural design, its components/units and their functions, and interaction.

Update the architectural design based on the sum of the accumulated analysis and comments.
Software Project Manager accepts architectural model and requirements allocation.

Outputs

The outputs from this process are listed below:
Documented and approved architectural design
Documented GUI design
Initial database scheme.

Exit Criteria

The exit criteria for this process are listed below:
Architectural model accepted and under CM in Software Development Library (SDL)
Requirements allocated to architectural components validated by the SQA Group
Documented GUI under CM.

Process Measurements

The following data shall be collected:
Earned Value Schedule

Event	% Complete
Start component level IDEF 0 diagrams	10
IDEF 0 at component level complete and associated algorithms defined	40
Each software component's SUs identified and structural relations defined	55
Architectural model trace to SRS complete	65
Start Formal Inspection of Architectural Model	75
Finish Formal Inspection and update of Architectural Model	90
CCB Baselines GUI Screens, initial database scheme, and Architectural Model (IDEF 0)	100

b. Track % build allocated requirements traceable to architectural model components

c. Track actual vs planned staff hours expended

d. Track actual vs planned costs expended

7.2.3 CSCI Test Planning Process

Purpose

The purpose of software test planning is to identify test resources and schedule, and identify tests cases and procedures necessary to the qualification of the developing software in a STP.

Roles and Responsibilities

The roles and responsibilities for this process are listed below:
Software Test and Evaluation Group performs analysis to create a strategy for qualification of system software.
SQA Group verifies process is performed.
IV&V Group validates traceability.
SCM Group performs CM of artifacts.
Entry Criteria
The entry criteria for this process are listed below:
SRS under CM
Build Plan with targeted requirements and acceptance criteria.
Inputs

The inputs for this process are listed below:
Approved SRS
Build Plan with documented acceptance criteria
Test requirements data base initialized.
Process Activity
The steps to complete this process are listed below:
Perform analysis of software performance requirements documented in the SRS and identified in Build Plan.

Define test levels for software testing, including general test requirements, test cases (e.g. stress, erroneous input, maximum capacity, and timing). Define the overall test concept and objectives supported by the defined tests.

Identify the environment in which tests will be conducted. Define plans for implementing and controlling the test environment.

Estimate the personnel and other resources required to support the test concept and objectives.
The Software Project Manager directs Software Test and Evaluation Manager to prepare, conduct, analyze data, and report the results of testing.

Trace software requirements allocated to the test cycle to defined test cases. Verify completeness of requirement's traceability.

Develop a general schedule of defined tests, including time for problem correction and retest.
Draft the STP, following the instructions of the assigned documentation standard.
Refine the test requirements database to create structure of test case and procedures with attribution on test characteristics.

The SQA and IV&V provide analysis and comment on the STP. Perform Formal Inspection to ensure that the STP meets the system requirements allocated to software and is in context with the architectural model and GUI designs. Revise draft STP to correct discrepancies and incorporate recommended changes. The LCCB baselines the STP placing it in the SDL.
Outputs

The outputs of this process are listed below:
Approved STP
Test requirements database with attributed test case entries.

Exit Criteria
The exit criteria for this process are listed below:
All SRS performance requirements allocated to defined test cases
STP approved and under CM.

Process Measurements
The following data shall be collected:
a. Earned Value Schedule

Event	% Complete
Start Analysis	10
Start Formal Inspection of STP	50
Finish Formal Inspection and update to STP	75
CCB Baselines STP in SDL	100

b. Track % build allocated testable requirements traceable to test cases identified in STP
c. Track actual vs planned staff hours expended
d. Track actual vs planned costs expended.

CSCI Detailed Design Process

Purpose

The purpose of Detailed Design is to incrementally define and describe the CSCI software components in terms of their SUs and the internal relationships of those SUs using detailed graphical representation.
If development of detailed design identifies deficiencies in architectural design or software requirements, the Software Design Team will submit requirement changes and/or update the developmental architectural model as required.

Roles and Responsibilities

The roles and responsibilities for this process are listed below:
Software Design Team performs analysis to create the detailed design of system software
SQA Group verifies process is performed
IV&V Group validates traceability
SCM Group performs CM of artifacts.

Entry Criteria

The entry criteria for this process are listed below:
Approved Software Architectural Design
Approved SRS
Approved GUI design.

Inputs

The inputs for this process are listed below.
Documented and approved architectural design
SRS, build plan, and attributed software requirements database
Documented GUI design
Initial database scheme.

Process Activity

The steps needed to complete this process are listed below:
Construct graphical representation for each software component detailing the interaction of its
SUs.
Add algorithm, data attributes, and operations to each SU.
Draft Software Detailed Design (SDD) document.
Provide traceability matrix from SUs to the parent requirement in the SRS and build plan.

The SQA Group and IV&V Group provide analysis and comment on the detailed design.
Conduct Formal Inspection of the SDD by component, each component's SUs and their functions and interaction. Repeat process several times until detailed design meets approval.
Update the SDD based on the sum of the accumulated analysis and comments.
The LCCB baselines the SDD placing it in the SDL.

Outputs

The output for this process is the SDD containing graphical representation of the detailed design, internal data, and algorithms definitions.

Exit Criteria

The exit criteria for this process is the software components and their SUs graphical design representation, internal data, algorithms defined, reviewed and approved.

Process Measurements

The following data shall be collected:

a. Earned Value Schedule

Event	% Complete
Start analysis	10
Graphical representation complete and associated algorithms defined	40
SU structural relations, algorithms, and interfaces defined	55
Draft SDD	65
Start Formal Inspection of SDD	75
Finish Formal Inspection of SDD	90
LCCB Baseline SDD	100

b. Track % build allocated requirements traceable to SUs described in the SDD
c. Track actual vs planned staff hours expended
d. Track actual vs planned costs expended.

Test Case Design Process

Purpose

The Software Test and Evaluation Group will design the set of test cases defined in the STP. The design will be in keeping with the overall test concept and objectives of the STP to verify the software meets all performance requirements allocated to the functional components of the architectural design.

Roles and Responsibilities

The roles and responsibilities for this process are listed below:
Software Test and Evaluation Group performs analysis to design the test cases
SQA Group verifies process is performed
IV&V Group validates traceability
SCM Group performs CM of developed artifacts.

Entry Criteria

The entry criteria for the process is that the STP is published and under CM.

The inputs for this process are listed below:
Base lined SRS
Base lined STP
Base lined Architectural Design.

Process Activity

The steps needed to complete this process are listed below:
Define the test case(s) identified in the draft STD, expressing the purpose and conditions associated with each test case and identify included test procedures. For each test case, perform the following activities:
Define the inputs (stimuli to the component under test) required to fulfill the test purpose.
Define the expected results (outputs from the component under test in response to inputs) required to fulfill the test purpose.
Define insertion and extraction methods to/from the component under test.
Identify points of data input/output and volumes of data.
Define evaluation criteria for test results analysis. Define ranges of values, capacities, and times for test pass/fail.
Identify test procedure suite (i.e., required discrete procedures).
Identify test environment configuration, interface drivers, database loaders, controllers/monitors, and other test tools to support test case purposes.
Trace SRS performance requirements to specific test procedures within the test cases.
The SQA Group and IV&V Group provide analysis and comment on the test case description.
Perform Formal Inspection to ensure that the test case design meets the system requirements allocated to functional component.
Revise draft test case descriptions to correct discrepancies and incorporate recommended changes.
The LCCB baselines the test case designs placing them in the SDL.
Outputs
The output from this process is the approved test case descriptions in STD format.

Exit Criteria

The exit criteria for this process are listed below:
Base lined test case descriptions
Functional component performance requirements allocated to test case procedures suite.

Process Measurements

The following data shall be collected:

a. Earned Value Schedule

Event	% Complete
Start Analysis	10
Start Technical Review of test case design	50
Finish Technical Review and update test case design	75
LCCB Baseline test case design	100

b. Track % build allocated testable requirements traceable to test procedures defined for the test cases.

c. Track actual vs planned staff hours expended.

d. Track actual vs planned costs expended.

Software Implementation and Test / Code and Unit Testing

The purpose of Software Implementation and Unit Testing is to implement and test the detailed design for SUs into new code, or as modifications of existing NDI code following documented programming style guidelines.

The following paragraphs describe Software Implementation and Unit Testing processes.

Purpose

The purpose of software implementation and unit testing is to create qualification test ready CSCI functional components by implementing and testing each component SUs as identified in the detailed design. This may involve developing new code and/or modifications of existing code, and following documented programming style guidelines.

Roles and Responsibilities

The roles and responsibilities for this process are listed below:
Software Production Team performs code/unit test and/or COTS selection
SQA Group verifies process is performed and validates SDFs
IV&V Group validates traceability
SCM Group performs CM of artifacts.

Entry Criteria

The entry criteria for this process are listed below:
SDD containing graphical representation of the detailed design, internal data, and algorithms definitions base lined.
Software units allocated to individual engineers for production and test.
SDF formats defined.
Programming style guides defined.
Engineering staff has completed required training on project processes, style guidelines, and tools.

The inputs for this process are listed below:

Base lined SRS and build plan
Base lined SDD
Approved GUI design
Initial database scheme.

Process Activity

The steps needed to complete this process are listed below:

Review requirements allocated to the SUs, the SDD, and programming style guidelines.

Maintain working source for assigned task within the SDF.

Compile software.

Perform tracing of each SU implementation to the parent software component in the SDD.

Review all inputs and outputs for each software unit. Identify any test drivers necessary to stimulate unit execution, provide inputs, and capture final outputs.

Identify all interfaces to other units and determine if other units are available or if unit stubs must be written.

Identify the various paths that may be taken by the unit (a McCabe complexity analysis tool may be used to determine the unit paths). Identify data variables and conditions that determine which paths are taken. Identify parameter limits for all input, output, and control parameters. Examine and list all error handling facilities of the units and error conditions.

Identify a set of unit tests to conduct the required testing (path testing, boundary condition testing, and input validation and syntax testing). There should be a test for each path through the unit, to test the upper and lower limits of all parameters, and to test all error conditions (pass and fail). Write step-by-step procedures for each test case.

Document the unit tests in the SDF.

Conduct technical review of each SUs implementation and unit tests. Resolve all comments.

Perform the test in accordance with the unit test plan.

Compare test results with expected results and SU acceptance criteria (e.g., 80% path test analysis successful). Document the results in the SDF. If discrepancies are found, attempt to determine whether the errors are associated with the software, test/test driver, or hardware. Rework as required.

The SQA Group and IV&V Group provide analysis and comment on the content of the SDFs.

The Software Production Team, upon successful completion of the unit test for a software component included SUs, promotes the software component for integration testing, and submits the software component for developmental CM in the SDL.

Document and report requirement and/or design errors/inconsistencies.

Update metrics information, including defect analysis database.

Outputs

The outputs from this process are listed below:

Current SDFs
Updated SDL (includes qualified SUs)
Defect database.

Exit Criteria
The exit criteria for this process are listed below:
Updated SDF
Successfully tested units
Metrics database updated with defect information from SU technical review and unit testing
SU's checked into SDL ready for integration.

Process Measurements
The following data shall be collected:
Earned Value Schedule
(Note: Earned value will be tracked against SDF contained SU set (i.e. a software component))

Event	Report % Complete
Start Code and Unit Test of component SUs	10
Complete Code and Unit Tests draft	50
Tracing of each SUs Unit Test to component allocated requirements complete	60
SU implementation technical review complete	70
Execution of Unit Test Complete, SU source code and Unit Tests updated	85
Component SDF validated by QA and IV&V	90
Component SU Code Checked into SDL	100

Track % actual number of SUs completing code and unit test against build's total number of SUs.

Track SLOC developed, reused, and modified to facilitate process modeling and estimation.
Collect for process defect analysis P/CRs, and their origin (i.e., code logic, design, requirements, etc).

Track actual vs planned staff hours expended.
Track actual vs planned costs expended.

8 – Unit Testing
The purpose of the unit testing addressed in Section 5.7.1 is to identify and correct as many internal logic errors as possible. Problems not uncovered by unit testing are, in general more difficult to isolate when uncovered at the CSCI level. Unit testing will be conducted throughout the implementation process, first as part of the initial development process and later as changes to the unit are made. Unit tests will be repeatable and may be conducted at any point in the implementation process in accordance with the approved unit test plan. For the purposes of unit testing, a

unit is an object. The goal for unit testing by developers is to perform path testing in which every affected branch is navigated in all possible directions at least once and every affected line of code is executed at least once. Unit test drivers and stubs will be developed as needed and will be placed under CM as part of the overall test utility. All unit test results will be recorded in the SDFs.

Test Case/Procedure Implementation

Purpose
The purpose of this process is to develop detailed steps for controlling tests, injecting inputs, recording results, and comparing actual to expected results. Document test case procedures in the final STD.

Roles and Responsibilities
The roles and responsibilities for this process are listed below:
Software Test and Evaluation Group performs analysis to implement the test case defined test procedures
SQA Group verifies process is performed
IV&V Group validates traceability
SCM Group performs CM of developed artifacts.

Entry Criteria
The entry criteria for this process are listed below:
Test cases designed
Software performance requirements allocated to test cases
GUI screens designed.
Inputs
The inputs for this process are listed below:
GUI screen and/or operator manual
Test case designs.

Process Activities
The steps needed to complete this process are listed below:
Review software GUI and/or operator manuals to identify methods of operator input, use of simulator/emulator tools, and software data recording.
Define and document detailed test procedure steps for providing inputs for test cases.
Prepare input data files to provide test stimuli. Prepare operator logs for the component under test as well as the test environment.
Define evaluation steps for conducting post-test analysis and comparing actual and expected test results.
Define and document the test procedures in the STD following the instructions of the assigned documentation standard.
(Recommended Option). Develop test harness with automated test procedures and test result analysis.
Trace specific test case procedures to requirements allocated to parent test case.

The SQA Group and IV&V Group provide analysis and comment on the test procedures.
Conduct a technical review to verify consistency of the STD test procedures (or test harness/procedures) with the test case descriptions, the STP, baselined requirements, and planning documents. Recommend revisions, as appropriate.
Revise draft STD (or test harness/procedures) to correct discrepancies and incorporate recommended changes to test case definitions and test procedures.
The LCCB baselines the test procedures in the STD placing them in the SDL.

Outputs
The outputs from this process are listed below:
Baselined STD containing test procedures
Test data files compiled.

Exit Criteria
The exit criteria for this process is that the STD is approved and under CM.

Process Measurements
The following data shall be collected:
a. Earned Value Schedule
(Note: Earned Value will be tracked against each Test Case Procedure suite)

Event	% Complete
Start Test Case Procedures development	10
Complete Test Case Procedures draft	50
Tracing of Test Case Procedures to test case allocated requirements complete	60
Test Procedure Technical Review complete	70
LCCB Baselines STD Test Procedures into SDL	100

b. Track total number of test procedures required to implement all test cases
c. Track % test procedures completed
d. Track actual vs planned staff hours expended
e. Track actual vs planned costs expended

9 - Software Integration and Testing
Purpose
The purpose of integration and test is to incrementally integrate SUs into larger software components, and components into a complete system. Testing is performed to validate each component's ability to meet its stated requirements and to ensure interoperability of the major software components.
Integration continues until all software components are integrated with the system-level hardware suite into a single functioning system.

Roles and Responsibilities
The roles and responsibilities for this process are listed below:
Software Test and Evaluation Group plans, integrates, and executes tests

Software Development Group performs updates in response to developmental P/CRs
SQA Group verifies process is performed and validates test results
IV&V Group validates traceability of successful tests runs
SCM Group performs configuration control of test reports, and updated project artifacts.

Entry Criteria
The entry criteria for this process are listed below:
STP
Build Plan and schedule
Any previous integration test reports
Approved architectural design
Approved STD with Test Procedures.

Inputs
The inputs for this process are listed below:
Updated SDL (includes qualified SUs)
Baselined STD (includes execution ready test cases with included test procedures)
Integration Test Report form
P/CR forms
Test drivers and test inspection tools as required by STD

Process Activities
The steps needed to complete this process are listed below:
Software Test and Evaluation Group reviews the Build Plan and schedule, the list of software units/components to be included in the build, and the results of any previous integration tests to determine the software and software fixes to be tested.
Determine a set of test cases to be used to test the build.
For the requirements identified for the build, develop an integration test plan. The integration test plan should identify the sequence of SU and component integration, traceability to the requirements to be validated by the test, test tools and drivers to be used, and the applicable test cases including procedures for conducting the test and analyzing the results. Document the integration test plan.
Perform a Technical Review of the integration test plan and resolve all comments.
Software Test and Evaluation Group places the integration test plan under developmental configuration control.
Integration Test Team requests and receives increment build, installs it in the integration test area.
Conduct the test in accordance with the integration test plan procedures. Record test results as they are observed.
Perform any required post test analysis or data reduction to determine pass/fail criteria as specified in the integration test plan.
Compare test results with expected results. If discrepancies are found, attempt to determine whether errors are associated with the software, test/test driver, or hardware.

Document all test results using the project integration test report form. This report should contain all data recorded from test tools, test results, and deviations from the test plan. Document any problems detected on a P/CR form. Document requirements satisfactorily tested in the test report.
Determine if another incremental build is required. Proceed with activity (e) above if needed. If not proceed to next step.
The SQA Group and IV&V Group provide analysis and comment on the test results.
Software Test and Evaluation Manager reviews the integration test report and any P/CRs for accuracy and thoroughness. File the test report in the project history file in the document library and submit a copy of the test report and P/CRs to the Software Development Manager for review and analysis.
LCCB baselines all updated developmental and test artifacts placing them in the SDL

Outputs
The outputs from this process are listed below:
Updated SDFs
Updated SDL (i.e., source code and test procedures)
Project History File updated with test results
P/CRs submitted for all detected problems.

Exit Criteria
The exit criteria for this process are listed below:
Integration test report reviewed and approved
All build allocated testable requirements have been successfully tested
CCB baselines SDL with updated source code and test procedures
Project History File updated with test results.

Process Measurements
The following data shall be collected:
Earned Value Schedule

Event	% Complete
Start Analysis	10
Integration test plan drafted	20
Integration test plan completes Technical Review	35
Initial increment test begins	50
Incremental integration testing completed	80
Test results validated by QA and IV&V	90
LCCB Baseline updated SDL artifacts	100

Track % build allocated testable requirements successfully completing integration test
(Note: Test Procedures trace to testable requirements. See Test Case Design)
Track SU successfully completing integration as a percent of total targeted for integration
Track open versus closed STRs
Track actual vs planned staff hours expended

Track actual vs planned costs expended.

9.1 System Acceptance Test

This section describes the approach to be followed for participating in system qualification testing. System acceptance testing will be conducted in accordance with the project SEMP. The software engineer(s) will be responsible for supporting the testing activities. These activities will include assisting with the debugging process as problems are uncovered and correcting any software related problems. Subtopics to be addressed include:

Independence in system acceptance testing
Testing on the target computer system
System acceptance test preparation
Performing system acceptance test dry runs
Performing system acceptance testing
Revision and retesting

9.2 Preparing for Software Use

The Delivery Team is responsible for correctly packaging the software consistent with the Build Plan and according to the BrEAD software documented processes.

Preparing the Executable Software
The executable software will be prepared for each user site, including any batch files, command files, data files, or other software files needed to install and operate the software on its target computer(s). The result is included in the executable deliverable portion of the Program Package (PP) as documented in the Build Plan.

Preparing Version Descriptions for User Sites
The Software Version Description (SVD) identifies and describes a version of a CSCI component or interim change (i.e., changes that occur between CSCI versions) to the previously released version. The SVD records data pertinent to the status and usage of a CSCI version or interim change. It is used to release CSCI versions or interim changes to the customer and will be included in the Program Package (PP).

Preparing User Manuals
User Manuals will be prepared by the Software Development Group and validated by the Software Test and Evaluation Group. The Software Librarian will baseline the User Manuals, and provides copies as part of the deliverable Program Package (PP).

Installation at User Sites
Installation and integration schedules must be developed and site surveys completed prior to development of the individual Program Packages (PP). The installation can begin at the completion of system development. All hardware and

software components must be assembled and tested in a lab environment prior to being shipped to the user site. The BrEAD software and hardware systems will then be shipped to the user site and installed and checked out by the designated Delivery Team. Each Delivery Team will identify needed training and prepare training materials. Training should be provided to users at the time of installation. Other assistance, such as user consultation, must be readily available after installation of each revision. Planning for the delivery of each build is contained in the Build Plan for that delivery.

9.3 Software Product Evaluation

The processes defined in the Software Quality Assurance Plan (SQAP), reference (g) address software product evaluations.

9.4 Software Quality Assurance

The Software Quality Assurance Plan (SQAP), reference (g),Software Quality Assurance Plan (SQAP) is the guiding document for the conduct of SQA within the BrEAD software

9.5 Corrective Action

The BrEAD architecture software will use the processes for data basing, tracking, and directing correction of P/CRs.

10 - Technical and Management Reviews

The purpose of technical and management reviews is to provide management with tracking and oversight of the progress of software development undertaken by the BrEAD architecture software and fulfillment of requirements. Timely technical and management reviews at the appropriate level of detail facilitate information reporting and interchange that tracks progress against plans, identify and resolve action items, and verify appropriate expenditure of assigned resources.

Joint Technical Reviews

The developer shall plan and participate in joint technical reviews at locations and dates proposed by the developer and approved by the acquirer. These reviews shall be attended by persons with technical knowledge of the software products to be reviewed. The reviews shall focus on in-process and final software products, rather than materials generated especially for the review. The reviews shall have the following objectives:

Review evolving software products, review and demonstrate proposed technical solutions; provide insight and obtain feedback on the technical effort; and surface and resolve technical issues.

Review project status and surface near and long-term risks regarding technical, cost, and schedule issues.

Arrive at agreed-upon mitigation strategies for identified risks, within the authority of those present.
Identify risks and issues to be raised at joint management reviews.
Ensure ongoing communication between acquirer and developer technical personnel.

Joint Management Reviews
The Software Project Manager shall plan and participate in joint management reviews at locations and dates approved by the Program Manager. These reviews shall be attended by persons with authority to make cost and schedule decisions. The reviews will be scheduled in the MS Project Plan for the BrEAD architecture software project. Keep management informed about project status, directions being taken, technical agreements reached, and overall status of evolving software products.

The objectives of management reviews are listed below:
Resolve issues that could not be resolved at joint technical reviews.
Arrive at agreed-upon mitigation strategies for near and long-term risks that could not be resolved at joint technical reviews.
Identify and resolve management-level issues and risks not raised at joint technical reviews.
Obtain commitments and acquirer approvals needed for timely accomplishment of the project.

Given below is a set of candidate joint management reviews that might be held during a software development project. There is no intent to require these reviews or to preclude alternatives or combinations of these reviews.
Software plan reviews. These reviews are held to resolve open issues regarding one or more of the items listed below:
Software Development Plan (SDP)
Software Test Plan (STP)
Operational concept reviews. These reviews are held to resolve open issues regarding the operational concept for a software system.
A system/subsystem requirement reviews. These reviews are held to resolve open issues regarding the specified requirements for a software system or subsystem.
System/subsystem design reviews. These reviews are held to resolve open issues regarding one or more of the items listed below:
The system or subsystem-wide design decisions
The architectural design of a software system or subsystem.
A software requirement reviews. These reviews are held to resolve open issues regarding the specified requirements for a CSCI.
Software design reviews. These reviews are held to resolve open issues regarding one or more of the items listed below:
The CSCI-wide design decisions
The architectural design of a CSCI
The detailed design of a CSCI or portion thereof (such as a database).

Test readiness reviews. These reviews are held to resolve open issues regarding one or more of the items listed below:
The status of the software test environment.
The test cases and procedures to be used for CSCI qualification testing or system qualification testing.
The status of the software to be tested.
Test results reviews. These reviews are held to resolve open issues regarding the results of CSCI qualification testing or system qualification testing.
Software usability reviews. These reviews are held to resolve open issues regarding one or more of the items listed below:
The readiness of the software for installation at user sites
The user and operator manuals
The software version descriptions
The status of installation preparations and activities.
Software supportability reviews. These reviews are held to resolve open issues regarding one or more of the items listed below:
The readiness of the software for transition to the support agency.
The software product specifications.
The software support manuals.
The software version descriptions.
The status of transition preparations and activities, including transition of the software development environment, if applicable.
Critical requirement reviews. These reviews are held to resolve open issues regarding the handling of critical requirements, such as those for safety, security, and privacy.

11 – Other Software Development Activities

These paragraphs describe the technical and management approach to coordinating and providing oversight of software development activities undertaken by the BrEAD. Timely technical and management oversight at the appropriate level is necessary to accomplish the activities listed below:
Track progress against plans
Identify and resolve problems
Identify and verify appropriate expenditure of assigned resources
Depict activities and relationships associated with planning and conducting integration
Consider other related development actions and schedules that could impact BrEAD system.
These paragraphs address those actions to be taken to maintain a broad management picture of BrEAD architecture software development.

Risk Management

Risk analysis and risk management deal with concerns of project personnel regarding the capability of the designed system to achieve program objectives of technical performance, schedule and cost. Risk analysis is an iterative process. It

attempts to identify what could go wrong, plan courses of action to mitigate the occurrence, and plan contingencies in the event problems arise. Risk analysis identifies potential problem areas, quantifies risks associated with these problems, assesses the effect of these risks, and generates alternative actions/mitigations to reduce risks and recover from their occurrence.
The major areas of risk for the software development will be managed following the process identified in the Risk Management Process.

Software Management Indicators
The BrEAD architecture management team will use software metrics for measurement of cost and schedule variance, productivity and quality. BrEAD architecture measurement program is defined in the Software Measurement Plan (SMP).

Security and Privacy
BrEAD architecture software development is anticipated to contain a limited amount of classified data. Development of classified software will occur only in properly cleared spaces utilizing Automated Information Systems (AIS)Automated Information Systems (AIS) approved computers. Transfer of classified programs between facilities shall utilize approved methods. The highest security level that should be encountered during software development is SECRET. Each software engineer or other software support person who requires access to classified material will be cleared to the appropriate level and supplied with an approved container in which classified material can be stored when not in use.

Subcontractor Management
The Software Project Manager will use the SEPO-developed CAPM Process for Software Contracts as a guide for contractor management. The Software Project Manager will augment the CAPM by initiating additional management actions described below.
Contractor personnel will participate in all development team activities including weekly status meetings, and informal and/or formal reviews as directed by Software Project Manager. Contractors will supply weekly measurements and monthly status reports on progress against their statement of work as called for in their individual Contract data requirements list (CDRLs). Contract Data Requirements List (CDRL).

When deemed necessary, the Software Project Manager will establish working groups made up of government and contractor personnel to address major project areas such as architecture, integration, and display.

The Software Project Manager will also establish an overall BrEAD software development, integration, and testing schedule. Included in this schedule will be all major project milestones that will enable a contractor to plan for delivery of their individual sections.

To manage and track BrEAD software status/progress, the Software Project Manager will direct the use of the software management tool, Microsoft Project.

Software Independent Verification and Validation (IV&V) Agents

Coordination will be provided by having the IV&V Group participate/monitor activities in each phase of the development process including attending both formal and informal program reviews. In addition, the IV&V Group will be provided on-line access to source code, test scripts and documentation, and will participate in the SCCB.

During all phases of software development, the IV&V Group will be able to initiate change requests to the evolving baseline or to add their review comments into the active action items to be resolved as a part of each end-of-phase review. These comments will be treated like any comment or change request initiated by a member of the project organization. Review comments are expected to be supplied in written form, or electronically in Microsoft Word, for subsequent tracking by the Software Project Manager in the same manner review actions items are handled. All comments will be adjudicated and responses provided back to the IV&V Manager describing the intended action to be taken. Change requests to change something in the developmental baseline, after the phase has been completed, will be entered into the configuration control system as described in SCMP and addressed like any other change request. The IV&V Group will be able to monitor the status of change requests they submitted directly from the SCM system.

Coordination with Associate Developers

The Software Project Manager and key staff members will meet weekly to clarify issues, obtain project status on deliverables, and elicit program comments. Updates will be provided and technical issues that affect the overall development effort will be discussed. In addition, the Software Project Manager will form sub-groups or committees to resolve complex technical and administrative issues.

To provide a documentation trail, E-mail will be used as the primary means of distributing data amongst participants.

Improvement of Project Processes

As the BrEAD architecture program matures, the SDP should undergo revision reflecting improvements to the processes. Quality and process improvement comes by analyzing and measuring the process in a structured, controlled manner and then changing the process. These improvements should be the product of lessons learned data, findings of audits, and process measurements collected and fed back into the processes.

The software development process described in this SDP generally follows MIL-STD-498. Where a process described in the SDP is deemed to be inadequate, cumbersome or non-responsive, any BrEAD architecture team member may submit a recommendation to change or improve that process.

When compliance with specific software policies, processes, forms, or templates would impose severe impacts on project cost, schedule, resources, or customer relations, the implementing developer may request a waiver/deviation of specific provisions. They must propose reasonable and detailed alternatives during project planning. Requests for waivers/deviations must cite specific provisions of policies, processes, forms, or templates for which waivers/deviations are being requested, the impact of compliance, and alternatives to be implemented by the BrEAD. The Department Software Engineering Process Group (SEPG), interfacing with SEPO, shall be responsible for evaluating and acting upon suggestions for improvements in policies, processes, procedures, waivers/deviations, and standard forms and templates referenced in this SDP. In addition, process metrics collected during program development and lessons learned from this and other projects will be used to update the BrEAD architecture processes. The Software Project Manager is authorized to approve SEPG recommended changes to the BrEAD processes described in the SDP.

The Software Project Manager is responsible for post mortem analysis of each BrEAD architecture build to identify candidate processes for improvement and for the submission of Project Data Forms (PDF) to the Organization Software Process Database (OSPD) maintained by SEPO.

Attachment 4 – Marketing

a. **Marketing ideas within the RMI:** BRAINS definition of the target audience for a small scale advertising campaign would be the people living in the direct proximity of the breadfruit. In many cases the individual would be responsible to processing of their own breadfruit and our advertising would simply be aimed at informing them of how to utilize breadfruit.
   - Advertise and put on breadfruit product demonstrations which include hands on demos of how to process the breadfruit as well as examples of food that can be made using breadfruit. During these demos either give away or sell basic tools that might be necessary for processing the breadfruit on a family scale.
   - Sponsor a breadfruit cooking competition with prizes to motivate participation. This will give the double benefit of getting people to talk about breadfruit and possibly enabling someone to discover a popular recipe containing breadfruit.
   - Advertise breadfruit as a way for locals to gain some control and stability to their family's food supply and finances. Advocate breadfruit to local, current and future farmers as a profitable and readily available crop.
   - Sell surplus (if any) breadfruit in their community (between families) markets in various forms.

Figure 211. Small Scale Marketing (Within the Marshal Islands)

b. **Marketing ideas outside the RMI:** For a Large scale marketing campaign BRAINS would not limit the target audience to only locals, instead BRAINS could advertise breadfruit as an alternative food to all types of people around the world.
   Marketing options:
   - Sponsor TV commercials and a website promoting the uses as breadfruit as a new and exotic food. Hire a marketing consulting business to develop a plan to advertise and promote breadfruit as a premium food product.
   - Emphasize different benefits of breadfruit depending on who our target audience is.

- An obvious method would be to advertise the great taste of breadfruit products. To do this, BRAINS will need a collection of popular easy to make recipes. A possibility for a way of collecting these recipes would be a website where users can upload their latest recipe and share feedback. Try to partnership with companies such as Betty Croker, General Mills, and most of the flour production companies within the US to use their work-force, leadership in the market, and resources to promote breadfruit flour as well.
- Another benefit of breadfruit that might be important to many consumers would be its health benefits. Advertising the nutritional benefits of breadfruit will help attract that certain market segment.
- The final benefit and possibly the most influential is the charitable side of breadfruit. Thus BRAINS can advertise to consumers the impact they can have on someone's life simply by purchasing breadfruit products, explaining to the population that every breadfruit product purchased provides money to the local farmers and process workers.

Figure 212. Large Scale Marketing (overseas)

Attachment 5 – Training

1.0 GENERAL INFORMATION

GENERAL INFORMATION

1.1 Introduction

This Training Plan defines the support activities, schedules, curriculum, methods and tools, and equipments required for system training. It also describes the trainer plan for conducting a training effort to fulfill the training requirements of the entire project

1.2 Scope and Purpose

This document describes the activities, schedules, curriculum, methods and tools used by Bright Robust Architects for the Institute of New Systems (BRAINS) to develop a Breadfruit Processing System and Breadfruit Enterprise Architecture Demonstration (BrEAD) for the National Green Revolution Foundation (NGRF).

The New Green Revolution Foundation (NGRF) is a non-profit foundation and has been doing research in the Marshall Islands on expanding the use of breadfruit as food source to feed the growing world population, particularly the Marshall Island. Therefore, the purpose of the NGRF is to write a proposal to raise sufficient funds to develop the Enterprise Architecture Demonstration (BrEAD) to do further research and development to develop new breadfruit food products, to promote those products worldwide, to facilitate the design of scalable breadfruit processing systems, to create a virtual collaborative knowledge sharing automation and communication system to share research and marketing information in real-time, and to seek funding from various sources, to continue the development of breadfruit food products and breadfruit processing technology development.

This document identifies the need for training in order to carry out successfully the program and enable the local population to use all the equipments proposed during the development of the system. It also presents the overall goals of the training, the instructional design strategy employed, the particular media and methods which will be used during the training, the methods which will be used to assess training effectiveness, and descriptions of the three major training deliverables: User Manuals, Video-Based Training, and on-the-job Training

1.3 System Overview

The BrEAD shall organize the collection of the breadfruit and inspect them before starting the processing phase. For our system, we choose to transform the breadfruit into flour using these steps:

Cleaning
Peeling
Slicing
Drying

Grinding
Screening

Some of these steps shall use electricity provide by wind turbine.

1.4 Project References

The following documents were use in the preparation of this training plan:
Poor man wind turbine manual
NGRF PBSS

1.5 Acronyms and Abbreviations

BrEAD
NGRF
PBSS
VBT

1.6 Points of Contact

For assistance or more information, please contact the system development engineer that are involved in the design and development of the BrEAD:

Name	Function	Contact
Danielle Eastberg	Program Manager	
Abdy Hachi	Entiprise Lead	
Gabriella Farias	Technical Lead	

1.7 General Training Prerequisites

This Training Plan is predicated upon certain assumptions which, although seldom discussed, are nevertheless critical to its fulfillment.
The first such assumption is the local population involvement in the analysis, design and development phases of training. Such participation will have a direct influence on the validity and reliability of training content, and is the most efficient way to ensure that the needs and special circumstances of the users are adequately and realistically represented.
The second assumption is that training implementation is a function of the overall Project Plan, which has been contractually agreed to. Future modifications in the Project Plan may therefore impact training. For example, it is presently assumed that the system will be launched on January 10, 2010. Projections of training facilities usage, equipment needs, etc. are based on this time schedule.
The training assumes that the level of education of the local population is very low in general. However, they have a good knowledge of the use of VHS with television.
The training assumes no prior system knowledge on the part of participants

2.0 Training Approach

TRAINING APPROACH

2.1 Training Requirements

The training has been designed to fulfill the following requirements:
The training shall help each user to gain an overall understanding of the scope and purpose of the BrEAD System
The training shall help each user to gain an overall understanding of the subsystem with which he/she works,
The training shall give hands-on experience in the use of the system functions that person uses to perform his/her job,
These are the functional, objective goals of training.
There is another set of goals which cannot be stated as succinctly - what might be called the subjective goals. For example, in addition to teaching screens and keystrokes, good training helps to "sell" a new system. It must overcome the anxieties of some regarding the use of automated systems in general, the resistance of others with respect to the use of this system as opposed to the one they're using now, and the concerns of all about "getting up to speed" quickly enough to make the whole enterprise feel worthwhile.
The training must not overwhelm. It is probably an impossible task to totally learn Subsystem A or Subsystem B from a single week of training. Rather than attempting to accomplish the impossible, then, the multiple curricula (courses of study) which will be created for the various training audiences will use an "empowering" approach: it is a goal of the training program to produce active learners rather than passive trainees
.

2.2 Roles and Responsibilities

The training shall be conducted by Bright Robust Architects for the Institute of New Systems (BRAINS). However, the materials and tools can be developed by an external entity that has a very good knowledge of the local population culture.

2.3 Techniques and Tools

Effective systems applications training often require the use of a variety of training tools, techniques and delivery methods. Specific training deliverables include a (paper-based) User Manual, a Video-Based Training (VBT)

User Manuals

The User Manuals are printed reference documents that contain step-by-step instructions for using the new subsystems.
Each manual contains:

- a title page,
- a list of illustrations,
- a table of contents,

- sufficient chapters to cover the subject material,
- a glossary (list of terms),
- an index,
- Appendices (to be distributed on a limited basis).

Video-Based Training (VBT)
VBT is a useful tool which can include tutorials, drill and practice, and units simulating the action of the new application subsystems. The integration of VBT into the on-the-job training provides a user-friendly, self-paced environment in which trainees will be able to practice skills. VBT units will be developed to present and assess the retention of the skills.

2.4 Schedule

As this training does not contain any lecture or classroom training, The schedule should be as comprehensive as possible and should be based on the availability of the various operators. It can be revised at later point to reflect any change in the project lifecycle.

3.0 EVALUATION

EVALUATION
Methods for evaluating training effectiveness from various perspectives are built into the training program in a range of different ways.
Months before full implementation of the BrEAD System, at least three groups of users on each of the component systems will receive training: the Acceptance Test group and personnel at the pilot site at Marshal Islands. Part of the acceptance testers' job is to provide feedback on the training materials (which will be in draft form at this point) and the training itself.
Based on the Acceptance Test experience, the training approach and materials will be refined, put in more final form, and tested again.
VBT, a major component of the training as described previously, has internal provisions for testing participants' degree of mastery of material presented, and issuing formal reports to trainers or management, in addition to its ongoing feedback to the learner. The Training Team will use this facility of the medium as yet another way of evaluating training effectiveness.

Attachment 6 – Human Systems Integration Plan

Human Factor Integration

The system in creation, BrEAD, deals mainly with human, right from the collection of the bread fruit to the processed final product. Therefore the Human Factors Integration is a very important aspect for the success of the system.
Its scope defines the relationship between all the activities of our processes and the Human Factors domains and then provides a systematic way to ensure that:

- The human role in the system is defined to optimize human performance in relation to the core system architecture
- Adequate human-equipment analyses and trade-off studies are performed
- Analysis and design support includes the environmental protection necessary to promote health and safety, and the capability for safe operation and maintenance of the core architecture
- System testing and evaluation is conducted to verify that users can safely and effectively operate, maintain and support equipment in its intended environment

Thus, BRAINS conducts Human Factors Engineering by given specific considerations to the following factors:

- Anthropometric factors
- Human sensory factors
- Physiological factors
- Psychological factors

A - Anthropometric factors

All our design requirements involving the human being take into account the physical dimensions of the human body. Both "structural" dimensions (when the body is fixed and in a static state) and "functional" dimensions (when the body is engaged in some physical activity and in a dynamic state) of the local population were considered.
In addition, we consider the possible variations for both males and females in our design.
For example, the Bread fruit processing Peeling subsystem for the family size is powered by a bicycle pedaling system that is adjustable in height and in speed to accommodate both males and females.

B – Human sensory factors

In this category BRAINS relates the HFI to the human sensory capacities, particularly sight or vision, hearing, feel or touch, smell, and so on.
All the breadfruit inspection subsystem interfaces were designed in regard to these factors.

C – Physiological factors

Under these factors BRAINS identify the effects of environmental stresses on the human body during the performance of tasks before engineering the system. In addition, the choice of the sites for collection and implementation were made to minimize their effects. Some typical elements of the physiological factors are:

- high and low temperatures, or temperature extremes,
- high humidity,
- high levels of vibration,
- high levels of noise, and
- large amounts of radiation or toxic substances

To varying degrees, these environmental effects will negatively impact on human performance; that is, physical fatigue will occur, motor response will be slower, mental processes will slow down, and the likelihood of error will increase.

D - Psychological factors

These factors are very important because though the system is designed here, it has to work in an environment completely different from ours. Therefore, BRAINS identifies under these factors all the elements that pertain to the users culture and mind; that is, the emotions, traits, attitudinal responses, and behavioral patterns as they relate to job performance because they generally are the source of the individual's motivations and expectations.

These factors weight heavily in the choice of the collection method of the bread fruit and the determination of the incentives to given out.

All the phases including planning and requirements specification, design and analysis, and test and evaluation of the BrEAD system development were impacted by the HFE. During the planning and requirements specification phases, HFE evaluates predecessor systems and operator tasks; analyzes needs; analyzes and allocates functions; and analyzes tasks and associated workload. In the design phase, HFE evaluates alternative designs using equipment mock-ups and software Prototypes, refines analysis of tasks and workload; and uses modeling tools such as human figure models to evaluate crew station and workplace design and operator procedures. For testing and evaluation, HFE confirms and physically demonstrate that the design and the human-machine interfaces, task sequences, task time and personnel qualifications meet HFE specification requirements; and identifies any undesirable design or procedural features

A detail HFE tasks that were accomplished during the various phases of the BrEAD development are captured in the table HFE – 1 below.

System development phases	HFE Task	Task description and application
Planning and requirements specification phases	Human Factors Program Plan	To develop a human factors program plan which identifies, integrates, and assists in the implementation of all management tasks applicable in fulfilling human factors engineering requirements. This plan includes a description of the human factors organization, organizational interfaces, a listing of tasks, task schedules and milestones, applicable policies and procedures, and projected resource requirements. This plan is tied directly to the System Engineering Management Plan (SEMP).
	Review and control of suppliers or subcontractors	To establish initial human factors requirements and to accomplish the necessary program review, evaluation, feedback, and control of component supplier/subcontractor program activities. Supplier program plans are developed in response to the requirements of the overall Human Factors Plan for the system.

	Human factors program reviews	To conduct periodic program and design reviews at the designated milestones; e.g. conceptual design review, system design reviews, equipment/software design reviews, and critical design review. The objective is to ensure that human factors requirements will be achieved.
design and analysis Phase	System analysis (mission analysis	To determine the overall capabilities and the performance requirements for the system and to develop appropriate mission scenarios identifying basic activity sequences This was accomplished as part of the system requirements definition process in conceptual design
	Functional analysis	To identify the major functions that the s stem is to perform (based on operational requirements), and to, develop functional flow block diagrams defining system design requirements in functional terms. This task must "track" the system-level functional analysis.
	Function allocation	To conduct trade-off studies, evaluate, and determine the resources required in accomplishing the functions identified through the Functional Analysis activity; i.e. determining the 'HOWs" (versus the 'WHATs"), particularly in situations where there are human-machine interfaces.
	Detailed operator task analysis	To evaluate functions that are to be accomplished by the human, and to establish a hierarchical breakdown to the lowest level where a human activity exists; i.**e. job** operation, duty, task, subtask, and task element. Personnel quantity and skill-level requirements are identified through analysis.
	Operational sequence diagrams	To identify the human-machine interfaces, and to develop a sequential flow of information, decisions, and actions through the generation of operational sequence diagrams
	Time line analysis	To select and evaluate critical task sequences, and to verify that the necessary events can be performed and that they are compatible in terms of allocated time; i.e. can the tasks be performed within the appropriate time allotted for accomplishing the mission
	Workload analysis	To evaluate human operator activities throughout a given scenario (or through a number **of** designated scenarios) to determine the workload level; e.g. the relationship between the maximum time allowed and the actual time for task performance
	Error analysis	To systematically determine the various ways in which errors can be made by the human and to make design recommendations to reduce the likelihood of such errors occurring in the future. This

		task is comparable to the reliability FMECA, except that the system/equipment failures are the result of human errors
	Safety analysis	To systematically evaluate, through cause-and-effect analysis, the effects of system/ equipment failures on safety. Although safety pertains to both personnel and equipment, the aspect of personnel safety is emphasized herein. This task was tied in directly with the reliability GECA and the Human Factors Error Analysis.
	Models and/or mock-ups	To develop a three-dimensional physical model or a mockup of the system (or a component thereof) to demonstrate human-machine interfaces, spatial relationships, equipment layouts. panel displays, accessibility provisions for maintenance, and **so** on
Test and Evaluation Phases	Training program Requirem ents	To plan and implement a formal training program. This includes the determination of personnel training requirements (quantity of personnel and the skill levels desired as an output), categories of training, training equipment, training data, training facilities, mockups and models, special training aids, and so on. The plan should include a description of the training organization, a **listing** tasks, task schedules and milestones, policies and procedures, and projected resource requirements.
	Personne l test and evaluatio n	To plan and implement a program to physically demonstrate human-machine interfaces, task sequences, times, personnel quantity and skill-level requirements, the adequacy of operating procedures the adequacy of personnel training, and so on. This test and evaluation activity is accomplished prior to entering production

Table HFE - 1 Human-factors integrating tasks

Attachment 7 – BRAINS Company Profile

Experience & About Us

BRAINS

Bright Robust Architects for the Institute of New Systems (BRAINS) has been a leading global systems engineering firm for 38 years. We engage in research, design, development, and integration of infrastructure systems and services. We are headquartered out of Minneapolis, MN, USA currently employ 10,200 people worldwide. We distinguish ourselves by our passion for originality and initiative. The following is a profile of our company:

BRAINS

Type	Public (NYSE: **BRN**)
Founded	1971
Headquarters	Minneapolis, MN, USA
Area Served	Worldwide
Key People	Sophia Sandoval (CEO) Bruce Tan (CFO) Charles Goldbloom (CTO)
Industry	Systems consulting, technology services, sourcing services
Products	Infrastructure frameworks, Data systems, Transportation systems, Business plans
Revenue	\$5 billion USD (2008)
Operating Income	\$6.25 million (2008)
Net Income	\$5.8 million (2008)
Total assets	\$4.2 billion (2008)
Total Equity	\$1.2 billion (2008)
Employees	10,200 (2008)
Website	www.BRAINS.org

Vision: *To embrace whole-systems thinking, turning extravagant ideas into achievable actions*

Over the years we have developed mature practices that have contributed to our success. We have a strong network of existing suppliers and manufacturers that we partner with to develop infrastructures of all types. We also house a Systems Integration Laboratory (SIL) that includes environmental test chambers (humidity, temperature, etc.), simulation software, and modeling capabilities.

Out past experience has given us a trusted formula to take a large-scale, complex business idea and transform it into a viable product. We work one-on-one with individuals and also with big businesses. We have dedicated, capable employees with the expertise to get a new idea off the ground.

How We Work

We pride ourselves on being a catalyst for unique ideas, and our business is run a similar unique fashion. Because systems engineering firms are not entirely common, we would like to explain in a little more detail how are company works and the processes we go through to create a deliverable.

Step 1: Individuals, businesses, and organizations come to BRAINS with an idea for a business plan

Step 2: BRAINS takes on full responsibility for implementing the business plan, using our existing resources. We utilize our operations, engineering, IT, finance, and marketing departments as appropriate to transform the business idea into life. We form our team around the specific needs of the business plan submitted. We deliver a complete implementation plan detailing the how, who, and what would need to happen to implement the idea.

Step 3: Depending on the specific business plan, we either deliver the implementation plan back to the customer or offer to run the business ourselves. In the first case the customer will own the research and will be given access and contact information to all the resources and companies needed to carry out their plan. In the latter case BRAINS will offer to carry out the plan and co-own the business.

Program Plan

We offer to take NGRF's request for proposal and set up a complete solution for creating a Breadfruit Enterprise Architecture Demonstration (BrEAD) in the Marshall Islands. Our existing resources include numerous partnerships with supply chain companies throughout the world. We have contacts with the preferred supply chain companies in each region.

We are well-versed in the DoDAF systems engineering processes, which provide us with a structured plan to manage complicated and large-scale projects. We have taken on the BrEAD proposal using this method. This ensures we develop our ideas in an iterative way, looking at the design from multiple views. When our design is finalized we also provide verification methods and a validation process to ensure we have met the needs of the customer entirely. If carried through to implementation, we employ lean six sigma methodologies to guarantee conservative cost profiles and to maximize efficiency.

We are confident that we have constructed the optimal solution using this method for the BrEAD. If chosen, we will utilize our finance department to manage accounting, our operations department to handle logistics and sourcing, and our marketing department to become a strong advocate for the breadfruit.

Something new to this program will be the materials and suppliers needed to create the wind turbines. We will be working with a new supplier local to the Marshall Islands, <insert supplier name here> to provide the materials. We also will be managing orders for wind turbine supplies for general purposes on the Marshall Islands once we establish our presence.

Case Studies

The following examples showcase some of past experience in systems consulting and implementation.

P.B. Loco used BRAINS to create their first peanut butter restaurant in 2003 which now has franchising opportunities across the nation and a complete online inventory

Elliot Bay Book Company out of Seattle used BRAINS to create a real-time inventory of books in stock and delivery tracking method for books purchased online.

The United Nations selected BRAINS to aid in the 2004 Tsunami Relief efforts. BRAINS provided logistics for temporary shelter and food distribution for victims.

Past Performance

The following is a list of some of our recent additions to our growing portfolio:

- Our outsourcing provides the Minnesota Department of Transportation (MNDOT) with their IT transit network
- We set up a temporary camp and provided transportation routes to aid in the 2004 Tsunami Relief efforts

- Our database of packaging companies allowed us to prepare and deploy $5 million dollars worth of food packages within a single month in the hurricane Katrina recovery efforts

- Our infrastructure capabilities were selected by Google to aid in *Project 10 to the 100th*, bringing to fruition a new water carrier for underdeveloped countries in 2006

- Our company landed the number 14 spot on Business Week's top 100 place to work in 2007, followed by the 11th spot in 2008

BRAINS has been lauded for our past performance in the press. The following are quotes from the article featuring BRAINS in *Business Weekly*, November 2008:

> *"I have never experienced such a seamless transition between my pie-in-the-sky idea to my small business. Thank you, BRAINS, for making this dream possible!"*
>
> *-Dan Zachmann, Chartt Industry founder*

> *"Wow! You guys really know your stuff! I would have never been able to find these resources on my own! My business is booming! Here's to living the good life!"*
>
> *-Chloe Black, IT & Beyond founder*

Attachment 8 - Cost Analysis

1) **Assumptions:** Since BRAINS have not carried out any experiments with Breadfruit Flour Production yet, the cost analysis involves several assumptions as listed below

a. **Plant Construction Phase:**

i. **Land for Processing Unit**: One plant will need half acre. (Including processing u storage, wind mill, facilities etc.)

ii. **Land for collection points:** Since there will be no storage required at collection po that will be just like a meeting place and arranged by chiefs

iii. One Engineer is needed full time during plant construction. (One Engineer per plants)

iv. **Time needed for construction:** approx. 15 days, however engineer's work will be for supervision(approx. 12 days; 96 man-hours; 48 hours/plant)

v. **Labor needed during plant construction:** 600 man-hours (5 people, 8 hours/day days)

vi. Wood for construction will be arranged locally, so long distance transportation nee for processing equipments only

vii. **Travelling Cost for BRAINSs' employees:** $ 3000 lump sum

b. **Breadfruit Production Phase:**

i. **Total Manpower for One Processing Unit:**

- Labor: 4 (1 each for peeler, slicer, dryer, grinder)
- Supervisor: 1 (full time. He will supervise & allocate work, coordinate mainten transportation arrangements, to & fro communication with NGRF); 1 / 6 Plan

ii. **Max. Output per day:** approximately 240 pounds (to get 240 lbs. output, weight of BF needed= 800 Pounds)

iii. **Working Hours:** Plant will operate for 6 days/week, 9 hours/day (considering short availability of BF)

iv. BRAINS will make arrangements of Radio Phones(for communication between pl NGRF will provide satellite phone for all long distance communication & data tra needs

c. **Rates:**

- Payback Period: 10 years
- Cost of Land ($): 4000/acre/year
- Transportation of Material ($): 2500 lump sum
- Labor Rate ($): 2 /hour
- Engineer's Hourly Rate ($): 18 /hour
- Plant Supervisor: salary($) 40000 /year
- Breadfruit buying Rate ($): 0.15 /lb
- Incentive ($): 0.1 /lb

2) Cost Calculations:

a. Capital Cost

Sr.No	Item description	Quantity	Unit	$per Unit	Total ($)
1	Dryer	2	Nos.	150	300
2	Grinder	1	Nos.	4900	4900
3	Windmills	6	Nos.	150	900
4	Backup power generator	1	Nos.	5000	5000
5	Peelers	4	Nos.	15	60
6	Slicers	4	Nos.	25	100
7	Storage Shelves	3	Nos.	150	450
8	Fire Extinguishers	2	Nos.	20	40
9	Construction Material (including wood)	1	lot	5000	5000
10	Electrical Installation & wiring	1	lump sum	1000	1000
11	Transportation (lump sum)	1	lump sum	2500	2500
12	Desktop Computer	1	Nos.	250	250
13	Labor for Construction	600	hours	2	1200
14	Engineer for supervision	48	hours	18	864
15	Wireless Radio Phone System	1	Unit	100	100
16	Travelling	1	lump sum	500	500
17					0
18					0
19					0
20					0
				Total ($)	**23,164.00**

- Annual Maintenance Contract ($): 8000 lump sum
- Transportation raw BF & BF flour ($): 12000 lump sum

b. Cost of Proposal

	Price ($)
Capital Cost for 1 Plant	23,164.00
Total Number of Plants	6
Total Cost for 6 Plants	138984.00

With other Miscellaneous charges	152882.40
Total Capital Cost (Engineering + Construction)	**153,000.00**
Base Fee	**100,000.00**
Total Price to be quoted in Part-A of Proposal	**$ 253,000.00**

c. Monthly Expenses

Workdays per month	26	days
Breadfruits collection per day	800	lbs
Hours of Operation/day	9	hours
labor	4	people
Daily Production	240	lbs
Monthly Production	6240	lbs

Sr.No	Description	Cost($)
1	Capital Cost Share/month	0
2	Land Lease Share/month	0
3	buying BF ($ 0.15/lb)	3120
4	Incentives to locals	0
5	transportation cost share/month	166.67
6	Maintenance cost share/month	111.11
7	labor (9 hours/day; 26 days; 4 people)	1872
8	supervisor's salary share/month	555.55
9	misc. expenses	100
10		
11		
12		
13		
	Total Monthly Cost	5925.33

Total Monthly Cost for Plant Operation ($)	5925.33	
Maximum Monthly Production	6240	lbs

Production Cost of BF flour/ lb	**$ 0.95**

d. Notes:

i. The cost analysis mentioned above does not include packaging cost, as that will vary according to size of package bag. That cost will get added to production cost.

ii. The expenses for marketing have been considered totally in NGRF's scope and they are not included in any of the categories mentioned above.

Acronyms List:

Term	Definition
BF Capture	Breadfruit Capture Team
BrEAD	Breadfruit Enterprise Architecture Demonstration
BRAINS	Bright Robust Architects of the Institute of New Systems
COTS	Commercial Off-The-Shelf
CPT	Cross product Team
DFM	Design for Manufacturing
DOD	Department of Defense
DOORS	Dynamic Object Oriented Requirements System
EIPT	Enterprise Integration Product Team
FIT	Fault-Isolation Test
FMEA	Failure, Modes, and Effects Analysis
FoS	Family of Systems
GB	Giga-Byte
I&T	Integration and Test
I/O	Input/Output
ILS	Integrated Logistic Support
IPT	Integrated product team
IMP	Integrated Master Plan
IMS	Integrated Master Schedule
MOE	Measure of effectiveness
MOP	Measure of performance
NGRF	New Green Revolution Foundation
NQMS	NGRF Quality Management System
OSA	Open Systems Architecture
PC	Personal Computer
RAM-T	Reliability, Availability, Maintainability - Testability
SDD	System Development and Demonstration
SDP	Software Development Plan
SE	Systems Engineering
SEMP	Systems Engineering Management Plan
SEP	Systems Engineering Process
SIL	System Integration Lab
SMP	Subcontract Management Plan
SOW	Statement of Work
SRR	System Requirements Review
SRS	Software Requirements Specification
SSD	Systems and Services Development

Term	Definition
SSDD	System/Subsystem Design Description
WBS	Work Breakdown Structure

Contributors

Volume Editors – Danielle Eastberg, Kyle Dahlstrom
Program and Schedule – Danielle Eastberg
Proposed SOW and CRDLS – Aditya Srynath
Integrated Master Plan – Danielle Eastberg
BRAINS Organizational Structure – Khuong Vu
System Node Operational View 2 – Edem Kavi
Integrated Master Schedule – Danielle Eastberg
Systems Engineering (SE) Management – Khuong Vu
Risk Management Plan – Subodh Deodhar
Subcontract Management – Will Tamayo
System Integration Lab – Edem Kavi
Software Development Plan – Edem Kavi
Annex C – CWBS – Leland Smith
Annex D – SOW – Aditya Srynath
Annex E – IMP –Danielle Eastberg
Annex F – IMS – Danielle Eastberg
Annex G – SEMP – Khuong Vu
Annex H – Software Development Plan – Edem Kavi
Attachment 4 – Marketing – Kyle Dahlstrom
Attachment 5 – Training – Edem Kavi
Attachment 6 – Human System Integration Plan – Edem Kavi
Attachment 7 – BRAINS Company Profile – Danielle Eastberg
Attachment 8 – Cost Analysis – Subodh Deodhar

Proposal 2: Solutions International Proposal

System Development and Demonstration of the Breadfruit Enterprise Demonstration (BrEAD) and Systems Design

Solutions International Proposal for the Breadfruit Enterprise Demonstration and System Design
Volume 1: Executive Summary

RFP NO: E00010-07-R0001
Copy 1 of 2
May 1, 2009

Prepared For:

New Green Revolution Foundation (NGRF)

Prepared By:

Solutions International
123 Pillsbury Drive
St. Paul, MN 55114

CAGE CODE SI4ME

TABLE OF CONTENTS

TABLE OF FIGURES

Offeror Summary Table

Solutions International's Offeror Summary Table not only identifies our subcontractors, but also identifies our partners required for program success. Our proposed solution relies on the cooperation of subcontractors, the Breadfruit Institute, and Local Aid Organizations at the proposed processing system sites.

Contractor Name	**Place of Performance**	**Brief Work Description and/or Program Responsibility**
CTI / University of St. Thomas Students (Subcontractor)	St. Paul, MN	Provide processing system equipment
Inmarsat (Subcontractor)	Headquarters in United Kingdom, with support in numerous countries	Provide satellite communication services
Onset (Subcontractor)	Cape Cod, Massachusetts	Provide support for data logger components
Breadfruit Institute (Teammate)	National Tropical Botanical Garden in Hawaii	Provide location and resources for our Systems Integration Lab (SIL)
Dr. Carolle (Teammate)	Haiti	Provide Point of Contacts in Haiti to facilitate our system integration and educate us on cultural norms.
Father Reiser Relief (Teammate)	Haiti	Provide Point of Contacts in Haiti to facilitate our system integration and educate us on cultural norms.

Technical

Our proposed solution creates a Breadfruit Enterprise Architecture Demonstration (BrEAD), provides a method for underdeveloped countries to utilize an unused food resource, and provides a venue to further research and development. Equally as important, our BrEAD architecture provides a solution to boost the local economy in these undeveloped countries. Solution International's goal is to provide feasible processing and Enterprise Architecture systems that can be incorporated in any region in the world.

Figure 1-213 illustrates our vision at the highest level. Our headquarters, also known as the Central Data Center (CDC), is located in St. Paul, MN. Our BrEAD implementation plan calls for the creation of Regional Centers wherever the breadfruit processing systems are installed. Multiple processing sites, located in local villages and communities, can report to one Regional Center. Finally, our SIL is located at the Breadfruit Institute in Hawaii.

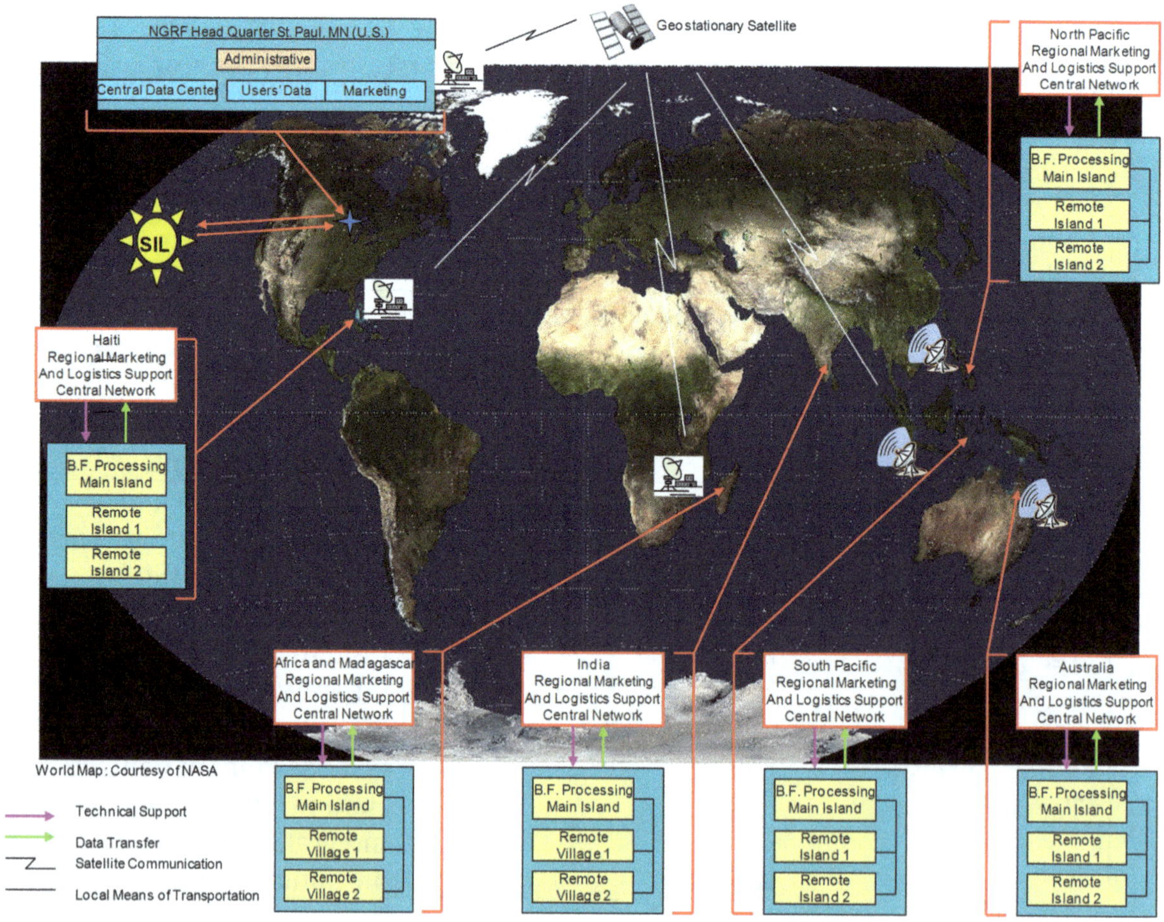

Figure 1-213: Solutions International's BrEAD System. *Our concept allows flexibility for each breadfruit region, providing a feasible and robust system.*

Our solution builds flexibility into our communication systems, allowing for the processing sites to be located at various distances from the Regional Center. The relationship between the Regional Center and the processing site is shown in **Figure 1-214**.

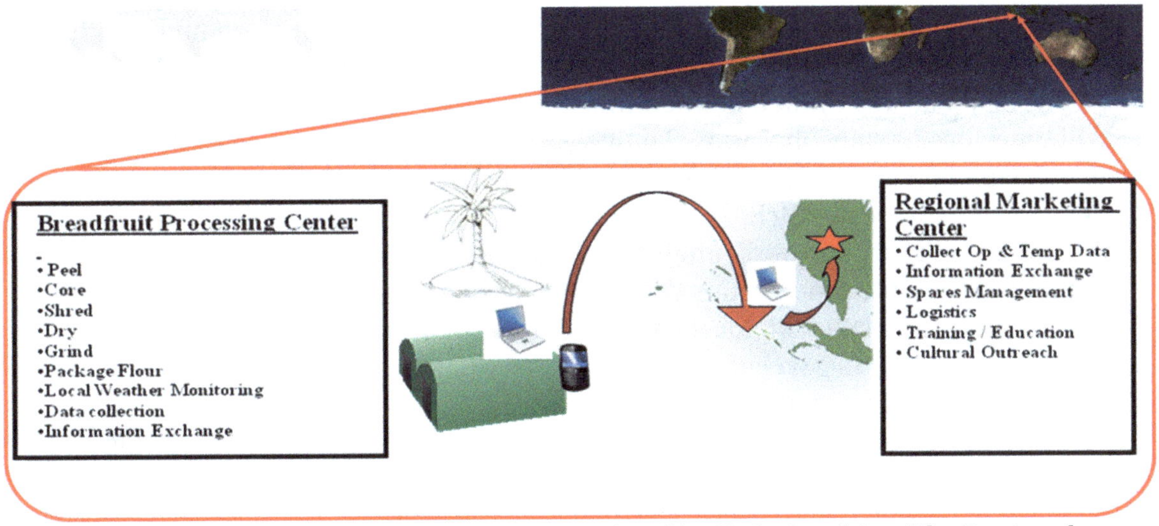

Figure 1-214: Regional Center and Processing Site Relationship. *The Regional Center and Processing Site have different focuses, but together they form the complete solution for breadfruit processing system integration in local communities.*

We propose to install a prototype processing system in a beta site, located in Haiti. The Regional Center will be located in Port Au Prince. Dr. Carolle and Father Reiser provided us with local contacts in Haiti to facilitate our installation. This prototype will be released prior to project completion, so that we can receive feedback on its function and operation. This allows us time to enhance the processing system prior to production.

In order to properly fund the Breadfruit Enterprise, Solutions International has designed a strategic marketing plan. This marketing plan contains four key areas in which Solutions International will spread the word about the cause and generate revenue for this non-profit venture. By using a website, brochures, targeting the gluten-free market, and using established resources to obtain government grants, Solutions International has positioned itself to be a prime selection for this contract.

Subsystem Descriptions

Our proposed system design is based on preliminary modeling and simulation we have already performed. Based on our initial research on available breadfruit quantities in Haiti, we are proposing a system that is capable of processing 450 breadfruit a day, with an expected yield of 295 kg of breadfruit flour per day. We are assuming the breadfruit harvest season is 3 weeks, but we would need processing system staff for about 2 months. By employing a minimum of 33 people to run the processing site for two months, we can produce enough flour to feed 47,700 people (assuming 0.5 lbs/person/day) or 130 people for an entire year.

Regional Center

The Regional Centers will communicate with and support the processing sites. These centers will act as communication, support, and coordination nodes in the BrEAD enterprise architecture. Since breadfruit harvesting is only possible for any given region during limited defined times throughout the year, the Regional Centers will therefore be fully operational only during those times when breadfruit processing systems are deployed and active. Rather than staffing the Regional Centers with dedicated BrEAD personnel all year round, local aid organizations will be tapped for part time personnel. In addition to being less expensive, part-time personnel from local aid groups have the following advantages:

- Familiarity with local culture, mores, local contacts
- Social and governmental contacts
- Familiarity with local languages
- Possible use or lease of existing facilities/communications equipment
- Possible integration with existing plans/goals of other aid groups

Beyond providing local relations, the Regional Center provides logistic support for the processing system. Spares for all processing system components are stored at the Regional Center. In addition, the Regional Center provides the option for processing systems to be leased or purchased by the local community. If the lease option is chosen, then following the breadfruit season, the local community can return the processing system to the Regional Center. This option allows flexibility in our design. A single processing system can be used in multiple regions of the world.

Following the breadfruit season, the Regional Center can be disassembled and put into safe storage. For efficient allocation of resources, the Regional Center equipment can be sent to another region for set up and operation there.

The Regional Center collects performance data from the processing data site and from the weather station system deployed nearby. The Regional Center passes this information on to the Central Data Center for processing.

Processing Data Site

The breadfruit processing system contains the equipment necessary to convert breadfruit into a sustainable resource (flour). Our processing system receives breadfruit, cores and peels the breadfruit, shreds and dries the breadfruit, grinds the breadfruit into flour, and finally packages the flour. In addition, we have proposed a method for managing the breadfruit waste.

One month prior to breadfruit harvesting season, we will construct the breadfruit processing site, train processing system staff, and advertise for fresh breadfruit. Breadfruit collection and breadfruit harvesting is outside the scope of the processing system. We expect locals (who may or may not be working for us) to harvest healthy, ripe breadfruit and deliver it to us for a fair price. Depending on the geographic location, this price may change. We will use our established local resources to decide exactly what this fair price is, based on the prevailing social and cultural norms. Our processing system provides an additional revenue source for the locals. By paying locals to bring us breadfruit and by paying the locals who work the processing system, we are not only creating an additional food source, but we are also boosting the local economy.

Since breadfruit spoils after only 48 hours and the breadfruit harvesting season is only a few weeks long, there is a hard time constraint on our processing system productivity. Because the time constraint is based on the fruit not spoiling (spoiling is related to water content), once the fruit has been dried out, the time constraint is no longer operative. Our design strategy is therefore focused on maximizing productivity by getting as much breadfruit to the drying stage as possible before proceeding to later steps.

After the breadfruit harvesting season, the processing system staff will grind and package all the dried, shredded breadfruit. This approach provides additional opportunities for the locals and, again, will boost the local economy. Once all the breadfruit is processed into flour, we will clean the processing system, pack the system up, and return the system to the Regional Center until the next breadfruit processing season.

Central Data Center

The Central Data Center, located at our Headquarters in St. Paul, Minnesota, collects all the data from the various Regional Centers, processes and analyzes the data, provides an avenue for researchers to use our data, and provides all the administrative support. Our Central Data Center will create a website that allows researchers, local governments, and local aid organizations to access our data. The Central Data Center has IT and software engineers that will maintain all the servers and develop the software. Our system has remote access capabilities for future software upgrades.

Systems Integration Lab (SIL)

Our SIL has a dual purpose: 1) Provide capabilities to model and simulate our system, verify our system requirements, and identify improvement opportunities for the processing system; 2) Produce breadfruit flour for sale to the gluten-free market. Since our SIL is located in Hawaii and partnered with the Breadfruit Institute, we will have access to their 200 different kinds of breadfruit. This allows us to robustly test our processing design as well as make it adaptable to the many different variations of breadfruit.

Because the Breadfruit Institute has over 200 different varieties of breadfruit, there is a constant supply of breadfruit and we are no longer constrained by a limited breadfruit season. This is not only valuable for our testing and verification, but also for our marketing strategy. Our proposed plan is to process the breadfruit into flour all year long and distribute in the US to the gluten-free market. There is a strong market for individuals who are allergic to gluten and as such cannot use wheat flour.

We will make our breadfruit flour available for purchase via our website maintained at the Central Data Center. The money we earn from our gluten-free sales will be used to further develop processing systems, to install processing systems in new communities, and to increase education about our mission and its benefits to communities in breadfruit producing regions.

As mentioned earlier, the SIL will perform modeling and simulation of the system and its performance. However, our SIL will also perform economic and cultural modeling of potential consumer markets, potential Regional Center sites, and potential processing system sites. Solutions International understands that each region will have its own cultural and social norms. In order for our proposed solution to be successful, we have to acknowledge and understand those cultural norms. For that purpose, our proposal includes the establishment and operation of a Cultural Norms Working Group.

System Development and Demonstration of the
Breadfruit Enterprise Demonstration (BrEAD) and Systems Design

Solutions International Proposal for the
Breadfruit Enterprise Demonstration and System Design
Volume 2: Technical
Book 1: Design Approach

RFP NO: E00010-07-R0001
Copy 1 of 2
May 4, 2009

Prepared For:

New Green Revolution Foundation (NGRF)

Prepared By:

Solutions International
123 Pillsbury Drive
St. Paul, MN 55114

CAGE CODE SI4ME

Table of Contents

Table of Figures

Design Approach

Third World countries lack means to diversify and increase their crops yield in order to meet the needs of their exponentially increasing population. Consequently, the death toll due to hunger especially among children is of all time high in these regions. Breadfruit trees are fast growing plants in favorable conditions with a remarkable high yield (350 to 1100 lbs fruit per tree per year in intensive cultivation). Although breadfruits can significantly help reduce hunger in these countries, the high poverty rate makes it more difficult for the population of this part of the world to obtain the basic infrastructure necessary to quickly process and/or extend breadfruits' shelf life. Tropical islands such as Marshall Islands and Haiti are among these countries.

To decrease the impact of hunger on the local populations, preliminary research work done by Dr Camille George's undergraduate students of the University of St Thomas, St Paul, led to the design of breadfruit processing system with the intent to extend the breadfruit's shelf life. The concept is to processing fresh breadfruit into flour, which has an extensive shelf life. The design approach of the Breadfruit Enterprise Architecture Demonstration (BrEAD) is to use preliminary research and prior breadfruit processing system prototypes to design an intuitive, reliable, modular and scalable breadfruit processing system. The breadfruit processing system's main functions will be to perform functions such as peeling, coring, shredding, drying, grinding into flour, and packaging. Our Enterprise Architecture system (including the processing system) will have the ability to collect and store local environmental and operational data. This data can be used by regional marketing research centers and to the New Green Revolution Foundation (NGRF) headquarter in St. Paul Minnesota. Our design approach offers options and flexibility, making it marketable to any region in the world.

Overall Design Approach

Because of the wide range of the system's capability highlighted in the PBSS requirements, the overall design approach of the Breadfruit Enterprise Architecture Demonstration (BrEAD) is based on:

- The V-diagram, **Figure 2.1-215**, of the DoD Architecture Framework to address the design of the individual systems such as modules of the Breadfruit Processing System
- BrEAD's overall design approach is also based on the DoDAF views linkage diagram, as shown in **Figure 2.1-216**, to explicitly address the Breadfruit Processing System operational and Technical needs, and the needs of the system as a whole in order to identify tasks and standards to be used.
- The DoD Architecture Development Process, **Figure 2.1-217**, to address a method for developing information technology that meets DoDAF requirements to support the breadfruit processing system. In that matter, SI follows each step to:
    a. Determine the intended use of the architecture
    b. Determine the overall system architecture scope
    c. Determine data required to support the architecture development
    d. Collect, organize and correlate systems architecture data
    e. Conduct analyses to support architecture objectives
    f. Document result in accordance with the architecture framework

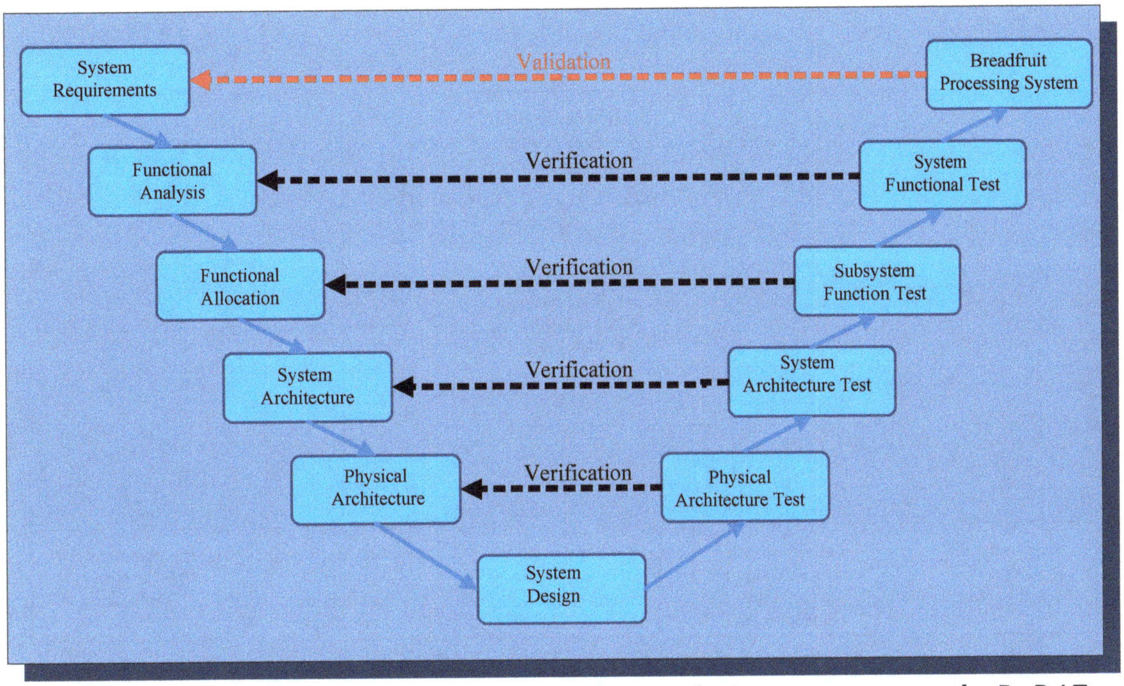

Figure 2.1-215: DoDAF V-Diagram. *Our system design encompasses the DoDAF methodology, proving our design to be robust.*

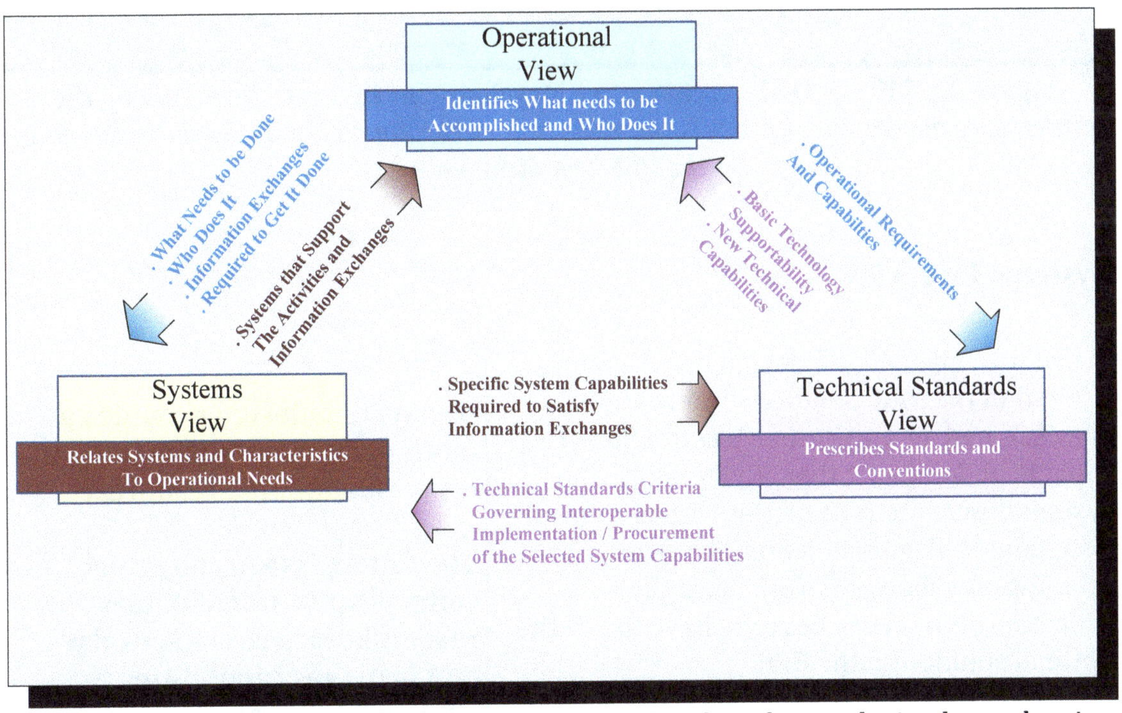

Figure 2.1-216: DoDAF 3-View Approach. *We developed our solution by evaluating all three views, creating a comprehensive solution.*

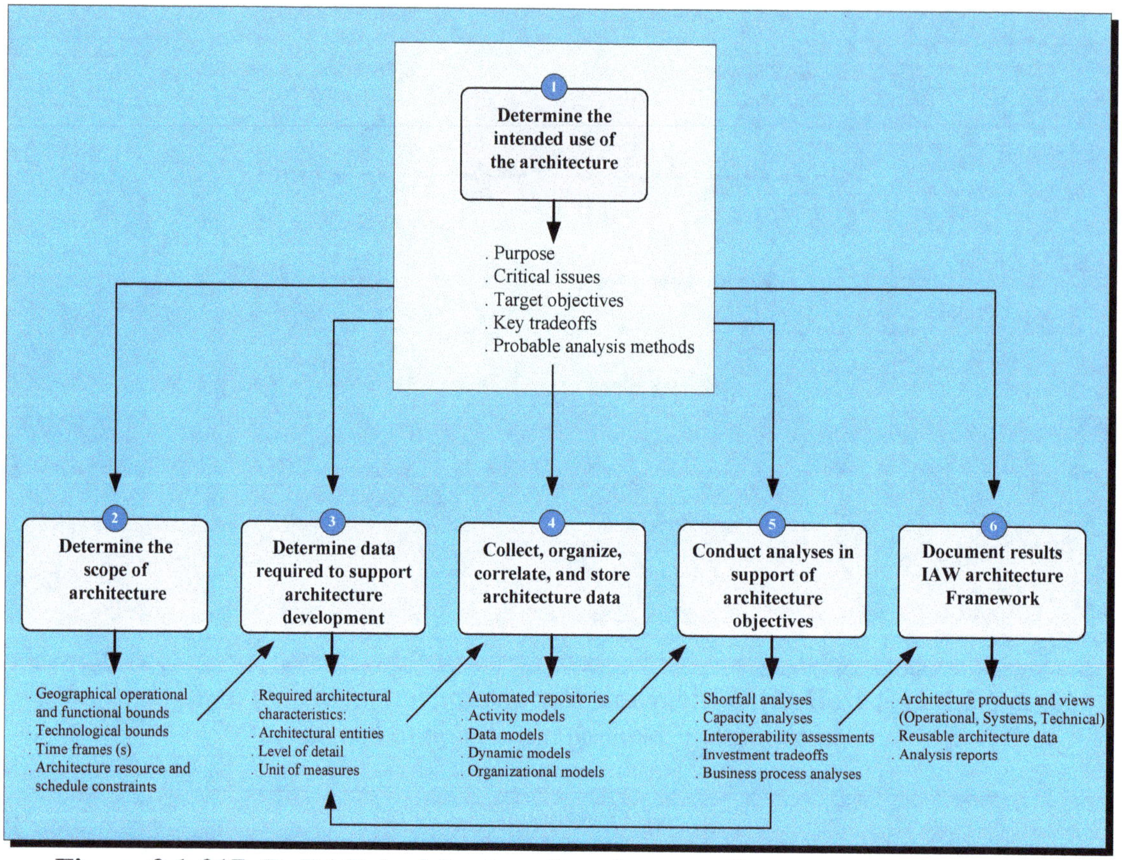

Figure 2.1-217: DoDAF Architecture Development Process. *By following the architecture development process, we were able to design a complete solution, meeting the PBSS requirements.*

Systems Overview

In order to meet all the design requirements laid out in the RFP, Solutions International has designed an Enterprise Architecture (EA) capable of doing just that. By breaking the system down into four distinct areas, Central Data Center, Regional Center, Systems Integration Laboratory (SIL), and the Processing Site, we believe we have found the optimal solution to the RFP.

Our system design provides a method for turning breadfruit into a sustainable resource.

The Enterprise Architecture collects operational data, allows researchers access to our data, and provides capabilities to evaluate cultural norms in potential installation sites.

Figure 2.1-218 shows the interactions between the different areas of the enterprise architecture. Each major subsystem is described in more detail in following sections.

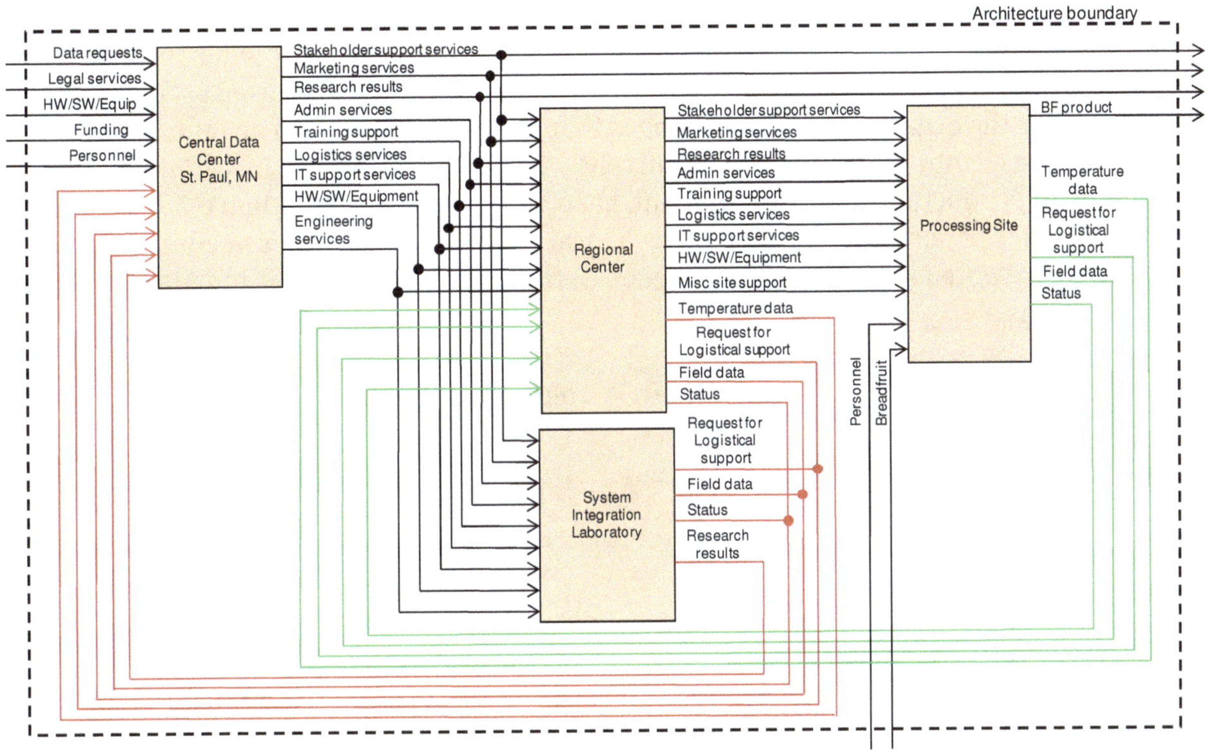

Figure 2.1-218: Breadfruit systems enterprise architecture. *Our Enterprise Architecture is defined, illustrating the relationships between the major subsystems.*

The Central Data Center, located in St. Paul, is the mainframe of the whole enterprise. Here is where the personnel, funding, hardware, software, equipment, legal services and the website are all run from. The central data center will provide the Regional Center and the SIL with everything that they need to run, day in and day out.

The Regional Center is the middle man of our enterprise architecture. The Regional Center will provide all the smaller processing sites with their equipment, hardware, software, and training needed to run. As well as providing the processing site with their needed items, the Regional Center will relay the information collected at each processing site back to the Central Data Center in a timely manner.

The Processing Sites are remote locations, where breadfruit is converted into flour. The processing systems can be either leased or purchased by the communities in which they are used. The Processing Site block, as shown in **Figure 2.1-218**, also contains the Weather Station, the BrEAD Measurement System, and the Communication System. These systems provide the Regional Center with the following data: operational performance and weather station data. This block also represents the communication system which provides means to communicate between the processing site, regional center, and central data center.

The SIL is Solutions Internationals research facility located in Hawaii. In addition to providing test, evaluation, modeling & simulation capabilities, it will have a medium scale processing system. The flour processed here will then be used as a marketing tool for the organization. The changes that are made at the SIL can later be integrated into the units that are currently out in the field or be incorporated into future productions of the equipment, hardware and software. **Figure 2.1-219** shows the types of personnel needed at each location. The specific roles needed at each area are noted in the figure but broken out in more detail in their individual sections.

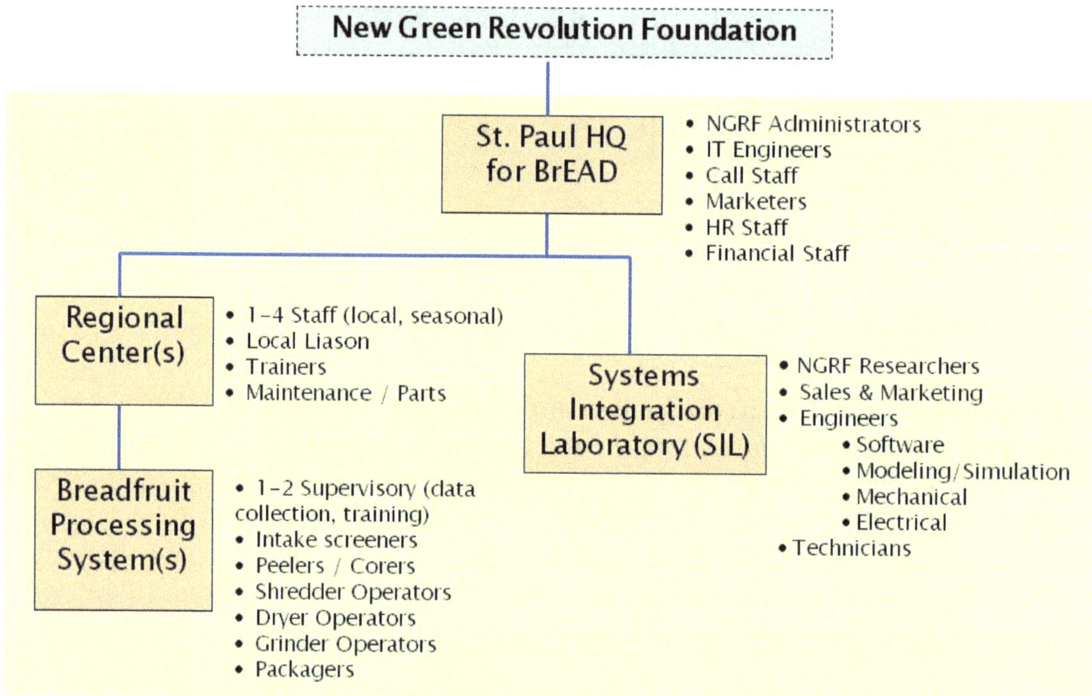

Figure 2.1-219: Enterprise Architecture Organizational Chart. *The breadfruit organizational chart shows the personnel needed in each area of the Enterprise Architecture.*

Breadfruit Processing Site

Our processing system has the capacity to process 450 breadfruit/day, yielding 295 kg of flour/day.

Our proposed solution includes a breadfruit processing system located at a Breadfruit Processing Site, where the Processing System converts fresh breadfruit into flour, collects performance data, and manages breadfruit waste. The Processing Site, primarily run by the local community, can be located in a village and/or near a breadfruit tree grove. It is divided into two major subsystems: Processing subsystem (converting breadfruit into flour) and the Waste Management System.

Each processing site will have a supervisor, who has overall responsibility for all processing system functions. This includes: processing equipment operation, safety of personnel, data collection, and waste management. **Figure 2.1-220** illustrates our anticipated organizational chart for the Processing Site. Initially, the supervisor will be a Solutions International employee, but as the processing site matures and acceptance grows in the region, the local region will take all responsibilities for the site.

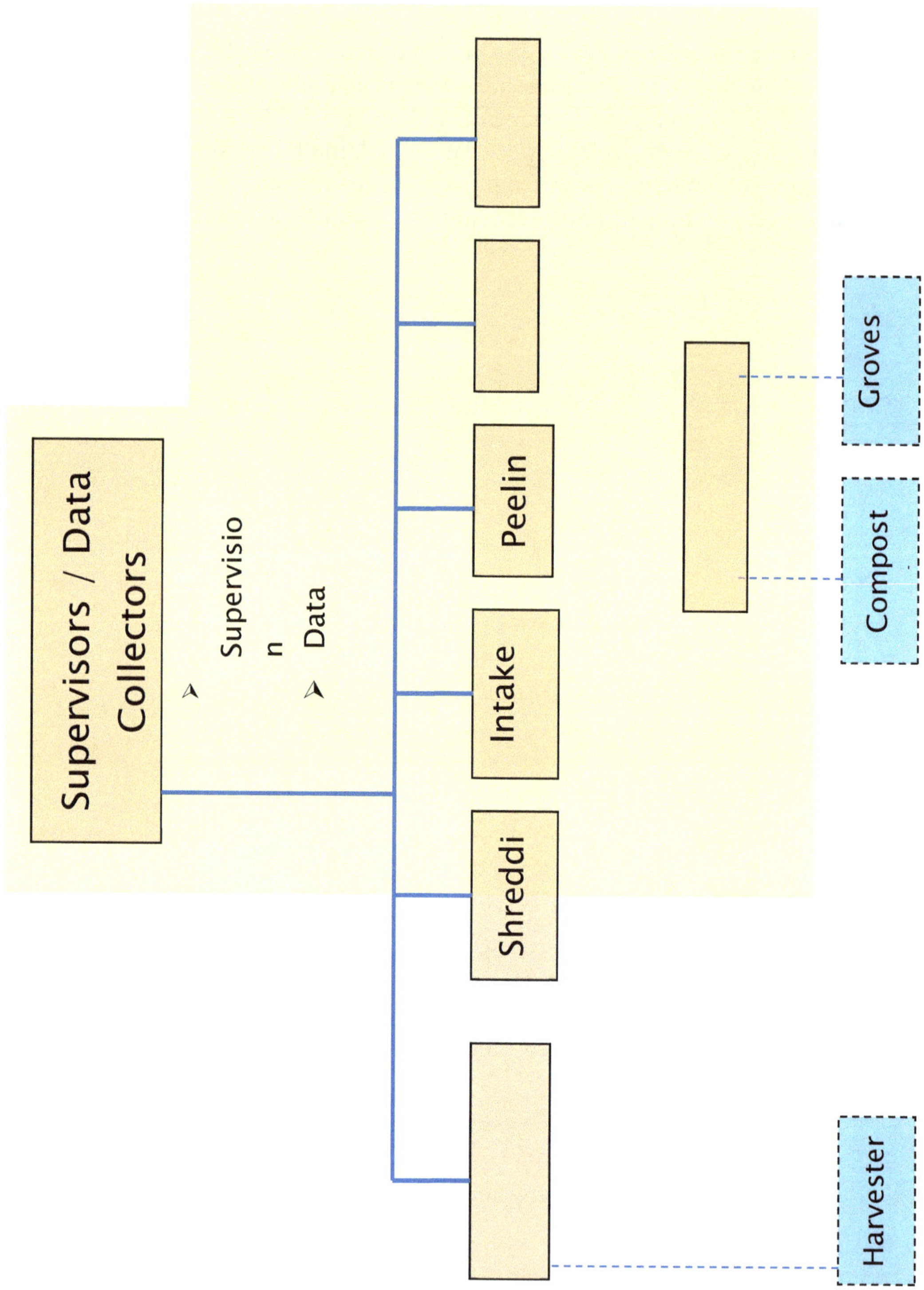

Figure 2.1-220: Processing Site Organizational Chart. *Our staff is defined for the processing site, where the blue boxes indicate activities not managed by the supervisor, but are necessary for proper breadfruit operation.*

The breadfruit processing system is only needed during breadfruit harvesting season, which is three to six weeks out of the year. Thus, our design allows the processing system to be mobile, enabling its use at multiple locations or to be transported to safe storage facilities. One month prior to breadfruit harvesting season, we will construct the breadfruit processing site, train processing system staff, and advertise for fresh breadfruit. Breadfruit collection and breadfruit harvesting is outside the scope of the processing system. We expect locals (who may or may not be working for us) to harvest healthy, ripe breadfruit and deliver it to us for a fair price. Depending on the geographic location, this price may change. We will use our established local resources to decide what exactly this fair price is based on the social and cultural norms. With our processing system design, we have added an additional revenue source for the locals. By paying locals to bring us breadfruit and locals who work the processing system, we are not only creating an additional food source, but we are also boosting the local economy.

Since breadfruit spoils after only 48 hours and the breadfruit harvesting season is only a few weeks long, we understand the importance of having a timing constraint on the processing system. Therefore, we have designed our processing system to dry 450 breadfruit per day, based on modeling calculations of a typical processing site in Haiti. Because the timing constraint is based on the fruit not spoiling, once the fruit has been dried out, the timing constraint is no longer active.

After the breadfruit harvesting season, the processing system staff will grind and package all the dried, shredded breadfruit. This approach provides additional opportunities for the locals and, again, will boost the local economy. Once all the breadfruit is processed into flour, we will clean the processing system, pack the system up, and return back to the Regional Center until the next breadfruit processing season. **Figure 2.1-221** illustrates the components of the processing system.

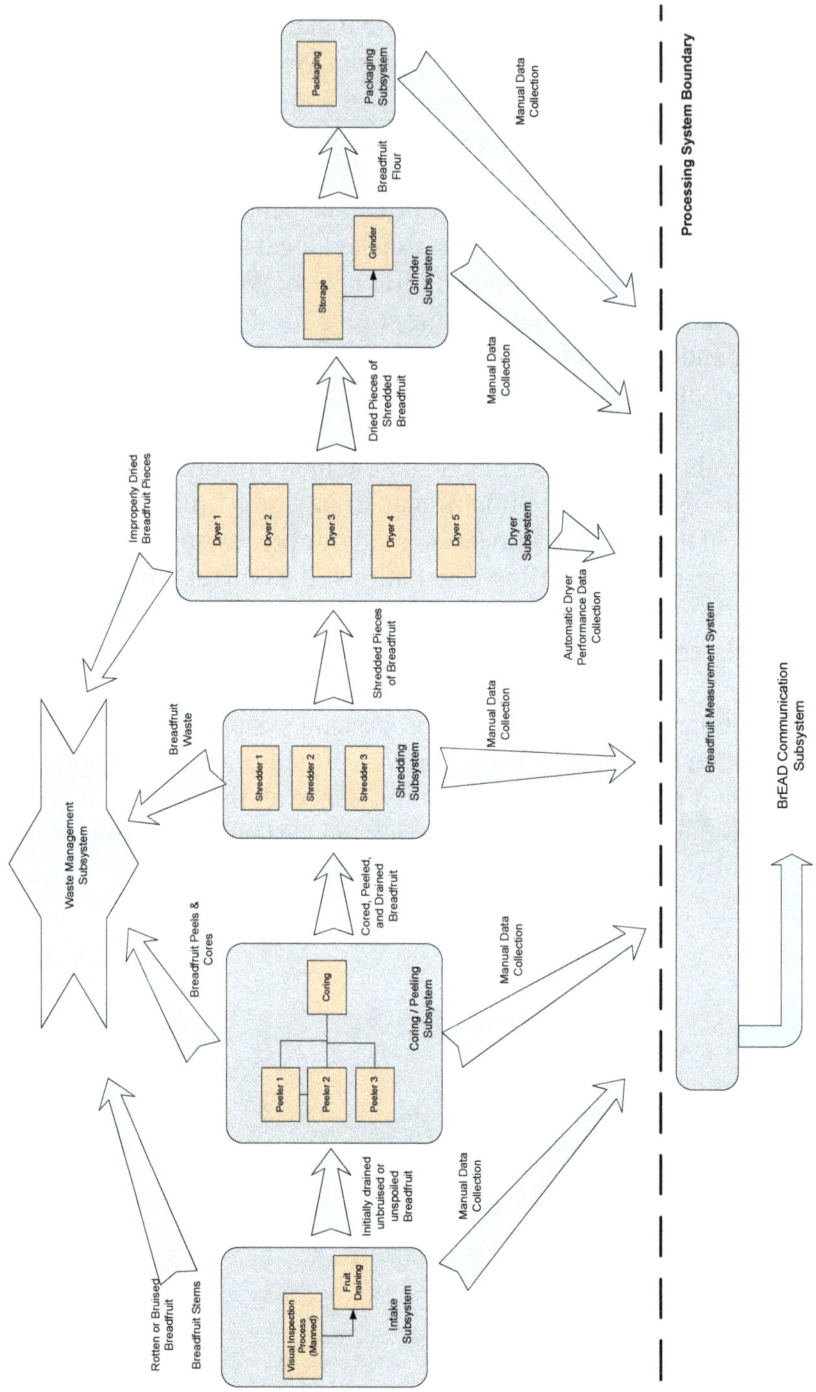

Figure 2.1-221: Breadfruit Processing System. *Our Processing System is simplistic and largely manually operated, allowing quick acceptance and integration into most cultures and geographic locations.*

Processing System

The processing system contains six subsystems, as shown in **Figure 2.1-221**: Intake subsystem, Coring/Peeling Subsystem, Shredder Subsystem, Dryer Subsystem, Grinder Subsystem, and Packaging Subsystem. Each processing system is maintained by the processing system supervisor. This supervisor reports to the Processing Site supervisor. The processing system supervisor has the following responsibilities:

- Hire local personnel for processing system operation
- Train personnel on system operation
- Provide maintenance and logistics field support
- Ensure the processing system operates smoothly

To determine the size of our workforce and the number of subsystems required, we modeled a daily timeline for the processing system. Assuming we have 450 breadfruit at the start of the day, we will need 3 coring/peeling stations, 3 shredders, and 5 dryers. With the breadfruit ready for draining at 5 am, it is possible that the first dryer load is ready by 12:15 pm. Staging these every 30 minutes will leave enough time in the heat of the day to dry all 450 breadfruit by 7:15 pm, as shown in **Figure 2.1-222**. More detail can be found in Annex A, section 4.3.

Time of Day	Action
5:00 AM	Cut off top of breadfruit
7:00:00 AM	450 breadfruit tops cut off
7:15:00 AM	Processing Begins
12:15:00 PM	Dryer #1 full - ready for drying
12:45:00 PM	Dryer #2 full - ready for drying
1:15:00 PM	Dryer #3 full - ready for drying
1:45:00 PM	Dryer #4 full - ready for drying
2:15:00 PM	Dryer #5 full - ready for drying

Figure 2.1-222: Breadfruit Processing System Timeline. *Our initial modeling determined 450 breadfruit can be processed daily, enabling our system to handle large quantities of breadfruit.*

The BrEAD Processing system is comprised of components assembled from Commercial Off The Shelf (COTS) materials and COTS components themselves. Our design leverages the work of University of St. Thomas engineering students Kara Torgerson, Ross McGruder, Thomas Mauritzen, Jason Emiliusen, Cody Benoy, Tony Caruso, Jake Thomas, Rachel Willenbring, and Aaron Brooks. Their team designed and enhanced previous designs of the processing system components. The designs are simple in nature, but provide an effective solution to meet the PBSS requirements. All components of the processing system are self-sustaining through use of solar or hand power. Due to these reasons, Solutions International chooses to use the designs as is, with the exception of the dryer. As discussed later, our approach enhances the current dryer design. As an additional note, our processing system can be altered with minimal change to use electric power or natural gas, when available. This feature would be a future enhancement.

The six subsystems of the processing system are discussed in detail below.

Intake Subsystem

The breadfruit enter the processing system through the Intake System. The subsystem contains the following components: a scale, counter, visual inspection guide, and a table. Our staff (the employed locals) receives the breadfruit from the local community. After our staff inspects each breadfruit and verifies it is not bruised or rotten, they compensate the person who delivered the fruit. The breadfruit is then passed on to be weighed using the scale. The quantity of breadfruit entering our system is counted and documented. Our staff cuts off the top or head of the breadfruit and turns them over to allow for internal fruit juices to seep out. The breadfruit must drain for a minimum of 2 hours on a table free of dirt to support them. Because of the large quantity of breadfruit, a fair number of tables will be needed to ensure enough Breadfruit can be 'de-headed' and begin this two hour draining process. Bug netting will then be placed over the draining fruit to ensure bugs do not get into the fruit. The rotten or bruised breadfruit and breadfruit stems are transferred to the waste management subsystem for proper disposal.

The intake system performance data consists of the metrics collected from the scale and counter. Since many choices in COTS scales and counters are available in local market places, we have not specified scales and counters at this time.

Coring/Peeling Subsystem

Once the breadfruit passes the intake subsystem, it enters the Coring/Peeling Subsystem. In the coring/peeling subsystem, our staff peels and cores the breadfruit. **Figure 2.1-223** illustrates the peeler options. The University of St. Thomas students anticipate the hand-crank peeler will be the final solution. However, in lieu of a hand-crank peeler, a COTS vegetable peeler is an effective alternative. We will provide both solutions to our staff for their convenience. After the fruit is peeled, our staff cuts the breadfruit into sections with a knife, removing the core. To meet the capacity of 450 breadfruit/day, we need 3 coring/peeling subsystems.

The breadfruit skins and cores are transferred to the waste management subsystem for proper disposal.

The coring/peeling performance data consists of the number of breadfruit peeled and cored during an hour. Our staff will use a counter to document how many breadfruit were processed during the hour.

Figure 2.1-223: Hand-Operated Peelers. *University of St. Thomas students identified two different hand-operated peelers, providing a self-contained, convenient, and adaptable peelers.*

Shredding Subsystem
Once the breadfruit is cored and sliced into sections, it enters the shredding subsystem. The sectioned breadfruit arrives to the shredder subsystem in a bucket. Prior to placing the sectioned breadfruit in the shredder, our staff weighs the bucket of breadfruit. Our staff then places the breadfruit sections into the opening, and the shredded pieces fall into the bucket located at the end of the shredder. After shredding the entire bucket of breadfruit, our staff weighs the bucket again to document the breadfruit loss in this process.

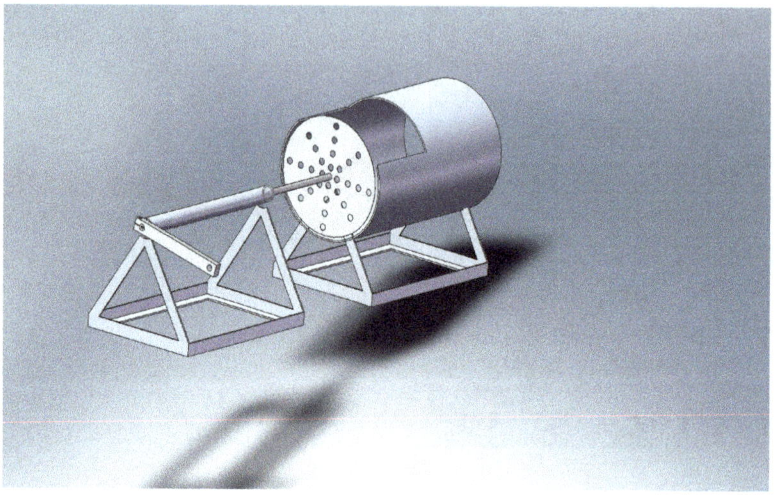

Using hand power, the shredder shreds the breadfruit down into approximately even sized prices of fruit. The shredder, shown in **Figure 2.1-224**, shreds the breadfruit into 2"x1/8"x1/2" pieces per PBSS requirements.

Figure 2.1-224: Hand-crank Shredder. *The horizontal shredder minimizes breadfruit waste, increasing the effectiveness of our processing system.*

The University of St. Thomas students built in safety guards and designed the shredder for ease of cleaning. The breadfruit is weighed on the same COTS scale described in the Intake Subsystem. To meet the capacity of 450 breadfruit/day, we need 3 shredding subsystems.

The shredder performance data consists of the weight of the breadfruit before entering and after leaving the shredder. The performance data also documents how many buckets were processed in an hour.

Dryer Subsystem
Once the bucket of shredded breadfruit leaves the shredding subsystem, it enters the dryer subsystem. The Dryer Subsystem dries the breadfruit shreds to less than 15% of total water weight. To prevent hardening, the dryer takes about 6 hours and should not operate above 105°F degrees. To meet the capacity of 450 breadfruit/day, our system contains 5 dryers.

Our team's dryer subsystem leverages previous dryer concepts from the University of St. Thomas teams and Compatible Technology International (CTI). Their design, shown in

Figure 2.1-225, is pyramid shaped and covered with plastic. The plastic traps the heat from the sun, providing an oven effect. Within the pyramid, there are twelve trains, each containing breadfruit shreds. The slit at the bottom of the dryer acts as an air inlet to the collector, which directs the warmer air into the dryer. As the air heats up, it exits out the top, creating air movement. This air movement is crucial to proper dryer operation.

Figure 2.1-225: CTI Dryer. *CTI initial dryer concept provides simplistic dryer approach.*

We evaluated the initial dryer design and consulted dehydrator experts at Commercial Dehydrator Services (CDS). From them, we learned that the most critical design criterion in the dryer was the air movement. By increasing air movement, the drying time and the chance of mold growth reduces. Thus our team designed enhancements to the initial dryer concept, as shown in **Figure 2.1-226**.

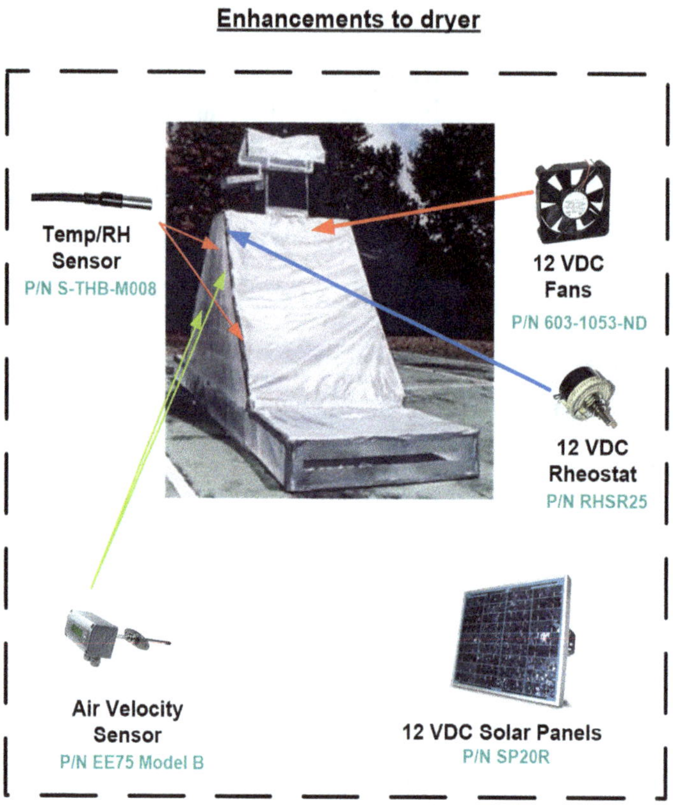

Figure 2.1-226: Enhanced Dryer. *Our enhancements to the initial dryer design increase the dryer operation and reduce risks of spoiled breadfruit.*

Our enhancements, as identified in **Figure 2.1-226**, include four solar-powered fans. These fans increase the air flow within the dryer by drawing in the air from the inlet at the bottom. An additional enhancement is tray position markings. These markings allow air to channel through all of the trays evenly.

Finally, our dryer design includes control capabilities of the fans via a rheostat. An automatic feedback loop controls the rheostat based on temperature/humidity and air velocity sensor data. **Figure 2.1-227** illustrates the electrical connection of the solar panel, the sensors, and data logger, which captures all the information. The data logger and intelecell are part of the BrEAD Monitoring System (Section 0), but are included in this illustration.

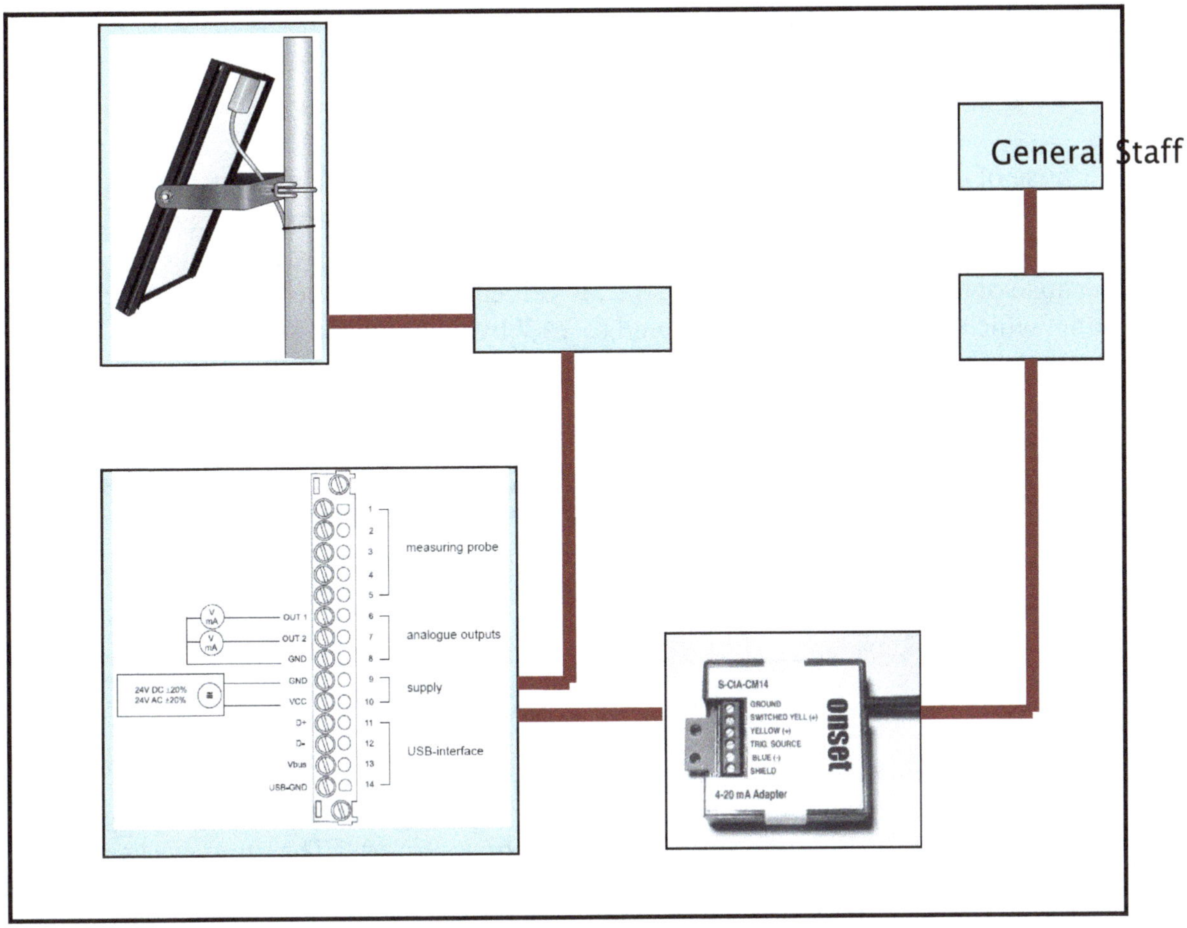

Figure 2.1-227: Air Velocity Sensor Powered by SP20R Solar Panel. *Our dryer design captures the internal environment and stores the information, allowing for real time adjustments to maximize dryer operation.*

The SP20R is a photovoltaic power source designed by Campbell Scientific. The SP20R can provide up to 20W peak power and 1.19A current by converting sunlight into direct current. Like any solar panel, the SP20R can source current on cloudy days but not at night. But because of its reduced power sourcing capability in cloudy, shaded, and dirty panel conditions, the SP20R is installed with the combination of a regulator and either a rechargeable or alkaline batteries to provide a sustainable power source.

EE75 series air velocity transmitters are designed by the Austrian company E and E Elektronik. The EE75 Model B is designed to measure a "wide range of velocities and temperature". The transmitter uses "hot film sensor based on cutting-edge thin film technology to ensure maximum sensitivity even at a low mass flow". The EE75 series operating temperature range of -40ºF to 248ºF is made possible because of its integrated temperature compensation that minimizes the temperature cross-sensitivity. The EE75 series can measure up to 80000ft/min air flow.
Accuracy:

- ± 0.03 m/s (6ft/min)
- ± 0.10 m/s (20ft/min) + 1% of measuring value
- ± 0.20 m/s (40ft/min) + 1% of measuring value

Because of the analog outputs option that offers the EE75 series transmitters, the BrEAD team connects the EE75 Model B transmitter to a hobo data logger through the S-CIA-CM14 analog input adapter for an automatic velocity data collection. Because of its power consumption, the air velocity sensor is powered by a 12V solar panel which voltage is in turn doubled to 24 V by a voltage booster

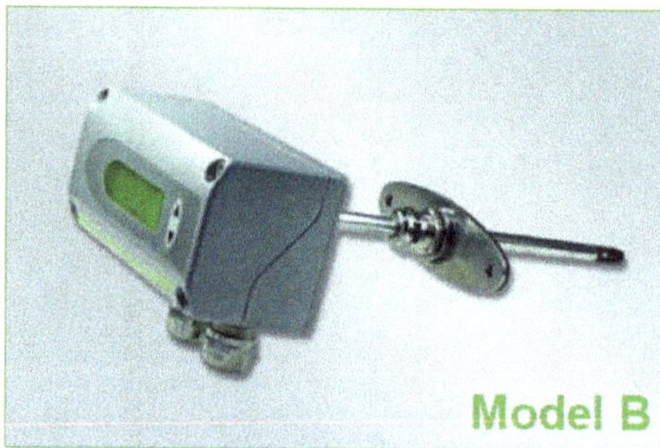

Figure 2.1-228: EE75 Model B, Air Velocity Transmitter. *Air velocity transmitter used to evaluate fan performance, providing the operators maximum control of the dryer.*

Grinder Subsystem

After the shredded breadfruit is dry, it is placed in a storage container for grinding. Once the breadfruit is dry, it is no longer at risk for becoming rotten. Therefore, we can store the breadfruit until after the breadfruit season for grinding. By waiting until later to grind the breadfruit, we will employ the local workforce longer.

Figure 2.1-223 illustrates the grinder as previously designed by University of St. Thomas students. The shredded breadfruit enters the opening is ground between the two cylinder stones shown. The flour falls into a bucket located below the two stones.

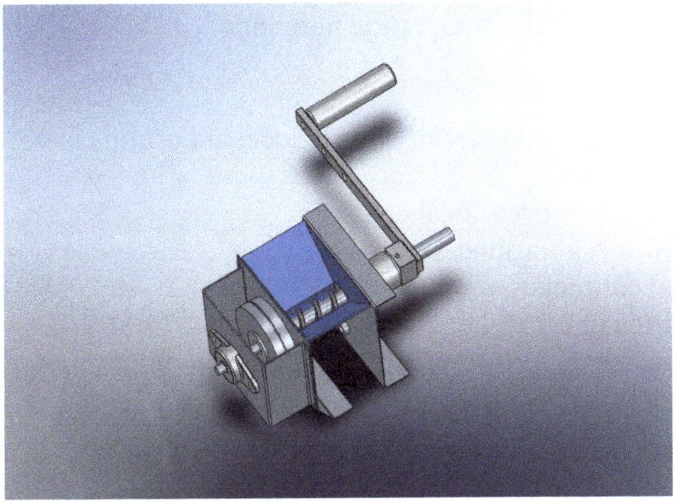

Figure 2.1-229: Hand-crank Grinder. *The Grinder is self-contained and convenient, which eliminates any constraints on the grinding system.*

Packaging Subsystem

Once the flour is in the buckets, it moves to the Packaging Subsystem. In the packaging subsystem, our staff packages the flour in 5 lb bags for individual sales and in 5 gallon buckets for commercial sales. The packaging material options are: plastic bags, plastic tubs, and paper packages. The material chosen will be native to the location of the Processing Site. Our only requirement is that the package prevents insects and moisture from infiltrating the flour.

Waste Management System

The Waste Management subsystem will be managed by the Processing Center supervisor. Local laws or customs may dictate what must be done with the waste. However, our team assumes that a compost pile will be allowed. All breadfruit waste will be stored in a compost pile, to be used later as fertilizer. Our staff will transport the waste on a daily basis to the compost pile to meet food processing sanitation standards.

Regional Center

The Regional Center acts as communication, support, and coordination nodes in the BrEAD enterprise architecture (see figure 2.1-4). Breadfruit harvesting is only possible for any given region during limited defined times throughout the year. The Regional Centers will therefore be fully operational only during those times when breadfruit processing systems are deployed and active. Rather than staffing the Regional Centers with dedicated BrEAD personnel all year

Our concept for Regional Centers creates a local platform to support and expand breadfruit processing in every area BrEAD serves. Local centers ensure an on-site presence for training, marketing, and logistics, and fast response to regional problems. And our flexible staffing strategy means the regional centers can be up and running at an initial low cost

round, local aid organizations will be tapped for part time personnel.

In addition to being less expensive, part-time personnel from local aid groups have the following advantages:

- Familiarity with local culture, mores, local contacts
- Familiarity with local work regulations, safety laws.
- Social and governmental contacts
- Familiarity with local languages
- Possible use or lease of existing facilities/communications equipment
- Possible integration with existing plans/goals of other aid groups

The major functions of the Regional Centers include,

- **Public Relations/Advocacy**
  - *Liaison to other aid organizations and local governments.* The Regional Center will be the de facto local representative for BrEAD, and by extension, the NGRF. The center will work with government officials to maintain acceptance and approval for additional processing sites. It will cooperate with other aid organizations to accomplish broad NGRF objectives.
  - *Marketing Outreach.* The Regional Center will host/participate in local events to promote the use of breadfruit and raise awareness if its potential.
  - *Site Search.* The Regional Center will work with BrEAD HQ in expanding the network of processing sites in each region. Locations for processing sites will be selected according to a weighted list of criteria that emphasize a combination of proximity to breadfruit groves, political stability and suitability of the host community. Likewise, locations for Regional Centers will be selected according to a list of criteria that emphasize proximity to potential breadfruit processing sites and access to local aid organizations. See **Figure 2.1-231** and **Figure 2.1-232** for our scoring strategy.
- **Data and Communications.** Regional Centers will collect operational data from each processing site and transmit that data to BrEAD Headquarters in St. Paul. They will also relay directives and data from St. Paul HQ, and coordinating information to each site, via electronic or other means.
- **Operations**
  - *Logistics Support.* The Regional Center will store processing systems as needed during the off season, depot spare parts, perform repairs and maintenance on the processing hardware.
  - *Transportation/Shipping.* As a coordination point, the regional centers will facilitate transportation of processing systems between sites, to/from the Regional Center, or to other regions.
  - *Training and Supervision.* The Regional Center can supervise BF processing and collect operational data from each site. Ideally however, all onsite tasks will ultimately be conducted by

owner/operators of the processing systems. To that end, the Regional Center will provide training for processing system operators, including manual operations, supervision, and data collection/transmission.

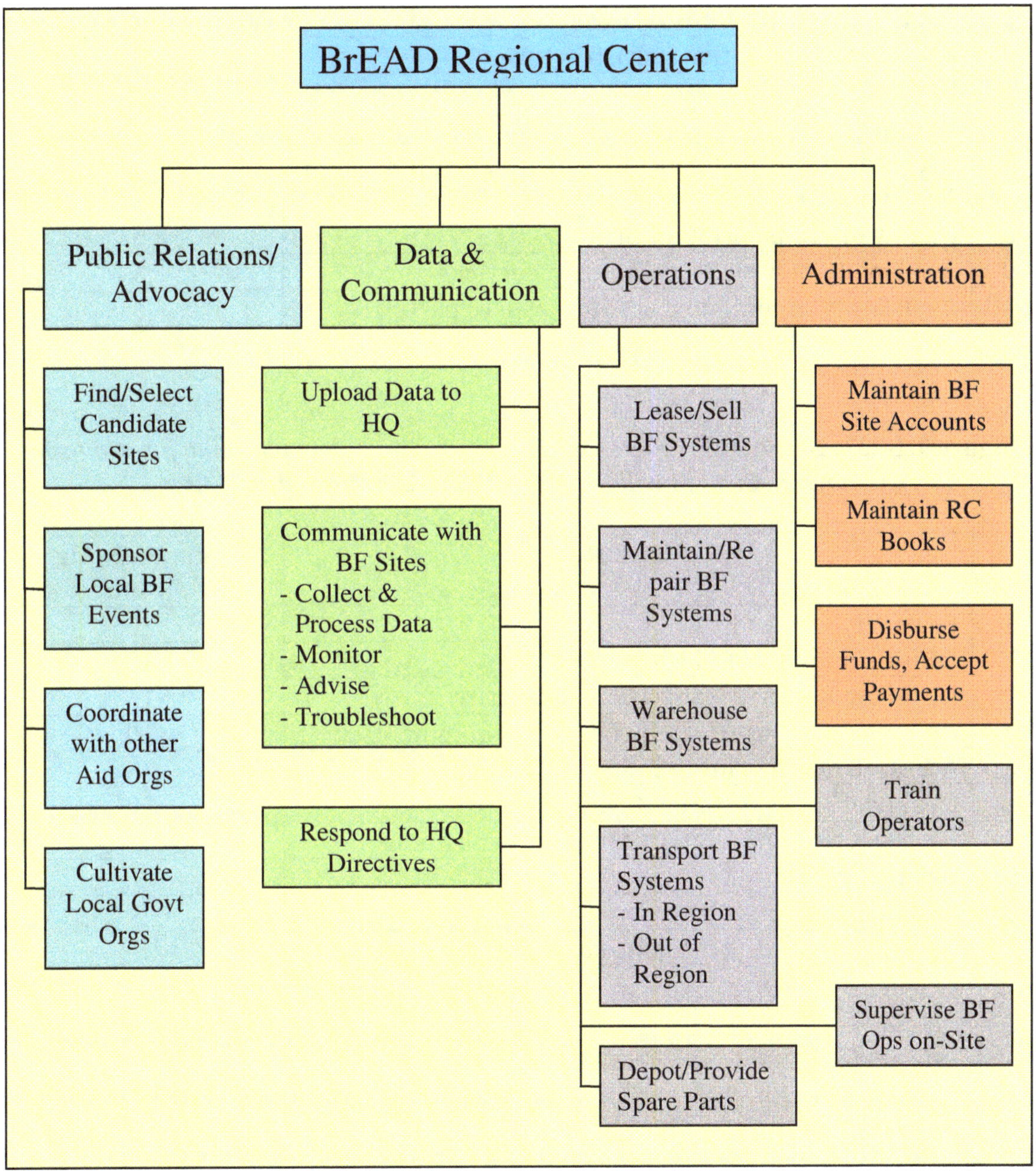

Figure 2.1-230: Regional Center Functions. *Our concept enables each center to develop the full breadfruit potential of its region.*

Figure 2.1-231: Regional Center Site Selection Strategy. *Potential new Regional Centers will be evaluated according to objective criteria.*

	Attribute	Max Possible Points
1	Proximity to potential breadfruit processing sites	20
2	Political stability/ Rule of Law	15
3	Local Aid Organization Infrastructure	15
4	Power Grid	10
5	Internet Access	10
6	Local government support for aid organizations	10
7	Telephone Grid	5
8	Quality of Transportation Grid	5
9	Availability of basic machine shop services	5
10	Availability of computer products/spare parts	5

Figure 2.1-232: Breadfruit Processing Site Selection. *When planning for expansion, Regional Centers will focus on the most promising sites first.*

	Attribute	Max Possible Points
1	Proximity to breadfruit	20
2	Political stability/ Rule of Law	15
3	Host Community Organizational Maturity	15
4	Ability to purchase/lease processing equipment	10
5	Access to markets for breadfruit products	10
6	Quality of Transportation Grid	10
7	Local government support for aid organizations	5
8	Telephone Grid	5
9	Power Grid	5
10	Availability of basic machine shop services	5

Central Data Center

> **Central Data Center provides administrative support and processes our operational data, which can be accessed by researches via our website.**

The Central Data Center, located in St. Paul, Minnesota, will be the headquarters for our Enterprise Architecture (EA). It will provide leadership and direction for all EA activities. The Central Data Center will communicate directly with the Regional Centers and the System Integration Laboratory and indirectly, through the Regional Centers, to the processing sites. It will be the repository for the data and research produced by our system.

It will be the access point, via website and call center, for our products. It will provide the following services for the BrEAD organization:

- Administrative
- Logistics
- Marketing
- Training support
- IT support
- Engineering
- Stakeholder support
- Research results
- Call center

The Field Data Database will reside at the Central Data Center. This database will be the repository for the field data, identified in Appendix A of the Performance Based Systems Specification (PBSS). In addition to field data, the weather data collected at the processing sites, will reside in this database. The output of the Field Data Database, will be the source of the data pertaining to processing breadfruit collected by the BrEAD for researchers and other stakeholders. Data from this database will be available 24 hours a day to stakeholders, and will be in the form of text files, spreadsheets, graphs, and reports. Access to the Field Data Database to stakeholders will be through a website which will also reside at the Central Data Center.

The Field Data Database will have the capability to have field data entered manually directly into the database, and to have field data received via internet connection, to be deposited in the database by automated means.

Figure 2.1-233depicts the Central Data Center functional organization and the primary services provided.

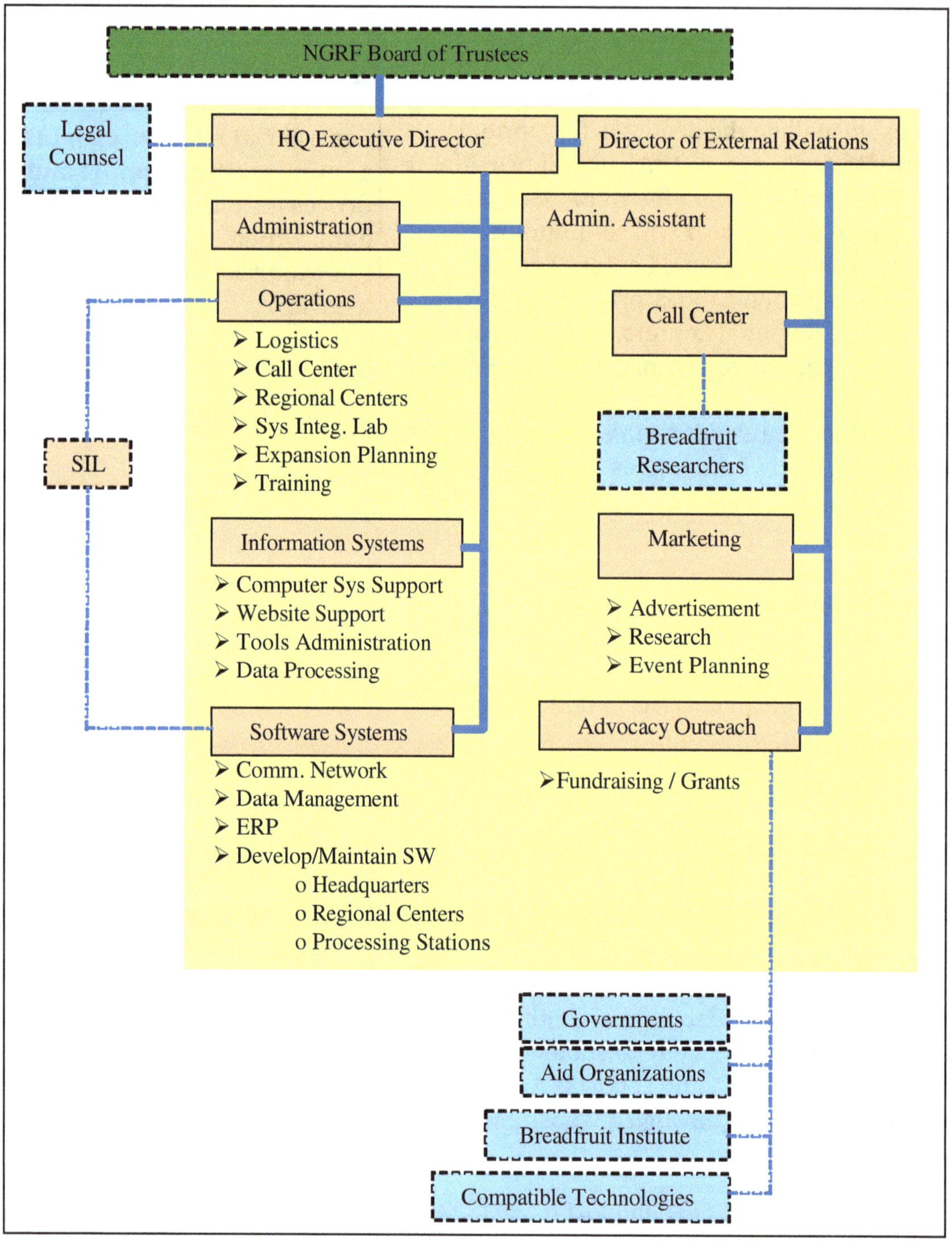

Figure 2.1-233: Central Data Center Functional Diagram. *The Central Data Center has many functions providing guidance and oversight to the other subsystems.*

Administrative support

Administrative services for the BrEAD will be provided at the Central Data Center. These services will include, but not be limited to,

- Management of BrEAD activities
- Grant administration.
- Human resources - hiring personnel, keeping personnel records,
- Accounting - payroll, purchasing, disbursements, financial records, financial reports.

Logistics support

Logistics services will be the source for equipment that is not purchased locally by the Regional Centers, the Systems Integration Lab, and the processing sites. And it will be the coordinator of funding for locally purchased equipment. Logistics services will coordinate the transportation to Regional Center and processing sites of personnel and equipment. It will also coordinate local resources as required.

Marketing support

Marketing services will provide advertising, research, and event planning for the BrEAD. Marketing will be the BrEAD advocacy outreach and will coordinate fund raising and grant applications to governments, companies that produce compatible technologies, the Breadfruit Institute, and aid organizations.

Training support

Training support will provide the writing, publication, and distribution of technical training materials in culturally compatible format to local users. It will provide training support resources to the Regional Centers for training processing site personnel. Also it will provide for any required training and training materials for Central Data Center and System Integration Laboratory personnel.

IT support

IT support services will specify computing hardware and COTS software products. This will include servers, PC computers, computer network hardware, firewall / virus protection software, engineering software, as well as application specific software. IT support services will also provide software/computer support to all BrEAD entities through the call center and problem report system.

Engineering support

Engineering services will create the custom software applications used by the BrEAD. This will include the personnel records database, website, the field data database, the website / field data database interface, automated field data input algorithm, database output report templates, the problem report database and any other non-COTS software needed by the BrEAD.

Stakeholder support

Stakeholders are defined as researchers, government or company officials, NGRF personnel, grant funding organizations, field personnel, BrEAD organization personnel and anyone else interested in BrEAD products. BrEAD products are the data and research results produced by the BrEAD and any other products marketed by the BrEAD.

Stakeholder support services will provide access to the field data through the website. It will provide support to interested parties in the form of database queries and data reports. It will respond to requests for technical information and support for field problems.

Research results

Results of research, simulation, and test analysis activities at the System Integration Lab will be made available to stakeholders via the website.

Call Center

A call center will be staffed during regular business hours at the Central Data Center, to provide support for field problems and requests for engineering or IT support. The call center will supplement the website as the means for stakeholders to access BrEAD data and support.

Communication Subsystem

Our communication system provides flexibility for implementation in areas with geographical barriers.

Automated data collection capability is one of the system requirements highlighted in the Performance Base System Specification (PBSS).

In order to meet the Breadfruit Enterprise Architecture Demonstration (BrEAD) data collection capability required in PBSS, Solutions International (SI) has developed two different communication schemes: utilizing a series of Intelecell transmitters or utilizing a laptop and Explorer 700 to communicate with satellite internet.

Although both communications schemes offer reliable and secure data transfer in remote and hostile geographical areas, the difference between the Intelecell based scheme and the Explorer700 based scheme is its distance capabilities. The Intelecell strategy can be used for distances of less than 40 miles between the weather station & regional center or processing system, and between the Regional Center and processing site. The Explorer 700 is a satellite based communication system, so it can be used for distances greater than 40 miles. This becomes important when a region contains scattered islands with more than 40 miles distance between them.

Intelecell is a long range wireless data logger capable of transmitting data (from the sensors connected to it) with no other supporting infrastructure. Intelecell is designed to be self-repairing self configuring and self-organizing in a mesh network topology and to be used unattended to monitor field data for a long period of time. Designed to acquire data from a wide variety of sensors, Intelecells can also be connected to Personal Computer through its dedicated serial port, as shown in **Figure 2.1-234**. Designed with ultra low power consumption in mind, Intelecells operate in a Sleep/Wakeup mode called "Doze Mode". In other words, the devices are asleep most of the time and wake periodically to listen, receive and send message to each other then go back to sleep. The operation between Intelecells is synchronized by a mother device named Base Station. Indeed, the Base Station initiates Intelecells data collection and handles the transfer of collected data to the server. The Base Station also connects Intelecells network to the Internet. The Base Station is one of the versatile attribute of Intelecells as it allows remote settings of the Intelecells. In addition to its remote programmability, Intelecells come standard with remote firmware upgrade and with a reliable, secure data transfer. Designed to be used in rugged environment and for long term, Intelecell is provided with a ten year lifetime rechargeable battery. The battery can be recharged using solar panel.

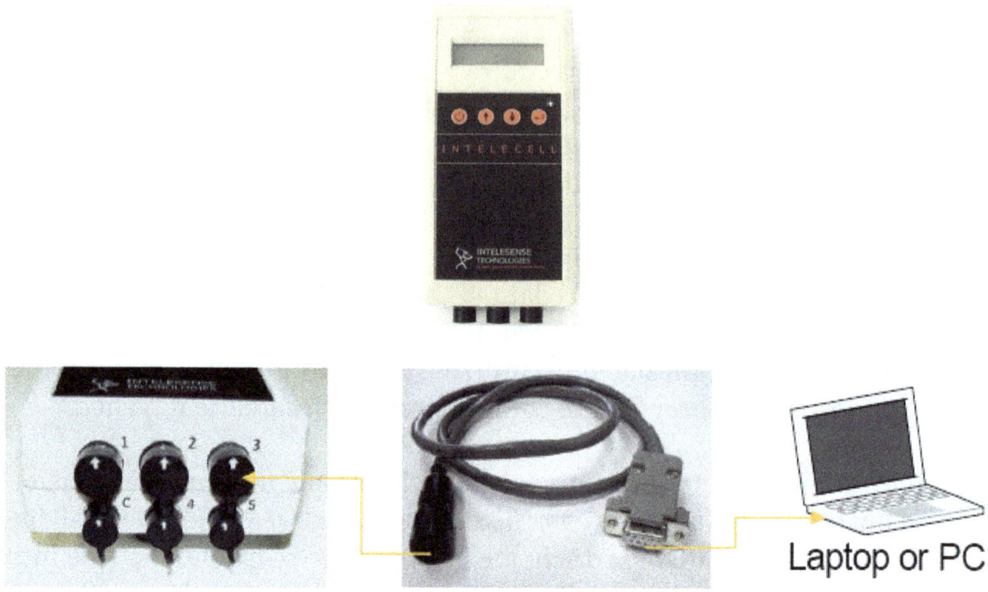

Figure 2.1-234: Intelecell and PC Interface Cable. *Interface between Intelecell and PC identified, proving a simple concept.*

The Explorer 700, as shown in **Figure 2.1-235**, is designed for voice and broadband data communications from anyplace on the planet earth via satellite. Designed to be operational in the most hostile environmental conditions, the Explorer 700 is capable of maintaining reliable communications in extreme weather conditions and less favorable geographical conditions. For these reasons, the explorer 700 is said to be rugged, durable and versatile. Among its multiple features, the Explorer 700 is capable of:

- Simultaneous voice and data communications over Broadband Global Area Network (BGAN)
- Full duplex, single or multi-user, up to 492 kbps
- Standard LAN, USB, Bluetooth and phone/fax ports
- ISDN compatible
- Built-in web server allowing the user to manage messages and calls, and customize the terminal to satisfy the user needs (Thrane and Thrane, Explorer 700, 2008).

Figure 2.1-235: Explorer 700. *Explorer 700 provides capabilities to transmit data over distances longer than 40 miles, allowing for EA integration into any region.*

The trade-off with the Explorer 700 is that a laptop is required for operation. The laptop connects to the explorer 700 and uploads the operational data (either processing system operational data or weather data).

However, the Intelecell is completely autonomous. The data logger from the weather station connects to the Intelecell and automatically transfers data to the Intelecell.

Weather Station Subsystem

To meet our PBSS requirement of collecting and providing hourly weather data, we have identified a weather station. The weather station is designed to frequently measure and collect geographical weather condition data. Our proposed weather station is a COTS component from Hobo: H21-001 Hobo Weather Station. This weather station automatically transfers the collected data to a data logger, the Hobo Microstation Data Logger

A legacy system of among Hobo data loggers, the H21-001 is a dependable 10, expandable to 15 channel weatherproof data logger specially designed for research-grade environmental data measurements. The H21-002 is a 4-channel smaller version of the H21-001.

Supported by Intelecells the multichannel capability of the H21-00X made it suitable for our design. It is used in our design to serve as bridge between the targeted object or environment and the Intelecells. As Intelecells support a limited range of sensors, the use of H21-00X gives more room for flexibility in the choice or the sensors that fit our application.

The Weather station contains the following sensors: Rain Gauge, Wind Sensor, Solar Radiation Sensor, and Temp/Humidity Sensor.

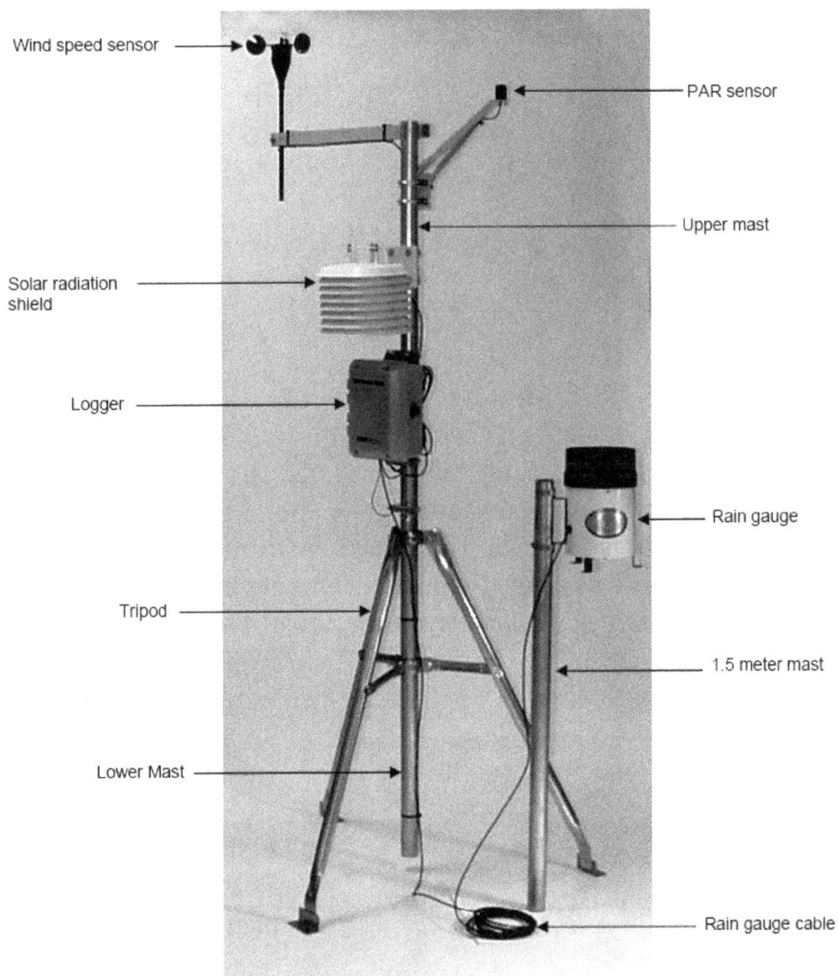

Figure 2.1-236: HOBO Weather Station (H21-002). *Not only does the HOBO Weather station have the required capabilities to meet the*

PBSS requirement, but also provides additional functionality.

S-RGA-M002 Rain Sensor

The S-RGA-M002 is a modular compatible Hobo Weather Station smart sensor designed to automatically communicate with the data logger with no additional configuration or programming.

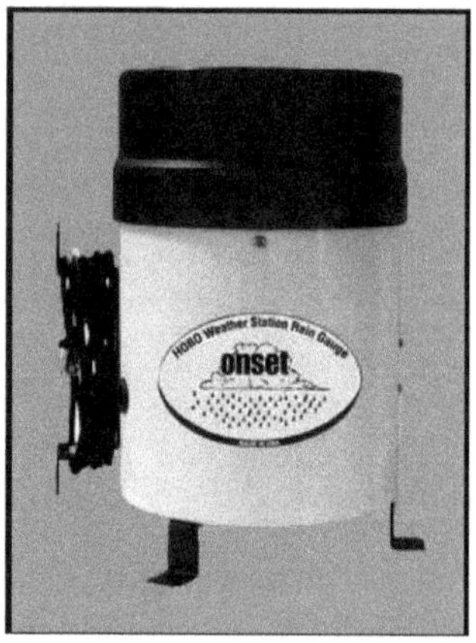

Figure 2.1-237: Rain Gauge Smart Sensor (S-RGA-M002). *Rain Sensor connects directly to data logger without addition configuration required.*

S-WCA-M003, Wind Sensor

The Onset smart wind sensor, S-WCA-M003, is designed to measure the speed and the direction of the wind of a specific environment. The smart wind sensor is capable of measuring the speed and direction of the highest wind gust in a 3-second time interval. The wind sensor measures wind speed/direction every three seconds and average the measurements before sending it to the data logger. Because of the logger's relatively large power consumption, the logging interval is usually set to be very large compared to the sampling interval. SI chooses the S-WCA-M003 for the high accuracy of its measurements.

Figure 2.1-238: S-WCA-M003, Wind Sensor. *The Wind Sensor collects wind speeds with high accuracy, providing highly reliable data for researchers.*

S-LIB-M003, Solar Radiation Sensor

The S-LIB-M003 solar radiation smart sensor is a light sensor designed to work with the H21 series data logger. It measures the light intensity over 300 to 1100 nm spectral range and reports the average measurement over the logging interval set by the user. In measurement averaging, multiple samples are taken and the average is calculated and fed into the logger at the logging time intervals. Solutions International collects the breadfruit processing environmental lighting data to feed researches on productivity increase and power equipment improvement.

Figure 2.1-239: S-LIB-M003 Solar Radiation Sensor Silicon Pyranometer. *The solar radiation sensor is an additional feature that Solutions International provides.*

S-THB-M008, Temperature/RH sensor

The S-THB-M008 is smart sensor design for temperature and Relative Humidity (RH) measurement. 0.36º F accuracy over temperature ranging from 32º F to 122º F and ± 2.5% accuracy over 10% to 90% Relative Humidity in addition to the operating temperature range (-40º F to 167º F) and the environment rating (0 - 100% RH), the S-THB-M0XX series is suitable for the breadfruit processing project. Designed to be compatible with all hobo data loggers, the S-THB-M008 is used by Solutions International to measure and record environmental data and to measure and record inside oven data during the dry process.

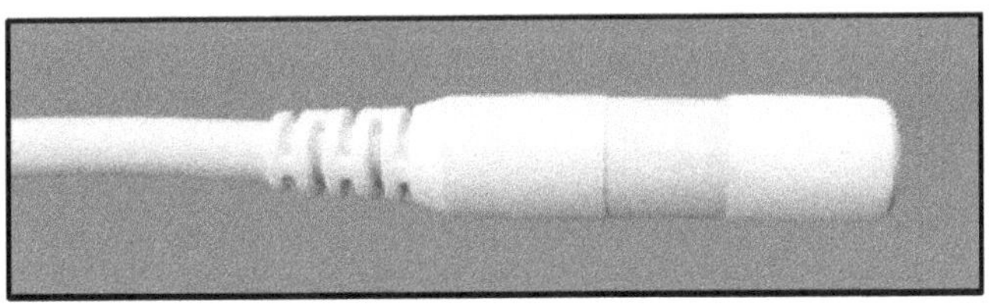

Figure 2.1-240: S-THB-M0XX Series, Temperature/RH sensor. *The temperature and humidity sensors are present on the dryer system as well as the weather station, demonstrating commonality within our EA architecture.*

BrEAD Monitoring Subsystem

The BrEAD monitoring system is essentially a combination of smart sensors, data loggers, and a secure wireless communication system explicitly set up to collect, store and transfer geographical environmental data in addition to the breadfruit processing operational data to regional centers and the NGRF headquarter in St. Paul, Minnesota.

Annex A identifies all the performance data collected during the processing system operation. Most of the information will be collected manually in the processing system, so forms are provided to each subsystem for proper data collection. Prior to the breadfruit season, the employed locals are trained on how to collect the performance data. The Processing Site supervisor transposes all the data from the collection forms to a GUI on the Processing site laptop on a daily basis. In addition, the dryer has sensors, which provide performance data to the data logger automatically. The Processing Site supervisor transfers the data from the data logger to the personal computer.

Once the data is uploaded into the Processing Site laptop, the Processing Site supervisor sends the information to the Regional Center via the communication system. Data transferred to regional centers is used for logistics; technical and marketing support while data transferred to NGRF headquarter in St. Paul is analyzed and used for product improvement and market development. Should any failures occur in the processing system, the supervisor is responsible for contacting the Regional Center via the satellite phone. Fault information is also maintained in the data collection forms.

The weather station operational data is collected via the HOBO data logger. The data logger transfers the logged data per measurement order set by the users to Intelecell device that in turn encrypts and send the data to the next Intelecell device (within 14 to 40 mile radius) until it reaches its final destination (data server), where the Intelecell is part of the communication system.

The BrEAD Monitoring System is shown in **Figure 2.1-241**, where the Intelecell is part of the communication system.

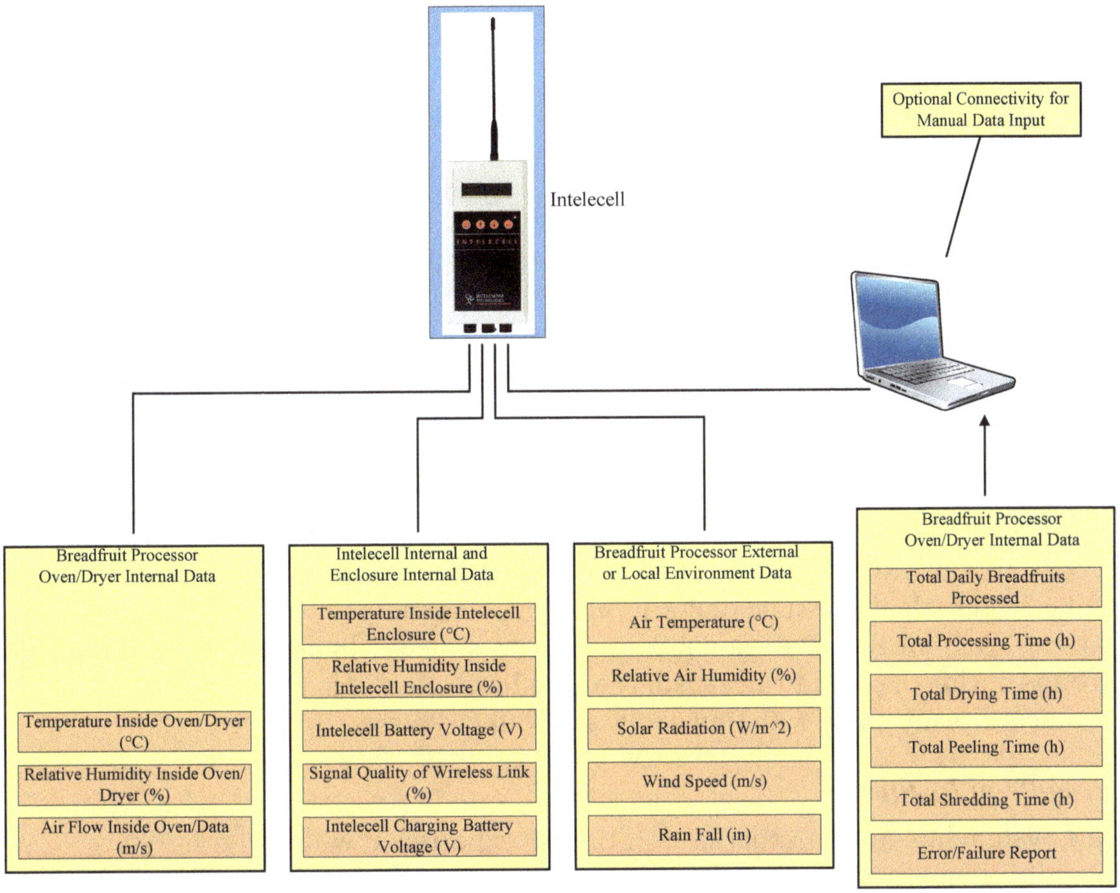

Figure 2.1-241: BrEAD Monitoring System. *The BrEAD Monitoring System collects all operational and temperature data, proving compliance to our PBSS requirements.*

External Systems

The BrEAD system must be capable of fulfilling all PBSS requirements. Our success depends on some external systems.

BrEAD is interactive with the world. Regional centers will work in cooperation with other aid organizations and local governments. The SIL will work with the Breadfruit Institute and seek markets for its products in Hawaii. Our breadfruit processing systems need cooperative groups to operate and protect them.

Regional Authorities

SI depends on the cooperation of local governments to allow and enable the administration and operation of breadfruit processing systems in their territories.

Local Aid Organizations

SI seeks to partner with other aid organizations, whose missions are to provide nutrition and promote prosperity. Other such groups can provide valuable insight on local culture and conditions. Additionally, our strategy calls for BrEAD Regional Centers to be staffed with personnel from local aid groups during the breadfruit season. For the Haiti site, SI plans to build on local contacts already made with Dr. Carolle and Father Reiser.

Breadfruit Institute

The System Integration Laboratory (SIL) will experiment with and process breadfruit into food products for sale to an upscale gluten-intolerant market. Our strategy depends on the Breadfruit Institute allowing us to access their groves in Hawaii. The SIL will depend on the Institute to provide access to breadfruit and possibly land for the SIL.

Satellites/Internet

The nodes in the BrEAD network, comprised of the Central Headquarters, System Integration Laboratory, and Regional Centers throughout the world will depend on the internet for communication. Much of this communication, especially in remote areas may depend on satellites as the transmitting medium (see 2.1.1.1.4 Communication Subsystem in Vol2, Book 1).

Proximity to Breadfruit

Breadfruit processing systems require proximity and easy access to groves of the breadfruit trees. These trees must be accessible and available to the processor host community.

Host Communities

Our strategy depends on viable local communities or groups to acquire, operate, safeguard and maintain the breadfruit processing systems during the harvesting season.

Electrical Infrastructure

Most of the BrEAD components (Headquarters, Regional Centers and SIL) depend on the availability of local electrical power. The breadfruit processing systems can operate in the absence of a local electrical grid.

Support System

> **BrEAD Architecture maximizes supportability and logistic support:**
>
> **Modular design limiting down-time**
> **Low maintenance required**
> **Spares located within two hours of processing site**

Solutions International understands the importance of an efficient and effective supportability approach. We designed our system so that minimal maintenance is necessary. Our Enterprise Architecture (EA) consists of the following subsystems: Central Data Center, Regional Center, Processing System, Weather Station, Communications System, and BrEAD Monitoring System. Our supportability approach incorporates Performance Based Logistics (PBL), Maintenance Planning and Condition Base Maintenance Plus (CBM+), as well as PBSS and SOO requirements. CBM+ is a conscious effort to shift equipment maintenance from unscheduled maintenance to preventive maintenance.

Our PBSS requirement for Mean Time To Repair (MTTR) is 15 minutes, so our general maintenance concept is designed to minimize impact on system operation. Therefore, all components of the BrEAD will be "remove and replace" maintenance. If a component fails, a spare will quickly be provided. Annex B provides the maintainability prediction of our BrEAD system.

The following sections provide specific PBL, CBM+, and logistics concepts for each subsystem.

Processing System Support

Because of the time constraint of the breadfruit season, our processing system cannot fail during breadfruit harvesting season. If a failure should occur, the whole processing system is shutdown and breadfruit may be wasted, if not processed in time. To avoid failure during the harvesting season (~ 2 months), the following preventative maintenance occurs prior to each harvesting season at the Regional Center:

- Sharpen all blades
- Visually inspect each piece of processing system equipment
- Oil all gears
- Refresh/replace all plastic in dryer

The processing system design has few electronic components, yielding a simple, maintainable system. Because the system is largely mechanical with few components, the processing system has maximized fault detection and fault isolation. For those electronic components (clicker, temperature sensor, fans), there is continuous human interaction, resulting in immediate fault detection.

As described in Section 0, our expected usage for the processing site laptop is only a few hours a day. Since we are not providing a self-contained electrical source for the processing site laptop, we will send five batteries with the laptop. At the end of the processing season, we expect all batteries will be returned to Regional Center for recharging.

After the harvesting season and when the processing systems are back at the Regional Center, we disassemble, clean, and dry each piece of the processing system equipment. To further the preventative maintenance, we will utilize the SIL to model the wearout life of each LRU. Per accepted PBL approaches, our system recommends replacement at 70% of the wearout life. If the 70% mark occurs during the harvesting season, the subject component will be replaced prior to the harvesting season.

Regional Center Support

The Regional Center is a combination of an office space and maintenance depot. The Regional Center also stores all the spare parts for the processing system. Acting as a call center, the Regional Center handles field questions and has access to the resources to fix processing site failures.

The office and data management equipment used at the Regional Center are COTS components. These COTS components have Built-in-Test (BIT) capabilities prior to purchase. As described in the Software Development Plan (Section 2.2.8), Solutions International provides custom software to read the COTS's BIT outputs and monitor the data management equipment. The regional center employees can detect office equipment failures continuously. Only the data management equipment is a necessity to BrEAD system operation.

The Regional Center will provide training and logistics support for both the processing system and weather station. For every hardware failure, the maintainer documents the time spent repairing the system and the cause of the failure. The maintainer sends this information to the central data center to be logged in the fault tracking database.

Weather Station Support

The Weather Station contains COTS equipment, which has its own internal built-in test (BIT) capabilities. The COTS weather station sends its BIT results in its data message to the regional center. The regional center then passes the data message on to the central data center. Both the regional and central data centers interpret the data center and log any failures. If a fault should occur on the weather station, personnel at the regional center will repair the weather station.

If a fault occurs preventing the weather station from communicating with the regional center, regional center personnel will check and repair, if necessary, the weather station. The weather station sends a fault and performance data message to the Regional Center every hour.

BrEAD Monitoring Subsystem

The BrEAD Monitoring System contains COTS equipment. As part of our subcontractor agreement, Onset will provide maintenance support for all data loggers used in our system. As part of our logistics concept, a spare data logger is kept at the Regional Center. Prior to each processing system and weather station installation, we will perform a functional performance test of the equipment. Each COTS component has internal built-in test (BIT) capabilities for debugging and isolation.

Communication Subsystem

The Communication System contains COTS equipment. As part of our subcontractor agreement, Inmarstat will provide maintenance support for all the satellite communication networks. If the communication subsystem fails, the performance data from both the Weather Station and the Processing System is still maintained. Each COTS component has internal built-in test (BIT) capabilities for debugging and isolation.

Central Data Center

The Central Data Center has remote access capabilities with the processing system laptop and regional center servers for diagnostics and periodic upgrades. The custom software on the regional/central data center tracks the faults from the COTS' BIT results.

Hardware and Software Integration Overview

Our proposed system is organized around four physical locations: a Central Data Center, Regional Center(s), breadfruit processing site(s), and a System Integration Lab (SIL). It is expected that these will not be co-located. Each has unique as well as common hardware (HW) / software (SW) needs as determined by the BrEAD functions performed at each location (**Figure 2.1-242**). This section will describe the HW/SW integration at each of the four locations. Note: there may be more than one Regional Center and processing site, but for the purpose of this section the following discussion will refer to one each of the four sites listed previously.

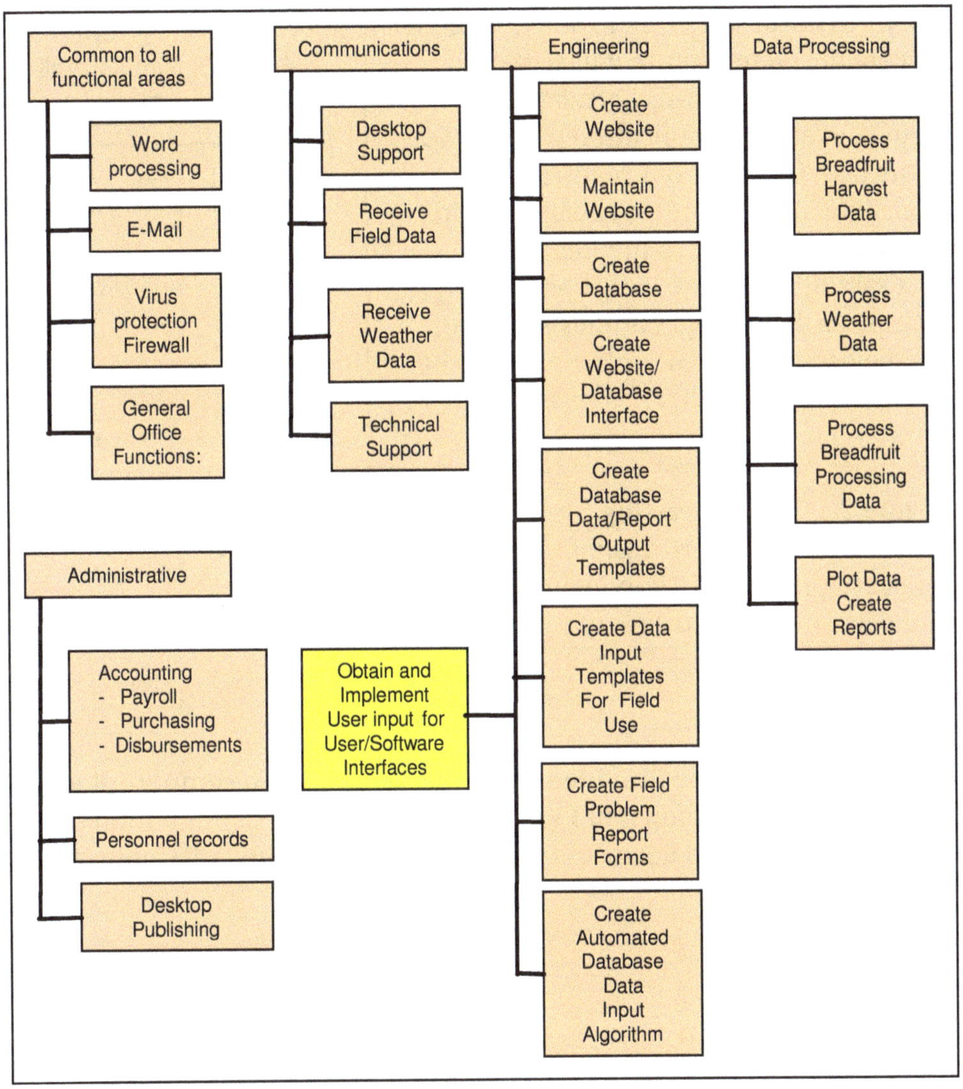

Figure 2.1-242: Software Functional Diagram. *Software functions defined, proving our system has multiple capabilities.*

Breadfruit Processing Site

The breadfruit processing site will be supplied with a laptop computer, the primary purpose of which is to provide a means of inputting breadfruit processing data into the BrEAD, as shown in **Figure 2.1-243**. This computer will run with the Microsoft Windows Professional operating system. It will have the Microsoft Office Professional suite of desktop software. It will be protected against unauthorized access by Symantec firewall and virus protection, and it will be configured to require a user ID and password to log on. A BrEAD Software Engineering developed application will provide a GUI to the user for input of breadfruit processing data into a text file. The application will also send this text file to the Regional Center via satellite internet. Local processing site user input will be sought and used in the design of the data input GUI for the purpose of making data input as intuitive as possible for the user.

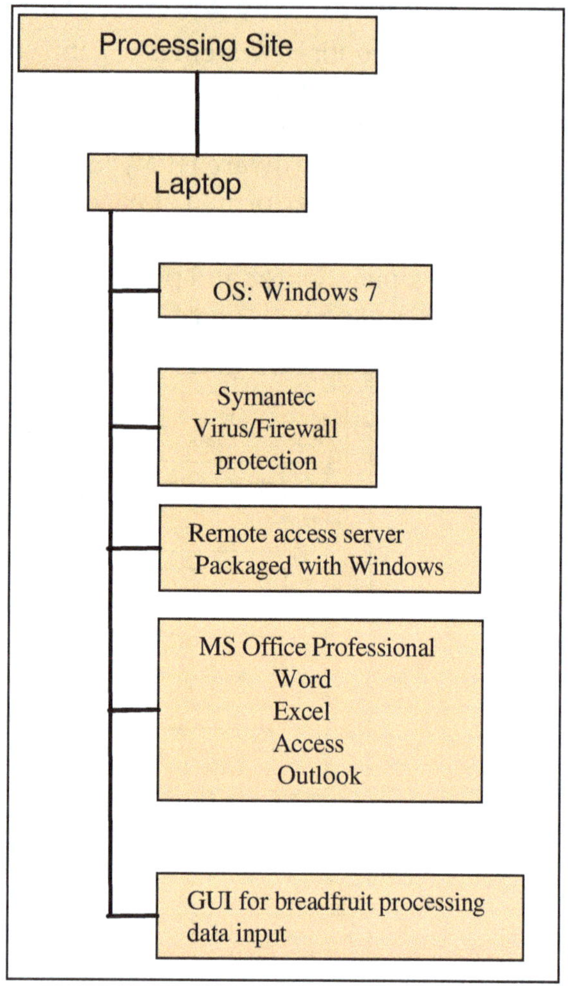

Figure 2.1-243: Processing Site Software / Hardware Integration. *Processing site software package is simplistic, allowing for quick development and ease of maintenance.*

Regional Center

The Regional Center will be provided with a server, a laptop computer for each user, and a wireless router, as shown in **Figure 2.1-244**. A 3 in 1 combination inkjet printer, copy machine, and fax machine will also be supplied. All items will be wireless network capable.

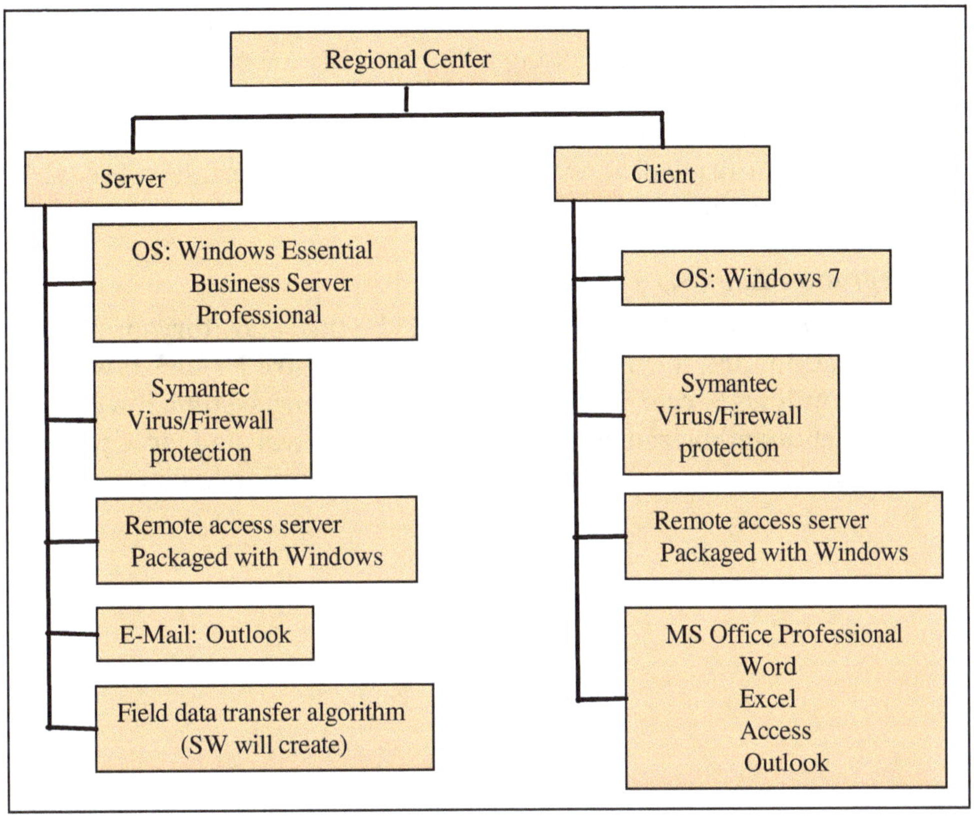

Figure 2.1-244: Regional Center Software / Software Integration. *Regional Center requires minimal custom software, proving to be a completely adaptable system.*

The primary function of the Regional Center server is to transfer data between the breadfruit processing site, the Central Data Center, and the System Integration Laboratory. It will run with the Microsoft Windows Server Standard Edition operating system. It will be protected against unauthorized access by Symantec firewall and virus protection, and it will be configured to require a user ID and password to log on.

An ACCESS database will be configured by Software Engineering for logging hardware and software Problem Reports (PR) from the processing site as well as at the Regional Center.

A BrEAD Software Engineering developed application will provide the functionality required for the transfer of data. Processing system data sent by the processing system site will be received, checked for validity, and sent to the Central Data Center via internet connection. This application will also receive, validate, and send to the Central Data Center, weather data sent from the processing site. Fault messages from the weather sensor system will be processed. PRs logged at the Regional Center will be forwarded to the Central Data Center.

The Regional Center will be provided with a laptop computer for each user for accessing the server and communicating with other BrEAD personnel. This computer will run with the Microsoft Windows Professional operating system. It will have the Microsoft Office Professional suite of office applications. It will be protected by Symantic firewall and virus protection software and will require a user ID and password to log in.

System Integration Lab (SIL)

The SIL will be provided with a server which will be the repository for multi-user software applications and the database, as shown in **Figure 2.1-245**. It will run with the Microsoft Windows Server Standard Edition operating system. It will be protected against unauthorized access by Symantec firewall and virus protection, and it will be configured to require a user ID and password to log on.

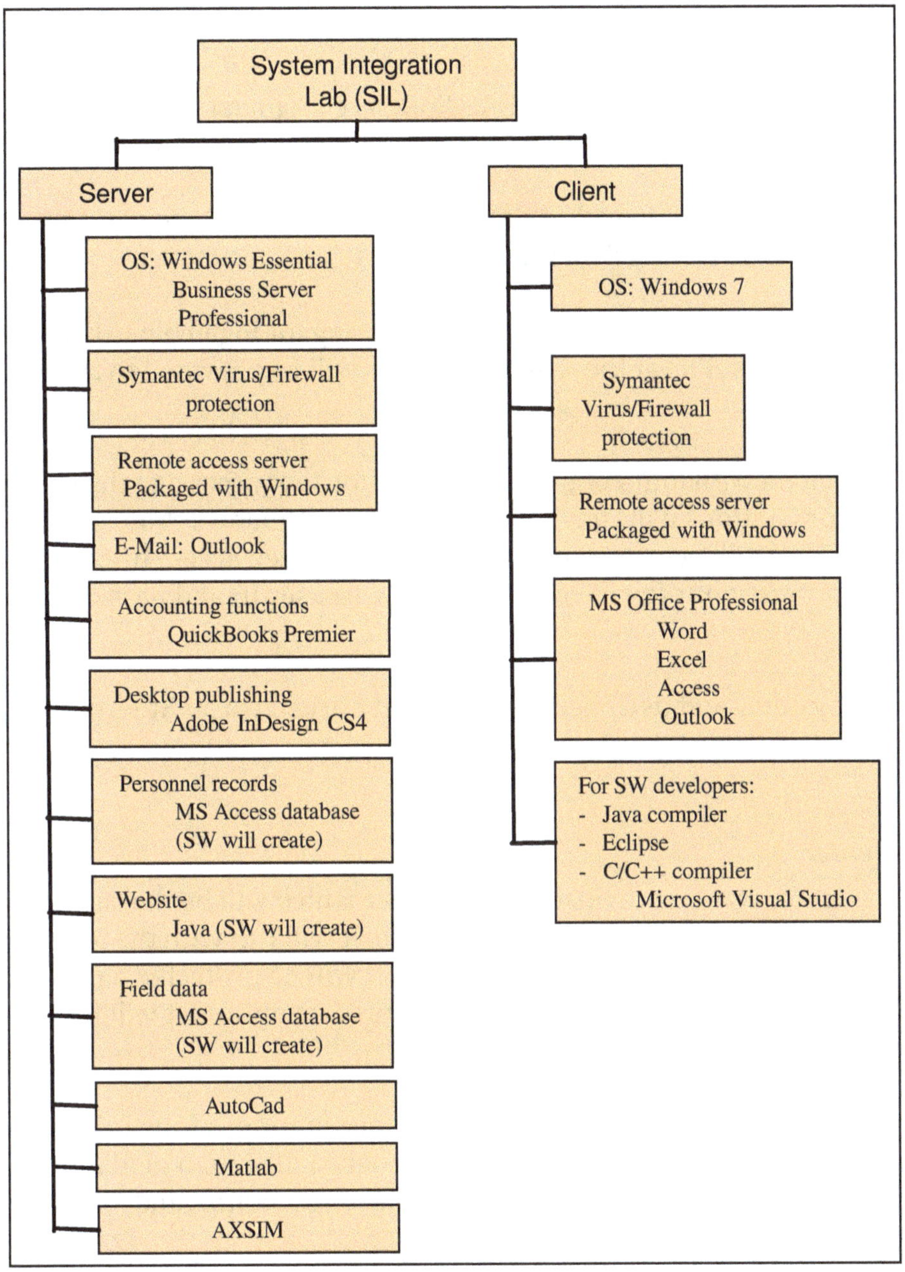

Figure 2.1-245: SIL Software / Hardware Integration. *SIL software packages and hardware components selected based on expected required capabilities of SIL, ensuring a swift installation.*

The multi-user software will be QuickBooks Premier accounting, Adobe InDesign CS4 desktop publishing, AutoCad computer aided design, and Matlab modeling software. Also, Sun Microsystems Java compiler, Eclipse text editor, and Microsoft Visual Studio C/C++ compiler will reside on the server.

The ACCESS database configured by the BrEAD Software Engineering for storing and processing field data at the Central Data Center will be provided to the SIL also.

The automated database input algorithm developed by Software Engineering at the Central Data Center will be installed on the SIL server.

The ACCESS database configured by Software Engineering for logging hardware and software Problem Reports (PR) at the Regional Center will be provided to the SIL also.

The SIL will be provided with four computer workstations for accessing the server software. These workstations will run with the Microsoft Windows Professional operating system. They will be protected against unauthorized access by Symantic firewall and virus protection software, and will require a user ID and password to log in.

The SIL computer systems and peripherals will be networked via an Ethernet network.

Central Data Center

The Central Data Center will be provided with a server which will be the repository for multi-user software applications and the database. It will run with the Microsoft Windows Server Standard Edition operating system. It will be protected against unauthorized access by Symantec firewall and virus protection, and it will be configured to require a user ID and password to log on.

The multi-user software will be QuickBooks Premier accounting, Adobe InDesign CS4 desktop publishing, AutoCad computer aided design, and Matlab modeling software. Also, Sun Microsystems Java compiler, Eclipse text editor, and Microsoft Visual Studio C/C++ compiler will reside on the server.

A BrEAD Software Engineering configured database will reside on the server, and will be configured for use by Human Resources for maintaining personnel records.

A BrEAD Software Engineering configured database will reside on the server, and will be configured for maintaining PRs from all areas of the BrEAD.

A BrEAD Software Engineering configured database will reside on the server, and will be configured for storing the field data received from the SIL and Regional Center. This database will provide a GUI for data input. It will provide reports and graphs of processed data for use by researchers and other interested stake holders.

Report and graph data output will be an ongoing effort as stakeholders changing needs for data are identified.

Figure 2.1-246: Central Data Center Software / Hardware Integration. *Central Data Center software contains mainly COTS software packages, allowing for quick integration.*

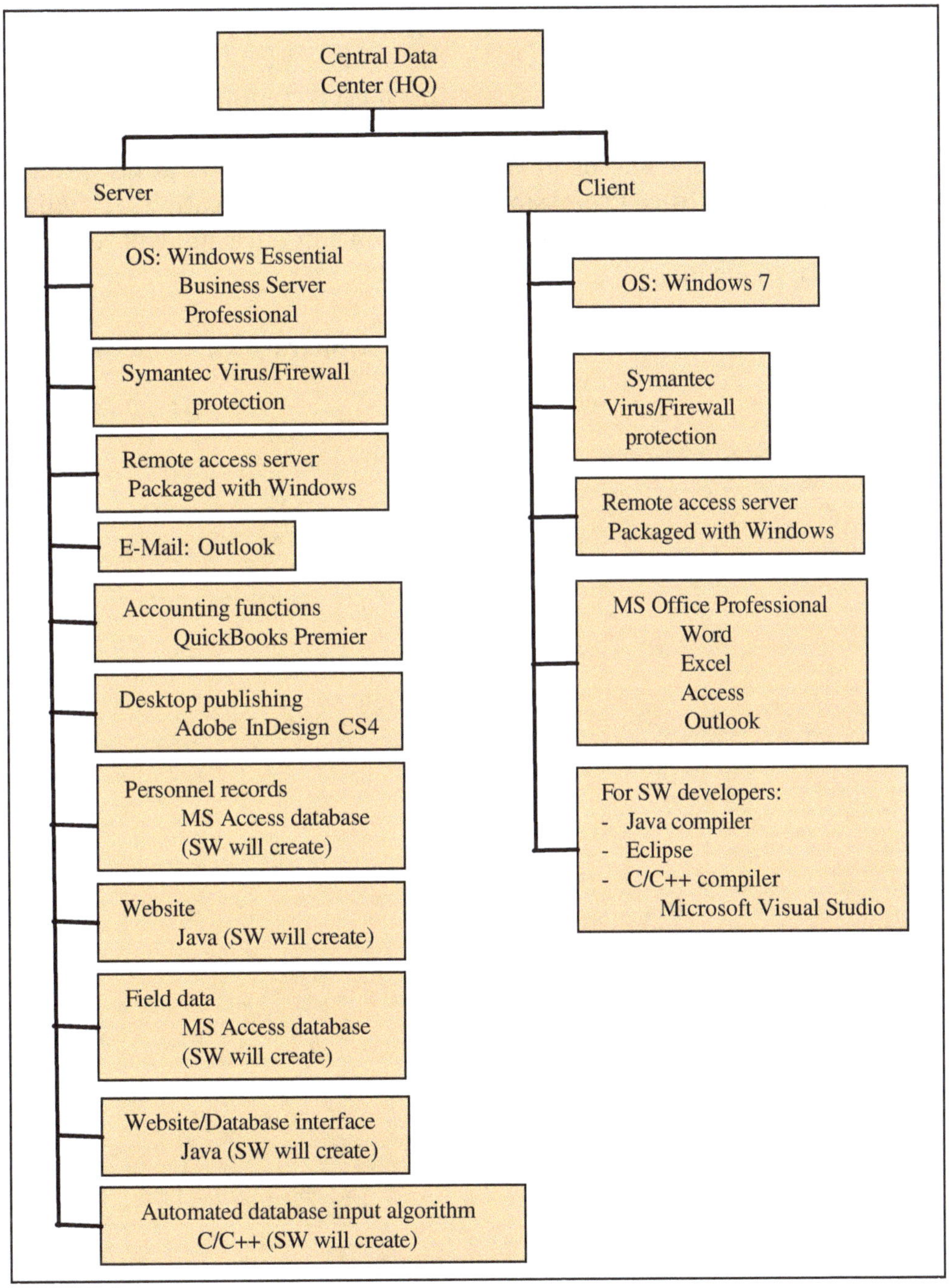

A BrEAD Software Engineering application will reside on the server that will automatically, without user effort, will receive field data from the internet and enter it into the data base.

A BrEAD Software Engineering website will be developed that will provide the internet interface to the BrEAD. It will contain general BrEAD information, e-mail access to key BrEAD personnel, access to BrEAD products (gluten free flower for example), and access to field data database reports, graphs, and raw data.

The Central Data Center will be provided with a laptop computer for each user for accessing the server and communicating with other BrEAD personnel. This computer will run with the Microsoft Windows Professional operating system. It will have the Microsoft Office Professional suite of office applications. It will be protected by Symantic firewall and virus protection software and will require a user ID and password to log in.

The Central Data Center computer systems and peripherals will be networked via an Ethernet network.

Performance Based Systems Specification (PBSS)

The SI PBSS is a comprehensive collection of measurable and verifiable requirements that define the SI Breadfruit Processing System and Enterprise Architecture. A total of 232 performance-based requirements make up the SI technical solution. The initial SI PBSS, included in Technical Attachment I, retains all 18 of the NGRF PBSS requirements as well as requirement statements included in the NGRF Statement of Objectives. The SI PBSS was formatted in compliance with Dept. of Defense Data Item Description DI-IPSC-81431A. This format was chosen because it is a well-established format used by Solutions International, our suppliers, and our partners.

The SI PBSS is a comprehensive collection of measurable and verifiable requirements that define the SI Breadfruit Processing System and Enterprise Architecture.

NGRF and other users will be solicited for feedback as the SI PBSS matures throughout the development effort.

SI PBSS changes will be summarized at all major customer reviews as identified in the IMS

The SI PBSS requirements have been categorized by major subsystems of the SI BrEAD, as shown in **Figure 2.1-247**. Consideration is given for important system engineering functions such as reliability, safety, and maintainability. Hardware and software interfaces, both internal and external, are also fully documented. The SI PBSS also contains a verification matrix which identifies and describes the expected method to verify that the design has met the requirement.

CATEGORY	# of REQTS
Enterprise Architecture - General	7
EA - Weather Station	7
EA - Measuring System	8
EA - Data Communication	6
EA - Regional Center	11
EA - Central Data Center	13
EA - SIL	14
EA - Adaptation requirements	4
EA - Safety	6
EA - Computer	13
EA - Interal Interfaces	21
EA - External Interfaces	7
EA - Reliability	5
EA - Maintenance	10
EA - Software	14
EA - Quality, Logistics, Security	3
TOTAL	149

CATEGORY	# of REQTS
Breadfruit Processing - General	17
BF Processing - Intake	6
BF Processing - Coring / Peeling	7
BF Processing - Shredder	6
BF Processing - Dryer	12
BF Processing - Grinder	3
BF Processing - Packaging / Storage	6
BF Processing - Waste Collection	5
BF Processing - Internal Interfaces	11
BF Processing - Reliability	5
BF Processing - Maintenance	5
TOTAL	83

Figure 2.1-247: PBSS Requirement Summary. *SI PBSS requirements encompass all essential elements of the SI Breadfruit Processing System and Enterprise Architecture.*
The SI PBSS and verification matrix will be updated periodically and delivered to NGRF under SOW 3.1.4.1 and (CDRL 0008). NGRF is an important stakeholder in the requirements development process and therefore will be solicited for feedback as the SI PBSS matures throughout the development effort. All changes to the SI PBSS will be negotiated with NGRF and summarized at all design reviews as shown in **Figure 2.1-248**.

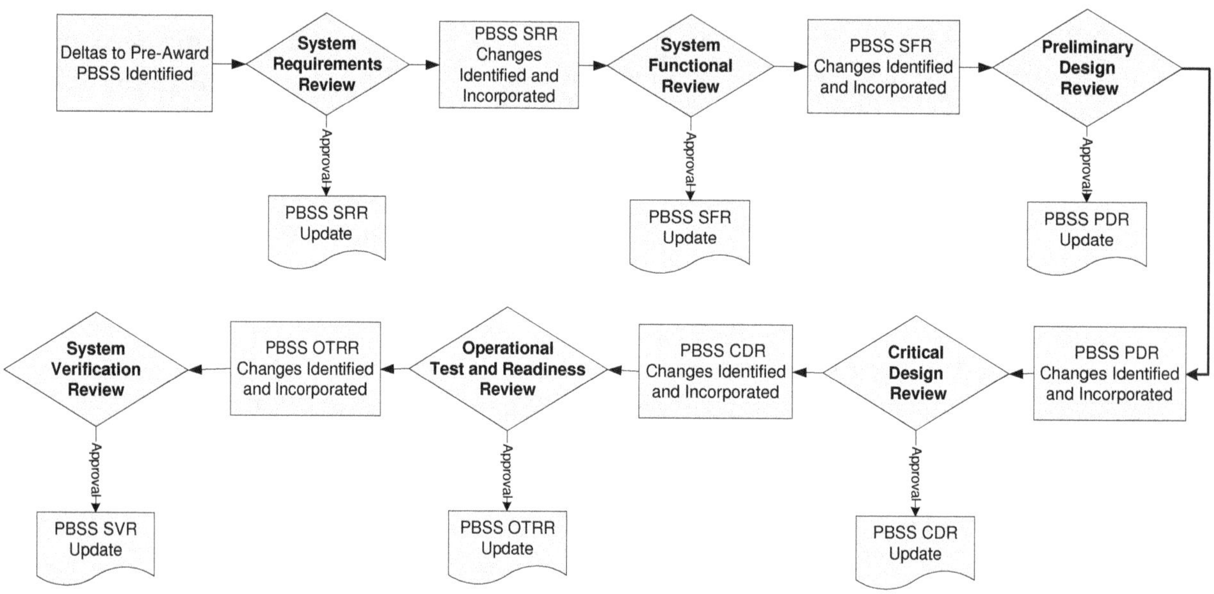

Figure 2.1-248: PBSS Maturity Flow Diagram. *NGRF and other users will have the opportunity to confirm validity and provide feedback on PBSS changes at all stakeholder review meetings.*

At the time of this proposal, there is only one NGRF PBSS requirement that we do not meet. We will provide a status briefing to the customer at the Systems Requirement Review to reach agreement on closing the gap between user need and current performance.

NGRF #10 The breadfruit processing system shall have a failure rate of one failure per 10,000 hours.

Current Estimate: Our current estimate for the breadfruit processing system is 1,400 hours based on hardware selected to meet all other requirements. The biggest contributor to our MTBF of 1,400 is the fan MTBF of 60,000 hours. In the reliability calculation, we modeled our system to be capable to process 450 breadfruit/day. This requires 5 dryers. Each dryer contains 4 fans, resulting in 20 fans in our system. The fan manufacturer's MTBF was calculated as the point at which 10 % of the fans fail. In fact, the fan failure mode is a wearout failure mode

CROSS DOCK ORDER

SON#: 83335744-848484LV

CUS#: NZOBZ36D

ORD#: SQAATP08

REF#: 5F663421

INGRAM DROP SHIP

FROM:

NASHVILLE
ONE INGRAM BLVD
LAVERGNE TN 370863629 US

TO: ORD#: NZOBZ36D

TRIFINITY PP C/O TRADE POINT
8505 100TH ST

PLEASANT PRAIRIE WI 531582226
United States

INGDS **REF#: 5F663421**

Ingram Drop Ship

PRO Number:

Bill of Lading: 38808710013155303

SON#: 83335744-848484LV	CUS#: SQAATP08
ORD#: NZOBZ36D	REF# 5F663421
Ship Date: 09/21/2026	Weight: 21.59
	69 OF 68

SSCC

(00) 0 0038808 716974916 7

and not a random failure (MTBF assumes random failure). Therefore, we are confident our processing system will still have the highest of reliability during the breadfruit harvesting season.

Proposed Resolution: The processing system will only be used during the breadfruit season, and then it will be returned to the depot for cleaning, refurbishing, and functional testing. Assuming the processing system will only be used for 8 weeks (12 hours, 7 days a week) our system only needs to have a failure rate of one failure per 672 hours. (MTBF assumes random failure rate). The SIL will model the wearout life of all the mechanical components and the fans to make sure that the components are replaced before a failure would occur.

Open Systems Architecture

Solutions International proposes an OSA design for the Breadfruit Enterprise Architecture Demonstration (BrEAD) EA and systems. We are applying the Modular Open Systems Architecture (MOSA) standards because the standards require us to evaluate our system against multiple design considerations. Additionally, given the fact that parts of this system will be deployed into multiple cultures with potentially different cultural norms, SI chooses the MOSA standard since MOSA's design approach is a global or 'total system' approach. Listed below are OSA & MOSA subsections and standards that were applied to the overall design of our BrEAD architecture. Our entire system both leverages and integrates COTS components. Each subsystem's design sections provide additional information on components or integration methodologies used.

Evolvability

The COTS materials or components integrated into the Enterprise Architecture (EA) will have the ability to incorporate technology advances. During the design process, we will provide close consideration of the industry our components and materials are originating from. SI will consider a particular industry's historical record with regard to changes or technological advancements. For situations, where we have a choice between two different technologies, we will perform a trade study (weighted scale or risk factor) to decide which technology to move forward with. Additionally, our technology selection will take into consideration other potential uses or other potential industry markets available.

By doing so, the components and materials of the EA will remain viable for the 'normal' life expectancy of at least 3 years for COTS components and materials. When obsolesce issues occur, the trade study performed initially will guide us in choosing a path forward. If additional functions or requirements are placed on our system, we will be able to confidently and quickly adapt to the new requirements since our components are limited in usage capabilities.

Finally, the application of this system demands an open design architecture, not a closed system architecture. The EA will be deployed around the world crossing multiple culture groups with multiple cultural norms. The opportunity exists for numerous local communities or countries to fabricate or develop some system components. By doing so, there is an immediate economic or social economic impact to the area and presents the possibility of reducing costs associated with shipping. Our success highly depends on 'buy-in' from local personnel and if we introduce closed components to our system, it would increase logistics and manufacturing costs while lowering our chance of success.

Scalability

In the event a larger system is needed, the hardware and software used in our EA natively supports 1 to 1000 users and has the ability to be scaled into a larger system. Mechanical components of the system are deployed 'readily' mechanically equipped for ease in installing motors or securing component. Furthermore, any additional design modifications or enhancements will require the ability to be 'scalable'. Upgrades to web interfaces and software upgrades to system components can be conducted without disruption to the rest of the system. The COTS hardware and software components used are Windows compatible which is a universally accepted operating system (OS) around the world. The communication protocols used in the system are those found in standard Information technology and/or communication networks such as Transmission Control Protocol / Internet Protocol (TCP/IP).

Portability

The EA software components are COTS-based utilities and OS, which are compatible with many different type of COTS hardware. The BrEAD processing system is comprised of COTS components found throughout the world or available in 'similar too' or 'like' manners. Description of any system component is available to any government body or interested party.

International safety, manufacturing, and environmental standards will be monitored throughout the design process and implementation. The system design is simple in nature, allowing it be used in multiple countries and multiple communities.

Modularity

Our software and hardware are COTS-based components, resulting in a modular system. OSs and utilities on components installed at the headquarters, SIL, a regional center, or a processing site can be mapped, transferred, or installed on other systems. No custom Application Programming Interfaces (API) are used. For example, Windows XP is the base OS on all processing center PC's. The GUI used in a particular region or deployment area will be developed in coordination with local personnel to maximize familiarity and effectiveness while reducing complexity. The BrEAD processing system components are designed with modularity in mind so they can be quickly disassembled and transferred from one community to the next. Conversely, the design of the system allows for components to be reassembled in short amount of time as well. The processing system is comprised of hardware that could be fashioned or modeled out of different materials found throughout the world.

Component changes, within the processing system, can be made during the time period when the processing system is active without impact the performance of the system. **Figure 2.1-249** depicts the processing system. Each subsystem component is independent of the next.

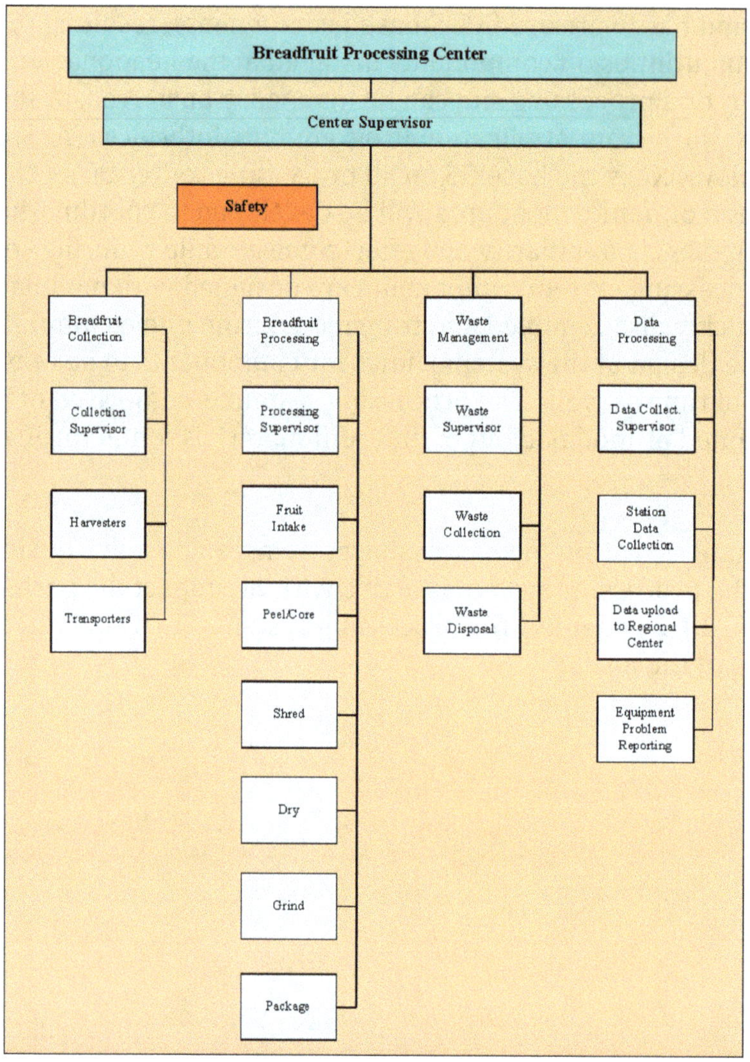

Figure 2.1-249: Processing System Components. *Each processing system component is independent, proving our system is modular.*

Not only are some of the BrEAD EA HW components partitioned into discrete, cohesive, and self-contained units, but so are the SW components. Similar to obsolescence concerns with the hardware, SI expects occasional support of software will be discontinued. Therefore, most of chosen software components have the ability to be quickly replaced with a similar OS or application.

Numerous COTS components and materials were considered for use in our system. SI chooses to leverage off of the work completed by University of St. Thomas engineering students, even through their design has limited field demonstration. A strong feature in the design of their processing system was the use of solar power. The availability of electric or natural gas cannot be counted on as a reliable and available source of energy because of the diversity of regions our system would be installed. Secondly, their design is simple. The processing system does not contain any specialized or custom components. The design has the capabilities of being fabricated here or at the host community. The combination of their field results and their design considerations provided enough confidence for us to move forward with their design.

Rationale for Modularization Choices

Solutions International has chosen a very modular system design to keep the system easily upgradeable and useful to many different types of people in many different cultures.

Employ Layered Modular Architecture

The system uses COTS hardware and software. These devices adhere to use open standard, industry accepted interfaces and can be sourced from different suppliers.

Maintainability

Solutions International deploys a simple but effective design that enables the user to process more food during the limited harvesting season. Additionally, it provides data to researchers and others about the effort ongoing throughout the world. The simple design of the system aids in operator's ease of use and maintenance. This reduces workload and increases productivity. Our system is operationally supported by both regional center and headquarters office. The Call Center within BrEAD EA is interoperable with the regional centers and the Solutions International's Software Integration Lab (SIL) making it easy for remote troubleshooting. Leased processing systems are available for low cost of maintenance. Refer to section 2.1.13 and Annex B for more information on BrEAD EA maintainability.

Figure 2.1-250 identifies some of the COTS components being integrated into the Central Data Center. Since this hardware is standard and commercially available all over the world, software like Windows XP or Microsoft Office products may be hosted on these systems. If problems would occur on a software or hardware basis, numerous resources worldwide exist to resolve the issue, instead of returning the equipment to the manufacturer.

Server	PowerEdge 2900 III	DELL
Network Switch	Powerconect 3524	DELL
Laptop	Latitude E6500	DELL
1 Printer/Fax/Scanner combo	OfficeJet Pro L7590	HP

Figure 2.1-250: Central Data Center COTS HW Components. *EA design utilizes COTS components, reducing the time required to perform maintenance.*

The System Change Requests (SCRs) system database will be used to track and manage any problem reported on hardware or software operation. A Change Review Board (CRB) co-chaired by the NGRF and SI personnel will triage SCRs and assign them to project engineers as appropriate for analysis.

The SCR will also be used to highlight potential weakness is the present design of the system along with identifying a potential area for improvement design in a technology insertion or a next generation improvement.

Technology Insertion and Refresh

COTS components and materials are used throughout the design of the entire system. The design of the system allows for any component to be replaced with minimum impact on the overall system. The OS and application software and utilities are upgradeable.

Figure 2.1-250 is a sample list components used in the system today. As technology moves forward and obsolescence issues come to life, these components have the ability to be replaced all at once or phased in over time. Hence, using COTS will minimize the impact this issue will have to the present of future design of the system. It is fully expected that COTS components can be inserted into any part of the system nearly anytime with either no or limited impact to the system.

Reusability

The majority of the BrEAD processing components can be re-used. For example, all pieces of station equipment like the peeler, shredder, grinder and the dryer may be returned to the regional center after the processing has been completed. Back at the regional center, the components of the peeler, shredder, and grinder may be sharpened and the plastic surrounding the dryer replaced. After these actions have been completed, the station equipment may be secured at the regional center or packaged up and shipped for use in another area of the world.

SW reusability will be maintained and used wherever possible keeping within the laws and regulation governing software license and possible re-use if permitted.

Life Cycle Supportability

SI's solution for the BrEAD EA entails MOSA and general Open Systems Architecture standards. The design of the system itself calls for materials and components that can be procured here or abroad. No custom methodology or 'specialized' software has been developed for any part of the system. The software applications that have been developed for use in the BrEAD EA could be re-mapped to another language in minimal time and at low risk.

Common life cycle support activities such as a System Change Requests (SCRs) database and a database with BrEAD EA call center tracking issues and problems.

Prior to any component being integrated into the system, a life cycle support analysis will be completed to gauge the expected level of lifetime support from anyone vendor and also to ensure no proprietary or vendor-unique technology exists in their product.

Traceability of System Requirements

Attachment 1 for a complete look at all system requirements and their associated traceability to BrEAD EA and systems

- The design of all data processing equipment at the BrEAD Regional center shall be intuitive to the user.
- The Breadfruit Enterprise Architecture Demonstration (BrEAD) shall provide an automated research, marketing, and user data collection capability for NGRF head quarters in St. Paul, Minnesota that will provide 24 hour per day data from researchers, marketers and users worldwide.
- The design of the breadfruit processing system shall be modular, such that any specific geographic requirements can be accommodated without breadfruit processing system redesign. Vehicle crossing time-Time stamp by software initiated by downstream loop.
- The breadfruit processing system human machine interface shall be intuitive; that is the human operation of the system shall require minimal documentation- ideally none.
- The breadfruit processing system shall be assembled to an operational state within TBD days after shipment to its operating location.
- The BrEAD Regional Center shall have operational resources for local troubleshooting of problems that may exist at the breadfruit processing sites.
- The BrEAD Regional Center shall have operational resources to transport equipment and other logistics supplies to other regional centers.
- The breadfruit processing system shall be scalable such that it can produce a minimum of 290 kg per day of breadfruit products for family use to TBD kg per day for industrial scale

System Development and Demonstration of the Breadfruit Enterprise Demonstration (BrEAD) and Systems Design

Solutions International Proposal for the Breadfruit Enterprise Demonstration and System Design

Volume 2: Technical

Book 2: Program and Schedule

RFP NO: E00010-07-R0001

Copy 1 of 2

May 4, 2009

Prepared For:

New Green Revolution Foundation (NGRF)

Prepared By:

Solutions International

123 Pillsbury Drive

St. Paul, MN 55114

CAGE CODE SI4ME

Table of Contents

Table of Figures

Program and Schedule

SI's Program and Schedule demonstrates a strong understanding of NGRF objectives and the means to accomplish them.

Our set of submissions is highly integrated and focused on the key elements crucial to meeting BrEAD's unique requirements.

SI has assembled a detailed set of integrated proposal artifacts that address the ways in which we propose to develop a successful BrEAD system. This set of data items is responsive to all NGRF objectives as delineated in the RFP and SOO. Collectively they illustrate what we plan to do to achieve program objectives and how we plan to do it. This set of submittals starts with a comprehensive SOW, developed directly from the RFP and SOO that lays out the objectives we plan to meet and the deliverables that we are committed to produce to ensure program success.

Our SOW relates directly to our IMP, which describes our roadmap to project completion in terms of major program events (reviews), the tasks tied to them, and their completion criteria. The IMS places the identified tasks and events in a timeline alongside supporting activities and program milestones. SI's approach and planning is cognizant of real world factors related to breadfruit and incorporates provisions that synchronize project events with the breadfruit season at the proposed processing sites.

With this proposal, SI has included a SEMP in Annex G. The SEMP describes IPT organizational structures and details how we will team with the NGRF to create the BrEAD system. SI's strengths in human factors engineering and modeling and simulation are explained and related to the accomplishment of program objectives. Project staffing, plans, tools, and work products are explained and interrelated in a timeline. Our plan for the integration of subcontractors is also discussed, and a

Software Development Plan describes our approach to creating and qualifying BrEAD software assets.

Finally, SI has produced a detailed and robust Risk Management Plan for this proposal, and a "Top Ten Risk List", to ensure that BrEAD is guided to successful completion. SI considers risk management a crucial component to ensure project success. We have defined detailed roles, responsibilities and processes for the identification, management and mitigation of program risk. Our mature risk procedures enable us to manage risk from day 1.

Proposed SOW and CDRLS

In Annex D, Solutions International has provided a detailed SOW that reflects all of the objectives stated in the NGRF provided SOO. The SOW will be coupled with the CDRL items for the appropriate task as a response to the SOO. The SOW addresses all the CDRLs, delineating the work task(s) required to develop each deliverable.

Integrated Master Plan

The Integrated Master Plan (IMP) called for in the Statement of Objectives (SOO) and detailed further in the Statement of Work (SOW), is an event-driven path that documents the events and tasks necessary to complete the work, while tying each task accomplishment to a key program event. The IMP is the foundation for the Integrated Master Schedule (IMS) and maps from the Contractual Work Breakdown Structure (CWBS). The IMP is provided in Annex E.

Two Integrated Project Teams (IPTs) will lead the design of the BrEAD EA, one team focused on the Enterprise Archictecture (EA) and the other on the Breadfruit processing system architecture. On each team, NGRF team members will combine with SI team members in order to help meet the project goals and milestones while sharing job insight and knowledge transfer.

SI prides itself in its Human Engineering and modeling expertise. The modeling and simulation work completed thus far has identified useful design constraints and requirements that will be needed in the final system design. Our work in human engineering is second to none since we've co-developed and teamed with over 50 other engineering organizations on various projects of all sizes.

The Integrated Master Schedule (IMS) depicts the key events and milestones, significant tasks, and success criteria for the BrEAD EA development, production, and deployment.

A detailed program plan describing SI's resources, practices, and capabilities and will be provided upon award of contract.

Integrated Master Schedule

By utilizing external resources and our integrated approach, Solutions International plans to deliver a completed BrEAD system on August 1, 2010.

The Project Schedule, referred to as the Integrated Master Schedule (IMS), is maintained in Microsoft Project. The Project Schedule is located in Annex F. The IMS is our program schedule, which identifies the technical tasks and activities necessary to complete the work effort scoped within the IMP. The BrEAD EA and systems IMS is based upon an estimate contract order award date of 6 June 09.

Schedule management is based on the Solutions International (SI) Earned Value Management System (EVMS) maintained by Program Management. The EVMS provide work completion status, critical path activities, and key program milestones.

Solutions International intends to complete and deliver the entire system NLT August 2010. In doing so, the system will be available to regions throughout the world where Breadfruit Harvesting occurs sometime in the late 3rd or 4th quarters of the year.

As a means of meeting this schedule, Solution's International has compiled a Risk Management Plan (RMP). This plan places a probability and subsequent priority to manage the particular risk during the development of the BrEAD EA. Risk mitigation plans are then incorporated into the IMS to identify program dependencies and schedule impacts.

The IMS has also identified the 'Critical tasks' crucial to the success of BrEAD EA. Our critical task is SI's Cultural Norms Working Group. Since we are planning to integrate our Enterprise Architecture in many different regions and culture, we need to understand the local cultural norms and their way of doing business. This group, containing Subject Matter Experts (SME), will clearly identify the processes in place and the 'steps' that will be used facilitate our system integration. The group will identify these processes for both multiple countries and for regions within a single country. SI's Cultural Norms Working Group plans to use our contacts with other non-profit or not-for-profit aid organizations as a starting point in developing this process.

Another critical task is the dryer design. The current breadfruit dryer design, as designed by the University of St. Thomas students, does not successfully dry large quantities of breadfruit. Once awarded the contract, SI's plans to use the St. Thomas design as a baseline and enhance the functionality of the design. If our proposed enhancement strategy is ineffective, we have provisioned for outsourcing the dryer design to an industry SME, such as Commercial Dehydrator Systems.

System Engineering (SE) Management

The methods and strategies by which SE management will be implemented for the BrEAD project is delineated in the System Engineering Management Plan (SEMP). The SEMP focuses on the technical planning, system engineering (SE) processes, personnel and tools that will be used to assure successful execution of the BrEAD project. It identifies and describes the project and technical reviews that support program management, and their place on the BrEAD event timeline. The SE management structure for BrEAD is detailed, as are vital SE activities, such as requirements analysis. The SEMP spells out the stakeholder involvement necessary for project success, and provides detail on some specific areas, such as Human Engineering, where specialized SI capabilities can make a difference in resolving technical challenges during BrEAD development.

SEMP Organization

The SEMP is organized into 5 sections,

1. Introduction – Executive Summary, Project Summary, Scope
2. Planning and Control – This section describes the organizational structure, personnel, events and planning tools that will be used to guide the BrEAD project to successful completion.
    a. Technical organization of the Integrated Project Teams (ITPs)
    b. Roles and responsibilities of IPT lead engineers and managers
    c. Sequence, timing, purpose and gating events of project reviews.
    d. Plans and specifications to be produced during BrEAD development.
    e. Document control procedures
    f. Technical Performance Measures tracking..
3. Key Technical and Management Processes – Describes the essential SE strategies that will be employed on the BrEAD project to manage requirements, risk, interfaces, configuration, changes, issues and problems. Discusses trade study and design analysis procedures that will be used to select amongst design alternatives.
4. Tools and Facilities – This section delineates the electronic tools that will be used to support system engineering and vital project processes for the BrEAD.
5. Engineering and Support Specialties – This section discusses how specialized engineering and support functions will be implemented during BrEAD development. Specialties discussed include human engineering, safety and security, logistics, reliability, maintainability, and supportability.

Risk Management Plan

Our comprehensive risk management plan prevents small issues from becoming emergencies. The risk management plan begins during proposal development and continues through project development and completion. With our risk management plan program, we are able to deliver a BrEAD system under the schedule constraints.

> **Solutions International's Risk Management Plan is robust, ensuring our project will complete on time and perform as expected.**
>
> **- Risk Management roles defined**
> **- Risk Management activities and schedule defined**
> **- Lessons learned from BrEAD and prior programs maintained and utilitized**

The Risk Management Board tracks each risk identified either in the proposal phase or during the development phase. At contract award, the risk list developed during the proposal phase will be used as a start. As new risks are identified through the IPT, they are added to the Risk List. Our team uses the Risk List to track potential risks. If a risk materializes, we will implement the contingency or mitigation plan immediately.

Roles and Responsibilities

The System Engineering Coordinating Teams for the project are described in the SEMP, Paragraph 2.1.1. Each IPT (BF Processing System and EA) has a system engineering team led by the Chief Systems Engineer (CSE). Each team also has a Risk Officer assigned. The roles and responsibilities for the CSE and Risk Officer are listed below:

Chief Systems Engineer	The CSE will assign a Risk Officer to the project and identify this individual on the project's organization chart The CSE, the Risk Officer, and other members of the Risk Management Board shall meet monthly to review the status of all risk mitigation efforts, review the exposure assessments for any new risk items, and redefine the project's Top Ten Risk List.
Risk Officer	The Risk Officer has the following responsibilities and authority: coordinate risk identification and analysis activities, maintain the project's risk list, and report risk resolution status to management.
Project Member Assigned a Risk	The Risk Officer will assign each newly identified risk to a project member, who will assess the exposure and probability for the risk factor and report the results of that analysis back to the Risk Officer. Assigned project members are also responsible for performing the steps of the mitigation plan and reporting progress to the Risk Officer biweekly.

Risk Documentation

Each risk will be maintained in Risk List for tracking. A risk data item, as defined below, identifies the minimum amount of information needed for each risk.

Risk List	The risk factors identified and managed for this project will be accumulated in a risk list, which is located in **Figure 2.2-256**. Once the contract is awarded, this list will be maintained on the IDE in an Excel spreadsheet. The ten risk items that currently have the highest estimated risk exposure are referred to as the project's Top Ten Risk List.
Risk Data Items	The following information will be stored for each project risk: Risk ID, classification (schedule, technical, or cost), description of risk, probability, impact, risk exposure (Green, Yellow, or Red), first indicator that risk is becoming a problem, mitigation approaches, owner, date due, contingency plan, author
Closing Risks	A risk item can be considered closed when the planned mitigation actions are completed and the risk exposure has been reduced to Green.

Risk Management Plan Process

Our risk management plan process contains the following steps: risk identification, risk analysis and prioritization, risk management planning, risk resolution, risk monitoring, and lessons learned. We maintain the lessons learned for future risks on this project, as well as risks on future programs. Our lessons learned from prior projects are readily available. With these prior lessons learned, our project execution will inherently have lower risk.

Our risk management plan requires that only risks with a risk exposure of yellow or red, need to identify a mitigation and contingency plan. A mitigation plan reduces the probability of occurrence whereas a contingency plan reduces the impact of the risk. Our risk assessment process assumes that green risks do not require a mitigation or contingency plan because the risk either has a low probability of occurrence or a low impact. However, if during the development phase, it appears a risk item's risk exposure should be elevated to yellow or red, the risk item will automatically make the Top Ten Risk List.

Risk Identification, Analysis, and Prioritization

The SEMP identifies the IPT structure that Solutions International uses. Each IPT will identify technical, schedule, and cost risks. Those who identify a risk will assign a probability (1 through 5) and impact (1 through5) to the risk. **Figure 1-213** and **Figure 2.2-252** define the probability and impact classification. These risks propagate to the Risk Management Board, which consists of members from the SE coordinating team for both EA and the Processing System. The Risk Officer reviews the risk classifications and adjusts, if necessary. The risks are sorted by descending risk exposure (Red, Yellow, and Green). **Figure 2.2-253** defines the Risk Exposure Matrix.

Risk Impact Classification			
ID	**Technical Performance**	**Schedule**	**Cost**
1	Minimal or no consequence to performance	Minimal or no impact <1 month	Minimal or no impact; <$1M
2	Minor reduction in performance or can be tolerated. Minor changes may be required	No impact on critical path. Able to meet non-critical path schedule slip of 1 to 2 months	Budget increase for SDD $1M-$25M; or production cost increase of $1M
3	Moderate reduction in performance or supportability workarounds are available. Moderate changes may be required	Potential non-critical path schedule slip of 2 to 3 months. Potential impact to critical path, recoverable. No impact to key dates	Budget increase for SDD $26M-$125M or production cost increase of $1M to $2M
4	Significant degradation in performance or major shortfall in workarounds may not be available or have significant consequences. Significant changes may be required	Potential non-critical path schedule slip of 4 to 5 months; potential schedule slip to critical path and is not recoverable; potential impact to key milestones	Budget increase for SDD $126M-$225M or production cost increase of $2M to $5M
5	Severe degradation in performance and cannot meet key threshold. No workarounds	Potential schedule slip of >6 months; critical path unrecoverable; cannot meet key program milestones	Budget increase for SDD >$225M or production cost increase of >$5M

Figure 2.2-251: Risk Impact Definition. *The impact definitions improve our ability to correctly categorize the risks.*

Risk Probability Classification	
ID	**Likelihood**
1	Not Likely
2	Low Likelihood
3	Likely
4	Highly Likely
5	Near Certainty

Figure 2.2-252: Risk Probability Definition. *The probability definitions improve our ability to correctly categorize the risks.*

Risk Exposure Matrix						
Probability	**5**	G	Y	R	R	R
	4	G	Y	Y	R	R
	3	G	Y	Y	Y	R
	2	G	G	G	Y	Y
	1	G	G	G	G	Y
		1	**2**	**3**	**4**	**5**
	Impact					

Figure 2.2-253: Risk Exposure Matrix. *The Risk Exposure Matrix provides guidance in risk management, making sure we correctly address each risk.*

Our integrated master plan (IMP) and integrated master schedule (IMS) incorporate schedule and budget contingencies defined during our risk analysis. During the proposal phase, an integrated risk team, including the planning/scheduling team, evaluates potential program risks. If risk has an exposed risk of yellow or red, our IMS accounts for the risk and its potential mitigation. Upon contract award, when an IPT identifies a risk, the Risk Officer engages the planning group if the risk affects the IMS or IMP. When we close a risk, we flag the risk and mitigation in the schedule. When schedule update occurs, we will incorporate the risk and mitigation changes.

Risk Management Planning

The Risk Officer assigns the top ten risks to individual project members for development and execution of a risk mitigation plan. Generally, these individual project members are those who identified the risk originally.

The individual who is assigned the risk recommends actions that will reduce either the probability of the risk materializing into a problem or the impact of the risk if it does. The assigned individual presents the proposed mitigation plan to the Risk Officer and Risk Management Board. The Risk Management Board and Risk Officer approve the mitigation plan.

The Risk Officer collates the mitigation plans for assigned risk items into a single list, which is the Top Ten Risk List. The Risk List also captures all other identified risks for future reference. The Risk Officer places the Risk List in a general location on the IDE so all individuals and program management team can access the risk list.

Risk Resolution
Each individual who is responsible for executing a risk mitigation plan carries out the mitigation activities.

Risk Monitoring
Bi-monthly, the Risk Management Board reviews the entire Risk List. The Risk Officer assigns each risk author to review the initial risk data item. The risk author documents if any parameters of the risk data item has changed, updates the mitigation plan if necessary, and closes the risk if possible.

The Risk Officer categorizes each risk data item after the bi-monthly review:

- New: A new risk is identified.
- Watch list: The risk exposure has increased
- Reduced Risk: The risk exposure has reduced
- No change: The risk data item has not changed from previous review
- Mitigation Impact: The risk mitigation/contingency plan has changed
- Closed: The risk has not materialized and can be closed.

The Risk Officer collects all the metrics for the entire Risk List and then presents a percentage for each category above. If the ratio of "Watch List" to "Closed" + "Reduced Risk" + "No Change" is greater than 0.5, then the Risk Management Board and Risk Officer will re-evaluate the risk management plan and perform a deep dive into the program execution.

Monthly, the Assigned Individuals (Top Ten Risk List) will report status of their open risks and mitigation plan. If the Risk Officer determines the assigned individual is not carrying out the mitigation effectively, the Risk Officer will escalate this risk to an appropriate level of management for visibility and action.

Lessons Learned
The Risk Officer enters risk-related lessons learned after each life cycle phase. The Risk Officer holds a 4-hour brainstorming session with all risk owners, IPT leads, the CSE, and the Program Manager. Risk Officer categorizes and documents the risks. The categories are as follows:

- Risk Rating
- Risk Exposure Matrix
- Mitigation
- Risk Review
- Risk Monitoring
- Schedule Risk
- Technical Risk
- Cost Risk

Schedule for Risk Management Activities

The following table identifies the when each risk management activity will occur and the frequency of its occurrence.

Risk Management Activities	Type of Activity	Schedule
Risk Identification	Proposal-identified risks	During proposal phase
Risk Identification	Risk Assessment Workshop	At Contract Award
Risk Identification	IPT Member Interviews	Each life cycle phase
Risk List	Risk Prioritization	1st : Week after Workshop 2nd on: Week following Risk List Review (bi-monthly)
Risk Tracking	Review status of Top Ten List	Monthly
Risk Monitoring	Entire Risk List Review	Bi-monthly
Lessons Learned	Brainstorming	Completion of each life cycle phase
Lessons Learned	Update lessons learned database	Week following brainstorming section.

Figure 2.2-254: Risk Management Activity Schedule. *By clearly defining when each program management activity will occur, robustness is built into our risk management plan.*

Risk Management Tools

The Risk List is contained in an Excel spreadsheet. The IDE contains the spreadsheet, where both NGRF and Solutions International individuals can access the spreadsheet. Assigned individuals and the Risk Officer will present the metrics described earlier in the risk management plan and risk item status using Microsoft power point. We will log and store all status presentations for reference in the future.

The lessons learned database is an Excel spreadsheet. The format is as described in the lessons learned section. Our risk management team (Risk Officer, CSE, and board members) will create the interview questionnaires in Microsoft Word.

BrEAD Program Risk Items

During the proposal phase, our team identified ten risks and proposed mitigation plans. **Figure 2.2-255** graphically depicts the risk item's risk exposure. We created a mitigation plan for nine of the risks, since only one risk has a "Green" risk exposure.

Figure 2.2-255: BrEAD Program Risks. *Solutions International is confident that we have completely and comprehensively addressed any risk in our proposal.*

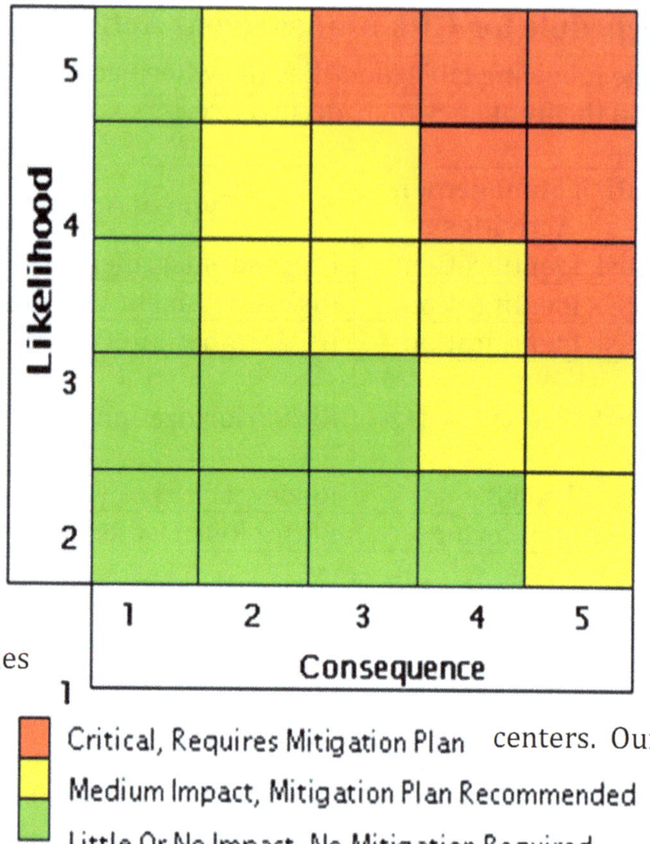

Our proposal team also evaluated potential risks, not only associated with internal systems, but also with external dependencies. Our success heavily relies on the partnership with local aid organizations and local communities in the areas we will install processing systems and regional centers. Our success also relies on the Breadfruit Institute in Hawaii, where we plan to install the SIL and subcontractor performance. As such, we identified the following risks associated with external dependencies:

- Locals aren't interested in the breadfruit flour
- Local Aid Organization not interested in working with us
- Locals do not bring us breadfruit to process
- CTI does not provide processing equipment to SIL or to prototype site
- Breadfruit Institute will not allow us to set up our SIL facility at their site and/or use their breadfruit

Figure 2.2-256 identifies all the technical and schedule risks known to date, each risk item's risk exposure, and each risk item's potential mitigation. All of these risks, with exception to risk items #4, #5, #7, and #9, are on the critical path. Items #4, #5, #7, and #9 are not on the critical path because they do not prevent Solutions International from meeting the schedule requirements or providing the BrEAD system as described in this proposal.

Our proposal team assumes a certain amount of uncertainty regarding the initial Risk List. As previously stated, this Risk List is based on information known to date. As our project develops, we will become more informed and will make the corrections necessary to our Risk List.

Figure 2.2-256: BrEAD Risk List. *Solutions International defines a first indicator for each risk and a mitigation for every yellow and red risk, providing confidence our project will perform as expected and be completed on time.*

Risk ID	Description	Prob.	Impact	Risk Exposure	First Indicator	Mitigation
1	Harvesters steal fruit from other people's trees, causing tension and fighting amongst the locals	3	1	Green	Complaints from locals	N/A
2	Harvesters bring in fruit that has been on the ground	3	3	Yellow	Visual Inspection at Intake	Create an incoming inspection process – Visual and physical inspection. Provide pictures to document if good or not.
3	Unknown how long fruit has been off the trees - could be longer than a couple days	3	3	Yellow	Visual Inspection at Intake	Create an incoming inspection process – Visual and physical inspection. Provide pictures to document if good or not. Ensure the processing system is less than 1 day

Risk ID	Description	Prob.	Impact	Risk Exposure	First Indicator	Mitigation
4	Manual processing system provides higher risk for human error, contamination, and health issues	4	4	Red	Complaints from locals	Use recommended health standards and practices for processing site operation and maintenance to avoid contamination. Set-up a visual check at each station
5	Won't be able to complete the processing system in time for proposed beta site	3	4	Yellow	PDR	Create a prototype processing system that can be delivered prior to project completion for testing and user feedback
6	External Dependency: Locals aren't interested in the breadfruit flour	3	4	Yellow	Cultural Modeling	Perform cultural modeling prior to installing a processing system. Investigate alternative uses for breadfruit flower: ie. animal feed
7	Manual performance data collection is tedious and locals won't accurately record the information.	4	3	Yellow	Prototype Processing system results	Define supervisor duties to review collected data. Automate the processing system and data collection.

Risk ID	Description	Prob.	Impact	Risk Exposure	First Indicator	Mitigation
8	External Dependency: Local Aid Organization not interested in working with us.	2	4	Yellow	Regional Center Development	Create a team to evaluate the proposed install location. This team will survey the location, environment, and government for a month to gather the necessary data.
9	Unable to meet the PBSS Reliability Requirement of 10,000 hrs for the processing system.	5	3	Red	Initial Reliability Prediction Report	Identify why requirement isn't met. Propose a new requirement (2,000 hrs)
10	Performance requirements determined from modeling and simulation could be incorrect.	3	3	Yellow	SIL and/or Prototype processing system	Update performance data based on verified modeling & simulation data
11	The dryer design does perform as anticipated: we can't control the temperature, the fans don't provide adequate air flow, or the dryer takes longer than expected to dry the fruit.	2	5	Yellow	SIL	Prior to releasing the prototype, test the dryer and verify it meets the dryer requirements. Utilize Commercial Dehydrator Services' expertise in dehydrator designs.
12	Subcontractor: CTI does not provide processing equipment to SIL or to prototype site	3	4	Yellow	Critical Design Review	Design our own processing system equipment

Risk ID	Description	Prob.	Impact	Risk Exposure	First Indicator	Mitigation
13	External Dependency: Breadfruit Institute will not allow us to set up our SIL facility at their site and/or use their breadfruit	3	4	Yellow	Preliminary Design Review	Change the SIL location to the Caribbean
14	Locals do not bring us breadfruit to process.	3	5	Red	Prototype processing system	Regional Center works with local government and community to promote this activity. Increase marketing in the local communities.

Test & Evaluation

Proactive T&E planning will ensure that the most efficient utilization of test articles, assets, and facilities are implemented to meet system test objectives within cost and schedule constraints.

The Systems Integration Laboratory located near Breadfruit Institute will be used to verify requirements and target the gluten-free market through M&S, breadfruit processing trials, and iterative prototyping

Test and Evaluation (T&E) is an important element in the development of the BrEAD Enterprise Architecture and breadfruit processing system. Our comprehensive test and evaluation approach provides feedback for potential design improvements, assesses risk, evaluates technical issues, identifies and analyzes data trends, and establishes confidence in the design before deploying the system operationally.

The BrEAD design will be verified against requirements throughout the development and operational phases of the program. We will generate a Test and Evaluation Master Plan (TEMP) (SOW 3.1.5 and CDRL 00018), which details the test objectives, articles, assets, facilities, and test schedule for verifying all requirements identified under demonstration and test methods. The Integrated Test Team will ensure that the

T&E approach makes most efficient use of all test articles, data processing assets, and test support equipment in the SIL.

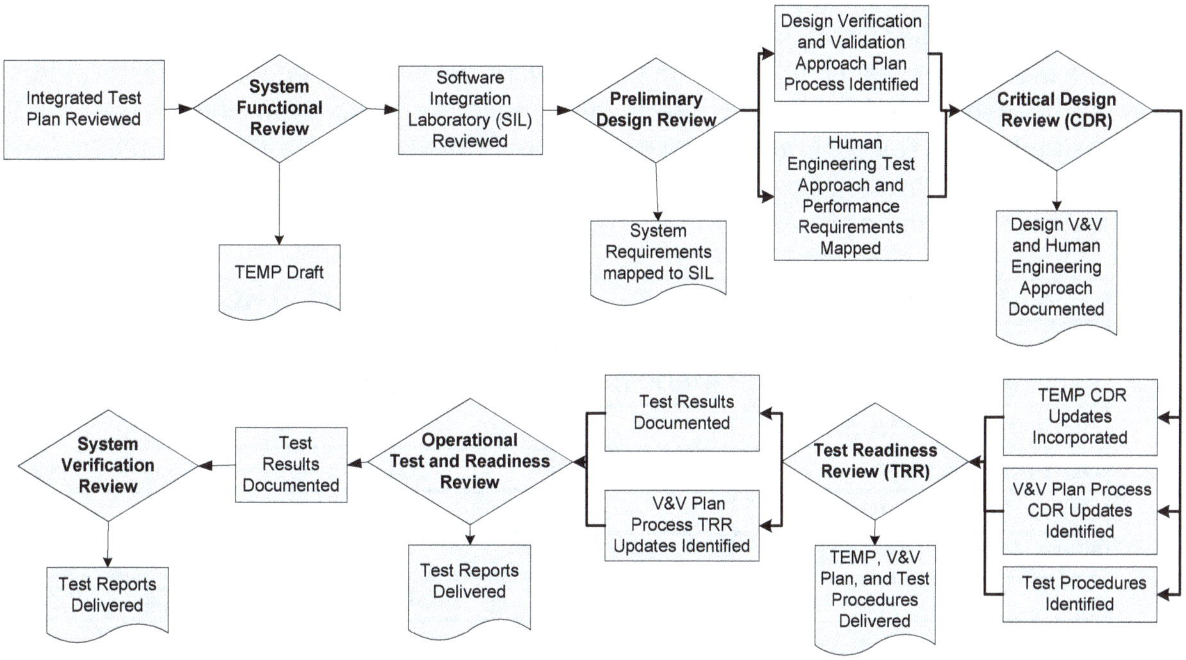

Figure 2.2-257: Test and Evaluation Process. *Test and Evaluation plans and reports will be presented at major customer reviews throughout the project lifecycle.*

Requirements Verification

The Solutions International requirement verification matrix included in Technical Attachment I details the methods of verification for each of the PBSS requirements. The method and expected program phase when the verification will be performed has been identified for each requirement. A majority of the requirements are verified late in the program because it requires considerable amount of time for specific hardware to be designed, built, and integrated into the system before system tests can be performed.

# of Requirements	VERIFICATION METHOD				
VERIFICATION PHASE	Inspection	Analysis	Demo	Test	TOTAL
Preliminary Design	43	0	4	0	**47**
Detailed Design	30	30	25	0	**85**
Operational Test	6	0	15	63	**84**
First Article Inspection	15	0	1	0	**16**
TOTAL	**94**	**30**	**45**	**63**	**232**

Figure 2.2-258: PBSS Requirements Verification Summary. *The SI BrEAD design will be verified against SI PBSS throughout the development and operational phases.*

T&E Metrics

T&E metrics have been developed to measure system test progress and effectiveness. The following metrics will ensure the T&E strategy is implemented to its maximum effectiveness and needs of the development program. Status of these metrics will be briefed at each major design and test readiness review.

a) % of Requirements compliance satisfied
b) % of Requirements verified/validated
c) % of Problem records / issues closed
d) % of Test plans / procedures completed

Systems Integration Laboratory

A System Integration Laboratory (SIL) will be the home for much of the system development, integration, validation and verification of the Enterprise Architecture and breadfruit processing system. The SIL will be located in Maui, Hawaii in order to leverage the technical contributions and resources at the Breadfruit Institute.

The Systems Integration Laboratory is an important multi-functional asset of the Enterprise Architecture. The SIL has the following purposes:

- **Initial System Proof-of-Concept** to demonstrate feasibility and to perform design trade studies of the breadfruit processing system and Enterprise Architecture.
- **Fully-Operational medium-scale breadfruit processing system** to generate breadfruit products for the gluten-free product market in Hawaii.
- **Developmental Test Asset** for subsystem unit testing, system integration testing, safety and human engineering evaluations, system assembly trials, and transportability assessments.
- **Configuration Baseline** of breadfruit processing hardware, models, and EA communication network throughout the product lifecycle.
- **Modeling, Simulation, and Prototype Laboratory** to improve BF processing methods via simulations and prototyping using inputs from worldwide collection of operational data from BF processing installations and info provided by researchers. Harvest, climate, cultural, economic, reliability, and ergonomic models will be developed to simulate the BrEAD operational environment to collect estimated performance data and assess technical risk before verification testing.
- **Operational Test Asset** to conduct experiments and simulations as requested by world-wide research organizations.

Details of the modeling, simulation, data and system performance estimates generated to date can be found in the System Performance Annex A of the proposal.

The physical architecture of the Systems Integration Laboratory, as shown in Figure 2.2-10, consists of integrated hardware and software to assist developers of the design, prototype conceptual designs, simulate performance, conduct operational trials, and perform test for verification of requirements.

The following hardware items expected in the SIL are:

- Dell PowerEdge 2900 III server (QTY 1)
- Dell Precision T5500 Workstation Desktop Computer (qty 6)
- Dell UltraSharp 2007FP 20.1-inch Flat panel LCD Monitor (QTY 6)
- HP Officejet Pro L7590 All-in-one Printer, Fax, Scanner, Copier (QTY 1)
- HP Thermal Inkjet
- Dell Powerconnect 3524 Ethernet switch (QTY 1)
- Dell Latitude E6500 Laptop computer
- HP DJ110 plus nr Designjet Printer/Plotter (QTY 1)
- Telephone (2)
- General office equipment as required (desks, chairs, file cabinets)

The following software items expected in the SIL are:

- OS: Windows Professional
- Accounting functions - QuickBooks Premier
- Desktop publishing - Adobe InDesign CS4
- Personnel records - Access database (SW will create)
- E-Mail - Outlook
- Virus/Firewall protection - Symantec
- Website - Java (SW will create)
- Field data - Access database (SW will create)
- Website/Database interface - Java (SW will create)
- Automated database input algorithm- C/C++ (SW will create)
- AutoCad (2 seats)
- Matlab

The physical architecture of the Systems Integration Laboratory, as shown in **Figure 2.2-259**, consists of integrated hardware and software to assist developers of the design, prototype conceptual designs, simulate performance, conduct operational trials, and perform test for verification of requirements.

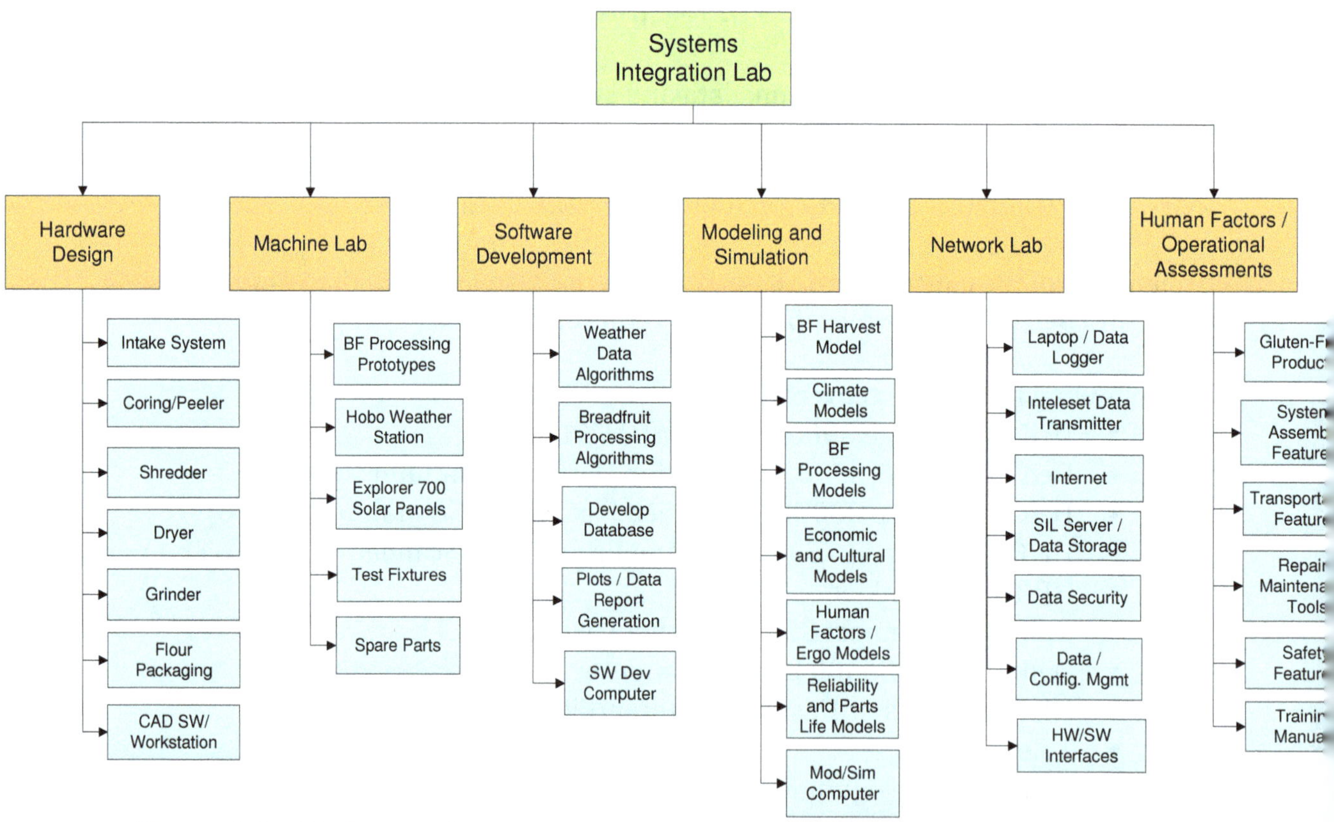

Figure 2.2-259: System Integration Laboratory (SIL) Physical Architecture. *All components of the Systems Integration Laboratory and its interfaces to the Enterprise Architecture have been identified, providing a comprehensive solution.*

An initial layout of the test and evaluation elements of the SIL is shown in **Figure 2.2-260**. The layout shown does not include office space for administration functions and other resource center functions which will also be co-located with the SIL. To perform some of the SIL activities will require some elements of the SIL to be located outside and in close proximity to the building. Upon award, the layout and functionality of the SIL will be reviewed with the customer to ensure all of the customer's needs are satisfied.

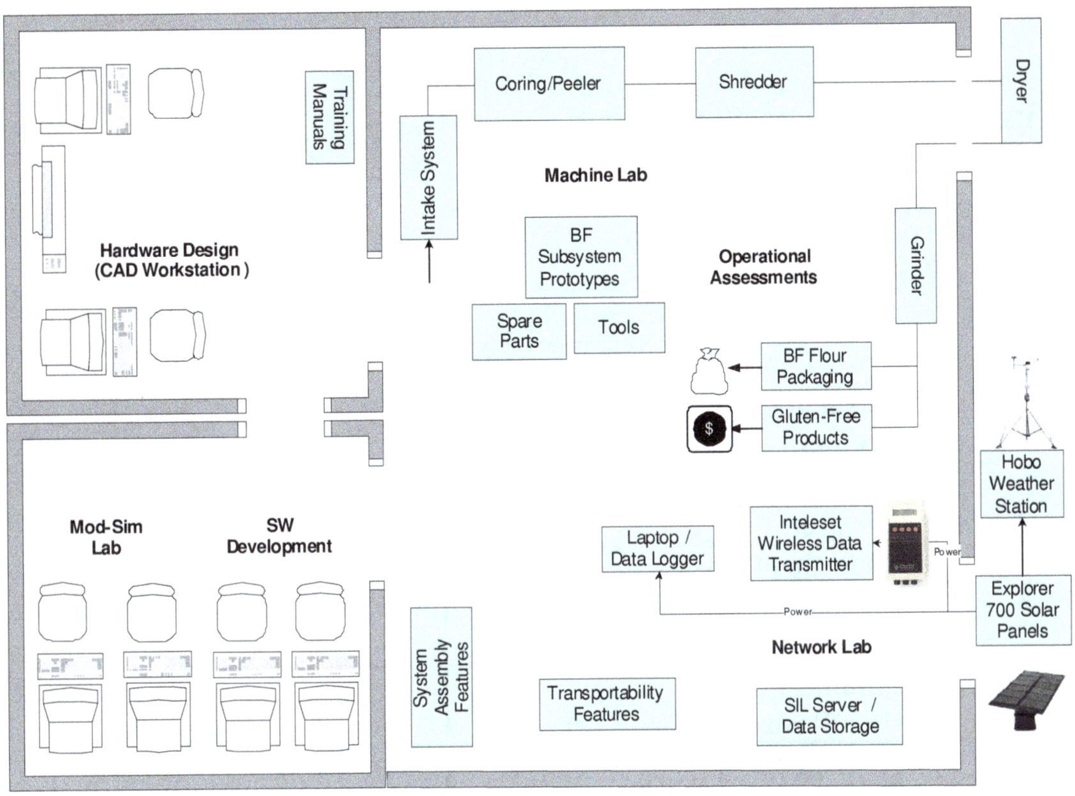

Figure 2.2-260: System Integration Laboratory (SIL) Layout. *The Systems Integration Laboratory is an important asset for design, testing, operational assessments, and marketing of the Breadfruit Processing System and Enterprise Architecture.*

Subcontract Management

Solutions International has identified subcontractors that will be crucial to our success. Their services will allow us to meet schedule deadlines and achieve a higher quality of work. These subcontractors are Compatible Technology (CTI), Onset and Inmarsat.

Through use of our subcontractors, we will provide quality products and support services.

Our Subcontract Management Plan (SMP) will guide us to efficiently manage our subcontractors.

Subcontract Management Plan

The Subcontract Management plan defines the relationship between Solutions International (SI) and its subcontractors.

Roles and Responsibilities

Functional Element	CORPORATION			
	Solutions International (SI)	CTI	ONSET	INMARSAT
Project Management	X			
Mechanical Design		X		
Communication System Design	X			
Data Logging Service			X	
Satellite Service				X
Training	X	X		
Technical Support	X	X	X	X

Figure 2.2-261: Corporations Functional Element Mapping. *The subcontractor strategy utilizes their expertise, allowing us to provide the complete solution.*

Since SI is the prime recipient of the project's funds, we have the responsibility to manage and make a good use of the project's funds. Subcontractors are sub-recipients. SI assumes the overall management of the project. Therefore, SI provides project direction to the subcontractors. The Project Director oversees the project and represents SI in product reviews and negotiations with subcontractors.

Project Management

Solutions International (SI) has overall responsibility to coordinate subcontractors' work activities, to effectively support the needs of subcontractors, and to establish a good working relationship with them.

Subcontractors Statement of Work (SOW)

The contract between Solutions International and CTI is limited to the design of peeling tools, coring equipments and grinders. Although CTI will still own the design right, CTI is required to design tools and equipments that meet PBSS requirements and at lower cost. Used in the past by University of St Thomas students for similar design, the new system requires that CTI modifies and/or improves the existing design by making critical functional parts of the system modular (electric power modules and non-power modules) and scalable. CTI agrees to provide technical training to local community on system assembly and repair.

As the majority of Solutions International's data communication system is automated, SI seeks the service of Onset for the repair of the damaged data loggers.

Inmarsat is chosen to provide the satellite communications service.

Terms and Conditions

By signing the contract, subcontractors agree to provide high quality service and will be held accountable for any low quality service they provide.

The contract between SI and Onset requires that Onset provide free and lifetime technical support for all equipments purchased from Onset. All Onset equipment is covered under a one year warranty. Any damaged equipment will be replaced at no cost during the warranty period. At any time, damaged equipments that may hold important data can be sent to Onset for the data retrieval for a $10.00 fee per damaged equipment.

The contract between BrEAD and Inmarsat is limited to the maintenance of a reliable and secure satellite communication system and a technical support for any satellite communication equipment purchased.

Risk Management

Annually renewable, subcontractors are accountable once the subcontract is in place.

Progress recording and Record Keeping

Solutions International is kept informed of the subcontractors' project progress. Subcontractors will provide SI copies of change and expense records. All subcontractors are required to keep their own records as well.

Milestone Reviews

Biweekly, Solutions International will conduct formal review on CTI work progress in the overall project schedule and request an update on the quality of work achieved to date.

Deliverable Review

Solutions International has authority to review, comment and identify deficiencies that can subsequently lead to a reject of the deliverable.

Deliverable Acceptance Form

Subcontractors are required to submit with deliverable, deliverable acceptance form.

Invoices and Payments

Subcontractors are required to submit invoices that meet Solutions International requirements for invoices that outline the terms and conditions of the contract that bounds SI and each individual subcontractor.

Software Development

In Annex H, Solutions International has provided a Software Development Plan (SDP). Solutions International recognizes the importance of robust, reliable software products and systems that meet user needs for the BrEAD activities. The development of quality software systems requires the use of documented processes for development, test, integration, and deployment of software systems. As such, the SDP identifies the plans to be used during the development of the BrEAD software products. The SDP itself is developed using the Software Development Plan Template, TM-SPP-02 V2.0, April 5, 2005 developed by the Systems Engineering Process Office of the Space and Naval Warfare Systems Center in San Diego, California. This template is tailored for the specific needs of the BrEAD software system.

There are two categories of software products specified for the BrEAD: COTS and BrEAD specific software developed by the BrEAD / Solutions International software IPT. The COTS software products are identified with the rationale for their selection. The majority of the SDP is devoted to identifying the processes and procedures to be used for the development, test, and maintenance of the BrEAD specific software that will be developed by the IPT.

To minimize cost and development time, Solutions International has specified the use of COTS software products wherever practical. To minimize compatibility issues, the Microsoft Windows Professional operating system is specified for use on all BrEAD computers and only Windows compatible software products will be purchased.

System Development and Demonstration of the Breadfruit Enterprise Demonstration (BrEAD) and Systems Design

Solutions International Proposal for the Breadfruit Enterprise Demonstration and System Design

Volume 2: Technical

Annex A: Systems Performance

RFP NO: E00010-07-R0001

Copy 1 of 2

May 4, 2009

Prepared For:

New Green Revolution Foundation (NGRF)

Prepared By:

Solutions International

123 Pillsbury Drive

St. Paul, MN 55114

CAGE CODE SI4ME

TABLE OF CONTENTS

Table of Figures

Introduction

System performance calculations have been performed to establish an initial performance baseline for the proposal. It is expected that modeling and simulation capabilities and early operational trials will be leveraged during the contract to improve the confidence and fidelity of this initial performance estimate.

Cultural and Economic Modeling

Cultural and Economic modeling can be an effective tool to reduce risk and to evaluate scenarios that might be encountered during operations. Because the initial system is to be implemented in Haiti it is important to account for situations that are unpredictable. A few examples are noted here.

Operator training - A majority of Haitians receive no formal education, and only a small minority are educated beyond primary school. The training they do receive is through rote memorization. Analysis, problem solving, and thinking "outside-of-the-box" if a problem arises may be normal for SIL engineers but may be difficult for Haitian operators.

Operator Health - Many Haitians suffer from poor health due to lack of physician availability and modern medicine. It may be unreasonable to expect the same person will be able to perform manual tasks consistently and reliably for the entire 3 week breadfruit season. Therefore operators may be required to be trained on more than one operation to fill in for those that are missing from day-to-day. A decreased level of performance could be expected.

Operator Labor Traditions - Although Haitians value community cooperation and usually have close extended-family ties the traditional relationships can be a detriment in team formulation. Men are usually expected to perform the difficult manual labor such as harvest crops and operate machinery while women are expected to perform the lighter food preparation work and sell food products at the market. Selecting work groups and assigning work assignments that do not respect their cultural norm may disrupt the morale of the team.

http://www.earthyfamily.com/H-Culture.htm

http://www.cal.org/co/haiti/htoc.html

During the development effort the cultural and economic modeling will continue. We plan to leverage the work of other charitable organizations such as AMURT Haiti who have had several community development projects in Haiti according to their website, http://www.amurthaiti.org/text.php?tf=grants.txt

Since 2004 AMURT has successfully realized nine integrated community development projects, receiving and managing a total of 2.1 Million USD. The NGO has consistently received high ratings from the major International Development Agencies in Haiti. AMURT is one of a few organizations out of hundreds of applicants to consistently qualify for financing by funders such as USAID and ACDI. The diversity of the projects it has realized in the last 3 years alone, and their positive impact on the lives of thousands of people attest to their effectiveness and connection to the local communities.

Technical Performance Measures (TPMs)

As documented in the SEMP, Technical Performance Measures will be established from BrEAD use cases to track critical performance parameters throughout the development of the BrEAD system. These TPMs are parameters that will impact the cost, performance, or schedule if they exceed critical values. SI will track these parameters, which are either directly measurable or derivable from modeling of the BrEAD subsystems, to ensure that operational objectives of the BrEAD are met.

The top-level TPMs for the BrEAD system are:

- Drying rate: 85% by weight of moisture removed within 6 hours
- Dryer reliability: MTBF = 1500 hours minimum
- Data Communication Range/Rate between communication cells
    - 40 miles maximum at 115.2 kbps
    - 15 miles maximum at 1.5 Mbps (with high gain antenna)
- Minimum breadfruit processing capacity: 290 kg/day for family use with prototype system

- Maximum breadfruit processing capacity: TBD for industrial-sized system

Breadfruit Processing System Performance

Breadfruit Processing Model

The breadfruit processing system is model in Figure 4.1-1. The functional steps are modeled in series. Multiple instances of subsystems are modeled in parallel.

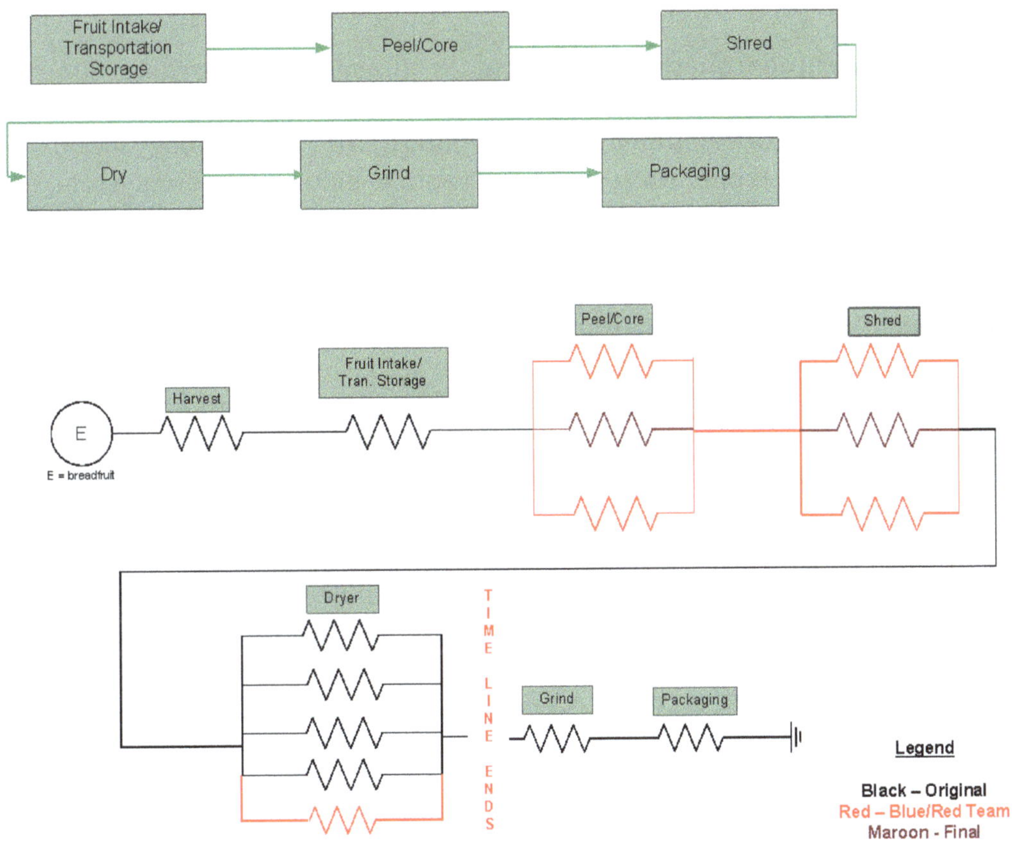

Figure 4.1-1: Breadfruit Processing Model

Breadfruit Processing Subsystem Rate Calculations

The components of the Breadfruit processing system are readily available COTS components or components fabricated from COTS materials. Modeling and simulation of the system was conducted because no performance information exists for these components with respect to their intended use.

The rate that Breadfruit could be processed through our system was modeled using a Pineapple. We chose a pineapple because it shares some of the same characteristics as breadfruit and was readily available for our modeling.

Two primary goals desired from the modeling of the pineapple was to; determine the daily rate the system could handle and secondly to see if the losses noted in the United Nations Industrial Development Organization (UNIDO) Technology Manual - Small Scale Fruit and Vegetable Processing and Products held true in our system.

First goal: our modeling determined that 450 breadfruit per day results in ~295 kg, or 650 lbs, of flour per day. We not only evaluated our throughput, but also evaluated the timeline and activities from one Breadfruit sub-system to the next. A timeline for each station was modeled to ensure the activities 'flowed' smoothly and to verify we didn't have any back-ups or stovepipe points in our system.

Second goal: the pineapple 'losses' we encountered as a result of processing the fruit were a bit higher than the percentages noted in the United Nations Industrial Development Organization (UNIDO) Technology Manual - Small Scale Fruit and Vegetable Processing and Products. However, our pineapple modeling used a common kitchen knife to peel and cut-up (shred), instead of using our proposed processing system peelers and shredders. This accounts for the increased losses that were incurred.

Processing waste loss estimate

The process began with following the guidelines established for processing fruits per the United Nations Industrial Development Organization (UNIDO) Technology Manual – Small Scale Fruit and Vegetable Processing and Products. The typical loss of a fruit's weight was notes as a result of processing activities like peeling, coring, etc. **Figure 4.2-1** (page 24 in the UNIDO) identifies the typical losses form each processing stage.

Stages in a Process	Typical Losses (%)
Washing Fruits/Vegetables	0-10
Sorting	5-50
Peeling	5-60
Slicing/dicing	5-10
Batch preparation/weighing	2-5
Drying	10-20
Packaging	5-10

Figure 4.2-1: Typical losses from each processing stage

To confirm these estimates, a readily available fruit in our geographical area with some similar characteristics to breadfruit was chosen to be modeled. SI chose a pineapple in this case. Additionally, for the purposes of this modeling, a standard kitchen knife and peeler that likely found in any kitchen were used instead of the actual breadfruit processing system components. The time taken to perform the task along with recording the loss of weight from one stage to the next was recorded during the process. **Figure 4.2-2** lists the Processing information recorded for the Pineapple. The modeling showed with the use of these 'simple' components, the processing system will be effective and able to manage their respective tasks thereby proving the overall capabilities of the system components that were modeled.

Modeling 1.96 kg (4 lbs 5 ounce Pineapple)				Loss %	Notes
	Weight	oz	kg		
Initial weight	4 lb 4 oz	69	1.96		
Top cut off	3 lb 10 5/8 oz	58.6	1.66	15	
Side skins cut off - fruit with core in	2 lb 1 1/4 oz	33.4	0.947	43	
Core out - Fruit only	1 lb 14 5/8 oz	30.5	0.865	9	
Side skins & core	1 lb 11 3/4 oz	27.5	0.78	10	Add .29 kg for top

Figure 4.2-2 Pineapple Model Processing Measurements

The next step was to cut the pineapple into slices representing the breadfruit shreds. The breadfruit shredder is designed to cut 2"x1/8"x1/2" shreds. Since it is difficult to slice a pineapple into 1/8th inch slices, a surface area equivalent 2"x1/2"x1/4" was chosen for the pineapple slice dimension. It would be expected that the shreds would lay on the dryer tray on the surface with the largest area. As can be seen in **Figure 4.2-3**, a total of 127 pieces of this dimension were obtained from a single pineapple.

Figure 4.2-3 Sliced Pineapple Model

Dryer Tray Sizing

Enhancements have been made to the dryer that was designed by the University of St. Thomas engineering students Jason Emiliusen, Thomas Mauritzen, Ross McGruder, and Kara Torgerson. However, ongoing enhancements based upon potential changes or possible re-design upon date of award of contract are TBD. Therefore, 12 ea 3’ x 2’ dryer trays were modeled. Each tray holds 6 ea 12” x 12” panels. In this arrangement, as shown in **Figure 4.2-4,** one tray can hold a total of 144 shreds when arranged as a row of 6 along with the 2" dimension and 24 along the ½ " dimension. Thus it is possible for each 2' x 3' tray to hold enough shreds produced from at least 6 breadfruit. Our dryer has a total of 12 trays so at least 72 breadfruit can be dried at one time.

Breadfruit Drying Tray

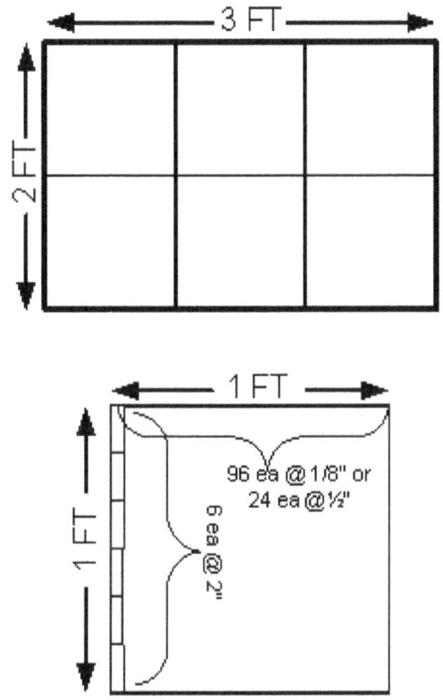

Breadfruit Shreds 2"x1/8"x1/2" (LxWxH)

Figure 4.2-4: Breadfruit Dryer Tray Model

Breadfruit Processing Timeline and Manpower Analysis

The total time and personnel needed to perform the entire operation from intake to drying is shown in **Figure 4.3-1**. Our timeline assumes a total of 450 breadfruit are available at the beginning of the day.

Step 1 - Intake - It is estimated that it will take 15 seconds per breadfruit to inspect the breadfruit for quality control, weigh each individual breadfruit, and log the data. A total of 5 people are expected for this operation. This process requires a total of 56.25 minutes for the entire 450 breadfruit.

Step 2 - Draining - It is estimated that it will take 2 hours to drain each breadfruit to reduce the initial moisture content in preparation for shredding. This will occur in parallel to the intake inspection process.

Step 3 Peeling and Coring - It is estimated that this process can be completed in 120 seconds per breadfruit. This process requires 10 operators at 3 coring / peeling stations. So 450 breadfruit can be processed in 300 minutes. (2 min/breadfruit *450 breadfruit / 3 stations)

Step 4 Shredding - This process is also estimated to be completed within 2 minutes per breadfruit and requires an additional 10 operators at 3 shredding stations. The operators will then place the shreds in one of 12 dryer trays. (2 min/breadfruit *450 breadfruit / 3 stations).

	Fruit Intake	Peel/Core (x3 Stations)	Shred (X3 Stations)
Tasks	Weighing, Counting fruit, Recording Info, Supervision	Peeling, Coring, Waste Collection, Supervision	Shredding, Weighing, Recording Information, Waste Collection, Supervision, Laying Shreds on drying trays
Time (Modeled) At Station (secs)	15	120	120
Personnel Needed	5	10	10
Total Time (min) (@ 450 Breadfruit)	56.25	300	300
Total Time (hrs) (@ 450 Breadfruit)	0.9375	5	5

Figure 4.3-1: Pre-Drying Processing time

Therefore with this number of operators and assuming all 450 breadfruit are ready to begin draining at 5 am, it is possible that the first dryer load is ready by 12:15 pm. Staging these every 30 minutes will leave enough time in the heat of the day to dry all 450 breadfruit by 7:15 pm.

Time of Day	Action
5:00 AM	Cut off top of breadfruit
7:00:00 AM	450 breadfruit tops cut off
7:15:00 AM	Processing Begins
12:15:00 PM	Dryer #1 full - ready for drying
12:45:00 PM	Dryer #2 full - ready for drying
1:15:00 PM	Dryer #3 full - ready for drying
1:45:00 PM	Dryer #4 full - ready for drying
2:15:00 PM	Dryer #5 full - ready for drying

Figure 4.3-2: System Timeline for 1 Day

An analysis of the manpower needed to operate our system outputting ~295 kg, or 650 lbs of flour per day concluded that 33 personnel would be needed with the majority of those personnel working an 8 hour day. Our results are largely based upon a mixture of analyzed and actual timelines recorded during modeling that was conducted. Due to potential differences in cultural norms, training, or changes that could occur before the final design of the system is completed, these findings are considered to preliminary manpower recommendations.

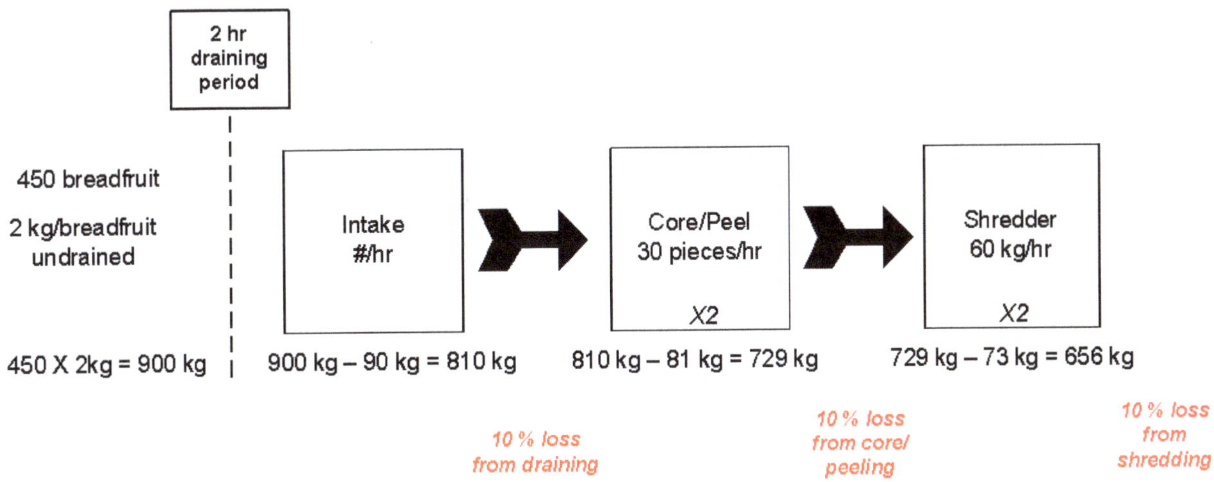

Figure 4.3-3: System Process for 1 Day

Breadfruit Processing System Modeling Performance

The following diagrams were generated to illustrate the integration of data parameters expected to be recorded, resources required to perform the operation, and post-processing calculations. Data to be measured and collected at the breadfruit processing sites are indicated in the dark green boxes. The items that can be used as performance parameters in Modeling and Simulation are highlighted in blue. Performance data as indicated in the yellow blocks will be post-processed in the SIL or regional center. The performance data shown in the bright green box represents current performance estimates.

Breadfruit Intake subsystem

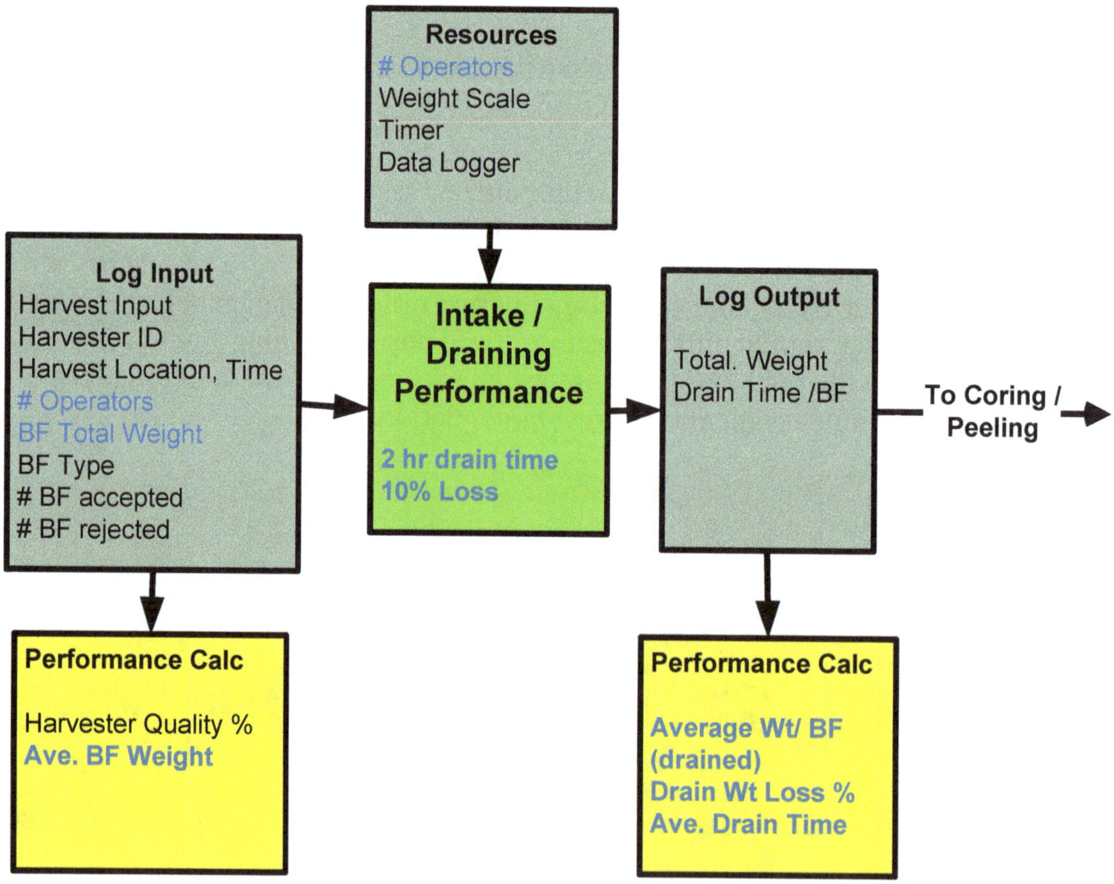

Figure 4.4-1: Breadfruit Draining Process Model

(Mod & Sim parameters indicated in blue highlights)

Breadfruit Coring and Peeling Subsystem

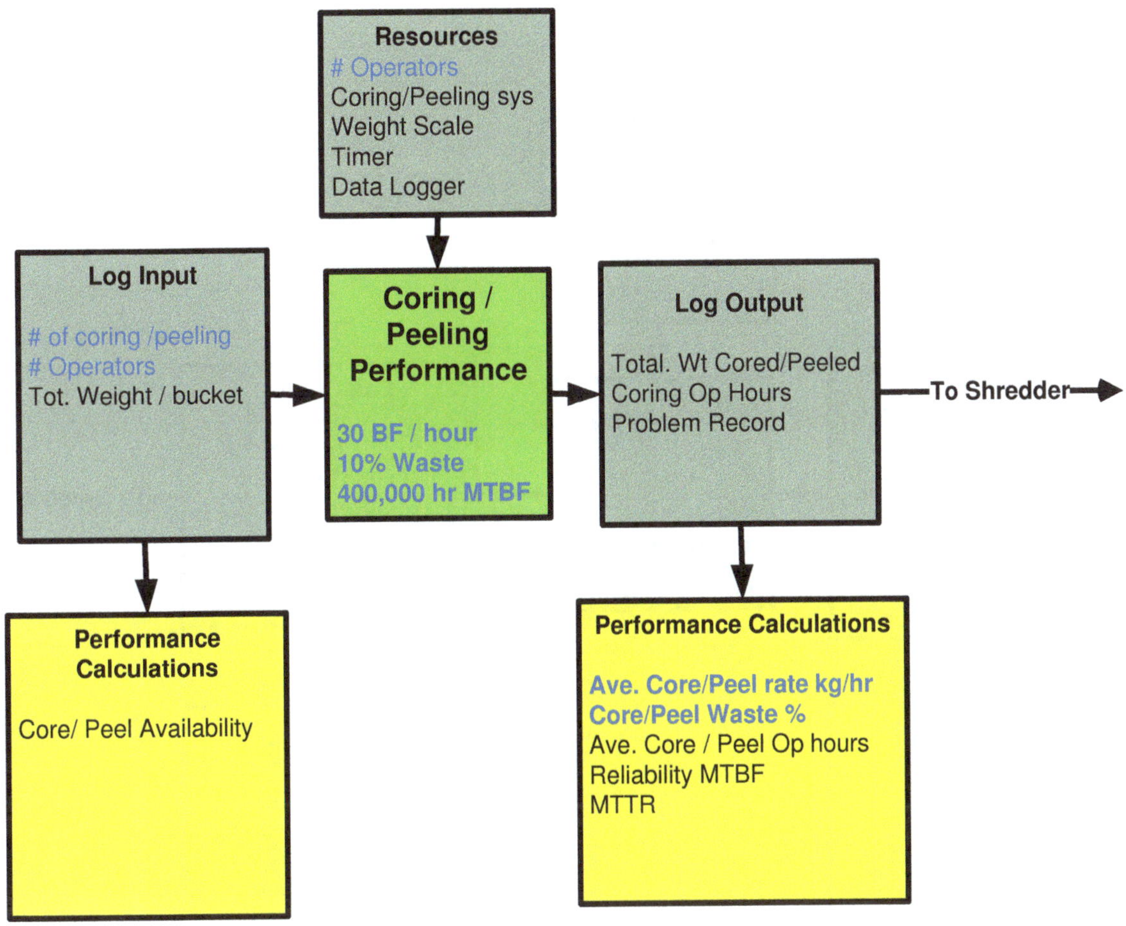

Figure 4.4-2: Breadfruit Coring / Peeling Process Model

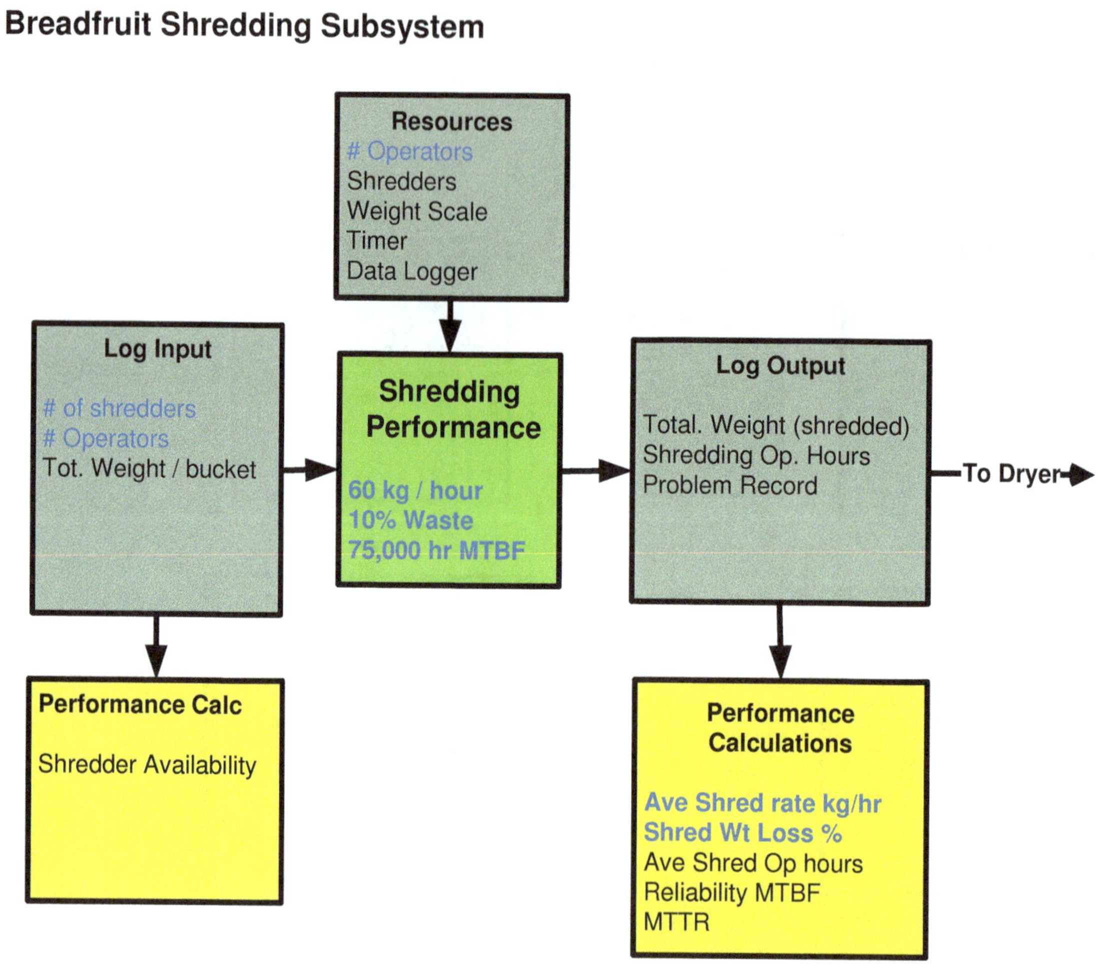

Figure 4.4-3: Breadfruit Shredding Process Model

Breadfruit Drying Subsystem

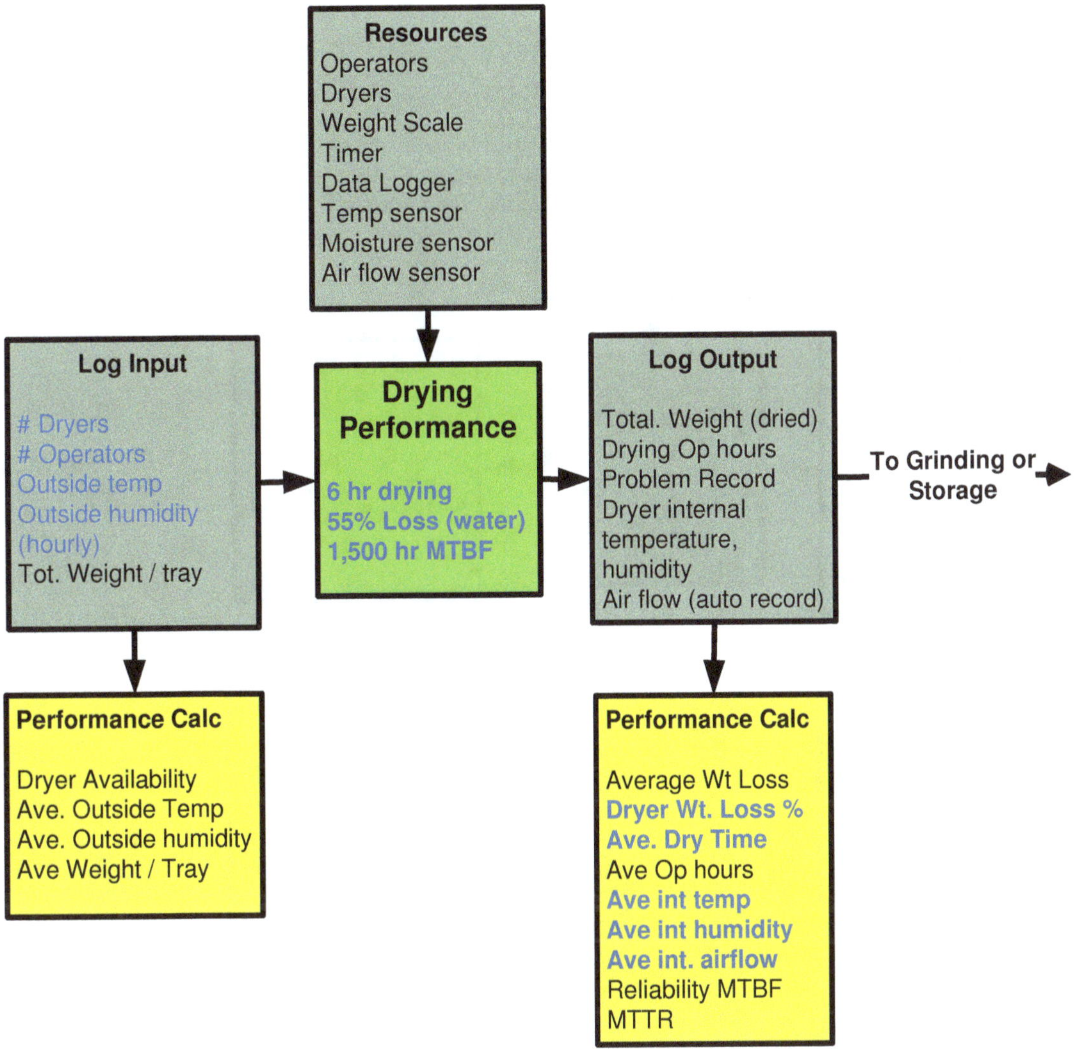

Figure 4.4-4: Breadfruit Drying Process Model

Breadfruit Grinding Subsystem

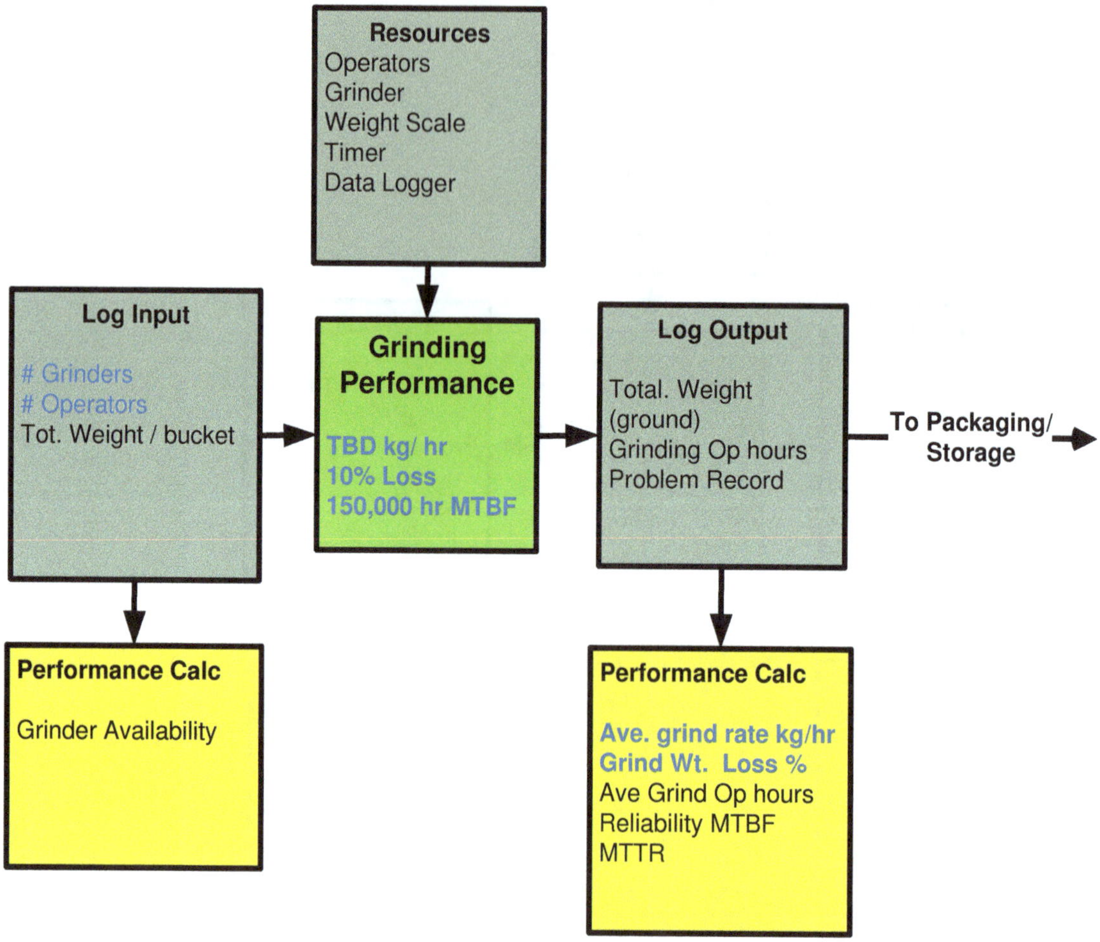

Figure 4.4-5: Breadfruit Grinding Process Model

Breadfruit Processing System Data Collection Form

The following forms have been generated to be made available for operational use at the breadfruit processing site or in the SIL. Each station has a dedicated data collection form (Intake, Coring/Peeling, Shredder, Dryer, Grinding, & Packaging). The form will be used to collect Performance Data & Failure Information. It is expected to be recorded manually on these sheets. At the end of the day the data will be entered manually into the data logger laptop computer.

Intake and Draining Subsystem	Units	Freq	Date	Time	Value	Measurement Device
Harvester ID	ID #	daily				ID list
When harvested (off-tree life)	date/time	daily				clock/calender
Processing system location ID	ID #					data logger
Total # of Breadfruit collected from harvesters	# of units	daily				counter
Total # of breadfruit rejected (bruised, rotten)	# of units	daily				counter
Individual Breadfruit weight	kg	per breadfruit				weight scale
Type of breadfruit	N/A	per breadfruit				visual aids
Total weight before draining	kg	per breadfruit				weight scale
Total weight after draining	kg	per breadfruit				weight scale

Coring / Peeling System	Units	Freq	Date	Time	Value	Measurement Device
Weight of breadfruit before coring / peeling process	kg	per bucket				weight scale and bucket
Weight of breadfruit after coring / peeling process	kg	per bucket				weight scale and bucket
Time to process 1 bucket of breadfruit	minutes	hourly				counter
# of operating hours per day	hours	daily				clock
Problem record ID	ID #	as reqd				data logger
Failed processing component	P/N	as reqd				data logger
Repairable / unrepairable	Y/N	as reqd				data logger
Time to repair	minutes	as reqd				timer

Shredder System	Units	Freq	Date	Time	Value	Measurement Device
Weight of breadfruit before shredding process	kg	per bucket				weight scale and bucket
Weight of breadfruit after shredding process	kg	per bucket				weight scale and bucket
Time to process 1 bucket of breadfruit	minutes	per bucket				counter
# of operating hours per day	hours	daily				clock
Problem record ID	ID #	as reqd				data logger
Failed processing component	P/N	as reqd				data logger
Repairable / unrepairable	Y/N	as reqd				data logger
Time to repair	minutes	as reqd				timer

Dryer / Oven SubSystem	Units	Freq	Date	Time	Value	Measurement Device
Weight of breadfruit before drying process	kg	start of drying				weight scale and tray
Weight of breadfruit after dryer process	kg	end of drying				weight scale and tray
Moisture content of breadfruit sample at start drying process	% water	start of drying				moisture sensor
Oven temperature at start of drying process	deg C	start of drying				thermometer
Internal air humidity level at start of drying process	% water	start of drying				air humidity sensor
Moisture content of breadfruit sample during process	% water	1 per hour				moisture sensor
Oven temperature during drying process	deg C	1 per hour				thermometer
Internal air humidity level during drying process	% water	1 per hour				air humidity sensor
Moisture content of breadfruit sample at end of drying process	% water	end of drying				moisture sensor
Oven temperature at the end during drying process	deg C	end of drying				thermometer
Internal air humidity level at the end of drying process	% water	end of drying				air humidity sensor
# of operating hours per day	hours	daily				clock
Problem record ID	ID #	as reqd				data logger
Failed processing component	P/N	as reqd				data logger
Repairable / unrepairable	Y/N	as reqd				data logger
Time to repair	minutes	as reqd				timer

Grinding SubSystem	Units	Freq	Date	Time	Value	Measurement Device
Weight of breadfruit before grinding process	kg	per bucket				weight scale and bucket
Weight of breadfruit after grinding process	kg	per bucket				weight scale and bucket
Grinding time per process	minutes	per process				timer
# of operating hours per day	hours	daily				clock
Problem record ID	ID #	as reqd				data logger
Failed processing component	P/N	as reqd				data logger
Repairable / unrepairable	Y/N	as reqd				data logger
Time to repair	minutes	as reqd				timer

Packaging SubSystem	Units	Freq	Date	Time	Value	Measurement Device
Weight of breadfruit before packaging process	kg	per bucket				weight scale and bucket
Weight of breadfruit after packaging process	kg	per package				weight scale and package
Packaging time per process	minutes	per process				timer
Package ID #	ID #	per package				labeler and data logger
# of operating hours per day	hours	daily				clock
Problem record ID	ID #	as reqd				data logger
Failed processing component	P/N	as reqd				data logger
Repairable / unrepairable	Y/N	as reqd				data logger
Time to repair	minutes	as reqd				timer

Table 4.5 -1 Breadfruit Processing data log forms

Regional and Central Data Center Performance Data

The following data will be processed at each regional center and transmitter to the central data center for collation of data world-wide.

Regional and Central Data Collection Description	Units	Frequency of calculation / measurement	Evaluated in SIL
Average Breadfruit flour produced	kg	1/day	N
System failure rate	#	1/day	Y
Mean Time to repair	minutes	1/day	Y
Harvester Quality %	%	1/day	N
Average # of Tools required to repair	#	1/day	Y
Ave. Drain Time per breadfruit	minutes	1/day	Y
Ave. Weight / Breadfruit Drained	kg/unit	1/day	Y
Ave. Drain Waste Loss %	kg/unit	1/day	Y
Total # of Breadfruit types cored/peeled	#	1/day	N
Coring / Peeling reliability	failures / operating hour	1/day	Y
# of coring / peeling operators	#	1/day	N
Ave. Coring / peeling rate	kg/hour	1/day	Y
Coring / Peeling effectiveness	kg/kg %	1/day	Y
Average Shredding rate	kg/ hour	1/day	Y
Average Shredding Waste Loss %	kg/kg %	1/ day	Y
Shredding reliability	failures / operating hour	1/ day	Y
Average Drying time	hours	1/day	Y
Drying reliability	failures / operating hour	1/ day	Y
Average Moisture content loss %	kg/kg %	1/day	Y
Average Grinding rate	kg / hour	1/day	Y
Grinding reliability	failures / operating hour	1/ day	Y
Average Packaging rate	#/ day	1/ day	Y
Packaging reliability	failures / operating hour	1/ day	Y
Location of Weather stations	lat / long	as required	N
Atmospheric temperature /accuracy	deg C	1/hour	Y
Atmospheric pressure / accuracy	Pa	1/hour	Y
Atmospheric humidity qnty / accuracy	%	1/hour	Y
Rainfall qnty / accuracy	cm	1/day	N
Average # of operating hours - coring	hours	1/day	Y
Average # of operating hours - shredding	hours	1/day	Y
Average # of operating hours - drying	hours	1/day	Y
Average # of operating hours - grinding	hours	1/day	Y
Average # of operating hours - packaging	hours	1/day	Y
Measurement system reliability	failures / operating hour	1/ day	Y
Average On-site to Regional data transfer rate	kB / sec	each occur.	N
Average Regional to Central data transfer rate	kB / sec	each occur.	N
Regional data collection reliability	failures / operating hour	1/ day	N
Central data collection reliability	failures / operating hour	1/ day	N

Regional and Central Data Collection Description	Units	Frequency of calculation / measurement	Evaluated in SIL
SIL reliability	failures / operating hour	1/ day	Y
# of injuries	#	1/day	Y
Injury severity / description		1/day	Y
Equipment causing injury		1/day	Y

System Development and Demonstration of the Breadfruit Enterprise Demonstration (BrEAD) and Systems Design

Solutions International Proposal for the Breadfruit Enterprise Demonstration and System Design

Volume 2: Technical

Annex B: Reliability, Maintainability, and Availability Report

RFP NO: E00010-07-R0001

Copy 1 of 2

May 4, 2009

Prepared For:

New Green Revolution Foundation (NGRF)

Prepared By:

Solutions International

123 Pillsbury Drive

St. Paul, MN 55114

CAGE CODE SI4ME

REVISION HISTORY

Revision	Comment	Date
RMA Rev. -	Preliminary delivery to NGRF	4 May 2009

TABLE OF CONTENTS

List of Tables

List of Figures

Introduction

This report documents results of the Breadfruit Enterprise Architecture Reliability, Maintainability & Availability (RMA) prediction. The report is submitted in response to RFP E00010-07R-0001.

Reliability Prediction Objective

The objective of the reliability analysis is to ensure that the system design meets reliability requirements stated in the PBSS. Those requirements are documented below:

[ID 024] The breadfruit processing system shall have a Mean Time Between Failure (MTBF) of no less than 1,400 hours.
[ID 046] The Coring / Peeling subsystem have a Mean Time Between Failure (MTBF) of no less than 400,000 hours
[ID 055] Breadfruit shredder shall have a Mean Time Between Failure (MTBF) of no less than 75,000 hours.
[ID 068] The breadfruit dryer subsystem shall have a Mean Time Between Failure (MTBF) of no less than 1500 hours.
[ID 074] The breadfruit grinder shall have a Mean Time Between Failure (MTBF) of no less than 150,000 hours.
[ID 099] The BrEAD Weather Station data logger shall have a Mean Time Between Failure (MTBF) of no less than 40,000 hours.
[ID 110] The BrEAD Measuring System shall have a MTBF of no less than 1,700 hours.
[ID 115] The BrEAD Communication shall have a MTBF of no less than 5,000 hours.

[ID 136] The BrEAD Regional center shall have a Mean Time Between Failure (MTBF) of no less than 5,000 hours.
[ID 157] The BrEAD Central Data Center shall have a Mean Time Between Failure (MTBF) of no less than 6,000 hours.

Maintainability Prediction Objective

The objective of the maintainability analysis is to ensure that the system design meets maintainability requirements stated in the PBSS. Those requirements are documented below:

[ID 115] The mean time to repair (MTTR) of the BrEAD Measuring System shall be no greater than 15 minutes.
[ID 123] The mean time to repair (MTTR) of the BrEAD Data Communication system shall be no greater than 15 minutes.
[ID 026] The mean time to repair (MTTR) of the breadfruit processing system shall be no greater than 15 minutes.
[ID 102] The mean time to repair (MTTR) of the BrEAD weather station shall be no greater than 15 minutes.

Availability Modeling Objective

The objective of the availability analysis is to ensure that the system design meets operational needs of the following PBSS requirements:

[ID 002] The Breadfruit Enterprise Architecture Demonstration (BrEAD) shall provide an automated research, marketing, and user data collection capability for NGRF head quarters in St. Paul, Minnesota that will provide 24 hour per day data from researchers, marketers and users worldwide.

Scope

The reliability/maintainability prediction contained within the analysis includes the entire BrEAD Architecture as defined by Solutions International. Within the BrEAD Architecture there are 6 major subsystems:

- Processing System
- Weather station System
- Communications System
- Regional Center
- Central Data Center
- BrEAD Measurement System (Performance Data)

These subsystems consist of new Commercial off the Shelf (COTS).

Applicable Documents

This analysis was accomplished using the following reliability and maintainability documents for guidance.

Document Number	Document Name
MIL-HDBK-217F Notice 2	Reliability Prediction of Electronic Equipment, December 2, 1998
Reliability Toolkit	The Rome Laboratory Reliability Engineer's Toolkit, April 1993
MIL-HDBK-472 Notice 1	Military Standardization Handbook Maintainability Predictions 12 January 1984

System Description

The subsystems that are included within the scope of this analysis are defined as: Processing System, Weather Station, Communication System, Regional Center, Central Data Center, and BrEAD Measurement System. Figure 262 illustrates the BrEAD Architecture.

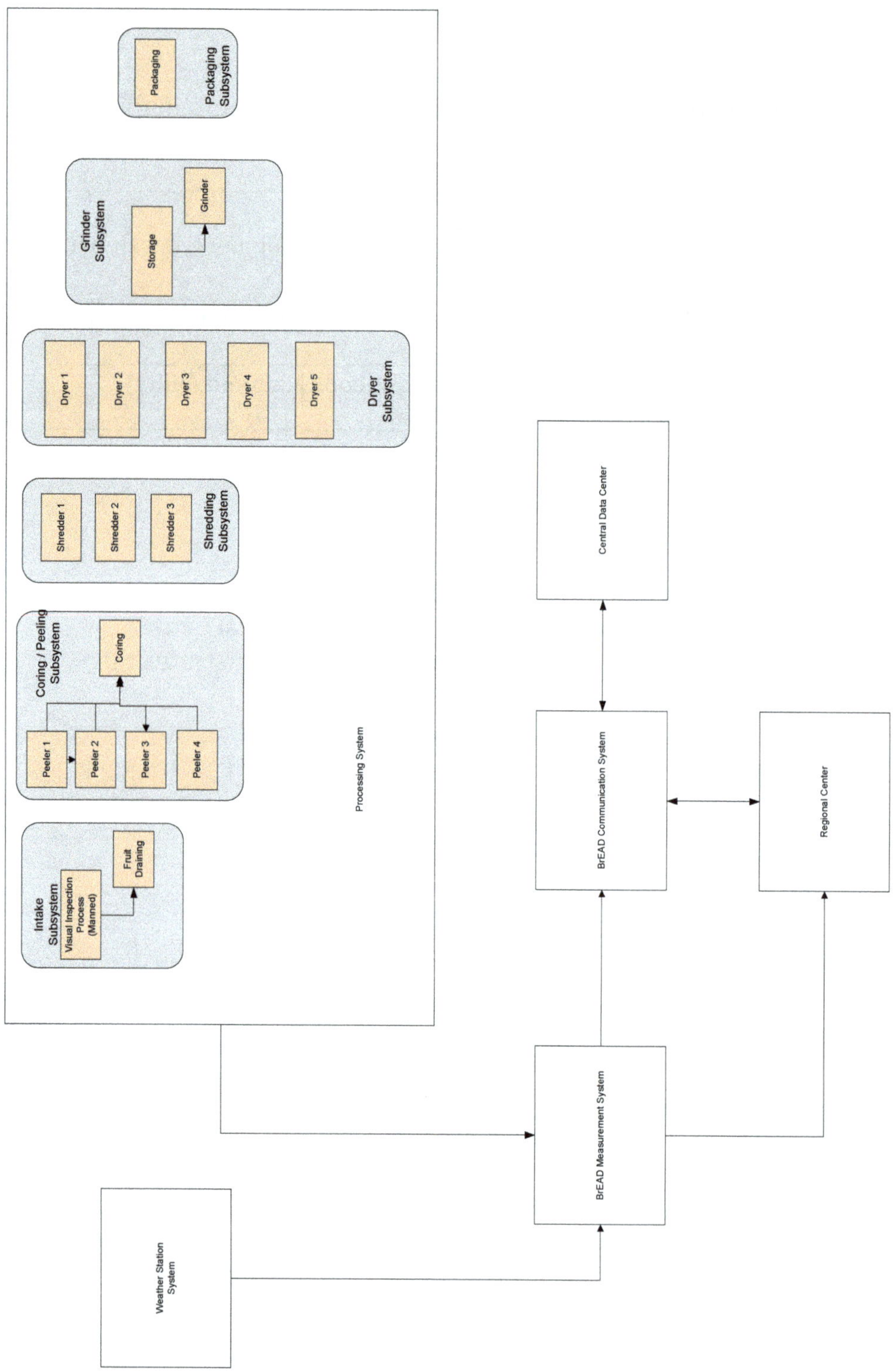

Figure 262: BrEAD Architecture

Processing System

The processing system contains six subsystems: Intake subsystem, Coring/Peeling Subsystem, Shredder Subsystem, Dryer Subsystem, Grinder Subsystem, and Packaging Subsystem. The intake subsystem weighs and counts the number of fruit entering the system. The Coring/Peeling subsystem consists of a peeler and corer. The shredder subsystem shreds the breadfruit into pieces. The dryer subsystem dries the shreds. The dryer consists of an automated feedback loop to control the dryer air flow. The Grinder subsystem grinds the dried breadfruit shreds into flour. The packaging subsystem packages the flour into sealed containers.

Weather Station

The Weather station collects environmental data, such as rain, temperature, relative humidity, wind speed, and solar radiation. All collected data is stored in the data logger.

Communication Station

The BrEAD architecture contains two different communication options: 1) a series of Intelecell transmitters or 2) Explorer 700. The communication system allows the weather station data to be transmitted to the regional center, the processing system to communicate with the regional center, and for the regional center to communicate with the central data center.

Regional Center

The Regional center contains standard office equipment, servers, and standard satellite communication devices. The regional center receives data from the processing site and the weather station and then passes it on the central data center.

Central Data Center

The central data center contains standard office equipment, servers, and standard internet communication devices. The central data center receives data from the regional centers and processes the data via the server and laptops.

BrEAD Measurement System

The BrEAD Measurement System automatically collects data from the weather station and from the dryer in the processing system. In addition, it contains a laptop at the processing system to manually collect performance data.

Reliability Modeling and Prediction

Reliability Requirements

The primary reliability requirement expressed in Technical attachment I - BrEAD PBSS is the following:

[ID 024] The breadfruit processing system shall have a Mean Time Between Failure (MTBF) of no less than 1,400 hours.
[ID 046] The Coring / Peeling subsystem have a Mean Time Between Failure (MTBF) of no less than 400,000 hours
[ID 055] Breadfruit shredder shall have a Mean Time Between Failure (MTBF) of no less than 75,000 hours.
[ID 068] The breadfruit dryer subsystem shall have a Mean Time Between Failure (MTBF) of no less than 1500 hours.
[ID 074] The breadfruit grinder shall have a Mean Time Between Failure (MTBF) of no less than 150,000 hours.
[ID 099] The BrEAD Weather Station data logger shall have a Mean Time Between Failure (MTBF) of no less than 40,000 hours.
[ID 110] The BrEAD Measuring System shall have a MTBF of no less than 1,700 hours.
[ID 115] The BrEAD Communication shall have a MTBF of no less than 5,000 hours.
[ID 136] The BrEAD Regional center shall have a Mean Time Between Failure (MTBF) of no less than 5,000 hours.
[ID 157] The BrEAD Central Data Center shall have a Mean Time Between Failure (MTBF) of no less than 6,000 hours.

There are no MTBF requirements associated for degraded mode operation or redundancy at the system level.

Assumptions and Ground Rules

The reliability prediction for all subsystem equipment uses a serial model in which all the hardware failure rates are summed to determine the top level system failure rate. If any one component of the system fails, the entire system is considered to have failed. The following two equations are used to calculate the overall system MTBF.

$$\lambda_{system} = \sum_{i=1}^{k} \lambda_{component}$$

$$MTBF_{system} = \frac{1}{\lambda_{system}}$$

Where:

λ = Failure rate expressed in Failure Per Million Hours (FPMH)
MTBF = Mean Time Between Failure of the system

The subsystem definition and equipment quantities reflect the contents as described in the Volume 2, Book 1. The reliability block diagram in Figures 2 through 7 show the subsystems modeled in the prediction.

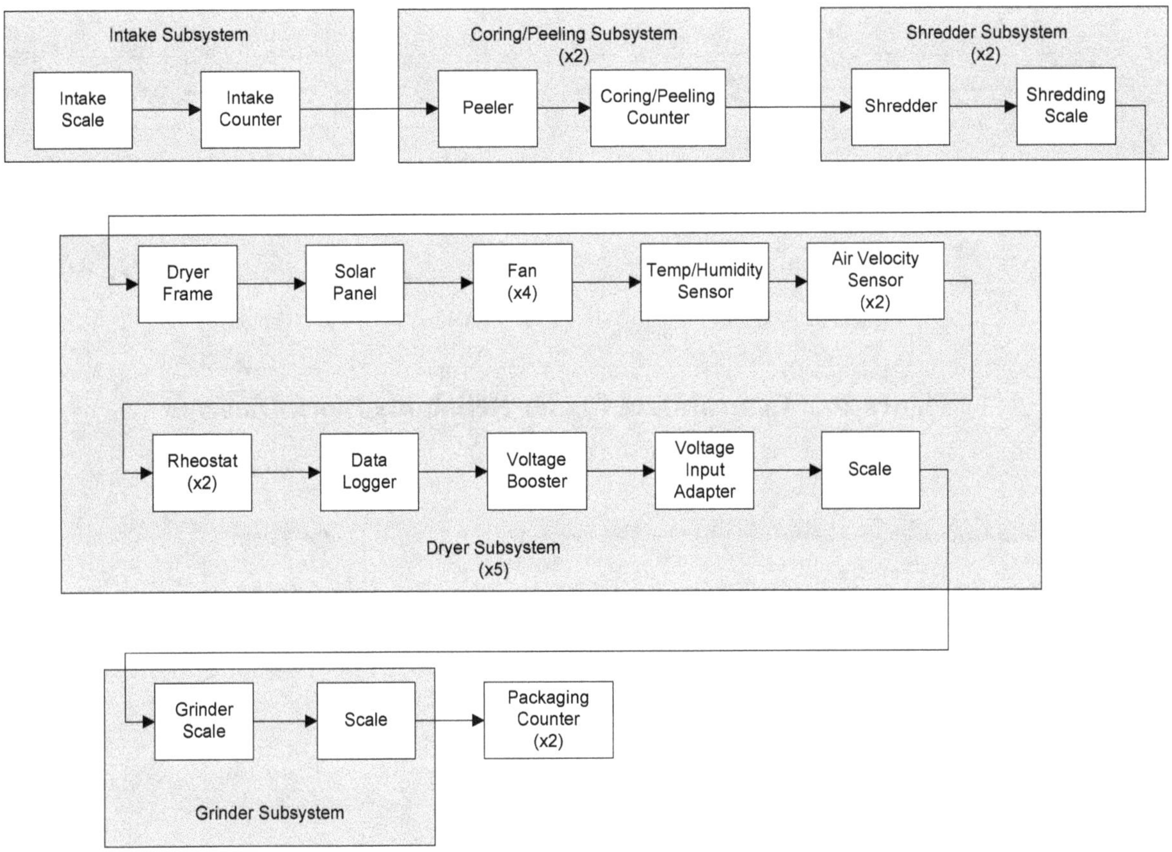

Figure 263: Processing System Reliability Block Diagram

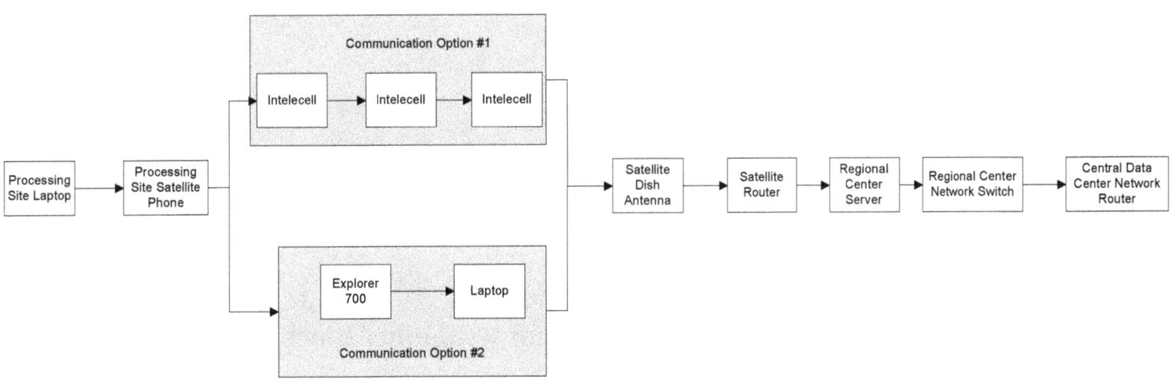

Figure 264: Communication Reliability Block Diagram

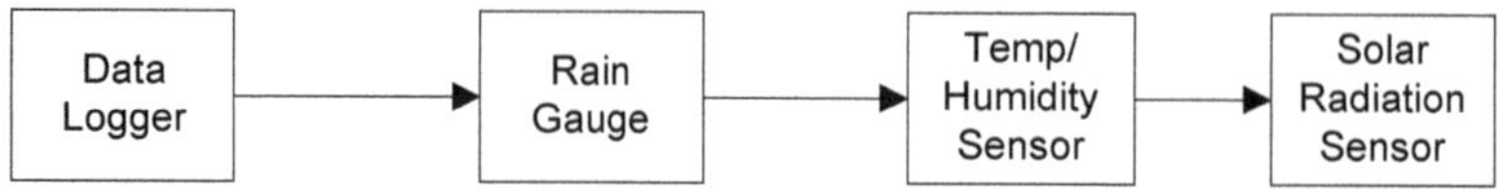

Figure 265: Weather Station Reliability Block Diagram

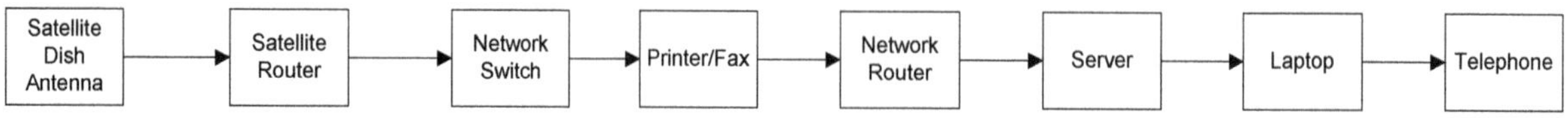

Figure 266: Regional Center Reliability Block Diagram

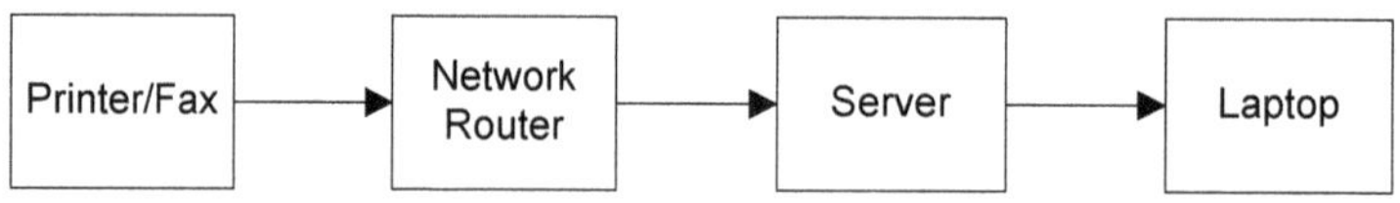

Figure 267: Central Data Center Reliability Block Diagram

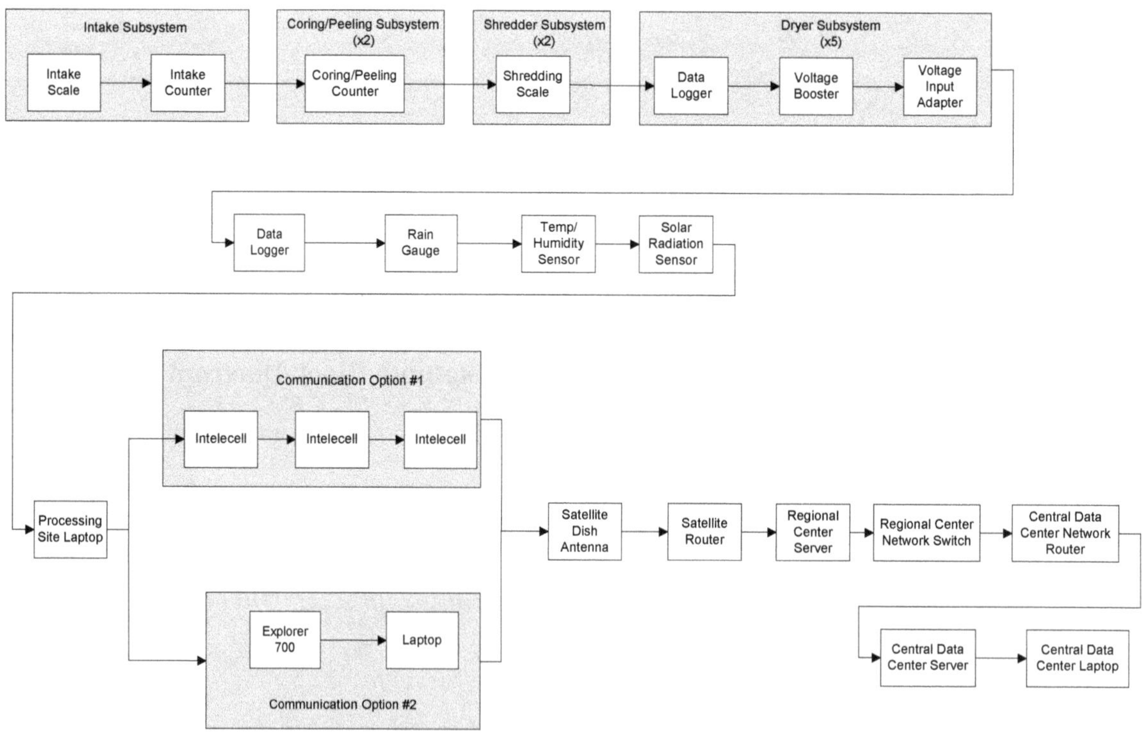

Figure 268: BrEAD Measurement Reliability Block Diagram

The failure rates for each of the components were drawn from the following sources:

- Supplier Data for most COTS equipment
- Engineering estimates based on equipment
- Reliability parts stressed analysis using MIL- HDBK-217F Notice 2 as a guide

Where predictions were performed using MIL-HDBK-217F Notice 2 as a guide, the following adjustments to the models were used to represent improvements in technology and manufacturing processes since the handbook was last updated in December of 1995.

Only the electromechanical components of the BrEAD architecture were modeled and included in the Reliability Prediction. The mechanical components were not included in the prediction because mechanical components typically have a wearout life and do not have random failures. Since the MTBF calculation is only valid for random failures, our mechanical components were not included.

Environmental Modification Factors

The BrEAD system reliability predictions are based on Ground Benign (GB) environments since this equipment will not be used in extreme temperature or vibration environments. For COTS equipment, vendor data was sometimes provided under Ground Fixed (GF) conditions. For these cases values were adjusted to an GB environment using Table 6.3.3-2 from The Rome Laboratory Reliability Engineer's Toolkit Commercial Practices Edition.

Summary of Reliability Prediction Results

Table 1 shows the predicted MTBF for each of the subsystem within the BrEAD architecture. All of the calculated MTBFs surpass the reliability requirements. No reliability critical items have been identified from this analysis.

Table 10: BrEAD Architecture MTBF

Subsystem Name	Predicted λ (FPMH)	Predicted MTBF (hrs)	Spec Limit (hours)
Coring/Peeling Subsystem	2.0	500,000	400,000
Shredder Subsystem	13.3	75,000	75,000
Dryer Subsystem	642.5	1,556	1,500
Grinder Subsystem	6.7	150,000	150,000
Total Processing System	674.2	1,483	1,400
Weather Station	24.5	40,816	40,000
Communication System (Intelecell)	190.8	5,240	5,000
Communication System (Explorer 700)	140.8	7,101	5,000
Regional Center	189.2	5,286	5,000
Central Data Center	150.0	6,667	6,000
BrEAD Measurement System (Intelecell)	536.2	1,865	1,700
BrEAD Measurement System (Explorer)	486.2	2,057	1,700

Maintainability Modeling and Prediction

Maintainability Requirements

The maintainability requirements expressed in Technical attachment I - BrEAD PBSS is the following:

[ID 115] The mean time to repair (MTTR) of the BrEAD Measuring System shall be no greater than 15 minutes.
[ID 123] The mean time to repair (MTTR) of the BrEAD Data Communication system shall be no greater than 15 minutes.
[ID 026] The mean time to repair (MTTR) of the breadfruit processing system shall be no greater than 15 minutes.
[ID 102] The mean time to repair (MTTR) of the BrEAD weather station shall be no greater than 15 minutes.

Assumptions and Ground Rules

The maintainability prediction for all subsystem equipment uses the standard MIL-HDBK-472 equation for calculating the overall system MTTR. The following represents the mathematical expression for the MTTR:

$$MTTR = \frac{\sum_{n=1}^{N} \lambda_n R_n}{\sum_{n=1}^{N} \lambda_n}$$

Where:

N = number of replaceable items (RI)

λ_n = The failure rate of the nth RI

R_n = The mean repair time of the nth RI

The MTTR and/or Mean Repair Times on the nth RI for each component is drawn from the following sources:

- Field Data
- Engineering estimates based on equipment similarity

All the equipment contained in the BrEAD architecture have a "remove and replace" maintenance approach. Therefore, if a processing system component fails, it will be removed, replaced with a new functioning component, and sent back to depot maintenance at the Regional Center.

Maintainability Summary of Results

Table 2 shows the predicted MTTR for the BrEAD processing systems, the BrEAD Measurement system, and the Weather station system. The MTTR analysis demonstrates that the BrEAD system can meet the specification requirements as summarized.

Table 11: Maintainability MTTR Summary

Subsystem Name	Spec Limit (Minutes)	Results (Minutes)
Processing system	15	**7.11**
Weather Station system	15	**10.10**
Communication system	15	Intelecell: **14.72** Explorer 700: **13.56**
Measurement System	15	Intelecell: **12.19** Explorer 700: **11.59**

Availability Modeling

Availability Requirements

The PBSS does not specific exact availability requirements, however high availability is required to meet the requirement below.

[ID 002] The Breadfruit Enterprise Architecture Demonstration (BrEAD) shall provide an automated research, marketing, and user data collection capability for NGRF head quarters in St. Paul, Minnesota that will provide 24 hour per day data from researchers, marketers and users worldwide.

In addition, the processing system needs to have a high availability because of the limited time it can be used. Our system cannot afford to have long periods of down time.

Availability Assumptions and Ground Rules

Availability is the probability that the system will be able to perform its task profile when required. Availability is primarily a function of how often failures occur or corrective/preventive maintenance is required (reliability), and then how quickly indicated or recorded failures can be confirmed and repaired or preventive maintenance performed (maintainability).

The following represents the mathematical expression for availability (A_0):

$$A_0 = \frac{MTBF}{MTBF+MTTR}$$

Availability Summary of Results

Table 3 shows the predicted availability for the BrEAD processing system and the BrEAD Measurement system. As demonstrated in the table our processing system high availability and will meet the system requirements.

Table 12: Availability Summary

Subsystem Name	Availability	Availability (%)
Processing system	0.999920108	**99.992%**
Measurement System (with Intelecell)	0.999874548	**99.987%**
Measurement System (with Explorer 700)	0.999884546	**99.988%**

Appendix A: Reliability and Maintainability Worksheets

Subsystem	Part Number	MFG	Qty	MTTR (hr)	Individual MTBF
Processing System					
Intake Subsystem					
Scale	TBD	TBD	1	0.03	150,000
Counter (Clicker)	TBD	TBD	1	0.03	1,000,000
				Subsystem MTBF	**130,435**
Coring / Peeling Subsystem					
Coring Device	TBD	St. Thomas/CTI	2	1	**N/A**
Counter (Clicker)	TBD	TBD	2	0.03	1,000,000
				Subsystem MTBF	**500,000**
Shredding Subsystem					
Horizontal Shredder	TBD	St. Thomas/CTI	2	1	**N/A**
Scale	TBD	TBD	2	0.03	150,000
				Subsystem MTBF	**75,000**
Drying Subsystem					
Dryer (Plastic & Frame)	TBD	CTI		2	**N/A**
Solar Panel (12 VDC)	SP20R	Campbell Scientific	5	0.25	60,000
Fans (12 VDC)	603-1053-ND	Delta	20	0.083	60,000
Temp / Humity Sensor	S-THB-M008	Hobo / Onset	5	0.25	500,000
Air Velocity Sensor	EE75 Model B	E & E Elektronik	10	0.083	500,000
Rheostat	RHSR25	Ohmite	10	0.083	10,000,000,000
Hobo Microstation Data Logger	H21-002	Hobo / Onset	5	0.25	80,000
Voltage Booster	TBD	TBD	5	0.083	250,000
Voltage Input Adapter	S-VIA-CM14	Hobo / Onset	5	0.083	50,000
Scale	TBD	TBD	2	0.03	150,000
				Subsystem MTBF	**1,556**
Grinding Subsystem					
Grinder	TBD	St. Thomas/CTI	1	0.083	**N/A**
Scale	TBD	TBD	1	0.03	150,000

Subsystem	Part Number	MFG	Qty	MTTR (hr)	Individual MTBF
				Subsystem MTBF	**150,000**
Packaging Subsystem					
Counter (Clicker)	TBD	TBD	2	0.03	1,000,000
				Subsystem MTBF	**500,000**
Monitoring Subsystem					
Laptop	Latitude E6500	DELL	1	0.03	30,000
Satellite Phone	GSP-1700	Globalstar	1	0.03	25,000
Weather Station					
Hobo Weatherstation Data Logger	H21-001	Hobo / Onset	1	0.25	80,000
Rain Gauge Smart Sensor	S-RGA-M002	Hobo / Onset	1	0.083	300,000
Wind Speed and Direction Smart Sensor	S-WCA-M003	Hobo / Onset	1	0.083	300,000
12-bit Temperature/RH Smart Sensor	S-THB-M008	Hobo / Onset	1	0.083	500,000
Solar Radiation Sensor	S-LIB-M003	Hobo / Onset	1	0.083	300,000
Communication					
Intelecell	InteleNet Starter Kit	Intelesense	3	0.3	40,000
Expolorer 700	Explorer 700	Thrane & Thrane	1	0.3	40,000
Regional Center					
Dish Antenna	Nova 1000	StarBand	1	0.2	80,000
Satellite Router	Model 484	StarBand	1	0.35	100,000
Network switch for the telephone	TBD	TBD	1	0.1	100,000
1 Printer/Fax/Scanner combo	OfficeJet Pro J8490	HP	1	0.1	25,000
Network Router	877W-G-A-K9	CISCO	1	0.35	100,000
Server	PowerEdge T100	DELL	1	0.3	30,000
Personal Computer	Latitude E6500	DELL	1	0.03	30,000
Telephone	TBD	TBD	1	0.03	25,000

Subsystem	Part Number	MFG	Qty	MTTR (hr)	Individual MTBF
				Subsystem MTBF	**5,286**
Central Data Center					
Server	PowerEdge 2900 III	DELL	2	0.45	30,000
Network Switch	Powerconect 3524	DELL	1	0.35	100,000
Laptop	Latitude E6500	DELL	1	0.03	30,000
1 Printer/Fax/Scanner combo	OfficeJet Pro L7590	HP	1	0.1	25,000
				Subsystem MTBF	**6,667**

System Development and Demonstration of the Breadfruit Enterprise Demonstration (BrEAD) and Systems Design

Solutions International Proposal for the Breadfruit Enterprise Demonstration and System Design

Volume 2: Technical

Annex C: Contractor Work Breakdown Structure (CWBS)

RFP NO: E00010-07-R0001

Copy 1 of 2

May 4, 2009

Prepared For:

New Green Revolution Foundation (NGRF)

Prepared By:

Solutions International

123 Pillsbury Drive

St. Paul, MN 55114

CAGE CODE SI4ME

CWBS

	Bread
1	EA

1.1 Contract Award
1.2 Task authorized
1.3 Milestones
1.3.1 Schedule Risk Assessment
1.3.2 Initial Baseline Review (IBR)
1.3.3 Program Management Review (PMR)
1.3.4 System Requirements Review (SRR)
1.3.5 System Functions/Design Readiness Review
1.3.6 Prelim Hazard Analysis Review
1.3.7 Preliminary Design Review/Software Readiness Review
1.3.8 System Hazard Analysis Review
1.3.9 Critical Design Review (CDR)/Software Verification and Validation Plan Reviews
1.3.10 Build (Prototype) Build Review
1.3.11 Test Readiness Review
1.3.12 Operational Test Readiness Review (OTRR)
1.3.13 System Verification Review
1.3.14 Production Readiness Review
1.3.15 Initial Operational Operability, Capability, and Support Review
1.4 CDRLS
1.4.1 0001 System Design Description
1.4.2 0002 Integrated Program Plan (IMP)
1.4.3 0003 Integrated Management Schedule (IMS)
1.4.4 0004 Integrated Test Plan
1.4.5 0005 Systems Engineering Management Plan (SEMP)
1.4.6 0005 Software Development Plan
1.4.7 0006 Validation and Verification Plan
1.4.8 0007 Statement Of Work
1.4.9 0008 Updated Performance Based Systems Specification
1.4.10 0009 Configuration Management Plan
1.4.11 00010 Data Management Plan
1.4.12 00011 System Operational Effectiveness Plan
1.4.13 00012 Quality Management Plan
1.4.14 00013 Modeling and Simulation Plan
1.4.15 00014 Human Engineering Task Analysis
1.4.16 00015 Human Engineering Design Approach
1.4.17 00016 Human Engineering Test Plan
1.4.18 00017 System Safety Management Plan (SSMP)
1.4.19 00018 Test and Evaluation Master Plan (TEMP)
1.4.20 00019 Contractors Performance Based Logistics Planning (PBL) Strategy Report
1.5 Other Project Plans
1.5.1 Risk Management Plan

1.5.2 Software Safety Plan
1.5.3 Software Program Measurement Plan
1.5.4 Software Quality Plan
1.5.5 Problem Reporting Process Plan
1.5.6 Support Equipment Program Plan
1.5.7 Training Plan
1.6 Program Completion
1.6.1 Level of Effort (LOE) Complete
1.7 BrEAD_EA_System
1.7.1 Program Management
1.7.1.1 Program Management Office (PMO)
1.7.1.2 Project Engineer (PE) - Bread EA
1.7.1.3 Project Engineer - BrEAD Processing Sys.
1.7.2 Contracts
1.7.3 Subcontracts
1.8 Business Management
1.8.1 Finance
1.8.2 Subcontracts
1.8.3 Deployments
1.9 Quality Assurance
1.10 Continuous Improvement - Working Groups
1.10.1 EA System Integration Working Group
1.10.2 Cultural Norms Working Group
1.10.3 Processing System Integration Group
1.10.4 Regional Center Working Group
1.10.5 Deployment/Scalability Working Group
1.10.6 Systems Safety Working Group
1.10.7 Test Plan Working Group
1.10.8 Systems Software Security/Safety Working Group
1.11 Program Panning and Control
1.11.1 Process and Environment
1.11.2 Integrated Master Plan (IMP) / Integrated Master Schedule (IMS)
1.12 Program Management Reviews
1.12.1 Integrated Baseline Review
1.12.1.1 Baseline Review Preparation
1.12.1.2 Baseline Review Execution
1.12.1.3 Post Baseline Review Updates
1.12.2 Program Management Reviews
1.12.2.1 Initial Program Management Review Preparation
1.12.2.2 Initial Program Management Review Conducted
1.12.2.3 Quarter 4 Program Management Review Preparation
1.12.2.4 Quarter 4 Program Management Review Conducted
1.12.2.5 Quarter 1 Program Management Review Preparation

1.12.2.6	Quarter 1 Program Management Review Conducted
1.12.2.7	Quarter 2 Program Management Review Preparation
1.12.2.8	Quarter 2 Program Management Review Conducted
1.12.2.9	Quarter 3 Program Management Review Preparation
1.12.2.10	Quarter 3 Program Management Review Conducted
1.12.2.11	Quarter 4 Program Management Review Preparation
1.12.2.12	Quarter 4 Program Management Review Conducted
1.12.3	Schedule Risk Assessment (SRA) Reviews
1.12.3.1	Initial SRA Preparation
1.12.3.2	Initial SRA Review Conducted
1.12.3.3	Quarter 4 SRA Preparation
1.12.3.4	Quarter 4 Baseline Review
1.12.3.5	Quarter 1 SRA Preparation
1.12.3.6	Quarter 1 Baseline Review
1.12.3.7	Quarter 2 SRA Preparation
1.12.3.8	Quarter 2 Baseline Review
1.12.3.9	Quarter 3 SRA Preparation
1.12.3.10	Quarter 3 Baseline Review
1.12.3.11	Quarter 4 SRA Preparation
1.12.3.12	Quarter 4 Baseline Review
1.12.4	Program Reviews
1.12.4.1	Systems Requirements Review (SRR)
1.12.4.1.1	Performance Based System Spec (PBSS) Prep
1.12.4.1.2	PBSS Review
1.12.4.1.3	Post PBSS Review
1.12.4.2	Software Requirements Specification
1.12.4.2.1	Software Requirements Spec prep
1.12.4.2.2	Software Requirements Spec review
1.12.4.2.3	Post Software Requirements Spec review
1.12.3.3	System Functions/Design Readiness Review
1.12.3.3.1	System Funct./Design Readiness Prep
1.12.3.3.2	System Funct./Design Readiness Execution
1.12.3.3.3	Post System Funct./Design Readiness Updates
1.12.3.4	Preliminary Hazard Analysis Review
1.12.3.4.1	Preliminary Hazard Analysis Review Prep
1.12.3.4.2	Preliminary Hazard Analysis Execution
1.12.3.4.3	Post Preliminary Hazard Analysis Updates
1.12.3.5	Design Reviews
1.12.3.5.1	Preliminary Design Review/Software Readiness Review
1.12.3.5.1.1	Preliminary Design Review/Software Readiness Review prep

1.12.3.5.1.2 Preliminary Design Review/Software Readiness Review execution
1.12.3.5.1.3 Post Preliminary Design Review/Software Readiness Review updates
1.12.3.5.2 System Hazard Analysis Review
1.12.3.5.2.1 System Hazard Analysis Review Prep
1.12.3.5.2.2 System Hazard Analysis Review Execution
1.12.3.5.2.3 Post System Hazard Analysis Review Updates
1.12.3.5.3 Critical Design Review (CDR)
1.12.3.5.3.1 CDR Prep
1.12.3.5.3.2 CDR Execution
1.12.3.5.3.3 Post CDR Updates
1.12.4 Software Verification & Validation Plan Review
1.12.4.1 Software Verification & Validation Plan Review Prep
1.12.4.2 Software Verification & Validation Plan Review Execution
1.12.4.3 Post Software Verification & Validation Plan Review Updates
1.12.5 Software Qualification Reviews (SQR)
1.12.5.1 SQR Prep
1.12.5.2 SQR Execution
1.12.5.3 Post SQR Updates
1.12.6 Prototype Build Reviews
1.12.6.1 Prototype Build Reviews Prep
1.12.6.2 Prototype Build Reviews Execution
1.12.6.3 Post Prototype Build Reviews Prep Reviews Updates
1.12.7 Test Readiness Review (TRR)
1.12.7.1 TRR Prep
1.12.7.2 TRR Execution
1.12.7.3 Post TRR Updates
1.12.8 Prototype Field Testing
1.12.8.1 Ship prototype system
1.12.8.2 Test & evaluation
1.12.8.3 System ship back to SI
1.12.9 Operational Test (SIL) and Readiness Reviews (OTRR)
1.12.9.1 OTRR Prep
1.12.9.2 OTRR Execution
1.12.9.3 Post OTRR Updates
1.12.10 System Verification Review (SVR)
1.12.10.1 SVR prep
1.12.10.2 SVR Execution
1.12.10.3 Post SVR Updates
1.12.11 Production Readiness Review (PRR)
1.12.11.1 PRR prep
1.12.11.2 PRR Execution

1.12.11.3	Post PRR Updates
1.13	Project Travel
1.14	Materials
1.15	Program Meetings
1.16	Systems Engineering - Bread_EA
1.16.1	Systems Management
1.16.1.1	Systems Management
1.16.2	Modeling and Simulation Management
1.16.2.1	Modeling and Simulation Management
1.16.3	Human Factors Management
1.16.3.1	Human Factors Management
1.16.4	Software Management
1.16.4.1	Software Management
1.16.5	Test Management
1.16.5.1	Test Management
1.16.6	Configuration and Data Management (CMDM)
1.16.6.1	Configuration and Data Management (CMDM)
1.16.7	Reliability, Maintainability, Supportability (RMS) Prognostics and Health Management (PHM)
1.16.7.1	Reliability, Maintainability, Supportability (RMS) Prognostics and Health Management (PHM)
1.16.8	Software Integration Lab (SIL) Management
1.16.8.1	Software Integration Lab (SIL) Management
1.16.8.2	Requirements Phase
1.16.8.3	System Design Phase
1.16.8.4	Detailed Design Phase
1.16.8.5	Prototype Phase
1.16.8.6	Integration Phase
1.16.8.7	Verification and Validation Phase
1.16.8.8	Production Phase
1.16.8.9	Transition Phase
1.16.8.10	BrEAD_EA Delivered
1.17	Systems Engineering - BrEAD Processing
1.17.1	Systems Management
1.17.1.1	Systems Management
1.17.2	Modeling and Simulation Management
1.17.2.1	Modeling and Simulation Management
1.17.3	Human Factors Management
1.17.3.1	Human Factors Management
1.17.4	Software Management
1.17.4.1	Software Management
1.17.5	Test Management
1.17.5.1	Test Management

System Development and Demonstration of the Breadfruit Enterprise Demonstration (BrEAD) and Systems Design

Solutions International Proposal for the Breadfruit Enterprise Demonstration and System Design
Volume 2: Technical
Annex D: Statement of Work (SOW)

RFP NO: E00010-07-R0001
Copy 1 of 2
May 4, 2009

Prepared For:

New Green Revolution Foundation (NGRF)

Prepared By:

Solutions International
123 Pillsbury Drive
St. Paul, MN 55114

CAGE CODE SI4ME

1.0 PROGRAM SCOPE

The New Green Revolution Foundation (NGRF) is a non-profit foundation has been doing research in several tropical island locations on expanding the used of breadfruit as a food source, and it believes that it could be a significant contribution to feeding a growing world population particularly for island nations. Currently breadfruit is an extremely perishable food source. It has a 48 hour shelf life after it is harvested from the breadfruit tree. The extremely short shelf life and lack of refrigeration on many of the islands has meant that the breadfruit has to be consumed locally in a very short time by the island populations. Preliminarily research results by the NGRF sponsored researchers has demonstrated that breadfruit can be processed into flour and other stable shelf life products using local technologies by the local population. The Breadfruit Enterprise Architecture Demonstration (BrEAD) will be developed to do further research and development to develop new breadfruit food products, to promote those products world wide, to facilitate the design of scalable breadfruit processing systems, to create a virtual collaborative knowledge sharing automation / communication system to share research and marketing information in real-time, and to seek funding from various sources, NGRFs and private organizations, to continue the development of breadfruit food products and breadfruit processing technology development.

2.0 PURPOSE

The purpose of the Breadfruit Enterprise Architecture Demonstration (BrEAD) System Development and Demonstration (SDD) acquisition phase is to: design, develop, fabricate, integrate, demonstrate, validate, test, and document the Breadfruit Enterprise Architecture (BrEAD) ensuring a scalable affordable, operationally supportable, tailorable breadfruit processing system, breadfruit product distribution system and a distributed virtual collaborative marketing and research communications system for future implementation.

This SOW is intended to provide enough information and detail to structure an executable program able to satisfy NGRF objectives. To this end, this SOW will specify critical program management, engineering and logistics processes considered necessary to meet these goals.

3.0 NGRF STRATEGIC ACQUISITION PHASE OBJECTIVES

Execution of the SDD Phase will demonstrate an operationally effective and suitable Breadfruit Enterprise Architecture Demonstration (BrEAD). The Breadfruit Enterprise Architecture Demonstration (BrEAD) at Initial Operating Capability IOC will be fully compliant with the PBSS(CDRL 0008) requirements to ensure an operationally effective and suitable report from Operational Test and Evaluation (OT&E). The Breadfruit Enterprise Architecture Demonstration (BrEAD) will provide a 24 hour research data collection, data analysis, and data distribution capabilities for every technical and marketing research facility in the NGRF organization. It will be interoperable with local, state and federal agencies and private organizations as deemed necessary to promote breadfruit processing systems ,provide products world wide, and products standards world wide.

The Breadfruit Enterprise Architecture Demonstration (BrEAD) will provide an integrated System of Systems (SoS) that will consist of the following: (1) scalable breadfruit processing systems, (2) virtual world-wide integrated (technical and marketing) data sharing (3) world-wide breadfruit processing system user information, (4) local weather conditions at breadfruit processing sites (5) a suite of communications with Line-Of-Sight (LOS) and Beyond-Line-Of Sight (BLOS) capabilities; (6) Local Data Processing Equipments and Software (LDPES) and NGRF Center Data Processing Equipments and Software located at the St. Paul, Minnesota world-wide head quarters of NGRF.

The following statement of work shall address the following objectives to produce the Breadfruit Enterprise Architecture Demonstration (BrEAD).

24. Obtain a Breadfruit Enterprise Architecture Demonstration (BrEAD) that meets or exceeds the PBSS requirements found in section 2.1.2;
25. Obtain a Breadfruit Enterprise Architecture Demonstration (BrEAD) that is economical to operate and maintain throughout its life cycle and ensure life cycle considerations are maximized throughout the design and integration process;
26. Obtain a Breadfruit Enterprise Architecture Demonstration (BrEAD) that is fully compliant with all applicable statutes and laws.
27. Obtain a Breadfruit Enterprise Architecture Demonstration (BrEAD) design that is implementable with Commercial-off-the-Shelf technology to support the program milestones and schedule;
28. Obtain a modular open systems architecture for hardware and software that supports efficient system upgrades and allows for interoperability with other systems. Obtain a design that maximizes the use of Commercial-Off-The-Shelf (COTS), Modified-COTS (MCOTS), NGRF-Off-The-Shelf (GOTS) items and Non-Developmental-Items. The open system architecture is defined in section 2.1.3;
29. Achieve effective integrated NGRF and Contractor product teams to execute the program. Integrated Product Teams (IPTs) will be the primary means used to gain insight and provide for timely exchange of information and data. The IPT structure is defined in section 2.2.4;
30. Employ a program management system based on the Integrated Master Plan/Integrated Master Schedule (IMP/IMS) that is able to provide accurate weekly cost, schedule and performance information to support Milestone C, IOC and throughout the life cycle of the program. The IMP and IMS are defined in sections 2.2.2 and 2.2.3 in the proposal respectively;
31. Employ an integrated NGRF/Contractor risk management system that actively identifies program risks and identifies mitigation steps for program risks. The risk management system is defined in section 2.2.4;
32. Adequately protect Critical Program Information (CPI) throughout the program's lifecycle;
33. Employ an integrated, comprehensive Configuration Manage(CM)(CDRL 0009) system that maintains consistency of the Breadfruit Enterprise Architecture Demonstration (BrEAD) performance, compliance of functional and physical attributes with requirements, tracking of design and operational information and optimizes System Operational Effectiveness (SOE) throughout the life of the system. This information can be found in section 2.1;

34. Minimize the amount of NGRF Furnished items (NGRF Furnished Information, Equipment, Property) ;
35. Employ a comprehensive Systems Engineering Process In Accordance With (IAW), the INCOSE Handbook,
36. Employ an Architecture that ensures timely, external data exchanges from the NGRF Command Center Data Processing Equipments and Software (CDEPS) to a wide Community Of Interest (COI) ;
37. Support applicable COI forums promulgated via joint NGRF/Contractor IPT;
38. Employ a Breadfruit Enterprise Architecture Demonstration (BrEAD) Modeling and Simulation (M&S) strategy, which includes an integrated environment of accredited M&S tools, simulation and stimulation environment, data and resources that enables collaboration across functional disciplines;
39. Employ a Breadfruit Enterprise Architecture Demonstration (BrEAD) Human System Integration strategy to efficiently and safely operate, maintain and support a Breadfruit Enterprise Architecture Demonstration (BrEAD).
40. Obtain a Breadfruit Enterprise Architecture Demonstration (BrEAD) design that incorporates sound human engineering design to optimize the Breadfruit Enterprise Architecture Demonstration (BrEAD) and maintenance user interfaces considering workload, manpower, mission and task effectiveness, integration, information management, situational awareness and data fusion;
41. Employs a design maximizing the use of automation and decision aids for operators and maintainers to reduce workload, enhance coherence and information management and safety;
42. Develop a system that is safe to operate, maintain, and transport and does not introduce hazards;
43. Design and deliver a NGRF Breadfruit Enterprise Architecture Demonstration (BrEAD) System Integration Laboratory (SIL) that will be integrated with other NGRF research sites, and are able to simulate and stimulate the Breadfruit Enterprise Architecture Demonstration (BrEAD); and support subsystem/system test, integration, Validation & Verification (V&V)(CDRL 0006), and system/software lifecycle sustainability;
44. Provide an integrated NGRF and Contractor test program, which demonstrates/validates the completeness of the design and development process, ensures design issues have been identified and solutions are available, and ensures compliance with Breadfruit Enterprise Architecture Demonstration (BrEAD) PBSS;
45. Develop Engineering Development Models (EDMs) to support developmental testing and the operational assessment of the Breadfruit Enterprise Architecture Demonstration (BrEAD);
46. Obtain a performance-based product support capability that:
    a. Provides a maintenance support concept based upon an effective Integrated Maintenance Concept (IMC);
24. Achieve an effective communication strategy that facilitates dialogue, ensures properly coordinated reviews, and audits, as well as the ability to assess progress towards meeting program requirements. The Contractor will

complete all the reviews, assessments and working groups requested by the customer to ensure appropriate entry and exit criteria are met.

4.3 **25. Create a System that is environmentally compliant with all applicable environmental statutes and laws;**

26. Create a design which maximizes the use of standard manufacturing concepts and industrial approved commercial best practices;

4.4

4.5 **3.1 Detailed Objectives**

4.6

4.7 **3.1.1 Schedule: It is an objective of the Breadfruit Enterprise Architecture Demonstration (BrEAD) program to achieve a December 29th 2010 IOC date or earlier. The Contractor will establish a program schedule that supports program milestones and allows adequate time for technical reviews, design, and test and evaluation to meet a December 29th 2010 IOC date.**

4.8

4.9 **3.1.2 Open Systems Architecture (OSA): The Breadfruit Enterprise Architecture Demonstration (BrEAD) will have Open Architecture designs and corresponding components. As part of this contract, the Contractor will define, document, and follow an open systems approach for using modular design and standards-based interfaces.**

3.1.3 Program Management: The Contractor will use appropriate management techniques to support program control and gain program insight. The Contractor will develop and implement an Integrated Master Plan (IMP)(CDRL 0002) and Integrated Master Schedule (IMS)(CDRL 0003). The Contractor will develop and implement an IMS and utilize the Contractor Work Breakdown Structure (CWBS). The Contractor will provide in the IMP; the structure of events, accomplishments and criteria required to design, develop, integrate, produce, demonstrate, test the capabilities and functionalities, and support the design to provide an operationally effective and operationally suitable system to meet the Breadfruit Enterprise Architecture Demonstration (BrEAD) requirements. The IMS will be part of the Breadfruit Enterprise Architecture Demonstration (BrEAD) SDD contract. The Contractor will detail the required work, to include subcontractor efforts, in the IMP and IMS. Working in concert with the NGRF, the Contractor and subcontractors will consider the following areas in tailoring a program that supports the NGRF's objectives.

3.1.3.1 Program Management Structure and System: The NGRF management objective is to take full advantage of integrated NGRF and Contractor product teams to execute the program. IPTs will be the primary means used to gain insight and provide for timely exchange of information and data. The IPTs will consist of NGRF, Contractor and subcontractor personnel. The Contractor will propose a portfolio of metrics and Technical Performance Measures (TPMs) to support the continual assessment of program progress against defined plans and schedules. The program management system will be based on an IMP/IMS that is able to provide accurate and timely schedule and performance information to support IOC. A detailed program management system will be established, implemented and controlled as documented in the IMP/IMS and the Systems Engineering Master Plan (SEMP)(CDRL 0005), including subcontractor management. The Contractor will maintain an IMP/IMS throughout the entire SDD contract and will provide access, licenses (if required) and training for IMP/IMS tools for the NGRF team. The system's detail will be such that the NGRF and Contractor can utilize schedule(s) for program planning, procurement planning, tracking, and execution of Breadfruit Enterprise Architecture Demonstration (BrEAD) SDD acquisition phase, decision/problem analysis, risk analysis and schedule reporting in accordance with the IMS. The IMS will contain the contract IMP events and milestones, accomplishments, criteria, required resources and activities from contract award to the completion of the contract in accordance with the IMS and IMP. IMS tasks will be directly traceable to control accounts included in the cost system, as collected by WBS and integrated with the Contract Performance Reports (CPRs). The Contractor will quantify the risk and provide optimistic, pessimistic, and most likely duration for each IMS activity and event. The Contractor will flow down the requirement of IMS to principal and/or critical subcontractors. It is the NGRF objective to minimize schedule risk during the program with respect to the IMP/IMS with particular emphasis on software development, SIL utilization, and test and evaluation including test asset allocation. The Contractor's schedule will have sufficient flexibility to accommodate for unforeseen technical issues without major impact to the IMP/IMS while maintaining a reasonable work and resource regimen. The Contractor will ensure schedule integration with its subcontractors and will ensure the validity of the subcontractor's schedule data including demonstrating effective methods for incorporating schedule data from subcontractors into the Contractor's IMS. The Contractor will participate in an IMS Schedule Risk Assessment (SRA) with the NGRF two weeks prior to the IBR. The Contractor will participate in a joint quarterly IMS-SRA following IBR continuing through the end of the contract period of performance with the NGRF in accordance with the IMS.

The Contractor will conduct quarterly Program Management Review (PMR) meetings at mutually agreed upon dates and locations. The PMR may be held in conjunction with the joint IMS-SRA review. During these reviews, the Contractor will present integrated cost, schedule, and technical performance status. Integrated Product Team leads or functional managers will include cost information in discussions of schedule status, technical performance, and risk, using earned value as an integrating tool. The following will be addressed: Cost/schedule trends, significant cost/schedule/technical variances, projected impacts, issues, quantified risk assessments, TPMs and corrective action plans. Additionally, the Contractor will present the status on personnel and staffing levels and compliance with the subcontractor management plan during the PMR.

3.1.3.2 Project Reporting:

3.1.3.2.1 Earned Value Management Systems (EVMS): The Contractor will establish, maintain, and use in the performance of this contract an integrated management system meeting the intent of the Industry Guidelines for Earned Value Management Systems (EVMS) ANSI/EIA-748-98 (reference DI-MGMT-81466 CPR; ANSI/EIA-748 (Latest Revision) and DI-MGMT-81650, IMS). The application of these concepts will provide for early indications of contract cost and schedule problems. Earned value assessments will correlate with technical achievement. A compliance review of the Contractor's EVMS will not be performed unless the NGRF program manager determines that it is necessary from IBR results or cost and schedule data quality assessments. As stated in the RFP, the contractor shall;

(a) Provide documentation that the Cognizant Federal Agency has determined that the proposed Earned Value Management System (EVMS) complies with the EVMS guidelines in ANSI/EIA Standard -748 (current version at time of solicitation).

tab (b) If the contractor proposes to use a system that has not been determined to be in compliance with the requirements of paragraph (a) of this provision, the contractor shall submit a comprehensive plan for compliance with the EVMS guidelines.

(1) The plan shall--

(i) Describe the EVMS the contractor intends to use in performance of the contracts;

(ii) Distinguish between the contractor's existing management system and modifications proposed to meet the guidelines;

(iii) Describe the management system and its application in terms of the EVMS guidelines;

(iv) Describe the proposed procedure for administration of the guidelines, as applied to subcontractors; and

(v) Provide documentation describing the process and results of any third-party or self-evaluation of the system's compliance with the EVMS guidelines.

(2) Provide information and assistance as required by the Contracting Officer to support review of the plan.

(3) The NGRF will review and approve the contractor's plan for an EVMS before contract award.

(4) The contractor's EVMS plan must provide milestones that indicate when the contractor anticipates that the EVM system will be compliant with the ANSI/EIA Standard -748 guidelines.

(c) Contractors shall identify the major subcontractors, or major subcontracted effort if major subcontractors have not been selected, planned for application of the guidelines. The prime Contractor and the NGRF shall agree to subcontractors selected for application of the EVMS guidelines.

3.1.3.2.2 Integrated Baseline Review: The Contractor will present its performance measurement baseline plan to the NGRF within six months of contract award, and subsequently, when required, following major changes to the baseline. The NGRF will verify during the IBR, and follow-on IBRs, when required, that the Contractor has established and maintains a reliable performance measurement baseline. The Contractor will ensure that the baseline includes the entire contract technical scope of work consistent with contract schedule requirements, and has adequate resources assigned. The Contractor will assure the NGRF that effective earned value methods are used to accurately assess contract cost, schedule, and technical performance. The Contractor will perform a self-assessment of the schedule risk for the IBR. The IBR will be used to achieve a mutual understanding of the baseline plan and schedule risk, and the underlying management processes used for planning and controlling the project. This review will be tailored to focus on risk areas identified by the NGRF.

3.1.3.3 Risk Management: The Contractor will implement a sound risk management system that facilitates risk identification and mitigates program risks. The risk management system will be integrated with program controls through the development of metrics to monitor program status, tied to the IMP/IMS. This system will be utilized to:

1) Perform timely and thorough assessment of cost, schedule and technical risk;
2) Identify and track all medium and high-risk areas;
3) Define, fund, and implement appropriate risk mitigation efforts and;
4) Provide for fall-back or alternate plans, including design off ramps in the event that the primary risk mitigation plans fail.

After contract award, the NGRF and the Contractor will form, operate and co-chair an integrated Risk Management Board, based on the description above.

3.1.3.4 Data Management:

The NGRF intends to minimize data costs by acquiring only the minimum essential data to support effective management of the Breadfruit Enterprise Architecture Demonstration (BrEAD) Program. Data will be managed via the use of electronic media and an IDE to improve data transfer, NGRF access, efficiency, and affordability. (CDRL 00010)

3.1.3.4.1 Integrated Digital Environment (IDE): Timely access to program data is of the utmost importance. The Contractor will use electronic media to improve data transfer efficiency and affordability. The IDE objective includes the utilization of a robust, cost-effective on-line digital data environment that allows the program acquisition and operational support activities, throughout the program's life-cycle, to digitally create, store, access, manipulate, share, and exchange programmatic and technical data. This data must be accessible to the Breadfruit Enterprise Architecture Demonstration (BrEAD) NGRF IPT at anytime from their workspaces (approximately 200 individuals). The Contractor will provide the maximum number of NGRF seats at no additional cost. The IDE will provide protection for program data in accordance with the Program Protection Plan (PPP). Strategy will include technical refresh provisions and any necessary accreditation activities. The Contractor will optimize the use of the IDE with its subcontractors and suppliers, as well as participating NGRF organizations. The Contractor will develop, update and maintain a Data Accession List (DAL) that identifies all data and documentation generated under the execution of the Breadfruit Enterprise Architecture Demonstration (BrEAD) effort.

The Contractor will define how the IDE will be used to facilitate the integration of the IMP, IMS, and WBS to provide the data necessary for the NGRF to assess the concept's ability to meet the requirements of the PBSS (verification matrix) and provide the necessary flight clearances for the flight test program. The IMS will include sufficient time for NGRF approval of data at significant events.

A list of the data to be provided on the IDE will be attached to the IDE implementation plan.

IDE process controls, in addition to use of business tools (such as Associate Contractor nondisclosure agreements), will apply and will use system-level and data-level Metadata tags for identification and protection of digital form data (such as for classified, Export Control, Proprietary, or Distribution Statements).

The NGRF will require continued use of the IDE beyond the SDD program. The Contractor will provide the NGRF with strategies to ensure continued use of the IDE

3.1.3.4.2 Contract Data Requirements List (CDRLs): The RFP includes a list of CDRLs expected to be included in the contract. The CDRL list is as follows:

CDRL 0001 System Design Description
CDRL 0002 Integrated Program Plan (IMP)
CDRL 0003 Integrated Management Schedule (IMS)
CDRL 0004 Integrated Test Plan
CDRL 0005 Systems Engineering Management Plan (SEMP)
CDRL 0005 Software Development Plan
CDRL 0006 Validation and Verification Plan
CDRL 0007 Statement of Work
CDRL 0008 Up dated Performance Based Systems Specification (PBSS)
CDRL 0009 Configuration Management Plan
CDRL 00010 Data Management Plan
CDRL 00011 System Operational Effectiveness Plan
CDRL 00012 Quality Management Plan
CDRL 00013 Modeling and Simulation Plan
CDRL 00014 Human Engineering Task Analysis
CDRL 00015 Human Engineering Design Approach
CDRL 00016 Human Engineering Test Plan
CDRL 00017 System Safety Management Plan (SSMP)
CDRL 00018 Test and Evaluation Master Plan (TEMP)
CDRL 00019 Contractors Performance Based Logistics Planning (PBL) Strategy Report

The Contractor may request additions and/or deletions which will be subject to NGRF approval.

3.1.3.5 Configuration Management: The Contractor will identify and utilize a configuration management approach which establishes and maintains consistency of the Breadfruit Enterprise Architecture Demonstration (BrEAD) performance, compliance of functional and physical attributes with requirements, and tracking of design and operational information throughout system life. The approach will also consider regulating life-cycle costs. The Contractor, as part of the IDE, will employ an integrated change management system that identifies, prioritizes, tracks, allocates and manages program, system and software requirement changes via a Change Review Board. The approach will consider all elements of the Breadfruit Enterprise Architecture Demonstration (BrEAD) program including data rights, product support, training infrastructure and will address any peculiar software/hardware items. Specifically, the approach will address product support development and implementation within a PBL support strategy and will include Product Support Integrator requirements, vendor integration and COTS/MCOTS/GFE/NDI integration and management. The approach will consider the establishment of the functional, allocated and product baselines; the establishment of these baselines will be included in the IMP and identified in the IMS.

3.1.3.6 Specifications and Standards: If the use of a NGRF specification or standard is not specified, the NGRF encourages the use of generally accepted commercial and industry specifications and standards. Foreign specifications and standards may be used if existing equipment from non-U.S. suppliers is being proposed.

3.1.3.7 Systems Operational Effectiveness: The Contractor will develop, and submit to the NGRF for review and comment, a SOE Plan that will establish and document the Contractor's process for implementing SOE(CDRL 00011). The plan will detail the following:

e) Integration of SOE with the Systems Engineering process including the process for managing SOE in trade studies and trade space;
f) Integration of SOE with the supportability and logistics planning activities;
g) Specific measures for establishing appropriate SOE goals and evaluating attainment thereof, including test and evaluation and validation and inclusion in the SIL
h) Modeling and simulation tools and processes.

3.1.3.8 Quality Management System: The Contractor will develop and maintain a Quality Management System (QMS)(CDRL 00012) that satisfies program objectives and meets the requirements of ANSI/ASQC-Q9001: 2000 or an equivalent QMS. The QMS procedures, planning, and all other documentation and data that comprise the QMS, will be made available to the NGRF for review. The NGRF may perform the inspections, verifications, and evaluations necessary to ascertain conformance to requirements and the adequacy of the implementing procedures. Third-party certification of the Contractor's QMS is not required. The Contractor will require subcontractors to maintain a QMS that achieves control of the quality, services and supplies provided.

3.1.4 Engineering

3.1.4.1 Systems Engineering: The Contractor will develop and update as necessary throughout the SDD phase a comprehensive SEMP that describes the Contractor's methodology for fully integrating the NGRF's requirements and PBSS. The SEMP will demonstrate an integrated systems engineering approach towards development of the requirements, configuration baseline definitions, system design, system integration PBL product support requirements, Open Architecture implementation, system architecture quality and concordance, Human Systems Integration (HSI) and logistics resources (maintenance personnel, support equipment and repair parts) that cumulatively provide the capabilities specified in the PBSS. Additionally, the Contractor will address how production capabilities and processes will be incorporated into the systems engineering approach. In particular, the Contractor will focus on a design engineering producibility program identifying and implementing the design and fabrication/assembly processes which will result in the system achieving its production performance, cost and schedule requirements. The Contractor will generate TPMs, and track and report the status of the TPMs. The Contractor will utilize a requirements traceability tool, (such as TRACE, Dynamic Object Oriented Requirements System (DOORS) or a compatible automated tool), throughout the design, development, production and verification phases of the program. The Contractor will provide sufficient access, licenses (if required) and training for the requirements traceability tool to the NGRF team. The traceability tool will include all levels of the system and subsystem requirements, derived requirements and links to the capabilities documents, and all testing requirements.

The Contractor will ensure HSI will be a fundamental driver for the design by implementing a strategy to coordinate and manage issues across the program concerning manpower, personnel, training, Human Factors Engineering (HFE), safety, occupational health hazards, personnel survivability, and habitability factors.

3.1.4.2 Modeling and Simulation (M&S): The Contractor will employ integrated M&S tools as appropriate in the Breadfruit Enterprise Architecture Demonstration (BrEAD) development and testing process as a means of optimizing the system design and demonstrating compliance with the PBSS(CDRL 00013). A key aspect of the Breadfruit Enterprise Architecture Demonstration (BrEAD) M&S strategy is an integrated environment of accredited M&S tools, data, and resources that enables collaboration across functional disciplines. M&S will integrate both simulation and stimulation capabilities. Development and utilization of this environment will place emphasis on open sharing of common tools/data. The Contractor will provide a detailed description of their proposed approach to the development and application of M&S assets in support of T&E, training, and other SDD development/engineering activities.

Modeling and simulation, including SIL M&S that includes Human-In-the-Loop (HIL), will leverage the Human Engineering Simulation Concept Report and human engineering related task analysis deliverables to characterize human interface functionality, as appropriate.

3.1.4.3 Human Factors Engineering (HFE): The Contractor will use human-centered HFE design principles and procedures, for all Breadfruit Enterprise Architecture Demonstration (BrEAD) systems, sub-systems and components to optimize operator and maintainer task performance, minimize operator and maintainer error, and minimize any safety, personnel health hazards, or environmental impacts. The Contractor will prepare and submit a Human Engineering Program Plan (HEPP)(CDRL 00015) to the NGRF for approval.

The Contractor will provide a detailed Human Engineering Task Analysis(CDRL 00014) which includes analysis of all functions of the processing system and the evaluation of the user interface of Breadfruit Enterprise Architecture Demonstration (BrEAD). This includes operators and maintainers and identification of critical tasks and operation in degraded modes. Information on each task will be provided to a level sufficient to identify operator and maintainer problem areas that can adversely affect mission accomplishment and to propose and evaluate corrective actions, including design solutions.

The Contractor will evaluate and test the Breadfruit Enterprise Architecture Demonstration (BrEAD) user-interface design, including implementation of controls and displays as specified in applicable design and requirements documents. All design decisions/validation related to Breadfruit Enterprise Architecture Demonstration (BrEAD) workstation configuration and layout, operator performance, controls and displays including information presentation and display management, decision aiding, adaptive system features and automation will be supported by Human Interface Laboratory (HIL) simulations. The Contractor will prepare and submit to the NGRF for approval a Human Engineering Simulation Concept/Report. Human Engineering activities will include demonstration and validation of Breadfruit Enterprise Architecture Demonstration (BrEAD) requirements for human performance, workload, personnel survivability, safety, and habitability. In addition to HIL simulations as required, the Contractor will perform evaluations and testing in functional operator workstations or mockups, and/or through computer models to verify overall system design.

The Contractor will use working groups, including NGRF working groups to assist in ensuring human engineering criteria as well as overall Breadfruit Enterprise Architecture Demonstration (BrEAD) human systems related performance requirements are met. The Contractor will develop and submit a Human Engineering Test Plan (HETP)(CDRL 00016) and a Human Engineering Test Report (HETR).

3.1.4.4 Software: The Contractor will define a software development approach appropriate for the computer software effort to be performed under this solicitation. This approach will be documented in a NGRF approved Software Development Plan (SDP)(CDRL 0005). The Contractor will follow this SDP for all computer software to be developed or maintained under this effort. The SDP will be updated as changes occur as a result of process improvement reviews and resubmitted for NGRF approval.

The SDP will contain the information defined in IEEE/EIA Std. 12207.1 section 5.2.1 (generic content) and the Plans or Procedures in Table 1 of IEEE/EIA Std. 12207.1. In all cases the level of detail will be sufficient to define all software development processes, activities and tasks to be conducted. Information provided must include, at a minimum, specific standards, methods, tools, actions, strategies, and responsibilities associated with development and qualification.

The Breadfruit Enterprise Architecture Demonstration (BrEAD) software will utilize a High Order Language (HOL), which follows ANSI, IEEE. The use of assembly language or low-level code will be restricted to processing time-constrained and memory-constrained functions. Standards will be established and utilized for embedding comments in source code per best industry practices. The comment standards for banners, headers, and special comments will be described in Contractor-approved standards or an equivalent methodology.

3.1.4.5 System Safety: The Contractor will ensure the Breadfruit Enterprise Architecture Demonstration (BrEAD) design is safe and efficient and conforms to current commercial standards for safety and human engineering. This effort will address hardware, software, and procedural safety and will eliminate or control hazards to personnel, the Breadfruit Enterprise Architecture Demonstration (BrEAD), other assets, and the environment to acceptable levels during all phases of the Breadfruit Enterprise Architecture Demonstration (BrEAD) lifecycle. To ensure a safe system design, the Contractor will develop and submit a System Safety Management Plan (SSMP)(CDRL 00017) for NGRF approval to indicate how the Contractor and their subcontractors plan to manage and implement a system safety program. The Contractor will develop and submit a System Safety Program Plan (SSPP) for NGRF approval that details how the Contractor plans to implement the Breadfruit Enterprise Architecture Demonstration (BrEAD) SSMP. Hazards will be identified in various hazard analyses (such as a Preliminary Hazard Analysis [PHA], System Hazard Analysis [SHA], and Operating and Support Hazard Analysis [O&SHA]) performed at various times during the design process commensurate with the maturity of the system design, with residual hazards/risks documented in Safety Assessment Reports prior to system test/fielding. The Breadfruit Enterprise Architecture Demonstration (BrEAD) will ensure against degradation or negation of safety features during operations, maintenance, storage, and shipping. The Breadfruit Enterprise Architecture Demonstration (BrEAD) will have fail-safe features with adequate redundancy in both hardware and software and be capable of being rendered safe during emergency or abnormal situations.

3.1.4.5.1 Software Safety: The objective is to ensure that the Breadfruit Enterprise Architecture Demonstration (BrEAD) software is safe. The Breadfruit Enterprise Architecture Demonstration (BrEAD) software will prevent or mitigate single point failures that could disable the entire system. Any software mitigation that is required to deal with hazards will be clearly identified in all software requirements documents and tools, architectural documentation or models as well as software design documentation and design tools. Hazards that could result from operator errors or operator input of improper navigation, control, payload operation, or environmental data will be mitigated by software monitoring and checks of operator inputs. The Breadfruit Enterprise Architecture Demonstration (BrEAD) software will perform appropriate safety checks to assure that dynamic updates of data do not adversely affect the health or safety of the Breadfruit Enterprise Architecture Demonstration (BrEAD).

3.1.4.5.2 Critical Safety Items: Breadfruit Enterprise Architecture Demonstration (BrEAD) Critical Safety Items (CSIs) and fatigue/fracture critical parts will be identified early in the design process through Failure Modes Effects and Criticality Analysis (FMECA) and tracked to ensure proper item management throughout the system lifecycle.

3.1.4.6 System Integration Laboratory (SIL): The Contractor will develop, deliver, install and maintain at a facility hardware, firmware, and software components for a NGRF Integration Test Facility to support the Breadfruit Enterprise Architecture Demonstration (BrEAD) Integrated Test (IT) and Operational Test (OT) program. The SIL will be used to verify and validate systems integration prior to deployment around the world.

3.1.4.7 Department of Defense Architecture Framework (DODAF): The Contractor will develop and deliver the Breadfruit Enterprise Architecture Demonstration (BrEAD) products in compliance with the current DODAF Volume I and II.

3.1.4.8 Design Decision Documentation. Utilize and maintain a Decision Database to document system-wide design decisions including decisions about the system's behavioral design and decisions affecting the selection and design of systems components and/or subsystems. Include the tradeoffs and other assessments, the decisions made, their justification, the actions taken including revised plans or changes to the baselines or functional architecture proposed or made as a result of the decisions. The decision database will be maintained over the life of the contract and will be readily available to and directly accessible by the NGRF. The Contractor will prepare and submit to the NGRF an archived image of the data in the Decision Database at the time of the close out of each technical review and audit.

3.1.5 Test & Evaluation (T&E): The Contractor, in conjunction with the NGRF, will develop and execute a detailed integrated NGRF and Contractor test program to demonstrate the completeness of the design and development process, to ensure design problems have been identified and solutions are available, to determine compliance with the Breadfruit Enterprise Architecture Demonstration (BrEAD) PBSS. Test planning will cover the Contractor testing, developmental testing, and operational testing.

The Integrated Test Team (ITT) will be responsible for developing the Integrated Test Plan (ITP)(CDRL 0004), approved by the cognizant authority per the Test and Evaluation Master Plan (TEMP)(CDRL 00018). The ITT should explore the most efficient use of existing and planned T&E resources including, but not limited to, systems integration labs, simulation and analysis tools, test ranges, telemetry and data gathering capabilities. A test V&V table will be provided to the NGRF to validate the PBSS compliance methodology. The NGRF will assess the test requirements, work with the Contractor if refinement is necessary and adjudicate any requests for test requirement removal or modification.

3.1.5.1 Problem Reporting
The NGRF will have insight into the Contractor's problem reporting database when system integration testing begins in the laboratory. The Contractor will ensure that full access to the problem reporting database is provided to the NGRF. The NGRF and Contractor will form, operate and co-chair a review board to prioritize problems for resolution.

3.1.6 Supportability: Breadfruit Enterprise Architecture Demonstration (BrEAD) will be designed to minimize scheduled and unscheduled maintenance, minimize the number of maintenance personnel, maximize SOE. The Contractor will determine and/or develop the necessary tools, support equipment, Logistics Management Information (LMI), technical documentation and skills to perform maintenance. The Contractor's supportability analysis and product development will reflect the following considerations:

d) The Breadfruit Enterprise Architecture Demonstration (BrEAD) will be supported using a performance based logistics concept. The performance based support system may be fully organic, fully commercial or a combination of both. The supply chain will maximize SOE by using a sense-and-respond process for identifying material requirements and will provide operating unit flexibility and self-sufficiency to mobilize, deploy and sustain;
e) The Breadfruit Enterprise Architecture Demonstration (BrEAD) will include a PHM subsystem that will facilitate reliability, maintainability and supportability requirements;
f) Minimize Total Ownership Cost;

3.1.6.1 Performance Based Logistics Planning: PBL(CDRL 00019) is an enabling factor for achieving a high level of SOE. Therefore, the NGRF intends to support the Breadfruit Enterprise Architecture Demonstration (BrEAD) through the application of a system-unique, PBL strategy. The Contractor will design, and report to the NGRF, a conceptual PBL strategy for use in supporting the Breadfruit Enterprise Architecture Demonstration (BrEAD). The strategy will optimize the Breadfruit Enterprise Architecture Demonstration (BrEAD) design for life cycle support and provide for continuous improvement in reliability, system readiness, TOC and product support processes. The NGRF may comment upon and approve the concept for use as a design-to support strategy or select portions of the proposed strategy. The Contractor will then utilize the approved strategy to influence the design and testing of the Breadfruit Enterprise Architecture Demonstration (BrEAD) subsystems and development of supportability products and data. The PBL strategy will comply with all relevant public laws, regulations and service preferences. The Contractors PBL Strategy Report will discuss the following at a minimum:

g) The Contractor's process for evaluating SOE impacts of the strategy;
h) The Contractor's process for influencing UAS design for enhanced supportability using the strategy, SDOE;
i) Assumptions used in analyzing and selecting a strategy;
j) NGRF actions necessary for implementing the strategy such as requests for waivers to policy or regulations;
k) Proposed performance metrics and incentives that may be used to achieve high levels of supportability and cost control;
l) Test and Evaluation and verification steps the Contractor will use to validate the strategy.

3.1.6.2 Maintenance Planning: The Contractor will incorporate processes for maintenance planning to reduce life-cycle cost, maintenance man-hours and system down time. The Contractor will develop a Logistic Management Information (LMI) database with appropriate LMI levels assigned, using a standardized LMI database software program capable of producing common output reports such as a Maintenance Plan, FMECA, Task Narrative, Provisioning Technical Documentation (PTD), Support Equipment Recommendation Data (SERD) and Training Requirements.

3.1.6.2.1 Integrated Maintenance Concept (IMC): The Contractor will maximize integration to a depth that maximizes SOE and cost effectiveness. In addition, the Contractor will implement maintenance planning concepts that take full advantage of SDOE and SOE to reduce life cycle costs.

3.1.6.3 Technical Manuals/Technical Data: The Contractor will prepare Breadfruit Enterprise Architecture Demonstration (BrEAD) technical manuals in the Interactive Electronic Technical Manuals (IETMs) format, and in accordance with a NGRF approved specification. Technical data will be accurate and in sufficient detail to support all planned levels of maintenance and operation. The technical data will include systems/functional description and theory of operation, maintenance and operating procedures, troubleshooting and testing procedures, fault isolation, Illustrated Parts Breakdown (IPB), Maintenance Requirement Cards (MRC), periodic maintenance and phased maintenance requirements, preventive maintenance, and corrective maintenance authorized for each level of maintenance. The Contractor will provide validated/verified technical data required for training, operations, testing, and maintenance. All updated publications and associated source data, XML files, schemas, Document Type Definition (DTD), business rules exchange, style sheets, and Format Output Specification Instance (FOSI) will be considered NGRF property at the conclusion of the contract.

3.1.6.4 Reliability, Maintainability, Supportability (RMS) Prognostics and Health Management (PHM) Engineering: The Contractor will maximize the inherent RMS and PHM capabilities of the Breadfruit Enterprise Architecture Demonstration (BrEAD) in a cost effective manner that will minimize O&S costs utilizing all RMS/PHM PBSS objectives as follows:

28. RMS/BIT/PHM Program Plan
29. RMS/BIT Organization and Structure
30. Schedule for RMS/BIT/PHM Tasks and Milestones
31. Reliability Modeling, Allocation and Predictions (e.g. RELEX ™, RAPTOR)
32. Maintainability Allocations, Analyses and Predictions
33. BIT and Testability Allocations, Analyses and Predictions
34. PHM Allocations, Analyses and Predictions
35. Failure Reporting, Analysis, and Corrective Action System (FRACAS)
36. Failure Review Board (FRB)
37. Fault Tree Analysis (FTA)
38. Failure Mode Effect and Criticality Analysis (FMECA)
39. Level of Repair Analysis (LORA)
40. Reliability Development/Growth Process and Tracking
41. Maintainability Demonstrations
42. BIT and Testability Demonstrations
43. Manufacturing Screening
44. RMS/BIT Testing and Tracking
45. PHM Program Plan
46. Reliability and BIT Performance Maturation
47. Parts Control/Application Program
48. Parts Derating Program
49. RMS/BIT Lessons Learned Incorporation
50. CBM+
51. Reliability Centered Maintenance (RCM) Analysis Plan
52. Maintenance Task Analysis (MTA)
53. Supportability Analysis
54. Maintenance Concept/Planning

The objective of PHM is to increase the survivability, reliability, maintainability, supportability, readiness and affordability of the Breadfruit Enterprise Architecture Demonstration (BrEAD) over the life cycle of the system. Specific sub systems and failure modes to be monitored by the PHM system will be determined by study of the system FMECAs and failure mode risk index, and a return on investment analysis of the benefit of versus the cost to implement a PHM system monitoring algorithm. Proposed failure modes to be monitored will be judged on effectiveness and maturity of the proposed monitoring algorithm relative to current and/or similar systems, and the potential benefit of new algorithms. The PHM system will allow for both hardware and software growth in order to address emerging failure modes over the life cycle of the program.

3.1.6.5 Unique Identification (UID): The Contractor will establish an effective serialized management program on all components and end items that are repairable, life-limited or time-controlled, configuration tracked items, and/or assigned logs and/or records. The Contractor will also be responsible for configuration management and associated data.

3.1.6.6 Support Equipment: The Contractor will ensure systems engineering processes are used to effectively trace supportability requirements to the Breadfruit Enterprise Architecture Demonstration (BrEAD) functional and physical architecture. The Contractor will develop and submit to the NGRF for review and comments, a Support Equipment Program Plan (SEPP). The SEPP will detail the Contractor's process for managing its SE development activities. At a minimum, the SEPP will detail how the Contractor plans to integrate SE engineering with the Systems Engineering process (unless adequately addressed in the SEMP), test and evaluation methods and LFp control and reduction efforts.

3.1.7 Training

3.1.7.1 Objective: The objective of the Breadfruit Enterprise Architecture Demonstration (BrEAD) training development program is to provide a Training System (TS) capable of producing operators and maintenance personnel in sufficient numbers to perform meet Integrated Test (IT) and OPEVAL requirements. The delivered TS will provide Cadre and Initial training system IOC date. The TS will include all hardware, software, courseware, technical and training documentation, and manuals necessary to train instructors, operators, and support personnel to full mission-ready proficiency. All documentation, Interactive Multi-media Instruction (IMI), instructional material, and training devices will maintain concurrency with Breadfruit Enterprise Architecture Demonstration (BrEAD) configuration.

3.1.8 Environment, Safety and Occupational Health (ESOH): The NGRF's objective is to reduce risks and Total Ownership Cost (TOC) by incorporating ESOH objectives into the systems engineering process, integrating elements of Contractor ESOH programs into the NGRF's ESOH programs, and ensuring continuous compliance with Federal, State, and local ESOH regulations. The Contractor will draft and submit to the NGRF, a Hazardous Material Management Program (HMMP) Plan. The HMMP Plan will detail the Contractor's management processes and controls, and outline requirements for periodic HMMP reports.

3.1.9 Transition to Production: The Contractor will maintain a management organization with the ability to coordinate all phases of the Breadfruit Enterprise Architecture Demonstration (BrEAD) transition to production while minimizing program risk.

3.1.10 Materials and Processes: The NGRF's objective is for all materials and processes considered for application in the Breadfruit Enterprise Architecture Demonstration (BrEAD) to be subjected to rigorous evaluation in a well-defined and documented design development test program. The principle objectives of the testing are to establish material system performance in the defined operational environments; identify, characterize, and optimize associated stable processes; verify methods used in the evaluation of materials, and establish the design. Design properties (in the appropriate chemical, thermal, and climatic environments) will be established during development testing to support transition and application of the material systems and processes into the Breadfruit Enterprise Architecture Demonstration (BrEAD). Environmentally conditioned tests will be performed at the development test level to meet design conditions. Whenever materials are proposed for which only a limited amount of data is available, the NGRF will be provided with sufficient background data so that a determination of the suitability of the material can be made. All testing and data used in establishing material system performance will be documented and submitted to the NGRF, including any data obtained from previous design development test programs.

3.1.10.1 Materials Producibility: The Contractor will develop and implement processes and procedures required to establish material systems maturity, repeatability, and integrity.

3.1.10.2 Material Repairability: Repair concepts and procedures will be developed to fully restore the structural, electrical or other applicable functional properties. Repair concepts and procedures will be demonstrated and validated. Use of previously qualified materials, processes, concepts, and procedures will be substantiated. Any studies or tests performed to develop and validate any repair materials, processes, concepts, and procedures will be documented in an Advanced Material Repair Development Program Plan. Results of developmental efforts will be documented in an Advanced Material Repair Development Report. Any existing data used to substantiate repair materials, processes, concepts, and procedures will also be documented.

3.1.11 Manufacturing: The Contractor will ensure implementation of processes and systems that consider manufacturing, quality and design functions in achieving a balanced product design which meets cost, schedule and performance requirements with acceptable risk. Appropriate practices for implementation should include production cost modeling; identification of key characteristics and processes; variability reduction; electronic simulation of the manufacturing environment; Cost As an Independent Variable (CAIV); cost/performance trade studies; manufacturing capability assessments/PRRs; product and process validation; and key supplier relationships.

3.1.11.1 Production Quality and Manufacturing Efficiency: The Contractor will implement processes and systems to assure program affordability through product quality and manufacturing efficiency. The following elements may be considered as appropriate practices for implementation: product improvement initiatives; variability reduction on product and process; manufacturing process control and continuous improvement; key supplier relationships.

The Contractor will identify drawings and process documents critical characteristics or critical processes. These critical characteristics will be inspectable and/or testable during the manufacturing/ fabrication/assembly processes. - END OF DOCUMENT -

System Development and Demonstration of the Breadfruit Enterprise Demonstration (BrEAD) and Systems Design

Solutions International Proposal for the Breadfruit Enterprise Demonstration and System Design

Volume 2: Technical

Annex E: Integrated Master Plan (IMP)

RFP NO: E00010-07-R0001

Copy 1 of 2

May 4, 2009

Prepared For:

New Green Revolution Foundation (NGRF)

Prepared By:

Solutions International

123 Pillsbury Drive

St. Paul, MN 55114

CAGE CODE SI4ME

	Integrated Master Plan (IMP)		
Activity #	Event	SOW REF	WBS REF
	IMP Accomplishment		
	IMP Criteria		
	IMP Task		
A	Event A - Risk Management Review	-	-
A01	Risk Management Planning Reviewed		
A01a	Post Award Risks Reviewed	3.1.3.3	1.5.1
A01b	Program Risks Identified	3.1.3.3	1.5.1
A01c	Risk Management Program Reviewed	3.1.3.3	1.5.1
A02	Risk Management Review Conducted		
A02a	Risk Mgmt Meeting Preparation	3.1.3.3	1.5.1
A02b	Risk Mgmt Meeting Conducted	3.1.3.3	1.5.1
A03b	Risk Mgmt Meeting Minutes and Action Items Generated	3.1.3.3	1.5.1
B	Event B - Initial Baseline Review (IBR)	-	-
B01	Management Planning Reviewed		
B01a	Program Organization Established	3.1.1	1.12.1.1
B01b	Initial Configuration Management Established	3.1.5	1.12.1.1
B01c	Program Schedule Reviewed	3.1.1.2	1.12.1.1
B01d	Risk Management Program Reviewed	3.1.3.3	1.12.1.1
B02	Baseline Design Reviewed	-	
B02a	Requirements Baseline Complete	3.1.3.4	1.12.1.2
B03	IBR Conducted	-	
B03a	IBR Meeting Preparation	3.1.3.1.1	1.12.1.1
B03b	IBR Meeting Conducted	3.1.3.1.2	1.12.1.2
B03c	IBR Meeting Minutes and Action Items Generated	3.1.3.1.2	1.12.1.3
C	Event C - Program Management Review (PMR)	-	-
C01	Program Management Baseline Completed	-	
C01a	IBR Deltas to Baseline Identified	3.1.3.1.2	1.12.2
C01b	IBR Deltas to Baseline Incorporated	3.1.3.1.2	1.12.2
C01c	Baseline Complete	3.1.3.1.2	1.12.2
C02	Program Performance Assessment Reviewed	-	
C02a	Earned Value Management System (EVMS) Reviewed	3.1.1.4.1	1.12.2
C02a01	Cost/schedule Trend Analysis Completed	3.1.3.2	1.12.2
C02a02	Significant Cost/Schedule/Technical Variances Identified	3.1.3.2	1.12.2
C02a03	Projected Impacts Identified	3.1.3.2	1.12.2
C02b	Cost and Software Data Reporting (CSDR) Reviewed	3.1.1.4.2	1.12.2
C02c	Contract Funds Status Reports (CFSR) Reviewed	3.1.1.4.3	1.12.2
C02d	Technical Performance Measures Analysis Completed	3.1.1.5	1.12.2
C02e	Corrective Action Plans Completed	3.1.3.2	1.12.2
C02f	Quantified Risk Assessments Analysis Completed	3.1.3.2	1.12.2
C03	PMR Conducted	-	
C03a	PMR Meeting Preparation	3.1.3.2	1.12.2
C03b	PMR Meeting Conducted	3.1.3.2	1.12.2
C03c	PMR Meeting Minutes and Action Items Generated	3.1.3.2	1.12.2
C08	Initial Hazardous Material Management Strategy Reviewed	-	1.12.2
C08a	Initial Hazardous Material Management Strategy Prep		1.12.2

	Integrated Master Plan (IMP)		
Activity #	**Event**	**SOW REF**	**WBS REF**
C08b	Initial Hazardous Material Management Strategy Conducted		1.12.2
C08c	Initial Hazardous Material Management Strategy Meeting Minutes and Action Items Generated		1.12.2
D	Event D - System Requirements Review (SRR)	-	-
D01	Performance Based System Specification (PBSS) Reviewed	-	
D01a	PBSS Deltas to Pre-Award PBSS Identified	3.1.3.4	1.12.4.1
D01b	PBSS Deltas to Pre-Award PBSS Incorporated	3.1.3.4	1.12.4.1
D02	Systems Engineering Management Plan (SEMP) Reviewed	-	
D02a	Significant Technical Variances Identified	3.2.1.1	1.12.4.1
D02b	Project Impacts Identified	3.2.1.1	1.12.4.1
D02c	Technical Updates Incorporated	3.2.1.1	1.12.4.1
D03	Software Requirements Specification (SRS) Reviewed	-	1.12.4.1
D04	SRR Conducted	-	
D04a	SRR Meeting Preparation	3.1.3.4	1.12.4.1
D04b	SRR Meeting Conducted	3.1.3.4	1.12.4.1
D04c	SRR Meeting Minutes and Action Items Generated	3.1.3.4	1.12.4.1
E	Event E - System Functional Review / Design Readiness Review	-	-
E01	SRR Approved	-	
E01a	SRR Updated	3.2.1.4	1.12.4.3
E01b	SRR Completed	3.1.3.4	1.12.4.3
E02	Performance Based System Specification (PBSS) Reviewed	-	
E02a	PBSS SRR Changes Identified	3.1.3.5	1.12.4.3
E02b	PBSS SRR Changes Incorporated	3.1.3.5	1.12.4.3
E03	Integrated Test Plan Reviewed	-	
E03a	Significant Technical Variances Identified	3.1.3.5	1.12.4.3
E03b	Technical Updates Incorporated	3.1.3.5	1.12.4.3
E04	Modeling and Simulation Plan Reviewed	-	
E04a	Modeling and Simulation Plan Developed	3.2.1.2	1.12.4.3
E05	Human Engineering Design Approach Reviewed	-	
E05a	Human Engineering Design Approach Documented	3.2.1.3	1.12.4.3
E05b	Design Approach and Performance Requirements Mapped	3.2.1.3	1.12.4.3
E07	System Safety Management Plan (SSPP) Reviewed	-	
E07a	System Safety Mgmt Approach Documented	3.2.1.5	1.12.4.3
E07b	SSPP and Industry Standards Mapped	3.2.1.5	1.12.4.3
E08	Systems Operational Efectiveness (SOE) Plan Reviewed		
E08a	System Operational Effectiuveness Approach Documented		
E08b	SOE Standards Mapped		
E09	System Functional Review / Design Readiness Review Conducted	-	
E09a	SFR / DRR Meeting Preparation	3.1.3.5	1.12.4.3
E09b	SFR / DRR Meeting Conducted	3.1.3.5	1.12.4.3
E09c	SFR / DRR Meeting Minutes and Action Items Generated	3.1.3.5	1.12.4.3
F	Event H - Software Development Plan	-	-
F01	Compliance to IEEE/EIA Std. 12207.1 (Table 1) Reviewed	-	1.12.4.2
F02	Software Safety Plan (SSP) Reviewed	-	
F02a	Software Safety Plan Approach Documented	3.1.3.5	1.12.4.2

	Integrated Master Plan (IMP)		
Activity #	Event	SOW REF	WBS REF
F03	Software Qualification Review Conducted	-	
F03a	Software Qualification Review Preparation	3.1.3.9	1.12.4.2
F03b	Software Qualification Review Conducted	3.1.3.9	1.12.4.2
F03c	Software Qualification Review Meeting Minutes and Action Items Generated	3.1.3.9	1.12.4.2
G	Event F - Preliminary Design Review / Software Readiness Review	-	-
G01	Performance Based System Specification (PBSS) Reviewed	-	
G01a	PBSS SFR / DRR Changes Identified	3.1.3.6	1.12.3.5.1
G01b	PBSS SFR / DRR Changes Incorporated	3.1.3.6	1.12.3.5.1
G02	System Design Reviewed	-	
G02a	System Design Documented	3.1.3.6	1.12.3.5.1
G02b	System Requirements Mapped	3.1.3.6	1.12.3.5.1
G03	Software Development Approach Documented	-	
G03a	Software Development Plan Reviewed	3.2.1.4	1.12.3.5.1
G03b	CMMI Software Processes Documented	3.2.1.4	1.12.3.5.1
G03c	Software Program Measurement Plan (SPMP) Reviewed	3.2.1.4	1.12.3.5.1
G03d	Software Quality Plan (SQP) Reviewed	3.2.1.4	1.12.3.5.1
G04	Software Integration Laboratory (SIL) Reviewed		
G04a	SIL Design Documented		
G04b	System Requirements mapped		
G05	Configuration Management Approach Documented	-	
G05a	Configuration Management Processes Documented	3.1.3.6	1.12.3.5.1
G05b	Configuration Management Plan Reviewed	3.1.3.6	1.12.3.5.1
G06	Preliminary Design Review / Software Readiness Review Conducted	-	
G06a	PDR / SRR Meeting Preparation	3.1.3.6	1.12.3.5.1
G06b	PDR / SRR Meeting Conducted	3.1.3.6	1.12.3.5.1
G06c	PDR / SRR Meeting Minutes and Action Items Generated	3.1.3.6	1.12.3.5.1
H	Event G - Critical Design Review (CDR) / Software Verification and Validation Plan Reviews	-	-
H01	Performance Based System Specification (PBSS) Reviewed	-	
H01a	PBSS PDR / SRR Changes Identified	3.1.3.7	1.12.3.5.3
H01b	PBSS PDR / SRR Changes Incorporated	3.1.3.7	1.12.3.5.3
H02	System Design Updates Reviewed	-	
H02a	System Design Updates from PDR / SRR Identified	3.1.3.7	1.12.3.5.3
H02b	System Design Updates from PDR / SRR Incorporated	3.1.3.7	1.12.3.5.3
H03	Systems Engineering Management Plan (SEMP) Reviewed	-	
H03a	Significant Technical Variances Identified	3.2.1.1	1.12.3.5.3
H03b	Project Impacts Identified	3.2.1.1	1.12.3.5.3
H03c	Technical Updates Incorporated	3.2.1.1	1.12.3.5.3
H04	Design Verification and Validation Approach Documented	-	
H04a	Verification and Validation Plan Process Identified	3.1.3.8	1.12.3.5.3
H04b	Verification and Validation Plan Reviewed	3.1.3.8	1.12.3.5.3
H04	Data Management Approach Documented	-	
H04a	Data Management Processes Documented	3.1.5	1.12.3.5.3

	Integrated Master Plan (IMP)		
Activity #	Event	SOW REF	WBS REF
H04b	Data Management Plan Reviewed	3.1.5	1.12.3.5.3
H05	Human Engineering Test Approach Reviewed	-	
H05a	Human Engineering Test Approach Documented	3.2.1.3	1.12.3.5.3
H05b	Test Approach and Performance Requirements Mapped	3.2.1.3	1.12.3.5.3
H06	Risks Reviewed	-	
H06a	Program Risks Identified	3.1.3.3	1.12.3.5.3
H06b	Current Risks Reviewed	3.1.3.3	1.12.3.5.3
H07	CDR / Software Verification and Validation Plan Review Conducted	-	
H07a	CDR Meeting Preparation	3.1.3.7	1.12.3.5.3
H07b	CDR Meeting Conducted	3.1.3.7	1.12.3.5.3
H07c	CDR Meeting Minutes and Action Items Generated	3.1.3.7	1.12.3.5.3
I	Event I - Build (Proto-type) Release Review (BRR)	-	-
I01	Initial Build Released	3.1.3.10	1.12.6
I02	Build Changes Reviewed	3.1.3.10	1.12.6
I03	Build Release Review Conducted	-	
I03a	BRR Preparation	3.1.3.10	1.12.6
I03b	BRR Conducted	3.1.3.10	1.12.6
I03c	BRR Meeting Minutes and Action Items Generated	3.1.3.10	1.12.6
J	Event J - Test Readiness Review (TRR)	-	-
J01	Integrated Test Plan Reviewed	-	
J01a	Significant Technical Variances Identified	3.1.3.11	1.12.7
J01b	Technical CDR Updates Incorporated	3.1.3.11	1.12.7
J02	Test and Evaluation Master Plan (TEMP) Documented	-	
J02a	TEMP Identified	3.2.1.6	1.12.7
J02b	TEMP Reviewed	3.2.1.6	1.12.7
J03	Design Verification and Validation Approach Documented	-	
J03a	Verification and Validation Plan Process CDR Updates Identified	3.1.3.11	1.12.7
J03b	Verification and Validation Plan Reviewed	3.1.3.11	1.12.7
J04	Human Engineering Test Approach Reviewed	-	
J04a	Human Engineering Test Approach CDR Updates Documented	3.2.1.3	1.12.7
J04b	Test Approach and Performance Requirements Mapped	3.2.1.3	1.12.7
J05	Risks Reviewed	-	
J05a	Program Risks Identified	3.1.3.3	1.12.7
J05b	Current Risks Reviewed	3.1.3.3	1.12.7
J06	Test Procedures Reviewed	-	
J06a	Test Procedures Identified	3.2.1.6	1.12.7
J06b	Test Procedures Documented	3.2.1.6	1.12.7
J07	Problem Reporting Approach Documented	-	
J07a	Problem Reporting Process Identified	3.1.4	1.12.7
J07b	Problem Reporting Process Plan Reviewed	3.1.4	1.12.7
J07	TRR Conducted	-	
J07a	TRR Preparation	3.1.3.11	1.12.7
J07b	TRR Conducted	3.1.3.11	1.12.7
J07c	TRR Meeting Minutes and Action Items Generated	3.1.3.11	1.12.7

	Integrated Master Plan (IMP)		
Activity #	Event	SOW REF	WBS REF
K	Event K - Operational Test and Readiness Review (OTRR)	-	-
K01	Design Verification and Validation Approach Documented	-	
K01a	Verification and Validation Plan Process TRR Updates Identified	3.1.3.12	1.12.9
K01b	Verification and Validation Plan Reviewed	3.1.3.12	1.12.9
K01c	Verification and Validation Plan Delivered	3.1.3.12	1.12.9
K02	Risks Reviewed	-	
K02a	Program Risks Identified	3.1.3.3	1.12.9
K02b	Current Risks Reviewed	3.1.3.3	1.12.9
K03	Test Procedures Reviewed	-	
K03a	Test ProcedureTRR Updates Identified	3.2.1.6	1.12.9
K03b	Test Procedures Documented	3.2.1.6	1.12.9
K04	Problem Reporting Approach Documented	-	
K04a	Problem Reporting Process TRR Updates Identified	3.1.4	1.12.9
K04b	Problem Reporting Process Plan Reviewed	3.1.4	1.12.9
K05	System Operational Effectiveness Approach Documented	-	
K05a	System Operational Effectiveness Process Identified	3.1.3.12	1.12.9
K05b	System Operational Effectiveness Plan Reviewed	3.1.3.12	1.12.9
K05c	System Operational Effectiveness Plan Delivered	3.1.3.12	1.12.9
K06	Test Results Reviewed	-	
K06a	Test Results Documented	3.1.3.12	1.12.9
K07	OTRR Conducted	-	
K07a	OTRR Preparation	3.1.3.12	1.12.9
K07b	OTRR Conducted	3.1.3.12	1.12.9
K07c	OTRR Meeting Minutes and Action Items Generated	3.1.3.12	1.12.9
L	Event L - System Verfication Review	-	-
L01	Test Results Reviewed	-	
L01a	Test Results Documented	3.1.3.13	1.12.10
L02	Problem Reports Reviewed	-	
L02a	Problem Reports Identified	3.1.4	1.12.10
L02b	Problem Reports Documented	3.1.4	1.12.10
L03c	Problem Reports Resolved	3.1.4	1.12.10
L04	Performance Based System Specification (PBSS) Reviewed	-	
L04a	PBSS OTRR Changes Identified	3.1.3.13	1.12.10
L04b	PBSS OTRR Changes Incorporated	3.1.3.13	1.12.10
L05	Risks Reviewed	-	
L05a	Program Risks Identified	3.1.1.3	1.12.10
L05b	Current Risks Reviewed	3.1.1.3	1.12.10
L06	System Verification Review Conducted	-	
L06a	System Verification Review Preparation	3.1.3.13	1.12.10
L06b	System Verification Review Conducted	3.1.3.13	1.12.10
L06c	System Verification Review Meeting Minutes and Action Items Generated	3.1.3.13	1.12.10
M	Event M - Production Readiness Review	-	-
M01	System Verification Review Approved	-	1.12.11
M02	First Article Build, Assembly, and Inspection Completed	-	
M02a	First Article Material Purchase and Build Completed	3.1.3.14	1.12.11

	Integrated Master Plan (IMP)		
Activity #	Event	SOW REF	WBS REF
M02b	First Article Assembly and Inspection/Test Completed	3.1.3.14	1.12.11
N	Event N - Initial Operational Capability Supportability Review	-	-
N01	Production Readiness Review Approved	-	1.12.8
N02	Support Equipment Program Plan (SEPP) Reviewed	-	
N02a	SEPP Process Identified	3.2.2.3	1.12.8
N02b	SEPP Documented	3.2.2.3	1.12.8
N02c	SEPP Delivered	3.2.2.3	1.12.8
N03	Contractors Performanced Based Logistics Planning (PBL) Strategy Reviewed	-	
N03a	Contractors PBL Identified	3.2.2.2	1.12.8
N03b	Contractors PBL Documented	3.2.2.2	1.12.8
N03c	Contractors PBL Delivered	3.2.2.2	1.12.8
N04	Logistics Management Information Database Reviewed	3.2.2.4	1.12.8
N05	Technical Manuals/ Interactive Electronic Technical Manauls (IETM) Delivered	3.2.2.5	1.12.8
N06	Training Plan Reviewed	3.2.2.6	1.12.8
N07	Transition Plan Reviewed	3.2.2.6	1.12.8

System Development and Demonstration of the Breadfruit Enterprise Demonstration (BrEAD) and Systems Design

Solutions International Proposal for the Breadfruit Enterprise Demonstration and System Design

Volume 2: Technical

Annex F: Integrated Master Schedule

RFP NO: E00010-07-R0001

Copy 1 of 2

May 4, 2009

Prepared For:

New Green Revolution Foundation (NGRF)

Prepared By:

Solutions International

123 Pillsbury Drive

St. Paul, MN 55114

CAGE CODE SI4ME

Table of Contents

Table of Figures

Introduction

1.1 Executive Summary

This Systems Engineering Management Plan (SEMP) describes the activities, processes and tools that will be used by the Breadfruit (BF) and Enterprise Architecture (EA) System Engineering (SE) teams to support the design, fabrication, and integration of the following BrEAD components:

- BF Processing System
- Enterprise Architecture
- Communications Network (NW)
- St. Paul Central Data Center (CDC)
- Regional Centers (RC)
- System Integration Laboratory (SIL)

The objective of the System Engineering effort is to assure successful development of the BrEAD by defining clear and accurate system requirements along with a stepwise definition and re-definition of all system design elements. System Engineering efforts include verifying compliance to the developed requirements and design. This SEMP is applicable to all System Engineering tasks to be performed in support of the BrEAD project. This document will be placed under change control upon its initial release.

1.2 Project Summary

The BrEAD will be a system composed of the components listed above in the Executive Summary. The purpose of the BrEAD project is to assist native populations around the world, living near breadfruit trees, to feed themselves more effectively with this natural resource. Any solution represented by the proposed BrEAD system, must be cognizant of the cultural, social, political and economic environment of the peoples to be assisted. The BrEAD system must be self expanding and include advocacy components for the use of breadfruit as a foodstuff.

Technical challenges include,

a) Optimization of the fruit drying processes for maximum speed
b) Containment of costs for the breadfruit processing equipment to within the range of impoverished peoples
c) Communication with breadfruit processing sites in remote locations and Regional Centers located in Third World countries.
d) Providing power for electronic components of processing system in "off the grid" areas.
e) Harvest and processing of breadfruit within a 2-day window.

1.3 Applicable Documents

- IEEE/EIA Std. 12207.1 Guide for Information Technology Software Life Cycle Processes

- DoD Directive 5000.59, "DoD Modeling and Simulation (M&S) Management," January 4, 1994
- ANSI/ASQC-Q9001-2000 Quality Management Systems - Requirements

Planning and Control

2.1 Technical Organization

Two Integrated Project Teams (IPTs) form the backbone of BrEAD Design and Development. One team is focused on the BF processing system, while the other will design and develop the Enterprise Architecture. On each team, a combination of SI and NGRF team members in a series of sub-teams are jointly responsible for project success. The SI team members are primarily responsible for project conformance to cost, schedule and technical performance, while the NGRF team members focus on job oversight and knowledge transfer (Figure 1).

The systems engineering process will involve coordinating the engineering efforts of the various subsystem teams as they design and implement the subsystems that will be integrated into a complete BF Processing System and a complete Enterprise Architecture (see Figure 2). Overall authority for the teams is held at the program management level by an SI Program Manager (PM) and an NGRF Executive Liason. These executives monitor and guide the project through the Chief System Engineer and the respective SE Coordinating Team of each IPT. The SE teams maintain requirements, manage interfaces, manage risk, and direct the work of the Subsystem Teams. They coordinate, schedule and chair the various committees responsible for key objectives. These include,

- Change Review Board
- Risk Management Board
- HSI Working Groups
- Interface Control Working Groups
- Test Planning Working Groups
- System Safety Working Groups

2.1.1 Systems Engineering Coordinating Teams

There is a separate SE Coordinating Team for each IPT. These teams are the principal system engineering functional groups for the BrEAD project. They constitute the interface between program management and the development and support groups on the project. They implement PM objectives, monitor and coordinate work activities, and follow project status (see Figure 3).

Each SE Coordinating Team is directed by a BrEAD Chief Systems Engineer (CSE). He is assisted and supported by a Project SE steering committee, comprised of System Engineers who manage risk, project changes, configuration, performance, and requirements on the project level as well as maintaining the BrEAD subsystem interfaces. Reporting to the Project SE sub-group are individual system engineers assigned to specific BrEAD subsystems, where they are responsible for requirements traceability, configuration management risks and subsystem performance. Subsystem-specific systems engineers or members of the SE Steering Group will also chair or be a member of most working groups, including the Interface Control, Interoperable/Net-Centric, Human Systems Integration (HSI), and System Software Safety Working Groups.

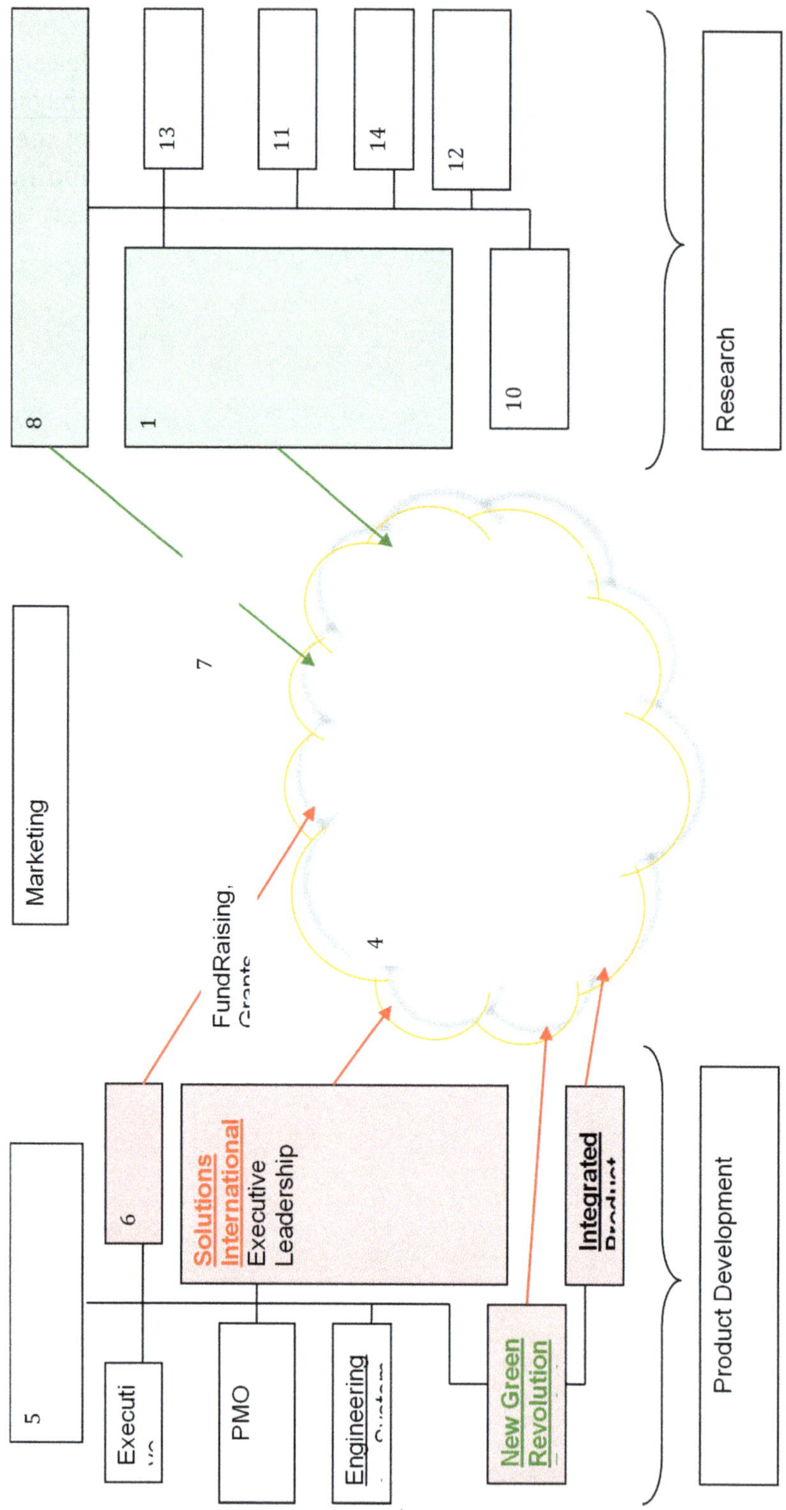

Figure 1- Integrated Project Teams (IPTs)

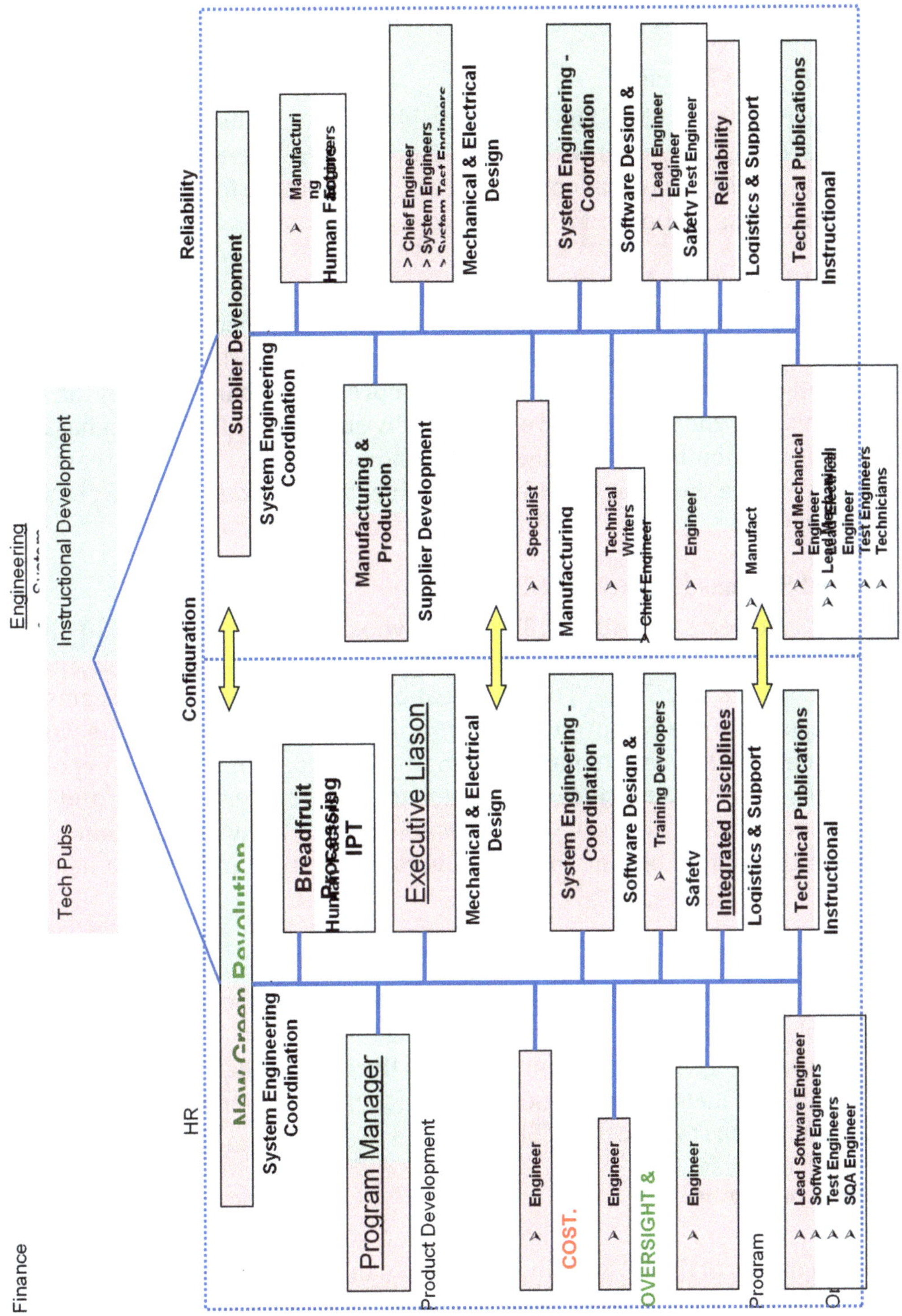

Figure 2- Integrated Project Teams

2.2 Technical Functions: IPT Roles and Responsibilities

2.2.1 SI Program Manager

The PM is part of the SI PMO. She is responsible for accomplishing the objectives of the BrEAD project. She works with the SE Coordinating Teams to manage schedule, resources, and risks, make tradeoffs, and ensure project performance. She reports on project progress to the SI PMO.

2.2.2 NGRF Executive Liason

The NGRF representative is responsible for monitoring project status, including progress, issues, and alignment with NGRF concerns. He works with the PM to manage changes to project requirements. He represents the interests of the NGRF Senior Council, to whom reports are periodically submitted. The Liason has a special responsibility to monitor knowledge and technology transfer to NGRF team members to ensure the ability of NGRF personnel to ultimately operate and extend the BrEAD system

2.2.3 Chief Systems Engineer (CSE)

There is a CSE for each BrEAD IPT. The CSEs, who are members of the SI PMO, report directly to the PM, and are responsible for the systems engineering activities of their SE Coordinating Team. The organizational structure of the SE Teams is shown in Figure 3. The CSEs provides technical leadership and direct the work of the subsystem development teams through the development and tracking of requirements, reserves, interface control documents, and performance and verification specifications. The CSEs are responsible for assuring that the subsystems are compatible and meet the overall objectives and requirements of the BrEAD.

2.2.4 Risk Officer

The Risk Officer is a member of the SE Coordinating Team Steering Committee and is responsible for administering the Risk Management Board. There is one Risk Officer for each integrated team, appointed by the CSE for that team. They coordinate all risk identification and analysis activities, maintain the team's Risk List, and report risk status to the CSE and PM.

2.2.5 System Engineer

The system engineers in each IPT belong to their respective SE Coordinating Team, either as members of the SE Steering Committee or dedicated to a BrEAD subsystem(s). The subsystem engineers report to the steering committee while the steering committee members report directly to the CSE.

Subsystem engineers monitor design progress with respect to satisfying BrEAD PBSS requirements. They maintain requirements traceability downward to design elements and test protocols and upward to higher level requirements or user needs.

They report project status and oversee risk mitigation strategies relative to their subsystem.

Steering committee members manage risks via the Risk Management Board, manage changes to requirement and design baselines via the Change Review Board, and manage interfaces for each subsystem via the Interface Control Working Groups. They manage project configuration via participation on the Change Review Board, which processes change requests for project baseline artifacts.

2.2.6 Subsystem Team Lead

The subsystem team lead for each of the BrEAD subsystems (e.g. SIL, NW) is responsible for planning and coordinating the work of individual functional group members to design, develop and test their particular subsystem. The team leads communicate status to the SE Coordinating Teams and are responsible for ensuring that system requirements and project requirements (such as for Configuration Management) are fulfilled with respect to their subsystem. They work with Function Group Leads and the SE Coordinating Team to schedule needed resources.

2.2.7 Functional Group Leads

The functional group lead for each functional group (e.g. Software, Human Factors) are responsible for allocating resources to each Subsystem Team. They enforce adherence to group procedures, manage standards and recordkeeping within the Group for all subsystems their group supports.

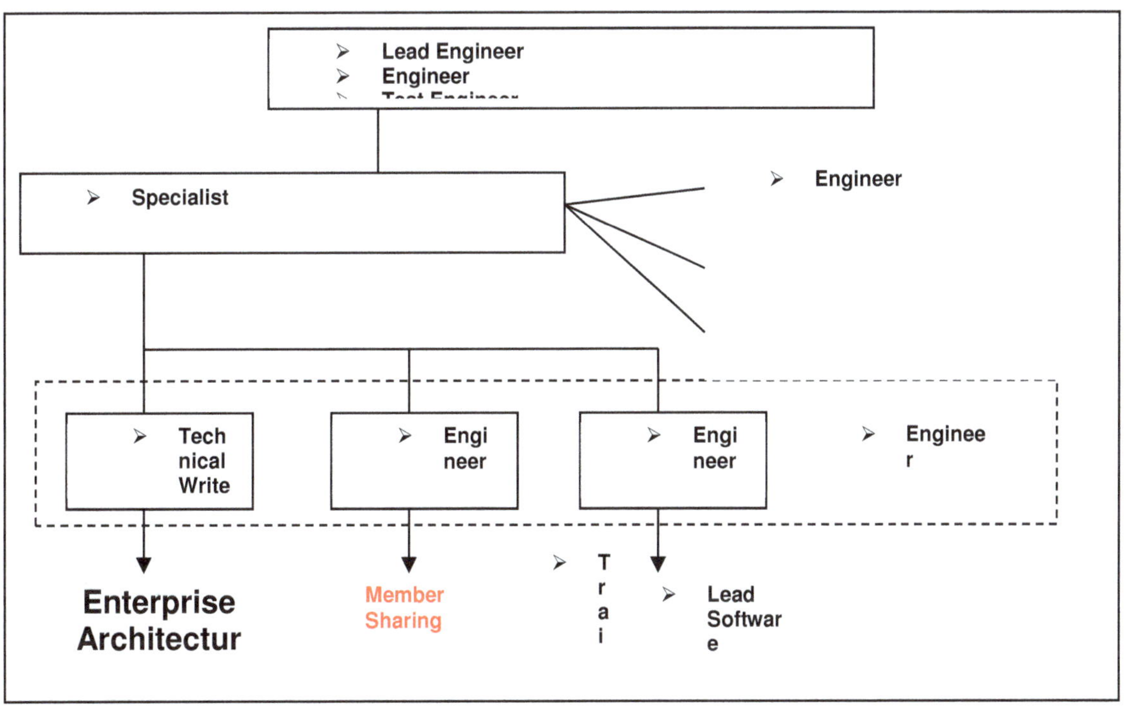

Figure 3– Structure of Systems Engineering Coordinating Team

2.3 Subcontractors

SI will engage the services of 3 principal subcontractors. These are Compatible Technologies Inc.(CTI), Onset and Inmarsat. See section 2.2.7 Subcontract Management in Volume 2, Book 2 of the main proposal). CTI as a contributing designer of non-COTS BrEAD components is required, and has agreed to maintain a Quality Management System to control the quality of services and products for BrEAD.

CTI will provide the designs for most of the breadfruit processing mechanical components. Building upon previous designs, CTI will refine these components to meet PBSS requirements for lower cost, modularity and scalability. CTI will work in conjunction with mechanical, electrical human factors and system engineers from the breadfruit processing system IPT. Their completed designs will be archived within BrEAD's document storage systems and BrEAD's requirements, risk, V&V, CM, and logistics processes like all other BrEAD designs. BrEAD test engineers will work with CTI engineers to create the test protocols used to verify and validate CTI designs.

Inmarsat will be a provider of satellite services for BrEAD's communication network. SI will engage Inmarsat on a long-term contract basis to provide bandwidth on their satellites, equipment to enable satellite communications, and maintenance and technical support for their provided equipment. Interfaces between BrEAD communication systems and Inmarsat equipment will be defined and maintained by engineers in the relevant BrEAD subsystems, based on Inmarsat-provided specifications.

The engagement of Onset is limited to the repair and long-term maintenance of COTS elements of the BrEAD communication network (i.e. Data Loggers). This will be reflected in BrEAD maintenance plans.

2.4 Principal Products (of IPTs)

The Breadfruit Processing and Enterprise Architecture IPTs will produce the BrEAD subsystems as listed in section 1.1 Executive Summary. In so doing, the IPTs will generate the following work products:

Risk List	Requirements Database	Technical Plans
Technical Specifications	Interface Control Drawings	Design Drawings
Software Applications	Trade Studies/Analyses	Project Issues Database
Logistics & Support Database	Issues Database	Test Protocols

2.5 Project and Technical Reviews

The SE Coordinating Teams are responsible for planning, scheduling and conducting technical and project reviews to promote communication and guidance within the BrEAD team and to provide status to and obtain feedback from Program Management and the NGRF. Reviews will include total system reviews of the entire BrEAD project, as well as major system reviews scoped at the level of the BF Processing and Enterprise Architecture IPTs. The reviews will encompass status reporting on the BrEAD technical and programmatic progress. Requirements, design data, analysis reports, budgets, schedules and project plans will be presented to Program Management and to attending members of the SI PMO, and executive leadership from both SI and NGRF. The review team will develop and present specific recommendations, actions, and concerns which will be tracked to resolution by the SE Coordinating Teams to ensure closure, and then re-presented as necessary at the next review.

These top-level reviews include the System Requirements Review (SRR), System Functional Review (SFR), Preliminary Design Review (PDR), and Critical Design Review (CDR). The time order of project reviews is depicted in Figure 4.

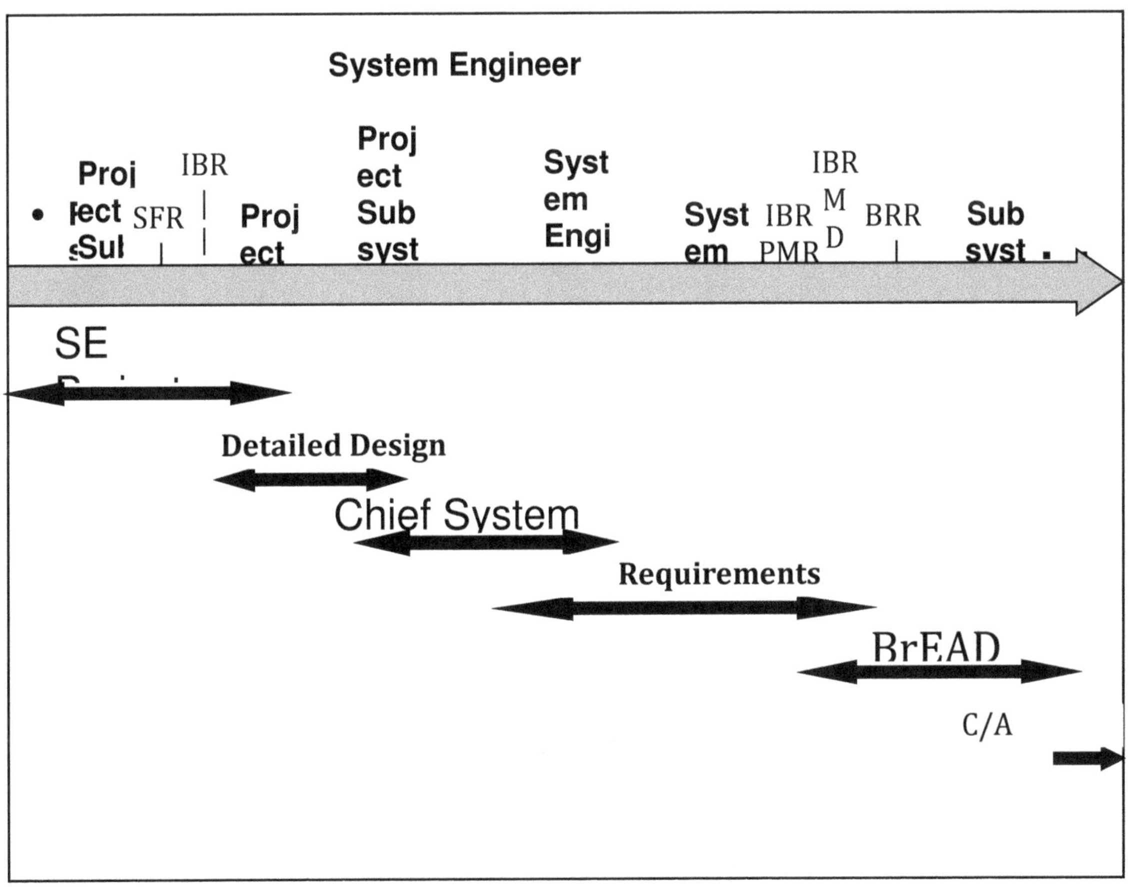

Figure 4- Event-based reviews throughout BrEAD

2.5.1 Integrated Baseline Review (IBR)

The initial BrEAD IBR will be held soon after Contract Award (CA). SI will present its initial Performance Measurement Baseline (PMB) to the NGRF so that project risks can be evaluated as well as management processes for control of budget, schedule and technical activities. Our PMB will establish the scope, schedule and budget targets for the project, and will include all artifacts necessary to assess the project's ability to meet contract milestones and deliver promised work packages. Project artifacts will include the SOW, WBS, and IMS.

Follow-on IBRs will be conducted concurrently with PDR and CDR, or whenever major program events warrant (at the discretion of the PM). The plan will be continually evaluated as changes to the baseline are made (modifications, restructuring, etc.). Each review will be tailored to focus on risk areas identified by the NGRF.

2.5.2 System Requirements Review (SRR)

The SRR occurs early in the Definition Phase of the project and will enable the NGRF and SI to reach mutual agreement on system requirement issues. Its purpose is to ensure that the intended use, performance, and effectiveness of the BrEAD will match its system and performance requirements. In this review, the draft system requirements are reviewed for completeness and necessity. All pending items will be assigned responsible project resources and due dates. The draft system architecture and external interfaces will also be reviewed at this time.

2.5.3 Program Management Review (PMR)

PMRs will be held at quarterly or more frequent intervals throughout the project to provide NGRF with status on project progress. Information presented will focus on the current status of cost, schedule and technical performance. Subsystem Team Leads will present the current state of the subsystems they are responsible for developing, in terms of schedule, risk, resources used, and dollars spent. The impact of shortfalls in critical project areas will be discussed, along with remediation planning. In order to synchronize with other important program milestones, PMRs will be held concurrently with the SFR, PDR, CDR, and quarterly thereafter.

2.5.4 System Functional Review (SFR)

The SFR is a technical review aimed at ensuring that requirements, design and the functional baseline has reached the point where BrEAD can move into the preliminary design phase. The functional baseline will be derived from the users' operational requirements for the program, which will be analyzed and broken down into a specification (SysRS?) that defines how these requirements are to be understood and approached from an engineering perspective, and implemented in terms of hardware and software. The SFR will be held near the end of the project Definition Phase, concurrently with the presentation of the DRR, an updated PBSS and an Integrated Test Plan.

The SFR will include assessment of the following items:

- Functional Analysis and Allocation of requirements to items below system level
- System Specification
- Design data defining the overall system
- Verification that design risks are acceptable for engineering development
- Verification that design selections have been optimized through trade study analyses
- Analyses, reports, Logistics support analysis data, and design documentation
- Technical Performance Measurement data and analysis,
- Associated enabling product plans (Acquisition Logistics and Systems Engineering plans)

2.5.5 Preliminary Design Review (PDR)

The Preliminary Design Review will occur at the end of the Preliminary Design Phase. It will be used to determine if the BrEAD project is ready to authorize the increase in manpower and cost concomitant with detailed design. Project deliverables to be presented at this review include the SDD, SDP, CMP, Modeling & Simulation Plan, hazard analyses, and human engineering deliverables such as the HETP and HEPP. .

The PDR will address the following items:

- Subsystem block and functional diagrams
- Equipment layouts and preliminary drawings
- Instrumentation, control, and diagnostic design approach
- Producibility and manufacturing considerations
- Preliminary parts lists
- Support system requirements and design approach
- Preliminary Development Specifications
- Modeling, test, and simulation data
- Software Development Plan
- Software requirements specifications (Preliminary Design)
- Risk and Mitigation strategy
- Interface control documents
- Design standardization and logistic considerations
- Trade and design studies
- Preliminary reliability, maintainability, and availability studies
- Transportation, packaging, and handling considerations
- Environmental, Health, and Safety analyses
- Quality Control Planning

- Test methodology
- Schedules
- Problems and Concerns

2.5.6 Critical Design Review (CDR)

The CDR occurs after the design is approximately 90% completed and is used to determine if the project is ready to proceed to the manufacture of the Breadfruit processing hardware and the coding of software. Each subsystem of BrEAD will be reviewed for cost, schedule, performance, risk, and compatibility with other subsystems. This review is focused on the acceptability of the detailed design, performance, and test characteristics of the design solution, and on the adequacy of the operation and support documents. Project deliverables to be presented at this review include the Software and System V&V Plans, the System Safety Management Plan, the SEMP, HETA, and DMP.

The following items will be addressed:

- Subsystem block and functional diagrams
- Drawing packages (assembly drawings and majority of remaining drawings)
- Final parts lists
- Final Development Specifications
- Design analysis and engineering test data
- Detailed software design, database design, interface design, firmware support, and computer resources integrated support documents
- Logistic support considerations
- Transportation, packaging, and handling
- Standardization
- Support equipment requirements
- Spares requirements
- Calibration requirements
- Risk: cost, schedule, and technical
- Plans for acquisition of parts, components, and materials needed for fabrication
- Integration and Test Plans
- Software Test Plans
- Design reliability and maintainability
- System safety
- Quality control plans
- Schedules
- Problems and Concerns

2.5.7 TRR (Test Readiness Review)

The TRR occurs prior to the start of testing of the integrated subsystems. The primary purpose of the TRR is to establish the readiness of the BrEAD systems for the various levels of test The TEMP will be presented at this review.

The following items will be addressed:

- Changes since the CDR
- Tests planned
- Test facilities
- Test configuration
- Test procedure status
- Staffing plans
- System performance during initial performance testing
- Significant Problem/Failure Reports (status and plans for closure)
- Risk: cost, schedule, and technical
- System safety
- Schedules
- Problems and Concerns

2.5.8 OTRR (Operational Test and Readiness Review)

The OTRR occurs after prototyping and prototype integration is complete. Its purpose is to establish the readiness of the BrEAD systems for operational test and to evaluate the operational test scenarios and protocols that have been planned.

2.5.9 SVR (System Verification Review)

The purpose of the SVR is to ensure that the BrEAD system can proceed into production. It assesses the Production Baseline to determine if it meets its functional requirements. The SVR establishes and verifies final product performance to ensure that the system will operate suitably and effectively. Generally this review is an audit trail from the Critical Design Review.

The SVR may be conducted concurrently with the Production Readiness Review.

2.5.10 PRR (Production Readiness Review)

The purpose of the PRR is to determine if the BrEAD design is ready for production and if production planning is sufficiently mature. This involves examining the production approach to asses the risk of production exceeding cost, schedule or performance targets. It also involves determining if the production system will implement all system requirements, and if traceability of final system requirements to the final production system is intact. The PRR will be tailored to the technical scope and risk associated with the system.

The following items will be addressed by the PRR:

- Quality Mgt System
- Readiness of the manufacturing processes
- Risk, Cost of Production, staffing
- Schedule Executability
- Production planning (facilities, tooling and test equipment capacity, process documentation)
- Adequacy of the System Product Baseline to support configuration management.

2.6 Systems Engineering and Design Specifications

The planned specification tree, Figure 5, shows the allocation of major BrEAD specifications to the functional units of the BrEAD system. The specification tree also represents the flow of requirements from the top-level mission requirements to increasingly detailed requirements for the subsystems. The specifications are grouped into levels within the context of the total BrEAD Project. The levels of specifications developed and controlled by the BrEAD Project are:
1 - Top-level performance-based requirements specifications
2 - Systems-level specifications. Changes to these specifications require approval of the Configuration Review Board (CRB) and NGRF.
3 – Subsystem requirements for the major subsystem elements. These requirements are the responsibility of the CSE for each IPT and their SE Coordinating Team. Changes to these specifications require approval of the CRB.
4 – Detailed designs describe individual parts of the BrEAD subsystems. They are the responsibility of the applicable Subsystem Team Lead and, once released, require approval of the responsible Subsystem Team Lead for any changes. See Figure 3.

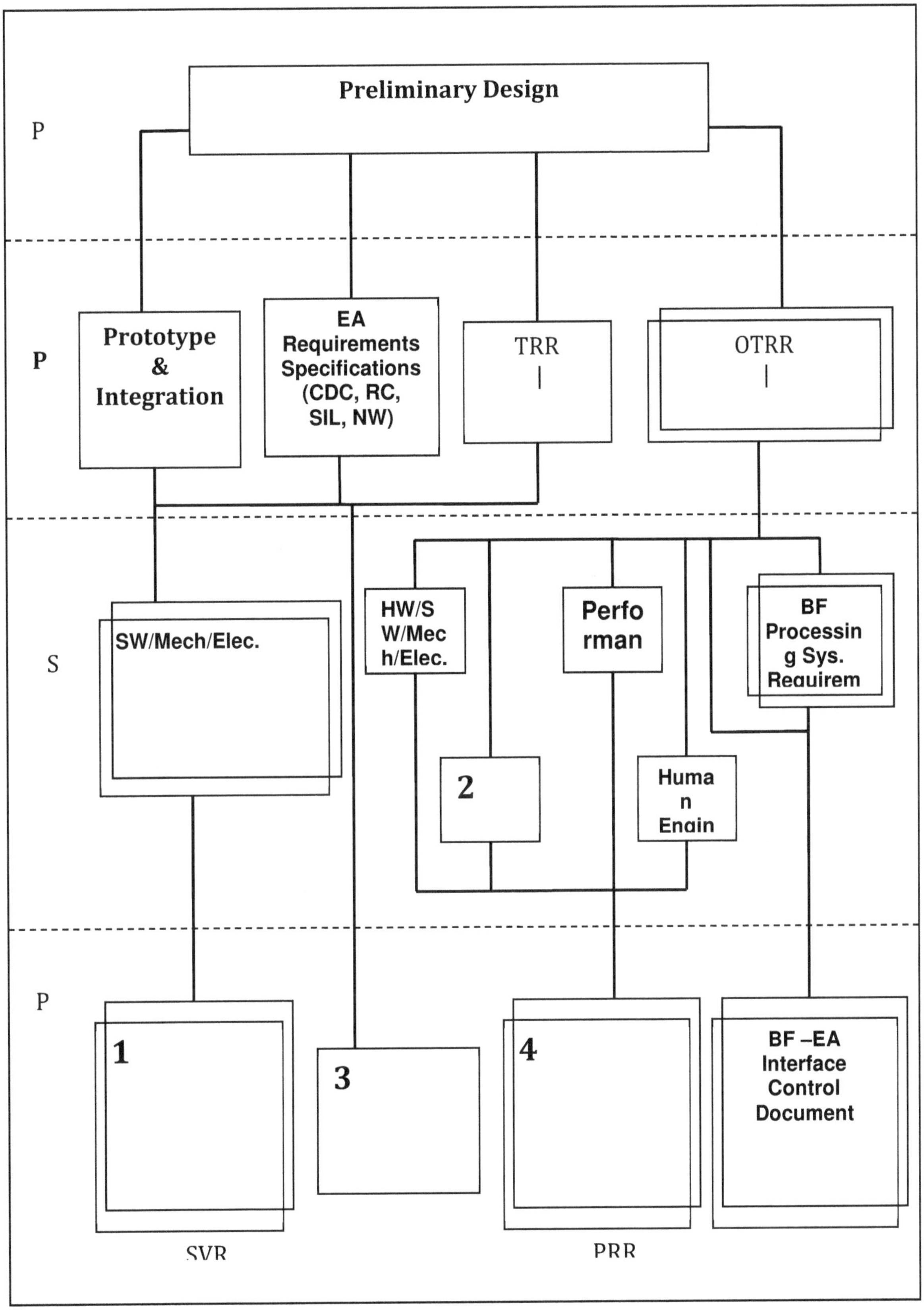
Preliminary Design
P
Prototype & Integration
EA Requirements Specifications (CDC, RC, SIL, NW)
TRR
OTRR
P
SW/Mech/Elec.
HW/S W/Mec h/Elec.
Perfo rman
BF Processin g Sys. Requirem
S
2
Huma n Engin
P
1
3
4
BF –EA Interface Control Document
SVR
PRR

Figure 5 - BrEAD Specification Tree

2.7 Project Plans

Detailed planning for the entire project will be recorded in a succession of plan documents covering all aspects of the BrEAD project. These plans serve to govern and direct all project activities to ensure the achievement of programmatic goals for cost, schedule and performance. Major plans include the Configuration Management Plan, Software Development Plan, Verification and Validation Plan, and Test and Evaluation Master Plan. Plan creation activities are provided for and scheduled in an event-based timeline in the IMP and IMS (see Annexes E and F). The time order of project plans is depicted in Figure 6.

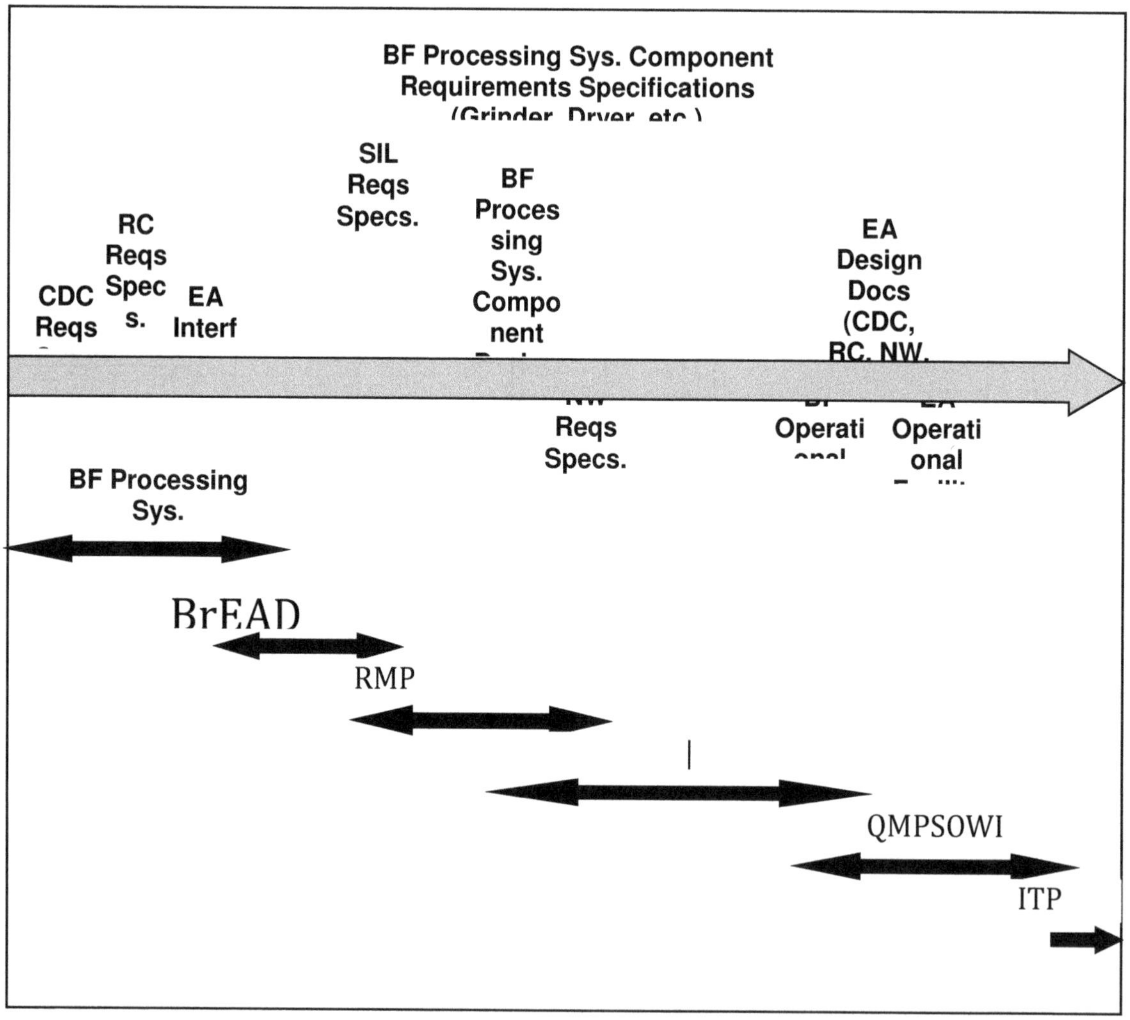

Figure 6- Detailed planning supports successful BrEAD development

2.7.1 RMP (Risk Management Plan)

Solutions International has already established a prototype risk plan, comprised of a Risk List, Probability Definition, Impact Classification scheme, and a Risk Matrix with mitigations identified. Upon contract award, that system will be expanded via a workshop with all ITP members. Risks will be examined frequently to prevent small issues from becoming large problems. The RMP will be presented at all major reviews beginning with the SFR. See section 3.2 of this SEMP and the Risk Management section of Vol. 2 Book 2 of the main proposal.

2.7.2 QMP (Quality Management Plan)

The QMP (CDRL00012) details the processes and procedures by which the quality of BrEAD work products will be assured. This plan will be created early in the project's definition phase by designated members of the SE Coordinating Teams, supported by quality engineers from the Reliability functional group. The plan will be compliant with ANSI/ASQC-Q9001-2000 and will clearly describe the processes to be used by NGRF to determine when quality goals have been reached.. The plan will be presented at SRR for approval by PM and NGRF.

2.7.3 SOW (Statement of Work)

The SOW contains the NGRF objectives that BrEAD will be designed to meet. The SOW was developed by the SI proposal team from the RFP and SOO. It is tied section by section to our IMP (see Annex D).

2.7.4 IMP (Integrated Master Plan)

The IMP (CDRL 0002) is our roadmap to producing a complete BrEAD system. It describes the project development process in terms of major program events (reviews), the tasks tied to them and their completion criteria. The IMP is provided in Annex E.

2.7.5 HEPP/HEDA (Human Engineering Program Plan)

SI's approach to the use of human engineering practices and techniques during BrEAD development will be detailed in the Human Engineering Product Plan (CDRL 00015). This plan will be created in the early stages of the BrEAD project by the Human Factors subteams, based on the Test and Evaluation Master Plan (TEMP) document and on inputs from NGRF members slated to become BrEAD users/operators, to help ensure that their requirements are met.

2.7.6 M&S (Modeling & Simulation Plan)

The foundation of SI's Modeling and Simulation Plan (CDRL 00013) for BrEAD is already mature, building upon a solid base of experience over the last 15 years (see section 5.1.3 HE Sim/Concept Report). The M&S that will be developed for BrEAD will detail a customized approach to applying our accredited (DoD Directive 5000.59) and proprietary MAXSIM toolset to BrEADs unique requirements (see 4.7 Modeling and Simulation Tools). Our M&S plan will be customized for BrEAD by members of the HF subteam for each IPT, starting on day one following contract award. The CSE from each ITP will appoint a principal plan contrubutor from their HF subteam. These authors, in concert with the HSIWG, will document a detailed and customized approach for the use of our M&S tools in developing and testing BrEAD subsystems. We will present it's final version for approval at PDR.

2.7.7 CMP (Configuration Management Plan)

The CMP (CDRL 0009) provides an overview of the organization, activities, overall tasks, and objectives of Configuration Management for the BrEAD. It addresses configuration item identification, change control and configuration audits. While configuration management is the overall responsibility of the PM, it will be administered for BrEAD by each CSE and their SE Coordinating Team. The CMP will be written as early as possible in the project Definition phase by designated Configuration Management Officers (CMOs) drawn from the SE Steering Groups. The CMOs will administer the plan, starting with definition of a functional baseline comprised initially of PBSS requirements. SI will present the CMP for final review and approval at PDR.

2.7.8 SDP (Software Development Plan)

The SDP (CDRL 0005) details the major activities, schedules, resources and milestones for developing the BrEAD Software. SI has provided a draft version of the SDP with this proposal (see Annex H). Our SDP contains information as defined in IEEE/EIA Std. 12207.1 section 5.2.1 and parts of IEEE/EIA Std. 12207.1. This plan will be updated during the project by a Software Development Manager appointed by the PM, with the support of the lead software engineers. It will be re-submitted for approval at major project milestones. SI will present the SDP for final review and approval at PDR.

2.7.9 HETA (Human Engineering Task Analysis)

SI's HE engineers will present the results of their task analyses in the HETA document at CDR. See section 5.1.2 for a description of HETA (CDRL 00014).

2.7.10 TEMP (Test & Evaluation Master Plan)

The TEMP (CDRL 00018) contains comprehensive planning for all T&E activities from component development through OT&E, and into production and acceptance. It provides a road map for integrated simulation, test, and evaluation plans, schedules, and resource requirements necessary to accomplish the T&E program objectives. It contains the required type and amount of test and their resource requirements. It will be the principal governing document for the V&VP (see Figure 7). The TEMP is produced jointly by the two CSEs in consultation with the PM and supported by SE Steering Committee members in charge of managing cost, schedule and resources. Work on the TEMP will begin at contract award, using the PBSS as its starting point and soliciting input from all stakeholders, especially future user/operators from NFGRF. The TEMP draft will be presented to, and reviewed by program management and NGRF representatives at major milestones such as the SFR, PDR, CDR, and TRR. At each review, feedback will be taken and incorporated into the next revision, along with changes driven by event or resource requirements changes. The TEMP will be completed near the end of the Prototype and Integration phase, and prior to the start of the V&V phase.

2.7.11 V&VP (Validation & Verification Plan)

The V&VP (CDRL 0006) will be written in accordance with the TEMP to satisfy the functional, operational and performance requirements of the BrEAD system. It details the projected implementation of all V&V protocols, at integration, subsystem and system levels. It will be the parent document for all ITPs, including the OT&EP, the HETP, and the ITPs for each BrEAD subsystem (see Figure 7). The V&V plan will be prepared jointly by the Systems Test Engineer for each SE Coordinating team, supported by the system engineers assigned to each BrEAD subsystem, and will be presented in the form of a general test approach at PDR, and as a completed document at CDR.

2.7.12 OT&EP (Operational Test & Evaluation Plan)

SI will create an OT&EP Plan (protocol) that describes how the integrated BrEAD system will be verified against system operational/performance requirements. It is anticipated that BrEAD will be operationally tested, both as a complete system comprised of the breadfruit processing system and EA, and also using operational scenarios involving only subsets of the entire system. This ITP will be developed by the Integrated Test Team in accordance with the V&VP and will be presented at the OT&RR, prior to use during the V&V phase of the project.

2.7.13 HETP (Human Engineering Test Plan)

The HETP (CDRL 00016) is an ITP that focuses on how the elements of BrEAD which interact and interoperate with humans will be tested (see section 5.1.4 HETP). SI members of the Human Factors subteam on each IPT, supported by the Test Plan Working Groups will generate an HETP for presentation at PDR. The final, approved HE test approach will be produced following CDR review.

2.7.14 ITP (Integrated Test Plan)

Integrated Test Plans (CDRL 0004) are the protocols developed to test specific elements of BrEAD under specific testing regimes. They are written in accordance with the V&VP and include protocols for operational tests, human engineering tests, and functional and performance testing (see Figure 7). The development of the Operational and Human Engineering ITPs are discussed in sections 2.5.15 and 2.5.16. Development of ITPs to verify the functional and performance requirements for each BrEAD subsystem are the responsibility of the Lead Test Engineer for each subsystem.

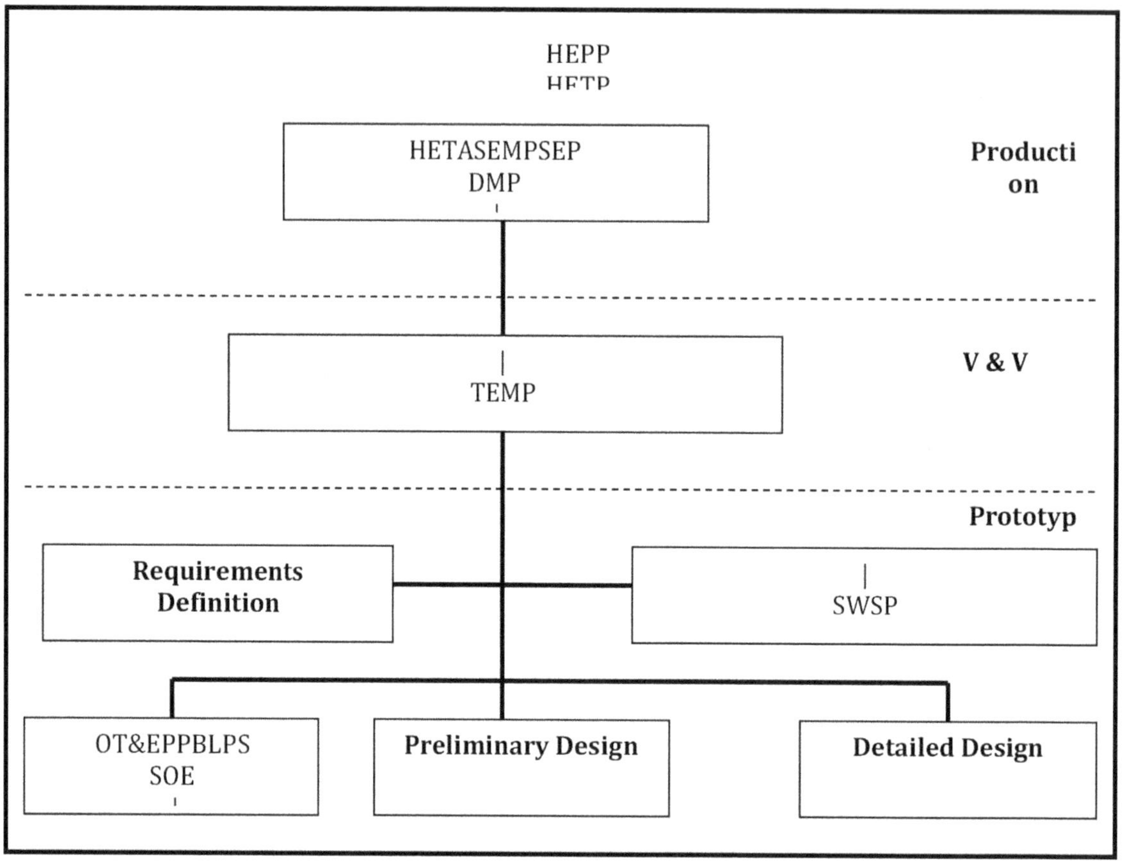

Figure 7- BrEAD Test Plan Hierarchy.

2.7.15 SOE (System Operational Effectiveness Plan)

The SOE plan (CDRL 00011) for BrEAD will delineate the timing, personnel and activities by which MOEs, MOPs and TPMs will be established and used in order to support system effectiveness (see section 2.9 Performance Control). SI's SOE Plan will be assembled by members of the Human Factors subteams, with the assistance of Logistics and Support engineers. The plan will be started early in the project in order to establish Metrics for operational effectiveness and suitability (MOEs & MOSs), supported by multiple Measures of Performance (MOPs) for the various BrEAD systems. The measures selected, and the success of the BrEAD project to meet established metrics will be assessed at SFR, PDR, and CDR. The SOE plan will be finalized at the start of the V&V phase.

2.8 Document Control

All project artifacts, including trade studies, standards, specifications, drawing packages, test protocols, and reports of every kind will be electronically stored in the project's Baseline Archives (see section 4.2). The storage systems will be web-based for remote access, and will have facilities for version control, revision labeling and electronic signatures. To control BrEAD project artifacts, SI will implement an internal document and drawing review and approval process summarized as follows:

a) Project artifacts will be separated into classes based on a graduated set of control requirements. Artifacts in a given class will have either no controls required, peer review required, or peer review and management approval required.
b) Artifacts with no required controls must be archived, but are under no other guidance.
c) Artifacts that require peer review may be revised and stored freely until peer review is performed. Following peer review, artifacts are deemed baselined. Subsequent changes to baselined revisions must be associated with a formal change request (SCR).
d) Following peer review, some documents may also require cover sheet electronic signatures from project authorities..
e) Peer reviews of specified artifact types will have specific peer review checklists and requirements for number and kind of reviewers. Peer review comments, and artifact updates based on the comments will be saved in an SCR and attached to the peer review record. The result of the review will be to approve the artifact, approve with changes, or reject for another review cycle.

2.9 Performance Control

Overall guidance of BrEAD system design towards desired operational characteristics will be implemented by an emphasis on system operational effectiveness (SOE). This is an approach to systems development that emphasizes a balance of system performance and availability, coupled with efficient maintenance and logistics. SI believes that maximum operational effectiveness for BrEAD will be achieved by ensuring that BrEAD has,

a) Properly trained personnel with well understood roles
b) Solid human factors design of BrEAD user interfaces
c) Efficient operational processes in place, continually evaluated
d) Sufficient maintenance and support

SI's SOE Plan will be created mainly by Human Factors engineers, assisted by engineers from the System, Support and Logistics disciplines. This reflects SI's depth in human engineering and our belief that a human-centric design approach yields maximum effectiveness. The focus of the plan will be to assure operational effectiveness in two ways. First, by coordinating the activities of training development, logistics & support planning (in this regard the SOE Plan will be a large input to, and subsume much of the responsibility for the Supportability Plan, Support Equipment Plan and Performance-Based Logistics (PBL) Plan). And secondly by deriving Measures of Effectiveness and Suitability (MOEs & MOSs), and from them the Measures of Performance (MOPs) by which the operational effectiveness of the BrEAD systems will be measured, and from which operational capabilities and requirements are derived.

SI will use a mature and successful methodology to derive and use these metrics:

a) Building upon the PBSS developed from the SOO during this proposal, and drawing upon interviews with future BrEAD user/operators, SI will create a set of typical use cases, describing how the BrEAD systems will be used.
b) Each use case will be translated into one or more MOEs and one or more MOSs which now serve as measures of BrEAD operational effectiveness.
c) The MOEs and MOSs from each use case will be translated into a set of performance requirements, bounded by specific MOPs. The requirements will go to enhance the PBSS and will be allocated to design elements by system engineers dedicated to each subsystem. The MOPs will also be decomposed and allocated to BrEAD subsystems and components.
d) From the list of MOPs, SI will select a limited set of critical technical performance measures. The TPMs provide a basis for determining how well BrEAD designs will be able to achieve their operational objectives and will be tracked during development for that purpose (see 2.8 Technical Performance Measures)..

2.10 Technical Performance Measures (TPMs)

In accordance with the SOE plan, SI will establish a set of TPMs from BrEAD use cases (see 2.7 Performance Control) to track critical performance parameters throughout the development of the BrEAD system. These TPMs are parameters that will impact the cost, performance, or schedule if they exceed critical values. SI will track these parameters, (which are either directly measurable or derivable from modeling of the BrEAD subsystems), to ensure that operational objectives of the BrEAD are met. The TPMs will be monitored and reported at each of the major technical reviews. The TPM report will include the current value and the threshold or "trigger point" for the current phase of development. The "trigger point" is the value which, if exceeded, triggers an automatic review of the entire subsystem or system by the appropriate SE Coordinating Team to assess impacts and corrective actions. The system-level metrics are flowed down and budgeted to the subsystems by the SE Coordinating Teams. The top-level TPMs for the BrEAD system are:

- Drying rate: 85% by weight of moisture removed within 6 hours
- Dryer reliability: MTBF = 1500 hours minimum
- Data Communication Range/Rate between communication cells
    - 40 miles maximum at 115.2 kbps
    - 15 miles maximum at 1.5 Mbps (with high gain antenna)
- Minimum breadfruit processing capacity: 290 kg/day for family use with prototype system
- Maximum breadfruit processing capacity: TBD for industrial-sized system

Key Technical and Management Processes

The SE Coordinating Teams provide status and other project information to the Program Manager. They manage configuration, risk, subsystem and internal interfaces, and track the progress of design and development throughout the course of the project. Their duties include the management of,

- Design Changes
- Action Items
- Requirements
- Risk
- System/subsystem Interfaces
- Configuration
- System Performance
- Project Status (monitoring and conducting reviews)

3.1 Requirements Management

The original eighteen requirements in the pre-award PBSS, along with requirements statements from the SOO have been used as the basis for an expanded set of requirements. This expanded set of 206 functional, operational and performance requirements have been allocated to proposed BrEAD systems and subsystems, and external interfaces have been identified. After contract award, this expanded PBSS will be presented to NGRF personnel, including future BrEAD user/operators for a further definition and understanding of customer needs. Requirements will be prioritized, conflicts resolved, gaps addressed, and ambiguities and redundancies removed. The resulting set of requirements will then be submitted to SRR for formal review.

Following SRR approval and incorporation of any changes, BrEAD design engineers will begin decomposing the requirements into more detailed requirements on lower levels of design. System engineers assigned to each subsystem will establish traceability between the requirements and design elements. They will monitor design activities to quickly identify potential deviations from requirements. .

The growing hierarchy of PBSS requirements will be reviewed frequently at project milestones such as the SFR, PDR, CDR, OTRR and SVR. At these reviews, walkthroughs of the PBSS will be held with the project team and project stakeholders to ensure the requirements are consistent, accurate, and complete.

Proposed changes to baselined requirements will be documented and submitted to the Change Review Board via the Issues Database.
Approved changes will be documented as SCRs in the Issues database and tracked to resolution by the appropriate SE Coordinating Team..

3.2 Risks Management

Risks will be managed to accomplish the following objectives:

- Identify the potential sources of risk and identify risk drivers.
- Quantify risks and assess their impacts on cost, schedule and performance.
- Determine the sensitivity of these risks to program, product and process assumptions, and the degree of correlation among the risks.
- Determine and evaluate alternative approaches to mitigate moderate and high risks.
- Take actions to avoid, control, assume or transfer each risk, and
- Ensure that risk is factored into decisions on selection of specification requirements and solution alternatives.

Risk management will be performed jointly by SI and NGRF personnel who will staff a Risk Management Board. The appointed Risk Officer chair the board. This board will meet shortly after Contract Award for the first time and perform a review of the risks already established in proposal phase. Risks will be re-evaluated for severity and likelihood and NGRF will participate in identifying any new risks. Mitigation strategies will be identified and owners and budgets assigned. Alternate plans will be established for situations where mitigations could fail to prevent the risk event from occurring.

Besides the regular Risk Management Board meetings (see Risk Management section Vol 2 Book 2 of main Proposal), formal Risk Management Reviews of the risks and their mitigations will be held concurrently with project milestones. These reviews will include evaluation of current mitigations, update of risk parameters, and recognition of risk retirements and any new risks.

3.3 Interface Control

Interface control drawings will be created to define and control both external interfaces and interactions between BrEAD subsystems. Interfaces based on initial BrEAD design concepts and early system requirements will be first introduced at SRR. As requirements become better defined and design progresses, these documents will be elaborated by engineers for the various BrEAD subsystems and re-presented for approval at PDR and CDR. Their development and the overall coordination between subsystems will be overseen by representatives from the SE Coordinating Teams, acting through the Interface Control Working Groups. After the interface documents are approved and baselined, changes to interfaces will require approval of the CRB.

3.4 Configuration Management

Throughout the entire BrEAD project, system configuration will be defined in a series of technical baselines, consisting of the approved documentation used to define the technical requirements, design, and interfaces. These baselines define the characteristics of the BrEAD system, subsystems, software and equipment at different development stages of the project. The technical baseline progresses from high-level requirements (Functional Baseline) to more detailed requirements, design drawings and specifications (Allocated Baseline) to complete "as-built" manufacturing drawings and specifications (Product Baseline). Specifications, interface control documents, and drawing packages are used to design and fabricate or procure items. These baseline documents will be organized in a hierarchy that provides design traceability to the lowest level. Once approved, these baseline documents are placed under configuration control.

While configuration management is the overall responsibility of the PM, it will be administered for BrEAD by the CSEs and their SE Coordinating Teams. All artifacts of the project baselines, as described above are stored in Baseline Archives (see Baseline Archives). Once baselined, project artifacts can only be updated via change requests submitted to the Issues Database. Systems engineering will exercise control of these changes via the Change Review Board (see Technical Issues Management).

3.5 Technical Issues Management

All problems and issues related to BrEAD design, development, test, or production will be documented as System Change Requests (SCRs) in the Issues Database. A Change Review Board (CRB) co-chaired by the NGRF and SI will triage SCRs and assign them to project engineers as appropriate for analysis. This analysis will further define the issue and the changes necessary to address it. If further studies or investigation is needed, this will also be indicated. The SCR then goes back to the CRB, who approve the analysis and assign the SCR to engineers as necessary to implement the provisions noted in the analysis. Each implementer will document in the SCR, their actions taken, and the precise revisions of updated work products. Likewise, the work product revisions will reference the SCR they address. In this way, problems and their resolution are cross-referenced. All implementations are then verified by independent reviewers before the CRB closes the SCR.

3.6 Change Management

All changes to baselined project artifacts will be documented in SCRs and submitted to the Issues Database (see Technical Issues Management).

3.7 Problems Resolution Process

See Technical Issues Management

3.8 Trade Studies Process

Throughout the requirements development and design phases of the BrEAD project, studies and analyses will be conducted to support decisions regarding requirements selection and system design alternatives. These trade studies may target performance drivers and constraints from limited resources, such as mass, power and dimension. Certain analyses may also address increased margins in capacity and reliability as well as ease of fabrication and testing.

The collection of these reports effectively documents the process of defining the BrEAD system and will be saved for future reference in the Design Decisions

Archive. As requirements are developed for the BrEAD systems and subsystems and recorded in the Requirements database, they will be cross-referenced to applicable analyses and trade studies in the Design Decisions Archive to provide traceability to the rationale for the requirement. By providing the original context in which a requirement was selected, any future reconsideration of the requirement can determine if the original constraints are still valid.

Trade studies conducted for BrEAD will follow a standardized process, which will be enforced via review by applicable system engineers. This process has been used successfully by SI on previous projects and may be summarized as follows:

a) State the Problem – Determine what project need must be fulfilled
b) Generate Ideas – Identify a broad range of options that satisfy the project need(s)
c) Candidate Solutions (brainstorming is encouraged)
d) Select Practicable Solutions based on project constraints and criteria. Constraints and criteria may be derived from current requirements, and from project stakeholders, especially NGRF.
e) Evaluate Candidates – objective weighting techniques are encouraged, based on technical, economic and operational perspectives.
f) Determine Optimal Solution (with rationale)

When a trade analysis is complete, it will be reviewed by the Chief Systems Engineer and selected system engineers from the applicable SE Coordinating Team. Following review, it will then be presented to the PM and NGRF liaison for final approval. No changes to design approach will be made without undergoing this scrutiny. Approved changes will be documented as SCRs in the Issues database and tracked to resolution.

Tools and Facilities

4.1 Requirements Database

In order to facilitate requirements tracking and traceability, a common requirements management software tool, such as DOORS®, will be implemented across the project. This software will allow the BrEAD Project to convert requirements documents into requirements databases, with each requirement receiving a unique identifier. Each requirement in the database can then be assigned a link to a higher-level requirement. As lower-level requirements are developed, imported into the database, and linked to the higher-level requirements, a structure evolves, allowing the flow down of requirements to be traced from the highest-level mission objectives and user needs to the lowest applicable level of component specifications and test protocols. The requirements database will include the following attributes for each requirement:

- Higher-level requirement satisfied

- Related documents (trade studies, system analyses, etc.)
- Priority
- Status
- Author
- Creation Date
- Change history
- Verification Method
- Verification protocol/test case

4.2 Baseline Archives

Multiple archives will be established to store artifacts from project baselines of differing file types such as MS-Word documents, PDF files, CAD drawings, etc.. These include project level documents such as the TEMP, and also design artifacts such as interface control drawings, software files, test protocols, schematics and parts lists. These archives will be mechanized by COTS version control systems such as (Windchill® or AccuRev®). The exact tool(s) to be used will be selected via trade study conducted soon after contract award.

4.3 Logistics Database

Logistics information will be stored in a database capable of producing reports on maintenance activities, support equipment, faults and failures. Following contract award, SI will conduct a trade study of Logistics tool vendors to determine the best system for BrEAD. Our previous experience with utilities such as the Eagle Logistics Toolset from Raytheon positions us to make an informed determination, with NGRF input.

4.4 Issues Database

To document, prioritize, track, report and assign problems for resolution and design issues, a common OTS tool will be deployed for the entire project. This software utility (e.g. JIRA®), will be used to track software bugs, HSI issues, project tasks and change requests in a database. The tool will allow all the tasks associated with the resolution of a given issue to be tracked and visible to all project stakeholders.

4.5 Design Decisions Archive

A project-wide document control system will be established to archive reports related to system-wide decisions of all kinds, including design decisions, system component selection; decisions resulting from technical reviews and trade studies, and architecture changes. The archive will be used to store documents related to the decision itself (with rationale), and also the discussions leading up to decisions, their justifications, and actions taken as a result of decisions reached. The archive will be available to all project personnel including NGRF staff. Artifacts stored in the archive will be cross-referenced to the applicable system requirement(s) to provide context and rationale for requirements (see Trade Studies Process).

4.6 Instructional Systems Development (ISD) Tools

In order to facilitate the development of training materials for BF Processing System operators and BrEAD staff, an Instructional Systems Development (ISD) tool, such as FlexTraining® will be utilized by the BrEAD training developers. This software will enable the rapid development of interactive computer-based training materials for the anticipated roles within the BrEAD system. The exact tool to be used will be chosen via a trade study early in the project.

4.7 Modeling and Simulation (M&S) Tools

SI has found wide scope for the use of modeling and simulation on its various contracts as techniques for reducing cost and risk, and achieving high levels of user satisfaction with minimal re-work. Over the years, our internal team of tool designers have developed a suite of tools (MAXSIM®) that have achieved international recognition, both from our customers (including the DoD), as well as from scientific bodies such as the Society for Modeling and Simulation International (SCS). For example, on the VanFlight project for the Department of Defense (DoD), our MAXSIM toolset was accredited under DoD Directive 5000.59 as acceptable for,

- simulating complex 3-D man/machine interactions in real time
- fast prototyping of GUI interfaces for optimization of task workload, fatigue, data absorption and other factors
- simulation of communication networks for speed/capacity estimation

In addition, our team of industrial designers have extensive experience in creating quick mockups in our model and paint shop from plywood, foamcore, ABS plastic and polyurethane foam for both industrial and military application. We anticipate great leverage for these in-house capabilities in the following areas of the BrEAD project:

- fast, inexpensive mockups of SIL and CDC workstations for proof of concept, to flesh out user requirements, and for operator training.
- GUI prototypes of CDC and SIL data systems for user requirements elaboration, training, test, and optimization of operational procedures.
- Simulation of the BrEAD communications architecture and modeling of its operational environment for design prototyping, design assurance, performance validation and test protocol development.
- Modeling/simulation of all BF processing stages for safety, throughput, ease of maintenance and general design optimization.

Engineering and Support Specialties

5.1 Human Factors Engineering

Human Systems Integration (HSI) is a process that ensures that designed products will interoperate with human capability in an optimal manner, and that designed systems are safe, efficient and operable by the personnel expected to interact with them. BrEAD HSI will achieve this by considering,

a) the manpower required to perform each task

b) the aptitudes, knowledge, skills and abilities of the people likely to be interacting with BrEAD systems.

c) the level of training necessary to perform each task

d) design features to maximize operability and effectiveness (human factors)

e) design features to maximize operator safety

To maximize the value of HSI contributions, HSI will be incorporated into the design process at its earliest stages by members of the Human Factors (HF) subteam on both IPTs. The SI HF subteam members provide expertise that includes design criteria, analysis and modeling tools, and measurement methods to help the system designers create cost-effective and efficient systems. Informed decisions about the human-machine interfaces of the BF Processing System and the human-software interfaces of utilities developed for the SIL, CDC and RC early in the design process will help ensure that human performance using the BrEAD system will be optimized. HF engineers will chair or be members of the Interface Control Working Groups, the System Safety Working Groups, and the System Software Safety Working Groups within each IPT to further facilitate integration of HE-based concerns. HF participation will continue during each succeeding acquisition phase. HF sub-team members will participate in all design Integrated Product Team (IPT) and program design reviews to ensure the primacy of human engineering-based considerations during the entire project.

HEPP/HEDA

5.1.1 HEPP/ HEDA (HE Program Plan and Design Approach)

SI's approach to the use of human engineering practices and techniques during BrEAD development will be detailed in the Human Engineering Product Plan (HEPP). This plan will be created in the early stages of the BrEAD project, based on the Test and Evaluation Master Plan (TEMP) document and on inputs from NGRF members slated to become BrEAD users/operators, to help ensure that their requirements are met. It will describe the entire human engineering effort in terms of its constituent elements and how they will be managed to integrate Human Engineering (HE) into the entire design process. It will describe necessary characteristics of the equipment and software systems that require human interaction to operate or maintain. It will be used as a planning guide and management tool to ensure that HE issues are appropriately and adequately accommodated during the project. And it will provide an HE audit trail by documenting program guidance, sources of applicable data, and acquisition decisions. The HEPP will be created before operational requirements for BrEAD are established and will be used to help formulate these requirements.

Because of the criticality of HE to the design, test and training processes, the HEPP will be maintained and updated throughout BrEAD development. It will also be a living document that will evolve as the BrEAD system develops. As part of this process, regular Human Systems Integration Working Group (HSIWG) meetings will be conducted; initially to work with NGRF user/operators of BrEAD to identify HSI issues and concerns. After that, members of the working group with CRB authority will monitor, plan, manage and coordinate the issues via the Issues Database to make sure they are properly dispositioned and reflected in the HEPP. The initial HEPP/HEDA document will be presented to the NGRF for approval at PDR.

5.1.2 HETA (HE Task Analysis)

SI HF engineers will perform task analysis on all defined BrEAD tasks, both those involving BF flour production, as well as engineering and data flow tasks associated with the BrEAD Enterprise Architecture. These analyses will be documented in a Human Engineering Task Analysis (HETA) report, and will provide a basis for making design concept decisions, identifying potential problem areas and will support later evaluations of BrEAD designs and equipment as they evolve; to ensure they meet human performance requirements and other human engineering criteria. HE task analyses will identify critical BrEAD tasks, modes of operation (including degraded modes), and will be used to verify that BrEAD design solutions meet human performance and safety needs. HETA data will also be used to help create training and instructional postings for operators/maintainers of the BF processing systems and BF Enterprise Architecture

In addition to task analysis, trade studies will be conducted both within the HSI domains applicable to this project (manpower, personnel, training, safety and occupational health, human factors) and between the HSI domains and BrEAD subsystems. The CSEs will integrate the system requirements from the HSI domains with each other, and also with the BrEAD subsystems. The results of these integration efforts will be reflected in updates to the requirements, objectives, and thresholds in BrEAD requirements documents.

5.1.3 HE Sim/Concept Report

SI is a leader in modeling and simulation of user tasks, both manual and computer-intensive. Several members of our HF staff worked originally at the Human Interaction Laboratories in Kanagawa, Japan and have extensive experience in "Human in the Loop" (HIL) simulations. Working with NGRF personnel, including potential operators of the BrEAD system, our experts will formulate a customized approach, using scale and full-size mockups, prototypes, 3-D anthropometric modeling tools, and Graphical User Interface simulations of BrEAD equipment and systems, for characterizing system/user interactions. Their approach will be documented in the HE Sim/Concept Report, and will be used to characterize the required functionality of the human-intensive portions of the BrEAD infrastructure (e.g. the BF processing system, including instrumentation, the weather station and data collection and communication devices, as well as software systems such as the BF data processing system and call center). Data derived from scenario-based HIL reenactments of BrEAD missions will be used to validate and demonstrate system and human performance and other human-based criteria, such as safety and workload. The data will support design decisions for workstation layout, presentation of data, and special features to aid the operator in decision making. Evaluations and testing with interface mockups or simulated interfaces will also be used to evaluate overall system design.

5.1.4 HETP (Human Engineering Test Plan)

Working in concert with the Test Plan Working Groups, and in accordance with the TEMP, SI HF engineers will generate a Human Engineering Test Plan (HETP). This plan will document the testing criteria and the operational testing that will demonstrate that each of the operator/hardware/software interactions identified as part of the BrEAD system can accomplish their intended functionality and meet human engineering requirements as specified. Development of this testing will be directed toward verifying that the system can be operated, maintained, supported, and controlled by user personnel in its intended operational environment with the intended training. This plan will be the principal means for validating the personnel/hardware/software interfaces, human performance, level of training, safety, workload, and criteria for personnel selection. The initial HETP will be presented for approval at PDR. The HETP will be finalized subsequent to the final TEMP version.

The HETP will form the core of the HE portion of the Initial Operational Test & Evaluation (IOT&E) phase of the project and will also specify criteria and protocols for design testing, integration testing and product demonstrations, including the evaluation of maintenance and support activities. After testing is concluded, a Human Engineering Test Report (HETR) records results and provide evidence that all human-based requirements for support, operations and maintenance have been met.

5.2 Logistics Support Planning

See 5.3 below

5.3 Reliability/Maintainability/Supportability/Availability Plans

SI will develop a comprehensive approach to maximizing the maintainability, supportability, reliability and availability of BrEAD subsystems and the system as a whole. This will be a performance-based approach, leveraging a sophisticated LMI database (see section 4.3 Logistics Database) capable of automating maintenance, and support reporting and generating support equipment recommendations and technical documentation. This approach will be detailed partly in the SOE plan (see section 2.9 Performance Control) with the remainder documented in the Support Equipment Plan (SEP), the Performance-Based Logistics Plan (PBLP) and Maintenance Task Analyses. These plans will be created by the Reliability, Logistics and Support teams and presented for review at CDR.

Appendix A. Acronym List

Term	Definition
BF	Breadfruit
BrEAD	Breadfruit Enterprise Architecture Demonstration
CDC	Central Data Center
CDR	Critical Design Review
CM	Configuration Management
CMP	Configuration Management Plan
CRB	Change Review Board
CSE	Chief Systems Engineer
DMP	Data Management Plan
EA	Enterprise Architecture
FMECA	Failure Modes and Effects Analysis
HE	Human Engineering
HEDA	Human Engineering Design Approach
HEPP	Human Engineering Program Plan
HETA	Human Engineering Task Analysis
HETP	Human Engineering Test Plan
HETR	Human Engineering Test Report
HE	Human Engineering
HF	Human Factors
HIL	Human in the Loop
HIS	Human Systems Integration
HSIWG	Human Systems Integration Working Group
HW	Hardware
ISD	Instructional Systems Development
IOT&E	Initial Operational Test & Evaluation
IMP	Integrated Master Plan
IPT	Integrated Project Team
LMI	Logistics Management Information
NGRF	New Green Revolution Foundation
NW	Communications Network
OT&EP	Operational Test and Evaluation Plan
PBLPS	Performance-Based Logistics Planning Strategy
PDR	Preliminary Design Review
PM	Program Manager
PMO	Project Management Office
PTD	Provisioning Technical Documentation
QA	Quality Assurance
QMP	Quality Management Plan
RMP	Risk Management Plan
SDP	Software Development Plan
SE	Systems Engineering

SECT	Systems Engineering Coordinating Team
SEMP	Systems Engineering Management Plan
SERD	Support Equipment Recommendations Data
SI	Solutions International
SIL	Systems Integration Laboratory
SOE	System Operational Effectiveness
SSMP	System Safety Management Plan
SW	Software
SWSP	Software Safety Plan
SFR	System Functional Review
SRR	System Requirements Review
SysRS	System Requirements Specification
T&E	Test & Evaluation
TEMP	Test and Evaluation Master Plan
TPM	Technical Performance Metric
V&V	Validation and Verification
V&VP	Validation and Verification Plan

System Development and Demonstration of the Breadfruit Enterprise Demonstration (BrEAD) and Systems Design

Solutions International Proposal for the Breadfruit Enterprise Demonstration and System Design

Volume 2: Technical

Annex H: Software Development Plan

RFP NO: E00010-07-R0001

Copy 1 of 2

May 1, 2009

Prepared For:

New Green Revolution Foundation (NGRF)

Prepared By:

Solutions International

123 Pillsbury Drive

St. Paul, MN 55114

CAGE CODE SI4ME

TABLE OF CONTENTS

SECTION 1 - Introduction

This Software Development Plan (SDP) establishes the plan for all aspects of software development, test, integration, purchase, implementation, and maintenance of the software resources needed for the Breadfruit Enterprise Architecture Demonstration (BrEAD). Updates to this SDP will address future BrEAD software upgrades.

The Software Development Plan Template, TM-SPP-02 V2.0 (refer to paragraph 2, Referenced Documents), is used as a template for this document and is tailored for the specific requirements of the BrEAD project.

Identification

The BrEAD software system falls under two general categories: Commercial Off The Shelf (COTS) purchased products and BrEAD specific software developed by the BrEAD / Solutions International software Integrated Product Team (IPT).

The COTS purchased software is listed below:

- Microsoft Windows Professional – operating system
- Microsoft Office Professional - desktop applications
- Microsoft Visual Studio – Java and C/C++ software development suite
- QuickBooks Premier – accounting application
- Adobe InDesign CS4 – desktop publishing application
- Symantic AntiVirus, Anti-Spyware, Endpoint Security, Firewall software
- AutoDesk Autocad – mechanical design CAD software
- MathWorks MATLAB – simulation and modeling software
- Telelogic DOORS – requirements management database

The developed software will be further delineated between database template creation and actual coded applications. Database templates will be developed for data input into the databases, and for extracting data from the databases.

Algorithms will be developed for automated input of field data into the Field Data Database, and for automated transfer of field data at the Regional Center between the processing sites and the Central Data Center.

System Overview

The Bread software system has several functions, of which the recording, transfer, storage, processing, and extraction of Field Data is primary. Field Data as used in this SDP refers to the weather data collected at the breadfruit processing sites and the breadfruit processing data, also collected at the processing sites. To accomplish this task a central database and several algorithms will be created.

In addition, there are general office functions, administrative, bookkeeping, accounting, personnel records, marketing, logistics, communication, and engineering functions. COTS products will be used to meet these needs. As will be detailed below, some of these will require the creation of database templates.

The BrEAD is organized around four physical locations: a Central Data Center, Regional Center(s), breadfruit processing site(s), and a System Integration Lab (SIL). It is expected that these will not be co-located. Each has unique as well as common software requirements. In addition, as stated previously, some of the software needs will be met by purchased COTS products, and some by software developed by the BrEAD Software Engineering IPT.

Commercial Off The Self (COTS) Software

Each PC computer within the BrEAD shall have the Microsoft Windows Professional operating system. Each PC computer shall have the Microsoft Office Professional office software suite. Each PC computer shall have Symantic AntiVirus, Anti-Spyware, Endpoint Security, and Firewall software.

The servers at the Central Data Center and at the SIL shall have additional COTS software:

- QuickBooks Premier accounting software
- Adobe InDesign CS4 desktop publishing software
- Sun Microsystems Java Enterprise compiler
- Eclipse text editor
- Microsoft Visual Studio Java and C/C++ software development suite

In addition, the server at the SIL shall have MATLAB, MAXSIM, and two seats of AutoCad.

Developed Software

The BrEAD Software Engineering IPT will develop software unique to the BrEAD needs.

Central Data Center

A Microsoft Access database will be used for the Human Resources department that will be utilized for personnel records. A Microsoft Access database will be used for the storage and processing of field and weather data collected at the processing sites. An algorithm will be used to receive field and weather data and enter it into the Field Data database automatically without user input. A website will serve as the interface between the BrEAD and the BrEAD stakeholders. An interface between the website and the database will provide access to the Field Data database for stakeholders.

Regional Center

An algorithm will accept field and weather data from the processing sites, format it for input into the Central Data Center Field Data database, and send it via Internet connection to the Central Data Center.

Processing Site

A GUI will be used for field data input, behind which a text file will be created, containing the entered data. The text file will be sent to the Regional Center.

System Integration Lab (SIL)

The same Field Data database and automated database input algorithm developed for the Central Data Center will also be installed at the SIL. An algorithm will be used for extracting data from the SIL database, formatting it, and sending it via Internet connection to the Central Data Center.

Document Overview

This SDP identifies applicable policies, requirements, and standards for the BrEAD software development. It defines schedules, organization, resources, and processes to be followed for all software activities necessary to accomplish the development. This SDP contains no privacy considerations pertaining to the BrEAD.

This document follows the outline of a Software Development Plan contained in the MIL-STD-498 Data Item Description for an SDP: DI-IPSC-8127. It also meets the content requirements for IEEE/EIA 12207.1 Development Process Plan (paragraph 6.5) and Project management plan (paragraph 6.11). These plans are tailored for the size and degree of complexity of this software system. The SDP is organized as follows:

Section 2 lists all documents referenced by this SDP and used during its preparation.
Section 3 provides an overview of the required work.
Section 4 describes the details of all software planning, design, development, reengineering, integration, test, evaluation, Software Configuration Management (SCM), product evaluation, Software Quality Assurance (SQA), and preparation for delivery activities.
Section 5 defines the project schedule and activity network.
Section 7 describes the project organization and the resources required to accomplish the work.

Relationship to Other Plans

This SDP, the documents listed below in section 2, the Integrated Management Schedule (IMS), the Systems Engineering Management Plan (SEMP), the Configuration Management Plan, Quality Management Plan, Test and Evaluation Master Plan (TEMP) serve as the guiding documents to develop the software for the BrEAD project.

SECTION 2 - Referenced Documents

- Software Development Plan Template, TM-SPP-02 V2.0, April 5, 2005: Systems Engineering Process Office; Space and Naval Warfare Systems Center San Diego; San Diego, CA.

- Data Item Description DI-IPSC-81427, Software Development Plan
- IEEE/EIA Standard 12207, Software Life Cycle Processes
- IEEE Std 1028, Software Reviews and Audits
- MIL-STD-498, Software Development and Documentation

SECTION 3 - Overview of Required Work

Requirements and Constraints

Requirements that are addressed by this plan are identified in the Performance Based System Specification (PBSS).

Program Schedule and Milestones

Major software development milestones are identified in the Integrated Management Schedule (IMS). Details of the schedule and milestones of development activities are contained in section 6 of this annex.

Software Development Activities

Field Data Database

A Field Data Database will be created. This database will reside on the Central Data Center server. This will be a Microsoft Access database. This database will be the repository for the field data as defined in Appendix A of the PBSS:

- Data Collected at Breadfruit Processing Site
- Data Processed at Breadfruit Regional and Central Data Site
- Data Processed at Breadfruit Central Data Site

This database will also be the repository for the weather data collected at the breadfruit processing sites.

The term Field Data, in this SDP, will refer to both the data collected and processed as listed above, and the weather data collected at the processing sites.

Microsoft Access templates will be developed for data extraction from the Field Data Database. It is anticipated that data output will be in the form of text files, spreadsheets, and graphs. However, stakeholder input will be sought and incorporated into the design as to the format and content of data output that is desired. It is anticipated, that the form of data output will evolve over the life of the program. As such this will be an ongoing effort.

Field Data will be entered into the database by automated means. However, as a backup, data may need to be entered manually. Thus Microsoft Access templates will be developed for manual data input of this data also.

The data in the database will be password protected against unauthorized entry or modification. Data Extraction will not be password protected so that any interested stakeholder may have access.

A copy of this database will also reside on the Systems Integration Lab (SIL) server. This database will contain Field Data for SIL activities. It is anticipated that data will be entered manually, however, the automated data input algorithm will be installed on the SIL server also to allow for data input from remote processing sites involved with SIL activities.

Automated Data Input Algorithm

An algorithm will be developed for automated data input into the Field Data Database. Field data will be received at the Central Data Center via internet connection. This algorithm will receive this data, validate it, and enter the data into the database. It will also send a message to the sender indicating successful receipt of the data. Field Data will be received from the Regional Centers and the Systems Integration Lab (SIL).

Central Data Center Website

A website will be created that will provide the internet interface to the BrEAD. It will reside on the Central Data Center server. It will provide general BrEAD organization information as determined by the BrEAD leadership. It will also provide read access to the Field Data Database, both raw data and the report, spreadsheet, and graphical output. Additional website content and features will be determined prior to the requirements review specified in the Integrated Management Schedule (IMS).

Website /Database Interface

An interface algorithm will be developed that will provide for the capability of the Field Data Database data to be viewed through the website, and for the database output, in its' various forms, to be viewed as well as downloaded through the website.

Personnel Database

A Microsoft Access database will be the repository of personnel records. This database will reside on the Central Data Center server. Microsoft Access templates will be developed for data input. Templates will also be created for data output as specified by the Human Resources department.

Problem Report Database

A Microsoft Access database will be developed for the repository of Problem Reports (PR). This database will contain technical problems discovered by users of the BrEAD software and hardware, including the breadfruit processing hardware. As with the other databases, templates will be developed for data input and output. This is where the information needed to adequately define and record the nature of the problem/defect will be stored. Tracking of progress on PR status and resolution will reside in this database.

A copy of this database will also reside on the SIL server for problems/defects associated with SIL activities.

Field Data Transfer Algorithm

An algorithm will be developed, the function of which is to receive Field Data from the breadfruit processing sites and forward it to the Central Data Center. This algorithm will reside on the Regional Center server. Data received will be checked for completeness and validity before forwarding to the Central Data Center. If incomplete or invalid data is received, error messages will be made to Regional Center system users. The erroneous data will be analyzed and PRs written.

When data is received, a message will be sent to the sender indicating successful receipt of the data. When data is forwarded to the Central Data Center, the message sent by the Central Data Center indicating successful receipt of data will be noted. If this message is not received within a specified time, the data will be resent. Each data package will be indexed so the receiver will know if it is getting duplicate data.

This algorithm will also validate weather data received from the breadfruit processing sites. It will also monitor the weather station heartbeat message to ensure the weather station is operating.
Note: a heartbeat message is a message sent on a regular periodic basis that tells the receiver the sending process is running.

Site Data Input GUI

A GUI interface will be created for entering the breadfruit processing site data defined in Appendix A of the PBSS. The data will be stored in a text file, which will be forwarded to the Regional Center. The data will be checked for validity as it is entered and the user will be notified if an attempt to enter invalid data is made. This application will reside on the processing site computer.

COTS Purchased Software

The BrEAD computers will have the COTS software installed as specified in section 1 of this SDP.

SECTION 4 – Plans for Performing Software Development Activities

Project Planning and Oversight

The software development process will be managed by a software project manager. His responsibility will be to plan and manage software development activities as specified below. He will also ensure that the processes described in the following paragraphs are documented in process specifications, and that these process specifications are implemented.

Software Development Process

This software development process is tailored from the guidelines of IEEE/EIA 12207 and Mil-Std-498. Software development is divided into four tasks. The first will consist of developing the Field Data Database and the site data input GUI. The second task will consist of developing the automated data input algorithm (paragraph 3.3.2) and the field data transfer algorithm (paragraph 3.3.7). The third task will consist of developing the Central Data Center website. The fourth task will consist of developing the personnel and problem report databases. Each of these tasks will be implemented in five phases.

Phase 1 - Software Requirements

Detailed software system requirements will be documented in a Software Requirements Specification (SRS) document. The SRS will be generated from a DOORS database that will be the repository for the software requirements. These requirements will specify system functionality, interfaces - software and user, database requirements, system security, data validation, data definition, and all other aspects of the software products. The requirements will be analyzed for traceability to source requirements (from the SOO, RFP, and BrEAD PBS) and testability. A Peer Review will be conducted prior to the Software Requirements Review in August of 2009 (as scheduled in the IMS). Upon successful completion of these reviews phase 2 will begin.

Phase 2 – Software Design

The software IPT will develop software solutions that satisfy the specifications of the requirements. Computer Software Configuration Items (CSCI) will be created as necessary to organize the software products into a logical functional organization. A Software Design Document (SDD) will be created that will document all aspects of the design of the software products. This will include design decisions as well as trade studies. All developed code will be developed using a high order language, Java or C/C++. Peer Reviews of the software design will be conducted. The Readiness Review will be conducted in November of 2009. Upon successful completion of these reviews phase 3 will begin.

Phase 3 – Software Unit Development, Test, and Integration

The software IPT will develop the executable code as defined in the SDD. Standards for embedded comments and file headers and banners will be developed, documented, and followed in code development. Code will be developed per best industry practices. Comments in code will be thorough and descriptive enough to make the code self-documenting.

Unit test procedures as well as unit test drivers will be developed and documented. Unit test will be conducted and the results documented. The software products will be integrated and Integration test procedures will be developed and executed and the test results documented. Peer Reviews of test plans and test results will be conducted. Peer Reviews will verify completeness of code coverage and expected test results. Upon successful completion of these reviews phase 4 will begin.

Phase 4 – System Qualification Test and Delivery

The software IPT will create system qualification test plans and procedures for each software product. Qualification test will verify each software requirement. The test plans and procedures will be Peer Reviewed and accepted prior to conducting qualification testing. The test procedures will be verified for requirements traceability, thoroughness of requirement coverage, testing of system robustness (error and fault handling). Upon successful completion of Peer Review, the Qualification test will be conducted. The following must be present for Qualification Test: the test conductor, a scribe to record test results, management, and the customer. As all software products are going to be used by BrEAD organization personnel, the customer in this case is defined as he user of the software product being tested. The test results will be recorded in the Qualification Test Report (QTP). Qualification Test is to be completed by March 2010. Upon successful completion of the Qualification test, the software products will be prepared for deployment and phase 5 will begin.

Phase 5 – Support of Installation and Use

The software IPT team will install the software products at the designated sites and conduct life cycle support. Life cycle support will include responding to requests for software upgrades and enhancements and Problem Reports (PR).

Configuration Management

As it is not anticipated that there will be a large number of code files, a simple system of configuration management will be implemented for managing software file revisions. A software configuration manager will establish a password protected directory in which he will store code and executable files. Each file will be identified with a file name and revision identifier. Executable files will also be identified with a revision identifier. He will maintain a spreadsheet listing each software file and it's revision history, which a brief description of changes made. This procedure will be documented in a process specification.

Quality Assurance

A Software Quality Assurance (SQA) process will be created and documented in a process specification that will ensure a high level of software quality. This process will include the reviews specified in the above paragraphs. It will specify a record keeping system to preserve the review and test results, and all other software documents specified in previous paragraphs. The SQA process will also specify procedures for identifying and implementing process improvements. The SQA department will be independent of the software IPT. They will not be involved in software development and will report up through a separate management chain than the software IPT.

Project Reviews

The purpose of the project reviews specified in section 4 is to ensure the quality of the software system and to provide management with a tracking mechanism to track development progress. Each review, as a minimum will have the product creator, two reviewers, a scribe to record review results, an SQA representative, and the development team lead present. The team lead may also serve as one of the reviewers. For the Qualification Test reviews a management representative must also be present.

Weekly status reviews will also be scheduled for team leads with management to provide status to management for the purpose of ensuring compliance with schedule, and to facilitate identification of issues that might cause schedule slip.

Corrective Action

A Problem Report (PR) system will be created for identifying software system defects and requested enhancements. This system will consist of a database for recording the defect/requested enhancement, a review process for determining the validity and seriousness of the PR, a change control board for approving/disapproving the PR, a process for implementing the PR, and a process for tracking progress on the PR.

Schedule

Software requirements development will start:	contract award (anticipated Jun 1, 2009)
Software requirements Peer Review:	July 1, 2009
Software requirement specification review:	August 2009 (refer to IMS)
Software top level design start:	August 2009
Software top level design Peer Review:	October 15, 2009
Software top level design readiness review:	November 2009 (refer to IMS)
Software detail design start:	November 2009
Software detail design Peer Review:	January 2, 2010
Software Unit test and Integration test procedure Peer Review:	January 20, 2010
Software Unit test results Peer Review:	January 25, 2010
Software Integration test results Peer Review:	January 30, 2010
Software Qualification test procedure Review:	February 15, 2010
Software Qualification test:	February 20, 2010
Software Qualification test results review:	March 2010 (refer to IMS)

System Development and Demonstration of the
Breadfruit Enterprise Demonstration (BrEAD) and Systems Design

Solutions International Proposal for the
Breadfruit Enterprise Demonstration and System Design
Volume 2: Technical
Annex I: Marketing Strategy

RFP NO: E00010-07-R0001
Copy 1 of 2
May 4, 2009

Prepared For:

New Green Revolution Foundation (NGRF)

Prepared By:

Solutions International
123 Pillsbury Drive
St. Paul, MN 55114

CAGE CODE SI4ME

TABLE OF CONTENTS

TABLE OF FIGURES

Marketing Strategy

In order to properly fund the Breadfruit Enterprise, Solutions International has designed a strategic marketing plan, as shown in **Figure 1-269**. This marketing plan contains four key areas in which Solutions International will spread the word about the cause and generate revenue for this non-profit venture. By using a website, brochures, targeting the gluten-free market, and using established resources to obtain government grants, Solutions International has positioned itself to be a prime selection for this contract

In Jamaica, they demonstrated the ability of a breadfruit orchard to generate a profit. With an 800 tree orchard they were able to generate $4 million in profit annually. Because of their success, Solutions International is confident that we can successfully productize a processing system that can help countries in need provide food for themselves. Using a good marketing strategy, Solutions International hopes to generate interest in this initiative worldwide.

Figure 1-269: Marketing Strategy. *Solutions International identifies 4 areas to focus marketing, showing our dedication to project success.*

Website

A website is a quick easy way for lots of information to be dispersed over the world in a very short amount of time. In order to try and connect with people all over the world, Solutions International will create a website. Since so many different things can be included in a website Solutions International plans to have four key areas listed on it. The first area is just an overview of what breadfruit is and what Solutions International is trying to do. The next item would be a place to actually buy the breadfruit flour. Earth friendly stores can purchase it to sell in their stores as well as everyday consumers and gluten-intolerant people. Researchers will be able to access all the different research data collected by the sites on the research section of the website. This portion of the website will be updated regularly with up-to-date information from around the world. Finally, since Solutions International is a non-profit organization, the final piece of website will include a section where donations can be given.

Brochure

A brochure describing our intention and mission would be a valuable tool in many areas of the world. This brochure would be used to increase interest in our processing system. Our initial brochure concept would contain the following information:

> *Breadfruit is a very abundant resource found in many of the world's tropical nations. Although it is very abundant, there is one major flaw with it, it rots within 48 hours of being off the tree. With this in mind it is very difficult for the natives of these countries to take advantage of this plentiful resource. Solutions International plans to help these natives change this. By providing affordable resources to nations around the world, Solutions International plans to save many lives. Through research, government grants, connections with local aid organizations and aggressive marketing plan, Solutions International plans to do just that.*

Gluten-free

Since flour will be produced as a by-product of the SIL in Hawaii, this excess flour can be sold rather than just being thrown away. Today, there are at least 110,000 Americans suffering from celiac disease, or gluten intolerance. Since breadfruit flour contains no gluten, it would be a perfect substitute for their culture to use. There are many different places where Solutions International can connect to this group. The first and probably most effective would be through the gluten free passport. This site includes a list of all kinds of different recipes and gluten free providers, but there are many other places where people can find resources. Solutions International has identified many of them including Livingwithout.com and Gluten Free Living Magazine. Included with all the advertisements would be a link to the website as well as Solutions International's mission statement.

Government Grants

The final way that Solutions International will try and market our product is through local aid organizations as well as through government organizations. Solutions International has identified two local aid organizations, Dr Carolle and Father Reiser's Relief. These organizations will help us with two key areas; cultural modeling and obtaining grant money. One of the risks that Solutions International has identified is the problem of being able to model the way the culture will react to the incorporation of their system. Both Dr. Carolle and Father have assured Solutions International they can provide a safe environment as well as enough information to model the culture. Given that Solutions International wins the contract, they will immediately pursue submitting proposals for a variety of different government grants utilizing these local aid organizations that have experience in this area.

System Development and Demonstration of the
Breadfruit Enterprise Demonstration (BrEAD) and Systems Design

Solutions International Proposal for the
Breadfruit Enterprise Demonstration and System Design
Volume 2: Technical
Annex J: Traceability Matrix and Acronym List

RFP NO: E00010-07-R0001
Copy 1 of 2
May 4, 2009

Prepared For:

New Green Revolution Foundation (NGRF)

Prepared By:

Solutions International
123 Pillsbury Drive
St. Paul, MN 55114

CAGE CODE SI4ME

TABLE OF CONTENTS

Traceability Matrix

The following traceability matrix can be used as a guide to evaluating our proposal. Our traceability matrix identifies where each SOO section and RFP-provided PBSS requirement is addressed in our proposal.

Requirements	Context	SOW Ref	SI PBSS Ref	Proposal Ref.
PBSS #1	BrEAD component of next green revolution	N/A	3.2.1	2.1
PBSS #1	BrEAD component of design	N/A	3.2.1	2.1
PBSS #1	BrEAD component of market	N/A	3.2.1	2.1
PBSS#1	BrEAD component of collect data	N/A	3.2.1	2.1
PBSS #1	BrEAD component of distribute and provide logistics support for breadfruit processing systems and SD for breadfruit processing equipment	N/A	3.2.1 (Enterprise Architecture)	2.1.1.3
PBSS #2	Provide an automated research, marketing, and user data collection capability for NGRF head quarters in St. Paul, Minnesota that will provide 24 hour per day data from researchers, marketers and users world wide	N/A	3.2.1 (Enterprise Architecture)	2.1.1.1

Requirements	Context	SOW Ref	SI PBSS Ref	Proposal Ref.
PBSS #3	The BrEAD shall provide organizational liaison to the Community of Interest (COI) and automated electronic connectivity for real time data interchange. The community of interest is comprised of the NGRF, government and private organizations that have or could have an interest in breadfruit products	N/A	3.2.14 (Central Data Center) 3.3.2 (External Interfaces)	2.1.1.1
PBSS #4	The initial baseline breadfruit systems design shall be based on the characterization of any prototype existing breadfruit processing systems. The characterization shall include, but not be limited to; operator usability, reliability, maintainability and manufacturability. The findings shall be documented in a requirements document and a product description	N/A	3.2.2 (Breadfruit Processing)	2.1.1.1
PBSS #5	The breadfruit processing system shall be operable in geographic regions that do not have an electrical power infrastructure and in geographic regions that have an electrical power infrastructure with minimal equipment differences	N/A	3.2.2 (Breadfruit Processing)	2.1.1.1

Requirements	Context	SOW Ref	SI PBSS Ref	Proposal Ref.
PBSS #6	The design of the breadfruit processing system shall be modular, such that any specific geographic requirements can be accommodated without breadfruit processing system redesign.	N/A	3.2.2.1 (BF Processing)	2.1.1.1
PBSS #7	The breadfruit processing system human machine interface shall be intuitive; that is the human operation of the system shall require minimal documentation- ideally none.	N/A	3.2.2.1 (BF Processing)	2.1.1.1
PBSS #8	Any design modifications that are indicated from operational data gathered from users of breadfruit processing systems in use in the field shall be analyzed and considered for incorporation as design modifications to future the breadfruit processing system designs	N/A	3.2.2.2 (BF Processing)	2.1.1.1
PBSS #9	The breadfruit processing system shall be easily repairable with a minimum of tools in the field	N/A	3.2.2.4 (Breadfruit Maintenance)	2.1.1.3
PBSS #10	The breadfruit processing system shall have a failure rate of one failure per 10,000 hours	N/A	3.2.2.3 (Breadfruit Reliability)	N/A
PBSS #11	The mean time to repair the breadfruit processing system shall be 15 minutes	N/A	3.2.2.4 (Breadfruit Maintenance)	N/A

Requirements	Context	SOW Ref	SI PBSS Ref	Proposal Ref.
PBSS #12	The breadfruit processing system shall extendible. Extendibility is the capability of the breadfruit processing system to produce more than one food product or food product component.	N/A	3.2.2.2 (BF Processing)	2.1.1.1
PBSS #13	The breadfruit processing system shall be scalable; the range of scalability to be considered is from family use to industrial scale use that could provide large quantities of breadfruit products. The range anticipated is from a few ponds per day per family to several tons per day for an industrial scale facility. The specific values shall be determined and documented in the proposal	N/A	3.2.2.2 (BF Processing)	2.1.1.1
PBSS #14	The BrEAD shall provide weather data gathered through out the world where breadfruit is used. It shall provide hourly temperature, pressure, humidity, and rain fall data to the Central Data Collection and Analysis Center of the NGRF located in St. Paul, Minnesota	N/A	3.2.11.1 (Weather Station)	2.1.1.1.4

Requirements	Context	SOW Ref	SI PBSS Ref	Proposal Ref.
PBSS #15	The BrEAD architecture shall have distributed local weather sensors, and data collection systems and shall be interoperable with the Central Data Collection and Analysis Center. The user data collection human machine interface at the data collection facilities shall be intuitive to the user	N/A	3.2.1 3.10.2	2.1.1.1.4
PBSS #16	The breadfruit processing system shall operate from 40 degrees Fahrenheit to 120 degrees Fahrenheit.	N/A	3.2.2.1 (BF Processing)	N/A
PBSS #17	The Systems Integration Laboratory (SIL) shall be used to evaluate candidate architectures for the BrEAD, the breadfruit processing systems, and the intuitive human interface capability desired.	N/A	3.2.15 (SIL)	2.2.6.2
PBSS #18	The breadfruit processing shall operate in 100% relative humidity	N/A	3.2.2.1 (BF Processing)	N/A
SOO Obj. #1	Obtain a Breadfruit Enterprise Architecture Demonstration (BrEAD) that meets or exceeds the PBSS requirements	3.0	All	2.1 (all)

Requirements	Context	SOW Ref	SI PBSS Ref	Proposal Ref.
SOO Obj. #2	Obtain a Breadfruit Enterprise Architecture Demonstration (BrEAD) that is economical to operate and maintain throughout its life cycle and ensure life cycle considerations are maximized throughout the design and integration process	3.0	3.2.2.3 3.2.2.4 3.2.14	2.1.1.3
SOO Obj. #3	Obtain a Breadfruit Enterprise Architecture Demonstration (BrEAD) that is fully compliant with all applicable statutes and laws	3.0	3.6 (Adaptation)	2.1 (all)
SOO Obj. #4	Obtain a Breadfruit Enterprise Architecture Demonstration (BrEAD) design that is implementable with Commercial-off-the-Shelf technology to support the program milestones and schedule	3.0	2.2 3.2.6.1 3.2.11.2	2.2.2 (IMP) 2.2.3 (IMS)
SOO Obj. #5	Obtain a modular open systems architecture for hardware and software that supports efficient system upgrades and allows for interoperability with other systems. Obtain a design that maximizes the use of Commercial-Off-The-Shelf (COTS), Modified-COTS (MCOTS), NGRF-Off-The-Shelf (GOTS) items and Non-Developmental-Item	3.0	3.6 and Several	2.1.3

Requirements	Context	SOW Ref	SI PBSS Ref	Proposal Ref.
SOO Obj. #6	Achieve effective integrated NGRF and Contractor product teams to execute the program. Integrated Product Teams (IPTs) will be the primary means used to gain insight and provide for timely exchange of information and data	3.0	N/A	2.2.4 (SEMP)
SOO Obj. #7	Employ a program management system based on the Integrated Master Plan/Integrated Master Schedule (IMP/IMS) that is able to provide accurate weekly cost, schedule and performance information to support Milestone C, IOC and throughout the life cycle of the program	3.0	N/A	2.2.2 (IMP) 2.2.3 (IMS)
SOO Obj. #8	Employ an integrated NGRF/Contractor risk management system that actively identifies program risks and identifies mitigation steps for program risks	3.0	N/A	2.2.5
SOO Obj. #9	Adequately protect Critical Program Information (CPI) throughout the program's lifecycle	3.0	3.2.14 (Central Data Center)	2.2.4 (SEMP)

Requirements	Context	SOW Ref	SI PBSS Ref	Proposal Ref.
SOO Obj. #10	Employ an integrated, comprehensive Configuration Management (CM) system that maintains consistency of the Breadfruit Enterprise Architecture Demonstration (BrEAD) performance, compliance of functional and physical attributes with requirements, tracking of design and operational information and optimizes System Operational Effectiveness (SOE) throughout the life of the system	3.0	3.2.14 (Central Data Center)	2.2.4 (SEMP)
SOO Obj. #11	Minimize the amount of NGRF Furnished items (NGRF Furnished Information, Equipment, Property)	3.0	N/A	N/A
SOO Obj. #12	Employ a comprehensive Systems Engineering Process In Accordance With (IAW), the INCOSE Handbook	3.0	N/A	2.2.4 (SEMP)
SOO Obj. #13	Employ an Architecture that ensures timely, external data exchanges from the NGRF Command Center Data Processing Equipments and Software (CDEPS) to a wide Community Of Interest (COI)	3.0	3.2.14 (Central Data Center) 3.3.2 (External Interfaces)	2.1.1.2
SOO Obj. #14	Support applicable COI forums promulgated via joint NGRF/Contractor IPT	3.0	3.3.2 (External Interfaces)	2.2.4 (SEMP)

Requirements	Context	SOW Ref	SI PBSS Ref	Proposal Ref.
SOO Obj. #15	Employ a Breadfruit Enterprise Architecture Demonstration (BrEAD) Modeling and Simulation (M&S) strategy, which includes an integrated environment of accredited M&S tools, simulation and stimulation environment, data and resources that enables collaboration across functional disciplines	3.0	3.2.15 (SIL)	2.2.6.2
SOO Obj. #16	Employ a Breadfruit Enterprise Architecture Demonstration (BrEAD) Human System Integration strategy to efficiently and safely operate, maintain and support a Breadfruit Enterprise Architecture Demonstration (BrEAD)	3.0	3.2.15 (SIL)	2.2.6.2
SOO Obj. #17	Obtain a Breadfruit Enterprise Architecture Demonstration (BrEAD) design that incorporates sound human engineering design to optimize the Breadfruit Enterprise Architecture Demonstration (BrEAD) and maintenance user interfaces considering workload, manpower, mission and task effectiveness, integration, information management, situational awareness and data fusion	3.0	3.2.1 (Enterprise Architecture) 3.2.2.1 (Breadfruit Processing)	2.1.1.3

Requirements	Context	SOW Ref	SI PBSS Ref	Proposal Ref.
SOO Obj. #18	Employs a design maximizing the use of automation and decision aids for operators and maintainers to reduce workload, enhance coherence and information management and safety.	3.0	3.2.1 (Enterprise Architecture) 3.2.2.1 (Breadfruit Processing)	2.1.1.3
SOO Obj. #19	Develop a system that is safe to operate, maintain, and transport and does not introduce hazards	3.0	3.7 (Safety)	2.1.1.3
SOO Obj. #20	Design and deliver a NGRF Breadfruit Enterprise Architecture Demonstration (BrEAD) System Integration Laboratory (SIL) that will be integrated with other NGRF research sites, and are able to simulate and stimulate the Breadfruit Enterprise Architecture Demonstration (BrEAD); and support subsystem/system test, integration, Validation & Verification (V&V), and system/software lifecycle sustainability	3.0	3.2.15 (SIL)	2.2.6

Requirements	Context	SOW Ref	SI PBSS Ref	Proposal Ref.
SOO Obj. #21	Provide an integrated NGRF and Contractor test program, which demonstrates/validates the completeness of the design and development process, ensures design issues have been identified and solutions are available, and ensures compliance with Breadfruit Enterprise Architecture Demonstration (BrEAD) PBSS	3.0	Verification matrix	2.2.6
SOO Obj. #22	Develop Engineering Development Models (EDMs) to support developmental testing and the operational assessment of the Breadfruit Enterprise Architecture Demonstration (BrEAD)	3.0	3.2.15 (SIL)	2.2.6.2
SOO Obj. #23	Obtain a performance-based product support capability that provides a maintenance support concept based upon an effective Integrated Maintenance Concept	3.0	3.2.13 (Regional), 3.2.14 (Central)	2.1.1.1.2

Requirements	Context	SOW Ref	SI PBSS Ref	Proposal Ref.
SOO Obj. #24	Achieve an effective communication strategy that facilitates dialogue, ensures properly coordinated reviews, and audits, as well as the ability to assess progress towards meeting program requirements. At a minimum, the Contractor and appropriate subcontractors will support the following reviews, assessments and working groups and ensure appropriate entry and exit criteria are met	3.0	N/A	2.2.4 (SEMP)
SOO Obj. #25	Obtain a System that is environmentally compliant with all applicable environmental statutes and laws	3.0	3.2.10	2.1.1.3
SOO Obj. #26	Obtain a design which maximizes the use of standard manufacturing concepts and industrial approved commercial best practices;	3.0	3.2	2.2.4 (SEMP)
SOO 3.1.2	Open Systems Architecture	3.1.2	3.2.14	2.1.3
SOO 3.1.3	Program Management	3.1.3	3.2.14	2.2.4 (SEMP)
SOO 3.1.3.1	Program Management Structure & System	3.1.3.1	N/A	2.2.4 (SEMP)
SOO 3.1.3.2.1	EVMS	3.1.3.2.1	3.2.14	
SOO 3.1.3.2.2	Cost and Software Data Reporting	3.1.3.2.2	3.2.14	
SOO 3.1.3.2.3	Contract Funds Status Reports	3.1.3.2.3	3.2.14	
SOO 3.1.3.2.4	Integrated Baseline Review	3.1.3.2.4	N/A	2.2.2 (IMP) 2.2.3 (IMS)

Requirements	Context	SOW Ref	SI PBSS Ref	Proposal Ref.
SOO 3.1.3.3	Risk Management	3.1.3.3	N/A	2.2.5
SOO 3.1.3.4	Data Management	3.1.3.4	3.2.14	2.2.4 (SEMP)
SOO 3.1.3.4.1	Integrated Digital Environment (IDE)	3.1.3.4.1	3.2.14	2.2.4 (SEMP)
SOO 3.1.3.5	Configuration Management	3.1.3.5	3.2.14	2.2.4 (SEMP)
SOO 3.1.3.6	Specifications and Standards	3.1.3.6	3.2.14	2.2.4 (SEMP)
SOO 3.1.3.7	Systems Operation Effectiveness	3.1.3.7	3.2.14	2.2.4 (SEMP)
SOO 3.1.3.8	Quality Management System	3.1.3.8	3.2.14	2.2.4 (SEMP)
SOO 3.1.4.1	Systems Engineering	3.1.4.1	3.2.14	2.2.4 (SEMP)
SOO 3.1.4.2	Modeling and Simulation (M&S)	3.1.4.2	3.2.15 (SIL)	2.2.6.2
SOO 3.1.4.3	Human Factors Engineering	3.1.4.3	3.2.15 (SIL)	2.2.6.2
SOO 3.1.4.4	Software	3.1.4.4	3.10.3 3.2.13, 3.2.14, 3.2.15 (SIL)	2.2.6.2, 2.2.8
SOO 3.1.4.5	System Safety	3.1.4.5	3.2.14	2.2.4 (SEMP)
SOO 3.1.4.5.1	Software Safety	3.1.4.5.1	3.2.14	2.2.4 (SEMP)
SOO 3.1.4.5.2	Critical Safety Items	3.1.4.5.2	3.2.14	2.2.4 (SEMP)
SOO 3.1.4.6	System Integration Laboratory (SIL)	3.1.4.6	3.2.15 (SIL)	2.2.6.2
SOO 3.1.4.7	DODAF	3.1.4.7	N/A	2.1.1
SOO 3.1.4.8	Design Decision Documentation	3.1.4.8	3.2.15 (SIL)	2.2.6.2
SOO 3.1.5	Test & Evaluation	3.1.5	3.2.15 (SIL)	2.2.6.2
SOO 3.1.5.1	Problem Reporting	3.1.5.1	3.2.13	2.1.1.1.2
SOO 3.1.6	Supportability	3.1.6	3.2.13	2.1.1.1.2
SOO 3.1.6.1	Performance Based Logistics Planning	3.1.6.1	3.2.13	2.1.1.1.2
SOO 3.1.6.2	Maintenance Planning	3.1.6.2	3.2.13	2.1.1.1.2
SOO 3.1.6.2.1	Integrated Maintenance Concept (IMC)	3.1.6.2.1	3.2.13	2.1.1.1.2
SOO 3.1.6.3	Technical Manuals/Technical Data	3.1.6.3	3.2.13	2.1.1.1.2

Requirements	Context	SOW Ref	SI PBSS Ref	Proposal Ref.
SOO 3.1.6.4	RM&S	3.1.6.4	3.2.15 (SIL)	2.2.6.2
SOO 3.1.6.5	Unique Identification	3.1.6.5	3.2.15 (SIL)	2.2.6.2
SOO 3.1.6.6	Support Equipment	3.1.6.6	3.2.13	2.1.1.1.2
SOO 3.1.7	Training	3.1.7	3.2.13	2.1.1.1.2
SOO 3.1.7.1	Objective	3.1.7.1	3.2.13	2.1.1.1.2
SOO 3.1.8	ESOH	3.1.8	3.2.14	2.2.4 (SEMP)
SOO 3.1.9	Transition to Production	3.1.9	3.2.14	2.2.4 (SEMP)
SOO 3.1.10	Materials & Processes	3.1.10	3.2.15 (SIL)	2.2.4 (SEMP)
SOO 3.1.10.1	Materials Producibility	3.1.10.1	3.2.14	2.2.4 (SEMP)
SOO 3.1.10.2	Material Repairability	3.1.10.2	3.2.15 (SIL)	2.2.4 (SEMP)
SOO 3.1.11	Manufacturing	3.1.11	3.2.14	2.2.4 (SEMP)
SOO 3.1.11.1	Production Quality and Manufacturing Efficiency.	3.1.11.1	3.2.14	2.2.4 (SEMP)

Acronym List

Acronym	Description
ANSI	American National Standards Institute
BLOS	Beyond Line of Sight
BrEAD	Bread Enterprise Architecture Demonstration
CAD	Computer-aided Design
CAIV	Cost As an Independent Variable
CDEPS	Command Center Data Processing Equipments and Software (CDEPS)
CDR	Critical Design Review
CDRL	Contract Data Requirements List
CMO	Configuration Management Officer
CMP	Configuration Management Plan
COTS	Commercial Off the Shelf
CPI	Critical Program Information
CPR	Contract Performance Requirements
CRB	Configuration Review Board
CWBS	Contactor Work Breakdown Structure
DMP	Data Management Plan
DoD	Department of Defense
DoDAF	Department of Defense Architecture Framework
EDM	Engineering Development Model
EIA	Electronic Industries Alliance
EVMS	Earned Value Management System
ESOH	Environmental, Safety, Occupational Health
FMECA	Failure Modes Effects Criticality Analysis
FTA	Fault Tree Analysis
GOTS	NGRF Commercial Off the Shelf
GB	Ground Benign
GPS	Global Positioning System
HEPP	Human Engineering Program Plan
HETP	Human Engineering Test Plan
HFE	Human Factors Engineering
HIL	Human Interface Library
HMMP	Hazardous Material Management Plan
HSI	Human Systems Integration

HISWG	Human Systems Integration Working Group
IAW	In Accordance With
IBR	***Integrated Baseline Review***
IDE	***Integrated Development Environment***
IEEE	***Institute of Electrical and Electronics Engineers***
IOC	Initial Operating Capability
IMI	Interactive Multi-media Instructions
IMC	Integrated Maintenance Concept
IMP	Integrated Master Plan
IMS	Integrated Master Schedule
IPT	Integrated Product Team
ISD	Instructional Systems Development
IT	Integrated Test
ITP	Integrated Test Plan
LDPES	Local Data Processing Equipment and Software
LMI	Logistics Management Information
LOS	Line of Sight
MCOTS	Modified Commercial Off the Shelf
MOE	Measure of Effectiveness
MOS	Measure of Suitability
M & S	Modeling and Simulation
MTBF	Mean Time Between Failure
MTTR	Mean Time To Repair
NGRF	New Green Revolution Foundation
OSA	Open Systems Architecture
OTRR	Operational Test and Readiness Review
PBSS	Performance Based System Specifications
PBL	***Performance Based Logistics***
PDR	***Preliminary Design Review***
PDF	***Portable Document Format***
PMO	Program Management Office
PMR	Program Management Review
PRR	Production Readiness Review
QMP	Quality Management Plan
QMS	Quality Management System
RFP	Request for Proposal

SCR	System Change Request
SDD	System Development Demonstration
SDP	Software Development Plan
SEMP	Software Engineering Management Plan
SEP	Support Equipment Plan
SIL	Software Integration Lab
SME	Subject Matter Expert
SOE	System Operational Effectiveness
SOS	System of Systems
SOW	Statement Of Work
SRA	Scheduled Risk Assessment
SRR	System Requirement Review
SVR	System Verification Review
SWSP	Software Safety Plan
TEMP	Test and Evaluation Master Plan
TBD	To Be Determined
TPM	Technical Performance Measure
TRR	Test Readiness Review
UID	Unique Identification
V & V	Validation & Verification

System Development and Demonstration of the

Breadfruit Enterprise Demonstration (BrEAD) and Systems Design

Solutions International Proposal for the

Breadfruit Enterprise Demonstration and System Design

Volume 2: Technical

Technical Attachment I: Performance-Based System Specification (PBSS) and Verification Matrix

RFP NO: E00010-07-R0001

CDRL: 0008

Copy 1 of 2

May 4, 2009

Prepared For:

New Green Revolution Foundation (NGRF)

Prepared By:

Solutions International

123 Pillsbury Drive

St. Paul, MN 55114

CAGE CODE SI4ME

TABLE OF CONTENTS

APPENDICES

1. Scope

This specification identifies the design, performance, and verification requirements of the Breadfruit Enterprise Architecture Demonstration (BrEAD) and Systems Design Phase.

Identification.

This performance based specification applies to all components and subsystems included in the Solutions International BrEAD and breadfruit processing system including a weather station, data communication system, Systems Integration Laboratory (SIL), regional and central data centers as described herein.

System overview.

The design approach of the Breadfruit Enterprise Architecture Demonstration (BrEAD) is to use the preliminary research work and prototypes to design an intuitive, reliable, modular and scalable breadfruit processing system that will have the ability to collect, temporarily store and transfer local environment and operation data to regional marketing research centers and to the New Green Revolution Foundation (NGRF) headquarter in St. Paul Minnesota. Moreover, the breadfruit processing system's main functions will be to perform functions such as peeling, coring, shredding, drying and grinding.

1.3 Document overview.

This document contains the performance based specification for use by Solutions International and its customers for the development and operational use of the BrEAD Enterprise Architecture and breadfruit processing system. The use of this document is limited to users identified in the contract.

Referenced documents.

Government Documents (Specifications, Standards, and Handbooks)

DoD Directive 5000.59, "DoD Modeling and Simulation (M&S) Management," January 4, 1994

MIL-HDBK-217F - Reliability Prediction of Electronic Equipment, December 2, 1998

Reliability Toolkit - The Rome Laboratory Reliability Engineer's Toolkit, April 1993

MIL-HDBK-472 Notice 1- Military Standardization Handbook Maintainability Predictions 12 January 1984

Commercial Standards

UNIDO (United Nations Industrial Development Organization) Technology Manual, Small-scale Fruit and Vegetable Processing and Products - Production methods, equipment and quality assurance practices.

ANSI/ISO/ASQ Q9001-2000: Quality Management Systems:Requirements

Institute of Electrical and Electronics Engineers (IEEE)/Electronic Industries Association (EIA) 12207 Series, IEEE and EIA, January 2008

IEEE/EIA Std. 12207.1 - Guide for Information Technology Software life cycle processes

Capability Maturity Model Integration (CMMI) Version 1.2

Underwriters Laboratory Standards for Safety

CE standards and British Standards such as;

> 2006/95/EC - Basic and safety principles for man-machine interface, marking and identification

98/37/EC - Portable grinding machines and Fans - Mechanical safety

EN 415-3:1999 Safety of packaging machines - Part 3: Form, fill and seal machines.

EN 454:2000 Food processing machinery - Planetary mixers - Safety and hygiene requirements.

Solutions International Documents

MAXSIM toolset

Requirements.

Required states and modes.

Data located at the Central Data Center will be available 24 hours a day.

The breadfruit processing system will be operational only during breadfruit processing harvesting seasons.

System capability requirements.

The BrEAD architecture shall employ a design maximizing the use of automation and decision aids for operators and maintainers to reduce workload, enhance coherence and information management and safety. The design shall maximize the use of standard manufacturing concepts and industrial approved commercial best practices.

Enterprise Architecture

[ID 001] The BrEAD shall be comprised of the Breadfruit Enterprise Architecture Demonstration (BrEAD) that forms the organizational structure that will: be an advocate for breadfruit as a component of the next green revolution, design, market, collect data, distribute and provide logistics support for breadfruit processing systems and the System Design (SD) for breadfruit processing equipment.

[ID 002] The Breadfruit Enterprise Architecture Demonstration (BrEAD) shall provide an automated research, marketing, and user data collection capability for

NGRF head quarters in St. Paul, Minnesota that will provide 24 hour per day data from researchers, marketers and users world wide.

[ID 003] The BrEAD shall provide organizational liaison to the Community of Interest (COI) and automated electronic connectivity for real time data interchange. The community of interest is comprised of the NGRF, government and private organizations that have or could have an interest in breadfruit products.

[ID 004] The BrEAD shall provide weather data gathered through out the world where breadfruit is used. It shall provide hourly temperature, pressure, humidity, and rain fall data to the Central Data Collection and Analysis Center of the NGRF located in St. Paul, Minnesota.

[ID 005] The BrEAD architecture shall have distributed local weather sensors that are interoperable with the Central Data Collection and Analysis Center.

[ID 006] The BrEAD architecture shall have distributed data collection systems that are interoperable with the Central Data Collection and Analysis Center.

[ID 007] The BrEAD architecture maintenance user interfaces shall make consider workload, manpower, mission and task effectiveness, integration, information management, situational awareness and data fusion in optimization of the design.

Breadfruit Processing System

[ID 008] The initial baseline breadfruit systems design shall be based on the characterization of any prototype existing breadfruit processing systems.

Breadfruit Processing System - Operational Design

[ID 009] The breadfruit processing system shall be operable in geographic regions that do not have an electrical power infrastructure.

[ID 010] The breadfruit processing system shall be operable in geographic regions that have an electrical power infrastructure with minimal equipment differences.

[ID 011] The design of the breadfruit processing system shall be modular and portable, such that any specific geographic requirements can be accommodated without breadfruit processing system redesign.

[ID 012] The breadfruit processing system human machine interface shall be intuitive; that is the human operation of the system shall require minimal documentation- ideally none.

[ID 013] The systems design of the breadfruit processing system shall be capable of accepting design modifications based on operational data gathered from users in the field.

[ID 014] Any design modifications that are indicated from operational data gathered from users of breadfruit processing systems in use in the field shall be analyzed and considered for incorporation as design modifications to future breadfruit processing systems.

[ID 015] The breadfruit processing system shall be capable of operating up to 100% humidity

[ID 016] The breadfruit processing system shall operate from 40 degrees Fahrenheit to 120 degrees Fahrenheit.

[ID 017] The breadfruit processing system shall be ruggedized such that it is capable of being assembled and fully operational after transport by air, rail, ship, and local delivery trucks over rough local roads.

[ID 018] The breadfruit processing system shall be "green" and as such shall make maximum use of sustainable energy sources, shall not generate carbon emissions, and shall not use pesticides.

[ID 019] The breadfruit processing system equipment design shall be allow for daily cleaning during non-operational time.

[ID 020] The breadfruit processing system shall employ a design maximizing the use of automation and decision aids for operators and maintainers to reduce workload, and enhance coherence and information management and safety.

[ID 021] The breadfruit processing system shall make use of reused and common components as much as feasible.

Breadfruit Processing System - Performance

[ID 022] The breadfruit processing system shall extendible such that it has the capability to produce more than one food product or food product component.

[ID 023] The breadfruit processing system shall be scalable such that it can produce a minimum of 290 kg per day of breadfruit products for family use to TBD kg per day for industrial scale.

[ID 024] The breadfruit processing system shall be assembled to an operational state within 3 days after shipment to its operating location.

Breadfruit Processing System - Reliability

[ID 025] The breadfruit processing system shall have a Mean Time Between Failure (MTBF) of no less than 1,400 hours.

Breadfruit Processing System - Maintainability

[ID 026] The breadfruit processing system shall be repairable in the field with a maximum of 2 tools.

[ID 027] The mean time to repair (MTTR) of the breadfruit processing system shall be no greater than 15 minutes.

[ID 028] Spare parts must be provided for any part or subsystem that cannot be repaired at the bread processing site.

[ID 029] Daily routine maintenance procedures shall be identified and made available at each breadfruit processing site.

[ID 030] Daily routine maintenance procedures shall consist of a minimum of the following procedures; sharpen blades, visually inspect equipment, oil gears, and refresh / replace plastic covering.

Breadfruit Processing System - Interfaces

[ID 031] The breadfruit processing system shall allow data transfer with the BrEAD data communications system.

Breadfruit Intake Subsystem

Breadfruit Intake Subsystem - Operational Design

[ID 032] The Breadfruit Intake subsystem shall not cause bruising to the received breadfruit.

[ID 033] The Breadfruit Intake subsystem shall protect the breadfruit from insects and rodents.

[ID 034] The Breadfruit Intake subsystem shall allow for visual quality inspection of individual breadfruit.

[ID 035] Visual guides and procedures shall be provided for the Breadfruit Intake subsystem.

Breadfruit Intake Subsystem - Performance

[ID 036] The Breadfruit Intake subsystem shall collect and temporarily store no less than 450 breadfruit or 900 kg of breadfruit at any time for a maximum of 24 hours.

[ID 037] The Breadfruit Intake subsystem shall allow draining of fluid from sliced breadfruit.

Breadfruit Intake Subsystem - Reliability

Not applicable.

Breadfruit Intake Subsystem Interfaces

[ID 038] The Breadfruit Intake subsystem shall allow easy manual transfer of breadfruit to the peeling and coring subsystem.

[ID 039] The breadfruit intake subsystem shall interface with the breadfruit measurement subsystem to record relevant data at this processing step as listed in Appendix A.

Breadfruit Peeling and Coring Subsystem

Breadfruit Peeling and Coring Subsystem - Operational Design

[ID 040] The breadfruit peeling and coring subsystem shall be designed for safe and efficient manual peeling of coring of breadfruit.

[ID 041] The breadfruit peeling and coring subsystem shall accommodate all breadfruit types and sizes in the region of operation.

[ID 042] The breadfruit peeling and coring subsystem shall weigh no more than 4.5 kg.

[ID 043] The breadfruit peeling and coring subsystem shall be capable of being manually operated.

Breadfruit Peeling and Coring Subsystem - Performance

[ID 044] The breadfruit peeling and coring subsystem shall process breadfruit to achieve a 90% efficiency by weight of usable breadfruit product from unprocessed breadfruit.

[ID 045] The breadfruit peeling / coring system shall be capable of coring / peeling up to 30 breadfruit / hour.

[ID 046] The breadfruit peeling and coring operational performance data shall be collected at the frequency stated in Appendix A.

[ID 047] The Coring / Peeling subsystem shall have a Mean Time Between Failure (MTBF) of no less than 400,000 hours

Breadfruit Peeling and Coring Subsystem - Interfaces

[ID 048] Breadfruit waste (unusable product) created from the coring / peeling system shall be manually transferred to the waste storage system.

[ID 049] The peeling and coring subsystem shall interface with the breadfruit measurement subsystem to record relevant data at this processing step as listed in Appendix A.

Breadfruit Shredder Subsystem

Breadfruit Shredder Subsystem - Operational Design

[ID 050] The breadfruit shredder must produce shreds of breadfruit that are of optimal size and shape for the breadfruit grinder (shreds approximately 1/2"x 1/8" x 2-3").

[ID 051] The breadfruit shredder shall not weigh more than 5 kg.

[ID 052] The breadfruit shredder subsystem shall be capable of being manually operated.

Breadfruit Shredder Subsystem - Performance

[ID 053] The breadfruit shredder shall be capable of shredding no less than 60 kg of breadfruit per hour.

[ID 054] The breadfruit shredding process shall be at least 90% effective by weight, in converting core breadfruit in usable shreds.

[ID 055] Breadfruit shredding operational performance data shall be collected at the frequency stated in data collection Appendix A.

Breadfruit Shredder Subsystem - Reliability

[ID 056] Breadfruit shredder shall have a Mean Time Between Failure (MTBF) of no less than 75,000 hours.

Breadfruit Shredder Subsystem - Interfaces

[ID 057] The breadfruit shredding subsystem shall interface with the breadfruit measurement subsystem to record relevant data at this processing step as listed in Appendix A:

Breadfruit Dryer / Oven

Breadfruit Dryer / Oven - Operational Design

[ID 058] The dryer design shall use COTS thermostat for monitoring of internal temperatures.

[ID 059] The dryer shall not weigh more than **TBD kg.**

[ID 060] The dryer shall not be greater than **TBD m^3.**

[ID 061] The dryer shall be UL/CE certified.

[ID 062] Breadfruit drying operational performance data shall be collected at the frequency stated in data collection Appendix A.

[ID 063] The breadfruit dryer subsystem shall be capable of being manually operated.

[ID 064] The breadfruit dryer shall utilize solar energy for generation of power.

Breadfruit Dryer / Oven - Performance

[ID 065] The breadfruit dryer / oven shall remove moisture from the shredded breadfruit product to less than 15% by weight moisture content level within 6 hours.

[ID 066] The breadfruit dryer / oven shall accommodate drying at least 140 kg of shredded breadfruit at any time.

[ID 067] Breadfruit drying operational performance data shall be collected at the frequency stated in Appendix A.

[ID 068] The dryer fans shall produce airflow at no less than 47 cubic feet per minute (CFM).

[ID 069] The breadfruit dryer / oven shall limit the internal temperature to no more than 105 degrees F.

Breadfruit Dryer / Oven - Reliability

[ID 070] The breadfruit dryer subsystem shall have a Mean Time Between Failure (MTBF) of no less than 1500 hours.

Breadfruit Dryer / Oven - Interfaces

[ID 071] The dryer operating electronics shall be capable of operating on 12V, 3.1 Amp, 62W power from the solar panels.

[ID 072] The breadfruit dryer / oven subsystem shall interface with the breadfruit measurement subsystem to record relevant data at this processing step as listed in Appendix A.

-

Breadfruit Grinder

Breadfruit Grinder - Operational Design

[ID 073] Breadfruit grinding operational performance data shall be collected at the frequency stated in data collection Appendix A.

[ID 074] The breadfruit grinder subsystem shall be capable of being manually operated.

Breadfruit Grinder - Performance

[ID 075] The breadfruit grinder shall convert shredded breadfruit to flour at a rate of TBD kg /hr.

Breadfruit Grinder - Reliability

[ID 076] The breadfruit grinder shall have a Mean Time Between Failure (MTBF) of no less than 150,000 hours.

Breadfruit Grinder - Interfaces

[ID 077] The breadfruit grinder subsystem shall interface with the breadfruit measurement subsystem to record relevant data at this processing step as listed in Appendix A.

Breadfruit Packaging and Storage Subsystem

Breadfruit Packaging and Storage - Operational Design

[ID 078] The packaging and storage subsystem shall be capable of storing up to 300 kg of dried breadfruit in shredded or ground form.

[ID 079] The packaging and storage subsystem shall protect the breadfruit product from humidity.

[ID 080] The packaging and storage subsystem shall protect the breadfruit product from contamination from by insects or small rodents.

[ID 081] Breadfruit packaging operational performance data shall be collected no less than 1 data recording per hour of operation.

[ID 082] The breadfruit packaging subsystem shall be capable of being manually operated.

Breadfruit Packaging and Storage - Performance

[ID 083] The packaging subsystem shall be capable of packaging up to 50 kg of finished breadfruit product in individual containers.

Breadfruit Packaging - Reliability

Not applicable

Breadfruit Packaging Subsystem- Interfaces

[ID 084] The breadfruit packaging subsystem shall interface with the breadfruit measurement subsystem to record relevant data at this processing step as listed in Appendix A.

[ID 085] The interface between the grinder and packaging system shall be not lose more than 5% during transfer from the grinder to the packaging system (95% effective).

Breadfruit Waste Collection Subsystem

The breadfruit waste collection subsystem will help ensure that the System is environmentally compliant with all applicable environmental statutes and laws.

Breadfruit Waste Collection Subsystem - Operational Design

[ID 086] The breadfruit waste collection subsystem shall accommodate easy removal of waste product.

[ID 087] The breadfruit waste collection subsystem shall accept waste water from cleansing and processing operations.

[ID 088] The breadfruit waste collection subsystem shall accommodate easy cleaning as frequently as 1 time per day.

[ID 089] The breadfruit waste collection subsystem shall protect waste product from access by insects and small rodents.

Breadfruit Waste Collection Subsystem - Performance

[ID 090] The breadfruit waste collection subsystem shall accommodate up to 600 kg of waste at any time.

Breadfruit Waste Collection Subsystem - Reliability

Not applicable

Breadfruit Waste Collection Subsystem - Interfaces

[ID 091] The breadfruit waste collection subsystem shall accommodate breadfruit waste from the intake/draining, peeling / coring, shredding, drying, and grinding subsystems.

BrEAD Weather Station

BrEAD Weather Station - Operational Design

[ID 092] The BrEAD Weather station shall be capable of recording hourly temperature, pressure, humidity, and rain fall data shall be installed in regions that breadfruit processing systems are being used.

[ID 093] The BrEAD Weather station shall have self-sustaining power.

[ID 094] The BrEAD Weather station shall be capable of performing autonomous BIT at hourly intervals.

BrEAD Weather Station - Performance

[ID 095] The BrEAD temperature sensors shall be capable of measuring temperatures from -40 to 100 deg C with an accuracy of +/- 0.2 deg C.

[ID 096] The BrEAD pressure sensors shall be capable of measuring atmospheric pressures from 660 mbar to 1070 mbar with an accuracy of +/- 1.5mbar.

[ID 097] The BrEAD rain fall collection subsystem shall be capable of measuring rain fall amounts up to 10 cm/hr with an accuracy of +/- 1%.

[ID 098] BrEAD humidity sensors shall be capable of measuring humidity levels from 0 to 100% with an accuracy of +/- 1 %.

[ID 099] The BrEAD solar radiation sensors shall be capable of measuring from solar radiation from 0 to 1280 W /m2 with an accuracy of ±10.0 W/m^2.

BrEAD Weather Station - Reliability

[ID 100] The BrEAD Weather Station data logger shall have a Mean Time Between Failure (MTBF) of no less than 40,000 hours.

BrEAD Weather Station - Maintainability

[ID 101] The BrEAD weather station shall be repairable in the field with a maximum of 10 tools.

[ID 102] The mean time to repair (MTTR) of the BrEAD weather station shall be no greater than 15 minutes.

BrEAD Weather Station - Interfaces

[ID 103] Each BrEAD weather station shall be capable of transmitting hourly temperature, pressure, humidity, and rainfall data to the to the data communication system.

[ID 104] The BrEAD weather station shall be capable of autonomously transmitting BIT results to the data communication system.

Breadfruit Measurement System

Breadfruit Measurement System - Operational Design

[ID 105] The breadfruit measurement system shall have machine interface to collect data.

[ID 106] The breadfruit measurement system shall receive power from rechargeable batteries.

Breadfruit Measurement System - Performance

[ID 107] A clock shall be available to record the operating time of the drying stage.

[ID 108] A timer/stop watch shall be used to record operating time for coring/peeling, shredder, and grinder subsystems with an accuracy of +/- 1 seconds.

[ID 109] The Breadfruit weight scales shall be capable of measuring the weight of individual breadfruit with an accuracy of +/- 50 grams.

[ID 110] Temperature sensors shall be used to record the internal temperature of the breadfruit oven.

[ID 111] Humidity sensors shall be used to record the internal humidity of the breadfruit oven from 0 to 100% humidity with an accuracy of +/- 1 %.

[ID 112] Moisture sensors shall be used to record the moisture content of the breadfruit oven.

Breadfruit Measurement System - Reliability

[ID 113] The BrEAD Measurement System shall have a MTBF of no less than 1,700 hours.

Breadfruit Measurement System - Maintainability

[ID 114] The BrEAD Measurement System shall be repairable in the field with a maximum of 10 tools.

[ID 115] The mean time to repair (MTTR) of the BrEAD Measurement System shall be no greater than 15 minutes.

Breadfruit Measurement System - Interfaces

[ID 116] All measurements made for the breadfruit processing system shall be performed manually and entered manually in the data logger for transmission to the regional or central data center.

BrEAD Data Communication System

BrEAD Data Communication System - Operational Design

[ID 117] The BrEAD data communication system shall be able to transmit operational data to regional center via internet or satellite communications.

[ID 118] The design of the BrEAD data communication system shall allow for technology insertion such that any component can be replaced with minimum impact on the overall system.

[ID 119] The design of the BrEAD data communication system shall allow for reconfiguration of components or subsystems in the event of communication failures.

BrEAD Data Communication System - Performance

[ID 120] The data communication system shall have a communication range of 40 miles between individual communication cells with a maximum data rate of 115.2 kbps and a communication range of 15 miles while transmitting data at 1.5 Mbps using high gain antenna.

BrEAD Data Communication System - Reliability

[ID 121] The BrEAD Data Communication system shall have a MTBF of no less than 5,000 hours.

BrEAD Data Communication System - Maintainability

[ID 122] The BrEAD Data Communication system shall be repairable in the field with a maximum of 10 tools.

[ID 123] The mean time to repair (MTTR) of the BrEAD Data Communication system shall be no greater than 15 minutes.

BrEAD Data Communication System - Interfaces

[ID 124] The BrEAD Data Communication Subsystem shall be capable of transmitting data listed in PBSS Appendix A from all breadfruit processing sites and weather stations in the region.

[ID 125] The BrEAD Data Communication Subsystem shall transmit all data listed in PBSS Appendix B from the regional data center to the Central Data Collection and Analysis Center.

[ID 126] The data communication system shall provide communications from the Regional Center and breadfruit processing sites.

[ID 127] The data communication system shall provide communications between the SIL and the Central Data Center.

[ID 128] The data communication system shall provide communications from the Headquarters to Regional Center.

BrEAD Regional Resource and Logistics Center

Breadfruit processing systems in each region will communicate with, and be supported by Regional Centers. These centers will act as communication, support, and coordination nodes in the BrEAD enterprise architecture. Breadfruit harvesting is only possible for any given region during limited defined times throughout the year. The Regional Centers will therefore be fully operational only during those times when breadfruit processing systems are deployed and active. Rather than staffing the Regional Centers with dedicated BrEAD personnel all year round, local aid organizations will be tapped for part time personnel.

[ID 129] The initial BrEAD Regional Center shall be located in Port Au Prince, Haiti

[ID 130] The BrEAD Regional Center shall have operational resources to advocate and market breadfruit products as a component of the next green revolution (NGRF) on a regional level.

[ID 131] The BrEAD Regional Center shall have operational resources to provide administration services such as expense and income accounting, budget, payroll, and resource and asset management.

[ID 132] The BrEAD Regional Center shall have operational resources to provide training to processing station operators on the operation of the breadfruit processing equipment.

[ID 133] The BrEAD Regional Center shall have operational resources to provide training to data processing operators on the collection of breadfruit processing data.

[ID 134] The BrEAD Regional Center shall have operational resources for local troubleshooting of problems that may exist at the breadfruit processing sites.

[ID 135] The BrEAD Regional Center shall have operational resources to assist HQ in identifying/facilitating expansion of new processing stations.

[ID 136] The BrEAD Regional Center shall have operational resources to receive equipment and other logistics supply shipments from headquarters.

[ID 137] The BrEAD Regional Center shall have operational resources to transport equipment and other logistics supplies to other regional centers.

[ID 138] Personnel at BrEAD Regional Center shall be able to speak efficiently both in English and in the predominant language native to that region.

[ID 139] Personnel at BrEAD Regional Center shall be knowledgeable in the breadfruit processing equipment such that they can assist in the initial installation, repair, and regular maintenance.

BrEAD Regional Center - Operational Design

[ID 140] The BrEAD Regional Center operational layouts and communications networks shall be designed such that they are entirely compatible and achieve maximize commonality with other regional centers within the EA.

[ID 141] The BrEAD Regional Center operational layouts and communications networks shall be designed such that they can be easily expanded to accommodate increased utilization of regional resources.

[ID 142] When not in use at the processing sites, the BrEAD Regional Center shall have a secure shelter for storing breadfruit processing equipment, measurement systems, and data loggers.

[ID 143] During breadfruit processing season, the BrEAD Regional center shall have a secure shelter for storing breadfruit processing packaging material (bags, boxes, shipping supplies, etc), breadfruit processing spare parts, data processing spare parts (laptop battery, cables, etc), and environmentally -controlled storage of undistributed breadfruit product.

[ID 144] The design of all data processing equipment at the BrEAD Regional center shall be intuitive to the user.

BrEAD Regional Center - Performance

Not Applicable.

BrEAD Regional Center - Reliability

[ID 145] The BrEAD Regional Center shall have a Mean Time Between Failure (MTBF) of no less than 5,000 hours.

BrEAD Regional Center - Interfaces

[ID 146] The BrEAD Regional Center shall have internet and satellite communications resources to collect and store all data listed in Appendix A and continuously transmitted by all on-site data processing systems and weather stations within the region.

[ID 147] The BrEAD Regional Center shall be capable of processing and transmitting all regional data as listed in Appendix B to the Central Data Center.

[ID 148] The BrEAD regional center shall be capable of remote access from the headquarters.

[ID 149] The BrEAD regional center shall interpret BIT results data from weather stations and log the fault mode.

[ID 150] The BrEAD regional center shall respond to weather station faults or loss of communication within one day.

Central Data Center

[ID 151] The BrEAD Central Data Center shall be located at the NGRF facilities in St. Paul, Minnesota.

[ID 152] The BrEAD Central Data Center shall be operable all year including breadfruit harvest seasons.

[ID 153] The BrEAD Central Data Center shall have operational resources to provide administration services such as expense and income accounting, human resources, budget, payroll, and resource and asset management.

[ID 154] The BrEAD Central Data Center shall have operational resources to provide Information Technology services such as computer systems support, data and network security, website support, SW tools administration, and data processing.

[ID 155] The BrEAD Central Data Center shall have operational resources to provide SW services such as intranet communications, data management, configuration management, and ERP.

[ID 156] The BrEAD Central Data Center shall have operational resources (computers, software, network) to collect, analyze, and store breadfruit harvest and breadfruit processing info.

[ID 157] The BrEAD Central Data Center shall have operational resources to collect, store, and process breadfruit processing logistics (e.g. repair, reliability) information.

[ID 158] The BrEAD Central Data Center shall have operational resources to procure or manufacture new breadfruit processing systems.

[ID 159] The BrEAD Central Data Center shall have operational resources to assemble or disassemble breadfruit processing equipment in prepare for transportation to the regional centers.

[ID 160] The BrEAD Central Data Center shall have operational resources to coordinate the implementation and deployment of processing systems across the regional centers to minimize the number of processing systems required for efficient operations.

[ID 161] The BrEAD Central Data Center shall have operational resources to plan, organize, set-up, and formally establish new regional centers as the EA expands worldwide.

Central Data Center - Operational Design

As a minimum, the Central Data Center is expected to consist of:

a) One PC-based computer system to act as a data server
b) One laptop computer
c) One All-in-one Printer, Fax, Scanner, Copier
d) One Ethernet switch

[ID 162] The BrEAD Central Data Center operational layout and communications network shall be designed such that is entirely compatible with all regional center interfaces within the EA.

[ID 163] The BrEAD Central Data Center operational layout and communications network shall be designed such that they it can be easily expanded to accommodate increased utilization of central data center resources.

Central Data Center - Performance

[ID 164] The BrEAD Central Data Center shall accept all data transmitted from regional data processing systems (as listed in PBSS Appendix B).

[ID 165] The BrEAD Central Data Center shall have the capacity to store 600GB of breadfruit data.

Central Data Center - Reliability

[ID 166] The BrEAD Central Data Center shall have a Mean Time Between Failure (MTBF) of no less than 6,000 hours.

Central Data Center - Interfaces

[ID 167] The BrEAD Central Data Center shall have operational resources to collect and store data continuously transmitted from all regional data processing systems within the EA as listed in Appendix B.

[ID 168] The BrEAD Central Data Center shall have operational resources to collect and store data continuously transmitted from the System Integration Laboratory as listed in Appendix B.

[ID 169] The BrEAD Central Data Center shall have operational resources to continuously transmit Central Data Center data as listed in Appendix C to the regional centers and the System Integration Laboratory.

[ID 170] The BrEAD Central Data Center shall be capable of remote access with the regional center and system integration laboratory.

BrEAD Systems Integration Laboratory (SIL)

A System Integration Laboratory (SIL) shall be established that is capable of integrating with other NGRF research sites, simulate and stimulate the Breadfruit Enterprise Architecture Demonstration (BrEAD), support subsystem/system test, integration, Validation & Verification (V&V), and system/software lifecycle sustainability. It will also be used to develop Engineering Development Models

(EDMs) to support developmental testing and the operational assessment of the Breadfruit Enterprise Architecture Demonstration (BrEAD).

[ID 171] The BrEAD Systems Integration Laboratory (SIL) shall be used to evaluate candidate architectures for the BrEAD, the breadfruit processing systems, and the intuitive human interface capability desired.

[ID 172] The BrEAD Systems Integration Laboratory (SIL) shall be used to demonstrate feasibility of BF processing and Enterprise Architecture by establishing an initial System Proof-of-Concept.

Systems Integration Laboratory - Operational Design

[ID 173] The BrEAD SIL shall have operational resources to advocate breadfruit products to an "upscale" gluten-intolerant market in Hawaii.

[ID 174] The BrEAD SIL shall have operational resources to serve as a developmental test-bed for subsystem/system unit test, integration test, verification, and validation.

[ID 175] The BrEAD SIL shall have operational resources to produce drawings and CAD models of the breadfruit processing hardware.

[ID 176] The BrEAD SIL shall have operational resources to maintain configuration baselines of breadfruit processing hardware, models, and EA communication network throughout product lifecycle.

[ID 177] The BrEAD SIL shall have operational resources to simulate breadfruit processing and communication throughout the EA in order to develop/optimize methods.

[ID 178] The BrEAD SIL shall have operational resources to simulate the EA communication network in order to develop/optimize its functions.

[ID 179] The BrEAD SIL shall have operational resources to simulate mechanical parts life of the breadfruit processing system.

[ID 180] The BrEAD SIL shall have operational resources to conduct human factors studies.

[ID 181] The BrEAD SIL shall have operational resources to conduct assembly / disassembly trials of the breadfruit processing system.

[ID 182] The BrEAD SIL shall have operational resources to conduct transportability studies.

[ID 183] The BrEAD SIL shall have operational resources to conduct repairability and maintainability studies of the breadfruit processing system.

[ID 184] The BrEAD SIL shall have operational resources to conduct safety evaluations of the breadfruit processing system.

[ID 185] The BrEAD SIL shall have operational resources to conduct experiments and simulations as requested for world-wide research organizations.

[ID 186] The BrEAD SIL shall have operational resources to perform economic modeling and cultural studies for future processing system sites.

Systems Integration Laboratory - Performance

Not applicable

Systems Integration Laboratory - Reliability

Not applicable

Systems Integration Laboratory - Interfaces

[ID 187] The BrEAD SIL shall have operational resources to access the breadfruit database server at the Central Data Center in order to gather operational data from world-wide breadfruit processing sites.

[ID 188] The BrEAD SIL shall have operational resources to transfer (i.e. upload) SIL-generated information to the database server at the Central Data Center.

System External Interface requirements.

System External Interface identification and diagrams.

TBD

System External Interfaces

[ID 189] The BrEAD Regional Center shall have operational resources for liaison with Headquarters, NGRF, local authorities, and local aid organizations. (NGOs, NFPs, etc.)

[ID 190] The BrEAD Central Data Center shall have operational resources to advocate and market breadfruit products as a component of the next green revolution (NGRF) on a worldwide level.

[ID 191] The BrEAD Central Data Center shall have operational resources (i.e. call center) for liaison with regional centers, NGRF, breadfruit researchers, SIL operators, and government aid organizations.

[ID 192] The regional center shall be provided an electrical infrastructure by host country.

System internal interface requirements.

All internal interfaces have been defined elsewhere in this document.

System internal data requirements.

All internal data requirements have been defined in this document.

Adaptation requirements.

[ID 193] The breadfruit processing system shall comply with all local, regional, and national regulations **of Port Au Prince, Haiti and** all surrounding areas within **TBD** miles

[ID 194] The breadfruit processing system shall be operable under the electrical infrastructure of Port Au Prince, Haiti and all surrounding areas within **TBD** miles.

[ID 195] The breadfruit processing system shall comply with all local, regional, and national regulations of Maui, Hawaii and all surrounding areas within **TBD** miles

[ID 196] The EA shall comply with Modular Open Systems Assessment (MOSA) or an equivalent open systems architecture standard.

Safety requirements.

[ID 197] The breadfruit processing system shall be designed to be safe to operate, maintain, and transport and does not introduce hazards.

[ID 198] The breadfruit shredder shall have a protective shield to prevent direct contact with the shredder's moving parts during shredding operations.

[ID 199] The breadfruit grinder shall have a protective shield to prevent direct contact with the grinder's moving parts during grinding operations.

[ID 200] The breadfruit processing electrical components shall be comply with grounding and all other applicable UL and CE requirements.

[ID 201] Safety incidents reports shall be issued daily.

[ID 202] Safety reports shall include description, severity, and component causing injury.

Security and privacy requirements.

[ID 203] The BrEAD shall insure protection of BrEAD data, software, and computer systems by a system of virus protection, firewalls, and passwords.

System environment requirements.

Described elsewhere in this document.

Computer resource requirements.

Computer hardware requirements.

[ID 204] The on-site data logger / laptop shall be capable of operating in conditions up to 100% humidity.

[ID 205] The data transmitter shall be capable of operating in conditions up to 100% humidity.

[ID 206] The on-site data logger / laptop shall be capable of operating within temperature ranges between 40 deg F and 120 deg F. (4.4 and 49 degrees C).

[ID 207] The data transmitter shall be capable of operating within temperature ranges between 40 deg F and 120 deg F. (4.4 and 49 degrees C).

[ID 208] The on-site data logger / laptop shall have a minimum of **TBD** memory, **TBD** data storage, and **TBD** Mhz processing speed.

[ID 209] The data transmitter shall be capable of transmitting data at a rate of least 100 kbps per second.

[ID 210] The regional server shall have a minimum of 2 GB memory, expandable to 8 GB memory, 600GB data storage, and 2.4GHz processing speed.

[ID 211] The central data server shall have a minimum of 8 GB memory expandable to 16 GB memory, 1800 GB data storage, and 1.86GHz processing speed.

Computer hardware resource utilization requirements

[ID 212] The BrEAD data logger shall be capable of recording and storing all data listed in Appendix A.

[ID 213] The human machine interface for the BrEAD data logger shall be intuitive to the user.

[ID 214] The human machine interface for the BrEAD data logger shall be in the language native to the user.

[ID 215] User manuals shall be included with each data logger.

[ID 216] The Input/Output (I/O) interfaces used on the computer hardware components shall use open standard, industry accepted interfaces.

Computer software requirements.

[ID 217] The Breadfruit Enterprise Architecture Demonstration (BrEAD) software shall utilize a High Order Language (HOL), which follows ANSI, IEEE.

[ID 218] The BrEAD software system shall maximize the use of Commercial-Off-The-Shelf (COTS) software products.

[ID 219] The BrEAD shall have an on-line digital data environment that allows the program acquisition and operational support activities, throughout the program's life-cycle, to digitally create, store, access, manipulate, share, and exchange programmatic and technical data.

[ID 220] The BrEAD shall provide access to the BrEAD Integrated Digital Environment (IDE) for approximately 200 individuals at anytime from their workspaces.

[ID 221] The BrEAD shall employ a configuration management system that identifies, prioritizes, tracks, allocates and manages software requirement changes via a Change Review Board.

[ID 222] The BrEAD software shall prevent or mitigate single point failures that could disable the entire system.

[ID 223] Hazards that could result from operator errors or input of invalid data shall be mitigated by software monitoring and checks of inputs.

[ID 224] The BrEAD software shall perform appropriate safety checks to assure that dynamic updates of data do not cause data corruption.

[ID 225] Hazards that could result from operator errors or input of invalid data shall be mitigated by software monitoring and checks of inputs.

[ID 226] The BrEAD software shall perform appropriate safety checks to assure that dynamic updates of data do not cause data corruption.

[ID 227] A Field Data Database shall be created for storing and processing weather and breadfruit processing data (Field Data) from the processing sites.

[ID 228] Field data, consisting of weather data and breadfruit processing data shall be recorded and preserved in a database.

[ID 229] When Field Data is received via the internet connection, it shall be input into the Field Data Database by automated means.

[ID 230] The Field Data Database shall be available for receipt and input of data 24 hours a day seven days a week.

Computer communications requirements.

Described elsewhere in this document.

System quality factors.

[ID 231] The enterprise shall meet ISO9001 requirements.

Design and construction constraints.

Described elsewhere in this document.

Personnel-related requirements.

Described elsewhere in this document.

Training-related requirements.

Described elsewhere in this document.

Logistics-related requirements.

[ID 232] Spares for the breadfruit processing system shall be available within 2 hours.

Other requirements.

Described elsewhere in this document.

Packaging requirements

Described elsewhere in this document.

Precedence and criticality of requirements.

Not applicable.

Qualification provisions.

Qualification will be performed by operational test and first article inspection. See details in the Validation / Verification Matrix included as Appendix D.

Requirements traceability.

Not applicable to system level requirements.

Notes.

Appendices.

A. Appendix A Data Collected at Breadfruit Processing Site

Intake and Draining Subsystem	Units	Freq	Measurement Device
Harvester ID	ID #	daily	ID list
When harvested (off-tree life)	date/time	daily	clock/calender
Processing system location ID	ID #		data logger
Total # of Breadfruit collected from harvesters	# of units	daily	counter
Total # of breadfruit rejected (bruised, rotten)	# of units	daily	counter
Individual Breadfruit weight	kg	per breadfruit	weight scale
Type of breadfruit	N/A	per breadfruit	visual aids
Total weight before draining	kg	per breadfruit	weight scale
Total weight after draining	kg	per breadfruit	weight scale

Coring / Peeling System	Units	Freq	Measurement Device
Weight of breadfruit before coring / peeling process	kg	per bucket	weight scale and bucket
Weight of breadfruit after coring / peeling process	kg	per bucket	weight scale and bucket
Time to process 1 bucket of breadfruit	minutes	hourly	counter
# of operating hours per day	hours	daily	clock
Problem record ID	ID #	as reqd	data logger
Failed processing component	P/N	as reqd	data logger
Repairable / unrepairable	Y/N	as reqd	data logger
Time to repair	minutes	as reqd	timer

Dryer / Oven SubSystem	Units	Freq	Measurement Device
Weight of breadfruit before drying process	kg	start of drying	weight scale and tray
Weight of breadfruit after dryer process	kg	end of drying	weight scale and tray
Moisture content of breadfruit sample at start drying process	% water	start of drying	moisture sensor
Oven temperature at start of drying process	deg C	start of drying	thermometer
Internal air humidity level at start of drying process	% water	start of drying	air humidity sensor
Moisture content of breadfruit sample during process	% water	1 per hour	moisture sensor
Oven temperature during drying process	deg C	1 per hour	thermometer
Internal air humidity level during drying process	% water	1 per hour	air humidity sensor
Moisture content of breadfruit sample at end of drying process	% water	end of drying	moisture sensor
Oven temperature at the end during drying process	deg C	end of drying	thermometer
Internal air humidity level at the end of drying process	% water	end of drying	air humidity sensor
# of operating hours per day	hours	daily	clock
Problem record ID	ID #	as reqd	data logger
Failed processing component	P/N	as reqd	data logger
Repairable / unrepairable	Y/N	as reqd	data logger
Time to repair	minutes	as reqd	timer

Grinding SubSystem	Units	Freq	Measurement Device
Weight of breadfruit before grinding process	kg	per bucket	weight scale and bucket
Weight of breadfruit after grinding process	kg	per bucket	weight scale and bucket
Grinding time per process	minutes	per process	timer
# of operating hours per day	hours	daily	clock
Problem record ID	ID #	as reqd	data logger
Failed processing component	P/N	as reqd	data logger
Repairable / unrepairable	Y/N	as reqd	data logger
Time to repair	minutes	as reqd	timer

Packaging SubSystem	Units	Freq	Measurement Device
Weight of breadfruit before packaging process	kg	per bucket	weight scale and bucket
Weight of breadfruit after packaging process	kg	per package	weight scale and package
Packaging time per process	minutes	per process	timer
Package ID #	ID #	per package	labeler and data logger
# of operating hours per day	hours	daily	clock
Problem record ID	ID #	as reqd	data logger
Failed processing component	P/N	as reqd	data logger
Repairable / unrepairable	Y/N	as reqd	data logger
Time to repair	minutes	as reqd	timer

B. Data Processed at Breadfruit Regional and Central Data Site

Regional and Central Data Collection Description	Units	Frequency of calculaton / measurement	Evaluated in SIL
Average Breadfruit flour produced	kg	1/day	N
System failure rate	#	1/day	Y
Mean Time to repair	minutes	1/day	Y
Harvester Quality %	%	1/day	N
Average # of Tools required to repair	#	1/day	Y
Ave. Drain Time per breadfruit	minutes	1/day	Y
Ave. Weight / Breadfruit Drained	kg/unit	1/day	Y
Ave. Drain Waste Loss %	kg/unit	1/day	Y
Total # of Breadfruit types cored/peeled	#	1/day	N
Coring / Peeling reliability	failures / operating hour	1/day	Y
# of coring / peeling operators	#	1/day	N
Ave. Coring / peeling rate	kg/hour	1/day	Y
Coring / Peeling effectiveness	kg/kg %	1/day	Y
Average Shredding rate	kg/ hour	1/day	Y
Average Shredding Waste Loss %	kg/kg %	1/ day	Y
Shredding reliability	failures / operating hour	1/ day	Y
Average Drying time	hours	1/day	Y
Drying reliability	failures / operating hour	1/ day	Y
Average Moisture content loss %	kg/kg %	1/day	Y
Average Grinding rate	kg / hour	1/day	Y
Grinding reliability	failures / operating hour	1/ day	Y
Average Packaging rate	#/ day	1/ day	Y
Packaging reliability	failures / operating hour	1/ day	Y
Location of Weather stations	lat / long	as required	N
Atmospheric temperature /accuracy	deg C	1/hour	Y
Atmospheric pressure / accuracy	Pa	1/hour	Y
Atmospheric humidity qnty / accuracy	%	1/hour	Y
Rainfall qnty / accuracy	cm	1/day	N
Average # of operating hours - coring	hours	1/day	Y
Average # of operating hours - shredding	hours	1/day	Y
Average # of operating hours - drying	hours	1/day	Y
Average # of operating hours - grinding	hours	1/day	Y
Average # of operating hours - packaging	hours	1/day	Y
Measurement system reliability	failures / operating hour	1/ day	Y
Average On-site to Regional data transfer rate	kB / sec	each occurence	N
Average Regional to Central data transfer rate	kB / sec	each occurence	N
Regional data collection reliability	failures / operating hour	1/ day	N
Central data collection reliability	failures / operating hour	1/ day	N
SIL reliability	failures / operating hour	1/ day	Y
# of injuries	#	1/day	Y
Injury severity / description		1/day	Y
Equipment causing injury		1/day	Y

C. Data Processed at Breadfruit Central Data Site

See Appendix B

D. Verification and Validation Matrix

Requirement	PBSS Req ID	Ref #	SOO Ref #	Insp ect	Anal ysis	De mo	Test
The BrEAD shall be comprised of the Breadfruit Enterprise Architecture Demonstration (BrEAD) that forms the organizational structure that will: be an advocate for breadfruit as a component of the next green revolution, design, market, collect data, distribute and provide logistics support for breadfruit processing systems and the System Design (SD) for breadfruit processing equipment.	1	1	N/A	X			
The Breadfruit Enterprise Architecture Demonstration (BrEAD) shall provide an automated research, marketing, and user data collection capability for NGRF head quarters in St. Paul, Minnesota that will provide **24 hour per day** data from researchers, marketers and users world wide.	2	2	N/A	X			

Requirement	PBSS Req ID	Ref #	SOO Ref #	Insp ect	Anal ysis	De mo	Test
The BrEAD shall provide organizational liaison to the Community of Interest (COI) and automated electronic connectivity for real time data interchange. The community of interest is comprised of the NGRF, government and private organizations that have or could have an interest in breadfruit products.	3	3	N/A	X			
The BrEAD shall provide weather data gathered through out the world where breadfruit is used. It shall provide hourly temperature, pressure, humidity, and rain fall data to the Central Data Collection and Analysis Center of the NGRF located in St. Paul, Minnesota.	4	14	N/A			X	
The BrEAD architecture shall have distributed local weather sensors that are interoperable with the Central Data Collection and Analysis Center.	5	15	N/A			X	
The BrEAD architecture shall have distributed data collection systems that are interoperable with the Central Data Collection and Analysis Center.	6	15	N/A			X	

Requirement	PBSS Req ID	Ref #	SOO Ref #	Inspect	Analysis	Demo	Test
The BrEAD architecture maintenance user interfaces shall make use and consider workload, manpower, mission and task effectiveness, integration, information management, situational awareness and data fusion in optimization of the design.	7	N/A	17			X	
The initial baseline breadfruit systems design shall be based on the characterization of any prototype existing breadfruit processing systems.	8	4	N/A	X			
The breadfruit processing system shall be operable in geographic regions that do not have an electrical power infrastructure.	9	5	N/A				X
The breadfruit processing system shall be operable in geographic regions that have an electrical power infrastructure with minimal equipment differences.	10	5	N/A				X
The design of the breadfruit processing system shall be modular, such that any specific geographic requirements can be accommodated without breadfruit processing system redesign.	11	6	5	X			

Requirement	PBSS Req ID	Ref #	SOO Ref #	Insp ect	Anal ysis	De mo	Test
The breadfruit processing system human machine interface shall be intuitive; that is the human operation of the system shall require minimal documentation- ideally none.	12	7	N/A			X	
The systems design of the breadfruit processing system shall be capable of accepting design modifications based on operational data gathered from users in the field.	13	8	N/A			X	
Any design modifications that are indicated from operational data gathered from users of breadfruit processing systems in use in the field shall be analyzed and considered for incorporation as design modifications to future breadfruit processing systems.	14	8	N/A			X	
The breadfruit processing system shall be capable of operating up to 100% humidity	15	18	N/A				X
The breadfruit processing system shall operate from 40 degrees Fahrenheit to 120 degrees Fahrenheit.	16	16	N/A				X
The breadfruit processing system shall be ruggedized such that it is capable of being assembled and fully operational after transport by air, rail, ship, and local delivery trucks over rough local roads.	17	N/A	19			X	

Requirement	PBSS Req ID	Ref #	SOO Ref #	Insp ect	Anal ysis	De mo	Test
The breadfruit processing system shall be "green" and as such shall make maximum use of sustainable energy sources, shall not generate carbon emissions, and shall not use pesticides.	18	N/A	N/A	X			
The breadfruit processing system equipment design shall allow for daily cleaning during non-operational time.	19	N/A	N/A			X	
The breadfruit processing system shall employ a design maximizing the use of automation and decision aids for operators and maintainers to reduce workload, and enhance coherence and information management and safety.	20	N/A	18	X			
The breadfruit processing system shall make use of reused and common components as much as feasible.	21	N/A	N/A	X			
The breadfruit processing system shall extendible such that it has the capability to produce more than one (TBR) food product or food product component.	22	12	N/A			X	
The breadfruit processing system shall be scalable such that it can produce a minimum of 290 kg per day of breadfruit products for family use to TBD kg per day for industrial scale.	23	13	N/A		X		

Requirement	PBSS Req ID	Ref #	SOO Ref #	Inspect	Analysis	Demo	Test
The breadfruit processing system shall be assembled to an operational state within TBD days after shipment to its operating location.	24	N/A	N/A			X	
The breadfruit processing system shall have a Mean Time Before Failure (MTBF) of no less than 2,000 hours.	25	10	N/A		X		
The breadfruit processing system shall be repairable in the field with a maximum of 2 tools.	26	9	N/A			X	
The mean time to repair (MTTR) of the breadfruit processing system shall be no greater than 15 minutes.	27	11	N/A		X		
Spare parts must be provided for any part or subsystem that cannot be repaired at the bread processing site.	28	N/A	N/A	X			
Daily routine maintenance procedures shall be identified and made available at each breadfruit processing site.	29	N/A	N/A	X			
Daily routine maintenance procedures shall consist of a minimum of the following procedures; sharpen blades, visually inspect equipment, oil gears, and refresh / replace plastic covering.	30	N/A	N/A	X			
The breadfruit processing system shall allow data transfer with the BrEAD data communications system.	31	N/A	N/A				X

Requirement	PBSS Req ID	Ref #	SOO Ref #	Inspect	Analysis	Demo	Test
The Breadfruit Intake subsystem shall not cause bruising to the received breadfruit.	32	N/A	N/A				X
The Breadfruit Intake subsystem shall protect the breadfruit from insects and rodents.	33	N/A	N/A	X			
The Breadfruit Intake subsystem shall allow for visual quality inspection of individual breadfruit.	34	N/A	N/A	X			
Visual guides and procedures shall be provided for the Breadfruit Intake subsystem.	35	N/A	N/A	X			
The Breadfruit Intake subsystem shall collect and temporarily store no less than 450 breadfruit or 900 kg of breadfruit at any time for a maximum of 24 hours.	36	N/A	N/A		X		
The Breadfruit Intake subsystem shall allow draining of fluid from sliced breadfruit.	37	N/A	N/A			X	
The Breadfruit Intake subsystem shall allow easy manual transfer of breadfruit to the peeling and coring subsystem.	38	N/A	N/A	X			
The breadfruit intake subsystem shall interface with the breadfruit measurement subsystem to record relevant data at this processing step as listed in Appendix A.	39	N/A	N/A				X

Requirement	PBSS Req ID	Ref #	SOO Ref #	Insp ect	Anal ysis	De mo	Test
The breadfruit peeling and coring subsystem shall be designed for safe and efficient manual peeling of coring of breadfruit.	40	N/A	N/A	X			
The breadfruit peeling and coring subsystem shall accommodate all breadfruit types and sizes in the region of operation.	41	N/A	N/A			X	
The breadfruit peeling and coring subsystem shall weigh no more than 4.5 kg.	42	N/A	N/A	X			
The breadfruit peeling and coring subsystem shall be capable of being manually operated.	43	N/A	N/A	X			
The breadfruit peeling and coring subsystem shall process breadfruit to achieve a 90% efficiency by weight of usable breadfruit product from unprocessed breadfruit.	44	N/A	N/A				X
The breadfruit peeling / coring system shall be capable of coring / peeling 30 breadfruit / minute.	45	N/A	N/A				X
The breadfruit peeling and coring operational performance data shall be collected at the frequency stated in Appendix A.	46	N/A	N/A				X
The Coring / Peeling subsystem shall have a Mean Time Before Failure (MTBF) of no less than 900,000 hours	47	N/A	N/A		X		

Requirement	PBSS Req ID	Ref #	SOO Ref #	Insp ect	Anal ysis	De mo	Test
Breadfruit waste (unusable product) created from the coring / peeling system shall be manually transferred to the waste storage system.	48	N/A	N/A			X	
The peeling and coring subsystem shall interface with the breadfruit measurement subsystem to record relevant data at this processing step as listed in Appendix A	49	N/A	N/A			X	
The breadfruit shredder must produce shreds of breadfruit that are of optimal size and shape for the breadfruit grinder. (shreds approximately 1/2"x 1/8" x 2-3")	50	N/A	N/A				X
The shredder shall not weigh more than 5 kg.	51	N/A	N/A	X			
The breadfruit shredder subsystem shall be capable of being manually operated.	52	N/A	N/A			X	
The breadfruit shredder shall be capable of shredding up to 60 kg of breadfruit per hour.	53	N/A	N/A				X
The breadfruit shredding process shall be at least 90% effective, by weight, in converting core breadfruit in usable shreds.	54	N/A	N/A				X
Breadfruit shredding operational performance data shall be collected at the frequency stated in Appendix A	55	N/A	N/A				X

Requirement	PBSS Req ID	Ref #	SOO Ref #	Inspect	Analysis	Demo	Test
Breadfruit shredder shall have a Mean Time Before Failure (MTBF) of no less than 150,000 hours.	56	N/A	N/A		X		
The breadfruit shredding subsystem shall interface with the breadfruit measurement subsystem to record relevant data at this processing step as listed in Appendix A.	57	N/A	N/A				X
The dryer design shall use COTS thermostat for monitoring of internal temperatures.	58	N/A	N/A	X			
The dryer shall not weigh more than TBD kg.	59	N/A	N/A	X			
The dryer shall not be greater than TBD m^3.	60	N/A	N/A	X			
The dryer shall be UL/CE certified.	61	N/A	N/A	X			
Breadfruit drying operational performance data shall be collected at the frequency stated in data collection annex.	62	N/A	N/A				X
The breadfruit dryer subsystem shall be capable of being manually operated.	63	N/A	N/A			X	
The breadfruit dryer shall utilize solar energy for generation of power.	64	N/A	N/A	X			
The breadfruit dryer / oven shall remove moisture from the shredded breadfruit product to less than 15% by weight moisture content level within 6 hours.	65	N/A	N/A				X

Requirement	PBSS Req ID	Ref #	SOO Ref #	Insp ect	Anal ysis	De mo	Test
The breadfruit dryer / oven shall accommodate drying no less than 140 kg of shredded breadfruit at any time.	66	N/A	N/A				X
Breadfruit drying operational performance data shall be collected at the frequency stated in Appendix A.	67	N/A	N/A				X
The dryer fans shall produce airflow at no less than 47 cubic feet per minute (CFM).	68	N/A	N/A				X
The breadfruit dryer / oven shall limit the internal temperature to no more than 105 degrees F.	69	N/A	N/A				X
The breadfruit dryer subsystem shall have a Mean Time Before Failure (MTBF) of no less than 1500 hours.	70	N/A	N/A		X		
The dryer shall be capable of receiving 24V, 3.1 Amp, 62W power from the breadfruit power generation system.	71	N/A	N/A				X
The breadfruit dryer / oven subsystem shall interface with the breadfruit measurement subsystem to record relevant data at this processing step as listed in Appendix A	72	N/A	N/A				X
Breadfruit grinding operational performance data shall be collected at the frequency stated in Appendix A.	73	N/A	N/A				X

Requirement	PBSS Req ID	Ref #	SOO Ref #	Insp ect	Anal ysis	De mo	Test
The breadfruit grinder subsystem shall be capable of being manually operated.	74	N/A	N/A			X	
The breadfruit grinder shall convert shredded breadfruit to flour at a rate of TBD kg /hr.	75	N/A	N/A				X
The breadfruit grinder shall have a Mean Time Before Failure (MTBF) of no less than 150,000 hours.	76	N/A	N/A		X		
The breadfruit grinder subsystem shall interface with the breadfruit measurement subsystem to record relevant data at this processing step:	77	N/A	N/A				X
The packaging and storage subsystem shall be capable of storing up to 300 kg of dried breadfruit in shredded or ground form.	78	N/A	N/A		X		
The packaging and storage subsystem shall protect the breadfruit product from humidity.	79	N/A	N/A				X
The packaging and storage subsystem shall protect the breadfruit product from contamination from by insects or small rodents.	80	N/A	N/A	X			
Breadfruit packaging operational performance data shall be collected no less than 1 data recording per hour of operation.	81	N/A	N/A				X
The breadfruit packaging subsystem shall be capable of being manually operated.	82	N/A	N/A			X	

Requirement	PBSS Req ID	Ref #	SOO Ref #	Insp ect	Anal ysis	De mo	Test
The packaging subsystem shall be capable of packaging up to 50 kg of finished breadfruit product in individual containers.	83	N/A	N/A				X
The breadfruit packaging subsystem shall interface with the breadfruit measurement subsystem to record relevant data at this processing step:	84	N/A	N/A				X
The interface between the grinder and packaging system shall be not lose more than 5% during transfer from the grinder to the packaging system (95% effective).	85	N/A	N/A				X
The breadfruit waste collection subsystem shall accommodate easy removal of waste product.	86	N/A	N/A			X	
The breadfruit waste collection subsystem shall accept waste water from cleansing and processing operations.	87	N/A	N/A			X	
The breadfruit waste collection subsystem shall accommodate easy cleaning as frequently as 1 time per day.	88	N/A	N/A			X	
The breadfruit waste collection subsystem shall protect waste product from access by insects and small rodents.	89	N/A	N/A			X	
The breadfruit waste collection subsystem shall accommodate up to 600 kg of waste at any time.	90	N/A	N/A		X		

Requirement	PBSS Req ID	Ref #	SOO Ref #	Insp ect	Anal ysis	De mo	Test
The breadfruit waste collection subsystem shall accommodate waste from the peeling / coring, shredding, drying, and grinding subsystems.	**91**	**N/A**	**N/A**			**X**	
Weather sensors capable of recording hourly temperature, pressure, humidity, and rain fall data shall be installed in regions that breadfruit processing systems are being used.	**92**	**N/A**	**N/A**				**X**
The BrEAD Weather station shall have self-sustaining power.	**93**	**N/A**	**N/A**	**X**			
The BrEAD Weather station shall be capable of performing autonomous BIT at hourly intervals.	**94**	**N/A**	**N/A**				**X**
The BrEAD temperature sensors shall be capable of measuring temperatures from -40 to 100 deg C with an accuracy of +/- 0.2 deg C.	**95**	**N/A**	**N/A**				**X**
The BrEAD pressure sensors shall be capable of measuring atmospheric pressures from 660mb to 1070 mb with an accuracy of +/- 1.5mbar.	**96**	**15**	**N/A**				**X**
The BrEAD rain fall collection subsystem shall be capable of measuring rain fall amounts up to 10 cm/hr with an accuracy of +/- 1%.	**97**	**N/A**	**N/A**				**X**
BrEAD humidity sensors shall be capable of measuring humidity levels from 0 to 100% with an accuracy of +/- 1 (TBR) %.	**98**	**N/A**	**N/A**				**X**

Requirement	PBSS Req ID	Ref #	SOO Ref #	Insp ect	Anal ysis	De mo	Test
The BrEAD solar radiation sensors shall be capable of measuring from solar radiation from 0 to 1280 W /m2 with an accuracy of ±10.0 W/m2 .	**99**	**N/A**	**N/A**				**X**
The BrEAD Weather Station data logger shall have a Mean Time Before Failure (MTBF) of no less than 40,000 hours.	**100**	**N/A**	**N/A**		**X**		
The BrEAD weather station shall be repairable in the field with a maximum of 10 tools.	**101**	**N/A**	**N/A**			**X**	
The mean time to repair (MTTR) of the BrEAD weather station shall be no greater than 15 minutes.	**102**	**N/A**	**N/A**		**X**		
Each BrEAD weather station shall be capable of transmitting hourly temperature, pressure, humidity, and rainfall data to the to the data communication system.	**103**	**N/A**	**N/A**				**X**
The BrEAD weather station shall be capable of autonomously transmitting BIT results to the data communication system.	**104**	**N/A**	**N/A**				**X**
The breadfruit measurement system shall have machine interface to collect data.	**105**	**N/A**	**N/A**	**X**			
The breadfruit measurement system shall be receive power from rechargeable batteries.	**106**	**N/A**	**N/A**	**X**			
A clock shall be available to record the operating time of the drying stage.	**107**	**N/A**	**N/A**	**X**			

Requirement	PBSS Req ID	Ref #	SOO Ref #	Inspect	Analysis	Demo	Test
A timer/stop watch shall be used to record operating time for coring/peeling, shredder, and grinder subsystems with an accuracy of +/- 1 seconds.	**108**	N/A	N/A	X			
The Breadfruit weight scales shall be capable of measuring the weight of individual breadfruit with an accuracy of +/- 50 grams.	**109**	N/A	N/A			X	
Temperature sensors shall be used to record the internal temperature of the breadfruit oven.	**110**	N/A	N/A	X			
Humidity sensors shall be used to record the internal humidity of the breadfruit oven from 0 to 100% humidity with an accuracy of +/- 1 %.	**111**	N/A	N/A				X
Moisture sensors shall be used to record the moisture content of the breadfruit oven.	**112**	N/A	N/A	X			
The BrEAD Measuring System shall have a MTBF of no less than 30,000 hours.	**113**	N/A	N/A		X		
The BrEAD Measurement System shall be repairable in the field with a maximum of 10 tools.	**114**	N/A	N/A			X	
The mean time to repair (MTTR) of the BrEAD Measuring System shall be no greater than 15 minutes.	**115**	N/A	N/A		X		

Requirement	PBSS Req ID	Ref #	SOO Ref #	Insp ect	Anal ysis	De mo	Test
All measurements made for the breadfruit processing system shall be performed manually and entered manually in the data logger for transmission to the regional or central data center.	**116**	**N/A**	**N/A**			**X**	
The BrEAD data communication system shall be able to transmit operational data to regional center via internet or satellite communications.	**117**	**N/A**	**N/A**				**X**
The design of the BrEAD data communication system shall allow for technology insertion such that any component can be replaced with minimum impact on the overall system.	**118**	**N/A**	**N/A**			**X**	
The design of the BrEAD data communication system shall allow for reconfiguration of components or subsystems in the event of communication	**119**	**N/A**	**N/A**			**X**	
The data communication system shall have a communication range of 40 miles between individual communication cells with a maximum data rate of 115.2 kbps and a communication range of 15 miles while transmitting data at 1.5 Mbps using high gain antenna.	**120**	**N/A**	**N/A**				**X**

Requirement	PBSS Req ID	Ref #	SOO Ref #	Insp ect	Anal ysis	De mo	Test
The BrEAD Communication shall have a MTBF of no less than 5,000 hours.	**121**	**N/A**	**N/A**		**X**		
The BrEAD Data Communication system shall be repairable in the field with a maximum of 10 tools.	**122**	**N/A**	**N/A**			**X**	
The mean time to repair (MTTR) of the BrEAD Data Communication system shall be no greater than 15 minutes.	**123**	**N/A**	**N/A**		**X**		
The BrEAD Data Communication Subsystem shall be capable of transmitting data listed in PBSS Appendix A from all breadfruit processing sites and weather stations in the region.	**124**	**N/A**	**N/A**				**X**
The BrEAD Data Communication Subsystem shall transmit all data listed in PBSS Appendix B from the regional data center to the Central Data Collection and Analysis Center.	**125**	**N/A**	**N/A**				**X**
The data communication system shall provide communications from the Regional Center and breadfruit processing sites.	**126**	**N/A**	**N/A**		**X**		
The data communication system shall provide communications from the Headquarters to Regional Center.	**127**	**N/A**	**N/A**		**X**		

Requirement	PBSS Req ID	Ref #	SOO Ref #	Insp ect	Anal ysis	De mo	Test
The data communication system shall provide communications between the Headquarters and Regional Center.	**128**	N/A	N/A		X		
The initial BrEAD Regional Center shall be located in Port Au Prince, Haiti	**129**	N/A	N/A	X			
The BrEAD Regional Center shall have operational resources to advocate and market breadfruit products as a component of the next green revolution (NGRF) on a regional level.	**130**	N/A	N/A	X			
The BrEAD Regional Center shall have operational resources to provide administration services such as expense and income accounting, budget, payroll, and resource and asset management.	**131**	N/A	N/A	X			
The BrEAD Regional Center shall have operational resources to provide training to processing station operators on the operation of the breadfruit processing equipment.	**132**	1	N/A	X			
The BrEAD Regional Center shall have operational resources to provide training to data processing operators on the collection of breadfruit processing data.	**133**	N/A	N/A	X			

Requirement	PBSS Req ID	Ref #	SOO Ref #	Insp ect	Anal ysis	De mo	Test
The BrEAD Regional Center shall have operational resources for local troubleshooting of problems that may exist at the breadfruit processing sites.	134	N/A	N/A	X			
The BrEAD Regional Center shall have operational resources to assist HQ in identifying/facilitating expansion of new processing stations.	135	N/A	N/A	X			
The BrEAD Regional Center shall have operational resources to receive equipment and other logistics supply shipments from headquarters.	136	N/A	N/A	X			
The BrEAD Regional Center shall have operational resources to transport equipment and other logistics supplies to other regional centers.	137	N/A	N/A	X			
Personnel at BrEAD Regional Center shall be able to speak efficiently both in English and in the predominant language native to that region.	138	N/A	N/A			X	
Personnel at BrEAD Regional Center shall be knowledgeable in the breadfruit processing equipment such that they can assist in the initial installation, repair, and regular maintenance.	139	N/A	N/A			X	

Requirement	PBSS Req ID	Ref #	SOO Ref #	Insp ect	Anal ysis	De mo	Test
The BrEAD Regional Center operational layouts and communications networks shall be designed such that they are entirely compatible and achieve maximize commonality with other regional centers within the EA.	**140**	**N/A**	**N/A**		**X**		
The BrEAD Regional Center operational layouts and communications networks shall be designed such that they can be easily expanded to accommodate increased utilization of regional resources.	**141**	**N/A**	**N/A**		**X**		
When not in use at the processing sites, the BrEAD Regional Center shall have a secure shelter for storing breadfruit processing equipment, measurement systems, and data loggers.	**142**	**N/A**	**N/A**	**X**			
During breadfruit processing season, the BrEAD Regional center shall have a secure shelter for storing breadfruit processing packaging material (bags, boxes, shipping supplies, etc), breadfruit processing spare parts, data processing spare parts (laptop battery, cables, etc), and environmentally -controlled storage of undistributed breadfruit product.	**143**	**N/A**	**N/A**	**X**			

Requirement	PBSS Req ID	Ref #	SOO Ref #	Insp ect	Anal ysis	De mo	Test
The design of all data processing equipment at the BrEAD Regional center shall be intuitive to the user.	**144**	N/A	N/A	X			
The BrEAD Regional center shall have a Mean Time Before Failure (MTBF) of no less than 30,000 hours.	**145**	N/A	N/A		X		
The BrEAD Regional Center shall have internet and satellite communications resources to collect and store all data listed in Appendix A and continuously transmitted by all on-site data processing systems and weather stations within the region.	**146**	N/A	N/A				X
The BrEAD Regional Center shall be capable of processing and transmitting all regional data as listed in Appendix B to the Central Data Center.	**147**	N/A	N/A				X
The BrEAD regional center shall be capable of remote access from the headquarters.	**148**	N/A	N/A				X
The BrEAD regional center shall interpret BIT results data from weather stations and log the fault mode.	**149**	N/A	N/A				X
The BrEAD regional center shall respond to weather station faults or loss of communication within one day.	**150**	N/A	N/A				X

Requirement	PBSS Req ID	Ref #	SOO Ref #	Insp ect	Anal ysis	De mo	Test
The Central Data Center shall be located at the NGRF facilities in St. Paul, Minnesota.	**151**	**N/A**	**N/A**	**X**			
The Central Data Center shall be operable all year including breadfruit harvest seasons.	**152**	**N/A**	**N/A**	**X**			
The BrEAD Central Data Center shall have operational resources to provide administration services such as expense and income accounting, human resources, budget, payroll, and resource and asset management.	**153**	**1**	**N/A**	**X**			
The BrEAD Central Data Center shall have operational resources to provide Information Technology services such as computer systems support, data and network security, website support, SW tools administration, and data processing.	**154**	**3**	**N/A**	**X**			
The BrEAD Central Data Center shall have operational resources to provide SW services such as intranet communications, data management, configuration management, and ERP.	**155**	**N/A**	**10**	**X**			

Requirement	PBSS Req ID	Ref #	SOO Ref #	Insp ect	Anal ysis	De mo	Test
The BrEAD Central Data Center shall have operational resources (computers, software, network) to collect, analyze, and store breadfruit harvest and breadfruit processing info.	156	N/A	N/A	X			
The BrEAD Central Data Center shall have operational resources to collect, store, and process breadfruit processing logistics (e.g. repair, reliability) information.	157	N/A	N/A	X			
The BrEAD Central Data Center shall have operational resources to procure or manufacture new breadfruit processing systems.	158	N/A	N/A	X			
The BrEAD Central Data Center shall have operational resources to assemble or disassemble breadfruit processing equipment in prepare for transportation to the regional centers.	159	N/A	N/A	X			
The BrEAD Central Data Center shall have operational resources to coordinate the implementation and deployment of processing systems across the regional centers to minimize the number of processing systems required for efficient operations.	160	N/A	N/A	X			

Requirement	PBSS Req ID	Ref #	SOO Ref #	Inspect	Analysis	Demo	Test
The BrEAD Central Data Center shall have operational resources to plan, organize, set-up, and formally establish new regional centers as the EA expands worldwide.	**161**	**N/A**	**N/A**	**X**			
The BrEAD Central Data Center operational layout and communications network shall be designed such that is entirely compatible with all regional center interfaces within the EA.	**162**	**N/A**	**N/A**	**X**			
The BrEAD Central Data Center operational layout and communications network shall be designed such that they it can be easily expanded to accommodate increased utilization of central data center resources.	**163**	**N/A**	**N/A**	**X**			
BrEAD Central Data Center shall accept all data transmitted from regional data processing systems (as listed in PBSS Appendix B).	**164**	**N/A**	**N/A**				**X**
The BrEAD Central Data Center shall have the capacity to store 600GB of breadfruit data.	**165**	**N/A**	**N/A**				**X**
The BrEAD Central Data Center shall have a Mean Time Before Failure (MTBF) of no less than 6,000 hours.	**166**	**N/A**	**N/A**		**X**		

Requirement	PBSS Req ID	Ref #	SOO Ref #	Insp ect	Anal ysis	De mo	Test
The BrEAD Regional Center shall have operational resources to collect and store data continuous transmitted from all regional data processing systems within the EA as listed in Appendix B.	167	N/A	N/A				X
The BrEAD Regional Center shall have operational resources to collect and store data continuous transmitted from the System Integration Laboratory as listed in Appendix B.	168	N/A	N/A				X
The BrEAD Regional Center shall have operational resources to continuously transmit Central Data Center data as listed in Appendix C to the regional centers and the System Integration Laboratory.	169	N/A	N/A				X
The BrEAD Central Data Center shall be capable of remote access with the regional center and system integration laboratory.	170	N/A	N/A				X
The BrEAD Systems Integration Laboratory (SIL) shall be used to evaluate candidate architectures for the BrEAD, the breadfruit processing systems, and the intuitive human interface capability desired.	171	N/A	16				X

Requirement	PBSS Req ID	Ref #	SOO Ref #	Insp ect	Anal ysis	De mo	Test
The BrEAD Systems Integration Laboratory (SIL) shall be used to demonstrate feasibility of BF processing and Enterprise Architecture by establishing an initial System Proof-of-Concept.	**172**	**N/A**	**N/A**			**X**	
The BrEAD SIL shall have operational resources to advocate breadfruit products to an "upscale" gluten-intolerant market in Hawaii.	**173**	**N/A**	**N/A**	**X**			
The BrEAD SIL shall have operational resources to serve as a developmental test-bed for subsystem/system unit test, integration test, verification, and validation.	**174**	**N/A**	**N/A**	**X**			
The BrEAD SIL shall have operational resources to produce drawings and CAD models of the breadfruit processing hardware.	**175**	**N/A**	**15, 22**	**X**			
The BrEAD SIL shall have operational resources to maintain configuration baselines of breadfruit processing hardware, models, and EA communication network throughout product lifecycle.	**176**	**N/A**	**10**	**X**			
The BrEAD SIL shall have operational resources to simulate breadfruit processing and communication throughout the EA in order to develop/optimize methods.	**177**	**N/A**	**15**	**X**			

Requirement	PBSS Req ID	Ref #	SOO Ref #	Insp ect	Anal ysis	De mo	Test
The BrEAD SIL shall have operational resources to simulate the EA communication network in order to develop/optimize its functions.	178	N/A	N/A	X			
The BrEAD SIL shall have operational resources to simulate mechanical parts life of the breadfruit processing system.	179	N/A	N/A	X			
The BrEAD SIL shall have operational resources to conduct human factors studies.	180	N/A	16,1 7	X			
The BrEAD SIL shall have operational resources to conduct assembly / disassembly trials of the breadfruit processing system.	181	N/A	N/A	X			
The BrEAD SIL shall have operational resources to conduct transportability studies.	182	N/A	N/A	X			
The BrEAD SIL shall have operational resources to conduct repairability and maintainability studies of the breadfruit processing system.	183	N/A	N/A	X			
The BrEAD SIL shall have operational resources to conduct safety evaluations of the breadfruit processing system.	184	N/A	N/A	X			
The BrEAD SIL shall have operational resources to conduct experiments and simulations as requested for world-wide research organizations.	185	N/A	N/A			X	

Requirement	PBSS Req ID	Ref #	SOO Ref #	Insp ect	Anal ysis	De mo	Test
The BrEAD SIL shall have operational resources to perform economic modeling and cultural studies for future processing system sites.	186	N/A	N/A	X			
The BrEAD SIL shall have operational resources to access the breadfruit database server at the Central Data Center in order to gather operational data from world-wide breadfruit processing sites.	187	N/A	N/A	X			
The BrEAD SIL shall have operational resources to transfer (i.e. upload) SIL-generated information to the database server at the Central Data Center.	188	N/A	N/A	X			
The BrEAD Regional Center shall have operational resources for liaison with Headquarters, NGRF, local authorities, and local aid organizations. (NGOs, NFPs, etc.)	189	N/A	13	X			
The BrEAD Central Data Center shall have operational resources to advocate and market breadfruit products as a component of the next green revolution (NGRF) on a worldwide level.	190	N/A	N/A	X			

Requirement	PBSS Req ID	Ref #	SOO Ref #	Insp ect	Anal ysis	De mo	Test
The BrEAD Central Data Center shall have operational resources (i.e. call center) for liaison with regional centers, NGRF, breadfruit researchers, SIL operators, and government aid organizations.	191	N/A	13	X			
The regional center shall be provided an electrical infrastructure by host country.	192	N/A	N/A	X			
The breadfruit processing system shall comply with all local, regional, and national regulations of Port Au Prince, Haiti and all surrounding areas within TBD miles	193	N/A	3		X		
The breadfruit processing system shall be operable under the electrical infrastructure of Port Au Prince, Haiti and all surrounding areas within TBD miles.	194	N/A	3		X		
The breadfruit processing system shall comply with all local, regional, and national regulations of Maui, Hawaii and all surrounding areas within TBD miles	195	N/A	3		X		
The EA shall comply with Modular Open Systems Assessment (MOSA) or an equivalent open systems architecture standard.	196	N/A	3,5	X			
The breadfruit processing system shall be designed to be safe to operate, maintain, and transport and does not introduce hazards.	197	N/A	19	X			

Requirement	PBSS Req ID	Ref #	SOO Ref #	Insp ect	Anal ysis	De mo	Test
The breadfruit shredder shall have a protective shield to prevent direct contact with the shredder's moving parts during shredding operations.	198	N/A	19			X	
The breadfruit grinder shall have a protective shield to prevent direct contact with the grinder's moving parts during grinding operations.	199	N/A	19			X	
The breadfruit processing electrical components shall be comply with grounding and all other applicable UL and CE requirements.	200	N/A	19	X			
Safety incidents reports shall be issued daily.	201	N/A	19			X	
Safety reports shall include description, severity, and component causing injury.	202	N/A	19	X			
The BrEAD shall insure protection of BrEAD data, software, and computer systems by a system of virus protection, firewalls, and passwords.	203	N/A	N/A	X			
The on-site data logger / laptop shall be capable of operating in conditions up to 100% humidity.	204	N/A	N/A				X
The data transmitter shall be capable of operating in conditions up to 100% humidity.	205	N/A	N/A				X
The on-site data logger / laptop shall be capable of operating within temperature ranges between 40 deg F and 120 deg F. (4.4 and 49 degrees C).	206	N/A	N/A				X

Requirement	PBSS Req ID	Ref #	SOO Ref #	Insp ect	Anal ysis	De mo	Test
The data transmitter shall be capable of operating within temperature ranges between 40 deg F and 120 deg F. (4.4 and 49 degrees C).	207	N/A	N/A				X
The on-site data logger / laptop shall have a minimum of TBD memory, TBD data storage, and TBD Mhz processing speed.	208	N/A	N/A	X			
The data transmitter shall be capable of transmitting data at a rate of least 100 kbps per second.	209	N/A	N/A				X
The regional server shall have a minimum of 2 GB memory, expandable to 8 GB memory, 600GB data storage, and 2.4GHz processing speed.	210	N/A	N/A	X			
The central data server shall have a minimum of 8 GB memory expandable to 16 GB memory, 1800 GB data storage, and 1.86GHz processing speed	211	N/A	N/A	X			
The BrEAD data logger shall be capable of recording and storing all data listed in Appendix A.	212	N/A	N/A				X
The human machine interface for the BrEAD data logger shall be intuitive to the user.	213	N/A	N/A			X	
The human machine interface for the BrEAD data logger shall be in the language native to the user.	214	N/A	N/A	X			
User manuals shall be included with each data logger.	215	N/A	N/A	X			

Requirement	PBSS Req ID	Ref #	SOO Ref #	Insp ect	Anal ysis	De mo	Test
The Input/Output (I/O) interfaces used on the computer hardware components shall use open standard, industry accepted interfaces.	**216**	N/A	N/A	X			
The Breadfruit Enterprise Architecture Demonstration (BrEAD) software shall utilize a High Order Language (HOL), which follows ANSI, IEEE.	**217**	N/A	N/A	X			
The BrEAD software system shall maximize the use of Commercial-Off-The-Shelf (COTS) software products.	**218**	N/A	N/A	X			
The BrEAD shall have an on-line digital data environment that allows the program acquisition and operational support activities, throughout the program's life-cycle, to digitally create, store, access, manipulate, share, and exchange programmatic and technical data.	**219**	N/A	N/A			X	
The BrEAD shall provide access to the BrEAD Integrated Digital Environment (IDE) for approximately 200 individuals at anytime from their workspaces	**220**	N/A	N/A	X			
The BrEAD shall employ a configuration management system that identifies, prioritizes, tracks, allocates and manages software requirement changes via a Change Review Board.	**221**	N/A	N/A	X			

Requirement	PBSS Req ID	Ref #	SOO Ref #	Insp ect	Anal ysis	De mo	Test
The BrEAD software shall prevent or mitigate single point failures that could disable the entire system.	222	N/A	N/A			X	
Hazards that could result from operator errors or input of invalid data shall be mitigated by software monitoring and checks of inputs.	223	N/A	N/A		X		
The BrEAD software shall perform appropriate safety checks to assure that dynamic updates of data do not cause data corruption.	224	N/A	N/A		X		
Hazards that could result from operator errors or input of invalid data shall be mitigated by software monitoring and checks of inputs.	225	N/A	N/A		X		
The BrEAD software shall perform appropriate safety checks to assure that dynamic updates of data do not cause data corruption.	226	N/A	N/A		X		
A Field Data Database shall be created for storing and processing weather and breadfruit processing data (Field Data) from the processing sites.	227	N/A	N/A	X			
Field data, consisting of weather data and breadfruit processing data shall be recorded and preserved in a database.	228	N/A	N/A	X			
When Field Data is received via the internet connection, it shall be input into the Field Data Database by automated means.	229	N/A	N/A			X	

Requirement	PBSS Req ID	Ref #	SOO Ref #	Inspect	Analysis	Demo	Test
The Field Data Database shall be available for receipt and input of data 24 hours a day seven days a week.	230	N/A	N/A			X	
The enterprise shall meet ISO9001 requirements	231	N/A	N/A	X			
Spares shall be available within 2 hours.	232	N/A	N/A			X	

Section	Author
Volume 1: Exec Summary	Christa Triebwasser
Volume 2, Book 1: Design Approach	
2.1	Oscar Kalipe
2.1.1	Oscar Kalipe
2.1.1.1	Brian Falk
2.1.1.1.1	Christa Triebwasser
2.1.1.1.2	Marc Loos
2.1.1.1.3	Jim Woodcock
2.1.1.1.4	Oscar Kalipe
2.1.1.1.5	Oscar Kalipe
2.1.1.1.6	Oscar Kalipe
2.1.1.2	Marc Loos / Oscar Kalipe
2.1.1.3	Christa Triebwasser
2.1.1.4	Jim Woodcock
2.1.2	Tom Marquette
2.1.3	Gary Wallenhorst
Annex A	Gary Wallenhorst / Tom Marquette
Annex B	Christa Triebwasser
Technical Attachment I	Tom Marquette
Volume 2, Book 2: Program	
2.2	Marc Loos
2.2.1	Brian Falk
2.2.2	Gary Wallenhorst
2.2.3	Gary Wallenhorst
2.2.4	Marc Loos
2.2.5	Christa Triebwasser
2.2.6	Tom Marquette
2.2.7	Oscar Kalipe
2.2.8	Jim Woodcock
Annex C	Brian Falk
Annex D	Brian Falk
Annex E	Gary Wallenhorst
Annex F	Gary Wallenhorst
Annex G	Marc Loos
Annex H	Jim Woodcock
Annex I	Brian Falk
Annex J	Gary Wallenhorst / Tom Marquette

www.ingramcontent.com/pod-product-compliance
Lightning Source LLC
LaVergne TN
LVHW062150200926
848484LV00007B/1895